HYDRODYNAMIC AND HYDROMAGNETIC STABILITY

BY

S. CHANDRASEKHAR

Morton D. Hull Distinguished Service Professor
University of Chicago

DOVER PUBLICATIONS, INC.
NEW YORK

Published in Canada by General Publishing Company, Ltd., 30 Lesmill Road, Don Mills, Toronto, Ontario.
Published in the United Kingdom by Constable and Company, Ltd., 10 Orange Street, London WC2H 7EG.

This Dover edition, first published in 1981, in an unabridged republication of the third (1970) printing of the original Clarendon Press edition, which appeared in 1961. This work originally belonged to the International Series of Monographs on Physics edited by W. Marshall and D. H. Wilkinson.

International Standard Book Number: 0-486-64071-X
Library of Congress Catalog Card Number: 80-69678

Manufactured in the United States of America
Dover Publications, Inc.
180 Varick Street
New York, N.Y. 10014

PREFACE

In this book I have tried to bring together into a coherent account what I have learnt of hydrodynamic and hydromagnetic stability, with my students and with my associates, during the past years.

I regret that even in a book of this size, I have not been able to do justice to many aspects of the subject which could claim as much right to attention as those that have received their due or ample share. Perhaps the most serious omission is the absence of any reference to viscous shear flow; but this is a large and a highly specialized field and I have not felt myself competent to write about it. This argument is indeed a general one: in the last analysis an author chooses to write about only those matters in which he has some confidence of his understanding.

In the writing of this book I have received co-operation and assistance, in generous measure, from so many that in order to be reasonably complete I have listed them separately. I should, however, like to mention here the extent of my obligation to Miss Donna D. Elbert: in a real sense this book is the outcome of our joint efforts over the years and without her part there would have been no substance.

S. C.

ACKNOWLEDGEMENTS

THE following persons provided copies for many of the illustrations in this book; and in several instances new computations or new experiments were necessary.

Dr. F. E. Bisshopp (Figs. 5, 6, 9, and 57)
Dr. R. J. Donnelly (Figs. 77, 78 *a*, *b*, and 79 *a*, *b*)
Drs. R. J. Donnelly and D. Fultz (Figs. 80, 82, 83, 84, 87, and 97)
Dr. J. R. D. Francis (Fig. 117)
Dr. D. Fultz (Figs. 20, 33, 35, and 64)
Mr. I. Goroff (Figs. 38 *a*, *b*)
Dr. B. Lehnert (Fig. 46)
Dr. Y. Nakagawa (Figs. 30, 31, 43, 44, 45, 51, 52, 53, and 54)
Dr. A. H. Nissan (Figs. 96 *a*, *b*)
Dr. W. H. Reid (Figs. 65, 66, 67, 88, 89, 99, 100, 108, and 113)
Drs. W. H. Reid and D. L. Harris (Figs. 2, 3, 4, and 8)
Dr. P. L. Silveston (Figs. 12 *a*, *b*, 13, 14, and 17 *a*, *b*)
Dr. S. K. Trehan (Figs. 130 and 131)
Dr. P. O. Vandervoort (Figs. 27, 28, 42, 71, 112, 114, 121, 135, and 136)
Dr. G. Veronis (Figs. 24 *a*, *b*, 25 *a*, *b*, and 26)
Mr. J. P. Wright (Fig. 41)

Permission was granted by the Editors of the following periodicals for the reproduction of some of the illustrations:

The Royal Society (Figs. 11, 15, 16, 32, 34, 44, 50, 55, 68, 85, and 86)
The Journal of Fluid Mechanics (Figs. 24 *a*, *b*, 25 *a*, *b*, and 26)
The Philosophical Magazine (Fig. 117)
Proceedings of the Physical Society (Figs. 133 and 134)

Mr. Norman R. Wolfe inked all of the line drawings in the book.

The following persons read various parts of the completed manuscript and made helpful suggestions.

Dr. W. H. Reid read Chapters I–IX and contributed new information included in Chapters II, VII, VIII, IX, and X
Dr. P. Ledoux read Chapters I–VI
Dr. M. N. Rosenbluth read Chapters IX–XII

The experimental investigations of Drs. D. Fultz, Y. Nakagawa, and R. J. Donnelly in close association with the theoretical developments have been a source of constant encouragement; much of the experimental work summarized in this book is theirs.

Dr. Norman R. Lebovitz read the entire book in proofs and helped to eliminate many errors.

Miss Donna D. Elbert carried out the relevant numerical calculations for most of the problems treated in this book; she is responsible for the numerical information included in all of the tables with the exception of Tables I–VI, X, XXIII–XXX, XXXVI–XXXIX, XLVII, XLIX, LXIV–LXVII, and LXX.

And I am grateful to all of these persons.

The theoretical investigations included in this book have in large part been supported by contracts with the Geophysics Research Directorate of the Air Force Cambridge Research Center, Air Research and Development Command (Contracts AF 19(604)–299 and AF 19(604)–2046).

And the experimental investigations carried out at the Enrico Fermi Institute for Nuclear Studies (University of Chicago) have been supported by contracts with the Office of Naval Research (Contracts N6ori-02056 and Nonr-2121(20)).

And finally, I am grateful to the Clarendon Press for bringing to this book that excellence of craftsmanship and typography which is characteristic of all of their work.

S. C.

CONTENTS

I

BASIC CONCEPTS

1. Introduction

THE equations of hydrodynamics, in spite of their complexity, allow some simple patterns of flow (such as between parallel planes, or rotating cylinders) as stationary solutions. These patterns of flow can, however, be realized only for certain ranges of the parameters characterizing them. Outside these ranges, they cannot be realized. The reason for this lies in their inherent *instability*, i.e. in their inability to sustain themselves against small perturbations to which any physical system is subject. It is in the differentiation of the stable from the unstable patterns of permissible flows that the problems of *hydrodynamic stability* originate.

In recent years the class of such problems of stability has been enlarged by the interest in hydrodynamic flows of electrically conducting fluids in the presence of magnetic fields. This is the domain of *hydromagnetics*; and there are problems of *hydromagnetic stability* even as there are problems of hydrodynamic stability.

This book is devoted to a consideration of some typical problems in hydrodynamic and hydromagnetic stability; and in this introductory chapter we shall formulate the basic principles and concepts of the subject.

2. Basic concepts

Suppose, then, that we have a hydrodynamic system which, in accordance with the equations governing it, is in a *stationary state*, i.e. in a state in which none of the variables describing it is a function of time. Let $X_1, X_2, \ldots, X_j$ be a set of parameters which define the system. These parameters will include geometrical parameters such as the dimensions of the system; parameters characterizing the velocity field which may prevail in the system; the magnitudes of the forces which may be acting on the system, such as pressure gradients, temperature gradients, magnetic fields and rotation; and others.

In considering the stability of such a system (with a given set of parameters $X_1, \ldots, X_j$) we essentially seek to determine the reaction of the system to small disturbances. Specifically, we ask: if the system is disturbed, will the disturbance gradually die down, or will the disturbance grow in amplitude in such a way that the system progressively departs from the initial state and never reverts to it? In the former case, we say that the system is *stable* with respect to the particular disturbance; and in the latter case, we say that it is *unstable*. Clearly, a system must be considered as unstable even if there is only *one* special mode of disturbance with respect to which it is unstable. And a system cannot be considered stable unless it is stable with respect to *every* possible disturbance to which it can be subject. In other words, stability must imply that there exists *no* mode of disturbance for which it is unstable.

If all initial states are classified as stable, or unstable, according to the criteria stated, then in the space of the parameters, $X_1,..., X_j$, the locus which separates the two classes of states defines the states of *marginal stability* of the system. By this definition, a *marginal state* is a state of *neutral stability*.

The locus of the marginal states in the $(X_1,..., X_j)$-space will be defined by an equation of the form

$$\Sigma(X_1,..., X_j) = 0. \tag{1}$$

The determination of this locus is one of the prime objects of an investigation on hydrodynamic stability.

In thinking of the stability of a hydrodynamic system, it is often convenient to suppose that all parameters of the system, save one, are kept constant while the chosen one is continuously varied. We shall then pass from stable to unstable states when the particular parameter we are varying takes a certain critical value. We may say that *instability sets in* at this value of the chosen parameter when all the others have their preassigned values.

States of marginal stability can be one of two kinds. The two kinds correspond to the two ways in which the amplitudes of a small disturbance can grow or be damped: they can grow (or be damped) aperiodically; or they can grow (or be damped) by oscillations of increasing (or decreasing) amplitude. In the former case, the transition from stability to instability takes place via a marginal state exhibiting a stationary pattern of motions. In the latter case, the transition takes place via a marginal state exhibiting oscillatory motions with a certain definite characteristic frequency.

If at the onset of instability a stationary pattern of motions prevails, then one says that the *principle of the exchange of stabilities* is valid and that instability sets in as stationary *cellular convection*, or *secondary flow*. (The significance of the qualification 'cellular' will become clear presently.) On the other hand, if at the onset of instability oscillatory motions prevail, then one says (following Eddington) that one has a case of *overstability*. Eddington explains this choice of the terminology as follows: 'In the usual kinds of instability, a slight displacement provokes restoring forces tending away from equilibrium; in overstability it provokes restoring forces so strong as to overshoot the corresponding position on the other side of equilibrium.'

In classifying marginal states into the two classes—stationary and oscillatory—we have supposed that we are dealing with *dissipative systems*. In non-dissipative, conservative, systems the situation is generally somewhat different. In these cases the stable states, when perturbed, execute undamped oscillations with certain definite characteristic frequencies; while in the unstable states small initial perturbations tend to grow exponentially with time; and the marginal states themselves are stationary.

3. The analysis in terms of normal modes

The mathematical treatment of a problem in stability generally proceeds along the following lines.

We start from an initial flow which represents a stationary state of the system. By supposing that the various physical variables describing the flow suffer small (infinitesimal) increments, we first obtain the equations governing these increments. In obtaining these equations from the relevant equations of motion, we neglect all products and powers (higher than the first) of the increments and retain only terms which are linear in them. The theory derived on the basis of such *linearized equations* is called the linear stability theory in contrast to *non-linear theories* which attempt to allow for the finite amplitudes of the perturbations. In this book we shall be concerned only with the linear stability theory which presupposes that the perturbations are infinitesimal; and this must be understood even if the qualification is not explicitly made every time.

As we have explained, stability means stability with respect to all possible (infinitesimal) disturbances. Accordingly, for an investigation on stability to be complete, it is necessary that the reaction of the system to all possible disturbances be examined. In practice, this is accomplished by expressing an arbitrary disturbance as a superposition of certain basic possible modes and examining the stability of the system with respect to each of these modes. In illustration of how one does this, consider a system confined between two parallel planes and in which the physical variables in the stationary state are functions only of the coordinate (z, say) normal to the planes. In this case, we may analyse an arbitrary disturbance in terms of two-dimensional periodic waves. Thus, if $A(x,y,z,t)$ represents a typical amplitude describing the disturbance, we expand it in the manner

$$A(x,y,z,t) = \int_{-\infty}^{+\infty}\int_{-\infty}^{+\infty} dk_x\, dk_y\, A_k(z,t)\exp[i(k_x x+k_y y)], \tag{2}$$

where
$$k = \sqrt{(k_x^2+k_y^2)} \tag{3}$$

is the wave number associated with the disturbance $A_k(z,t)$. Since the perturbation equations are linear, the reaction of the system to a general disturbance can be determined if we know the reaction of the system to disturbances of all assigned wave numbers. In particular, the stability of the system will depend on its stability to disturbances of *all* wave numbers; and instability will follow from its instability to disturbances of even *one* wave number. From this discussion it is clear that for 'plane problems' of the kind considered, the marginal state will be of neutral stability for disturbances of one particular wave number (k_c, say) and will be stable for disturbances of all other wave numbers. At the onset of instability, therefore, motions (stationary or oscillatory, as the case may be) will be manifested which will belong to a particular wave number (namely, k_c); they will accordingly present a cellular pattern, the horizontal dimension of the cell being determined by the wavelength $\lambda_c = 2\pi/k_c$.

In problems with other geometries, the disturbance will have to be analysed correspondingly. Thus, if the stationary state has symmetry about an axis, we may expand the disturbance in the manner

$$A(\varpi,z,\theta,t) = \sum_{m=-\infty}^{+\infty}\int_{-\infty}^{+\infty} A_{m,k}(\varpi,t)e^{i(kz+m\theta)}\,dk \tag{4}$$

(where z is measured along the axis of symmetry, ϖ is the distance from the axis, and θ is the azimuthal angle), and investigate the stability of the system with respect to all modes distinguished by k and m. Similarly, if the initial state has spherical symmetry, we may expand the disturbance in spherical harmonics, $Y_l^m(\vartheta,\varphi)$, in the manner

$$A(r,\vartheta,\varphi) = \sum_{l=0}^{\infty}\sum_{m=-l}^{+l} Y_l^m(\vartheta,\varphi)A_l^m(r,t) \tag{5}$$

(where r denotes the distance from the origin, and ϑ and φ are the spherical polar angles), and investigate the stability of the system with respect to all modes distinguished by m and l.

The essential point of the foregoing remarks is that in all cases the disturbance must be expanded in terms of some suitable set of *normal modes* which must be *complete* for such an expansion to be possible. Let the various modes appropriate to a particular problem be distinguished by the symbol $\mathbf{k}$. In practice, several parameters may be needed to distinguish the different modes; and the symbol $\mathbf{k}$ is assumed to represent all the parameters that may be needed. With this understanding, we may write (again, symbolically)

$$A(\mathbf{r},t) = \int A_{\mathbf{k}}(\mathbf{r},t)\,d\mathbf{k}. \tag{6}$$

The equations governing the general (infinitesimal) perturbations can be specialized for the normal modes. We then eliminate the dependence on time by seeking solutions of the form

$$A_{\mathbf{k}}(\mathbf{r},t) = A_{\mathbf{k}}(\mathbf{r})e^{p_{\mathbf{k}}t}, \tag{7}$$

where $p_{\mathbf{k}}$ is a constant to be determined. The subscript $\mathbf{k}$ has been attached to p to emphasize that the value of this constant will be different for the different normal modes distinguished by $\mathbf{k}$.

With the dependence on time separated in this manner, the perturbation equations will involve $p_{\mathbf{k}}$ as a parameter. And solutions of these last equations must be sought which satisfy certain necessary boundary conditions (such as, that there be no slip at a 'rigid boundary', or that there be no tangential viscous stresses at a 'free boundary'). In general the equations will not allow non-trivial solutions (i.e. not vanishing everywhere) for any arbitrarily assigned $p_{\mathbf{k}}$. Indeed, the requirement that the equations allow non-trivial solutions satisfying the various boundary conditions leads directly to a *characteristic value problem* for $p_{\mathbf{k}}$. The problem is thus reduced to determining $p_{\mathbf{k}}$ for the various modes. In general the characteristic values for $p_{\mathbf{k}}$ will be complex:

$$p_{\mathbf{k}} = p_{\mathbf{k}}^{(r)} + ip_{\mathbf{k}}^{(i)}, \tag{8}$$

where $p_{\mathbf{k}}^{(r)}$ and $p_{\mathbf{k}}^{(i)}$ will depend, apart from $\mathbf{k}$, on the parameters $X_1,..., X_j$ of the basic flow. The condition for stability is that $p_{\mathbf{k}}^{(r)}$ be negative for all $\mathbf{k}$. The states which will be of neutral stability with respect to disturbances belonging to $\mathbf{k}$ will be characterized by

$$p_{\mathbf{k}}^{(r)}(X_1,..., X_j) = 0. \tag{9}$$

This is a condition on the parameters $X_1,..., X_j$; and it defines a locus,

$$\Sigma_{\mathbf{k}}(X_1,..., X_j) = 0, \tag{10}$$

in the $(X_1,..., X_j)$-space; this locus separates states which are stable from those which are unstable with respect to disturbances belonging to the particular mode **k**. It follows from this that *the locus* $\Sigma(X_1,..., X_j)$ *separating the regions of complete stability from the regions of instability in the* $(X_1,..., X_j)$*-space is the envelope of the loci* $\Sigma_{\mathbf{k}}$; *further, that when the system becomes unstable as we cross* Σ *at some particular point, the mode of disturbance which will be manifested* (*at onset*) *will be the one whose locus* $\Sigma_{\mathbf{k}}$ *touches* Σ *at that point.*

A further observation might be made. The distinction between the two kinds of marginal states (stationary and oscillatory) made in § 2 corresponds to whether or not the imaginary part, $p_{\mathbf{k}}^{(i)}$ of $p_{\mathbf{k}}$, vanishes when the real part, $p_{\mathbf{k}}^{(r)}$ of $p_{\mathbf{k}}$, does. If $p_{\mathbf{k}}^{(r)} = 0$ implies that $p_{\mathbf{k}}^{(i)} = 0$ for every **k**, then the principle of the exchange of stabilities will be valid; otherwise, we will have overstability at least when instability sets in as certain modes. If overstability occurs, then $p_{\mathbf{k}}^{(i)}$ will have determinate values on $\Sigma_{\mathbf{k}}$. Accordingly, in case of overstability, the theory must tell us not only what the mode of disturbance is that will be manifested at onset of instability, but also the characteristic frequency of the oscillations.

4. Non-dimensional numbers

As we have explained, the solution of a problem in stability generally centres on the determination of the locus Σ of the marginal states in the space of the parameters $(X_1,..., X_j)$ characterizing the initial flow. In expressing the results of the theory, it will often be convenient to combine the various parameters into certain non-dimensional combinations or *numbers*. The most famous of such numbers is the *Reynolds number* R which is a combination of the form

$$\mathrm{R} = Lv/\nu, \tag{11}$$

where L is a typical dimension of the system, v is a measure of the velocities which prevail in the stationary flow, and ν is the kinematic viscosity. However, particular problems may require the definition of special numbers. Thus, in the problem of the stability of viscous flow between two rotating coaxial cylinders, the parameters are the radii R_1 and R_2 of the two cylinders, the angular velocities Ω_1 and Ω_2 of their rotations, and the kinematic viscosity ν. The non-dimensional combinations of these parameters in terms of which it is convenient to discuss this problem are (see Chapter VII):

$$\eta = \frac{R_1}{R_2}, \quad \mu = \frac{\Omega_2}{\Omega_1}, \quad \text{and} \quad T = \frac{4\Omega_1^2 R_1^4}{\nu^2}\frac{(1-\mu)(1-\mu/\eta^2)}{(1-\eta^2)^2}. \tag{12}$$

The combination $$4\Omega^2 L^4/\nu^2$$

of an angular velocity Ω, a linear dimension L, and the kinematic viscosity ν, which occurs in the definition of T, always appears when we are dealing with rotating systems; it is called the *Taylor number*. Similarly, in the problem of the thermal stability of a horizontal layer of fluid heated from below, the combinations which occur are:

$$R = \frac{g\alpha\beta}{\kappa\nu} d^4 \quad \text{and} \quad \mathfrak{p} = \nu/\kappa, \tag{13}$$

where g is the acceleration due to gravity, d is the depth of the layer, β is the adverse temperature gradient which is maintained, and α, κ, and ν are the coefficients of volume expansion, thermometric conductivity, and kinematic viscosity, respectively; R as defined here is called the *Rayleigh number*; and $\mathfrak{p}$ is the *Prandtl number*. And when we are dealing with an electrically conducting fluid in which a magnetic field H prevails, a 'number' of frequent occurrence is

$$Q = \frac{\mu^2 H^2 \sigma}{\rho\nu} d^2, \tag{14}$$

where μ denotes the magnetic permeability, ρ is the density, σ is the coefficient of electrical conductivity, and d is a linear dimension.

It should be pointed out that there is nothing unique in the way these various numbers are defined; they happen to be the ones which have been chosen; and in some sense they occur most naturally in certain types of problems.

BIBLIOGRAPHICAL NOTES

The only book which is devoted exclusively to problems in hydrodynamic and hydromagnetic stability is:

1. C. C. Lin, *The Theory of Hydrodynamic Stability*, Cambridge, England, 1955.

A major portion of Lin's book treats problems which are related to viscous shear flows and the Orr–Sommerfeld equation: topics which on account of their very special character will not be considered in this book.

Problems in hydromagnetic stability are also considered in:

2. T. G. Cowling, *Magnetohydrodynamics*, Interscience Tracts on Physics and Astronomy, No. 4, Interscience Publishers, Inc., New York, 1957; see particularly chapter 4.

Among other books on related subjects, we may refer to:

3. H. Alfvén, *Cosmical Electrodynamics*, International Series of Monographs on Physics, Oxford, England, 1950.

4. Lyman Spitzer, Jr., *Physics of Fully Ionized Gases*, Interscience Tracts on Physics and Astronomy, No. 3, Interscience Publishers, Inc., New York, 1956.

5. J. W. Dungey, *Cosmic Electrodynamics*, Cambridge Monographs on Mechanics and Applied Mathematics, Cambridge, England, 1958.

The following review articles bear on topics treated in this book:

6. S. LUNDQUIST, 'Studies in magneto-hydrodynamics', *Arkiv för Fysik*, **5,** 297–347 (1952).

7. T. G. COWLING, 'Solar electrodynamics', *The Sun*, chapter 8, edited by G. P. Kuiper, University of Chicago Press, Chicago, 1953.

8. W. M. ELSASSER, 'Hydromagnetism. I. A review,' *American J. Physics*, **23,** 590–609 (1955); 'Hydromagnetism. II. A review,' ibid. **24,** 85–110 (1956).

9. G. H. A. COLE, 'Some aspects of magnetohydrodynamics,' *Advances in Physics*, **5,** 452–497 (1956).

This book represents an expansion of the point of view taken in:

10. S. CHANDRASEKHAR, 'Problems of stability in hydrodynamics and hydromagnetics', *Monthly Notices Roy. Astron. Soc. London*, **113,** 667–78 (1953).

11. S. CHANDRASEKHAR, 'Thermal convection', *Daedalus*, **86,** 323–39 (1957).

§ 2. The quotation from Eddington is from:

12. A. S. EDDINGTON, *The Internal Constitution of the Stars*, p. 201, Cambridge, England, 1926.

II

THE THERMAL INSTABILITY OF A LAYER OF FLUID HEATED FROM BELOW

1. THE BÉNARD PROBLEM

5. Introduction

THE problem of the onset of thermal instability in horizontal layers of fluid heated from below is well suited to illustrate the many facets, mathematical and physical, of the general theory of hydrodynamic stability. If we enlarge the problem to include the effects of rotation and magnetic field, we can also exhibit the conflicting tendencies to which a fluid can be subject; and many striking aspects of fluid behaviour, disclosed by the mathematical theory in these connexions, have been experimentally demonstrated. For these reasons, we have selected the general topic of the instability of layers of fluid heated from below for a detailed treatment; and Chapters II–V are devoted to this problem.

In this chapter we shall consider the simplest problem in the absence of rotation and magnetic field; the effects of these will be considered in the following chapters.

6. The nature of the physical problem

Consider then a horizontal layer of fluid in which an adverse temperature gradient is maintained by heating the underside. The temperature gradient thus maintained is qualified as adverse since, on account of thermal expansion, the fluid at the bottom will be lighter than the fluid at the top; and this is a top-heavy arrangement which is potentially unstable. Because of this latter instability there will be a tendency on the part of the fluid to redistribute itself and remedy the weakness in its arrangement. However, this natural tendency on the part of the fluid will be inhibited by its own viscosity. In other words, we expect that the adverse temperature gradient which is maintained must exceed a certain value before the instability can manifest itself.

The earliest experiments to demonstrate in a definitive manner the onset of thermal instability in fluids are those of Bénard in 1900, though the phenomenon of thermal convection itself had been recognized earlier by Count Rumford (1797) and James Thomson (1882). In § 18 we shall describe the experiments of Bénard and others. It will suffice to summarize here the principal facts established by them. They are: *first*, a certain critical adverse temperature gradient must be exceeded before instability can set in; and *second*, the motions which ensue on surpassing the critical temperature gradient have a stationary cellular character. What actually happens at the onset of instability is that the layer of fluid resolves itself into a number of cells; and if the experiment is performed with sufficient care, the cells become equal and they align themselves to form a regular hexagonal pattern. Fig. 1 (which is a reproduction of one of Bénard's early photographs) illustrates the phenomenon.

The theoretical foundations for a correct interpretation of the foregoing facts were laid by Lord Rayleigh in a fundamental paper. Rayleigh showed that what decides the stability, or otherwise, of a layer of fluid heated from below is the numerical value of the non-dimensional parameter,

$$R = \frac{g\alpha\beta}{\kappa\nu} d^4, \tag{1}$$

where g denotes the acceleration due to gravity, d the depth of the layer, β $(= |dT/dz|)$ the uniform adverse temperature gradient which is maintained, and α, κ, and ν are the coefficients of volume expansion, thermometric conductivity, and kinematic viscosity, respectively; R as defined here is called the *Rayleigh number*. Rayleigh further showed that instability must set in when R exceeds a certain critical value R_c; and that when R just exceeds R_c, a stationary pattern of motions must come to prevail. The principal theoretical question is, therefore: how is one to determine R_c? This chapter is largely devoted to answering this question.

7. The basic hydrodynamic equations

For a general treatment of the problem of thermal instability described in § 6, we need the equations governing the hydrodynamical flow of a viscous fluid of varying density and temperature. In this section we shall obtain these basic equations.

Consider then a fluid in which the density ρ is a function of position x_j $(j = 1, 2, 3)$. Let u_j $(j = 1, 2, 3)$ denote the components of the velocity. In writing the various equations we shall use the notation of Cartesian tensors with the usual summation convention.

(a) The equation of continuity

We have

$$\frac{\partial \rho}{\partial t} + \frac{\partial}{\partial x_j}(\rho u_j) = 0. \tag{2}$$

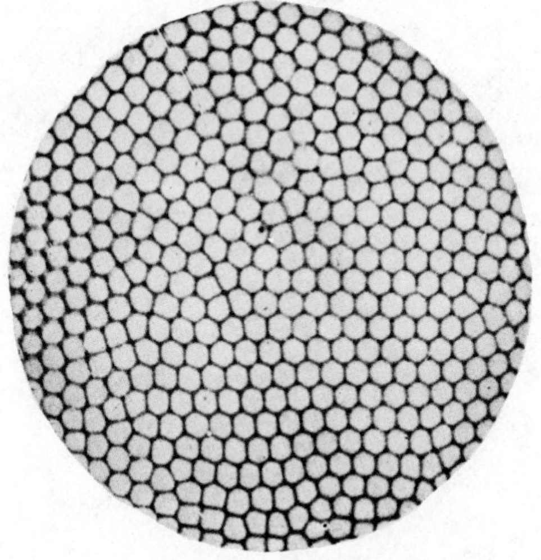

Fig. 1. Bénard cells in spermaceti. A reproduction of one of Bénard's original photographs.

This equation expresses the *conservation of mass*: for, by integrating the equation over an arbitrary volume V, we obtain

$$\frac{\partial}{\partial t}\int\limits_V \rho\, d\tau = -\int\limits_V \frac{\partial}{\partial x_j}(\rho u_j)\, d\tau, \tag{3}$$

where $d\tau$ $(= dx_1\, dx_2\, dx_3)$ is the element of volume. By Gauss's theorem, the volume integral on the right-hand side is equal to an integral over the bounding surface S of V; thus,

$$\frac{\partial}{\partial t}\int\limits_V \rho\, d\tau = -\int\limits_S \rho u_j\, dS_j, \tag{4}$$

where dS_j represents a vector normal to the element of surface dS and whose absolute magnitude is equal to dS. [The vector dS_j is dual to the antisymmetric tensor dS_{lm} whose components are equal to the projected area of dS on the coordinate planes.] Equation (4) expresses the fact that the rate of change of the mass contained in a fixed volume of the fluid is given by the rate at which the fluid flows out of it across the boundary S. We could, indeed, have started with equation (4) as the expression of the conservation of mass and retraced the arguments to derive equation (2).

An alternative form of equation (2) which we shall find useful is

$$\frac{\partial \rho}{\partial t} + u_j \frac{\partial \rho}{\partial x_j} = -\rho \frac{\partial u_j}{\partial x_j}. \tag{5}$$

The quantity on the left-hand side of equation (5) is clearly the total derivative of ρ with respect to the time.

For an incompressible fluid, the equation of continuity reduces to

$$\frac{\partial u_j}{\partial x_j} = 0; \tag{6}$$

in this case, the velocity field is, therefore, *solenoidal*.

(*b*) *The equations of motion*

Before we can write down the equations of motion, it is necessary to know the *stress* P_{ij} acting in the direction of x_j per unit area on an element of surface normal to x_i. This stress must be related to the *rate of increase of strain* in the fluid; the latter is given by

$$e_{ij} = \frac{1}{2}\left(\frac{\partial u_i}{\partial x_j} + \frac{\partial u_j}{\partial x_i}\right). \tag{7}$$

A basic assumption of fluid dynamics is that P_{ij} *and* e_{ij} *are related linearly*; thus,

$$P_{ij} = \varpi_{ij} + q_{ij;kl}\, e_{kl}, \tag{8}$$

where ϖ_{ij} is a symmetric tensor to which P_{ij} tends in the limit $e_{ij} = 0$, and $q_{ij;kl}$ is a tensor of the fourth rank. For an isotropic fluid, the form of the relation (8) must be invariant to arbitrary rotations and translations of the coordinate system. This requires that ϖ_{ij} and $q_{ij;kl}$ are *isotropic tensors*; they must, therefore, have the forms

$$\varpi_{ij} = -p\delta_{ij} \tag{9}$$

and

$$q_{ij;kl} = \lambda\delta_{ij}\delta_{kl}+\mu(\delta_{ik}\delta_{jl}+\delta_{il}\delta_{jk}), \tag{10}$$

where p, λ, and μ are (arbitrary) scalar functions of x_i. Inserting the expressions (9) and (10) in equation (8), we have

$$P_{ij} = -p\delta_{ij}+2\mu e_{ij}+\lambda\delta_{ij}e_{kk}. \tag{11}$$

It is convenient to define p as the isotropic pressure at x_i when there is no strain; then,

$$P_{ii} = -3p = -3p+2\mu e_{ii}+3\lambda e_{kk}. \tag{12}$$

Hence, with the definition of p adopted,

$$\lambda = -\tfrac{2}{3}\mu \tag{13}$$

and

$$P_{ij} = -p\delta_{ij}+2\mu e_{ij}-\tfrac{2}{3}\mu\delta_{ij}e_{kk}. \tag{14}$$

The coefficient μ which occurs in this equation is the *coefficient of viscosity*; it can be an arbitrary function of position.

The terms in (14) which are proportional to μ define the stresses due to viscosity; denoting this by p_{ij}, we have

$$p_{ij} = \mu\left(\frac{\partial u_i}{\partial x_j}+\frac{\partial u_j}{\partial x_i}\right)-\tfrac{2}{3}\mu\frac{\partial u_k}{\partial x_k}\delta_{ij}. \tag{15}$$

For an incompressible fluid, the viscous tensor has the simpler form (cf. equation (6))

$$p_{ij} = \mu\left(\frac{\partial u_i}{\partial x_j}+\frac{\partial u_j}{\partial x_i}\right). \tag{16}$$

In terms of the stresses P_{ij} we can write down the hydrodynamical equations of motion. We have

$$\rho\frac{\partial u_i}{\partial t}+\rho u_j\frac{\partial u_i}{\partial x_j} = \rho X_i+\frac{\partial P_{ij}}{\partial x_j}, \tag{17}$$

where X_i is the ith component of whatever external force may be acting on the fluid. Substituting for P_{ij} in this equation, we have

$$\rho\frac{\partial u_i}{\partial t}+\rho u_j\frac{\partial u_i}{\partial x_j} = \rho X_i-\frac{\partial p}{\partial x_i}+\frac{\partial}{\partial x_j}\left\{\mu\left(\frac{\partial u_i}{\partial x_j}+\frac{\partial u_j}{\partial x_i}\right)-\tfrac{2}{3}\mu\frac{\partial u_k}{\partial x_k}\right\}. \tag{18}$$

For an incompressible fluid in which μ is constant, equation (18) simplifies to

$$\rho\frac{\partial u_i}{\partial t}+\rho u_j\frac{\partial u_i}{\partial x_j} = \rho X_i-\frac{\partial p}{\partial x_i}+\mu\nabla^2 u_i. \tag{19}$$

This is the original form of the *Navier–Stokes equations*.

Equation (17) expresses the *conservation of momentum*. This can be seen by integrating the equation over an arbitrary volume V. We find

$$\int\limits_V \left(\rho\frac{\partial u_i}{\partial t}+\rho u_j\frac{\partial u_i}{\partial x_j}\right) d\tau = \int\limits_V \rho X_i\, d\tau+\int\limits_V \frac{\partial P_{ij}}{\partial x_j}\, d\tau. \tag{20}$$

By integrating by parts the second integral on the left-hand side and transforming the integral over P_{ij} on the right-hand side by Gauss's theorem, we obtain

$$\int\limits_V \left\{\rho\frac{\partial u_i}{\partial t}-u_i\frac{\partial}{\partial x_j}(\rho u_j)\right\} d\tau+\int\limits_S \rho u_i u_j\, dS_j = \int\limits_V \rho X_i\, d\tau+\int\limits_S P_{ij}\, dS_j. \tag{21}$$

By making use of the equation of continuity, we find that the first of the two integrals on the left-hand side of (21) is

$$\int\limits_V \left(\rho\frac{\partial u_i}{\partial t}+u_i\frac{\partial \rho}{\partial t}\right) d\tau = \frac{\partial}{\partial t}\int\limits_V \rho u_i\, d\tau. \tag{22}$$

Equation (21) can accordingly be written as

$$\frac{\partial}{\partial t}\int\limits_V \rho u_i\, d\tau = \int\limits_V \rho X_i\, d\tau+\int\limits_S P_{ij}\, dS_j-\int\limits_S \rho u_i u_j\, dS_j. \tag{23}$$

This equation expresses the fact that the rate of change of the momentum contained in a fixed volume V of the fluid is equal to the volume integral of the external forces acting on the elements of the fluid plus the surface integral of the normal stresses acting on the bounding surface S of V minus the rate at which momentum flows out of V across the boundaries of V by the motions prevailing on S. It is clear that by expressing the law of the conservation of momentum in the form (23), we can, conversely, derive the equation of motion (17).

(c) *The rate of viscous dissipation*

By multiplying equation (17) by u_i (and, of course, summing over i) and integrating over a volume V, we obtain

$$\frac{1}{2}\int\limits_V \rho\frac{\partial}{\partial t}u_i^2\, d\tau+\frac{1}{2}\int\limits_V \rho u_j\frac{\partial}{\partial x_j}u_i^2\, d\tau = \int\limits_V \rho u_i X_i\, d\tau+\int\limits_V u_i\frac{\partial P_{ij}}{\partial x_j}\, d\tau. \tag{24}$$

By a sequence of transformations similar to those used in passing from equation (20) to equation (23), we find

$$\frac{1}{2}\frac{\partial}{\partial t}\int\limits_V \rho u_i^2\,d\tau$$

$$=\int\limits_V \rho u_i X_i\,d\tau+\int\limits_S u_i P_{ij}\,dS_j-\tfrac{1}{2}\int\limits_S \rho u_i^2 u_j\,dS_j-\int\limits_V P_{ij}\frac{\partial u_i}{\partial x_j}\,d\tau. \quad (25)$$

This equation gives the rate at which the kinetic energy contained in a volume V of the fluid changes; and we observe that this is the sum of four terms. The first three terms represent, respectively, the rate at which the external forces do work on the elements of the fluid in V; the rate at which the stresses P_{ij} do work on the surface bounding V; and finally, the rate at which energy actually flows out of V across S.

It remains to interpret the last term on the right-hand side of equation (25); this is an integral over the volume of the quantity

$$-P_{ij}\frac{\partial u_i}{\partial x_j}. \quad (26)$$

Using the definitions of e_{ij} and P_{ij} (equations (7) and (14)), we can write

$$P_{ij}\frac{\partial u_i}{\partial x_j}=P_{ij}\,e_{ij}=(-p\delta_{ij}+2\mu e_{ij}-\tfrac{2}{3}\mu\delta_{ij}\,e_{kk})e_{ij}$$

$$=-pe_{jj}+2\mu e_{ij}^2-\tfrac{2}{3}\mu(e_{jj})^2. \quad (27)$$

The first term on the right-hand side is

$$-pe_{jj}=-p\frac{\partial u_j}{\partial x_j}=\frac{p}{\rho}\left(\frac{\partial\rho}{\partial t}+u_j\frac{\partial\rho}{\partial x_j}\right)=\frac{p}{\rho}\frac{d\rho}{dt}; \quad (28)$$

the volume integral of this quantity represents, therefore, the increase in the internal energy due to the compression which the fluid experiences. The remaining term,

$$\Phi=2\mu e_{ij}^2-\tfrac{2}{3}\mu(e_{jj})^2, \quad (29)$$

in (27) must, therefore, represent the rate at which energy is dissipated, irreversibly, by viscosity in each element of volume of the fluid. Consistent with this statement, Φ is indeed positive definite: for, it can be readily verified that an equivalent form for Φ is

$$\Phi=4\mu(e_{12}^2+e_{23}^2+e_{31}^2)+\tfrac{2}{3}\mu[(e_{11}-e_{22})^2+(e_{22}-e_{33})^2+(e_{33}-e_{11})^2]. \quad (30)$$

For an incompressible fluid $e_{jj}=0$, and the corresponding expression for Φ is

$$\Phi=2\mu e_{ij}^2. \quad (31)$$

(d) The equation of heat conduction

We have seen that the laws of conservation of mass and of momentum lead to the equations of continuity and motion, respectively. It remains to express the law of the *conservation of energy*. As we shall presently see, this leads to the equation of heat conduction in the form required for our present purposes.

Apart from an additive constant, the energy ϵ per unit mass of the fluid can be written as

$$\epsilon = \tfrac{1}{2}u_i^2 + c_V T, \tag{32}$$

where c_V is the specific heat at constant volume and T is the temperature. Counting the gains and losses of energy which occur in a volume V of the fluid, per unit time, we have

$\dfrac{\partial}{\partial t}\displaystyle\int_V \rho\epsilon\, d\tau =$ rate at which work is done on the boundary S of V by the stresses $P_{ij}+$

$+$rate at which work is done on each element of the fluid inside V by the external forces—

—rate at which energy in the form of heat is *conducted* across S—

—rate at which energy is *convected* across S by the prevailing mass motions

$$= \int_S u_i P_{ij}\, dS_j + \int_V \rho u_i X_i\, d\tau + \int_S k\frac{\partial T}{\partial x_j}\, dS_j - \int_S \rho\epsilon u_j\, dS_j, \tag{33}$$

where k is the coefficient of heat conduction. Making use of equations (25), (27), (28), and (29), we can rewrite the first term on the right-hand side of equation (33) as

$$\int_S u_i P_{ij}\, dS_j = \frac{1}{2}\frac{\partial}{\partial t}\int_V \rho u_i^2\, d\tau + \tfrac{1}{2}\int_S \rho u_i^2 u_j\, dS_j - \int_V \rho u_i X_i\, d\tau - \\ - \int_V p\frac{\partial u_j}{\partial x_j}\, d\tau + \int_V \Phi\, d\tau; \tag{34}$$

also, alternative forms of the third and the fourth terms are

$$\int_S k\frac{\partial T}{\partial x_j}\, dS_j = \int_V \frac{\partial}{\partial x_j}\left(k\frac{\partial T}{\partial x_j}\right) d\tau \tag{35}$$

and

$$-\int_S \rho\epsilon u_j\, dS_j = -\int_S \rho(\tfrac{1}{2}u_i^2 + c_V T)u_j\, dS_j \\ = -\frac{1}{2}\int_S \rho u_i^2 u_j\, dS_j - \int_V \frac{\partial}{\partial x_j}(\rho u_j c_V T)\, d\tau. \tag{36}$$

Combining the foregoing equations, we have

$$\int\limits_V \frac{\partial}{\partial t}(\rho c_V T)\,d\tau = \int\limits_V \frac{\partial}{\partial x_j}\left(k\frac{\partial T}{\partial x_j}\right)d\tau - \int\limits_V p\frac{\partial u_j}{\partial x_j}\,d\tau + \int\limits_V \Phi\,d\tau - \int\limits_V \frac{\partial}{\partial x_j}(\rho c_V T u_j)\,d\tau. \quad (37)$$

Since this equation must hold for every volume V, we must have

$$\frac{\partial}{\partial t}(\rho c_V T)+\frac{\partial}{\partial x_j}(\rho c_V T u_j) = \frac{\partial}{\partial x_j}\left(k\frac{\partial T}{\partial x_j}\right)-p\frac{\partial u_j}{\partial x_j}+\Phi. \quad (38)$$

Making use of the equation of continuity, we can simplify the foregoing equation to the form

$$\rho\frac{\partial}{\partial t}(c_V T)+\rho u_j\frac{\partial}{\partial x_j}(c_V T) = \frac{\partial}{\partial x_j}\left(k\frac{\partial T}{\partial x_j}\right)-p\frac{\partial u_j}{\partial x_j}+\Phi. \quad (39)$$

Equations (2), (14), (17), (29), and (39) are the basic hydrodynamic equations. They must be supplemented by an equation of state. For substances with which we shall be principally concerned, we can write

$$\rho = \rho_0[1-\alpha(T-T_0)], \quad (40)$$

where α is the coefficient of volume expansion and T_0 is the temperature at which $\rho = \rho_0$.

8. The Boussinesq approximation

In deriving the hydrodynamical equations in the preceding section, no assumptions were made regarding the constancy, or otherwise, of the various coefficients (μ, c_V, α, and k) which were introduced. The equations which were derived are, therefore, of quite general validity. However, as was first pointed out by Boussinesq, there are many situations of practical occurrence in which the basic equations can be simplified considerably. These situations occur when the variability in the density and in the various coefficients is due to variations in the temperature of only moderate amounts. The origin of the simplifications in these cases is due to the smallness of the coefficient of volume expansion: for gases and liquids such as we shall be mostly concerned with, α is in the range 10^{-3} to 10^{-4}. For variations in temperature not exceeding 10° (say), the variations in the density are at most 1 per cent. The variations in the other coefficients (consequent to variations in density of the amounts stated) must be of the same order. Variations of this small amount can, in general, be ignored. But there is one important exception: the variability of ρ in the term ρX_i in the equation of motion cannot be ignored; this is because the acceleration resulting from $\delta\rho X_i = \alpha\Delta T X_i$ (where ΔT is a measure of the variations in temperature which occur) can be quite large; larger than, for example, the acceleration due to the inertial term $u_j\,\partial u_i/\partial x_j$ in the equation of motion. Accordingly, we may treat ρ as a constant in all terms in the equations of motion except the one in the external force. This is the Boussinesq approximation. We shall verify in § 9 that for the particular problem of thermal instability we need not make the Boussinesq approximation: for, we shall show that in this case the general equations of § 7 lead to the same perturbation equations. Nevertheless, the equations which follow on the Boussinesq approximation are of interest in themselves; and they provide also the basis for further developments in the non-linear domain.

On the basis of the foregoing remarks, we first replace the equation of continuity (5) by

$$\frac{\partial u_j}{\partial x_j} = 0; \tag{41}$$

for, the terms on the left-hand side of (5) are of order α compared with those on the right-hand side. With this condition on u_j, the expression for the viscous stress-tensor is

$$p_{ij} = \mu\left(\frac{\partial u_i}{\partial x_j} + \frac{\partial u_j}{\partial x_i}\right), \tag{42}$$

where for the same reason we may treat μ as a constant. In the frame of these approximations, the equation of motion (17) becomes

$$\frac{\partial u_i}{\partial t} + u_j\frac{\partial u_i}{\partial x_j} = -\frac{1}{\rho_0}\frac{\partial p}{\partial x_i} + \left(1+\frac{\delta\rho}{\rho_0}\right)X_i + \nu\nabla^2 u_i, \tag{43}$$

where $\nu\ (=\mu/\rho_0)$ denotes the *kinematic viscosity*, ρ_0 is the density at some properly chosen mean temperature T_0, and

$$\delta\rho = -\rho_0\,\alpha(T-T_0). \tag{44}$$

Considering next the equation of heat conduction (39), we can treat c_V and k as constants and take them outside the differentiation signs; and we can ignore also the term $-p\,\mathrm{div}\,\mathbf{u}$ on the right-hand side. The term in the viscous dissipation Φ can also be ignored: for, according to equations (43) and (44), the prevailing velocities are of the order $[\alpha\Delta T|\mathbf{X}|d]^{\frac{1}{2}}$, where d is a measure of the linear dimensions of a system. Consequently, the term in Φ is of the order

$$\mu\alpha|\mathbf{X}|d/k \tag{45}$$

relative to the term arising from the conduction of heat; and this ratio for ordinary liquids (such as water and mercury) is 10^{-7} or 10^{-8} for

$d \sim 1$ cm and $|\mathbf{X}| \sim g$ (the acceleration due to gravity). Under these circumstances, the equation of heat conduction (39) reduces to

$$\frac{\partial T}{\partial t}+u_j\frac{\partial T}{\partial x_j} = \kappa\nabla^2 T, \tag{46}$$

where κ $(= k/\rho_0 c_V)$ is the *coefficient of thermometric conductivity.*

Equations (41), (43), (44), and (46) are the basic equations in the Boussinesq approximation.

9. The perturbation equations

Consider an infinite horizontal layer of fluid in which a steady adverse temperature gradient is maintained; further, let there be no motions. The initial state is, therefore, one in which

$$u_j \equiv 0 \quad \text{and} \quad T \equiv T(\lambda_j x_j), \tag{47}$$

where $\boldsymbol{\lambda} = (0, 0, 1)$ is a unit vector in the direction of the vertical.

When no motions are present, the hydrodynamical equations require only that the pressure distribution is governed by the equation (cf. equation (18))

$$\frac{\partial p}{\partial x_i} = \rho X_i = -g\rho\lambda_i, \tag{48}$$

where

$$\rho = \rho_0[1+\alpha(T_0-T)] \tag{49}$$

and ρ_0 and T_0 are the density and the temperature at the lower boundary ($z = \lambda_j x_j = 0$, say). The temperature distribution is governed by the equation

$$\nabla^2 T = 0. \tag{50}$$

In writing equation (39), under conditions of a steady state in the form (50), we have assumed that the coefficient of heat conduction k is a constant independent of T; this assumption is justified in our present connexion.

The solution of equation (50) appropriate to the problem on hand is

$$T = T_0-\beta\lambda_j x_j; \tag{51}$$

β is the adverse temperature gradient which is maintained. The corresponding distribution of the density is given by

$$\rho = \rho_0(1+\alpha\beta\lambda_j x_j). \tag{52}$$

With this expression for ρ, equation (48) can be integrated to give

$$p = p_0-g\rho_0(\lambda_i x_i+\tfrac{1}{2}\alpha\beta\lambda_i\lambda_j x_i x_j). \tag{53}$$

Let the initial state described by equations (47), (51), (52), and (53) be slightly perturbed. Let u_j denote the velocity in the perturbed state; and let the altered temperature distribution be

$$T' = T_0 - \beta\lambda_j x_j + \theta. \tag{54}$$

Finally, let δp denote the change in the pressure distribution. We shall now obtain the linearized form of the equations of motion which govern this perturbed state.

We shall first consider the problem on the basis of Boussinesq's approximation. By ignoring terms of the second and higher orders in the perturbations, equations (43) and (46) give (see equation (62) below)

$$\frac{\partial u_i}{\partial t} = -\frac{\partial}{\partial x_i}\left(\frac{\delta p}{\rho_0}\right) + g\alpha\theta\lambda_i + \nu\nabla^2 u_i \tag{55}$$

and

$$\frac{\partial \theta}{\partial t} = \beta\lambda_j u_j + \kappa\nabla^2\theta; \tag{56}$$

and the velocity field remains, of course, solenoidal:

$$\frac{\partial u_i}{\partial x_i} = 0. \tag{57}$$

It is of interest to verify that we obtain the same equations (55)–(57) from the general equations without first making the Boussinesq approximation. The general equations of motion are:

$$\frac{\partial \rho}{\partial t} + u_j\frac{\partial \rho}{\partial x_j} = -\rho\frac{\partial u_j}{\partial x_j}, \tag{58}$$

$$\frac{\partial T}{\partial t} + u_j\frac{\partial T}{\partial x_j} = \kappa\nabla^2 T - \frac{p}{\rho c_V}\frac{\partial u_j}{\partial x_j} + \frac{1}{\rho c_V}\Phi, \tag{59}$$

and

$$\rho\frac{\partial u_i}{\partial t} + \rho u_j\frac{\partial u_i}{\partial x_j} = -\frac{\partial p}{\partial x_i} + \frac{\partial}{\partial x_j}\left\{\mu\left(\frac{\partial u_i}{\partial x_j} + \frac{\partial u_j}{\partial x_i}\right) - \tfrac{2}{3}\mu\frac{\partial u_j}{\partial x_j}\right\} - g\rho\lambda_i. \tag{60}$$

In the unperturbed state

$$\frac{\partial \rho}{\partial x_j} = \lambda_j\rho_0\alpha\beta; \tag{61}$$

and the change in the density $\delta\rho$ caused by the perturbation θ in the temperature is given by

$$\delta\rho = -\alpha\rho\theta = -\alpha\rho_0(1 + \alpha\beta\lambda_j x_j)\theta. \tag{62}$$

Consequently, while the first-order terms on the left-hand side of equation (58) occur with the factor α, the corresponding terms on the right-hand side do not have this factor; and since $\alpha \sim 10^{-3} - 10^{-4}$, we may neglect the term $\lambda_j u_j \alpha\beta$ on the left-hand side and *deduce* the solenoidal character of $\mathbf{u}$. Making use of this fact and remembering that the dissipation term Φ is of the second order in $\mathbf{u}$ in any case, we obtain for θ the equation

$$\frac{\partial \theta}{\partial t} = \beta \lambda_j u_j + \kappa \nabla^2 \theta; \tag{63}$$

this is the same equation as (56).

Considering next equation (60), we first observe that we may treat μ as a constant; for, the variations in μ must be of order $\alpha \delta\rho$ and we may ignore variations of this order when they occur multiplied by u_j. Thus we obtain the linearized equation

$$\frac{\partial u_i}{\partial t} = -\frac{1}{\rho} \frac{\partial}{\partial x_i} \delta p + \frac{\mu}{\rho} \nabla^2 u_i + g\alpha\theta\lambda_i; \tag{64}$$

and in this equation we may clearly write ρ_0 in place of ρ. We thus recover the equations which were obtained earlier on the Boussinesq approximation.

Returning to equations (55)–(57), we eliminate the term $\delta p/\rho_0$ in (55) by applying the operator,

$$\text{curl}_k = \epsilon_{ijk} \frac{\partial}{\partial x_j}, \tag{65}$$

to the k-component of the equation. Letting

$$\omega_i = \epsilon_{ijk} \frac{\partial u_k}{\partial x_j} \tag{66}$$

denote the *vorticity*, we have the equation

$$\frac{\partial \omega_i}{\partial t} = g\alpha\epsilon_{ijk} \frac{\partial \theta}{\partial x_j} \lambda_k + \nu \nabla^2 \omega_i. \tag{67}$$

Taking the curl of this equation once again, we have

$$\frac{\partial}{\partial t} \epsilon_{ijk} \frac{\partial \omega_k}{\partial x_j} = g\alpha \epsilon_{ijk} \epsilon_{klm} \frac{\partial^2 \theta}{\partial x_l \partial x_j} \lambda_m + \nu \nabla^2 \epsilon_{ijk} \frac{\partial \omega_k}{\partial x_j}. \tag{68}$$

Making use of the identity

$$\epsilon_{ijk} \epsilon_{klm} = \delta_{il} \delta_{jm} - \delta_{im} \delta_{jl}, \tag{69}$$

we find

$$\epsilon_{ijk} \frac{\partial \omega_k}{\partial x_j} = \epsilon_{ijk} \epsilon_{klm} \frac{\partial^2 u_m}{\partial x_j \partial x_l} = \frac{\partial}{\partial x_i} \left(\frac{\partial u_j}{\partial x_j} \right) - \nabla^2 u_i = -\nabla^2 u_i. \tag{70}$$

Similarly,
$$\epsilon_{ijk} \epsilon_{klm} \frac{\partial^2 \theta}{\partial x_j \partial x_l} \lambda_m = \lambda_j \frac{\partial^2 \theta}{\partial x_j \partial x_i} - \lambda_i \nabla^2 \theta. \tag{71}$$

Thus, equation (68) becomes

$$\frac{\partial}{\partial t} \nabla^2 u_i = g\alpha \left(\lambda_i \nabla^2 \theta - \lambda_j \frac{\partial^2 \theta}{\partial x_i \partial x_j} \right) + \nu \nabla^4 u_i. \tag{72}$$

Now multiplying equations (67) and (72) by λ_i, we get

$$\frac{\partial \zeta}{\partial t} = \nu \nabla^2 \zeta \tag{73}$$

and

$$\frac{\partial}{\partial t} \nabla^2 w = g\alpha\left(\frac{\partial^2 \theta}{\partial x^2} + \frac{\partial^2 \theta}{\partial y^2}\right) + \nu \nabla^4 w, \tag{74}$$

where

$$\zeta = \lambda_j \omega_j \quad \text{and} \quad w = \lambda_j u_j \tag{75}$$

are the z-components of the vorticity and the velocity. We also have the equation (cf. equation (56))

$$\frac{\partial \theta}{\partial t} = \beta w + \kappa \nabla^2 \theta. \tag{76}$$

Equations (73), (74), and (76) are the required perturbation equations. We must seek solutions of these equations which satisfy certain boundary conditions. We shall now formulate these conditions.

(a) The boundary conditions

The fluid is confined between the planes $z = 0$ and $z = d$; on these two planes certain boundary conditions must be satisfied. Regardless of the nature of these bounding surfaces, we must require

$$\theta = 0 \quad \text{and} \quad w = 0 \qquad \text{for} \qquad z = 0 \quad \text{and} \quad d; \tag{77}$$

for, the surfaces $z = 0$ and d are maintained at constant temperatures and as such they can suffer no change; and it is also clear that the normal component of the velocity must vanish on these surfaces. There are two further boundary conditions which, however, depend on the nature of the surfaces at $z = 0$ and d.

We shall distinguish two kinds of bounding surfaces: *rigid surfaces* on which no slip occurs and *free surfaces* on which no tangential stresses act.

Consider first a rigid surface. The condition that no slip occurs on this surface implies that not only w, but also the horizontal components of the velocity, u and v, vanish. Thus

$$u = 0 \quad \text{and} \quad v = 0 \quad \text{in addition to} \quad w = 0 \quad \text{on a rigid surface.} \tag{78}$$

Since this condition must be satisfied for all x and y on the surface, it follows from the equation of continuity,

$$\frac{\partial u}{\partial x} + \frac{\partial v}{\partial y} + \frac{\partial w}{\partial z} = 0, \tag{79}$$

that

$$\frac{\partial w}{\partial z} = 0 \quad \text{on a rigid surface.} \tag{80}$$

The conditions on a free surface are

$$P_{xz} = P_{yz} = 0. \tag{81}$$

Since the isotropic term $-p\delta_{ij}$ has no transverse component, the condition (81) is equivalent to the vanishing of the components p_{xz} and p_{yz} of the viscous stress tensor:†

$$p_{xz} = \mu\left(\frac{\partial u}{\partial z}+\frac{\partial w}{\partial x}\right) \quad \text{and} \quad p_{yz} = \mu\left(\frac{\partial v}{\partial z}+\frac{\partial w}{\partial y}\right). \tag{82}$$

Since w vanishes (for all x and y) on the bounding surface, it follows from (82) that

$$\frac{\partial u}{\partial z} = \frac{\partial v}{\partial z} = 0 \quad \text{on a free surface.} \tag{83}$$

From the equation of continuity (79) differentiated with respect to z, we conclude that

$$\frac{\partial^2 w}{\partial z^2} = 0 \quad \text{on a free surface.} \tag{84}$$

The boundary conditions on the normal component of the vorticity ζ can be deduced from the foregoing. Since

$$\zeta = \frac{\partial v}{\partial x}-\frac{\partial u}{\partial y}, \tag{85}$$

it follows from equations (78) and (83) that

$$\zeta = 0 \quad \text{on a rigid surface} \tag{86}$$

and

$$\frac{\partial \zeta}{\partial z} = 0 \quad \text{on a free surface.} \tag{87}$$

10. The analysis into normal modes

According to the general ideas described in Chapter I (§ 3), we must analyse an arbitrary disturbance into a complete set of normal modes and examine the stability of each of these modes, individually. For the problem on hand, the analysis can be made in terms of two-dimensional periodic waves of assigned wave numbers. Thus, we ascribe to all quantities describing the perturbation a dependence on x, y, and t of the form

$$\exp[i(k_x x+k_y y)+pt], \tag{88}$$

where

$$k = \sqrt{(k_x^2+k_y^2)} \tag{89}$$

is the wave number of the disturbance and p is a constant (which can be complex). As we have explained in § 3, the solution of the stability problem requires the specification of the states, for each k, which are characterized by the real part of p being zero.

† It will be observed that we are applying the conditions (81) on the undisplaced surface; in so doing we are ignoring, for example, the effects arising from the existence of gravity waves (see Chapter X, § 94 *e*). If one should want to include such effects, then the conditions (81) should be applied to the surface as displaced by the perturbation.

In accordance with the remarks in the preceding paragraph, we suppose that the perturbations θ, w, and ζ have the forms:

$$\begin{aligned} w &= W(z)\exp[i(k_x x+k_y y)+pt], \\ \theta &= \Theta(z)\exp[i(k_x x+k_y y)+pt], \\ \zeta &= Z(z)\exp[i(k_x x+k_y y)+pt]. \end{aligned} \tag{90}$$

For functions with this dependence on x, y, and t,

$$\frac{\partial}{\partial t} = p, \quad \frac{\partial^2}{\partial x^2}+\frac{\partial^2}{\partial y^2} = -k^2, \quad \text{and} \quad \nabla^2 = \frac{d^2}{dz^2}-k^2; \tag{91}$$

and equations (73), (74), and (76) become

$$p\left(\frac{d^2}{dz^2}-k^2\right)W = -g\alpha k^2\Theta+\nu\left(\frac{d^2}{dz^2}-k^2\right)^2 W, \tag{92}$$

$$p\Theta = \beta W+\kappa\left(\frac{d^2}{dz^2}-k^2\right)\Theta, \tag{93}$$

and

$$pZ = \nu\left(\frac{d^2}{dz^2}-k^2\right)Z. \tag{94}$$

Solutions of these equations must be sought which satisfy the boundary conditions

$$\Theta = 0 \text{ and } W = 0 \quad \text{for} \quad z = 0 \text{ and } d, \tag{95}$$

and

$$\left.\begin{aligned} &Z = 0 \text{ and } \frac{dW}{dz} = 0 \quad \text{on a rigid surface} \\ &\frac{dZ}{dz} = 0 \text{ and } \frac{d^2W}{dz^2} = 0 \quad \text{on a free surface} \end{aligned}\right\}. \tag{96}$$

(the second line of (96) is preceded by "and")

It is convenient to discuss equations (92)–(94) in non-dimensional variables. Choose the units

$$[L] = d \quad \text{and} \quad [T] = d^2/\nu \tag{97}$$

and let

$$a = kd \quad \text{and} \quad \sigma = pd^2/\nu \tag{98}$$

be the wave number and the 'time constant' in these units. We shall, however, let x, y, z stand for the coordinates in the new unit of length d. Equations (92) and (93) become

$$(D^2-a^2)(D^2-a^2-\sigma)W = \left(\frac{g\alpha}{\nu}d^2\right)a^2\Theta \tag{99}$$

and

$$(D^2-a^2-\mathfrak{p}\sigma)\Theta = -\left(\frac{\beta}{\kappa}d^2\right)W, \tag{100}$$

where $D = d/dz$ and $\mathfrak{p}$ $(= \nu/\kappa)$ is the Prandtl number. [Observe that

W and Θ have their usual dimensions: these are not measured in the units which follow from (97).] The associated boundary conditions are

$$\Theta = 0, \quad W = 0 \quad \text{for} \quad z = 0 \quad \text{and} \quad 1, \tag{101}$$

and

$$DW = 0 \text{ for } z = 0 \text{ and } 1 \quad \text{if both bounding surfaces are rigid} \tag{102}$$

or

$$DW = 0 \text{ for } z = 0 \quad \text{and} \quad D^2W = 0 \text{ for } z = 1 \tag{103}$$

if the bottom surface is rigid and the top surface is free.

By eliminating Θ between the equations (99) and (100), we obtain

$$(D^2-a^2)(D^2-a^2-\sigma)(D^2-a^2-\mathfrak{p}\sigma)W = -Ra^2W, \tag{104}$$

where

$$R = \frac{g\alpha\beta}{\kappa\nu}d^4 \tag{105}$$

is the Rayleigh number. An identical equation governs Θ.

(*a*) *The solutions for the horizontal components of the velocity*

Once we have obtained the proper solutions of equations (99) and (100) we can complete the solution of the problem by determining the horizontal components of the velocity as follows.

Express u and v in terms of two functions ϕ and ψ in the manner

$$u = \frac{\partial\phi}{\partial x} - \frac{\partial\psi}{\partial y} \quad \text{and} \quad v = \frac{\partial\phi}{\partial y} + \frac{\partial\psi}{\partial x}. \tag{106}$$

Then

$$-\frac{\partial w}{\partial z} = \frac{\partial u}{\partial x} + \frac{\partial v}{\partial y} = \frac{\partial^2\phi}{\partial x^2} + \frac{\partial^2\phi}{\partial y^2} = -a^2\phi \tag{107}$$

and

$$d\zeta = \frac{\partial v}{\partial x} - \frac{\partial u}{\partial y} = \frac{\partial^2\psi}{\partial x^2} + \frac{\partial^2\psi}{\partial y^2} = -a^2\psi. \tag{108}$$

Hence

$$\phi = \frac{1}{a^2}\frac{\partial w}{\partial z} \quad \text{and} \quad \psi = -\frac{d}{a^2}\zeta, \tag{109}$$

and we can write

$$u = \frac{1}{a^2}\left(\frac{\partial^2 w}{\partial x \partial z} + d\frac{\partial\zeta}{\partial y}\right) = \frac{i}{a^2}(a_x DW + a_y dZ)\exp[i(a_x x + a_y y) + \sigma t] \tag{110}$$

and

$$v = \frac{1}{a^2}\left(\frac{\partial^2 w}{\partial y \partial z} - d\frac{\partial\zeta}{\partial x}\right) = \frac{i}{a^2}(a_y DW - a_x dZ)\exp[i(a_x x + a_y y) + \sigma t]. \tag{111}$$

The foregoing equations relating u and v with the normal components of the velocity and the vorticity are quite general: they are not restricted to this particular problem.

11. The principle of the exchange of stabilities

We shall now show that, for the problem under discussion, the principle of the exchange of stabilities is valid, i.e. σ is real and the marginal states are characterized by $\sigma = 0$.

Let
$$G = (D^2-a^2)W \tag{112}$$
and
$$F = (D^2-a^2)(D^2-a^2-\sigma)W = (D^2-a^2-\sigma)G. \tag{113}$$
According to equation (99), the condition $\Theta = 0$ on the bounding surfaces is equivalent to
$$F = 0 \quad \text{for} \quad z = 0 \quad \text{and} \quad 1. \tag{114}$$
In terms of F, the equation satisfied by W is
$$(D^2-a^2-\mathfrak{p}\sigma)F = -Ra^2W. \tag{115}$$
We shall now prove that σ is real for all positive R.

Multiply equation (115) by F^* (the complex conjugate of F) and integrate over the range of z. We have
$$\int_0^1 F^*(D^2-a^2-\mathfrak{p}\sigma)F\,dz = -Ra^2\int_0^1 F^*W\,dz. \tag{116}$$
By an integration by parts we obtain
$$\int_0^1 F^*D^2F\,dz = -\int_0^1 |DF|^2\,dz, \tag{117}$$
since the integrated part vanishes on account of the boundary condition (114). Hence
$$\int_0^1 \{|DF|^2+(a^2+\mathfrak{p}\sigma)|F|^2\}\,dz = Ra^2\int_0^1 WF^*\,dz. \tag{118}$$
Consider the integral on the right-hand side. We have
$$\int_0^1 WF^*\,dz = \int_0^1 W(D^2-a^2-\sigma^*)G^*\,dz$$
$$= \int_0^1 WD^2G^*\,dz-(a^2+\sigma^*)\int_0^1 WG^*\,dz. \tag{119}$$
By successive integrations by parts, we obtain
$$\int_0^1 WD^2G^*\,dz = -\int_0^1 DWDG^*\,dz = \int_0^1 G^*D^2W\,dz, \tag{120}$$
the integrated part vanishing each time on account of the boundary conditions $W = 0$ and *either* $DW = 0$ *or* $G^* = (D^2-a^2)W^* = 0$

(depending on whether the particular bounding surface is rigid or free). Thus

$$\int_0^1 WF^*\,dz = \int_0^1 G^*\{(D^2-a^2)W-\sigma^*W\}\,dz$$

$$= \int_0^1 |G|^2\,dz-\sigma^*\int_0^1 W(D^2-a^2)W^*\,dz. \tag{121}$$

Again, by an integration by parts, we have

$$\int_0^1 WD^2W^*\,dz = -\int_0^1 |DW|^2\,dz. \tag{122}$$

Combining equations (118), (121), and (122), we obtain

$$\int_0^1 \{|DF|^2+a^2|F|^2+\mathfrak{p}\sigma|F|^2\}\,dz-Ra^2\int_0^1 \{|G|^2+\sigma^*[|DW|^2+a^2|W|^2]\}\,dz = 0. \tag{123}$$

The real and the imaginary parts of this equation must vanish separately; and the vanishing of the imaginary part gives

$$\text{im}(\sigma)\left\{\mathfrak{p}\int_0^1 |F|^2\,dz+Ra^2\int_0^1 [|DW|^2+a^2|W|^2]\,dz\right\} = 0. \tag{124}$$

But the quantity inside the curly brackets is positive definite for $R > 0$. Hence

$$\text{im}(\sigma) = 0. \tag{125}$$

This establishes that σ is real for $R > 0$ and that the principle of the exchange of stabilities is valid for this problem.

12. The equations governing the marginal state and the reduction to a characteristic value problem

Since σ is real for all positive Rayleigh numbers (i.e. for all adverse temperature gradients), it follows that the transition from stability to instability must occur via a stationary state. The equations governing the marginal state are therefore to be obtained by setting $\sigma = 0$ in the relevant equations. Thus, we have (cf. equations (99) and (100))

$$(D^2-a^2)^2W = +\left(\frac{g\alpha}{\nu}d^2\right)a^2\Theta \tag{126}$$

and

$$(D^2-a^2)\Theta = -\left(\frac{\beta}{\kappa}d^2\right)W. \tag{127}$$

Eliminating Θ between these equations, we obtain

$$(D^2-a^2)^3W = -Ra^2W. \tag{128}$$

Solutions of this equation must be sought which satisfy the boundary conditions

$$\left.\begin{array}{l} W = 0, \quad (D^2-a^2)^2W = 0 \quad \text{for} \quad z = 0 \quad \text{and} \quad 1, \\ \text{and } \textit{either } DW \textit{ or } D^2W = 0 \text{ depending on the nature of the} \\ \text{bounding surfaces at } z = 0 \text{ and } 1. \end{array}\right\} \tag{129}$$

Alternatively, we can eliminate W between the equations (126) and (127) and obtain

$$(D^2-a^2)^3\Theta = -Ra^2\Theta, \tag{130}$$

together with the boundary conditions

$$\left.\begin{array}{l} \Theta = 0, \quad (D^2-a^2)\Theta = 0 \quad \text{for} \quad z = 0 \quad \text{and} \quad 1, \\ \text{and } \textit{either } D(D^2-a^2)\Theta \textit{ or } D^2(D^2-a^2)\Theta = 0 \text{ depending on the} \\ \text{nature of the bounding surfaces at } z = 0 \text{ and } 1. \end{array}\right\} \tag{131}$$

In either of the two foregoing formulations (equations (128) and (129) *or* equations (130) and (131)) we have a differential equation of order six and we have to satisfy six boundary conditions, three at $z = 0$ and three at $z = 1$; this we cannot in general do. Only for particular values of R (for given a^2) will the problem allow a non-zero solution. We have thus a *characteristic value problem* for R.

The manner in which the solution for the stability problem is to be carried out is clear. For a given a^2 we must determine the lowest characteristic value for R; the minimum with respect to a^2 of the characteristic values so obtained is the critical Rayleigh number at which instability will manifest itself.

13. The variational principles

We shall now show that the solution of the characteristic value problems formulated in § 12 can be expressed in terms of variational principles. The first of these is due to Pellew and Southwell.

(a) *The first variational principle*

Consider first the characteristic value problem expressed in terms of W. Letting

$$F = (D^2-a^2)^2W = (D^2-a^2)G, \tag{132}$$

we have

$$(D^2-a^2)F = -Ra^2W; \tag{133}$$

and the boundary conditions are:

$$W = 0 \text{ and } F = 0 \text{ for } z = 0 \text{ and } 1; \text{ and } \textit{either } DW = 0 \textit{ or } D^2W = 0 \tag{134}$$

depending on the nature of the bounding surfaces at $z = 0$ and 1.

Let R_j be a characteristic value; and let the solution belonging to R_j

be distinguished by a subscript j. Multiplying the equation satisfied by W_j by F_i (belonging to a different characteristic value R_i) and integrating over the range of z, we have

$$\int_0^1 F_i(D^2-a^2)F_j\,dz = -R_j a^2 \int_0^1 W_j(D^2-a^2)G_i\,dz. \tag{135}$$

By a sequence of integrations by parts similar to that followed in § 11, we obtain from equation (135) the result (cf. equation (123))

$$\int_0^1 (DF_i\,DF_j+a^2F_i\,F_j)\,dz = R_j a^2 \int_0^1 G_i\,G_j\,dz. \tag{136}$$

Interchanging i and j in this last equation, we have

$$\int_0^1 (DF_j\,DF_i+a^2F_j\,F_i)\,dz = R_i a^2 \int_0^1 G_j\,G_i\,dz. \tag{137}$$

From equations (136) and (137) it is apparent that

$$\int_0^1 G_i\,G_j\,dz = 0 \quad \text{if} \quad i \neq j. \tag{138}$$

The functions G_j belonging to the different characteristic values are, therefore, orthogonal.

When $i = j$, equation (136) gives

$$R_j a^2 \int_0^1 G_j^2\,dz = \int_0^1 [(DF_j)^2+a^2F_j^2]\,dz, \tag{139}$$

which expresses R_j as the ratio of two positive definite integrals. Indeed, we shall show that R given by the formula

$$R = \frac{\int_0^1 [(DF)^2+a^2F^2]\,dz}{a^2\int_0^1 G^2\,dz} = \frac{I_1}{a^2I_2} \quad \text{(say)} \tag{140}$$

has a stationary property when the quantities on the right-hand side are evaluated in terms of the true characteristic functions.

To prove the stationary property, let R be evaluated in accordance with equation (140) in terms of a function W which is arbitrary except for the requirement (apart from boundedness and continuity) that it satisfy the boundary conditions (134). Let δR be the change in R when

W is subjected to a small variation δW which is again compatible with the boundary conditions on W; i.e.

$\delta W = 0$ and $\delta F = 0$ for $z = 0$ and 1; and *either* $D\delta W = 0$ *or* $D^2\delta W = 0$ depending on the nature of the bounding surfaces at $z = 0$ and 1. (141)

According to equation (140),

$$\delta R = \frac{1}{a^2 I_2}\left(\delta I_1 - \frac{I_1}{I_2}\delta I_2\right) = \frac{1}{a^2 I_2}(\delta I_1 - Ra^2\delta I_2), \tag{142}$$

where

$$\delta I_1 = 2\int_0^1 (DFD\delta F + a^2 F\delta F)\,dz \tag{143}$$

and

$$\delta I_2 = 2\int_0^1 [(D^2-a^2)W][(D^2-a^2)\,\delta W]\,dz \tag{144}$$

are the variations in I_1 and I_2 consequent to the variation δW in W. After a sequence of integrations by parts, we readily find

$$\delta I_1 = -2\int_0^1 \delta F(D^2-a^2)F\,dz \tag{145}$$

and

$$\delta I_2 = 2\int_0^1 W(D^2-a^2)^2\,\delta F\,dz = 2\int_0^1 W\,\delta F\,dz. \tag{146}$$

Thus

$$\delta R = -\frac{2}{a^2 I_2}\int_0^1 \delta F\{(D^2-a^2)F + Ra^2 W\}\,dz. \tag{147}$$

From (147) it follows that $\delta R = 0$ *if*

$$(D^2-a^2)F = -Ra^2 W; \tag{148}$$

and, conversely, if $\delta R = 0$ *for any arbitrary* $\delta F\,[= (D^2-a^2)^2\delta W]$ *compatible with the boundary conditions of the problem, then equation* (148) *must hold and the function* F *in terms of which* R *was initially calculated must have been a solution of the characteristic value problem.*

The foregoing arguments establish the stationary property of the characteristic values when they are regarded as given by equation (140). We shall now show, further, that the lowest characteristic value of R is, indeed, a true minimum.

We have seen that the functions G_j form an orthogonal set. We shall suppose that they are also normalized so that

$$\int_0^1 G_i\,G_j\,dz = \delta_{ij}. \tag{149}$$

Let

$$G = (D^2-a^2)W, \tag{150}$$

where W is an arbitrary, continuous, bounded function which satisfies the boundary conditions of the problem. We shall suppose that G can be expanded in terms of the basic set of functions G_j; thus,

$$G = \sum_{j=1}^{\infty} A_j G_j, \tag{151}$$

where
$$A_j = \int_0^1 G G_j \, dz. \tag{152}$$

Suppose G is normalized: then

$$1 = \int_0^1 G^2 \, dz = \sum_{j=1}^{\infty} \sum_{k=1}^{\infty} A_j A_k \int_0^1 G_j G_k \, dz = \sum_{j=1}^{\infty} A_j^2. \tag{153}$$

Associated with the expansion (151) for G, we have

$$W = \sum_{j=1}^{\infty} A_j W_j \tag{154}$$

and
$$F = \sum_{j=1}^{\infty} A_j (D^2-a^2)^2 W_j = \sum_{j=1}^{\infty} A_j (D^2-a^2) G_j. \tag{155}$$

From equation (155) it follows that

$$(D^2-a^2)F = \sum_{j=1}^{\infty} A_j (D^2-a^2)^3 W_j = -a^2 \sum_{j=1}^{\infty} A_j R_j W_j. \tag{156}$$

Multiply this last equation by F and integrate over the range of z. We obtain

$$\begin{aligned}
\int_0^1 F(D^2-a^2)F \, dz &= -a^2 \sum_{j=1}^{\infty} A_j R_j \int_0^1 W_j F \, dz \\
&= -a^2 \sum_{j=1}^{\infty} A_j R_j \left\{ \sum_{k=1}^{\infty} A_k \int_0^1 W_j (D^2-a^2)^2 W_k \, dz \right\} \\
&= -a^2 \sum_{j=1}^{\infty} \sum_{k=1}^{\infty} A_j A_k R_j \int_0^1 (D^2-a^2) W_j (D^2-a^2) W_k \, dz \\
&= -a^2 \sum_{j=1}^{\infty} \sum_{k=1}^{\infty} A_j A_k R_j \int_0^1 G_j G_k \, dz \\
&= -a^2 \sum_{j=1}^{\infty} A_j^2 R_j.
\end{aligned} \tag{157}$$

Thus

$$\int_0^1 F(D^2-a^2)F \, dz = -\int_0^1 [(DF)^2 + a^2 F^2] \, dz = -a^2 \sum_{j=1}^{\infty} A_j^2 R_j, \tag{158}$$

or, making use of (153), we can write

$$\int_0^1 [(DF)^2+a^2F^2]\,dz-a^2R_1 = a^2\Big\{\sum_{j=1}^{\infty} A_j^2\,R_j-R_1\sum_{j=1}^{\infty} A_j^2\Big\}$$

$$= a^2\sum_{j=2}^{\infty} A_j^2(R_j-R_1). \tag{159}$$

It is clear that the last summation in (159) cannot be negative. Hence

$$R_1 a^2 \leqslant \int_0^1 [(DF)^2+a^2F^2]\,dz, \tag{160}$$

where the equality sign can hold if and only if $A_j = 0$ for all j greater than 1. *This proves that the quantity on the right-hand side of* (160) *attains its true minimum when F belongs to R_1.*

(*b*) *The second variational principle*

There is a second variational principle which one can obtain by considering the characteristic value problem expressed in terms of Θ.

Letting
$$G = (D^2-a^2)\Theta, \tag{161}$$

we write the equation governing Θ in the form

$$(D^2-a^2)^2G = -Ra^2\Theta; \tag{162}$$

and the boundary conditions are

$\Theta = 0$ and $G = 0$ for $z = 0$ and 1; and *either* $DG = 0$ *or* $D^2G = 0$ (163)

depending on the nature of the bounding surfaces at $z = 0$ and 1.

Let the solution belonging to a particular characteristic value R_j be distinguished by a subscript j. Multiplying the equation satisfied by Θ_j by G_i (belonging to R_i) and integrating over the range of z, we obtain

$$\int_0^1 G_i(D^2-a^2)^2G_j\,dz = -R_j\,a^2\int_0^1 \Theta_j(D^2-a^2)\Theta_i\,dz. \tag{164}$$

By an integration by parts, the right-hand side becomes

$$R_j\,a^2\int_0^1 [(D\Theta_i)(D\Theta_j)+a^2\Theta_i\Theta_j]\,dz; \tag{165}$$

the integrated parts vanish on account of the boundary condition $\Theta = 0$ for $z = 0$ and 1. Similarly, we obtain after successive integrations by parts

$$\int_0^1 G_i D^2(D^2-a^2)G_j\,dz = -\int_0^1 DG_i\,D(D^2-a^2)G_j\,dz$$

$$= +\int_0^1 D^2G_i\,(D^2-a^2)G_j\,dz; \tag{166}$$

again the integrated parts vanish on account of the boundary conditions on G. Thus, we obtain

$$R_j a^2 \int_0^1 [(D\Theta_i)(D\Theta_j)+a^2\Theta_i\Theta_j]\,dz = \int_0^1 [(D^2-a^2)G_i][(D^2-a^2)G_j]\,dz. \tag{167}$$

From (167) it follows that

$$\int_0^1 [(D^2-a^2)G_i][(D^2-a^2)G_j]\,dz = 0 \quad \text{if } i \neq j; \tag{168}$$

and that (when $i = j$)

$$R_j = \frac{\int_0^1 [(D^2-a^2)G_j]^2\,dz}{a^2 \int_0^1 [(D\Theta_j)^2+a^2\Theta_j^2]\,dz}. \tag{169}$$

Again it can be shown that R given by the foregoing formula has a stationary property and that the true minimum of the quantity on the right-hand side is the lowest characteristic value of R.

14. The thermodynamic significance of the variational principle

We have seen that the Rayleigh number, at which disturbances of an assigned wave number become unstable, is the minimum value which a certain ratio of two positive definite integrals can attain. We shall now show that the quantity which is thus minimized has a simple physical meaning.

First we observe that when marginal conditions prevail, the z-component of the vorticity vanishes. This follows from equation (94); for, when $\sigma = 0$,

$$(D^2-a^2)Z = 0; \tag{170}$$

and this equation does not allow a non-zero solution which satisfies the boundary conditions (cf. equation (96)) on Z. [The vanishing of Z under stationary conditions follows even more generally from equation (73); for, when $\partial\zeta/\partial t = 0$, $\nabla^2\zeta = 0$ and as is well known, Laplace's equation does not allow non-zero solutions which (or, the normal derivatives of which) vanish on a closed boundary.] Hence, in this case, the solutions for the horizontal components of the velocity given by equations (110) and (111) become

$$u = \frac{1}{a^2}\frac{\partial^2 w}{\partial z \partial x} \quad \text{and} \quad v = \frac{1}{a^2}\frac{\partial^2 w}{\partial z \partial y}. \tag{171}$$

To be specific, let
$$w = W(z)\cos a_x x \cos a_y y, \tag{172}$$
where
$$a^2 = a_x^2+a_y^2. \tag{173}$$

Then

$$u = -\frac{DW}{a^2} a_x \sin a_x x \cos a_y y; \qquad v = -\frac{DW}{a^2} a_y \cos a_x x \sin a_y y. \quad (174)$$

Now consider the average rate of viscous dissipation of energy by a unit (vertical) column of the fluid. This is given by†

$$\epsilon_\nu = -\frac{\rho\nu}{d^2} \int_0^1 \{\langle w\nabla^2 w\rangle + \langle u\nabla^2 u\rangle + \langle v\nabla^2 v\rangle\}\, dz$$

$$= -\frac{\rho\nu}{d^2} \int_0^1 \{\langle w(D^2-a^2)w\rangle + \langle u(D^2-a^2)u\rangle + \langle v(D^2-a^2)v\rangle\}\, dz, \quad (175)$$

where angular brackets signify that the quantity enclosed is averaged over the horizontal plane. For w, u, and v given by equations (172) and (174), we have

$$\epsilon_\nu = -\frac{\rho\nu}{4d^2} \int_0^1 \left\{W(D^2-a^2)W + \frac{a_x^2}{a^4} DW(D^2-a^2)DW + \right.$$

$$\left. + \frac{a_y^2}{a^4} DW(D^2-a^2)DW \right\} dz$$

$$= -\frac{\rho\nu}{4a^2d^2} \int_0^1 \{a^2 W(D^2-a^2)W + DW(D^2-a^2)DW\}\, dz. \quad (176)$$

Integrating by parts the second term on the right-hand side, we obtain (cf. equation (132))

$$\epsilon_\nu = \frac{\rho\nu}{4a^2d^2} \int_0^1 [(D^2-a^2)W]^2\, dz = \frac{\rho\nu}{4a^2d^2} \int_0^1 G^2\, dz. \quad (177)$$

Consider next the average rate at which energy is released in a unit column of the fluid by the buoyancy force $g\delta\rho$ $(= -g\alpha\rho\theta)$ acting on the fluid. This is given by

$$\epsilon_g = \rho g\alpha \int_0^1 \langle \theta w\rangle\, dz. \quad (178)$$

Making use of equation (76) (and remembering that in this equation length is *not* measured in the unit d) we can rewrite (178) in the form

$$\epsilon_g = -\frac{\rho g\alpha\kappa}{\beta d^2} \int_0^1 \langle \theta\nabla^2\theta\rangle\, dz = -\frac{\rho g\alpha\kappa}{4\beta d^2} \int_0^1 \Theta(D^2-a^2)\Theta\, dz. \quad (179)$$

† Note that in this equation (as in all equations after (98)) length is measured in the unit d; this is the origin of the factor $1/d^2$ in this equation.

After an integration by parts, we have

$$\epsilon_g = \frac{\rho g\alpha\kappa}{4\beta d^2}\int_0^1 [(D\Theta)^2+a^2\Theta^2]\,dz. \tag{180}$$

On the other hand, according to equations (126) and (132)

$$\Theta = \frac{\nu}{g\alpha a^2 d^2}(D^2-a^2)^2 W = \frac{\nu}{g\alpha a^2 d^2}F. \tag{181}$$

In terms of F, the expression for ϵ_g becomes

$$\epsilon_g = \frac{\rho\kappa\nu^2}{4g\alpha\beta a^4 d^6}\int_0^1 [(DF)^2+a^2F^2]\,dz. \tag{182}$$

In a steady state, the kinetic energy dissipated by viscosity must be balanced by the internal energy released by the buoyancy force (see Appendix I). We must therefore require that

$$\epsilon_\nu = \epsilon_g. \tag{183}$$

Equations (177) and (182) now give

$$\frac{\rho\nu}{4a^2d^2}\int_0^1 G^2\,dz = \frac{\rho\kappa\nu^2}{4g\alpha\beta a^4 d^6}\int_0^1 [(DF)^2+a^2F^2]\,dz; \tag{184}$$

or, alternatively,

$$R = \frac{g\alpha\beta}{\kappa\nu}d^4 = \frac{\int_0^1 [(DF)^2+a^2F^2]\,dz}{a^2\int_0^1 G^2\,dz}. \tag{185}$$

This is the same as equation (140) which led to the first of the two variational principles considered in § 13. We may therefore state the physical content of this variational principle as follows.

Instability occurs at the minimum temperature gradient at which a balance can be steadily maintained between the kinetic energy dissipated by viscosity and the internal energy released by the buoyancy force.

15. Exact solutions of the characteristic value problem

We now return to the characteristic value problem formulated in § 12. We shall obtain the solution for the following three cases.

(*a*) Both the bounding surfaces are free, in which case

$$W = (D^2-a^2)^2W = 0 \quad\text{and}\quad D^2W = 0 \text{ for } z = 0 \text{ and } 1. \tag{186}$$

(*b*) Both the bounding surfaces are rigid, in which case

$$W = (D^2-a^2)^2W = 0 \quad\text{and}\quad DW = 0 \text{ for } z = 0 \text{ and } 1. \tag{187}$$

(*c*) One of the bounding surfaces (say, $z = 0$) is rigid and the other bounding surface ($z = 1$) is free, in which case

$$W = (D^2-a^2)^2W = 0 \text{ for } z = 0 \text{ and } 1; \text{ and } DW = 0 \text{ for } z = 0 \text{ and } D^2W = 0 \text{ for } z = 1. \tag{188}$$

From the point of view of realizability in the laboratory for precise quantitative experiments, case (*b*) is, of course, of the greatest interest. Case (*c*) is of interest when the top surface is kept free for visual observations (as in Bénard's original experiments). Case (*a*) (which was first considered by Rayleigh) can be realized, if at all, only under very artificial conditions (e.g. by having the liquid layer floating on top of a somewhat heavier liquid); nevertheless, the case is of theoretical interest because it allows an explicit solution and clearly this has some advantages.

(*a*) *The solution for two free boundaries*

In this case the boundary conditions (186) require

$$W = D^2W = D^4W = 0 \quad \text{for } z = 0 \text{ and } 1; \tag{189}$$

from the equation satisfied by W (namely (128)), it now follows that $D^6W = 0$ for $z = 0$ and 1. From equation (128) differentiated twice with respect to z, we next conclude that $D^8W = 0$ is also zero for $z = 0$ and 1. By further differentiations of equation (128), we can conclude, successively, that all the even derivatives of W vanish on the boundaries. Thus

$$D^{(2m)}W = 0 \quad \text{for } z = 0 \quad \text{and } 1 \quad \text{and} \quad m = 1, 2,\dots. \tag{190}$$

From this it follows that the required solutions must be

$$W = A \sin n\pi z \quad (n = 1, 2,\dots), \tag{191}$$

where A is a constant and n is an integer. Substitution of this solution in equation (128) leads to the *characteristic equation*

$$R = (n^2\pi^2+a^2)^3/a^2. \tag{192}$$

For a given a^2, the lowest value of R occurs when $n = 1$; then

$$R = \frac{(\pi^2+a^2)^3}{a^2}. \tag{193}$$

The meaning of this last relation is this: for all Rayleigh numbers less than that given by (193), disturbances with the wave number a will be stable; these disturbances will become marginally stable when the Rayleigh number equals the value given by (193); and when the Rayleigh number exceeds the value given by (193), the same disturbances will be unstable. The critical Rayleigh number for the onset of instability is, therefore, determined by the condition

$$\frac{\partial R}{\partial a^2} = 3\frac{(\pi^2+a^2)^2}{a^2} - \frac{(\pi^2+a^2)^3}{a^4} = 0, \tag{194}$$

or,

$$3a^2 = \pi^2+a^2 \quad \text{or} \quad a^2 = \pi^2/2. \tag{195}$$

The corresponding value of R is

$$R_c = \frac{(\frac{3}{2}\pi^2)^3}{\frac{1}{2}\pi^2} = \tfrac{27}{4}\pi^4 = 657{\cdot}5; \tag{196}$$

and the disturbances which will be manifest at marginal stability will be characterized by the wavelength

$$\lambda = \frac{2\pi}{k} = \frac{2\pi}{a}d = 2^{\frac{3}{2}}d. \tag{197}$$

(b) The solution for two rigid boundaries

In view of the symmetry of this problem with respect to the two bounding planes, it will be convenient to translate the origin of z to be midway between the two planes. Then, the fluid will be confined between $z = \pm\frac{1}{2}$; and we have to seek solutions of the equation

$$(D^2-a^2)^3W = -Ra^2W \tag{198}$$

which satisfy the boundary conditions

$$W = DW = (D^2-a^2)^2W = 0 \quad \text{for} \quad z = \pm\tfrac{1}{2}. \tag{199}$$

We may first observe that it follows from the evenness of the operator $(D^2-a^2)^3$, and the identity of the boundary conditions which have to be satisfied at $z = \pm\frac{1}{2}$, that the proper solutions of equation (198) fall into two non-combining groups of even and odd solutions. From general considerations, it is clear that the lowest 'state' will be even with no nodes while the first 'excited state' will be odd with a node at $z = 0$.

It is evident that the general solution of equation (198) can be expressed as a superposition of solutions of the form

$$W = e^{\pm qz}, \tag{200}$$

where q^2 is a root of the equation

$$(q^2-a^2)^3 = -Ra^2. \tag{201}$$

Letting

$$Ra^2 = \tau^3a^6, \tag{202}$$

we find that the roots of equation (201) are given by

$$q^2 = -a^2(\tau-1) \quad \text{and} \quad q^2 = a^2[1+\tfrac{1}{2}\tau(1\pm i\sqrt{3})]; \tag{203}$$

or, taking the square roots, we have the six roots†

$$\pm iq_0, \quad \pm q, \quad \text{and} \quad \pm q^*, \tag{204}$$

where

$$\begin{gathered} q_0 = a(\tau-1)^{\frac{1}{2}}, \\ \left.\begin{aligned} \text{re}(q) &= q_1 = a\{\tfrac{1}{2}\sqrt{(1+\tau+\tau^2)}+\tfrac{1}{2}(1+\tfrac{1}{2}\tau)\}^{\frac{1}{2}} \\ \text{im}(q) &= q_2 = a\{\tfrac{1}{2}\sqrt{(1+\tau+\tau^2)}-\tfrac{1}{2}(1+\tfrac{1}{2}\tau)\}^{\frac{1}{2}} \end{aligned}\right\}. \end{gathered} \tag{205}$$

From (203) we readily obtain the following relations which we shall find useful:

$$(q_0^2+a^2)^2 = a^4\tau^2$$

and

$$(q^2-a^2)^2 = \tfrac{1}{2}a^4\tau^2(-1\pm i\sqrt{3}). \tag{206}$$

(i) *The even solutions*

Considering first the even solutions, we can clearly write

$$W = A_0 \cos q_0 z + A \cosh qz + A^* \cosh q^*z, \tag{207}$$

where A_0 and A are constants, the latter being complex. We must now impose the boundary conditions (199) on this solution.

From (207) we obtain

$$DW = -A_0 q_0 \sin q_0 z + Aq \sinh qz + A^*q^* \sinh q^*z, \tag{208}$$

and

$$(D^2-a^2)^2W = A_0(q_0^2+a^2)^2 \cos q_0 z + A(q^2-a^2)^2 \cosh qz + \\ + A^*(q^{*2}-a^2)^2 \cosh q^*z. \tag{209}$$

Using the relations (206), we can rewrite (209) in the form

$$(D^2-a^2)^2W = \tfrac{1}{2}a^4\tau^2\{2A_0 \cos q_0 z + (i\sqrt{3}-1)A \cosh qz - \\ -(i\sqrt{3}+1)A^* \cosh q^*z\}. \tag{210}$$

The boundary conditions (199), therefore, require

$$\left|\begin{matrix} \cos\frac{1}{2}q_0 & \cosh\frac{1}{2}q & \cosh\frac{1}{2}q^* \\ -q_0 \sin\frac{1}{2}q_0 & q\sinh\frac{1}{2}q & q^*\sinh\frac{1}{2}q^* \\ \cos\frac{1}{2}q_0 & \frac{1}{2}(i\sqrt{3}-1)\cosh\frac{1}{2}q & -\frac{1}{2}(i\sqrt{3}+1)\cosh\frac{1}{2}q^* \end{matrix}\right\|\begin{matrix} A_0 \\ A \\ A^* \end{matrix}\right| = 0. \tag{211}$$

For a non-vanishing solution, the determinant of the matrix in (211) must vanish. This condition is

$$\left\|\begin{matrix} 1 & 1 & 1 \\ -q_0 \tan\frac{1}{2}q_0 & q\tanh\frac{1}{2}q & q^*\tanh\frac{1}{2}q^* \\ 1 & \frac{1}{2}(i\sqrt{3}-1) & -\frac{1}{2}(i\sqrt{3}+1) \end{matrix}\right\| = 0. \tag{212}$$

† An asterisk to a quantity denotes its complex conjugate.

Subtracting the first row from the third row and dividing the result by $-\sqrt{3}/2$, we get

$$\begin{Vmatrix} 1 & 1 & 1 \\ -q_0 \tan \tfrac{1}{2}q_0 & q \tanh \tfrac{1}{2}q & q^* \tanh \tfrac{1}{2}q^* \\ 0 & \sqrt{3}-i & \sqrt{3}+i \end{Vmatrix} = 0. \tag{213}$$

Expanding this last determinant, we have

$$\text{im}\{(\sqrt{3}+i)q \tanh \tfrac{1}{2}q\}+q_0 \tan \tfrac{1}{2}q_0 = 0. \tag{214}$$

TABLE I

The exact characteristic values for the first even and odd modes of instability

	Even solution		*Odd solution*	
a	R	$a(\tau-1)^{\frac{1}{2}}$	R	$a(\tau-1)^{\frac{1}{2}}$
0·0	∞	4·143514	∞	7·332130
1·0	5854·48	4·125906	163127·6	7·323920
2·0	2177·41	4·071204	47005·6	7·299920
3·0	1711·28	3·985000	26146·6	7·262095
3·117	1707·762	3·973639	24982	7·2569
4·0	1879·26	3·885334	19684·6	7·213707
5·0	2439·32	3·789634	17731·5	7·158774
5·365	..	..	17610·39	7·137877
6·0	3417·98	3·706519	17933·0	7·101236
7·0	4918·54	3·637524	19575·8	7·044260
8·0	7084·51	3·581053	22461·5	6·989981
∞	∞	3·141593	∞	6·283185

We can write this equation, alternatively, in the form

$$-q_0 \tan \tfrac{1}{2}q_0 = \text{im}\left\{(\sqrt{3}+i)(q_1+iq_2)\frac{\sinh q_1+i \sin q_2}{\cosh q_1+\cos q_2}\right\}. \tag{215}$$

Simplifying the right-hand side of equation (215), we have

$$-q_0 \tan \tfrac{1}{2}q_0 = \frac{(q_1+q_2\sqrt{3})\sinh q_1+(q_1\sqrt{3}-q_2)\sin q_2}{\cosh q_1+\cos q_2}, \tag{216}$$

where it may be recalled that q_0, q_1, and q_2 are defined in equations (205).

Equation (216) is a transcendental equation relating a and $\tau\;(=\sqrt[3]{(R/a^4)})$, and it must be solved numerically by trial and error. The method is to determine τ for a given a; the corresponding characteristic value of R then follows from equation (202). In this way the solution of this problem was first accomplished by Pellew and Southwell, though the exact solution had been obtained earlier by a somewhat different (but equivalent) method by Low. However, the definitive treatment of the problem is due to Reid and Harris; their results are given in Table I and illustrated in Fig. 2. It will be observed that the function attains its minimum at

$$a = 3{\cdot}117, \text{ where } R = 1707{\cdot}762. \tag{217}$$

The corresponding solutions for W and $F(\propto \Theta)$ are (with A_0 in (207) set equal to 1):

$$W = \cos q_0 z - 0{\cdot}06151664 \cosh q_1 z \cos q_2 z + 0{\cdot}10388700 \sinh q_1 z \sin q_2 z, \tag{218}$$

$$(a^2R)^{-\frac{2}{3}}F = \cos q_0 z + 0{\cdot}12072710 \cosh q_1 z \cos q_2 z + {} + 0{\cdot}001331473 \sinh q_1 z \sin q_2 z,$$

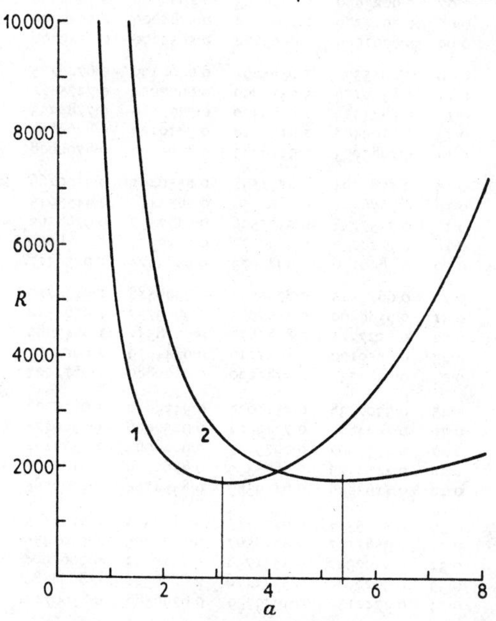

FIG. 2. The Rayleigh numbers at which instability sets in for disturbances of different wave numbers a for the first even (curve labelled 1) and the first odd (curve labelled 2) mode.

where $q_0 = 3{\cdot}973639; \quad q_1 = 5{\cdot}195214; \quad q_2 = 2{\cdot}126096.$ (219)

The solutions W and F are given in Table II and illustrated in Fig. 3.

(ii) *The odd solutions*

Considering next the odd solutions, we now have (instead of (207))

$$W = A_0 \sin q_0 z + A \sinh qz + A^* \sinh q^* z. \tag{220}$$

The characteristic determinant which follows from this solution is

Table II

The solutions for W and $(a^2R)^{-\frac{1}{3}}F$ for the first even and odd modes of instability

z	W_e	$(a^2R)^{-\frac{1}{3}}F_e$	W_o	$(a^2R)^{-\frac{1}{3}}F_o$
0·00	0·9384834	1·1207271	0·0000000	0·0000000
0·01	0·9377396	1·1200748	0·0698930	0·0728662
0·02	0·9355102	1·1181191	0·1394197	0·1453850
0·03	0·9318004	1·1148639	0·2082157	0·2172109
0·04	0·9266188	1·1103153	0·2759205	0·2880024
0·05	0·9199780	1·1044823	0·3421794	0·3574235
0·06	0·9118936	1·0973761	0·4066453	0·4251459
0·07	0·9023850	1·0890103	0·4689809	0·4908506
0·08	0·8914750	1·0794012	0·5288600	0·5542300
0·09	0·8791900	1·0685672	0·5859696	0·6149890
0·10	0·8655597	1·0565291	0·6400120	0·6728475
0·11	0·8506176	1·0433099	0·6907054	0·7275412
0·12	0·8344005	1·0289350	0·7377866	0·7788233
0·13	0·8169488	1·0134318	0·7810122	0·8264663
0·14	0·7983063	0·9968295	0·8201596	0·8702628
0·15	0·7785204	0·9791598	0·8550291	0·9100266
0·16	0·7576421	0·9604558	0·8854448	0·9455945
0·17	0·7357256	0·9407526	0·9112557	0·9768262
0·18	0·7128290	0·9200870	0·9323374	1·0036059
0·19	0·6890136	0·8984973	0·9485924	1·0258427
0·20	0·6643445	0·8760233	0·9599518	1·0434710
0·21	0·6388900	0·8527060	0·9663757	1·0564510
0·22	0·6127224	0·8285876	0·9678542	1·0647689
0·23	0·5859170	0·8037116	0·9644079	1·0684367
0·24	0·5585532	0·7781220	0·9560889	1·0674925
0·25	0·5307138	0·7518638	0·9429810	1·0619998
0·26	0·5024849	0·7249824	0·9252006	1·0520474
0·27	0·4739567	0·6975237	0·9028964	1·0377482
0·28	0·4452229	0·6695339	0·8762509	1·0192390
0·29	0·4163808	0·6410589	0·8454794	0·9966785
0·30	0·3875314	0·6121447	0·8108314	0·9702473
0·31	0·3587797	0·5828367	0·7725905	0·9401455
0·32	0·3302342	0·5531798	0·7310741	0·9065908
0·33	0·3020074	0·5232179	0·6866346	0·8698180
0·34	0·2742159	0·4929939	0·6396586	0·8300749
0·35	0·2469799	0·4625492	0·5905684	0·7876218
0·36	0·2204238	0·4319239	0·5398210	0·7427274
0·37	0·1946762	0·4011558	0·4879092	0·6956670
0·38	0·1698697	0·3702806	0·4353619	0·6467188
0·39	0·1461412	0·3393317	0·3827446	0·5961605
0·40	0·1236321	0·3083394	0·3306597	0·5442657
0·41	0·1024879	0·2773310	0·2797474	0·4913001
0·42	0·0828590	0·2463303	0·2306868	0·4375168
0·43	0·0649003	0·2153572	0·1841964	0·3831516
0·44	0·0487714	0·1844274	0·1410353	0·3284183
0·45	0·0346369	0·1535520	0·1020048	0·2735031
0·46	0·0226663	0·1227369	0·0679496	0·2185585
0·47	0·0130345	0·0919828	0·0397596	0·1636974
0·48	0·0059215	0·0612842	0·0183718	0·1089859
0·49	0·0015130	0·0306295	0·0047726	0·0544365
0·50	0·0000000	0·0000000	0·0000000	0·0000000

(cf. equation (213))

$$\begin{Vmatrix} 1 & 1 & 1 \\ q_0 \cot \tfrac{1}{2}q_0 & q \coth \tfrac{1}{2}q & q^* \coth \tfrac{1}{2}q^* \\ 0 & \sqrt{3}-i & \sqrt{3}+i \end{Vmatrix} = 0, \tag{221}$$

and this leads to the equation

$$q_0 \cot \tfrac{1}{2}q_0 = \operatorname{im}\left\{(\sqrt{3}+i)(q_1+iq_2)\frac{\sinh q_1 - i\sin q_2}{\cosh q_1 - \cos q_2}\right\}. \tag{222}$$

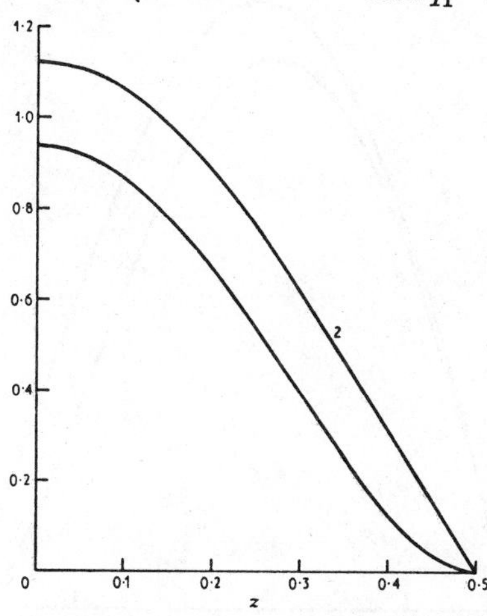

FIG. 3. The proper solutions for W (curve 1) and $(a^2R)^{-\frac{2}{3}}F$ (curve 2) for the state of marginal stability for the case when both bounding surfaces are rigid.

On simplifying this last equation, we obtain

$$q_0 \cot \tfrac{1}{2}q_0 = \frac{(q_1+q_2\sqrt{3})\sinh q_1 - (q_1\sqrt{3}-q_2)\sin q_2}{\cosh q_1 - \cos q_2}. \tag{223}$$

The results of Reid and Harris derived on the basis of the solution (223) are included in Tables I and II; and the solutions for W and F for the lowest odd mode are (see also Fig. 4):

$$W = \sin q_0 z - 0{\cdot}01707389 \sinh q_1 z \cos q_2 z + 0{\cdot}00345645 \cosh q_1 z \sin q_2 z,$$

$$(a^2R)^{-\frac{2}{3}}F = \sin q_0 z + 0{\cdot}01153032 \sinh q_1 z \cos q_2 z + {}$$
$$+ 0{\cdot}01305820 \cosh q_1 z \sin q_2 z, \tag{224}$$

where $$q_0 = 7{\cdot}137877; \quad q_1 = 9{\cdot}110819; \quad q_2 = 3{\cdot}789330. \tag{224$'$}$$

The first excited mode for the value of a $(= 3{\cdot}117)$ at which the lowest

Rayleigh number is obtained is also of interest (cf. Chapter VII, 71 (*d*)). The corresponding solutions for W and Θ are

$$W = \sin q_0 z - 0{\cdot}045302 \sinh q_1 z \cos q_2 z + 0{\cdot}012173 \cosh q_1 z \sin q_2 z,$$

$$(a^2R)^{-\frac{2}{3}}F = \sin q_0 z + 0{\cdot}033193 \sinh q_1 z \cos q_2 z + \\ + 0{\cdot}033146 \cosh q_1 z \sin q_2 z, \quad (225)$$

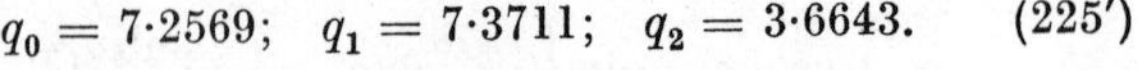

where $$q_0 = 7{\cdot}2569; \quad q_1 = 7{\cdot}3711; \quad q_2 = 3{\cdot}6643. \quad (225')$$

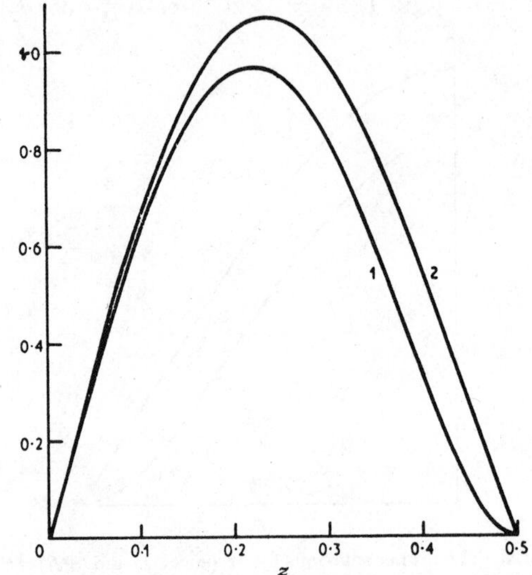

FIG. 4. The proper solutions for W (curve 1) and $(a^2R)^{-\frac{2}{3}}F$ (curve 2) for the state of marginal stability for the case when one bounding surface is rigid and the other is free.

(*c*) *The solution for one rigid and one free boundary*

The solution for the case when the top surface (for example) is free and the bottom surface is rigid can be deduced from the results for the first odd mode given in the last subsection: for it is evident that this odd solution satisfies at $z = 0$ the boundary conditions appropriate for a free surface, i.e. W, D^2W, and $(D^2-a^2)^2W$ all vanish at $z = 0$. Accordingly, an odd solution for the case of two rigid boundaries, applicable to a cell depth d, also provides a solution for the present case applicable, however, to a cell depth $\frac{1}{2}d$. When a and R are defined (as usual) with respect to the depth of the layer, we can infer from the results given in Table II that for the present case

$$a = 5{\cdot}365/2 = 2{\cdot}682_5 \quad \text{and} \quad R_c = 17610{\cdot}39/16 = 1100{\cdot}65. \quad (226)$$

(*d*) *Summary of the results for the three cases*

The results for the three cases we have considered are summarized in Table III.

TABLE III

The parameters characterizing the marginal state for three cases

Nature of the bounding surfaces	R_c	a	$2\pi/a$
Both free	657·511	2·2214	2·828
Both rigid	1707·762	3·117	2·016
One rigid and one free	1100·65	2·682	2·342

16. The cell patterns

We have seen that at the onset of instability the disturbances which will be manifested will be characterized by a particular wave number. Nevertheless, the pattern which the convection cells will exhibit is completely unspecified. This arises from the fact that a given wave vector **a** can be resolved into two orthogonal components in infinitely many ways; and, moreover, the waves corresponding to differing resolutions can be superposed with arbitrary amplitudes and phases. The present theory is unable to distinguish between these infinitely many possibilities. However, if we should specify from considerations of symmetry, or otherwise, the kind of cell patterns that can prevail, then the theory can describe the details of the resulting cell structures. Thus, it may be reasonable to argue that since there are no points or directions in the horizontal plane which are preferred, the entire layer in the marginal state must be tessellated into regular polygons with the cell-walls being surfaces of symmetry. Such complete symmetry will require that the polygons be either equilateral triangles, squares, or regular hexagons.† The structures of the cells in these special cases are worth examining.

First, it may be useful to state precisely what we mean when we say that a certain periodic cell pattern is presented. We mean that there is a unit cell which repeats itself regularly; that the walls of the unit cell are vertical and are surfaces of symmetry; that on the cell-walls the normal gradient of the vertical velocity vanishes; and, finally, that the cells occur contiguously.

† If n is the number of the sides of the regular polygon, then the angle at a vertex $(= \pi(1-2/n))$ must divide 2π an integral number of times. Hence we must have $1-2/n = 2/m$, where m is an integer; this relation can be satisfied only for $n = 3$, 4, and 6, when $m = 6$, 4, and 3, respectively.

The condition that the normal gradient of vertical velocity vanishes on the cell-walls requires

$$(\mathbf{n}\,.\,\nabla_{\perp})w = 0, \tag{227}$$

where $\mathbf{n}$ is a unit vector normal to the cell-walls and

$$\nabla_{\perp} = \left(\frac{\partial}{\partial x}, \frac{\partial}{\partial y}, 0\right). \tag{228}$$

For the problem under consideration, the condition (227) is equivalent to requiring that the component of the horizontal velocity normal to the cell-wall vanishes. To see this, we recall that according to equations (91) and (171)

$$w = F(x,y)W(z) \quad (\nabla_{\perp}^2 F = -a^2 F), \tag{229}$$

and
$$u = \frac{1}{a^2}\frac{\partial^2 w}{\partial x \partial z} \quad \text{and} \quad v = \frac{1}{a^2}\frac{\partial^2 w}{\partial y \partial z}. \tag{230}$$

Therefore,
$$\nabla_{\perp} w = W\nabla_{\perp} F, \qquad \mathbf{u}_{\perp} = \frac{DW}{a^2}\nabla_{\perp} F, \tag{231}$$

and
$$\mathbf{u}_{\perp} = \frac{1}{a^2}\frac{DW}{W}\nabla_{\perp} w. \tag{232}$$

The vanishing of $\nabla_{\perp} w$ along a certain direction implies the vanishing of $\mathbf{u}_{\perp}$ along the same direction; and conversely.

(*a*) *Rolls*

A particularly simple pattern occurs when all the quantities depend on only one of the two horizontal coordinates, say x. In this case the cells are infinitely elongated and it is more appropriate to call them *rolls*.

Let
$$w = W(z)\cos\frac{2\pi}{L}x, \tag{233}$$

where L $(= 2\pi/a)$ is a constant. Corresponding to this solution for w,

$$u = -\frac{DW}{a^2}\frac{2\pi}{L}\sin\frac{2\pi}{L}x \quad \text{and} \quad v = 0. \tag{234}$$

There are, therefore, no horizontal motions along the direction of the rolls; also
$$u = 0 \quad \text{for} \quad x = \pm(n+\tfrac{1}{2})L \text{ and } \pm nL, \tag{235}$$

where n is an integer. For values of x given in (235), the gradient of w normal to the roll is zero; the width of the roll is therefore L; it is the same as the wavelength of the disturbance.

(*b*) *Rectangular and square cells*

Consider the solution

$$w = W(z)\cos\frac{2\pi}{L_x}x\cos\frac{2\pi}{L_y}y, \tag{236}$$

where L_x and L_y are the wavelengths of the disturbance in the x- and the y-directions; they are related to a by

$$a^2 = 4\pi^2\left(\frac{1}{L_x^2}+\frac{1}{L_y^2}\right). \tag{237}$$

According to the solution (236),

$$\begin{aligned} w &= W(z) \text{ at the points, } (\pm nL_x, 0),\ [\pm(n+\tfrac{1}{2})L_x, \pm\tfrac{1}{2}L_y], \text{ etc.,} \\ w &= -W(z) \text{ at the points, } [\pm(n+\tfrac{1}{2})L_x, 0],\ (\pm nL_x, \pm\tfrac{1}{2}L_y), \text{ etc.,} \end{aligned} \tag{238}$$

and $\quad w = 0$ along the lines, $x = \pm\frac{1}{4}L_x, \pm\frac{3}{4}L_x$, etc.; and

$$y = \pm\tfrac{1}{4}L_y,\ \pm\tfrac{3}{4}L_y, \text{ etc.,} \tag{239}$$

where n is an integer.

The horizontal components of the velocity are given by

$$u = -\frac{DW}{a^2}\frac{2\pi}{L_x}\sin\frac{2\pi}{L_x}x\cos\frac{2\pi}{L_y}y,$$

$$v = -\frac{DW}{a^2}\frac{2\pi}{L_y}\cos\frac{2\pi}{L_x}x\sin\frac{2\pi}{L_y}y. \tag{240}$$

Hence, $\quad u = 0$ along the lines $x = 0, \pm\frac{1}{2}L_x, \pm L_x$, etc.; and

$$y = \pm\tfrac{1}{4}L_y,\ \pm\tfrac{3}{4}L_y, \text{ etc.,} \tag{241}$$

and $\quad v = 0$ along the lines $y = 0, \pm\frac{1}{2}L_y, \pm L_y$, etc.; and

$$x = \pm\tfrac{1}{4}L_x,\ \pm\tfrac{3}{4}L_x, \text{ etc.}$$

From the proportionality of $\mathbf{u}_\perp$ and $\nabla_\perp w$, it follows that

$$\partial w/\partial x = 0 \text{ along the lines } u \text{ is zero,}$$

and

$$\partial w/\partial y = 0 \text{ along the lines } v \text{ is zero.} \tag{242}$$

The cells are, therefore, rectangles with sides of lengths L_x and L_y; and the pattern of the motions in the cell, centred at $x = 0$ and $y = 0$, and bounded by $x = \pm\frac{1}{2}L_x$ and $y = \pm\frac{1}{2}L_y$, is repeated in all the others.

From the expressions for u and v given in (240), we find

$$L_x u \pm L_y v = -\frac{2\pi}{a^2}DW\sin 2\pi\left(\frac{x}{L_x}\pm\frac{x}{L_y}\right); \tag{243}$$

and it follows from this equation that

$$L_x u + L_y v = 0 \quad \text{for} \quad x/L_x + y/L_y = 0 \quad \text{and} \quad \pm\tfrac{1}{2}, \tag{244}$$

and

$$L_x u - L_y v = 0 \quad \text{for} \quad x/L_x - y/L_y = 0 \quad \text{and} \quad \pm\tfrac{1}{2}; \tag{245}$$

in other words, the streamlines intersect a principal diagonal in directions which are normal to the other.

The streamlines of the motion in the horizontal plane are determined by the equation

$$\frac{dx}{dy} = \frac{u}{v} = \frac{L_y}{L_x}\frac{[\sin(2\pi x/L_x)][\cos(2\pi y/L_y)]}{[\cos(2\pi x/L_x)][\sin(2\pi y/L_y)]}. \tag{246}$$

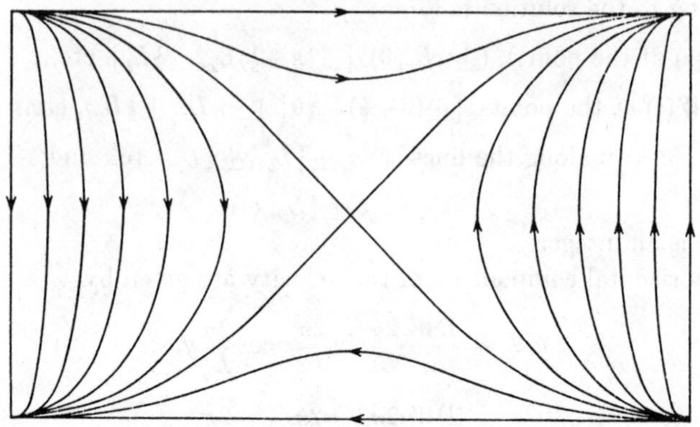

FIG. 5. The streamlines in the horizontal plane for a rectangular cell.

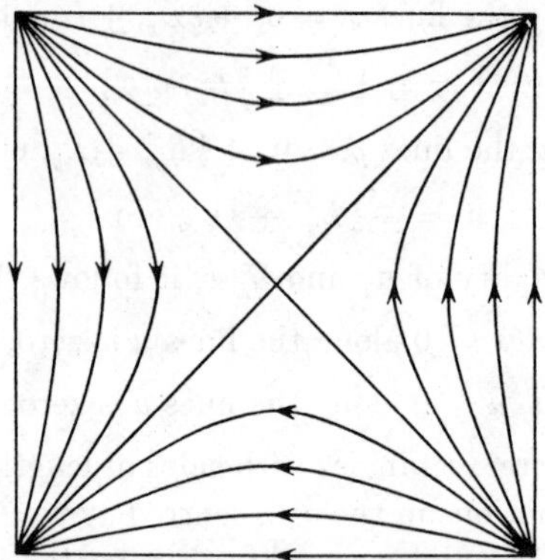

FIG. 6. The streamlines in the horizontal plane for a square cell.

On integrating this equation, we obtain

$$\sin\frac{2\pi x}{L_x} = \text{constant}\left(\sin\frac{2\pi y}{L_y}\right)^{L_y^2/L_x^2}. \tag{247}$$

The streamlines derived from this equation are illustrated in Fig. 5 for the case $L_y = \sqrt{3}L_x/2$.

The solution for the case of square cells can be obtained from the foregoing by setting $L_x = L_y$. The corresponding streamlines of the motion are illustrated in Fig. 6.

(c) Hexagonal cells

The solution for the hexagonal pattern was discovered by Christopherson. His solution is

$$w = \tfrac{1}{3}W(z)\left\{2\cos\frac{2\pi}{L\sqrt{3}}x\cos\frac{2\pi}{3L}y+\cos\frac{4\pi}{3L}y\right\}. \tag{248}$$

Alternative forms of this solution are

$$w = \tfrac{1}{3}W(z)\left\{\cos\frac{4\pi}{3L}\left(\frac{\sqrt{3}}{2}x+\tfrac{1}{2}y\right)+\cos\frac{4\pi}{3L}\left(\frac{\sqrt{3}}{2}x-\tfrac{1}{2}y\right)+\cos\frac{4\pi}{3L}y\right\} \tag{249}$$

and

$$w = \tfrac{1}{3}W(z)\left\{4\cos\frac{2\pi}{3L}\left(\frac{\sqrt{3}}{2}x+\tfrac{1}{2}y\right)\cos\frac{2\pi}{3L}\left(\frac{\sqrt{3}}{2}x-\tfrac{1}{2}y\right)\cos\frac{2\pi}{3L}y-1\right\}. \tag{250}$$

We shall presently show that L measures the side of the hexagon. But first we shall verify that the solution (248) does, indeed, represent a periodic disturbance of a given total wave number a. We have

$$\begin{aligned}\nabla_\perp^2 w &= -\tfrac{1}{3}W(z)\left\{2\left(\frac{2\pi}{3L}\right)^2(3+1)\cos\frac{2\pi}{L\sqrt{3}}x\cos\frac{2\pi}{3L}y+\left(\frac{4\pi}{3L}\right)^2\cos\frac{4\pi}{3L}y\right\}\\ &= -\left(\frac{4\pi}{3L}\right)^2 w.\end{aligned} \tag{251}$$

Hence,

$$a = 4\pi/3L. \tag{252}$$

The basic hexagonal symmetry of the solution expressed by (249) becomes apparent when we write

$$x = \varpi\cos\theta \quad\text{and}\quad y = \varpi\sin\theta. \tag{253}$$

Then

$$\frac{\sqrt{3}}{2}x\pm\tfrac{1}{2}y = \varpi\sin(60^\circ\pm\theta), \tag{254}$$

and we can rewrite equation (249) in the form

$$w = \tfrac{1}{3}W(z)\left\{\cos\left[\frac{4\pi\varpi}{3L}\sin(\theta+60^\circ)\right]+\cos\left[\frac{4\pi\varpi}{3L}\sin(\theta+120^\circ)\right]+\right.$$
$$\left.+\cos\left[\frac{4\pi\varpi}{3L}\sin\theta\right]\right\}, \tag{255}$$

from which it is apparent that

$$w(\varpi,\theta) \equiv w(\varpi,\theta+60^\circ). \tag{256}$$

In addition to this invariance for rotation by 60° about the origin, the solution exhibits periodicity in the x- and y-directions as well; thus

$$w(x+nL\sqrt{3},\ y+3mL) \equiv w(x,y), \tag{257}$$

where n and m are (arbitrary) integers. The wavelength of the periodicity in the y-direction is seen to be $\sqrt{3}$ times the wavelength in the x-direction.

We next observe that according to equation (250)

$$w(0) = W(z); \tag{258}$$

also, along the three pairs of lines,

$$y = \pm\tfrac{3}{4}L, \quad x\sqrt{3}+y = \pm\tfrac{3}{2}L, \quad x\sqrt{3}-y = \pm\tfrac{3}{2}L, \tag{259}$$

which describes the hexagon $GHIJKL$ (see Fig. 7a),

$$w = -\tfrac{1}{3}W(z); \tag{260}$$

and at the six points,

$$\left(\frac{\sqrt{3}}{2}L,\ \pm\tfrac{1}{2}L\right), \quad (0,\ \pm L), \quad \left(-\frac{\sqrt{3}}{2}L,\ \pm\tfrac{1}{2}L\right), \tag{261}$$

which are the vertices of the hexagon $ABCDEF$,

$$w = -\tfrac{1}{2}W(z). \tag{262}$$

The horizontal components of the velocity are given by

$$u = \frac{1}{a^2}\frac{\partial^2 w}{\partial x \partial z} = -\frac{DW}{3a^2}\frac{4\pi}{L\sqrt{3}}\sin\frac{2\pi}{L\sqrt{3}}x\cos\frac{2\pi}{3L}y, \tag{263}$$

and $$v = \frac{1}{a^2}\frac{\partial^2 w}{\partial y \partial z} = -\frac{DW}{3a^2}\frac{4\pi}{3L}\left(\cos\frac{2\pi}{L\sqrt{3}}x+2\cos\frac{2\pi}{3L}y\right)\sin\frac{2\pi}{3L}y.$$

From these equations it follows that

$$u = 0 \quad \text{for} \quad x = 0, \quad x = \pm\frac{\sqrt{3}}{2}L \quad \text{and} \quad y = \pm\tfrac{3}{4}L,$$

and $$v = 0 \quad \text{for} \quad y = 0 \quad \text{and} \quad y = \pm\tfrac{3}{2}L. \tag{264}$$

There is thus no flow *across* OD ($u = 0$ for $x = 0$), BC ($u = 0$ for $x = L\sqrt{3}/2$), and OI ($v = 0$ for $y = 0$). From the invariance of the solution for rotations by 60°, we conclude that there is no flow across *any* of the sides of the six equilateral triangles which form the hexagon $ABCDEF$, nor across any of the perpendiculars drawn from the centre of the hexagon to the sides. From the proportionality of $\mathbf{u}_\perp$ and $\nabla_\perp w$, it follows that the walls of the prisms, whose sections are the six equilateral triangles, are surfaces of symmetry. Again, from the fact that the flow is normal to KJ ($u = 0$ for $y = 3L/4$), we conclude that there is no flow *along* the sides of the inscribed hexagon $GHIJKL$; and this is consistent with the fact that w is constant along these same sides.

In Fig. 7b the flow pattern in the horizontal plane is illustrated.

The nature of the flow in the vertical planes is somewhat more complicated. However, in the planes $x = 0$ and $y = 0$ they are two-dimensional and can be described simply. Thus

$$\left.\begin{aligned} u &= 0 \\ v &= -\frac{DW}{3a^2}\frac{4\pi}{3L}\left(1+2\cos\frac{2\pi}{3L}y\right)\sin\frac{2\pi}{3L}y \\ w &= \tfrac{1}{3}W\left(2\cos\frac{2\pi}{3L}y+\cos\frac{4\pi}{3L}y\right) \end{aligned}\right\} (x = 0), \tag{265}$$

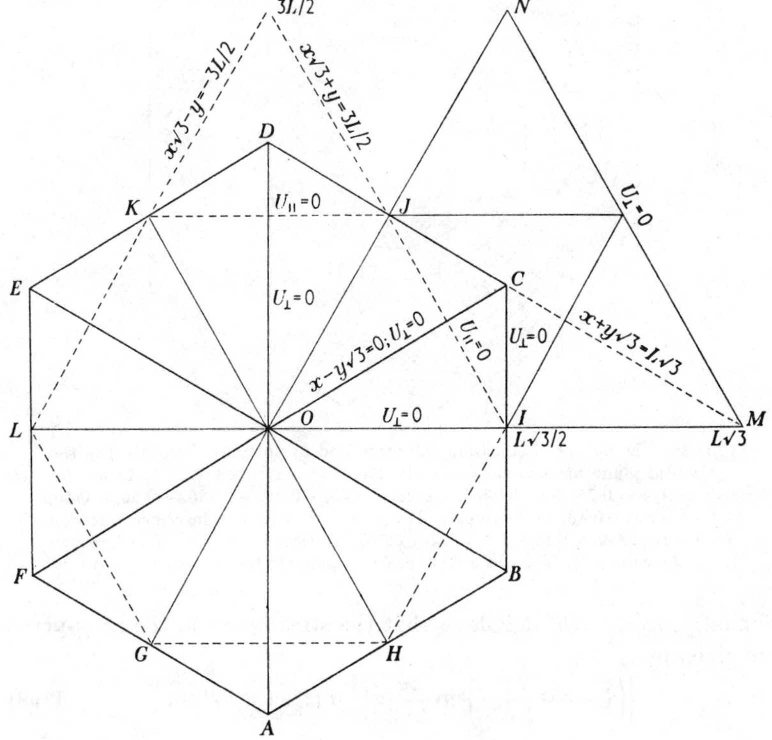

FIG. 7*a*. The geometry of the hexagonal cells.

and

$$\left.\begin{aligned} u &= -\frac{DW}{3a^2}\frac{4\pi}{L\sqrt{3}}\sin\frac{2\pi}{L\sqrt{3}}x \\ v &= 0 \\ w &= \tfrac{1}{3}W\left(2\cos\frac{2\pi}{L\sqrt{3}}x+1\right) \end{aligned}\right\} (y = 0). \tag{266}$$

For v and w given by (265), the equation, $dy/v = dz/w$, governing the streamlines in the (y,z)-plane can be readily integrated to give

$$\left\{\frac{1+2\cos(2\pi y/3L)}{1+\cos(2\pi y/3L)}\sin\frac{2\pi}{3L}y\right\}^2 W(z) = \text{constant.} \tag{267}$$

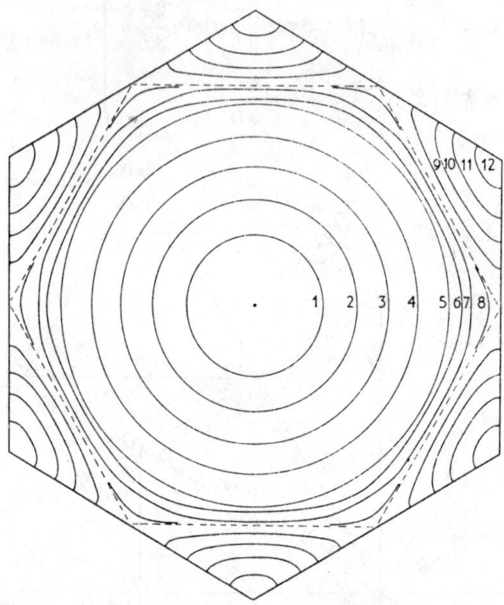

FIG. 7*b*. The curves of constant w (normalized to unity at the centre) in the horizontal plane for a hexagonal cell. The curves labelled 1, 2,..., 12 are for values of $w = 0{\cdot}75$, $0{\cdot}50$, $0{\cdot}25$, 0, $-0{\cdot}20$, $-0{\cdot}25$, $-0{\cdot}30$, $-0{\cdot}325$, $-0{\cdot}36$, $-0{\cdot}40$, $-0{\cdot}44$, and $-0{\cdot}48$, respectively. The value of w on the inscribed hexagon (shown by broken lines) is $-\frac{1}{3}$; and at the vertices of the principal hexagon it has the value $-\frac{1}{2}$. The streamlines are orthogonal to the curves of constant w.

Similarly, from (266) it follows that the streamlines in the (x,z)-plane are given by

$$\left\{\left(1-\cos\frac{2\pi}{L\surd 3}x\right)\sin\frac{2\pi}{L\surd 3}x\right\}^{\frac{2}{3}} W(z) = \text{constant.} \tag{268}$$

The streamlines of the flow derived from equations (267) and (268) are illustrated in Fig. 8.

(*d*) *Triangular cells*

This case does not need a separate discussion. The pattern which we considered as hexagonal can be considered, equally, as triangular. The unit cell is then represented by the equilateral triangle OMN (see Fig. 7*a*) whose sides are of length $L\surd 3$. Along the sides of this triangle there are no transverse motions; and at its vertices w takes the same value (which is twice that at the centroid C and of the opposite sign).

(e) *More general cell patterns*

A generalization of Christopherson's solution was discovered by Bisshopp. Bisshopp's solution appears to give the most general cell

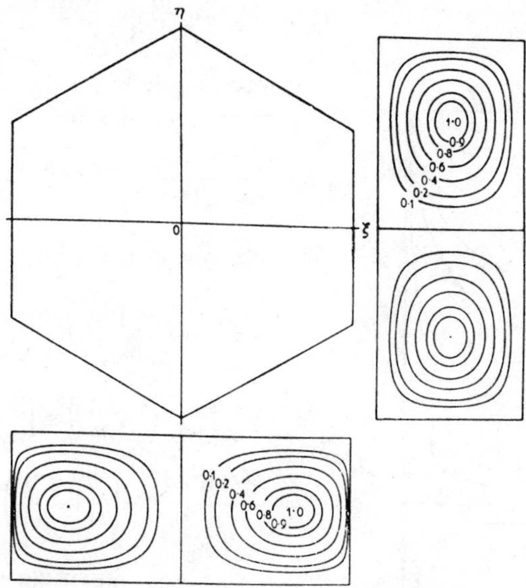

FIG. 8. The streamlines in a hexagonal cell in the planes of symmetry at the onset of instability for the first even mode. The streamlines (normalized to unity at the maxima) are computed in accordance with equations (267) (the patterns on the right) and (268) (the patterns at the bottom); the curves are labelled by the values of the constants to which they refer.

patterns which exhibit definite periodicities in the x- and y-directions and are characterized by a given total wave number.

Starting with a term having the periodicity of the assigned wave number a in a particular direction (say, the y-direction), we write the solution in the form

$$w = W(z)\left\{A\cos\left[a\left(1-\frac{1}{m^2}\right)^{\frac{1}{2}}x\right]\cos\frac{a}{m}y+\cos a(y-y_0)\right\}, \qquad (269)$$

where m is an integer greater than 1 and A and y_0 are constants. The solution (269) is constructed so as to satisfy two conditions: *first*, any periodic term which we may add to $\cos a(y-y_0)$ has a wave number in the y-direction which is a divisor of a (for, only then can we include the periodicity already implied by $\cos a(y-y_0)$); and *second* the added term has a periodicity in the x-direction of a wave number such as to make the total wave number a. These conditions are manifestly satisfied by the solution (269). Thus

$$\nabla_\perp^2 \cos\left[a\left(1-\frac{1}{m^2}\right)^{\frac{1}{2}}x\right]\cos\frac{a}{m}y = -a^2\cos\left[a\left(1-\frac{1}{m^2}\right)^{\frac{1}{2}}x\right]\cos\frac{a}{m}y, \quad (270)$$

and

$$\nabla_\perp^2 w = -a^2 w \quad (271)$$

as required.

The wavelengths of the disturbance in the x- and the y-directions are

$$\lambda_x = \frac{2\pi}{a}\frac{m}{\sqrt{(m^2-1)}} \quad \text{and} \quad \lambda_y = \frac{2\pi}{a}m. \quad (272)$$

The components of the horizontal velocity are

$$u = -\frac{DW}{a}\left(1-\frac{1}{m^2}\right)^{\frac{1}{2}}A \times \sin\left[a\left(1-\frac{1}{m^2}\right)^{\frac{1}{2}}x\right]\cos\frac{a}{m}y, \quad (273)$$

and

$$v = -\frac{DW}{a}\left\{\frac{A}{m}\cos\left[a\left(1-\frac{1}{m^2}\right)^{\frac{1}{2}}x\right]\sin\frac{a}{m}y + \sin a(y-y_0)\right\}. \quad (274)$$

The velocities transverse to the lines

$$x = \pm\tfrac{1}{2}n\lambda_x \quad \text{and} \quad y = \pm\tfrac{1}{2}n\lambda_y \quad (275)$$

(where n is an integer) will vanish if $y_0 = 0$. Accordingly, the solution

$$w = W(z)\left\{A\cos\left[a\left(1-\frac{1}{m^2}\right)^{\frac{1}{2}}x\right]\cos\frac{a}{m}y + \cos ay\right\} \quad (276)$$

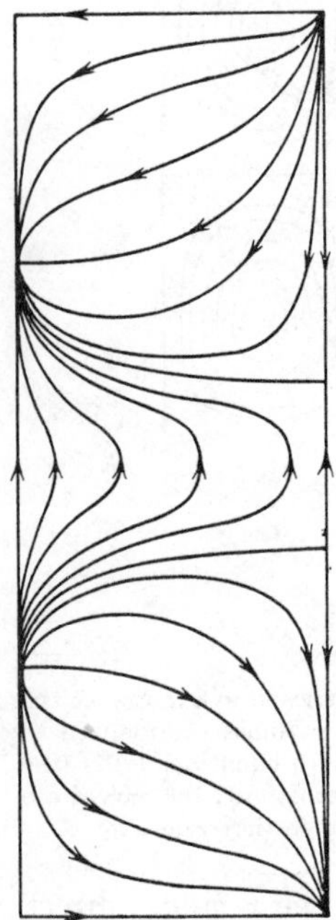

Fig. 9. A complex cell pattern derived from the solution $w = \frac{1}{3}W_0\{\cos[\frac{1}{3}k(x\sqrt{8}+y)] + \cos[\frac{1}{3}k(x\sqrt{8}-y)] + \cos ky\}$.

describes a cell pattern in which the unit cell is a rectangle with sides λ_x and λ_y along the two directions. However, the motions inside the unit cells are far more complicated than in the simple rectangular cells considered in § (b).

Finally, it may be noted that we recover Christopherson's solution (248) as a special case of (276) by setting $A = 2$ and $m = 2$. An example of these more general cell patterns is shown in Fig. 9.

17. The variational solution

In § 15 we have seen how the characteristic value problem underlying the determination of the critical Rayleigh number for the onset of instability can be solved exactly. In this section we shall describe an alternative method based on the variational principles of § 13. While this is necessarily an approximate method, it appears that by a judicious procedure, the method can lead to results of high precision. This is of some importance since most of the other problems we shall encounter do not easily lend themselves to an exact solution; and variational methods appear to be the only practicable ones. It is therefore fortunate that there is this one case in which the results of the approximate method can be compared with those of an exact calculation and some idea gained as to the accuracy which can be attained in the solution of similar problems by the variational method.

Consider then the problem appropriate to the case when the liquid is confined between two rigid boundaries at $z = \pm\frac{1}{2}$. The problem then is to solve the equations

$$(D^2-a^2)^2 W = F \tag{277}$$

and

$$(D^2-a^2)F = -Ra^2 W, \tag{278}$$

together with the boundary conditions

$$F = 0 \quad \text{and} \quad W = DW = 0 \quad \text{for} \quad z = \pm\tfrac{1}{2}. \tag{279}$$

As we have already pointed out in § 15, the solutions of equations (277) and (278) satisfying the boundary conditions (279) are either even or odd. Consider the even solutions. Then F is even and since it is required to vanish at $z = \pm\frac{1}{2}$, we can expand it in a cosine series in the form

$$F = \sum_m A_m \cos[(2m+1)\pi z]. \tag{280}$$

The summation over m in (280) may be assumed to run from zero to infinity; but this is not necessary since we shall consider the coefficients A_m as variational parameters (see below).

With F given by (280), we *determine* W as a solution of the equation

$$(D^2-a^2)^2 W = \sum_m A_m \cos[(2m+1)\pi z] \tag{281}$$

which satisfies the remaining boundary conditions on W. In view of the linearity of equation (281), we can express W as a sum in the form

$$W = \sum_m A_m W_m(z), \tag{282}$$

where W_m is a solution of the equation

$$(D^2-a^2)^2 W_m = \cos[(2m+1)\pi z] \tag{283}$$

which satisfies the boundary conditions

$$W_m = DW_m = 0 \quad \text{for} \quad z = \pm\tfrac{1}{2}; \tag{284}$$

since equation (283) is of the fourth order, such a solution can be found and is in fact uniquely determined. We shall presently obtain the solution explicitly.

The variational principle formulated in § 13 is equivalent to minimizing

$$\int_{-\frac{1}{2}}^{+\frac{1}{2}} [(DF)^2+a^2F^2]\,dz, \tag{285}$$

for such variations of F which preserve the constancy of

$$\int_{-\frac{1}{2}}^{+\frac{1}{2}} [(D^2-a^2)W]^2\,dz. \tag{286}$$

Introducing Ra^2 as a Lagrangian undetermined multiplier, we can equally minimize

$$J = \int_{-\frac{1}{2}}^{+\frac{1}{2}} [(DF)^2+a^2F^2]\,dz - Ra^2 \int_{-\frac{1}{2}}^{+\frac{1}{2}} [(D^2-a^2)W]^2\,dz, \tag{287}$$

for entirely arbitrary variations of F. When we choose for F a *trial function* with a certain number of parameters, the method is to minimize with respect to those parameters. Thus, if the expansion (280) for F is considered as a trial function, then the coefficients of the expansion A_m play the role of the variational parameters and the minimization of J is with respect to them.

When F and W satisfy the boundary conditions required of them and are related as in equation (277), an alternative form for J is

$$J = -\int_{-\frac{1}{2}}^{+\frac{1}{2}} F(D^2-a^2)F\,dz - Ra^2 \int_{-\frac{1}{2}}^{+\frac{1}{2}} WF\,dz. \tag{288}$$

We shall now evaluate the integrals on the right-hand side of (288) when F and W are given by equations (280) and (282). We have

$$-\int_{-\frac{1}{2}}^{+\frac{1}{2}} F(D^2-a^2)F\,dz = \int_{-\frac{1}{2}}^{+\frac{1}{2}} dz \sum_m A_m \cos[(2m+1)\pi z]\times$$

$$\times \sum_n A_n[(2n+1)^2\pi^2+a^2]\cos[(2n+1)\pi z] = \frac{1}{2}\sum_m \frac{A_m^2}{\gamma_{2m+1}}, \tag{289}$$

where

$$\gamma_{2m+1} = \frac{1}{(2m+1)^2\pi^2+a^2}. \tag{290}$$

Next, considering the second integral on the right-hand side of (288), we have

$$\int_{-\frac{1}{2}}^{+\frac{1}{2}} WF\,dz = \sum_n A_n \int_{-\frac{1}{2}}^{+\frac{1}{2}} dz\cos[(2n+1)\pi z]\sum_m A_m W_m(z). \tag{291}$$

Letting
$$(n|m) = \int_{-\frac{1}{2}}^{+\frac{1}{2}} \{\cos[(2n+1)\pi z]\}W_m(z)\,dz, \tag{292}$$

we can write
$$\int_{-\frac{1}{2}}^{+\frac{1}{2}} WF\,dz = \sum_n\sum_m A_n(n|m)A_m. \tag{293}$$

The matrix $(n|m)$ which we have defined in (292) is symmetric: for,

$$\begin{aligned}(n|m) &= \int_{-\frac{1}{2}}^{+\frac{1}{2}} W_m(z)\cos[(2n+1)\pi z]\,dz \\ &= \int_{-\frac{1}{2}}^{+\frac{1}{2}} [(D^2-a^2)^2W_n(z)]W_m(z)\,dz \\ &= \int_{-\frac{1}{2}}^{+\frac{1}{2}} [(D^2-a^2)W_n(z)][(D^2-a^2)W_m(z)]\,dz,\end{aligned} \tag{294}$$

and the symmetry is apparent.

Returning to equation (288) and making use of equations (289) and (293), we have

$$J = \frac{1}{2}\sum_m \frac{A_m^2}{\gamma_{2m+1}} - Ra^2\sum_m\sum_n A_n(n|m)A_m; \tag{295}$$

and we must minimize this expression with respect to the parameters A_n. Therefore, requiring

$$\frac{\partial J}{\partial A_n} = 0 \quad (n = 0, 1,...), \tag{296}$$

we obtain

$$\sum_m \left\{\frac{\delta_{mn}}{2a^2\gamma_{2m+1}R} - (n|m)\right\}A_m = 0 \quad (n = 0, 1,...). \tag{297}$$

This represents a system of linear homogeneous equations for the coefficients A_n. For a non-zero solution of these equations to exist, the determinant of the system must vanish. We thus obtain the condition

$$\left\|\frac{\delta_{mn}}{a^2R\gamma_{2m+1}} - 2(n|m)\right\| = 0 \tag{298}$$

which is in the form of a *characteristic equation* of a matrix.

Restricting ourselves to the first n coefficients $A_1, A_2,..., A_n$ in (280) and setting the others equal to zero is the same as considering a trial function for F with n parameters. The minimization of J is then with respect to these n parameters; and the order of the resulting 'secular' determinant (298) will also be n.

While we have derived the condition (298) as a consequence of the variational principle, it is of some significance to observe that it follows also from equation (278) by equating the Fourier coefficients of either side. Thus substituting for F and W in accordance with equations (280) and (282) in equation (278), we have

$$\sum_m \frac{A_m}{\gamma_{2m+1}} \cos[(2m+1)\pi z] = Ra^2 \sum_m A_m W_m(z). \tag{299}$$

Multiplying this equation by $\cos[(2n+1)\pi z]$ and integrating over the range of z, we obtain

$$\frac{1}{2}\frac{A_n}{\gamma_{2n+1}} = Ra^2 \sum_m (n|m)A_m \quad (n = 0, 1, 2,...); \tag{300}$$

this represents the same set of equations as (297) and leads to the same secular determinant (298). The present method of deriving the condition (298) is applicable even if a variational principle does not underlie the problem. But the existence of a variational principle ensures that by keeping more and more terms in the Fourier expansion (280) for F and solving the secular determinant (298) for R, we approach the true characteristic value *monotonically from above*.

We now return to equation (283) to determine the explicit form of the solution for W_m and of the matrix $(n|m)$.

The general solution of equation (283) which is even in z is

$$W_m = P_m \cosh az + Q_m z \sinh az + \gamma_{2m+1}^2 \cos[(2m+1)\pi z], \tag{301}$$

where γ_{2m+1} is defined in (290) and P_m and Q_m are constants of integration which are to be determined by the boundary conditions (284). These latter conditions give

$$P_m \cosh \tfrac{1}{2}a + \tfrac{1}{2}Q_m \sinh \tfrac{1}{2}a = 0$$

and

$$P_m a \sinh \tfrac{1}{2}a + Q_m(\tfrac{1}{2}a \cosh \tfrac{1}{2}a + \sinh \tfrac{1}{2}a) = (2m+1)\pi\gamma_{2m+1}^2 (-1)^m. \tag{302}$$

Solving these equations for P_m and Q_m, we find

$$P_m = (-1)^{m+1} \frac{(2m+1)\pi\gamma_{2m+1}^2}{a+\sinh a} \sinh \tfrac{1}{2}a,$$

and

$$Q_m = (-1)^m \frac{2(2m+1)\pi\gamma_{2m+1}^2}{a+\sinh a} \cosh \tfrac{1}{2}a. \tag{303}$$

The matrix element $(n|m)$ is now given by

$$(n|m) = \int_{-\frac{1}{2}}^{+\frac{1}{2}} \{P_m \cosh az + Q_m z \sinh az + \gamma_{2m+1}^2 \cos[(2m+1)\pi z]\} \times \\ \times \cos[(2n+1)\pi z]\, dz \quad (304)$$

or,

$$(n|m) = \tfrac{1}{2}\gamma_{2m+1}^2 \delta_{mn} + P_m \int_{-\frac{1}{2}}^{+\frac{1}{2}} \cosh az \cos[(2n+1)\pi z]\, dz + \\ + Q_m \int_{-\frac{1}{2}}^{+\frac{1}{2}} z \sinh az \cos[(2n+1)\pi z]\, dz. \quad (305)$$

The integrals on the right-hand side of (305) are readily evaluated. We find

$$\int_{-\frac{1}{2}}^{+\frac{1}{2}} \cosh az \cos[(2n+1)\pi z]\, dz = 2(-1)^n (2n+1)\pi \gamma_{2n+1} \cosh \tfrac{1}{2}a,$$

$$\int_{-\frac{1}{2}}^{+\frac{1}{2}} z \sinh az \cos[(2n+1)\pi z]\, dz \\ = (-1)^n (2n+1)\pi \gamma_{2n+1} (\sinh \tfrac{1}{2}a - 4a\gamma_{2n+1} \cosh \tfrac{1}{2}a). \quad (306)$$

We thus have

$$(n|m) = \tfrac{1}{2}\gamma_{2m+1}^2 \delta_{mn} + (2n+1)\pi \gamma_{2n+1} (-1)^n \times \\ \times \{2P_m \cosh \tfrac{1}{2}a + Q_m (\sinh \tfrac{1}{2}a - 4a\gamma_{2n+1} \cosh \tfrac{1}{2}a\}. \quad (307)$$

Substituting for P_m and Q_m from (303) in the foregoing equation and simplifying, we find

$$(n|m) = \tfrac{1}{2}\gamma_{2m+1}^2 \delta_{mn} - 8a(-1)^{m+n}(2n+1)(2m+1)\pi^2 \gamma_{2n+1}^2 \gamma_{2m+1}^2 \frac{\cosh^2 \frac{1}{2}a}{\sinh a + a}. \quad (308)$$

This is symmetrical in n and m as required.

The explicit form of the secular determinant (298) is, therefore,

$$\left\| \left(\frac{1}{a^2 R \gamma_{2m+1}} - \gamma_{2m+1}^2 \right) \delta_{mn} + \\ + (-1)^{n+m} 16 a\pi^2 \frac{\cosh^2 \frac{1}{2}a}{\sinh a + a} (2n+1)(2m+1) \gamma_{2n+1}^2 \gamma_{2m+1}^2 \right\| = 0. \quad (309)$$

A first approximation to R will be given by setting the (0, 0) element of the secular matrix equal to zero and ignoring all the others. This corresponds to the choice of $\cos \pi z$ as a trial function for F, i.e. one with *no* variational parameters. The corresponding result is

$$\frac{1}{a^2 R \gamma_1} = \gamma_1^2 - 16 a\pi^2 \gamma_1^4 \frac{\cosh^2 \frac{1}{2}a}{\sinh a + a}, \quad (310)$$

or, since $\gamma_1 = (\pi^2+a^2)^{-1}$,

$$R = \frac{(\pi^2+a^2)^3}{a^2\{1-16a\pi^2\cosh^2\frac{1}{2}a/[(\pi^2+a^2)^2(\sinh a+a)]\}}. \tag{311}$$

This formula gives

$$R = 1715{\cdot}08 \quad \text{for} \quad a = 3{\cdot}117; \tag{312}$$

this should be contrasted with the result (1707·76) of the exact calculation for the same value of a. Thus, even with no variational parameters,

TABLE IV

Rayleigh numbers derived by the variational method

(i) *The even mode*

$a = 3{\cdot}117$

First approximation	$R = 1715{\cdot}080$	
Second approximation	$R = 1707{\cdot}938$	$A_2/A_1 = +0{\cdot}028973$
Third approximation	$R = 1707{\cdot}775$	$A_2/A_1 = +0{\cdot}028963$; $A_3/A_1 = -0{\cdot}002694$
Exact value	$R = 1707{\cdot}76$	

(ii) *The odd mode*

$a = 5{\cdot}365$

First approximation	$R = 17803{\cdot}24$	
Second approximation	$R = 17621{\cdot}74$	$A_2/A_1 = +0{\cdot}06304$
Third approximation	$R = 17611{\cdot}84$	$A_2/A_1 = +0{\cdot}062945$; $A_3/A_1 = -0{\cdot}010088$
Exact value	$R = 17610{\cdot}39$	

the method achieves an accuracy somewhat better than $\frac{1}{2}$ per cent. This accuracy is attained largely on account of the fact that in the process of applying the variational method, we *solved* the fourth-order differential equation (277) relating F and W, i.e. in effect we solved 'two-thirds' of the problem exactly!

By including additional terms in the expansion for F, we can naturally reach higher precision in the deduced values of R. This is illustrated in Table IV.

The odd mode (in terms of which we found in § 15 the solution for the case of one rigid and one free boundary) can also be obtained by the variational method. Thus, we now assume for F an expansion of the form

$$F = \sum_m A_m \sin 2m\pi z. \tag{313}$$

The corresponding solution for W_m is

$$W_m = P_m \sinh az + Q_m z\cosh az + \gamma_{2m}^2 \sin 2m\pi z, \tag{314}$$

where P_m and Q_m are constants of integration and

$$\gamma_{2m} = \frac{1}{4m^2\pi^2+a^2}. \tag{315}$$

The boundary conditions determine P_m and Q_m. We find

$$P_m = (-1)^m \frac{2m\pi\gamma_{2m}^2}{\sinh a - a} \cosh \tfrac{1}{2}a,$$

$$Q_m = (-1)^{m+1} \frac{4m\pi\gamma_{2m}^2}{\sinh a - a} \sinh \tfrac{1}{2}a. \tag{316}$$

The matrix element $(n|m)$ is now given by

$$(n|m) = \int\limits_{-\frac{1}{2}}^{+\frac{1}{2}} \sin 2n\pi z\{P_m \sinh az + Q_m z\cosh az + \gamma_{2m}^2 \sin 2m\pi z\}\, dz. \tag{317}$$

On evaluating the integrals, we find

$$(n|m) = \tfrac{1}{2}\gamma_{2m}^2\, \delta_{mn} - 32(-1)^{m+n} amn\pi^2 \gamma_{2m}^2\, \gamma_{2n}^2 \frac{\sinh^2\frac{1}{2}a}{\sinh a - a}; \tag{318}$$

and the secular determinant takes the form

$$\left\|\left(\frac{1}{a^2R\gamma_{2m}} - \gamma_{2m}^2\right)\delta_{mn} + (-1)^{m+n} 64a\pi^2 \frac{\sinh^2\frac{1}{2}a}{\sinh a - a}\, mn\gamma_{2m}^2\, \gamma_{2n}^2\right\| = 0. \tag{319}$$

The first approximation to R will be given by setting the (1, 1) element of the secular matrix equal to zero and ignoring all the others. Thus we obtain

$$R = \frac{1}{a^2\gamma_2^3[1-64a\pi^2\gamma_2^2 \sinh^2\frac{1}{2}a/(\sinh a - a)]}, \tag{320}$$

or, substituting for γ_2, we have

$$R = \frac{(4\pi^2+a^2)^3}{a^2\{1-64a\pi^2 \sinh^2 \frac{1}{2}a/[(4\pi^2+a^2)^2(\sinh a - a)]\}}. \tag{321}$$

This formula gives

$$R = 17803{\cdot}24 \quad \text{for} \quad a = 5{\cdot}365; \tag{322}$$

this should be contrasted with the result 17610·39 of the exact calculation. The results obtained by including further terms in the expansion of F are given in Table IV. Again, we observe the high precision which can be reached by the variational method.

18. Experiments on the onset of thermal instability in fluids

In this section we shall give an account of some of the experimental work on the onset of thermal instability in fluids which bears on the theoretical developments.

As we stated at the outset, and as the subtitle to this chapter indicates, the experiments which stimulated this field of inquiry are those of

Bénard. Indeed, it was to account for the 'interesting results obtained by Bénard's careful and skilful experiments' that Rayleigh undertook his theoretical investigation: the quotation is in fact the opening remark in Rayleigh's paper.

(*a*) *Bénard's experiments*

Bénard carried out his experiments on very thin layers of fluid, about a millimetre in depth, or less, standing on a levelled metallic plate

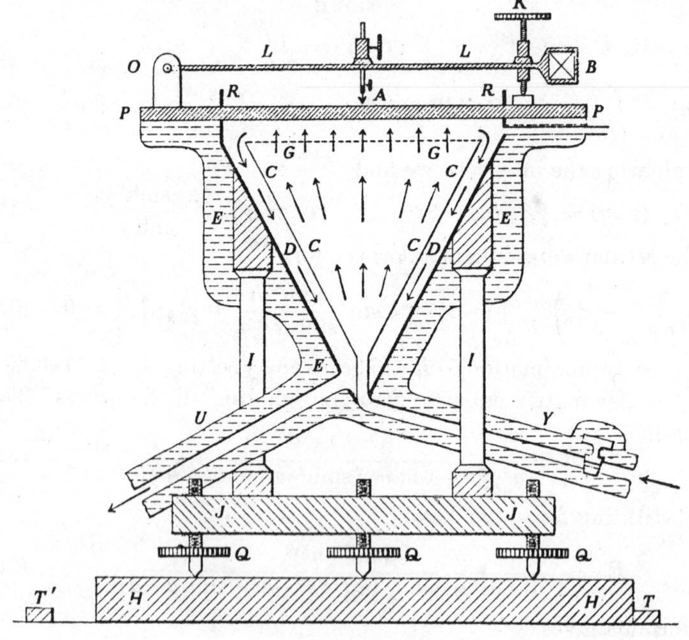

Fig. 10. The original apparatus of Bénard.

maintained at a constant temperature (see Fig. 10). The upper surface was usually free and being in contact with the air was at a lower temperature. Bénard experimented with several liquids of differing physical constants. He was particularly interested in the role of viscosity; and as liquids of high viscosity he used melted spermaceti and paraffin. In all cases, Bénard found that when the temperature of the lower surface was gradually increased, at a certain instant, the layer became reticulated and revealed its dissection into cells. He further noticed that there were motions inside the cells: of ascension at the centre, and of descension at the boundaries with the adjoining cells. Bénard distinguished two phases in the succeeding development of the cellular pattern: an initial phase of short duration in which the cells acquire a moderate degree of regularity and become convex polygons with four to seven sides and vertical walls; and a second phase of relative permanence in which the cells all become equal, hexagonal, and properly aligned. Fig. 1, which is a reproduction of one of Bénard's original photographs, illustrates the remarkable regularity which can be attained. Bénard himself was so interested in observing the emergence of the second phase and the associated motions that he was not very explicit regarding the conditions necessary for the onset of instability. However, in a later analysis of his early observations, he states that they are in qualitative agreement with the requirements of Rayleigh's criterion.

Some doubts have recently been raised whether, in view of the thinness of the layers of fluid with which Bénard experimented, his observations cannot all be accounted for in terms of effects arising from surface tension. As Bénard himself pointed out quite explicitly in his early papers, effects of surface tension are undoubtedly present; but they do not appear to be primary in most of his experiments.

(*b*) *The Schmidt–Milverton principle for detecting the onset of thermal instability*

Of the vast literature on experiments relating to thermal convection, we shall refer only to that which bears directly on the detection of the onset of thermal instability, and the determination of the critical Rayleigh number. The earliest of the experiments in this latter category appear to be those of Schmidt and Milverton. In their experiments, Schmidt and Milverton incorporated a principle for the detection of the onset of thermal instability which is so direct and simple that it has served as the basis for all later experiments in this subject. In view of the importance of this principle, we shall explain its origin and the manner of its use.

Suppose that a difference of temperature ΔT $(= T_2 - T_1;\ T_2 > T_1)$ is maintained between the two sides of a layer of fluid of depth d. In a steady state, this implies that there is a constant outward flux of heat from the surface at the higher temperature. Let $\mathcal{Q}$ denote this flux of heat: it measures the quantity of heat (in calories) emerging from the surface, per unit area and per unit time. If the flow of heat across the fluid layer is entirely by conduction, then clearly

$$\mathcal{Q} = k\Delta T/d, \tag{323}$$

where k is the coefficient of heat conduction. Therefore, so long as conduction prevails as the sole mechanism of heat transport, the flux of heat which must be provided to sustain a difference of temperature ΔT between the two sides increases linearly with ΔT with a constant of proportionality which is k/d. This particular linear relation must break down at the onset of thermal instability when other modes of heat transport become operative, and the effective thermal conductivity increases beyond its static value k.

Now a steady emergent flux of heat from a surface can be maintained only by supplying an equivalent amount of energy by some external agency. In practice, the required energy is, generally, supplied electrically by passing the necessary amount of current through heating coils suitably placed below the surface. The energy thus supplied can be conveniently measured by the square of the current C times the resistance R of the circuit, or, equivalently, by the product of the current and the applied electromotive force E. If we suppose that the electrical energy which is supplied contributes to the emergent heat flux without any extraneous loss, then,

$$EC = C^2R = 4{\cdot}1854\mathcal{Q}\times\text{area}, \tag{324}$$

where the current is measured in amperes, the resistance in ohms, the electromotive force in volts, and the quantity of heat in calories. (The factor 4·1854 is the mechanical equivalent of heat in joules.) Combining equations (323) and (324), we have in the conductive régime,

$$EC = C^2R = 4{\cdot}1854k\frac{\Delta T}{d}\times\text{area}. \tag{325}$$

Therefore, in the absence of heat losses, and in the conductive régime, the square of the current in the heating coils, necessary to maintain a difference in temperature ΔT between the two sides, increases linearly with ΔT with a constant of proportionality which is determinate. The linear relation which thus obtains will break down at the onset of instability. Hence, in an $(C^2, \Delta T)$-plot, we must observe a break when ΔT surpasses the value at which instability sets in. Such a break should be discernible, even if the entire energy equivalent of C^2R does not contribute to $\mathcal{Q}$: for, if we suppose that only a fraction q of C^2R contributes to $\mathcal{Q}$, it is unlikely that the variation of q, within the small range of C^2 in which instability occurs, will be such as to mask the discontinuity which will otherwise be present. Fig. 11 shows some $(C^2, \Delta T)$-plots made by Schmidt and Milverton: the plots clearly exhibit the breaks which occur at the onset of instability.

In practice, the electrical energy dissipated by the heating coils will not all contribute to the heat flux $\mathcal{Q}$: some of it will be lost by conduction and by radiation to the surrounding materials and air at lower temperatures. In a carefully designed experiment (see § (c) below) these losses can be reduced and allowed for with sufficient accuracy. A reliable quantitative measure of the heat flux $\mathcal{Q}$ can then be made without reference to the manner of the heat transfer in the intervening fluid. It

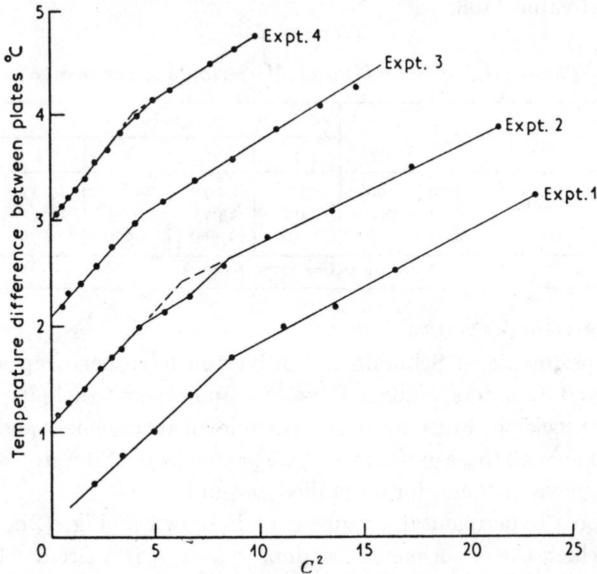

FIG. 11. The plots represent the square of the heating current C^2 required to maintain a given difference in temperature in a layer of water confined between two plates. The break in the curves indicates the onset of instability and illustrates the Schmidt–Milverton principle. The critical Rayleigh numbers derived from these plots are given in Table V.

will be convenient to express the flux $\mathcal{Q}$, measured in this way, in the unit $k\Delta T/d$ (which will be its value in the conductive régime); $\mathcal{Q}$ expressed in the unit $k\Delta T/d$ is often called the Nusselt number:

$$\text{Nu} = \mathcal{Q}/(k\Delta T/d). \tag{326}$$

One can then plot the Nusselt number against the Rayleigh number,

$$R = \frac{g\alpha\beta}{\kappa\nu}d^4 = \frac{g\alpha\,\Delta T}{\kappa\nu}d^3. \tag{327}$$

In such a plot, we must observe that the Nusselt number begins to assume values greater than unity, quite abruptly, at some value of R; the value of R at which this happens is the critical Rayleigh number for the onset of instability.

Schmidt and Milverton applied their principle to determine the critical Rayleigh number for the onset of thermal instability in horizontal layers of water confined between two rigid planes. The values they derived from their experimental results (exhibited in the plots of Fig. 11) are listed in Table V. It will be observed that the value for the critical Rayleigh number they obtain is in satisfactory agreement with the theoretical value 1708.

TABLE V

The results of Schmidt and Milverton's experiments

	Experiment			
	1	2	3	4
Temperature of upper plate	23·2°	22·75°	19·80°	17·63°
Temperature of lower plate	24·9°	23·75°	20·85°	18·47°
R_c	1,970	1,580	1,850	1,670
R_c (mean) = 1,770 ± 140				

(*c*) *The precision experiments of Silveston*

The experiments of Schmidt and Milverton have been repeated by Schmidt and Saunders, Malkus, Silveston, and others to achieve greater range and precision. Since the basic principles of arrangement and design are the same in all these experiments, we have selected Silveston's, as the most extensive of them, for a detailed account.

Silveston's experimental arrangement is shown in Fig. 12*a*, *b*. The fluid on which the experiments are done is at *a*; it is a circular layer of diameter 19·8 cm and of variable depth. The fluid layer is confined between two decks of plates *b* and *c*, and a surrounding Plexiglass cylinder *d*. The lower deck *b* consists of a pair of bolted copper plates between which is sandwiched a heating coil *e*. The latter is a chromium-nickel tape encased in mica and insulated from the metallic plates by asbestos sheets. The upper deck *c* consists also of a pair of bolted copper plates between which is engrooved a two-way spiral canal *f*. Through this canal, cold water from a constant temperature reservoir is allowed to circulate. The water enters the canal from the outside, spirals inward by one of the arms, and then spirals outward by the other arm. The small change in temperature which the water will suffer, as it circulates, is largely compensated—as far as its effects on maintaining a constant temperature of *c* is concerned—by the fact that the outflowing and the inflowing water is adjacent in the two arms. The entry and the exit for the circulating water are at places indicated by *g*.

The two decks *b* and *c* are separated by carefully machined blocks *h*

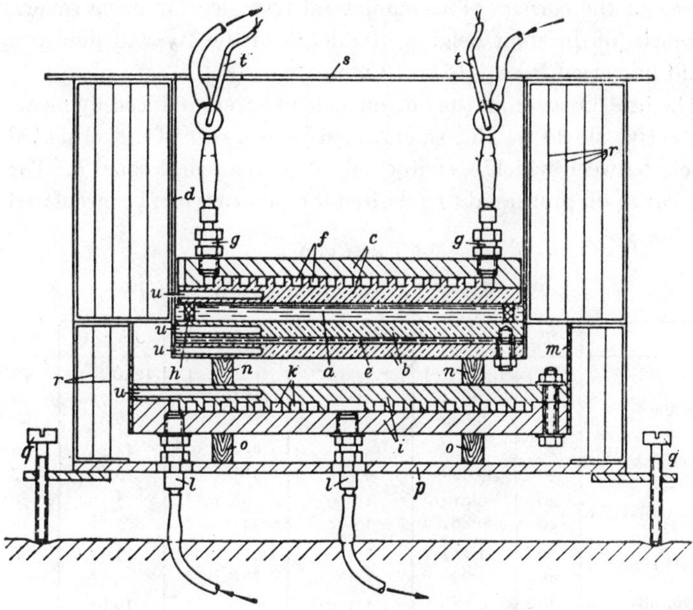

Fig. 12*a*. Silveston's experimental arrangement for measurements of heat transfer.

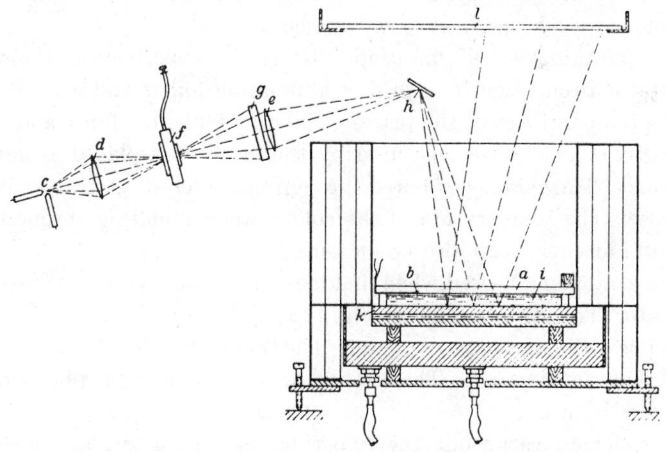

Fig. 12*b*. Silveston's experimental arrangement for optical observations.

placed at the corners of an equilateral triangle. By using several sets of blocks of different heights, the depth of the layer of fluid confined could be varied from 1·45 to 13 mm.

The heat losses from the bottom side of *b* were reduced by means of a protective shield *i*. This shield consists of a pair of nickel-plated iron sheets between which is engrooved (as in *c*) a spiral canal *k*. Through this canal, oil preheated to a desired temperature can be circulated. By

TABLE VI

The physical constants of certain liquids

	T	ρ	$c_v \times 10^{-7}$	ν	$\kappa \times 10^4$	$\alpha \times 10^3$
	(° C)	(gm/cm³)	(ergs/gm °C)	(cm²/sec)	(cm²/sec)	(°C)⁻¹
Heptane	20	0·684	2·22	$6{\cdot}16 \times 10^{-3}$	8·75	12·4
	40	0·666	2·26	$5{\cdot}11 \times 10^{-3}$	8·81	12·8
Water	20	0·998	4·19	$1{\cdot}006 \times 10^{-2}$	14·33	2·0
	40	0·992	4·18	$6{\cdot}58 \times 10^{-3}$	15·11	3·8
Silicon oil AK 3	20	0·912	1·609	$3{\cdot}20 \times 10^{-2}$	7·79	10·6
	40	0·892	1·668	$2{\cdot}32 \times 10^{-2}$	7·56	11·5
Ethylene glycol	20	1·113	2·38	$1{\cdot}915 \times 10^{-1}$	9·42	6·4
	40	1·099	2·48	$8{\cdot}79 \times 10^{-2}$	9·42	6·5
Silicon oil AK 350	20	0·980	1·496	$4{\cdot}67 \times 10^{-2}$	10·61	9·21
	40	0·962	1·542	$3{\cdot}20 \times 10^{-2}$	10·36	9·44

arranging that the shield is at nearly the same temperature as the bottom side of *b*, the heat losses can be reduced.

To maintain steady conditions during the experiments, the whole apparatus is enclosed by a grid of aluminium foils *r* and a roof *s*.

The temperatures of the plates *b* and *c* confining the fluid, and of the protective shield *i* are measured by thermocouples placed at suitable locations. The channels *u* bored through the different plates are for this purpose. The temperature of the cold water circulating through *f*, at exit and at entry, can also be measured.

The arrangement used when making optical observations is shown in Fig. 12 *b*. The upper deck (*c* in Fig. 12 *a*) is then replaced by a reflecting glass plate *a* of the same diameter. The thermocouple *b* is pasted to the glass plate inside a slit. The rest of the arrangement for photographic purposes is clear.

In order to carry out the experiments over as wide a range of parameters as possible, Silveston used, besides water, four other liquids. The physical constants of these liquids are listed in Table VI. The very high viscosity of the silicon oil (AK 350) is, particularly, to be noted.

The experiment consists, principally, in determining the emergent flux of heat $\mathcal{Q}$ as a function of the difference in temperature between the two sides which it is able to sustain. In a steady state, the required flux of heat is given directly by the electrical energy supplied to the heating element after proper allowance has been made for the heat-losses.

The heat losses were determined by raising the temperatures of the components *b*, *c*, and *i* to a value which was in excess of the room temperature by a known amount; the energy dissipated by the heating element, under these conditions, is a direct measure of the heat-loss. In this way the heat-losses which occur, as a function of the difference in temperature between the plate *b* and the surrounding air, were determined. In addition to these losses, allowance must also be made for the losses by conduction through the Plexiglass cylinder and the various other supports. These additional losses were estimated from the known heat conductivities of the materials.

The electrical energy dissipated by the heating element, in watts, is given by the product of the current, in amperes, and the applied electromotive force, in volts. By subtracting from this energy the heat-losses determined in the manner we have described, Silveston was able to deduce the value of the heat flux $\mathcal{Q}$ in absolute measure.

In carrying out the experiments, several precautions were taken. For example, care was taken to ensure that the liquid, before it was admitted into the apparatus, was free of occluded air: this was accomplished by heating the liquid to a temperature of 80°–100° C and keeping it at this temperature for about an hour. The liquid so treated was introduced into the apparatus to a level such as just to be in contact with the lower surface of *c*. The current through the heating element and the flow of cold water through the canal *f* were adjusted so that the desired difference of temperature between *b* and *c* was attained and then maintained. Measurements were made only after the conditions had remained steady for at least an hour. To allow for small fluctuations, the temperatures of the principal elements (namely *b*, *c*, and *i*) were recorded, first in one sequence and then in the reverse sequence. The room temperature, the temperature of the circulating cold water (at exit and at entry), the current, and the voltage, were all read at the beginning, during the middle, and at the end of the principal temperature measurements.

A check on the consistency of the heat measurements is obtained if one can account for all of the electrical energy supplied in terms of the heat carried away by the circulating water,† the heat lost from the top

† This heat is given by the rate of flow of the water times the difference in its temperature at entry and at exit.

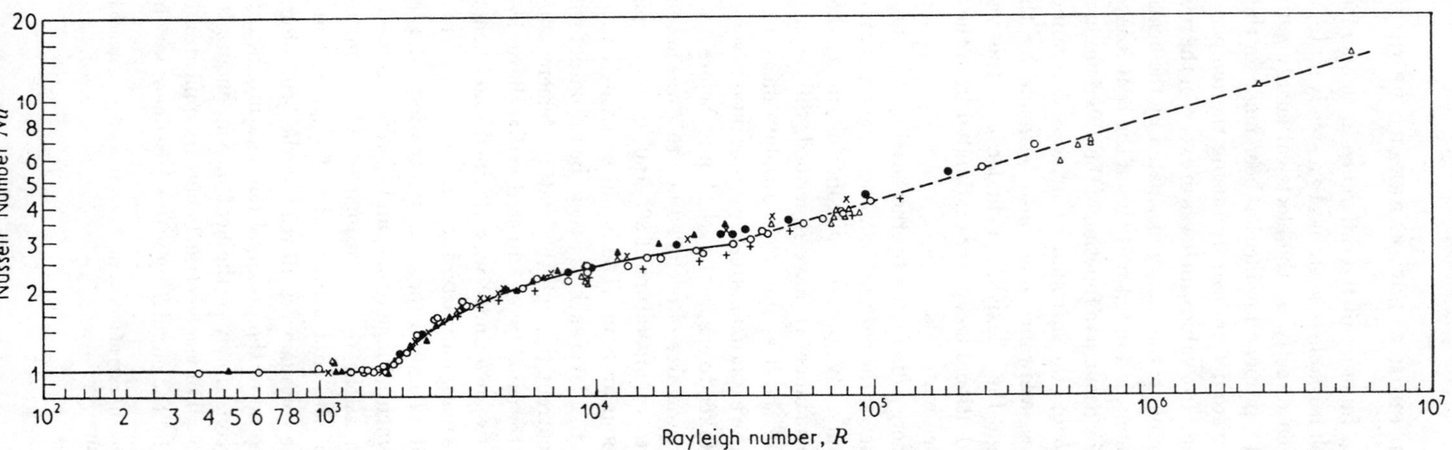

FIG. 13. Silveston's experimental results on the heat transfer in various liquids (○ water; + heptane; × ethylene glycol; ● silicone oil AK 3; ▲ silicone oil AK 350; △ air data of Mull and Reiher). The Nusselt number is plotted against the Rayleigh number.

surface of c, the walls, the shield, etc. Silveston found that in his experiments the check could always be made within the errors of measurement.

The final results of Silveston's experiments are shown in Fig. 13. The fact that the results of the measurements made under widely different conditions and with liquids with widely varying physical parameters all consistently fall on the same average curve when plotted in terms of

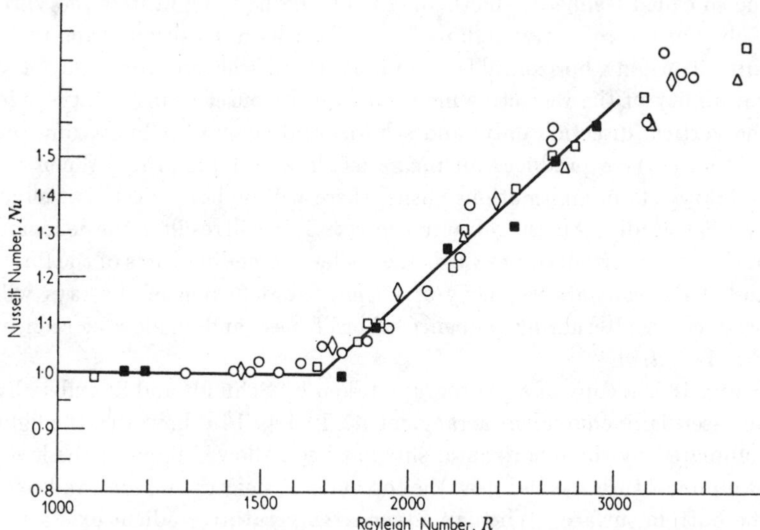

FIG. 14. Silveston's experimental results in the neighbourhood of instability in various liquids (■ silicone oil AK 350; ◇ silicone oil AK 3; □ ethylene glycol; △ heptane; o water). These data consistently indicate that in all of these liquids instability sets in at the Rayleigh number 1700±51 (to be compared with the theoretical value 1708).

the non-dimensional Nusselt and Rayleigh numbers, is a remarkable testimony to the reliability and the precision of the experiments.

From an examination of the results obtained for Rayleigh numbers in the range 10^3–10^4 (see Fig. 14), Silveston derives for the critical Rayleigh number for the onset of instability the value

$$R_c \text{ (experimental)} = 1700 \pm 51. \tag{328}$$

This is in very good accord with the theoretical value 1708.

(d) *Observations by optical methods*

The emergence of a cellular pattern of motions at instability, and its persistence afterwards, are phenomena which lend themselves to observations by optical methods and photographic means. Bénard's own observations fall in this category; and Fig. 1 is an example. The phenomena have since been investigated by more refined optical methods. Figs. 15, 16, and 17 are illustrations taken from these later investigations. We shall briefly describe what these illustrations represent and how they were obtained.

Fig. 15 is a copy of a photograph taken by Schmidt and Milverton by the so-called 'schlieren' method. In this method one utilizes the variability of the refractive index of a medium with density and temperature. Thus, in a horizontal layer of fluid which is heated from below, the variability of the refractive index, before the onset of instability, is in the vertical direction only; and a horizontal beam of light grazing the bottom surface produces an image as shown in Fig. 15*a*. When instability sets in and motions ensue, there will be horizontal, as well as vertical, gradients in the refractive index. This will result in the deflexion of the rays horizontally towards the cooler descending parts of the fluid; and if the cells are regularly arranged, the deflexion of the rays will produce an alternating sequence of brightness and shadow as seen in Fig. 15*c*, *d*.

Fig. 16 is a copy of a photograph taken by Schmidt and Saunders by an essentially equivalent arrangement. In Fig. 16*a*, however, the light collimated by three horizontal slits has been allowed to enter the layer of liquid at three levels: near the top surface, near the middle, and near the bottom surface. When an adverse temperature gradient exists and instability prevails, the photograph shows that the light passing through the middle slit is practically undeflected, while the light from the top and the bottom slits is deflected upwards, the deflexion varying periodically with distance across the width of the photograph. It will be noticed that the minima of the vertical deflexion of light from the top slit occur at the same places as the maxima of the vertical deflexion of light from the bottom slit; these are the places where the cooled liquid is descending. The absence of any appreciable vertical deflexion of light from the middle slit is an indication that in this plane, vertical temperature gradients are everywhere negligible. In Fig. 16*b*, *c*, the light has been admitted to the whole gap between the planes (very much as in Fig. 15). The characteristic features are the bright patches which are marked both in *b* and *c*. In Fig. 16*a* there are eleven cells; and in Fig. 16*b*, *c* there are twenty-two cells. These numbers of cells occur within a horizontal length corresponding to 22·9 cm; the width of the cells are, accordingly, 2·1 cm and 1·0 cm respectively, i.e. approximately twice the distance beween the plates. Schmidt and Saunders, in fact, found

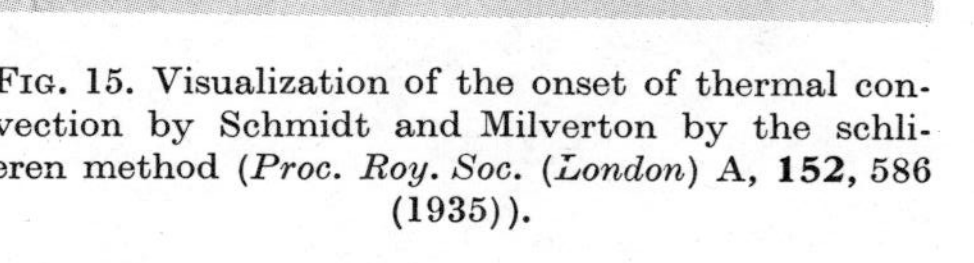

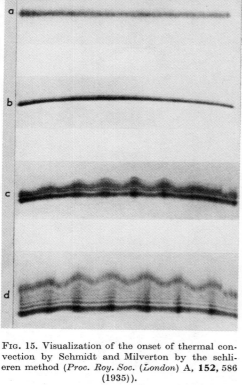

FIG. 15. Visualization of the onset of thermal convection by Schmidt and Milverton by the schlieren method (*Proc. Roy. Soc.* (*London*) A, **152**, 586 (1935)).

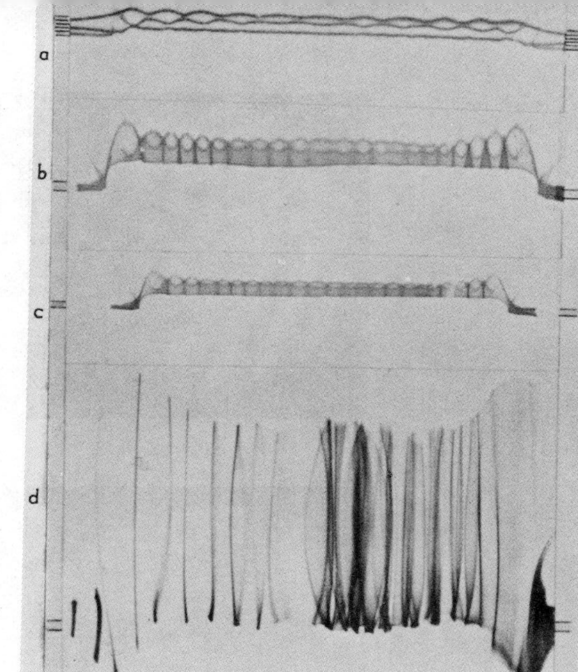

FIG. 16. Visualization of the onset of thermal convection by an optical arrangement by Schmidt and Saunders (*Proc. Roy. Soc.* (*London*) A, **165**, 216 (1938)). The data to which the illustrations refer are given below:

(*a*) $d = 1{\cdot}1$ cm, $R = 12{,}000$, $\Delta T = 0{\cdot}55°$ C
(*b*) $d = 0{\cdot}5$ cm, $R = 3{,}500$, $\Delta T = 1{\cdot}7°$ C
(*c*) $d = 0{\cdot}5$ cm, $R = 3{,}500$- $\Delta T = 1{\cdot}7°$ C
(*d*) $d = 1{\cdot}1$ cm, $R = 130{,}000$, $\Delta T = 4{\cdot}0°$ C

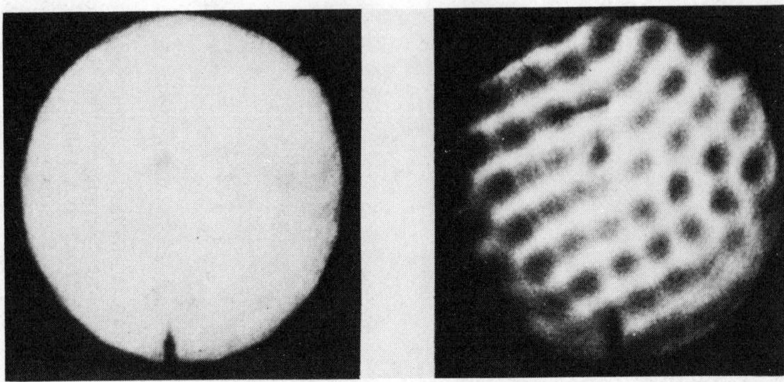

FIG. 17*a*. Visualization of the onset of thermal convection by Silveston. The photograph on the left was obtained for a Rayleigh number 1,500 while the photograph on the right was obtained for a Rayleigh number 1,800. The depth of the layer in these experiments was 7 mm.

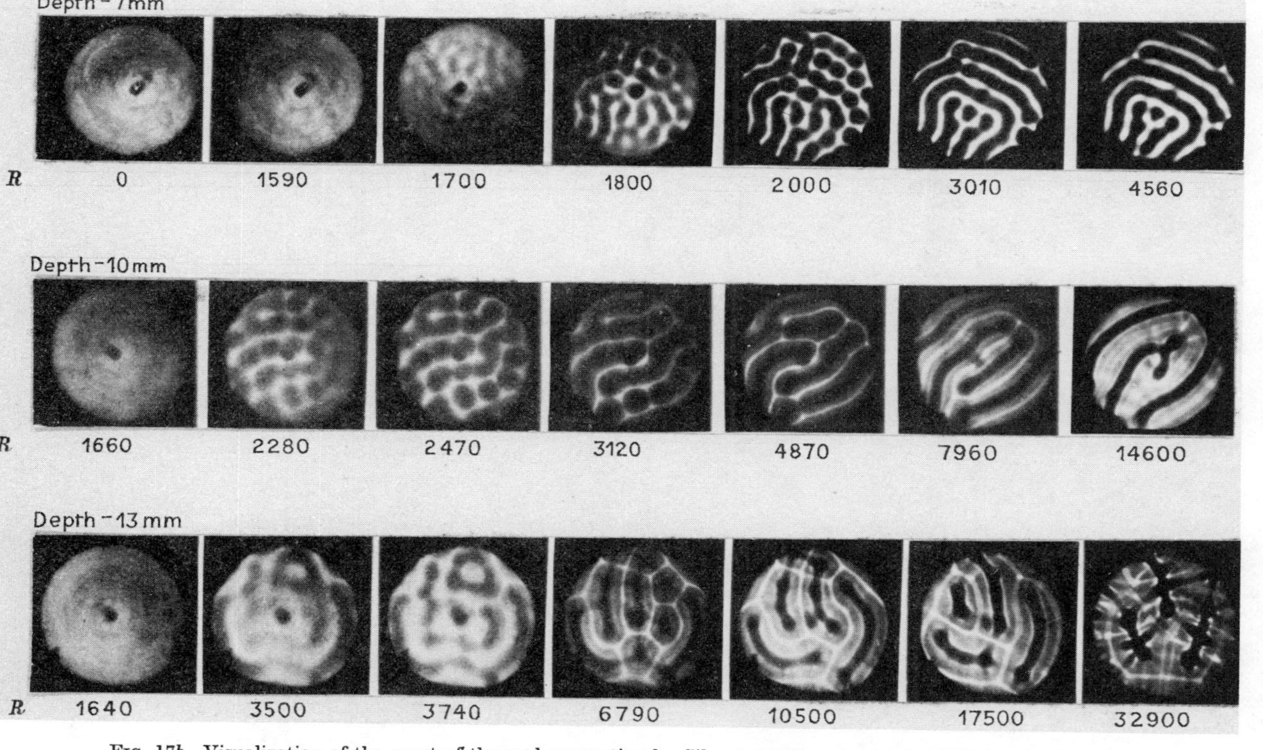

FIG. 17*b*. Visualization of the onset of thermal convection by Silveston: photographs for different depths and increasing Rayleigh numbers.

that whenever the cellular motion was well established, the width of the cell was about twice the depth of the layer.

Fig. 16*b*, *c* illustrate conditions which are substantially beyond marginal stability; nevertheless, it appears that the basic cellular pattern has not suffered any appreciable changes since marginal conditions. The situation is, however, quite different in Fig. 16*d*: here the Rayleigh number is 130,000; and it appears that now the cellular pattern has been completely replaced by random turbulent motions. (The photograph illustrated in Fig. 16*d* was taken somewhat differently from the others: in the left half of the photograph, the light entering the layer was limited by eleven vertical slits while on the right half, no vertical slits were used and the vertical lines on the image are due to refraction in the fluid.)

Finally, Fig. 17 is a copy of a photograph taken by Silveston with the experimental arrangement shown in Fig. 12*b*. The emergence of the cell pattern at the critical Rayleigh number is apparent; and its relative stability far beyond the marginal conditions is particularly striking.

BIBLIOGRAPHICAL NOTES

The phenomenon of thermal convection in fluids, as we now understand it, was discovered by Count Rumford:

1. COUNT RUMFORD, 'Of the propagation of heat in fluids', *Complete Works*, **1**, 239, American Academy of Arts and Sciences, Boston, 1870.

An interesting account of Rumford's discovery with relevant extracts from his writings will be found in:

2. S. C. BROWN, 'Count Rumford discovers thermal convection', *Daedalus*, **86**, 340–3 (1957).

In reference 2, Brown points out that the term 'convection' was first introduced in science by William Prout:

3. W. PROUT, *Bridgewater Treatises*, **8**, 65, edited by W. Pickering, London, 1834.

The following extract from Prout is taken from reference 2:

> 'There is at present no single term in our language employed to denote this mode of propagation of heat; but we venture to propose for that purpose the term *convection* (*convectio*, a carrying or converging), which not only expresses the leading facts, but also accords very well with the two other terms (conduction and radiation).'

The fact that convection in horizontal layers of fluid is often accompanied by a cellular pattern of motions seems to have been first observed by James Thomson:

4. J. THOMSON, 'On a changing tesselated structure in certain liquids', *Proc. Phil. Soc. Glasgow*, 13, 464–8 (1882).

The following extract is from reference 4:

> 'The occurrence of the phenomena seems to be associated essentially with cooling of the liquid at its surface where exposed to the air, when the main body of the liquid is at a temperature somewhat above that assumed by a thin superficial film. A very slight excess of temperature in the body of the fluid above that of the surrounding air is sufficient to institute the tesselated changing structure. . . . The motions now described and explained as occurring in the soapy water and other liquids when showing the tesselated structure he [Thomson] thus believes constitute one special case of what is often called convective circulation. He points out, however, that this case, with a thin film cooled at the surface while the great body of the liquid is at a somewhat higher temperature, presents essential distinctions from the case in which convective circulation is caused by heat applied at the bottom of the vessel. The surface phenomena and the actual motions throughout the body of the liquid are very different in the two cases, as may easily be seen by a little consideration, both theoretical and observational.'

The first quantitative experiments on the onset of thermal instability and the recognition of the role of viscosity in the phenomenon are due to H. Bénard:

5. H. Bénard, 'Les Tourbillons cellulaires dans une nappe liquide', *Revue générale des Sciences pures et appliquées*, **11,** 1261–71 and 1309–28 (1900).
6. ——, 'Les Tourbillons cellulaires dans une nappe liquide transportant de la chaleur par convection en régime permanent', *Annales de Chimie et de Physique*, **23,** 62–144 (1901).

On the theoretical side, the fundamental paper is that of Lord Rayleigh:

7. Lord Rayleigh, 'On convective currents in a horizontal layer of fluid when the higher temperature is on the under side', *Phil. Mag.* **32,** 529–46 (1916); also *Scientific Papers*, **6,** 432–46, Cambridge, England, 1920.

The paper by Lord Rayleigh is included in an anthology of related papers edited by Saltzmann:

8. B. Saltzmann, 'The general circulation as a problem in thermal convection; A collection of classical and modern theoretical papers', *Scientific Report* No. 1, *General Circulation Project*, Department of Meteorology, Massachusetts Institute of Technology, 1958.

A collection of papers containing some beautiful illustrations of convection phenomena in Nature is:

9. *Convection Patterns in the Atmosphere and Ocean*, edited by R. W. Miner, *Annals of the New York Academy of Sciences*, **48,** 1947.

References to the Bénard convection cells from the standpoint of meteorology will be found in:

10. D. Brunt, *Physical and Dynamical Meteorology*, 219–21, Cambridge, England, 1939.
11. O. G. Sutton, *Micrometeorology*, 119–25, McGraw-Hill Book Co. Inc., New York, 1953.

And a general account of the subject treated in this and the following chapters is given in:

12. S. Chandrasekhar, 'Thermal convection', *Daedalus*, **86,** 323–39 (1957).

§ 7. A useful reference for the foundations of hydrodynamics is:

13. S. Goldstein, *Modern Developments in Fluid Dynamics*, **1**, 90–101 and **2**, 601–9, Oxford, England, 1938.

See also:

14. W. F. Cope, 'The equations of hydrodynamics in a very general form', *Aeronautical Research Committee Reports and Memoranda*, No. 1903, 1–6, (1942).

For the notation of Cartesian tensors see:

15. H. Jeffreys, *Cartesian Tensors*, Cambridge, England, 1931.

16. H. and B. S. Jeffreys, *Methods of Mathematical Physics*, chapter 3, Cambridge, England, 1956.

§ 8. The approximation introduced in this section is due to:

17. J. Boussinesq, *Théorie Analytique de la Chaleur*, **2**, 172, Gauthier–Villars, Paris, 1903.

§§ 9 and 10. As we have stated, the foundations of the subject were laid by Lord Rayleigh in reference 7. The theory, from a somewhat more general point of view and leading to equation (104), is due to Jeffreys:

18. H. Jeffreys, 'The stability of a layer of fluid heated below', *Phil. Mag.* **2**, 833–44 (1926).

19. —— 'Some cases of instability in fluid motion', *Proc. Roy. Soc.* (*London*) A, **118**, 195–208 (1928).

See also:

20. H. Jeffreys, 'The instability of a compressible fluid heated below', *Proc. Camb. Phil. Soc.* **26**, 170–2 (1930).

In reference 20, Jeffreys shows that the criterion for the onset of instability derived for incompressible fluids can, under certain conditions, be used for compressible fluids if β is interpreted as the difference between the adiabatic temperature gradient and the prevailing temperature gradient.

§ 11. The validity of the principle of the exchange of stabilities for this problem was proved by Rayleigh (reference 7) for the case of two free boundaries. The proof in the general case is due to Pellew and Southwell:

21. A. Pellew and R. V. Southwell, 'On maintained convective motion in a fluid heated from below', *Proc. Roy. Soc.* (*London*) A, **176**, 312–43 (1940).

§ 13. The first of the two variational principles considered in this section is due to Pellew and Southwell (reference 21); for the second of these see:

22. S. Chandrasekhar, 'On characteristic value problems in high order differential equations which arise in studies on hydrodynamic and hydromagnetic stability', *American Math. Monthly*, **61**, 32–45 (1954).

The fact that the variational principle gives the true minimum does not seem to have been proved before.

§ 14. The thermodynamic significance of the variational principle was first considered explicitly though in the context of a somewhat different problem by:

23. H. Jeffreys, 'The thermodynamics of thermal instability in liquids', *Quart. J. Mech. Appl. Math.* **9**, 1–5 (1956).

Less explicitly, it is contained in:

24. W. V. R. Malkus, 'The heat transport and spectrum of thermal turbulence', *Proc. Roy. Soc.* (*London*) A, **225,** 196–212 (1954).

A complete discussion of the underlying principle including sources of dissipation, besides viscosity, and allowing for overstability will be found in:

25. S. Chandrasekhar, 'The thermodynamics of thermal instability in liquids', *Max Planck Festschrift 1958*, 103–14, Veb Deutscher Verlag der Wissenschaften, Berlin, 1958.

§ 15. The exact solution of the basic characteristic value problem was first accomplished by:

26. A. R. Low, 'On the criterion for stability of a layer of viscous fluid heated from below', *Proc. Roy. Soc.* (*London*) A, **125,** 180–95 (1929).

By a somewhat different method it was also carried out by Pellew and Southwell (reference 21). The method followed in the text is essentially that of Pellew and Southwell. The definitive treatment of the problem is due to:

27. W. H. Reid and D. L. Harris, 'Some further results on the Bénard problem', *The Physics of Fluids*, **1,** 102–10 (1958).

28. —— —— 'Streamlines in Bénard convection cells', ibid. **2,** 716–17 (1959).

§ 16. The solution of the equation

$$\frac{\partial^2\phi}{\partial x^2}+\frac{\partial^2\phi}{\partial y^2} = -a^2\phi$$

appropriate for a hexagonal cell pattern was discovered by:

29. D. G. Christopherson, 'Note on the vibration of membranes', *Quart. J. of Math.* (Oxford Series), **11,** 63–65 (1940).

The generalization of Christopherson's solution given in § (e) is due to:

30. F. E. Bisshopp, 'On two-dimensional cell patterns', *J. Math. Analysis and Applications*, **1,** 373–85 (1960).

§ 17. Variational solutions have been given by Pellew and Southwell (reference 21) and Reid and Harris (27).

§ 18. The references for the experimental work described in this section are:

31. H. Bénard and D. Avsec, 'Travaux récents sur les tourbillons cellulaires et les tourbillons en bandes; applications a l'astrophysique et a la météorologie', *Le Journal de physique et le radium*, **9,** 486–500 (1938).

32. R. J. Schmidt and S. W. Milverton, 'On the instability of a fluid when heated from below', *Proc. Roy. Soc.* (*London*) **A, 152,** 586–94 (1935).

33. —— and O. A. Saunders, 'On the motion of a fluid heated from below', ibid. **165,** 216–28 (1938).

34. O. A. Saunders, M. Fishenden, and H. D. Mansion, 'Some measurements of convection by an optical method', *Engineering*, **139,** 483–5 (1935).

35. P. L. Silveston, 'Wärmedurchgang in waagerechten Flüssigkeitsschichten', Part 1, *Forsch. Ing. Wes.* **24,** 29–32 and 59–69 (1958).

See also:

36. W. V. R. Malkus, 'Discrete transitions in turbulent convection', *Proc. Roy. Soc.* (*London*) A, **225,** 185–95 (1954).

In reference 31, Bénard's earlier work (references 5 and 6) is summarized and applications to astrophysical and meteorological phenomena are considered. The role of surface tension in Bénard's experiments is discussed by:

37. J. R. A. PEARSON, 'On convection cells induced by surface tension', *J. Fluid Mech.* **4,** 489–500 (1958).

The 'Schmidt–Milverton' principle is described in reference 32. And the experimental arrangement used in securing the photographs illustrated in Fig. 15 is described in reference 34.

Matters closely related to the subject-matter of this chapter, but not explicitly considered, are treated in:

38. H. TIPPELSKIRCH, 'Weitere Konvektionsversuche: der Nachweis der Ringzellen und ihrer Verallgemeinerung', *Beiträge zur Physik der Atmosphäre,* **32,** 2–22 (1959).

39. —— 'Über Konvektionszellen, insbesondere im flüssigen Schwefel', ibid. **29,** 37–54 (1956).

40. J. ZIEREP, 'Über die Bevorzugung der Sechseckzellen bei Konvektionsströmungen über einer gleichmässig erwärmten Grundfläche', ibid. **31,** 31–39 (1958).

41. —— 'Zur Theorie der Zellularkonvektion III', ibid. **32,** 23–33 (1959).

42. —— 'Über rotationssymmetrische Zellularkonvektionsströmungen', *Z. für angewandte Mathematik und Mechanik,* **38,** 1–4 (1958).

Some progress has recently been achieved in developing theories which will be applicable to Rayleigh numbers slightly in excess of R_c and when the amplitudes are finite. The following references may be noted in this connexion:

43. W. V. R. MALKUS and G. VERONIS, 'Finite amplitude cellular convection', *J. Fluid Mech.*, **4,** 225–60 (1958).

44. L. P. GOR'KOV, 'Stationary convection in a plane liquid layer near the critical heat transfer point', *Soviet Physics, JETP,* **6,** 311–15 (1958).

See also Appendix I.

III

THE THERMAL INSTABILITY OF A LAYER OF FLUID HEATED FROM BELOW

2. THE EFFECT OF ROTATION

19. Introduction

In this chapter we shall investigate the effect of rotation on the simple problem of thermal instability considered in the last chapter. It will appear that rotation introduces a number of new elements into the problem; and some of its consequences are, at first sight, unexpected: the role of viscosity is, for example, inverted. The origin of this and other consequences of rotation can be traced to certain general theorems, relating to vorticity, in the dynamics of rotating fluids. For this reason, we shall preface the investigation of the stability by a discussion of these general theorems.

20. The theorems of Helmholtz and Kelvin

Consider an incompressible, inviscid, fluid. The equations governing it are:

$$\frac{\partial u_i}{\partial t}+u_j\frac{\partial u_i}{\partial x_j}=-\frac{\partial}{\partial x_i}\left(\frac{p}{\rho}+V\right) \tag{1}$$

and†

$$\frac{\partial u_i}{\partial x_i}=0, \tag{2}$$

where it has been assumed that the external forces are derivable from a potential V. The vorticity $\boldsymbol{\omega}$ is, of course, defined by

$$\boldsymbol{\omega}=\operatorname{curl}\mathbf{u} \quad\text{or}\quad \omega_i=\epsilon_{ijk}\frac{\partial u_k}{\partial x_j}. \tag{3}$$

A curve drawn from point to point in the fluid, so that its direction is everywhere that of the instantaneous direction of $\boldsymbol{\omega}$, is called a *vortex*

† The principal theorems to be established are not restricted to incompressible fluids: they apply to compressible fluids, also, if p is a function of ρ only. In the latter case $(\operatorname{grad} p)/\rho$ can be written as $\operatorname{grad}\chi$ where $\chi=\int dp/\rho$; and the quantity on the left-hand side of equation (1) continues to be the gradient of a scalar function. It is on this last circumstance that the principal theorems to be established depend; we shall not, in particular, have occasion to use the equation of continuity (2).

line. The differential equations determining it are

$$\frac{dx_1}{\omega_1} = \frac{dx_2}{\omega_2} = \frac{dx_3}{\omega_3}. \tag{4}$$

If through every point of a small closed curve we draw the corresponding vortex line, we obtain a tube which is called a *vortex tube.*

Since
$$\operatorname{div} \boldsymbol{\omega} = 0, \tag{5}$$
it follows from Gauss's theorem that the integral of the normal component of $\boldsymbol{\omega}$ over any closed surface is zero:

$$\oint_S \boldsymbol{\omega}.d\mathbf{S} = 0 \quad (S \text{ any closed surface}). \tag{6}$$

Apply this result to the closed surface bounding an element of fluid contained in a section of the vortex tube (see Fig. 18). The sides of the vortex tube clearly do not contribute to the surface integral and equation (6), in this case, gives

$$\int_{S_1} \boldsymbol{\omega}.d\mathbf{S}_1 = \int_{S_2} \boldsymbol{\omega}.d\mathbf{S}_2. \tag{7}$$

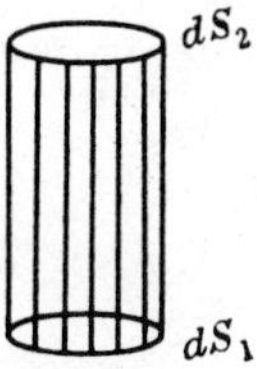

dS_2

dS_1

FIG. 18. A vortex tube.

The flux of vorticity across any section of a vortex tube is the same; it, therefore, represents a characteristic of the tube.

If we consider a vortex tube of infinitesimal cross-section, the flux across a section dS normal to itself is $|\boldsymbol{\omega}|dS$. By the theorem we have just proved, $|\boldsymbol{\omega}|dS$ is constant along the tube; it cannot therefore terminate at any point in the fluid. *Vortex lines must either be closed, or terminate on the boundaries.*

We shall now obtain an equation of motion for the vorticity. First, we observe that

$$\begin{aligned}\epsilon_{ijk} u_j \omega_k &= \epsilon_{ijk} u_j \epsilon_{klm} \frac{\partial u_m}{\partial x_l} \\ &= (\delta_{il}\delta_{jm} - \delta_{im}\delta_{jl}) u_j \frac{\partial u_m}{\partial x_l} = u_j \frac{\partial u_j}{\partial x_i} - u_j \frac{\partial u_i}{\partial x_j},\end{aligned} \tag{8}$$

or
$$u_j \frac{\partial u_i}{\partial x_j} = \frac{1}{2}\frac{\partial}{\partial x_i}|\mathbf{u}|^2 - \epsilon_{ijk} u_j \omega_k. \tag{9}$$

Using this relation in (1), we have

$$\frac{\partial u_i}{\partial t} - \epsilon_{ijk} u_j \omega_k = -\frac{\partial}{\partial x_i}\left(\frac{p}{\rho} + \tfrac{1}{2}|\mathbf{u}|^2 + V\right). \tag{10}$$

Letting
$$\varpi = p/\rho + \tfrac{1}{2}|\mathbf{u}|^2 + V, \tag{11}$$
we can write
$$\frac{\partial \mathbf{u}}{\partial t} - \mathbf{u} \times \boldsymbol{\omega} = \operatorname{grad} \varpi. \tag{12}$$

Taking the curl of this equation, we obtain

$$\frac{\partial \boldsymbol{\omega}}{\partial t} - \text{curl}(\mathbf{u} \times \boldsymbol{\omega}) = 0; \tag{13}$$

this is the required equation of motion for $\boldsymbol{\omega}$. The principal theorem on vorticity follows from this equation.

Consider a surface S enclosed by a simple closed contour C. Let $d\mathbf{S}$ be an element of this surface. Multiplying equation (13) scalarly by $d\mathbf{S}$, and integrating over S, we obtain

$$\int_S \frac{\partial \boldsymbol{\omega}}{\partial t} . d\mathbf{S} - \int_S \text{curl}(\mathbf{u} \times \boldsymbol{\omega}) . d\mathbf{S} = 0. \tag{14}$$

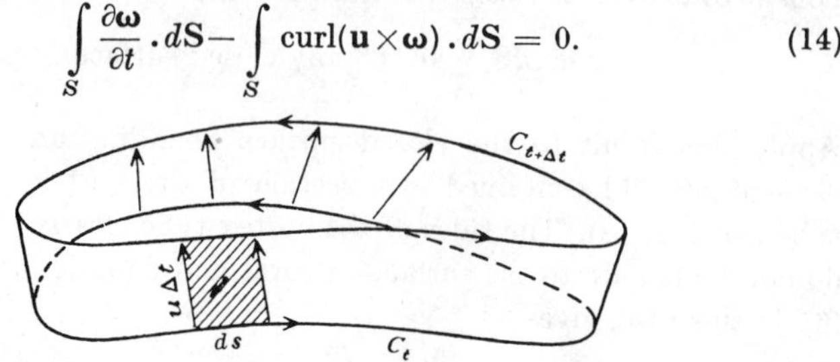

FIG. 19. Illustrating the Helmholtz–Kelvin theorem.

Transforming the second of these integrals by Stokes's theorem, we have

$$\int_S \frac{\partial \boldsymbol{\omega}}{\partial t} . d\mathbf{S} + \int_C \boldsymbol{\omega} . (\mathbf{u} \times d\mathbf{s}) = 0, \tag{15}$$

where $d\mathbf{s}$ is an element of arc of the contour C. The physical meaning of this last equation becomes clearer if we multiply the equation by an infinitesimal interval of time Δt, and then interpret the different terms. We have

$$\Delta t \int_{S_t} \frac{\partial \boldsymbol{\omega}_t}{\partial t} . d\mathbf{S} + \int_{C_t} \boldsymbol{\omega}_t . (\mathbf{u}\Delta t \times d\mathbf{s}) = 0, \tag{16}$$

where we have inserted a subscript t to the various quantities to emphasize that they are the instantaneous values at the time indicated (see Fig. 19). Now $\mathbf{u}\Delta t \times d\mathbf{s}$ is the area swept out by the fluid elements along $d\mathbf{s}$ in a time Δt. Accordingly, the integral along C_t in (16) is, in reality, *a surface integral* over the elementary strip connecting the contour C_t with the contour $C_{t+\Delta t}$ to which C_t will be carried in the time Δt. Applying the result (6) to the closed surface consisting of S_t (bounded by C_t), $S_{t+\Delta t}$ (bounded by $C_{t+\Delta t}$), and the strip connecting C_t and $C_{t+\Delta t}$, we have†

$$\int_{C_t} \boldsymbol{\omega}_t . (\mathbf{u}\Delta t \times d\mathbf{s}) = \int_{S_{t+\Delta t}} \boldsymbol{\omega}_t . d\mathbf{S} - \int_{S_t} \boldsymbol{\omega}_t . d\mathbf{S}. \tag{17}$$

† Note that the validity of this equation depends on $\text{div}\, \boldsymbol{\omega} = 0$.

Using this in equation (16) ,we obtain

$$\Delta t \int\limits_{S_t} \frac{\partial \boldsymbol{\omega}_t}{\partial t}.d\mathbf{S}+ \int\limits_{S_{t+\Delta t}} \boldsymbol{\omega}_t.d\mathbf{S} = \int\limits_{S_t} \boldsymbol{\omega}_t.d\mathbf{S}. \tag{18}$$

With an error of order $(\Delta t)^2$, we can replace the integral over S_t on the left-hand side by an integral over $S_{t+\Delta t}$; we thus find,

$$\int\limits_{S_{t+\Delta t}} \left(\boldsymbol{\omega}_t+\Delta t\frac{\partial \boldsymbol{\omega}_t}{\partial t}\right).d\mathbf{S} = \int\limits_{S_t} \boldsymbol{\omega}_t.d\mathbf{S}+O(\Delta t^2), \tag{19}$$

or

$$\int\limits_{S_{t+\Delta t}} \boldsymbol{\omega}_{t+\Delta t}.d\mathbf{S} = \int\limits_{S_t} \boldsymbol{\omega}_t.d\mathbf{S}+O(\Delta t^2). \tag{20}$$

Passing to the limit $\Delta t = 0$, we obtain

$$\frac{d}{dt}\int\limits_{S_t} \boldsymbol{\omega}.d\mathbf{S} = 0. \tag{21}$$

We have thus proved: *The integral of the normal component of* $\boldsymbol{\omega}$ *over any surface S bounded by a closed curve remains constant as we follow the surface S with the motion of the fluid elements constituting it*:

$$\int\limits_{S_t} \boldsymbol{\omega}_t.d\mathbf{S} = \text{constant}. \tag{22}$$

This is the principal theorem on vorticity due to Helmholtz and Kelvin. We may state it also in the form: *the strength of a vortex tube is an integral of the equations of motion.*

Transforming (22) by Stokes's theorem, we have

$$\int\limits_{S_t} \boldsymbol{\omega}_t.d\mathbf{S} = \int\limits_{S_t} \operatorname{curl}\mathbf{u}_t.d\mathbf{S} = \int\limits_{C_t} \mathbf{u}_t.d\mathbf{s} = \text{constant}. \tag{23}$$

The integral of $\mathbf{u}$ along any closed curve is the *circulation* along the curve. By (23), *the circulation along any closed curve C remains constant as we follow the curve C with the motion of the fluid elements constituting it.*

It is customary to derive the constancy of the circulation along a closed curve first, and then deduce from it the constancy of the flux of the vorticity across a surface. We have followed a reverse procedure to make comparisons with the motions of the magnetic lines of force in hydromagnetics (Chapter IV, § 38 (*b*)).

A further important fact is that vortex lines retain their character as vortex lines as we follow the fluid motion. This follows from equation (13) written somewhat differently. Since

$$\epsilon_{ijk}\frac{\partial}{\partial x_j}\epsilon_{klm}u_l\omega_m = (\delta_{il}\delta_{jm}-\delta_{im}\delta_{jl})\frac{\partial}{\partial x_j}u_l\omega_m = \frac{\partial}{\partial x_j}(u_i\omega_j-u_j\omega_i), \tag{24}$$

we can write
$$\frac{\partial \omega_i}{\partial t}+\frac{\partial}{\partial x_j}(u_j\omega_i-u_i\omega_j) = 0. \tag{25}$$

Using the solenoidal character† of $\mathbf{u}$ and $\boldsymbol{\omega}$, we can rewrite this last equation in the form
$$\frac{\partial \omega_i}{\partial t}+u_j\frac{\partial \omega_i}{\partial x_j} = \omega_j\frac{\partial u_i}{\partial x_j}, \tag{26}$$

or
$$\frac{\partial \boldsymbol{\omega}}{\partial t}+(\mathbf{u}\,.\,\text{grad})\boldsymbol{\omega} = \frac{d\boldsymbol{\omega}}{dt} = (\boldsymbol{\omega}\,.\,\text{grad})\mathbf{u}. \tag{27}$$

Consider now a vortex line. Let A and B be two adjacent points on it. Since the element of arc joining AB is in the direction of $\boldsymbol{\omega}$,
$$\overrightarrow{AB} = \delta q\boldsymbol{\omega}, \tag{28}$$
where δq is an infinitesimal constant. The velocities at A and B differ by the amount
$$\mathbf{u}_B-\mathbf{u}_A = (\overrightarrow{AB}\,.\,\text{grad})\mathbf{u}_A = \delta q(\boldsymbol{\omega}\,.\,\text{grad})\mathbf{u}_A; \tag{29}$$
or, making use of equation (27), we have
$$\mathbf{u}_B-\mathbf{u}_A = \delta q\frac{d\boldsymbol{\omega}}{dt}. \tag{30}$$

During a time Δt, the points A and B will suffer displacements of amounts $\mathbf{u}_A\Delta t$ and $\mathbf{u}_B\Delta t$, respectively. If the displaced positions of A and B are A' and B', then
$$\overrightarrow{A'B'} = \overrightarrow{AB}+(\mathbf{u}_B-\mathbf{u}_A)\Delta t = \delta q\left(\boldsymbol{\omega}+\frac{d\boldsymbol{\omega}}{dt}\Delta t\right) = \delta q\,\boldsymbol{\omega}_{t+\Delta t}. \tag{31}$$

In other words, $\overrightarrow{A'B'}$ is tangential to the vortex line through A' at time $t+\Delta t$. *A vortex line, therefore, consists of the same fluid elements*: *it moves with the fluid like a material substance.* We may if we like say: *the vortex lines are permanently attached to the fluid.* If a vortex line should disappear in a real fluid, it can only be because of dissipation by viscosity.

21. The equations of hydrodynamics in a rotating frame of reference

Consider a fluid in rotation about some fixed axis with a constant angular velocity Ω. It will be convenient to describe the motions which occur in it as they will appear to an observer at rest in a frame rotating about the same axis and with the same angular velocity. In such a

† We are using the solenoidal character of $\mathbf{u}$ here for the first time in this section.

rotating frame of reference, quantities which will be recognized as velocities and accelerations by an observer at rest in it are not the same as the velocities and accelerations as recognized by an observer at rest with respect to the fixed inertial frame. To avoid the resulting ambiguities and to fix meanings, consider explicitly the inertial frame (ξ, η, ζ) with respect to which the chosen coordinate system (x, y, z) is rotating with the angular velocity Ω about the ζ-axis. Also, the z- and the ζ-axes will be assumed to coincide. The transformation relating the two coordinate systems is

$$\begin{aligned} x &= +\xi\cos\Omega t + \eta\sin\Omega t, \\ y &= -\xi\sin\Omega t + \eta\cos\Omega t, \\ z &= +\zeta. \end{aligned} \tag{32}$$

A vector $\mathbf{q}$, with components q_ξ, q_η, and q_ζ in the inertial frame, will have the following components along the instantaneous directions of the axes of the rotating frame:

$$\begin{aligned} q_x^{(0)} &= +q_\xi\cos\Omega t + q_\eta\sin\Omega t, \\ q_y^{(0)} &= -q_\xi\sin\Omega t + q_\eta\cos\Omega t, \\ q_z^{(0)} &= +q_\zeta. \end{aligned} \tag{33}$$

We have distinguished the components of the vector obtained by this transformation by a superscript (0) because they are not necessarily the same quantities which will be recognized by the observer, at rest in the rotating frame, as having the physical meaning of $\mathbf{q}$. The superscript (0) will distinguish then the quantities which will have the same *absolute*, as distinct from *relative*, meanings. The reason for this distinction will become clear when we consider the quantities which have the meanings of velocity and acceleration in the rotating frame and their relations to the 'absolute' velocity and acceleration.

By differentiating equation (32) with respect to t, we obtain

$$\frac{dx}{dt} = \left(+\frac{d\xi}{dt}\cos\Omega t + \frac{d\eta}{dt}\sin\Omega t\right) - \Omega(\xi\sin\Omega t - \eta\cos\Omega t), \tag{34}$$

$$\frac{dy}{dt} = \left(-\frac{d\xi}{dt}\sin\Omega t + \frac{d\eta}{dt}\cos\Omega t\right) - \Omega(\xi\cos\Omega t + \eta\sin\Omega t). \tag{35}$$

Clearly, dx/dt and dy/dt are the quantities which will be recognized by the observer at rest in the rotating frame as the components u_x and u_y of the velocities of the fluid element along the x- and the y-directions; whereas the quantities grouped together in the first brackets on the right-hand sides of equations (34) and (35) are the components along the instantaneous directions of x and y of what *is* the velocity in the inertial frame; they are the quantities which must be distinguished by the superscript (0). Equations (34) and (35) relate u_x and u_y with $u_x^{(0)}$ and $u_y^{(0)}$; we have

$$u_x = u_x^{(0)}+\Omega y; \quad u_y = u_y^{(0)}-\Omega x; \quad u_z = u_z^{(0)}. \tag{36}$$

These equations can be combined into the single vector equation

$$\mathbf{u} = \mathbf{u}^{(0)}-\boldsymbol{\Omega}\times\mathbf{r}. \tag{37}$$

Now differentiating equations (34) and (35) once again with respect to t, we obtain

$$\begin{aligned} \frac{d^2x}{dt^2} &= \left(+\frac{d^2\xi}{dt^2}\cos\Omega t+\frac{d^2\eta}{dt^2}\sin\Omega t\right)+2\Omega\left(-\frac{d\xi}{dt}\sin\Omega t+\frac{d\eta}{dt}\cos\Omega t\right)-\Omega^2 x, \\ \frac{d^2y}{dt^2} &= \left(-\frac{d^2\xi}{dt^2}\sin\Omega t+\frac{d^2\eta}{dt^2}\cos\Omega t\right)+2\Omega\left(-\frac{d\xi}{dt}\cos\Omega t-\frac{d\eta}{dt}\sin\Omega t\right)-\Omega^2 y. \end{aligned} \tag{38}$$

The equations are equivalent to

$$\begin{aligned} \frac{du_x}{dt} &= \left(\frac{du_x^{(0)}}{dt}\right)^{(0)}+2\Omega u_y^{(0)}-\Omega^2 x, \\ \frac{du_y}{dt} &= \left(\frac{du_y^{(0)}}{dt}\right)^{(0)}-2\Omega u_x^{(0)}-\Omega^2 y. \end{aligned} \tag{39}$$

Substituting for $u_x^{(0)}$ and $u_y^{(0)}$ from (36) in the foregoing equations, we obtain

$$\begin{aligned} \left(\frac{du_x^{(0)}}{dt}\right)^{(0)} &= \frac{du_x}{dt}-2\Omega u_y-\Omega^2 x, \\ \left(\frac{du_y^{(0)}}{dt}\right)^{(0)} &= \frac{du_y}{dt}+2\Omega u_x-\Omega^2 y, \\ \left(\frac{du_z^{(0)}}{dt}\right)^{(0)} &= \frac{du_z}{dt}. \end{aligned} \tag{40}$$

These equations can be combined into the single vector equation

$$\left(\frac{d\mathbf{u}^{(0)}}{dt}\right)^{(0)} = \frac{d\mathbf{u}}{dt}+2\boldsymbol{\Omega}\times\mathbf{u}-\tfrac{1}{2}\operatorname{grad}(|\boldsymbol{\Omega}\times\mathbf{r}|^2). \tag{41}$$

The term $2\boldsymbol{\Omega}\times\mathbf{u}$ in this equation represents the Coriolis acceleration and the term $-\frac{1}{2}\operatorname{grad}(|\boldsymbol{\Omega}\times\mathbf{r}|^2)$ represents the centrifugal force.

For differentiations with respect to the space coordinates, there are no complicating considerations such as we have just explained for differentiations with respect to time leading to velocities and accelerations. Accordingly, the standard hydrodynamical equation (equation II (17))

$$\frac{du_i}{dt} = X_i+\frac{1}{\rho}\frac{\partial P_{ij}}{\partial x_j} \tag{42}$$

(where X_i is the ith component of whatever external forces may be acting on the fluid and P_{ij} is the stress tensor), in the rotating frame of reference, becomes

$$\frac{du_i}{dt} = \frac{\partial u_i}{\partial t}+u_j\frac{\partial u_i}{\partial x_j} = X_i+\frac{1}{\rho}\frac{\partial P_{ij}}{\partial x_j}+\frac{\partial}{\partial x_i}(\tfrac{1}{2}|\mathbf{\Omega}\times\mathbf{r}|^2)+2\epsilon_{ijk}u_j\Omega_k. \quad (43)$$

The equation of continuity and the equation of heat conduction remain unaffected.

For an incompressible fluid, the equations of motion take the explicit form (equation II (19))

$$\frac{\partial u_i}{\partial t}+u_j\frac{\partial u_i}{\partial x_j} = X_i-\frac{\partial}{\partial x_i}\left(\frac{p}{\rho}-\tfrac{1}{2}|\mathbf{\Omega}\times\mathbf{r}|^2\right)+\nu\nabla^2 u_i+2\epsilon_{ijk}u_j\Omega_k. \quad (44)$$

The equation of continuity in the form (2) continues to hold.

22. The Taylor–Proudman theorem

Consider the equation of motion (44) for an inviscid fluid in case the external forces are derivable from a potential V. We then have

$$\frac{\partial u_i}{\partial t}+u_j\frac{\partial u_i}{\partial x_j} = -\frac{\partial}{\partial x_i}\left(\frac{p}{\rho}-\tfrac{1}{2}|\mathbf{\Omega}\times\mathbf{r}|^2+V\right)+2\epsilon_{ijk}u_j\Omega_k. \quad (45)$$

Making use of the identity (9), we can write

$$\frac{\partial u_i}{\partial t}-\epsilon_{ijk}u_j\omega_k = -\frac{\partial \varpi}{\partial x_i}+2\epsilon_{ijk}u_j\Omega_k, \quad (46)$$

where now
$$\varpi = p/\rho+\tfrac{1}{2}|\mathbf{u}|^2+V-\tfrac{1}{2}|\mathbf{\Omega}\times\mathbf{r}|^2. \quad (47)$$

Alternatively, we can also write

$$\frac{\partial \mathbf{u}}{\partial t}-\mathbf{u}\times(\boldsymbol{\omega}+2\mathbf{\Omega}) = -\operatorname{grad}\varpi; \quad (48)$$

and taking the curl of this equation, we have

$$\frac{\partial \boldsymbol{\omega}}{\partial t}-\operatorname{curl}[\mathbf{u}\times(\boldsymbol{\omega}+2\mathbf{\Omega})] = 0. \quad (49)$$

Since $\mathbf{\Omega}$ is a constant vector, it is clear from a comparison of equations (13) and (49) that *in a rotating frame of reference* $\boldsymbol{\omega}+2\mathbf{\Omega}$ *plays the role of* $\boldsymbol{\omega}$ *in an inertial frame.* We can, therefore, conclude that now

$$\int_S (\boldsymbol{\omega}+2\mathbf{\Omega})\,.\,d\mathbf{S} = \text{constant} \quad (50)$$

as we follow a surface S (enclosed by a simple closed curve C) with the motion in the rotating frame. An equivalent statement is:

The circulation round a closed contour C bounding a surface $S+$

$$+2\Omega \times (\textit{the projection of the area of } S \textit{ on a plane perpendicular to } \mathbf{\Omega}) = \textit{constant}. \quad (51)$$

In this form, the theorem appears to have been first stated by V. Bjerknes.

When the motions are slow and steady, the theorem asserts:

The projection of the area of S on a plane perpendicular to $\mathbf{\Omega}$ is a constant. (52)

What this implies for the motions themselves can be inferred more directly from equation (49). In a state of steady, slow motions when the squares of the velocity components can be neglected, equation (49) requires that

$$\operatorname{curl}(\mathbf{u} \times \mathbf{\Omega}) = 0, \quad (53)$$

or (cf. equation (24))

$$\frac{\partial}{\partial x_j}(u_i\,\Omega_j - u_j\,\Omega_i) = 0. \quad (54)$$

Since $\mathbf{\Omega}$ is a constant vector and $\mathbf{u}$ is solenoidal, equation (54) reduces to

$$\Omega_j \frac{\partial u_i}{\partial x_j} = 0. \quad (55)$$

The motions cannot, therefore, vary in the direction of $\mathbf{\Omega}$. In other words, *all steady slow motions in a rotating inviscid fluid are necessarily two-dimensional.* This is the form in which the Taylor–Proudman theorem is generally stated; equation (50) is, however, a more complete statement.

What the condition (55) implies is this: the fact that the motions transverse to $\mathbf{\Omega}$ cannot vary in the direction of $\mathbf{\Omega}$ means that any two fluid elements which are initially on a line parallel to $\mathbf{\Omega}$ will always remain on that line; and the fact that motions in the direction of $\mathbf{\Omega}$ cannot also vary along this direction means that two fluid elements which are initially a certain distance apart (in the direction of $\mathbf{\Omega}$) will always remain at this distance apart. This predicted behaviour of slow, steady motions in a rotating fluid was strikingly demonstrated by Taylor as follows.

Water in a tank was first made to rotate steadily as a solid body. A small motion was then communicated to it and a few drops of coloured liquid (or ink) were inserted. The slow movement of the fluid then drew this coloured portion of the fluid into sheets; and these sheets always remained parallel to the axis of rotation. Taylor noted: 'The accuracy with which they remained parallel to the axis of rotation is quite extraordinary.' These experiments of Taylor have been repeated by Fultz; one of his photographs illustrating this phenomenon is reproduced in Fig. 20.

23. The propagation of waves in a rotating fluid

We shall conclude this excursion into general hydrodynamical theorems by showing that the equations of motion in a rotating fluid allow periodic solutions representing the propagation of waves.

In the absence of external forces, or in the presence of forces which are derivable from a potential, the equations of motion governing a viscous incompressible fluid take the form (cf. equation (44))

$$\frac{\partial u_i}{\partial t}+u_j\frac{\partial u_i}{\partial x_j} = -\frac{\partial \varpi}{\partial x_i}+\nu\nabla^2 u_i+2\epsilon_{ijk}\,u_j\,\Omega_k. \tag{56}$$

We seek solutions of this equation with a space-time dependence given by

$$e^{i(k_j x_j+pt)}. \tag{57}$$

For solutions having this space-time dependence,

$$\frac{\partial}{\partial t} = ip, \quad \frac{\partial}{\partial x_j} = ik_j, \quad \text{and} \quad \nabla^2 = -k^2; \tag{58}$$

and equation (56) gives

$$(ip+\nu k^2)u_i+ik_j\,u_j\,u_i = -ik_i\varpi+2\epsilon_{ijk}\,u_j\Omega_k; \tag{59}$$

while the equation of continuity gives

$$k_j\,u_j = 0. \tag{60}$$

This last equation implies that the waves are *transverse*.

In view of (60), equation (59) becomes

$$nu_i+k_i\varpi+2i\epsilon_{ijk}\,u_j\Omega_k = 0, \tag{61}$$

where we have written $$n = p-i\nu k^2. \tag{62}$$

Multiplying equation (61) scalarly by u_i, k_i, and Ω_i, we obtain, respectively,

$$\begin{aligned} nu_i^2 &= 0,\\ k^2\varpi+2i\epsilon_{ijk}\,k_i\,u_j\Omega_k &= 0,\\ nu_i\Omega_i+\varpi k_i\Omega_i &= 0. \end{aligned} \tag{63}$$

It is now convenient to choose the orientation of the coordinate system such that $$\mathbf{\Omega} = (\Omega_x, 0, \Omega_z) \quad \text{and} \quad \mathbf{k} = (0, 0, k); \tag{64}$$

this clearly implies no loss of generality. With this choice of the coordinate system, equation (60) gives

$$ku_z = 0 \quad \text{or} \quad u_z = 0. \tag{65}$$

Equations (63) now give

$$n(u_x^2+u_y^2) = 0, \tag{66}$$

$$2i\Omega_x u_y = +k\varpi, \tag{67}$$

$$n\Omega_x u_x = -k\varpi\Omega_z. \tag{68}$$

These equations enable us to express the amplitudes u_y and ϖ in terms of u_x; we find

$$u_y = +i\frac{n}{2\Omega_z}u_x; \quad \varpi = -\frac{n}{k}\frac{\Omega_x}{\Omega_z}u_x; \quad u_z = 0. \tag{69}$$

Substituting the first of these relations in equation (66), we obtain

$$n\left(1-\frac{n^2}{4\Omega_z^2}\right)u_x^2 = 0. \tag{70}$$

From this equation it follows that n is real; and ignoring the root $n = 0$, we have

$$n = \pm 2\Omega_z = \pm 2\Omega\cos\vartheta, \tag{71}$$

where ϑ is the inclination of the direction of wave-propagation to the direction of $\mathbf{\Omega}$. The gyration frequency p is given by (cf. equation (62))

$$p = n+i\nu k^2 = \pm 2\Omega\cos\vartheta+i\nu k^2. \tag{72}$$

The waves are, therefore, damped; and the damping constant is νk^2.

With n given by equation (72), the expressions for the amplitudes become

$$u_x = \pm iu_y, \qquad \varpi = \mp\frac{2\Omega}{k}u_x\sin\vartheta, \qquad u_z = 0. \tag{73}$$

The waves are, therefore, transverse, circularly polarized, and *damped.*

In the limit of zero viscosity, the velocity of the propagation of the waves is given by

$$V = \frac{p}{k} = \pm\frac{2\Omega}{k}\cos\vartheta. \tag{74}$$

This is the *phase velocity*. The *group velocity* is given by

$$\frac{\partial p}{\partial k_i} = \pm 2\frac{\partial}{\partial k_i}\left(\frac{\Omega_j k_j}{k}\right) = \pm 2\left(\frac{\Omega_i}{k}-\frac{\Omega_j k_j}{k^3}k_i\right), \tag{75}$$

or

$$\frac{\partial p}{\partial \mathbf{k}} = \pm\frac{2}{k^3}\mathbf{k}\times(\mathbf{\Omega}\times\mathbf{k}). \tag{76}$$

Finally, it should be noted that the foregoing solutions representing the propagation of waves are not limited to infinitesimal amplitudes.

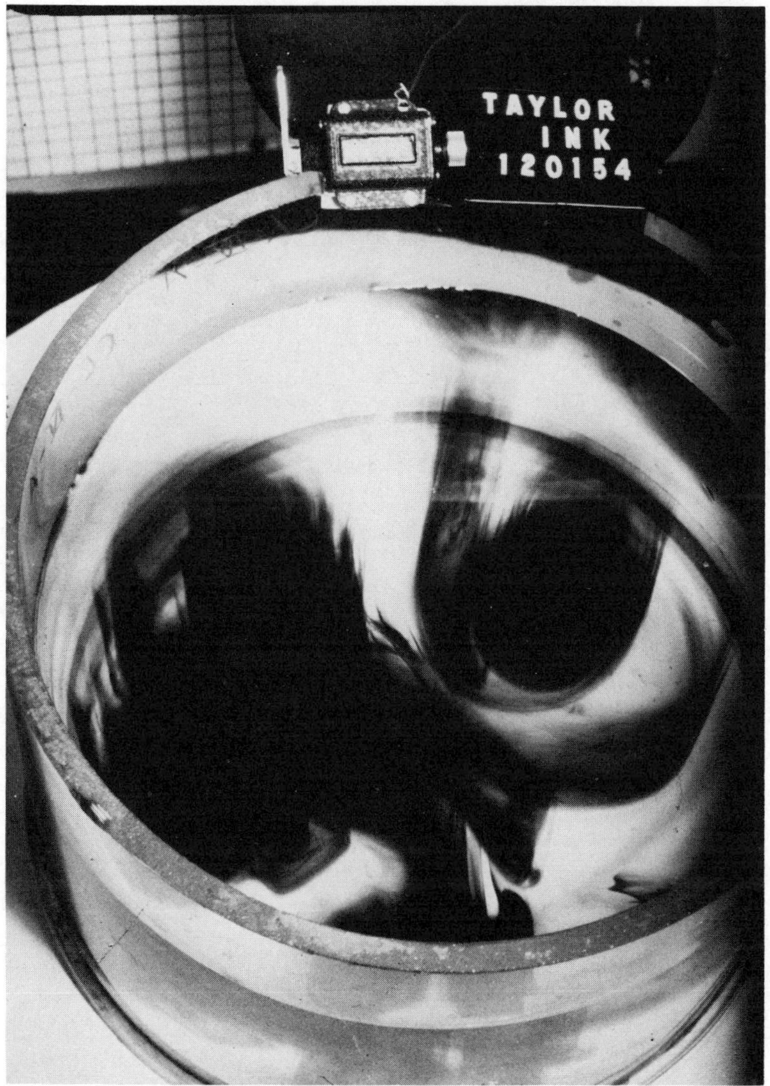

FIG. 20. Illustrating the Taylor–Proudman theorem in a rotating cylinder of water. In the experiment, ink was poured into the rotating cylinder and stirred $1\frac{1}{2}$ min later by dipping a stationary rod for 15 sec. Note the thinness of many of the filaments into which the ink is drawn in accordance with the Taylor–Proudman theorem.

24. The problem of thermal instability in a rotating fluid: general considerations

From the theorems proved in §§ 22 and 23 it is apparent that rotation should have a profound effect on the onset of thermal instability. For convection implies that motions which occur have, necessarily, a three-dimensional character; this the Taylor–Proudman theorem expressly forbids for an inviscid fluid so long as the non-linear terms in the equations of motion are neglected and the motions are steady. The first of these two conditions is certainly met in a linear stability theory; and the second obtains for steady convection. Therefore, in contrast to non-rotating fluids, an inviscid fluid in rotation should be expected to be thermally stable for *all* adverse temperature gradients. Indeed, only in the presence of viscosity can thermal instability arise; for only then can the Taylor–Proudman theorem be violated.

There is another factor to remember. While the Taylor–Proudman theorem forbids any variation of the velocity in the direction of $\mathbf{\Omega}$, it does not forbid oscillatory motions. As we have seen in § 23, the propagation of transverse waves is possible under the same circumstances. For this reason, we may anticipate that overstability as a means of developing instability may, under certain conditions, play a decisive role. We shall see that this is indeed the case.

25. The perturbation equations

Consider, then, an infinite horizontal layer of fluid which is kept rotating at a constant rate. Let $\mathbf{\Omega}$ denote the angular velocity of rotation. As we have shown explicitly in Chapter II, § 9 for problems of the type we are considering, it will suffice to work with the relevant equations of motion and heat conduction in the Boussinesq approximation. The only additional factors which we have to allow for now are the effects of Coriolis acceleration and centrifugal force in the equations of motion. Thus, in view of equation (41), we replace equation II (43) by

$$\frac{\partial u_i}{\partial t}+u_j\frac{\partial u_i}{\partial x_j}=-\frac{\partial}{\partial x_i}\left(\frac{p}{\rho_0}-\tfrac{1}{2}|\mathbf{\Omega}\times\mathbf{r}|^2\right)+\left(1+\frac{\delta\rho}{\rho_0}\right)X_i+ +\nu\nabla^2u_i+2\epsilon_{ijk}\,u_j\,\Omega_k. \quad (77)$$

The remaining equations (equations II (41), (44), and (46)) are unaffected.

We envisage now an initial state in which a steady adverse temperature gradient β is maintained and there are no motions. The characterization of this initial state differs in no respect from that described in Chapter II, § 9; and the equations governing small perturbations can be obtained in exactly the same way. Thus, equation II (55) is replaced by

$$\frac{\partial u_i}{\partial t} = -\frac{\partial}{\partial x_i}\left(\frac{\delta p}{\rho_0}\right) + g\alpha\theta\lambda_i + \nu\nabla^2 u_i + 2\epsilon_{ijk}\, u_j \Omega_k, \tag{78}$$

where $\boldsymbol{\lambda} = (0, 0, 1)$ is a unit vector in the direction of the vertical. We also have (equations II (56) and (57))

$$\frac{\partial \theta}{\partial t} = \beta\lambda_j\, u_j + \kappa\nabla^2\theta \tag{79}$$

and

$$\frac{\partial u_i}{\partial x_i} = 0. \tag{80}$$

As in Chapter II, § 9, we can eliminate the term in $\delta p/\rho_0$ in equation (78) by applying the operator $\epsilon_{ijk}\partial/\partial x_j$ to the k-component of the equation; since (cf. equation (24))

$$\epsilon_{ijk}\frac{\partial}{\partial x_j}\epsilon_{klm}\, u_l \Omega_m = \frac{\partial}{\partial x_j}(u_i\Omega_j - u_j\Omega_i) = \Omega_j\frac{\partial u_i}{\partial x_j}, \tag{81}$$

the result is (cf. equation II (67))

$$\frac{\partial \omega_i}{\partial t} = g\alpha\epsilon_{ijk}\frac{\partial \theta}{\partial x_j}\lambda_k + \nu\nabla^2\omega_i + 2\Omega_j\frac{\partial u_i}{\partial x_j}. \tag{82}$$

Taking the curl of this equation once again, we obtain (cf equation II (72))

$$\frac{\partial}{\partial t}(\nabla^2 u_i) = g\alpha\left(\lambda_i\,\nabla^2\theta - \lambda_j\frac{\partial^2\theta}{\partial x_i\,\partial x_j}\right) + \nu\nabla^4 u_i - 2\Omega_j\frac{\partial \omega_i}{\partial x_j}. \tag{83}$$

Now multiplying equations (82) and (83) by λ_i, we get

$$\frac{\partial \zeta}{\partial t} = \nu\nabla^2\zeta + 2\Omega_j\frac{\partial w}{\partial x_j} \tag{84}$$

and

$$\frac{\partial}{\partial t}\nabla^2 w = g\alpha\left(\frac{\partial^2\theta}{\partial x^2} + \frac{\partial^2\theta}{\partial y^2}\right) + \nu\nabla^4 w - 2\Omega_j\frac{\partial \zeta}{\partial x_j}, \tag{85}$$

where w and ζ are the z-components of the velocity and the vorticity, respectively.

Most of our discussions relating to equations (84) and (85) will be restricted to the case when $\mathbf{\Omega}$ and $\mathbf{g}$ are parallel; the case when they are not will be briefly considered in § 34.

When the axis of rotation coincides with the vertical, the relevant equations are

$$\frac{\partial \theta}{\partial t} = \beta w + \kappa\nabla^2\theta, \tag{86}$$

$$\frac{\partial \zeta}{\partial t} = 2\Omega\frac{\partial w}{\partial z} + \nu\nabla^2\zeta, \tag{87}$$

and
$$\frac{\partial}{\partial t}\nabla^2 w = g\alpha\left(\frac{\partial^2\theta}{\partial x^2}+\frac{\partial^2\theta}{\partial y^2}\right)+\nu\nabla^4 w-2\Omega\frac{\partial\zeta}{\partial z}; \tag{88}$$

and we have to seek solutions of these equations which satisfy the boundary conditions described in Chapter II, § 9 (*a*).

(*a*) *The analysis into normal modes*

Following the procedure in Chapter II, § 10, we analyse the perturbations w, θ, and ζ into two-dimensional periodic waves; and considering disturbances characterized by a particular wave number k, we suppose that w, θ, and ζ have the forms assumed in equation II (90). Equations (86)–(88) then become

$$p\Theta = \beta W+\kappa\left(\frac{d^2}{dz^2}-k^2\right)\Theta, \tag{89}$$

$$pZ = 2\Omega\frac{dW}{dz}+\nu\left(\frac{d^2}{dz^2}-k^2\right)Z, \tag{90}$$

$$p\left(\frac{d^2}{dz^2}-k^2\right)W = -g\alpha k^2\Theta+\nu\left(\frac{d^2}{dz^2}-k^2\right)^2 W-2\Omega\frac{dZ}{dz}. \tag{91}$$

Measuring lengths in the unit d and letting

$$a = kd, \quad \sigma = pd^2/\nu, \quad \text{and} \quad \mathfrak{p} = \nu/\kappa, \tag{92}$$

we can reduce equations (89)–(91) to the forms

$$(D^2-a^2-\mathfrak{p}\sigma)\Theta = -\left(\frac{\beta}{\kappa}d^2\right)W, \tag{93}$$

$$(D^2-a^2-\sigma)Z = -\left(\frac{2\Omega}{\nu}d\right)DW, \tag{94}$$

$$(D^2-a^2)(D^2-a^2-\sigma)W-\left(\frac{2\Omega}{\nu}d^3\right)DZ = \left(\frac{g\alpha}{\nu}d^2\right)a^2\Theta, \tag{95}$$

where $D = d/dz$ (z being measured in the new unit). Solutions of equations (93)–(95) must be sought which satisfy the boundary conditions given in equations II (95) and (96).

Once we have obtained the proper solutions of the foregoing equations, we can complete the solution by determining the horizontal components of the velocity by making use of the general relations derived in Chapter II, § 10 (*a*).

26. The case when instability sets in as stationary convection. A variational principle

We shall find that in contrast to the simple Bénard problem, the principle of the exchange of stabilities is not valid, generally, when rotation is present. However, it is not possible to state in simple analytical terms the necessary and sufficient conditions for its validity; it will appear that they are best stated in terms of the solution for the onset of instability as stationary convection. For this reason, we shall first establish the conditions for this latter type of instability to occur before deciding whether instability will in fact occur in this way.

When instability sets in as ordinary convection, the marginal state will be characterized by $\sigma = 0$, and the basic equations reduce to (cf. equations (93)–(95))

$$(D^2-a^2)\Theta = -\left(\frac{\beta}{\kappa} d^2\right)W, \tag{96}$$

$$(D^2-a^2)Z = -\left(\frac{2\Omega}{\nu} d\right)DW, \tag{97}$$

and

$$(D^2-a^2)^2W-\left(\frac{2\Omega}{\nu} d^3\right)DZ = \left(\frac{g\alpha}{\nu} d^2\right)a^2\Theta. \tag{98}$$

We can eliminate Z and Θ from the last equation by operating (D^2-a^2) on it. We find:

$$(D^2-a^2)^3W+TD^2W = -Ra^2W, \tag{99}$$

where R is the Rayleigh number as usually defined, and

$$T = \frac{4\Omega^2}{\nu^2} d^4 \tag{100}$$

is the Taylor number.

The boundary conditions with respect to which we must seek solutions of the foregoing equations are (cf. equations II (95) and (96))

$$W = 0 \quad \text{and} \quad \Theta = 0 \quad \text{for } z = 0 \text{ and } 1, \tag{101}$$

and

$$\left.\begin{aligned} &\textit{either} \quad Z = 0 \quad \text{and} \quad DW = 0 \quad \text{(on a rigid surface)} \\ &\textit{or} \quad DZ = 0 \quad \text{and} \quad D^2W = 0 \quad \text{(on a free surface)} \end{aligned}\right\}. \tag{102}$$

In view of equation (98) we can replace the conditions (101) by

$$W = 0 \quad \text{and} \quad (D^2-a^2)^2W-\left(\frac{2\Omega}{\nu} d^3\right)DZ = 0 \quad \text{for } z = 0 \text{ and } 1. \tag{103}$$

Since the boundary conditions (103) involve Z, it is clear that we cannot treat equation (99) independently of equation (97) for Z; the order of the system we are effectively dealing with is, therefore, eight *not* six.

Equations (97) and (99), together with the boundary conditions (102) and (103), clearly constitute a characteristic value problem for R for given a^2 and T; and the problem of determining the critical Rayleigh number for the onset of instability as stationary convection reduces to the following.

We must first determine the lowest characteristic value of R, through the solution of equations (97), (99), (102), and (103), as a function of a^2 for a given T and then find the minimum of this function; and the minimum so determined is the critical Rayleigh number for the onset of stationary convection for the given T.

(a) *A variational principle*

As in the case of the simple Bénard problem, we can formulate the present characteristic value problem also in terms of a variational principle.

Letting
$$F = (D^2-a^2)^2W-\left(\frac{2\Omega}{\nu}d^3\right)DZ, \tag{104}$$

we can rewrite the differential equations governing W and Z in the forms

$$(D^2-a^2)F = -Ra^2W \tag{105}$$

and
$$(D^2-a^2)Z = -\left(\frac{2\Omega}{\nu}d\right)DW. \tag{106}$$

The boundary conditions (103) require that

$$W = F = 0 \quad \text{for } z = 0 \text{ and } 1. \tag{107}$$

Now multiply equation (105) by F and integrate over the range of z. After an integration by parts, the left-hand side of the equation gives

$$\int_0^1 F(D^2-a^2)F\,dz = -\int_0^1 [(DF)^2+a^2F^2]\,dz, \tag{108}$$

the integrated part vanishing on account of the boundary condition on F; and the right-hand side of the equation requires us to consider

$$\int_0^1 WF\,dz = \int_0^1 W(D^2-a^2)^2W\,dz-\frac{2\Omega}{\nu}d^3\int_0^1 W\,DZ\,dz. \tag{109}$$

After two integrations by parts, the first of the two integrals on the right-hand side of (109) becomes (cf. equation II (120))

$$\int_0^1 W(D^2-a^2)^2W\,dz = \int_0^1 [(D^2-a^2)W]^2\,dz, \tag{110}$$

while the second, after an integration by parts, gives (cf. equations (103) and (106))

$$-\frac{2\Omega}{\nu}d^3\int_0^1 W\,DZ\,dz = \frac{2\Omega}{\nu}d^3\int_0^1 Z\,DW\,dz = -d^2\int_0^1 Z(D^2-a^2)Z\,dz. \tag{111}$$

Remembering that on a bounding surface either Z or DZ vanishes, we obtain after a further integration by parts

$$-\frac{2\Omega}{\nu}d^3\int_0^1 W\,DZ\,dz = d^2\int_0^1 [(DZ)^2+a^2Z^2]\,dz. \tag{112}$$

Combining equations (109), (110), and (112), we have

$$\int_0^1 WF\,dz = \int_0^1 [(D^2-a^2)W]^2\,dz+d^2\int_0^1 [(DZ)^2+a^2Z^2]\,dz. \tag{113}$$

The result of multiplying equation (105) by F and integrating over z is

$$R = \frac{\int_0^1 [(DF)^2+a^2F^2]\,dz}{a^2\int_0^1 \{[(D^2-a^2)W]^2+d^2[(DZ)^2+a^2Z^2]\}\,dz} = \frac{I_1}{a^2I_2}\ \text{(say)}. \tag{114}$$

This formula expresses R as a ratio of two positive definite integrals.

Consider now the effect on R of variations δW and δZ in W and Z compatible with the boundary conditions on W and Z. To the first order, we have

$$\delta R = \frac{1}{a^2I_2}\left(\delta I_1-\frac{I_1}{I_2}\delta I_2\right) = \frac{1}{a^2I_2}(\delta I_1-Ra^2\delta I_2), \tag{115}$$

where δI_1 and δI_2 are the corresponding variations in I_1 and I_2:

$$\delta I_1 = 2\int_0^1 [(DF)(D\delta F)+a^2F\,\delta F]\,dz,$$

$$\delta I_2 = 2\int_0^1 \{[(D^2-a^2)W][(D^2-a^2)\,\delta W]+ +d^2[(DZ)(D\,\delta Z)+a^2Z\,\delta Z]\}\,dz. \tag{116}$$

Making use of the boundary conditions which F, δF, W, δW, Z, and δZ satisfy, we can reduce the expressions for δI_1 and δI_2 by one or more integrations by parts. Thus

$$\delta I_1 = -2\int \delta F(D^2-a^2)F\,dz, \tag{117}$$

and

$$\delta I_2 = 2\int_0^1 W(D^2-a^2)^2\,\delta W - 2d^2\int_0^1 \delta Z(D^2-a^2)Z\,dz$$

$$= 2\int_0^1 W(D^2-a^2)^2\,\delta W + \frac{4\Omega}{\nu}\,d^3\int_0^1 \delta Z\,DW\,dz$$

$$= 2\int_0^1 W\left\{(D^2-a^2)^2\,\delta W - \left(\frac{2\Omega}{\nu}\,d^3\right)D\,\delta Z\right\}dz$$

$$= 2\int_0^1 W\,\delta F\,dz. \tag{118}$$

Now combining equations (115), (117), and (118), we have

$$\delta R = -\frac{2}{a^2 I_2}\int_0^1 \delta F\{(D^2-a^2)F + Ra^2W\}\,dz, \tag{119}$$

where it may be recalled that equation (104) defining F and equation (106) relating Z and W have been explicitly used in the reductions.

From equation (119) it follows that

$$\delta R = 0 \quad \text{if} \quad (D^2-a^2)F = -Ra^2W; \tag{120}$$

and conversely if $\delta R = 0$ *for any arbitrary variation,*

$$\delta F = (D^2-a^2)^2\delta W - (2\Omega d^3/\nu)D\,\delta Z,$$

compatible with the boundary conditions of the problem, then equation (120) *must hold and the function* W *in terms of which* R *was initially calculated must have been a solution of the characteristic value problem.*

The foregoing arguments establish the stationary property of the characteristic values when they are regarded as given by equation (114). And it can be shown, by a procedure exactly analogous to that used in connexion with the first variational principle for the simple Bénard problem (Chapter II, § 13 (*a*)), that the lowest characteristic value of R is, indeed, a true minimum of the functional (114). We are thus enabled to formulate the following variational procedure for solving equations (97) and (99) (for any assigned a^2 and T) and satisfying the boundary conditions of the problem.

Assume for F an expansion involving one or more parameters A_k which vanishes for $z = 0$ and 1. With the chosen form of F, determine W and Z as solutions of the equations

$$(D^2-a^2)^2W - \left(\frac{2\Omega}{\nu}\,d^3\right)DZ = F \tag{121}$$

and
$$(D^2-a^2)Z = -\left(\frac{2\Omega}{\nu}d\right)DW \tag{122}$$

which satisfy a total of six boundary conditions on W and Z at $z = 0$ and 1. Since equations (121) and (122) are together of order six, there will be just enough constants of integration in the general solution to satisfy all the boundary conditions. Having solved for W and Z in this manner, evaluate R according to the formula (114) and minimize it with respect to the parameters A_k. In this way, we shall obtain the 'best' value of R for the chosen form of F. We shall see in the following section that even with the simplest trial function for F, we can reach quite high precision in the deduced values of R.

Finally, it may be noted that according to equations (121) and (122), the equation for W that must be solved when using the variational method is
$$(D^2-a^2)^3W+TD^2W = (D^2-a^2)F, \tag{123}$$
where T is the Taylor number.

27. Solutions for the case when instability sets in as stationary convection

We shall obtain the solution for the following three cases: when both the bounding surfaces are free; when both the bounding surfaces are rigid; and when one bounding surface is rigid and the other is free.

(a) *The solution for two free boundaries*

In this case the boundary conditions (102) and (103) require
$$W = D^2W = D^4W = 0 \quad \text{and} \quad DZ = 0 \text{ for } z = 0 \text{ and } 1. \tag{124}$$
From the equation satisfied by W (namely (99)), it follows that $D^6W = 0$ for $z = 0$ and 1. By differentiating equation (99) even numbers of times, we can successively conclude that all the even derivatives of W must vanish for $z = 0$ and 1. The proper solutions for W must, therefore, be
$$W = A\sin n\pi z, \tag{125}$$
where A is a constant and n is an integer. The corresponding solution for Z is
$$Z = A\left(\frac{2\Omega}{\nu}d\right)\frac{n\pi}{n^2\pi^2+a^2}\cos n\pi z. \tag{126}$$

Substitution of the solution (125) in equation (99) leads to the characteristic equation
$$R = \frac{1}{a^2}[(n^2\pi^2+a^2)^3+n^2\pi^2T]. \tag{127}$$

For a given a^2, the lowest characteristic value for R occurs when $n = 1$; then,

$$R = \frac{1}{a^2}[(\pi^2+a^2)^3+\pi^2 T]; \tag{128}$$

and this equation has to be interpreted in the same way as equation II (193) was interpreted in Chapter II, § 15 (*a*).

Letting
$$a^2 = \pi^2 x, \tag{129}$$

we can rewrite equation (128) in the form

$$R = \pi^4 \frac{1}{x}\left[(1+x)^3+\frac{T}{\pi^4}\right]. \tag{130}$$

TABLE VII

Critical Rayleigh numbers and wave numbers of the unstable modes at marginal stability for the onset of stationary convection when both bounding surfaces are free

T	a_c	R_c	T	a_c	R_c
0	2·233	$6·575\times10^2$	3×10^5	10·45	$4·257\times10^4$
10	2·270	$6·771\times10^2$	10^6	12·86	$9·222\times10^4$
10^2	2·594	$8·263\times10^2$	10^7	19·02	$4·147\times10^5$
5×10^2	3·278	$1·275\times10^3$	10^8	28·02	$1·897\times10^6$
10^3	3·710	$1·676\times10^3$	10^9	41·20	$8·746\times10^6$
2×10^3	4·221	$2·299\times10^3$	10^{10}	60·52	$4·047\times10^7$
5×10^3	5·011	$3·670\times10^3$	10^{11}	88·87	$1·876\times10^8$
10^4	5·698	$5·377\times10^3$	10^{12}	130·46	$8·701\times10^8$
3×10^4	6·961	$1·021\times10^4$	10^{13}	191·51	$4·037\times10^9$
10^5	8·626	$2·131\times10^4$			

As a function of x, R given by equation (130) attains its minimum when

$$2x^3+3x^2 = 1+T/\pi^4. \tag{131}$$

With x determined as the root of this cubic equation, equation (130) will give the required critical Rayleigh number R_c. Values of R_c determined in this fashion for various values of T are given in Table VII; the wave numbers characterizing the marginal states are also given. The results are further illustrated in Figs. 21 and 22. The inhibiting effect of rotation on the onset of instability is apparent from these results.

For T/π^4 sufficiently large, the required root of equation (131) tends to

$$x_{\min} \to \left(\frac{T}{2\pi^4}\right)^{\frac{1}{3}} \quad (T \to \infty). \tag{132}$$

The corresponding asymptotic behaviours of R_c and $a_{\min}$ are:

$$\left.\begin{aligned} R_c &\to 3\pi^4\left(\frac{T}{2\pi^4}\right)^{\frac{2}{3}} = 8·6956T^{\frac{2}{3}} \\ \text{and}\qquad a_{\min} &\to (\tfrac{1}{2}\pi^2 T)^{\frac{1}{6}} = 1·3048T^{\frac{1}{6}} \end{aligned}\right\} \quad (T \to \infty). \tag{133}$$

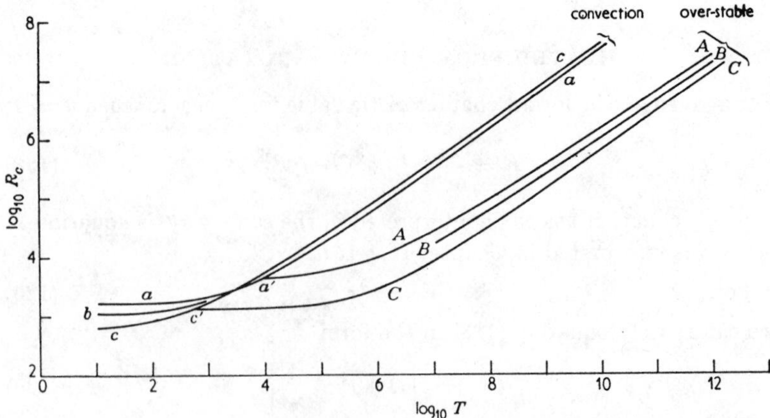

FIG. 21. The (R_c, T)-relations for the three cases (i) both bounding surfaces rigid, (ii) one bounding surface rigid and the other free, and (iii) both bounding surfaces free; the curves labelled aa, b, and cc are the relations for the onset of ordinary cellular convection for the three cases, respectively. The curves labelled $a'AA$, BB, and $c'CC$ are the corresponding relations for the onset of overstability for $\mathfrak{p} = 0{\cdot}025$. At a' (respectively c') we have a change from one type of instability to another as T increases.

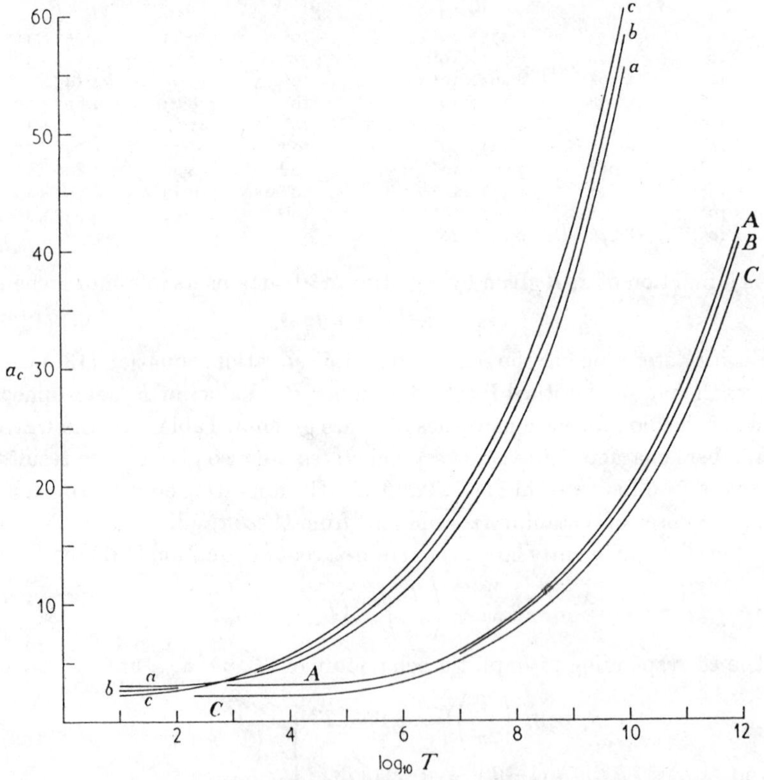

FIG. 22. The (a_c, T)-relations for the three cases (i) both bounding surfaces rigid, (ii) one bounding surface rigid and the other free, and (iii) both bounding surfaces free; the curves labelled aa, bb, and cc are the relations for the onset of ordinary cellular convection for the three cases, respectively. The curves labelled AA, B, and CC are the corresponding relations for the onset of overstability for $\mathfrak{p} = 0{\cdot}025$.

Substituting for R and T in accordance with their definitions, we find that the formula which determines the critical temperature gradient for the onset of stationary convection, when $T \to \infty$, is

$$g\alpha\beta_c \to 21{\cdot}911\left(\frac{\Omega}{d}\right)^{\frac{4}{3}}\kappa\nu^{-\frac{1}{3}} \quad (\Omega \to \infty;\text{ or } \nu \to 0); \tag{134}$$

this should be contrasted with the formula,

$$g\alpha\beta_c = \text{constant } \kappa\nu d^{-4}, \tag{135}$$

valid in the absence of rotation. The dependence of β_c on an inverse power of ν, for Ω fixed and $\nu \to 0$, implies that an *inviscid, ideal fluid, in rotation, is stable with respect to the onset of stationary convection for all adverse temperature gradients.* This is clearly a consequence of the Taylor–Proudman theorem.

(*b*) *The solution for two rigid boundaries*

We shall obtain the solution for this case by the variational method.

In view of the symmetry of this problem with respect to the bounding planes, we shall find it convenient to translate the origin of z to be midway between the two planes. The fluid will then be confined between $z = \pm\frac{1}{2}$ and we shall have to seek solutions of equations (120)–(123) which satisfy the boundary conditions

$$F = W = DW = Z = 0 \quad \text{for} \quad z = \pm\tfrac{1}{2}. \tag{136}$$

It is apparent from the equations and the boundary conditions that the proper solutions of the problem fall into two non-combining groups; these consist of solutions which are even in W and odd in Z and solutions which are odd in W and even in Z; we shall call these the even and the odd solutions, respectively. It is also clear that the lowest characteristic value of R will occur among the even solutions; we shall accordingly consider such solutions.

Since F is assumed to be even and is required to vanish at $z = \pm\frac{1}{2}$, we can expand it in a cosine series in the form

$$F = \sum_m A_m \cos[(2m+1)\pi z], \tag{137}$$

where the summation over m may be assumed to run from zero to infinity; but this is not necessary since we shall consider the coefficients A_m as variational parameters.

With the chosen form for F, the equation to be solved for W is (cf. equation (123))

$$[(D^2-a^2)^3+TD^2]W = -\sum_m A_m c_{2m+1}\cos[(2m+1)\pi z], \tag{138}$$

where

$$c_{2m+1} = (2m+1)^2\pi^2+a^2. \tag{139}$$

In view of the linearity of equation (138), we can express W and Z as sums of the form

$$W = \sum_m A_m W_m \quad \text{and} \quad Z = \sum_m A_m Z_m, \tag{140}$$

where W_m and Z_m are solutions of the equations

$$(D^2-a^2)^3 W_m + TD^2 W_m = -c_{2m+1}\cos[(2m+1)\pi z] \tag{141}$$

and

$$(D^2-a^2)Z_m = -\left(\frac{2\Omega}{\nu}d\right) DW_m. \tag{142}$$

The variational principle formulated in § 26 (*a*) is equivalent to minimizing

$$\int_{-\frac{1}{2}}^{+\frac{1}{2}} [(DF)^2+a^2F^2]\,dz \tag{143}$$

for such variations of the parameters (the A_m's in the present connexion) which preserve the constancy of

$$\int_{-\frac{1}{2}}^{+\frac{1}{2}} WF\,dz. \tag{144}$$

By arguments similar to those used in Chapter II, § 17, it can be shown, by the use of a Lagrangian undetermined multiplier, that the application of this principle is exactly equivalent to solving a secular determinant which will arise from a Fourier analysis of the equation

$$(D^2-a^2)\sum_m A_m \cos[(2m+1)\pi z] = -Ra^2 \sum_m A_m W_m. \tag{145}$$

Returning to equation (141), we observe that the general solution of this equation, which is even, can be expressed in the form

$$W_m = c_{2m+1}\gamma_{2m+1}\cos[(2m+1)\pi z] + \sum_{j=1}^{3} B_j^{(m)} \cosh q_j z, \tag{146}$$

where

$$\frac{1}{\gamma_{2m+1}} = [(2m+1)^2\pi^2+a^2]^3+(2m+1)^2\pi^2 T = c_{2m+1}^3+(2m+1)^2\pi^2 T, \tag{147}$$

the $B_j^{(m)}$'s $(j = 1, 2, 3)$ are constants of integration, and the q_j^2's $(j = 1, 2, 3)$ are the three roots of the cubic equation

$$(q^2-a^2)^3+Tq^2 = 0. \tag{148}$$

An identity which readily follows from equation (148) is

$$\frac{1}{\gamma_{2m+1}} = \prod_{j=1}^{3} [(2m+1)^2\pi^2+q_j^2]; \tag{149}$$

for,

$$(q^2-a^2)^3+Tq^2 = \prod_{j=1}^{3} (q^2-q_j^2) \tag{150}$$

is, in reality, an identity.

The solution for Z_m corresponding to the solution (146) for W_m is

$$Z_m = -\left(\frac{2\Omega}{\nu}d\right)\left\{(2m+1)\pi\gamma_{2m+1}\sin[(2m+1)\pi z]+\right.$$
$$\left.+\sum_{j=1}^{3} B_j^{(m)}\frac{q_j}{x_j}\sinh q_j z\right\}, \quad (151)$$

where

$$x_j = q_j^2-a^2. \quad (152)$$

The boundary conditions $W = DW = Z = 0$ for $z = \pm\frac{1}{2}$ require

$$\sum_{j=1}^{3} B_j^{(m)}\cosh\tfrac{1}{2}q_j = 0,$$
$$\sum_{j=1}^{3} B_j^{(m)}q_j\sinh\tfrac{1}{2}q_j = (-1)^m(2m+1)\pi c_{2m+1}\gamma_{2m+1},$$
$$\sum_{j=1}^{3} B_j^{(m)}\frac{q_j}{x_j}\sinh\tfrac{1}{2}q_j = (-1)^{m+1}(2m+1)\pi\gamma_{2m+1}. \quad (153)$$

On solving these equations, we find that

$$B_1^{(m)} = (-1)^m(2m+1)\pi\gamma_{2m+1}\left\{q_3\left(1+\frac{c_{2m+1}}{x_3}\right)\coth\tfrac{1}{2}q_2-\right.$$
$$\left.-q_2\left(1+\frac{c_{2m+1}}{x_2}\right)\coth\tfrac{1}{2}q_3\right\}\Delta\,\mathrm{cosech}\,\tfrac{1}{2}q_1, \quad (154)$$

where

$$\frac{1}{\Delta} = \frac{q_2q_3}{x_2x_3}(x_3-x_2)\coth\tfrac{1}{2}q_1+\frac{q_3q_1}{x_3x_1}(x_1-x_3)\coth\tfrac{1}{2}q_2+\frac{q_1q_2}{x_1x_2}(x_2-x_1)\coth\tfrac{1}{2}q_3, \quad (155)$$

and $B_2^{(m)}$ and $B_3^{(m)}$ are given by similar expressions which can be obtained by cyclically permuting the q_j's and the x_j's in the solution (154).

The characteristic equation is now obtained from (cf. equations (145) and (146))

$$\sum_m A_m c_{2m+1}\cos[(2m+1)\pi z]$$
$$= Ra^2\sum_m A_m\left\{c_{2m+1}\gamma_{2m+1}\cos[(2m+1)\pi z]+\sum_{j=1}^{3} B_j^{(m)}\cosh q_j z\right\}. \quad (156)$$

Multiplying this equation by $\cos[(2n+1)\pi z]$ and integrating over the range of z, we obtain

$$\tfrac{1}{2}c_{2n+1}A_n = Ra^2\left\{\tfrac{1}{2}c_{2n+1}\gamma_{2n+1}A_n+\sum_m (n|m)A_m\right\} \quad (n = 0,1,2,\ldots), \quad (157)$$

where

$$(n|m) = \sum_{j=1}^{3} B_j^{(m)}\int_{-\frac{1}{2}}^{+\frac{1}{2}}\cosh q_j z\cos[(2n+1)\pi z]\,dz$$
$$= 2(2n+1)\pi(-1)^n\sum_{j=1}^{3}\frac{B_j^{(m)}\cosh\frac{1}{2}q_j}{(2n+1)^2\pi^2+q_j^2}. \quad (158)$$

Equation (158) provides a set of linear homogeneous equations for the constants A_m; this same set of equations ensures that expression (143) is a minimum for all variations of the A_m's which preserve the constancy of (144). In this latter formulation Ra^2 is the Lagrangian undetermined multiplier.

The determinant of the system of equations represented by (157) must vanish; and this provides the characteristic equation for R. Thus,

$$\left\|\tfrac{1}{2}c_{2n+1}\left(\frac{1}{Ra^2}-\gamma_{2n+1}\right)\delta_{nm}-(n|m)\right\| = 0. \tag{159}$$

Substituting for the $B_j^{(m)}$'s in accordance with equation (154) and making use of the identity (149), we find after some minor reductions that the expression for $(n|m)$ given in (158) becomes

$$\begin{aligned}(n|m) = {} & 2(-1)^{m+n}(2n+1)(2m+1)\pi^2\gamma_{2n+1}\gamma_{2m+1}\Delta\prod_{j=1}^{3}\coth\tfrac{1}{2}q_j\times\\ & \times\left\{q_1(q_3^2-q_2^2)[(2n+1)^2\pi^2+q_1^2]\left(1+\frac{c_{2m+1}}{x_1}\right)\tanh\tfrac{1}{2}q_1+\right.\\ & +q_2(q_1^2-q_3^2)[(2n+1)^2\pi^2+q_2^2]\left(1+\frac{c_{2m+1}}{x_2}\right)\tanh\tfrac{1}{2}q_2+\\ & \left.+q_3(q_2^2-q_1^2)[(2n+1)^2\pi^2+q_3^2]\left(1+\frac{c_{2m+1}}{x_3}\right)\tanh\tfrac{1}{2}q_3\right\}.\end{aligned} \tag{160}$$

Recalling the definitions of c_{2n+1} and x_j (equations (139) and (152)), we can rewrite the foregoing in the form

$$\begin{aligned}(n|m) = {} & 2(-1)^{m+n}(2n+1)(2m+1)\pi^2\gamma_{2n+1}\gamma_{2m+1}\Delta\prod_{j=1}^{3}\coth\tfrac{1}{2}q_j\times\\ & \times\left\{(x_3-x_2)(c_{2n+1}+x_1)\left(1+\frac{c_{2m+1}}{x_1}\right)q_1\tanh\tfrac{1}{2}q_1+\right.\\ & +(x_1-x_3)(c_{2n+1}+x_2)\left(1+\frac{c_{2m+1}}{x_2}\right)q_2\tanh\tfrac{1}{2}q_2+\\ & \left.+(x_2-x_1)(c_{2n+1}+x_3)\left(1+\frac{c_{2m+1}}{x_3}\right)q_3\tanh\tfrac{1}{2}q_3\right\}.\end{aligned} \tag{161}$$

This expression for the matrix element $(n|m)$ is manifestly symmetric in n and m: a fact which reflects the basic self-adjoint character of the underlying problem.

The expression for $(n|m)$ can be further reduced. First, we observe that according to equations (148) and (152) the x_j's are the roots of the equation

$$x^3+Tx+Ta^2 = 0. \tag{162}$$

This equation allows one real root and a pair of complex conjugate roots; let these be†

$$x \quad \text{and} \quad X \pm iY. \tag{163}$$

Denote the corresponding roots of q by

$$q = \sqrt{(a^2+x)} \quad \text{and} \quad \alpha_1 \pm i\alpha_2 = \sqrt{(a^2+X \pm iY)}. \tag{164}$$

On substituting for the x_j's and the q_j's in accordance with these definitions, we find that the expression (161) for $(n|m)$ can be reduced to the form

$$(n|m) = 2(-1)^{m+n+1}(2n+1)(2m+1)\pi^2\gamma_{2n+1}\gamma_{2m+1}\Phi_0^{(e)} \times$$
$$\times \left\{ Y(c_{2n+1}+x)\left(1+\frac{c_{2m+1}}{x}\right)q\tanh\tfrac{1}{2}q + \right.$$
$$\left. + \frac{c_{2n+1}c_{2m+1}}{X^2+Y^2}\Phi_1^{(e)} - (c_{2n+1}+c_{2m+1})\Phi_2^{(e)} + \Phi_3^{(e)} \right\}, \tag{165}$$

where $\Phi_0^{(e)},..., \Phi_3'^{(e)}$ are functions of q, α_1, α_2, x, X, and Y defined as follows. Let

$$\phi_1^{(e)} = \frac{\alpha_1 \sinh\alpha_1 - \alpha_2 \sin\alpha_2}{\cosh\alpha_1 + \cos\alpha_2}, \qquad \psi_1^{(e)} = \frac{\alpha_2 \sinh\alpha_1 + \alpha_1 \sin\alpha_2}{\cosh\alpha_1 + \cos\alpha_2},$$
$$\phi_2^{(e)} = \frac{\alpha_1 \sinh\alpha_1 - \alpha_2 \sin\alpha_2}{\cosh\alpha_1 - \cos\alpha_2}, \qquad \psi_2^{(e)} = \frac{\alpha_2 \sinh\alpha_1 + \alpha_1 \sin\alpha_2}{\cosh\alpha_1 - \cos\alpha_2}, \tag{166}$$

then

$$\Phi_0^{(e)} = \frac{\coth\tfrac{1}{2}q(\sinh^2\alpha_1+\sin^2\alpha_2)(\cosh\alpha_1-\cos\alpha_2)^{-2}}{q\left(\dfrac{Y\phi_2^{(e)}-X\psi_2^{(e)}}{X^2+Y^2}+\dfrac{\psi_2^{(e)}}{x}\right)-\dfrac{\alpha_1^2+\alpha_2^2}{X^2+Y^2}Y\coth\tfrac{1}{2}q},$$
$$\Phi_1^{(e)} = (X^2-Y^2-xX)\psi_1^{(e)} - Y(2X-x)\phi_1^{(e)}, \tag{167}$$
$$\Phi_2^{(e)} = Y\phi_1^{(e)} - (X-x)\psi_1^{(e)},$$
$$\Phi_3^{(e)} = (X^2+Y^2-xX)\psi_1^{(e)} - xY\phi_1^{(e)}.$$

By solving the determinantal equation (159) by including successively more rows and columns, we can evaluate the required characteristic values of R with increasing precision. The results of such calculations are summarized in Table VIII and illustrated in Figs. 21 and 22 (see p. 96). It will be seen that in no case is it really necessary to go higher than the second approximation.

(c) *The solution for one rigid and one free boundary*

In this case the conditions to be satisfied on the two bounding surfaces are different. However, the solution for this case can be reduced to

† It can be directly verified that

$$x = -2X \quad \text{and} \quad Y^2 = 3X^2+T.$$

I am indebted to Miss Donna Elbert for pointing out these relations.

that of case (*b*) by considering odd (instead of even) solutions for W. For it is clear that an odd solution for W satisfying the boundary conditions appropriate to case (*b*) vanishes at $z = 0$; therefore, at $z = 0$ the boundary conditions on W appropriate to a free surface are satisfied. Associated with an odd solution for W is an even solution for Z; therefore, at $z = 0$, DZ will vanish which is the further condition to be satisfied on a free surface. Consequently, a solution with W odd and Z even, suitable for case (*b*) and applicable to a cell depth d, provides a solution

TABLE VIII

Critical Rayleigh numbers and related constants for the case when both bounding surfaces are rigid and the onset of instability is as stationary convection

		R_c			Second approximation	Third approximation	
T	a_c	First approximation	Second approximation	Third approximation	A_2/A_1	A_2/A_1	A_3/A_1
10	3·10	$1{\cdot}720\times10^3$	$1{\cdot}7130\times10^3$		+0·02884		
100	3·15	$1{\cdot}764\times10^3$	$1{\cdot}7566\times10^3$		+0·02885		
500	3·30	$1{\cdot}948\times10^3$	$1{\cdot}9405\times10^3$	$1{\cdot}9403\times10^3$	+0·02849	+0·02849	−0·00291
1,000	3·50	$2{\cdot}159\times10^3$	$2{\cdot}1517\times10^3$		+0·02818		
2,000	3·75	$2{\cdot}538\times10^3$	$2{\cdot}5305\times10^3$		+0·02686		
5,000	4·25	$3{\cdot}476\times10^3$	$3{\cdot}4692\times10^3$	$3{\cdot}4686\times10^3$	+0·02255	+0·02253	−0·00415
10,000	4·80	$4{\cdot}717\times10^3$	$4{\cdot}7131\times10^3$		+0·01634		
30,000	5·80	$8{\cdot}326\times10^3$	$8{\cdot}3264\times10^3$		−0·00071		
10^5	7·20	$1{\cdot}674\times10^4$	$1{\cdot}6721\times10^4$	$1{\cdot}6721\times10^4$	−0·02493	−0·02495	−0·00197
10^6	10·80	$7{\cdot}159\times10^4$	$7{\cdot}1132\times10^4$		−0·06653		
10^8	24·5	$1{\cdot}545\times10^6$	$1{\cdot}5313\times10^6$		−0·09364		
10^{10}	55·5	$3{\cdot}482\times10^7$	$3{\cdot}4636\times10^7$	$3{\cdot}4574\times10^7$	−0·07731	−0·07798	+0·04193

for case (*c*) applicable to a cell depth $\frac{1}{2}d$ and Rayleigh and Taylor numbers which are sixteen times smaller.

We consider then the odd solutions appropriate to case (*b*) and obtain them once again by the variational method.

We now expand F in a sine series of the form

$$F = \sum_m A_m \sin 2mz \tag{168}$$

and follow the procedure outlined in § (*b*) above. We find that the corresponding solutions for W_m and z_m are

$$W_m = c_{2m}\gamma_{2m}\sin 2m\pi z + \sum_{j=1}^{3} B_j^{(m)} \sinh q_j z \tag{169}$$

and

$$Z_m = -\left(\frac{2\Omega}{\nu}d\right)\left\{-2m\pi\gamma_{2m}\cos 2m\pi z + \sum_{j=1}^{3} B_j^{(m)}\frac{q_j}{x_j}\cosh q_j z\right\}, \tag{170}$$

where q_j and x_j have the same meanings as in § (*b*) and c_{2m} and γ_{2m} are defined as in equations (139) and (147) with $2m$ replacing $(2m+1)$. The constants of integration $B_j^{(m)}$ are similarly determined by the boundary conditions.

Substituting for F and W in accordance with equations (168) and (169) in the equation relating them (namely, equation (105)), we now have (cf. equation (156))

$$\sum_m A_m c_{2m} \sin 2m\pi z = Ra^2 \sum_m A_m \Big\{c_{2m}\gamma_{2m} \sin 2m\pi z + \sum_{j=1}^{3} B_j^{(m)} \sinh q_j z\Big\}. \tag{171}$$

Multiplying this equation by $\sin 2n\pi z$ and integrating over the range of z, we obtain

$$\tfrac{1}{2}c_{2n} A_n = Ra^2\Big\{\tfrac{1}{2}c_{2n}\gamma_{2n} A_n + \sum_m (n|m)A_m\Big\}, \tag{172}$$

where

$$(n|m) = \sum_{j=1}^{3} B_j^{(m)} \int_{-\frac{1}{2}}^{+\frac{1}{2}} \sinh q_j z \sin 2n\pi z \, dz$$

$$= 4n\pi(-1)^n \sum_{j=1}^{3} \frac{B_j^{(m)} \sinh \frac{1}{2}q_j}{4n^2\pi^2+q_j^2}; \tag{173}$$

and this leads to the characteristic equation

$$\left\|\tfrac{1}{2}c_{2n}\left(\frac{1}{Ra^2}-\gamma_{2n}\right)\delta_{nm}-(n|m)\right\| = 0. \tag{174}$$

The explicit expression for the matrix element $(n|m)$ is found to be (cf. equation (165))

$$(n|m) = 8(-1)^{m+n+1} nm\pi^2 \gamma_{2n}\gamma_{2m} \Phi_0^{(o)} \times$$

$$\times \Big\{Y(c_{2n}+x)\left(1+\frac{c_{2m}}{x}\right) q \coth \tfrac{1}{2}q +$$

$$+\frac{c_{2n}c_{2m}}{X^2+Y^2}\Phi_1^{(o)} - (c_{2n}+c_{2m})\Phi_2^{(o)} + \Phi_3^{(o)}\Big\}, \tag{175}$$

where x, q, X, and Y have the same meanings as in § (*b*); and analogous to equations (166) and (167), we now have

$$\phi_1^{(o)} = \frac{\alpha_1 \sinh \alpha_1 + \alpha_2 \sin \alpha_2}{\cosh \alpha_1 - \cos \alpha_2}, \qquad \psi_1^{(o)} = \frac{\alpha_2 \sinh \alpha_1 - \alpha_1 \sin \alpha_2}{\cosh \alpha_1 - \cos \alpha_2},$$

$$\phi_2^{(o)} = \frac{\alpha_1 \sinh \alpha_1 + \alpha_2 \sin \alpha_2}{\cosh \alpha_1 + \cos \alpha_2}, \qquad \psi_2^{(o)} = \frac{\alpha_2 \sinh \alpha_1 - \alpha_1 \sin \alpha_2}{\cosh \alpha_1 + \cos \alpha_2},$$

$$\Phi_0^{(o)} = \frac{\tanh \frac{1}{2}q(\sinh^2\alpha_1 + \sin^2\alpha_2)(\cosh \alpha_1 + \cos \alpha_2)^{-2}}{q\left(\dfrac{Y\phi_2^{(o)} - X\psi_2^{(o)}}{X^2+Y^2} + \dfrac{\psi_2^{(o)}}{x}\right) - \dfrac{\alpha_1^2+\alpha_2^2}{X^2+Y^2} Y \tanh \frac{1}{2}q},$$

$$\Phi_1^{(o)} = (X^2-Y^2-xX)\psi_1^{(o)}-Y(2X-x)\phi_1^{(o)},$$
$$\Phi_2^{(o)} = Y\phi_1^{(o)}-(X-x)\psi_1^{(o)},$$
$$\Phi_3^{(o)} = (X^2+Y^2-xX)\psi_1^{(o)}-xY\phi_1^{(o)}. \tag{176}$$

The secular equation (174) has been solved in various approximations. The results of such calculations are summarized in Table IX and illustrated in Figs. 21 and 22 (see p. 96).

(d) The origin of the $T^{2/3}$-law

From the results for the three sets of boundary conditions illustrated in Figs. 21 and 22, it is apparent that all three cases exhibit the same general features. This is particularly true of the asymptotic dependence on T of the critical Rayleigh number and the associated wave number. For the case of two free boundaries, the laws

$$R_c \to \text{constant } T^{2/3} \quad \text{and} \quad a \to \text{constant } T^{1/6} \tag{177}$$

follow directly from the solution of the characteristic value problem. From the results of the calculations for the two other cases, it appears that the same power laws hold for them also though the constants of proportionality seem to depend slightly, but definitely, on the boundary conditions.

By going back to the original differential equations, we shall try to locate the common origin of the $T^{2/3}$- and the $T^{1/6}$-laws.

Consider the equation

$$(D^2-a^2)^3W+TD^2W = -Ra^2W. \tag{178}$$

When $T \to \infty$, we expect that for R in the neighbourhood of R_c, a also will tend to infinity. But we do not expect that the solution for W, in this limit, will show any special behaviour which will make any of its higher derivatives become 'disproportionately' large. The solution for the case of two free boundaries, as well as the progression of the values of the coefficients A_m in the variational solutions for the other two cases (see Tables VIII and IX), support this latter expectation. Accordingly, keeping only the terms in T and in the highest power of a in equation (178), as relevant in the limit $T \to \infty$, we have

$$TD^2W = -(Ra^2-a^6)W. \tag{179}$$

This is an equation of the second order in W. Consequently, we cannot satisfy all the boundary conditions of the problem; we can satisfy only two of the six boundary conditions. It would appear that the vanishing of W on the boundaries is one condition which we must satisfy on all

Table IX

Critical Rayleigh numbers and related constants for the case when one bounding surface is rigid and the other is free and the onset of instability is as stationary convection

T	a_c	R_c			Second approximation	Third approximation	
		First approximation	Second approximation	Third approximation	A_2/A_1	A_2/A_1	A_3/A_1
6·25	2·68	$1{\cdot}120\times10^3$	$1{\cdot}1085\times10^3$		+0·06289		
$3{\cdot}125\times10^1$	2·70	$1{\cdot}148\times10^3$	$1{\cdot}1365\times10^3$	$1{\cdot}1359\times10^3$	+0·06246	+0·06235	−0·00101
$6{\cdot}250\times10^1$	2·79	$1{\cdot}181\times10^3$	$1{\cdot}1695\times10^3$		+0·06235		
$1{\cdot}875\times10^2$	2·975	$1{\cdot}303\times10^3$	$1{\cdot}2917\times10^3$		+0·06093		
$6{\cdot}250\times10^2$	3·40	$1{\cdot}650\times10^3$	$1{\cdot}6387\times10^3$	$1{\cdot}6376\times10^3$	+0·05578	+0·05559	−0·01198
$1{\cdot}875\times10^3$	4·00	$2{\cdot}369\times10^3$	$2{\cdot}3603\times10^3$		+0·04349		
$6{\cdot}250\times10^3$	4·925	$4{\cdot}050\times10^3$	$4{\cdot}0477\times10^3$		+0·01953		
$1{\cdot}875\times10^4$	6·00	$7{\cdot}230\times10^3$	$7{\cdot}2291\times10^3$		−0·00801		
$6{\cdot}250\times10^4$	7·425	$1{\cdot}453\times10^4$	$1{\cdot}4511\times10^4$	$1{\cdot}4510\times10^4$	−0·03702	−0·03701	+0·00491
$1{\cdot}875\times10^5$	9·00	$2{\cdot}850\times10^4$	$2{\cdot}8412\times10^4$		−0·05796		
$6{\cdot}250\times10^5$	11·05	$6{\cdot}117\times10^4$	$6{\cdot}0874\times10^4$		−0·07423		
10^6	12·00	$8{\cdot}281\times10^4$	$8{\cdot}2382\times10^4$	$8{\cdot}2267\times10^4$	−0·07834	−0·07907	+0·03261
10^8	26·55	$1{\cdot}730\times10^6$	$1{\cdot}7208\times10^6$	$1{\cdot}7173\times10^6$	−0·08461	−0·08604	+0·04638
10^{10}	58·25	$3{\cdot}774\times10^7$	$3{\cdot}7619\times10^7$	$3{\cdot}7570\times10^7$	−0·06739	−0·06842	+0·03876

accounts. The solution of equation (179) which vanishes for $z = 0$ and 1 is

$$W = A\sin n\pi z, \tag{180}$$

where n is an integer. The corresponding lowest characteristic value of R is

$$R = \frac{1}{a^2}(a^6+\pi^2 T). \tag{181}$$

As a function of a, R given by this equation attains its minimum when

$$4a^3 - \frac{2}{a^3}\pi^2 T = 0, \tag{182}$$

or

$$a = (\tfrac{1}{2}\pi^2 T)^{\frac{1}{6}}; \tag{183}$$

and the associated value of R is

$$R = 3(\tfrac{1}{2}\pi^2 T)^{\frac{2}{3}}. \tag{184}$$

These results agree with those given in equation (133).

In one sense, the foregoing arguments do not go much beyond the discussion for the case of two free boundaries. But it does suggest that the origin of the $T^{\frac{2}{3}}$- and the $T^{\frac{1}{6}}$-laws probably lies in the effective lowering of the order of equation (178) with a corresponding reduction in the number of boundary conditions that need be satisfied as $T \to \infty$. It is worth pointing out that the preceding discussion is not 'fine' enough to account for the differences in the constants of proportionality in the asymptotic relations which the variational calculations strongly suggest.

28. The motions in the horizontal planes and the cell patterns at the onset of instability as stationary convection

We now turn to a description of the motions in the horizontal plane and of the cell patterns which can emerge at the onset of instability as stationary convection.

In terms of the solutions for w and ζ, the components of the velocity in the horizontal plane are given by (equations II (110) and (111))

$$u = \frac{1}{a^2}\left(\frac{\partial^2 w}{\partial x\partial z}+d\frac{\partial\zeta}{\partial y}\right), \qquad v = \frac{1}{a^2}\left(\frac{\partial^2 w}{\partial y\partial z}-d\frac{\partial\zeta}{\partial x}\right). \tag{185}$$

To be specific, let $\quad w = W(z)\cos a_x x\cos a_y y$

and $\quad \zeta = Z(z)\cos a_x x\cos a_y y. \qquad (186)$

Equations (185) then give

$$u = -\frac{1}{a^2}(a_x DW\sin a_x x\cos a_y y+a_y dZ\cos a_x x\sin a_y y),$$

$$v = -\frac{1}{a^2}(a_y DW\cos a_x x\sin a_y y-a_x dZ\sin a_x x\cos a_y y). \tag{187}$$

Since (cf. equation (122))

$$(D^2-a^2)Z = -\left(\frac{2\Omega d}{\nu}\right)DW, \tag{188}$$

we can also write

$$u = \left(\frac{\nu}{2\Omega d}\right)\frac{1}{a^2}\{(a_x \sin a_x x \cos a_y y)(D^2-a^2)Z - \\ -(a_y \cos a_x x \sin a_y y)Z\sqrt{T}\},$$

$$v = \left(\frac{\nu}{2\Omega d}\right)\frac{1}{a^2}\{(a_y \cos a_x x \sin a_y y)(D^2-a^2)Z + \\ +(a_x \sin a_x x \cos a_y y)Z\sqrt{T}\}. \tag{189}$$

From these expressions for u and v we find

$$u^2+v^2 = \left(\frac{\nu}{2\Omega d}\right)^2\frac{1}{a^4}(a_x^2 \sin^2 a_x x \cos^2 a_y y + a_y^2 \cos^2 a_x x \sin^2 a_y y)\times \\ \times\{[(D^2-a^2)Z]^2+TZ^2\}. \tag{190}$$

Therefore, for the average over a horizontal plane, we have

$$\langle u^2+v^2\rangle = \frac{1}{4a^2}\left(\frac{\nu}{2\Omega d}\right)^2\{[(D^2-a^2)Z]^2+TZ^2\}. \tag{191}$$

For the case of two free boundaries, the solutions for W and Z appropriate for the lowest mode can be written as (cf. equation (126))

$$W = W_0 \sin \pi z$$

and

$$Z = W_0\left(\frac{2\Omega d}{\nu}\right)\frac{\pi}{\pi^2+a^2}\cos \pi z. \tag{192}$$

A somewhat surprising result emerges when the foregoing solution for Z is used in equation (191). We find

$$\langle u^2+v^2\rangle = \tfrac{1}{4}W_0^2\,\frac{\pi^2}{a^2(\pi^2+a^2)^2}\{(\pi^2+a^2)^2+T\}\cos^2 \pi z. \tag{193}$$

Letting $a^2 = \pi^2 x$ (as in § 27 (a), equation (129)), we have

$$\langle u^2+v^2\rangle = \tfrac{1}{4}W_0^2\,\frac{(1+x)^2+T/\pi^4}{x(1+x)^2}\cos^2 \pi z. \tag{194}$$

On the other hand, the value of x at the onset of instability is related to T by equation (131); using this equation to eliminate T from (194), we find that we are left with

$$\langle u^2+v^2\rangle = \tfrac{1}{2}W_0^2 \cos^2 \pi z, \tag{195}$$

a result which is independent of T. Thus, *for a layer of fluid confined between two free boundaries, the ratio, of the mean kinetic energies of the motions in the horizontal plane and in the vertical direction at the onset of stationary convection, is independent of rotation.*

For other boundary conditions, the invariance expressed by equation (195) is asymptotically true for $T \to \infty$: for, according to the discussion in § 27 (*d*), the solution in the general case tends to that for two free boundaries as $T \to \infty$.

We now turn to the cell patterns. These have been described and beautifully illustrated by Veronis. The following account largely derives from his work.

(*a*) *Rolls, rectangles, and squares*

The case of rectangular cells (of which rolls and squares are special cases) can be derived from the solutions for u and v given in (189). For the case of two free boundaries (to which the present discussion on the cell patterns will be limited), we can insert for Z its solution from (192). We thus obtain the explicit formulae

$$u = -\frac{\pi}{a^2}\left(a_x \sin a_x x \cos a_y y + \frac{\sqrt{T}}{\pi^2+a^2} a_y \cos a_x x \sin a_y y\right) W_0 \cos \pi z,$$

$$v = -\frac{\pi}{a^2}\left(a_y \cos a_x x \sin a_y y - \frac{\sqrt{T}}{\pi^2+a^2} a_x \sin a_x x \cos a_y y\right) W_0 \cos \pi z. \qquad (196)$$

The corresponding solution for the vertical velocity is given by

$$w = W_0 \cos a_x x \cos a_y y \sin \pi z. \qquad (197)$$

The case of rolls is of particular simplicity. The solution for this case can be obtained by setting $a_x = a$ and $a_y = 0$ in equations (196) and (197); thus

$$u = -\frac{\pi}{a} W_0 \sin ax \cos \pi z,$$

$$v = \sqrt{T} \frac{\pi}{a(\pi^2+a^2)} W_0 \sin ax \cos \pi z, \qquad (198)$$

and

$$w = W_0 \cos ax \sin \pi z. \qquad (199)$$

When $T = 0$, there are no motions in the y-direction and the streamlines are confined to planes normal to the direction of the rolls. When the system is rotated, the Coriolis acceleration induces longitudinal motions.

Since in the present case

$$\frac{v}{u} = -\frac{\sqrt{T}}{\pi^2+a^2} = \text{constant}, \tag{200}$$

the streamlines are again confined to planes; but these planes are inclined to the x-axis (see Fig. 23 b, c). Under these circumstances we can define a wave number a_s which will describe the motions in the oblique planes containing the streamlines; it is given by

$$a_s = a\cos(\tan^{-1}v/u) = \frac{a}{\sqrt{(1+v^2/u^2)}}. \tag{201}$$

Substituting for v/u from (200), we obtain

$$a_s^2 = \frac{a^2(\pi^2+a^2)^2}{(\pi^2+a^2)^2+T}. \tag{202}$$

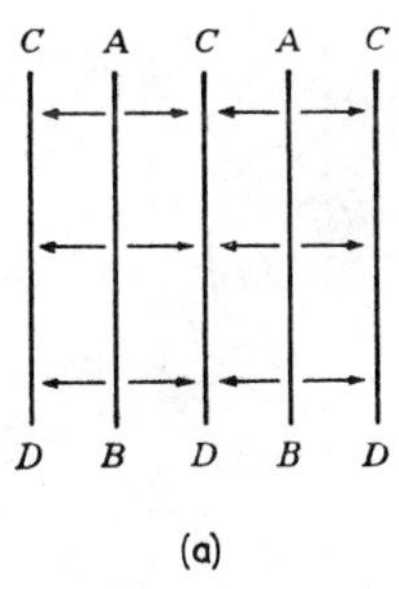

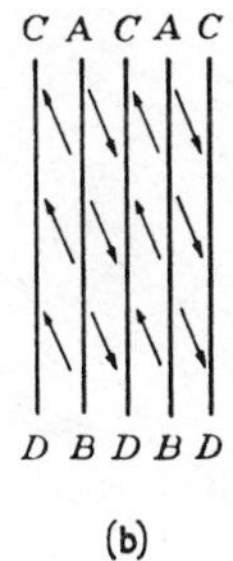

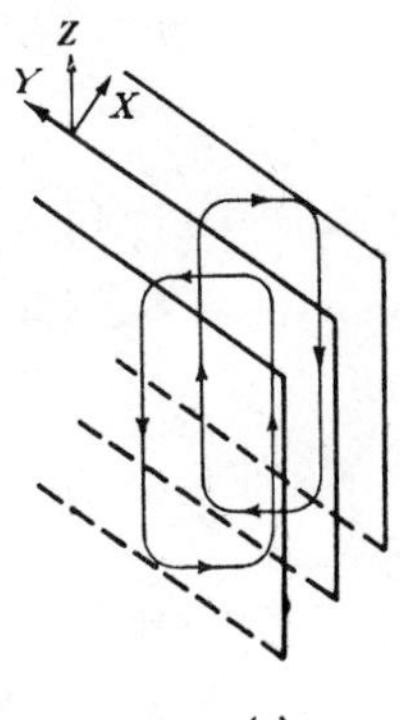

FIG. 23. (a) A top view of two-dimensional rolls in a non-rotating fluid. (b) The same view in a system rotating counterclockwise. (c) A perspective view of the particle motions in a roll in the rotating case. The arrows indicate the direction of particle motions.

The quantity on the right-hand side of this equation is, apart from a numerical factor, the reciprocal of the expression which we have already shown to be independent of T (cf. the arguments leading from equation (193) to (195)); it has the value $\frac{1}{2}\pi^2$. Thus,

$$a_s^2 = \tfrac{1}{2}\pi^2; \tag{203}$$

but this is the same as the formula which gives the wave number of the cell in the absence of rotation (see equation II (195)). Hence, *the wavelength of the roll measured in the plane containing the streamlines is independent of rotation*. This remarkable result is due to Veronis.

Considering next the case of square cells, we obtain the corresponding solutions by setting $a_x = a_y = a/\sqrt{2}$ in (196). We obtain

$$u = -\frac{\pi}{a\sqrt{2}}\left(\sin\frac{ax}{\sqrt{2}}\cos\frac{ay}{\sqrt{2}}+\frac{\sqrt{T}}{\pi^2+a^2}\cos\frac{ax}{\sqrt{2}}\sin\frac{ay}{\sqrt{2}}\right)W_0\cos\pi z,$$

$$v = -\frac{\pi}{a\sqrt{2}}\left(\cos\frac{ax}{\sqrt{2}}\sin\frac{ay}{\sqrt{2}}-\frac{\sqrt{T}}{\pi^2+a^2}\sin\frac{ax}{\sqrt{2}}\cos\frac{ay}{\sqrt{2}}\right)W_0\cos\pi z. \qquad (204)$$

The streamlines in the absence of rotation in such square cells have been

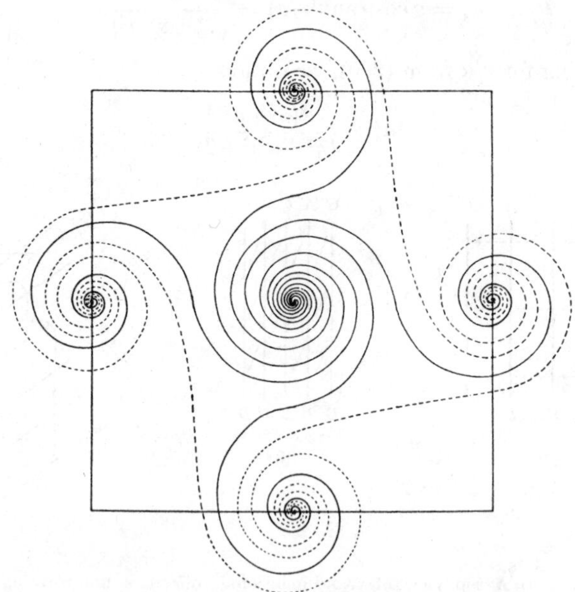

FIG. 24*a*. A square cell in a rotating fluid. The particle motions are from the centre outward along the spiral curves. The dashed curves form the boundary of the square cell.

illustrated in Chapter II (Fig. 6, p. 46). The dashed curves in Fig. 24 *a* correspond to the sides of the square in Fig. 6. In the presence of rotation, the Coriolis acceleration causes the fluid to turn as it moves towards or outwards from the centre; motions along the curves of constant w, as well as transverse to them, occur. Fig. 24 *b* is a perspective of the complete trajectory as sketched by Veronis. The drawing illustrates how a fluid element starting at the centre of a cell spirals upwards in a clockwise direction (for a counter-clockwise rotation $\mathbf{\Omega}$), and as it approaches the top it crosses over towards a corner of the cell and starts spiralling downwards in a counter-clockwise direction. When it arrives at the mid-plane ($z = \frac{1}{2}$), it reverses its sense of spiralling and proceeds downwards, and as it reaches the bottom it crosses back towards the centre of the cell, and starts spiralling upwards towards the centre in a clockwise direction. The reversal of the rotation at the mid-plane is associated with the circumstance that here DW and, therefore also, $\nabla_\perp \cdot \mathbf{u}_\perp$ change sign.

The square cells shown in Fig. 24 *b* correspond to the basic geometry

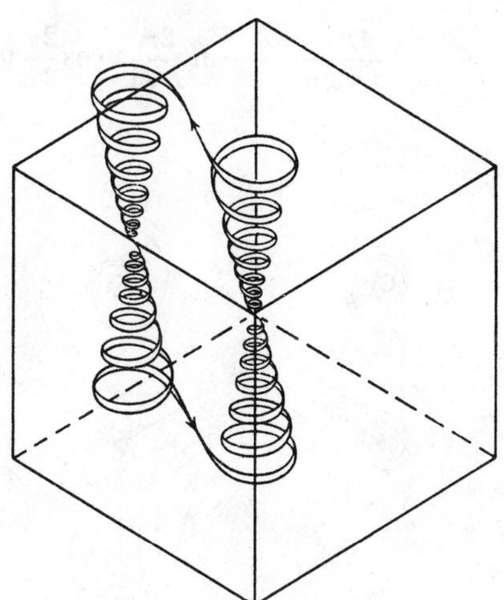

FIG. 24*b*. A perspective sketch of the path of a fluid particle in a square cell.

of the vertical velocity. The distortion of the cell consequent on the rotation is not shown.

(*b*) *Hexagons*

Christopherson's solution for a hexagonal pattern and appropriate for two free boundaries is (cf. equation II (248))

$$w = \frac{1}{3}\left\{2\cos\frac{2\pi}{L\sqrt{3}}x\cos\frac{2\pi}{3L}y+\cos\frac{4\pi}{3L}y\right\}W_0\sin\pi z. \tag{205}$$

The corresponding solution for the vertical component of the vorticity is

$$\zeta = \frac{1}{3}\left(\frac{2\Omega d}{\nu}\right)\frac{\pi}{\pi^2+a^2}\left\{2\cos\frac{2\pi}{L\sqrt{3}}x\cos\frac{2\pi}{3L}y+\cos\frac{4\pi}{3L}y\right\}W_0\cos\pi z. \tag{206}$$

Substituting for w and ζ in accordance with these solutions in (187) we obtain

$$u = -\frac{\pi}{3a^2}\left\{\frac{4\pi}{L\surd 3}\sin\frac{2\pi}{L\surd 3}x\cos\frac{2\pi}{3L}y+\right.$$

$$\left.+\frac{4\pi}{3L}\frac{\surd T}{\pi^2+a^2}\left(\cos\frac{2\pi}{L\surd 3}x+2\cos\frac{2\pi}{3L}y\right)\sin\frac{2\pi}{3L}y\right\}W_0\cos\pi z,$$

$$v = -\frac{\pi}{3a^2}\left\{\frac{4\pi}{3L}\left(\cos\frac{2\pi}{L\surd 3}x+2\cos\frac{2\pi}{3L}y\right)\sin\frac{2\pi}{3L}y-\right.$$

$$\left.-\frac{4\pi}{L\surd 3}\frac{\surd T}{\pi^2+a^2}\sin\frac{2\pi}{L\surd 3}x\cos\frac{2\pi}{3L}y\right\}W_0\cos\pi z. \quad (207)$$

Fig. 25*a*. A top view of seven rotating hexagonal cells. The fluid particles follow spiral paths from the centre toward the corners. The dashed lines form the boundaries of the centre cell.

Fig. 25 *a* shows the streamlines as they would appear at the top. This pattern should be contrasted with that illustrated in Fig. 7 *b* (p. 50) for a non-rotating system. The dashed spirals form the boundary of the central cell. It will be observed that the source at the centre of the cell provides for the flow downwards at the six corners; and each corner receives fluid from three adjacent cells. In this last respect it is not different from the non-rotating case.

A perspective drawing of the complete trajectory is shown in Fig. 25 *b*. And finally, in Fig. 26 we illustrate the manner in which the cell is distorted by the rotation.

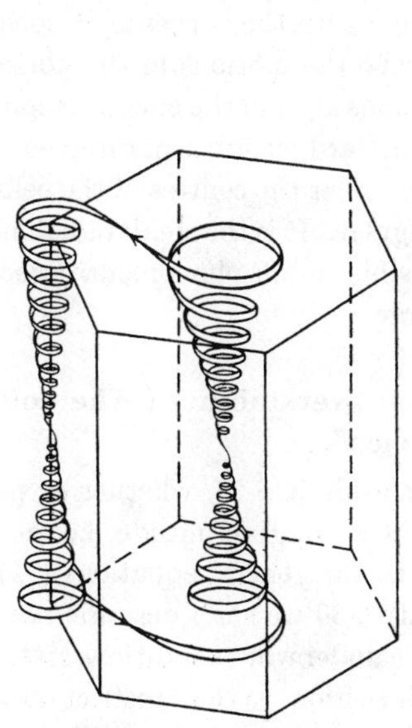

FIG. 25*b*. A perspective sketch of the path of a fluid particle in a hexagonal cell.

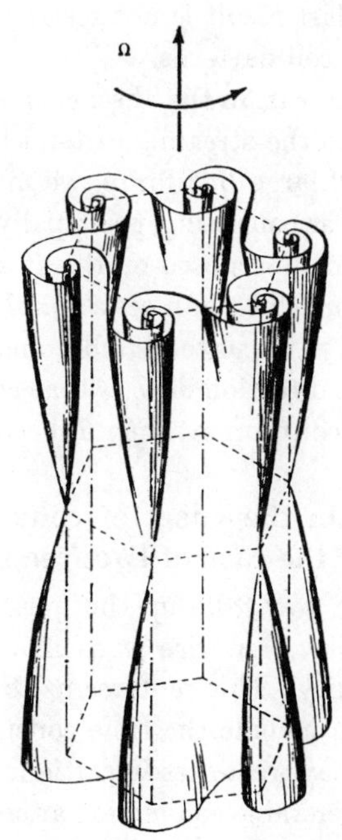

FIG. 26. A perspective sketch of a hexagonal cell as it is distorted by the rotation of the fluid.

(*c*) *The limiting behaviour of the streamlines for* $T \to \infty$

It is clear from the foregoing illustrations of the cell patterns that as T increases, the streamlines become increasingly close wound spirals. In the limit $T \to \infty$, when convection in fact ceases, the streamlines become closed curves. Thus, for T sufficiently large, the dominant terms in the solution (196) for u and v are, for example,

$$u \to -\frac{\pi\sqrt{T}}{a^2(\pi^2+a^2)} a_y W_0 \cos a_x x \sin a_y y \cos \pi z,$$

$$v \to +\frac{\pi\sqrt{T}}{a^2(\pi^2+a^2)} a_x W_0 \sin a_x x \cos a_y y \cos \pi z. \tag{208}$$

The corresponding equation for the streamlines is

$$\cos a_x x \cos a_y y = \text{constant} \tag{209}$$

or, $$w = \text{constant} \quad (\text{for } z = \text{constant}). \tag{210}$$

This last result is not restricted to rectangular cells: it applies equally to all cell patterns.

Whereas, in the absence of rotation $\mathbf{u}_\perp$ is parallel to $\nabla_\perp w$, in the limit $T \to \infty$ the streamlines tend to coincide with the curves $w =$ constant. For T large but finite, we may describe the motions in the horizontal plane as consisting, principally, of motions around the curves of constant w; and superposed on this is a slow outward, or inward, directed radial motion which induces a spiral pattern. Near the centres of the cells, the spiral patterns tend to become equiangular. It is the small radial motion in the direction of $\nabla_\perp w$ that is responsible for all phenomena associated with convection when T becomes large.

29. On the onset of convection as overstability. The solution for the case of two free boundaries

We now take up the question, set aside in § 26, whether or not instability can arise as oscillations of increasing amplitude, i.e. as overstability. This requires us to return to the general equations (93)–(95) which include the time constant σ. In § 30 we shall describe how best one may treat these equations to decide under what conditions stationary or overstable convection arise. In this section we shall restrict ourselves to the case of two free boundaries; for in this case the problem can be solved by elementary methods and it suggests how one may treat the general case.

By applying the operator $(D^2-a^2-\sigma)(D^2-a^2-\mathfrak{p}\sigma)$ to equation (95), we can eliminate Z and Θ and obtain

$$(D^2-a^2-\mathfrak{p}\sigma)[(D^2-a^2-\sigma)^2(D^2-a^2)+TD^2]W = -Ra^2(D^2-a^2-\sigma)W. \tag{211}$$

As in § 27 (a), we can show that in this case also the proper solution for W belonging to the lowest mode is

$$W = W_0 \sin \pi z. \tag{212}$$

Substituting this solution for W in equation (211), we obtain the characteristic equation

$$(\pi^2+a^2+\mathfrak{p}\sigma)[(\pi^2+a^2+\sigma)^2(\pi^2+a^2)+\pi^2 T] = Ra^2(\pi^2+a^2+\sigma), \tag{213}$$

where it must be remembered that σ can be complex. Letting

$$x = \frac{a^2}{\pi^2}, \quad i\sigma_1 = \frac{\sigma}{\pi^2}, \quad R_1 = \frac{R}{\pi^4}, \quad \text{and} \quad T_1 = \frac{T}{\pi^4}, \tag{214}$$

we can rewrite equation (213) in the form

$$R_1 = \frac{1}{x}(1+x+i\mathfrak{p}\sigma_1)\left\{(1+x)(1+x+i\sigma_1)+\frac{T_1}{1+x+i\sigma_1}\right\}. \tag{215}$$

It is apparent from equation (215) that for an arbitrarily assigned σ_1, R_1 will be complex. But the physical meaning of R_1 requires it to be real. Consequently, the condition that R_1 be real implies a relation between the real and the imaginary parts of σ_1. Since our principal interest is to specify the critical Rayleigh number for the onset of instability via a state of purely oscillatory motions, we shall in the first instance suppose that σ_1 in equation (215) is real and seek the conditions for such solutions to exist. This will suffice to answer the principal question as to when instability will set in as stationary convection and when as overstable oscillations. To establish that the criteria so obtained are both necessary and sufficient, we must consider the characteristic equation (215) more generally, and this we will do in § (*a*) below.

We shall assume, then, that in equation (215) σ_1 is real. On collecting the real and the imaginary parts of the expression on the right-hand side of equation (215), we have

$$R_1 = \frac{1+x}{x}\left\{(1+x)^2-\mathfrak{p}\sigma_1^2+\frac{T_1}{1+x}\,\frac{(1+x)^2+\mathfrak{p}\sigma_1^2}{(1+x)^2+\sigma_1^2}+\right.$$
$$\left.+i\sigma_1\left[(1+x)(1+\mathfrak{p})-T_1\frac{(1-\mathfrak{p})}{(1+x)^2+\sigma_1^2}\right]\right\}. \tag{216}$$

The real and the imaginary parts of this equation must vanish separately. We thus obtain the pair of equations:

$$R_1 = \frac{1+x}{x}\left\{(1+x)^2-\mathfrak{p}\sigma_1^2+\frac{T_1}{1+x}\,\frac{(1+x)^2+\mathfrak{p}\sigma_1^2}{(1+x)^2+\sigma_1^2}\right\} \tag{217}$$

and

$$(1+x)(1+\mathfrak{p}) = T_1\frac{(1-\mathfrak{p})}{(1+x)^2+\sigma_1^2}. \tag{218}$$

A relation which follows from equation (218) is

$$\frac{T_1}{1+x}\,\frac{(1+x)^2+\mathfrak{p}\sigma_1^2}{(1+x)^2+\sigma_1^2} = \frac{T_1}{1+x}-\frac{T_1}{1+x}\,\frac{(1-\mathfrak{p})\sigma_1^2}{(1+x)^2+\sigma_1^2} = \frac{T_1}{1+x}-(1+\mathfrak{p})\sigma_1^2. \tag{219}$$

Using this relation in (217), we have

$$R = \frac{1}{x}[(1+x)^3+T_1-(1+x)(1+2\mathfrak{p})\sigma_1^2]. \tag{220}$$

Next, solving equation (218) for σ_1^2, we find

$$\sigma_1^2 = \frac{T_1}{1+x}\frac{1-\mathfrak{p}}{1+\mathfrak{p}}-(1+x)^2. \tag{221}$$

Using this expression for σ_1^2 in (220), we obtain on further simplification the result

$$R_1 = 2(1+\mathfrak{p})\frac{1}{x}\left[(1+x)^3+\frac{\mathfrak{p}^2}{(1+\mathfrak{p})^2}T_1\right]. \tag{222}$$

Equations (221) and (222) are the equations which must be satisfied if overstability is to occur for a wave number corresponding to x and a Taylor number corresponding to T_1.

One conclusion which equation (221) enables us to draw, at once, is that solutions describing overstability cannot occur if

$$\frac{T_1}{1+x}\frac{1-\mathfrak{p}}{1+\mathfrak{p}} < 1; \tag{223}$$

for σ_1^2 would then be negative, contrary to hypothesis. Clearly, (223) will, *a fortiori*, be the case if $\mathfrak{p} > 1$. Accordingly, *for* $\mathfrak{p} > 1$, *overstability cannot occur and the principle of the exchange of stabilities is valid.*

For overstability to be at all possible, $\mathfrak{p}$ must be less than one; and even when this is the case, we shall obtain real frequencies, σ_1, only if

$$T_1 > \frac{1+\mathfrak{p}}{1-\mathfrak{p}}(1+x)^3. \tag{224}$$

For a given T_1, overstable solutions are therefore possible only for $x < x_*$ where x_* is such that

$$(1+x_*)^3 = T_1\frac{1-\mathfrak{p}}{1+\mathfrak{p}}. \tag{225}$$

When $x = x_*$, $\sigma_1^2 = 0$ and (cf. equation (220))

$$R_1 = \frac{1}{x_*}[(1+x_*)^3+T_1]; \tag{226}$$

and this is, as one should expect, the value of R_1 at which stationary convection will occur for a wave number corresponding to x_* (see equation (130)). For $x > x_*$, overstability is not possible for the given $\mathfrak{p}$ and T_1; and the onset of instability as stationary convection remains the only possibility. For $x < x_*$, overstability is possible; and the question of discriminating between the two manners of instability arises. In making the discrimination, we shall suppose that, other things being equal, that manner of instability will occur that allows a solution for a lower Rayleigh number. While this is intuitively obvious, it requires proof; this is furnished in § (*a*) below.

Consider the (x, R_1)-plane (see Fig. 27). In this plane, we have first the curve,

$$R_1^{(c)} = \frac{1}{x}[(1+x)^3+T_1], \tag{227}$$

which defines the locus of states which are marginal with respect to stationary convection. The overstable solutions branch off from this

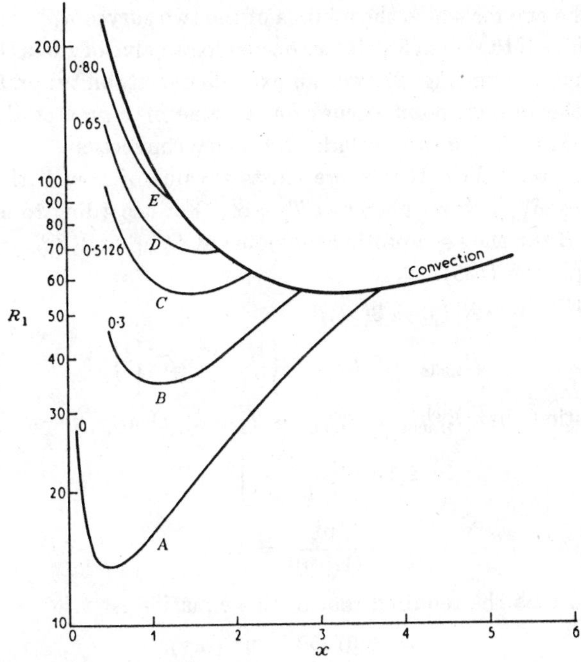

FIG. 27. The disposition of the curves for marginal stability in the (x, R_1)-plane for a Taylor number, $T = 10^4$, and for various values of the Prandtl number $\mathfrak{p}$. The curve labelled 'convection' applies for all values of $\mathfrak{p}$ and defines the locus $R_1^{(c)}$ (equation 227). The remaining curves are the overstable loci $R_1^{(o)}$ for values of $\mathfrak{p}$ by which they are labelled. The minimum for the curve C ($\mathfrak{p} = 0{\cdot}5126$) and for the convection curve occurs for the same value of R_1.

locus at the point x_* (see equation (226)); and for $x < x_*$ they are described by (cf. equations (220) and (222))

$$R_1^{(o)} = R_1^{(c)} - (1+2\mathfrak{p})\sigma_1^2\frac{1+x}{x} = 2(1+\mathfrak{p})\frac{1}{x}\left[(1+x)^3+\frac{\mathfrak{p}^2}{(1+\mathfrak{p})^2}T_1\right], \tag{228}$$

where the first form shows that $R_1^{(o)}$ (when this branch of the solution exists) is *always* less than $R_1^{(c)}$. Depending on $\mathfrak{p}$ and T_1, the branch point x_* can occur either before, or after, the point, $x_{\min}^{(c)}$, at which $R_1^{(c)}$ attains its minimum. If $x_* > x^{(c)}_{\min}$, then it is clear that for all $x < x_*$, overstability is the preferred manner of instability. If, however, $x_* < x^{(c)}_{\min}$, there are several possibilities: these are shown by the different curves labelled by B, C, D, and E in Fig. 27. It is apparent from this figure that the case which will distinguish whether, with increasing R_1 (for a given T_1), overstability or stationary convection will first manifest itself, is the one for which the minima of the two curves $R_1^{(c)}$ and $R_1^{(o)}$ are equal. Thus, if the branch point x_* occurs for a value of x less than what corresponds to C in Fig. 27, we can exclude overstability; on the other hand, if the branch point occurs for a value of x greater than what corresponds to C, we can exclude stationary convection.

We shall now show that there exists a value of $\mathfrak{p}$ such that $R^{(o)}_{1,\min}$ approaches $R^{(c)}_{1,\min}$ from above as $T_1 \to \infty$. For according to equations (227) and (228), the asymptotic behaviours of $R^{(c)}_{1,\min}$ and $R^{(o)}_{1,\min}$ are given by (cf. equation (133))

$$R^{(c)}_{1,\min} \to 3(\tfrac{1}{2}T_1)^{\frac{2}{3}},$$

$$R^{(o)}_{1,\min} \to 2(1+\mathfrak{p})\left\{3\left[\frac{1}{2}\frac{\mathfrak{p}^2}{(1+\mathfrak{p})^2}T_1\right]^{\frac{2}{3}}\right\}. \tag{229}$$

The condition that $R^{(o)}_{1,\min} \to R^{(c)}_{1,\min}$ as $T_1 \to \infty$, clearly, requires that

$$2(1+\mathfrak{p})\left[\frac{\mathfrak{p}^2}{(1+\mathfrak{p})^2}\right]^{\frac{2}{3}} = 1, \tag{230}$$

or

$$2\frac{\mathfrak{p}^{\frac{4}{3}}}{(1+\mathfrak{p})^{\frac{1}{3}}} = 1. \tag{231}$$

It is found that the required root of this equation is

$$\mathfrak{p} = 0{\cdot}67659 = \mathfrak{p}^* \text{ (say)}. \tag{232}$$

From the monotonic dependence of the minimum of the curve,

$$y = \frac{1}{x}[(1+x)^3+X], \tag{233}$$

on X, we conclude that for $\mathfrak{p} > \mathfrak{p}^*$, $R^{(c)}_{1,\min} < R^{(o)}_{1,\min}$ for all T_1. Hence, for $1 > \mathfrak{p} > \mathfrak{p}^*$, the situations indicated by D or E will prevail for all T_1. Therefore, *for* $\mathfrak{p} > \mathfrak{p}^*$, *instability will always manifest itself, first, as stationary convection.*

For $\mathfrak{p} < \mathfrak{p}^*$, the situation indicated by C occurs for a finite T_1, say $T_1^{(\mathfrak{p})}$. For $T_1 < T_1^{(\mathfrak{p})}$, the disposition of the curves $R_1^{(o)}$, with respect to $R_1^{(c)}$, is as shown by the curves D and E in Fig. 27; when $T_1 = T_1^{(\mathfrak{p})}$, the situation is as shown by the curve C; and for $T_1 > T_1^{(\mathfrak{p})}$ the disposition of the curves $R_1^{(o)}$, with respect to $R_1^{(c)}$, is as shown by the curves B and A. We conclude, then, that *for* $\mathfrak{p} < \mathfrak{p}^*$ *there exists a* $T_1^{(\mathfrak{p})}$ *such that for*

$T_1 \leqslant T_1^{(\mathfrak{p})}$, the onset of instability will be as stationary convection, while for $T_1 > T_1^{(\mathfrak{p})}$ it will be as overstability.

There is no simple formula which gives $T_1^{(\mathfrak{p})}$ as a function of $\mathfrak{p}$: it is simply determined by the condition that $R_{1,\min}^{(c)}$ and $R_{1,\min}^{(o)}$ are equal for

TABLE X

Taylor numbers above which instability occurs as overstability

$\mathfrak{p}$	$T^{(\mathfrak{p})}$	R	$\mathfrak{p}$	$T^{(\mathfrak{p})}$	R
0·0	548	1315	0·55	18,870	7748
0·1	728	1471	0·60	68,150	16,790
0·2	990	1669	0·63	$2{\cdot}588\times10^5$	$3{\cdot}876\times10^4$
0·4	3163	2890	0·65	$1{\cdot}223\times10^6$	$1{\cdot}050\times10^5$
0·5	8505	4910	0·6766	∞	∞

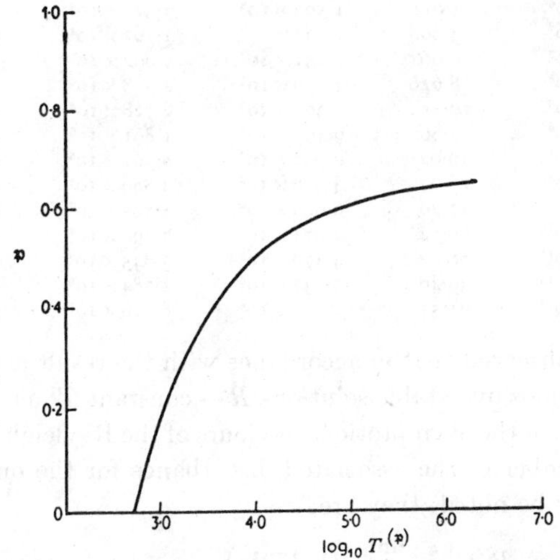

FIG. 28. The variation of $T^{(\mathfrak{p})}$ with the Prandtl number $\mathfrak{p}$.

$T_1 = T_1^{(\mathfrak{p})}$. Values of $T^{(\mathfrak{p})}$ determined by this condition are given in Table X for a few values of $\mathfrak{p}$; and Fig. 28 shows the division of the $(\mathfrak{p}, T)$-plane into regions where the manner of instability at onset is one or the other; the part of this plane where overstability can occur for Rayleigh numbers higher than are required for the onset of instability is bounded by $\mathfrak{p} = 1$ and the locus $T^{(\mathfrak{p})}$.

For $\mathfrak{p} < \mathfrak{p}^*$, the complete (R_c, T)-relation can be deduced from the results given in Table VII by a simple transformation of scales: for, according to equation (222), we have only to interpret T in Table VII as now signifying $\mathfrak{p}^2 T/(1+\mathfrak{p})^2$ and multiply the values under R_c by $2(1+\mathfrak{p})$. The (R_c, T)-relations derived in this manner are illustrated in Fig. 29. For later comparisons with results derived for other boundary conditions, the numerical form of the relation for $\mathfrak{p} = 0{\cdot}025$ is given in Table XI.

TABLE XI

Critical Rayleigh numbers and related constants for the onset of overstability in case both bounding surfaces are free and $\mathfrak{p} = 0{\cdot}025$

T	a_c	σ	R_c	p/Ω
0	2·233	—	$1{\cdot}348\times10^{3}$	imaginary
$1{\cdot}681\times10^{4}$	2·270	$1{\cdot}014\times10^{2}$	$1{\cdot}388\times10^{3}$	1·564
$1{\cdot}681\times10^{5}$	2·594	$3{\cdot}079\times10^{2}$	$1{\cdot}694\times10^{3}$	1·502
$8{\cdot}405\times10^{5}$	3·278	$6{\cdot}184\times10^{2}$	$2{\cdot}613\times10^{3}$	1·349
$1{\cdot}681\times10^{6}$	3·710	$8{\cdot}168\times10^{2}$	$3{\cdot}436\times10^{3}$	1·260
$3{\cdot}362\times10^{6}$	4·220	$1{\cdot}067\times10^{3}$	$4{\cdot}713\times10^{3}$	1·164
$8{\cdot}405\times10^{6}$	5·011	$1{\cdot}503\times10^{3}$	$7{\cdot}523\times10^{3}$	1·037
$1{\cdot}681\times10^{7}$	5·698	$1{\cdot}932\times10^{3}$	$1{\cdot}102\times10^{4}$	0·9424
$5{\cdot}043\times10^{7}$	6·961	$2{\cdot}851\times10^{3}$	$2{\cdot}092\times10^{4}$	0·8029
$1{\cdot}681\times10^{8}$	8·626	$4{\cdot}330\times10^{3}$	$4{\cdot}368\times10^{4}$	0·6680
$5{\cdot}043\times10^{8}$	10·45	$6{\cdot}308\times10^{3}$	$8{\cdot}728\times10^{4}$	0·5618
$1{\cdot}681\times10^{9}$	12·86	$9{\cdot}497\times10^{3}$	$1{\cdot}891\times10^{5}$	0·4633
$1{\cdot}681\times10^{10}$	19·02	$2{\cdot}062\times10^{4}$	$8{\cdot}501\times10^{5}$	0·3181
$1{\cdot}681\times10^{11}$	28·02	$4{\cdot}458\times10^{4}$	$3{\cdot}889\times10^{6}$	0·2175
$1{\cdot}681\times10^{12}$	41·20	$9{\cdot}620\times10^{4}$	$1{\cdot}793\times10^{7}$	0·1484
$1{\cdot}681\times10^{13}$	60·52	$2{\cdot}074\times10^{5}$	$8{\cdot}296\times10^{7}$	0·1012
$1{\cdot}681\times10^{14}$	88·87	$4{\cdot}470\times10^{5}$	$3{\cdot}845\times10^{8}$	0·0690
$1{\cdot}681\times10^{15}$	130·46	$9{\cdot}632\times10^{5}$	$1{\cdot}784\times10^{9}$	0·0470
$1{\cdot}681\times10^{16}$	191·51	$2{\cdot}075\times10^{6}$	$8{\cdot}276\times10^{9}$	0·0320

It will be observed that in accordance with the result given in (229), along the various overstable solutions $R \to \text{constant } T^{\frac{2}{3}}$ as $T \to \infty$. The explicit forms of the asymptotic behaviours of the Rayleigh number and the wave number of the associated disturbance for the onset of overstability may be noted; they are:

$$\left.\begin{aligned} R_c^{(o)} &\to \frac{6\mathfrak{p}^{\frac{4}{3}}}{(1+\mathfrak{p})^{\frac{1}{3}}}(\tfrac{1}{2}\pi^2 T)^{\frac{2}{3}} \\ a_c^{(o)} &\to \left(\frac{\mathfrak{p}}{1+\mathfrak{p}}\right)^{\frac{1}{3}}(\tfrac{1}{2}\pi^2 T)^{\frac{1}{6}} \end{aligned}\right\} \quad (\mathfrak{p}^2 T \to \infty). \tag{234}$$

The corresponding behaviour of $|\sigma| = \pi^2\sigma_1$ can be deduced from equation (221). We find

$$|\sigma| \to \frac{(2-3\mathfrak{p}^2)^{\frac{1}{2}}}{[\mathfrak{p}(1+\mathfrak{p})^2]^{\frac{1}{3}}}(\tfrac{1}{2}\pi^2 T)^{\frac{1}{3}} \quad (\mathfrak{p}^2 T \to \infty), \tag{235}$$

where it may be noted that according to the definition of σ (see equation (92)),

$$\frac{|p|}{\Omega} = \frac{2|\sigma|}{\sqrt{T}} = \frac{2|\sigma_1|}{\sqrt{T_1}}. \tag{236}$$

A further relation which follows from the foregoing is

$$|p|a_c^{(o)} \to 2\pi\Omega\frac{(1-1{\cdot}5\mathfrak{p}^2)^{\frac{1}{2}}}{1+\mathfrak{p}} \quad (\mathfrak{p}^2T \to \infty). \tag{237}$$

It should be noted that the asymptotic relations (234), (235), and (237) are valid only for $\mathfrak{p}^2T \to \infty$, *not* for $T \to \infty$, a distinction which is important when $\mathfrak{p} \to 0$ (see § 32).

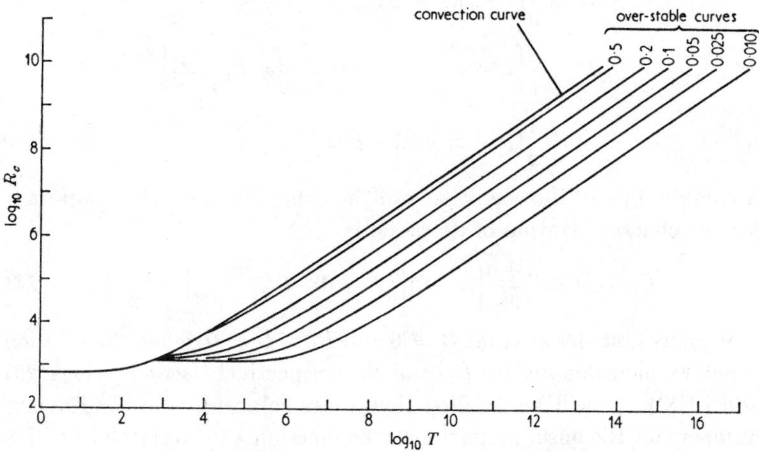

FIG. 29. The (R_c, T)-relations for a rotating horizontal layer of fluid heated below. The curves have been derived for the case when both bounding surfaces are free. The curve labelled 'convection curve' is the (R_c, T)-relation for the onset of ordinary cellular convection. The remaining curves are the corresponding relations for the onset of overstability. The values of $\mathfrak{p}$ to which the various curves refer are shown at the top of each curve. It will be seen that for each value of $\mathfrak{p} < 0{\cdot}677$, the instability sets in as ordinary cellular convection for T less than a certain $T^{(\mathfrak{p})}$ while it sets in as overstability for $T > T^{(\mathfrak{p})}$.

(*a*) *The nature of the roots of the characteristic equation* (215)

The preceding discussion of the dependence of the manner of instability on $\mathfrak{p}$ and T was carried out in terms of the explicit solutions for two special cases of equation (215): the case when $\sigma_1 = 0$ and the marginal state is a stationary one; and the case when σ_1 is real and the marginal state is an oscillatory one. It was further assumed that the manner of instability which will occur at onset is the one which allows a solution for a lower Rayleigh number. It was pointed out that this assumption requires justification; and this we shall now provide.

In considering the roots of equation (215), generally, we shall find it convenient to replace $i\sigma_1$ by σ_1 so that overstability now corresponds to σ_1 being a pure imaginary number. With this replacement, the equation we have to consider may be written as

$$(1+x+\mathfrak{p}\sigma_1)\left[(1+x+\sigma_1)^2+\frac{T_1}{1+x}\right] = R_1(1+x+\sigma_1)\frac{x}{1+x}. \tag{238}$$

This is a cubic equation for σ_1; its explicit form is

$$\sigma_1^3+B\sigma_1^2+C\sigma_1+D = 0, \tag{239}$$

where

$$B = \frac{1}{\mathfrak{p}}(1+x)(1+2\mathfrak{p}),$$

$$C = \frac{1}{\mathfrak{p}}\left[(2+\mathfrak{p})(1+x)^2+\mathfrak{p}\frac{T_1}{1+x}-R_1\frac{x}{1+x}\right],$$

$$D = \frac{1}{\mathfrak{p}}[(1+x)^3+T_1-R_1 x]. \tag{240}$$

A combination of the coefficients of the cubic (239), which is important for the characterization of its roots, is

$$BC-D = \frac{1+\mathfrak{p}}{\mathfrak{p}^2}\left\{(2+\mathfrak{p})\left[(1+x)^3+T_1\frac{\mathfrak{p}^2}{(1+\mathfrak{p})^2}\right]-R_1 x\right\}. \tag{241}$$

We may first observe that $D = 0$ and $BC-D = 0$ define the solutions which we have denoted by $R_1^{(c)}$ and $R_1^{(o)}$, respectively (see equations (227) and (228)). It will be recalled that these solutions give the Rayleigh numbers for the onset of stationary convection and overstability. This is consistent with equation (239): for $\sigma_1 = 0$ is clearly a root of the equation if $D = 0$; while, if σ_1 should be purely imaginary, then on equating the real and the imaginary parts of equation (239) separately, we should obtain $|\sigma_1|^2 = C = D/B$.

Now when $R_1 = 0$, C, D, and $BC-D$, are all positive; and B is, of course, always positive. When $D > 0$, equation (239) clearly allows at least one real root < 0; denote this root by $-d$. When equation (239) allows a real negative root (such as $-d$), we can factorize it in the form

$$(\sigma_1^2+2b\sigma_1+c)(\sigma_1+d) = 0. \tag{242}$$

By comparison with equation (239), we obtain the relations

$$B = 2b+d, \quad C = 2bd+c, \quad \text{and} \quad D = cd. \tag{243}$$

From these relations it follows that

$$BC-D = 2b(2bd+c+d^2) = 2b(C+d^2); \tag{244}$$

therefore,

$$b = \frac{BC-D}{2(C+d^2)} \quad \text{and} \quad c = \frac{D}{d}. \tag{245}$$

As we have noted, when $R_1 = 0$, $BC-D > 0$ and $D > 0$; therefore

$$b > 0 \quad \text{and} \quad c > 0 \quad (\text{when } R_1 \to 0). \tag{246}$$

The roots of equation (242), besides $-d$, are

$$\sigma_1 = -b \pm \sqrt{(b^2-c)}; \tag{247}$$

by (246), these two roots have also negative real parts. The state $R_1 = 0$ is, therefore, *absolutely* stable since $\text{re}(\sigma_1) < 0$ for all three roots.

Consider what happens when R_1 increases: D and $BC-D$ decrease and one of them may become zero. It is evident that if a real root of (242) becomes zero, $D \to 0$, since, either $c \to 0$ or $d \to 0$; while if the real part of a pair of complex roots becomes zero $b \to 0$, i.e. $BC-D \to 0$ (since $C > 0$, $C+d^2$ is necessarily positive). Therefore, as R_1 increases, *either* $D \to 0$ (in which case a real root tends to zero) *or* $BC-D \to 0$ (in which case the real part of a pair of complex roots tends to zero); and depending on whichever happens first, we shall have the onset of instability as stationary convection or as overstable oscillations. But this is exactly the premise on which our earlier discussion was based; it is now justified.

30. On a method of discriminating the character of the marginal state. A general variational principle

The discussion in the preceding section was restricted to the case of two free boundaries for which it was possible to obtain the required characteristic equation in an explicit form. We now return to the general case when this is not possible.

Replacing σ by $i\sigma$ in equations (93)–(95) and letting

$$F = (D^2-a^2)(D^2-a^2-i\sigma)W - \left(\frac{2\Omega}{\nu}d^3\right)DZ, \tag{248}$$

we have (cf. equations (105) and (106))

$$(D^2-a^2-i\sigma)Z = -\left(\frac{2\Omega}{\nu}d\right)DW \tag{249}$$

and

$$(D^2-a^2-i\mathfrak{p}\sigma)F = -Ra^2W. \tag{250}$$

Equations (248) and (249) can be combined to give

$$[(D^2-a^2-i\sigma)^2(D^2-a^2)+TD^2]W = (D^2-a^2-i\sigma)F. \tag{251}$$

Solutions of equations (248)–(250) must be sought which satisfy the boundary conditions

$$\left.\begin{aligned} W = F = 0 \quad &\text{for} \quad z = 0 \text{ and } 1, \\ \text{and} \quad \textit{either } DW = 0 \quad \text{and} \quad Z = 0 \quad &\text{(on a rigid surface)}, \\ \textit{or } D^2W = 0 \quad \text{and} \quad DZ = 0 \quad &\text{(on a free surface)}. \end{aligned}\right\} \tag{252}$$

There are eight boundary conditions and the requirement that a solution of equations (248)–(250) satisfy these conditions will determine for a given a^2 and $i\sigma$ (which can be real or complex) a sequence of possible values for R. These characteristic values of R will in general be complex; and the requirement that R be real implies a relation between the real and the imaginary parts of σ. We are not, however, interested in the relationship which may exist between the real and the imaginary parts of σ and the other parameters of the problem. We are interested only in the marginal state and its characterization. From the discussion of the case of two free boundaries in § 29, it is clear that it will suffice for our present purposes to determine the critical Rayleigh number for the onset of instability as stationary convection, and as overstability, for a given Taylor number; that which gives the lower Rayleigh number is the one that will be manifested first as we gradually increase the Rayleigh number beyond zero. The case $\sigma = 0$ has already been treated in §§ 26 and 27. It remains to treat the case when σ in equations (248)–(251) is real. The solution of this latter problem resolves itself, as we shall presently see, to solving a *double characteristic value problem.*

For an assigned a^2, σ (assumed to be real) is to be determined by the condition that R is real. In general, there will be a sequence of possible values of σ (for any given a^2 and T) which will make R real. We are, however, interested only in the particular σ which will give the lowest positive value for R. Let $R_0(a^2)$ and $\sigma_0(a^2)$ denote the corresponding values of R and σ. The meaning to be attached to these values is this: as the Rayleigh number is gradually increased, a disturbance in the horizontal plane characterized by the wave number a (in the unit $1/d$) first becomes unstable by overstability when it reaches the value $R_0(a^2)$; and $\sigma_0(a^2)$ is the frequency (in the unit ν/d^2) of the oscillations which are set up in the marginal state. To determine the critical Rayleigh number for the onset of overstability, we must determine the minimum of the function $R_0(a^2)$. We should then compare the minimum so obtained with the critical Rayleigh number for the onset of stationary convection; and that which is lower is the one that will prevail and be manifested at the onset of instability.

It should be apparent from the foregoing algorism that an exact solution of the double characteristic value problem will be difficult if not impracticable. However, a variational method can be devised which makes the problem feasible.

(*a*) *The variational principle*

Equations (248)–(250) can be treated in the same manner as equations (104)–(106) were treated in § 26 (*a*). Thus, by multiplying equation (250) by F and integrating over the range of z, we obtain after an integration by parts the equation

$$\int_0^1 [(DF)^2+(a^2+ip\sigma)F^2]\,dz = Ra^2\int_0^1 WF\,dz. \tag{253}$$

The right-hand side of this equation requires us to consider

$$\int_0^1 WF\,dz = \int_0^1 W\left\{(D^2-a^2)^2W-i\sigma(D^2-a^2)W-\left(\frac{2\Omega}{\nu}d^3\right)DZ\right\}dz. \tag{254}$$

After several integrations by parts we find that

$$\int_0^1 WF\,dz = \int_0^1 \{[(D^2-a^2)W]^2+i\sigma[(DW)^2+a^2W^2]\}\,dz+ \\ +\left(\frac{2\Omega}{\nu}d^3\right)\int_0^1 Z\,DW\,dz, \tag{255}$$

the integrated parts always vanishing on account of the boundary conditions. Now making use of equation (249), we find after further integrations by parts that

$$\int_0^1 WF\,dz = \int_0^1 \{[(D^2-a^2)W]^2+d^2[(DZ)^2+a^2Z^2]\}\,dz+ \\ +i\sigma\int_0^1 \{(DW)^2+a^2W^2+d^2Z^2\}\,dz. \tag{256}$$

Thus, the result of multiplying equation (250) by F and integrating over the range of z is

$$R = \frac{\int_0^1 [(DF)^2+a^2F^2+ip\sigma F^2]\,dz}{a^2\int_0^1 \{[(D^2-a^2)W]^2+d^2[(DZ)^2+a^2Z^2]+i\sigma[(DW)^2+a^2W^2+d^2Z^2]\}\,dz} \\ = \frac{I_1}{a^2I_2}\text{ (say)}. \tag{257}$$

Now it can be shown (exactly as in § 26 (a)) that the variation δR in R given by equation (257) due to variations δW and δZ in W and Z compatible only with the boundary conditions on W, Z, and F is given by

$$\delta R = -\frac{2}{a^2I_2}\int_0^1 \delta F\{(D^2-a^2-ip\sigma)F+Ra^2W\}\,dz. \tag{258}$$

Consequently, *if* $\delta R = 0$ *for all small arbitrary variations* δF, *then*

$$(D^2-a^2-i\mathfrak{p}\sigma)F = -Ra^2W \tag{259}$$

and conversely. In other words, the characteristic values of R, for assigned σ, a^2, and T, have an *extremal* character; but unlike the case considered in § 26 (a), they do not have a minimal character: indeed, they could not since they can be complex.

31. The onset of convection as overstability: the solution for other boundary conditions

The solution of the double characteristic value problem formulated in § 30 can be obtained by the variational method by a procedure exactly analogous to that used in §§ 27 (b) and (c). Thus, for obtaining the solution for the case of two rigid boundaries, we expand F in a cosine series, as †

$$F = \sum_m A_m \cos[(2m+1)\pi z], \tag{260}$$

and express W and Z as sums in the manner,

$$W = \sum_m A_m W_m \quad \text{and} \quad Z = \sum_m A_m Z_m. \tag{261}$$

In view of equation (251), W_m is now a solution of the equation

$$[(D^2-a^2-i\sigma)^2(D^2-a^2)+TD^2]W_m = -c_{2m+1}\cos[(2m+1)\pi z], \tag{262}$$

where
$$c_{2m+1} = (2m+1)^2\pi^2+a^2+i\sigma. \tag{263}$$

The solution of equation (262) appropriate to the problem on hand is

$$W_m = c_{2m+1}\gamma_{2m+1}\cos[(2m+1)\pi z]+\sum_{j=1}^{3} B_j^{(m)}\cosh q_j z, \tag{264}$$

where
$$\frac{1}{\gamma_{2m+1}} = c_{2m+1}^2[(2m+1)^2\pi^2+a^2]+(2m+1)^2\pi^2 T, \tag{265}$$

the $B_j^{(m)}$'s ($j = 1, 2, 3$) are constants of integration; and q_j^2's ($j = 1, 2, 3$) are the roots of cubic equation,

$$(q^2-a^2-i\sigma)^2(q^2-a^2)+Tq^2 = 0. \tag{266}$$

The corresponding solution for Z_m is

$$Z_m = -\left(\frac{2\Omega}{\nu}d\right)\left\{(2m+1)\pi\gamma_{2m+1}\sin[(2m+1)\pi z]+\sum_{j=1}^{3} B_j^{(m)}\frac{q_j}{x_j}\sinh q_j z\right\}, \tag{267}$$

where
$$x_j = q_j^2-a^2-i\sigma \quad (j = 1, 2, 3). \tag{268}$$

Comparing the solutions (264) and (267) with the solutions (146) and (151) obtained in § 27 (a), we observe that they are identical except for

† The origin of z has now been displaced so that its limits are $\pm\frac{1}{2}$.

the different definitions of q_j, x_j, c_{2m+1}, and γ_{2m+1}. With these same redefinitions of the constants, the solutions for $B_j^{(m)}$ given in equations (154) and (155) apply equally well to the present case.

Substituting the solution for W we have obtained in equation (250), we now have

$$\sum_m A_m[(2m+1)^2\pi^2+a^2+i\mathfrak{p}\sigma]\cos[(2m+1)\pi z]$$
$$= Ra^2\sum_m A_m\Big\{c_{2m+1}\gamma_{2m+1}\cos[(2m+1)\pi z]+\sum_{j=1}^{3} B_j^{(m)}\cosh q_j z\Big\}. \quad (269)$$

Equation (269) is identical in form with equation (156). The analysis following equation (156) up to, and inclusive of, equation (161) now applies.† In particular, we have the secular determinant:

$$\left\|\frac{1}{2}\left\{\frac{(2n+1)^2\pi^2+a^2+i\mathfrak{p}\sigma}{Ra^2}-c_{2n+1}\gamma_{2n+1}\right\}\delta_{mn}-(n|m)\right\| = 0, \quad (270)$$

where the matrix element $(n|m)$ is given by equation (161). The various quantities (such as q_j, x_j, c_{2n+1}, and γ_{2n+1}) appearing in equation (161) have the meanings now assigned to them.

The matrix $(n|m)$ is symmetrical in n and m. However, since the elements are complex, the symmetry does not ensure the reality of its characteristic roots; indeed, they will in general be complex.

In the first approximation in which we retain only the (0, 0)-element of the secular matrix, the characteristic equation for R becomes

$$R = \frac{1}{a^2}\frac{\pi^2+a^2+i\mathfrak{p}\sigma}{c_1\gamma_1+2(0|0)}, \quad (271)$$

where (cf. equation (161))

$$(0|0) = 2\pi^2\gamma_1^2\Delta\prod_{j=1}^{3}\coth\tfrac{1}{2}q_j\Big\{\frac{1}{x_1}(x_3-x_2)(c_1+x_1)^2q_1\tanh\tfrac{1}{2}q_1+$$
$$+\frac{1}{x_2}(x_1-x_3)(c_1+x_2)^2q_2\tanh\tfrac{1}{2}q_2+\frac{1}{x_3}(x_2-x_1)(c_1+x_3)^2q_3\tanh\tfrac{1}{2}q_3\Big\}. \quad (272)$$

It may be recalled here that Δ in (272) is defined as in equation (155) but with the present meanings for the various quantities.

From our previous experience with variational methods, like the present, we may be confident that already the first approximation should give the required characteristic values to an accuracy of 1 or 2 per cent.

We shall now present the results of certain calculations based on equations (271) and (272). The value

$$\mathfrak{p} = 0{\cdot}025 \quad (273)$$

† Equations following (161) do not apply since these depend on the particular definitions of the roots q_j, etc.

was used in these calculations; this is approximately the value of $\mathfrak{p}$ for mercury at ordinary room temperatures. Some details regarding the method by which the (R_c, T)-relation was derived may be given.

For a chosen value of a, R was evaluated in accordance with equations (271) and (272) for various assigned values of σ; and the value of σ for which R is real was deduced by interpolation. Thus for $T = 10^{10}$ and $a = 19{\cdot}8$ it was found that:

$$\left.\begin{aligned} R &= 1{\cdot}983\times 10^6 - 5{\cdot}882\times 10^5 i \text{ for } \sigma = 1{\cdot}420\times 10^4;\\ R &= 1{\cdot}322\times 10^6 + 9{\cdot}510\times 10^4 i \text{ for } \sigma = 1{\cdot}500\times 10^4;\\ R &= 1{\cdot}414\times 10^6 + 0{\cdot}329\times 10^4 i \text{ for } \sigma = 1{\cdot}489\times 10^4.\end{aligned}\right\} \quad (274)$$

TABLE XII

Critical Rayleigh numbers and related constants for the onset of overstability in case both bounding surfaces are rigid and $\mathfrak{p} = 0{\cdot}025$

T	a_c	σ	R_c	p/Ω
10^4	3·08	$4{\cdot}45\times 10^1$	$4{\cdot}39\times 10^3$	0·8902
10^6	4·09	$5{\cdot}82\times 10^2$	$9{\cdot}51\times 10^3$	1·1646
5×10^7	8·10	$2{\cdot}43\times 10^3$	$6{\cdot}29\times 10^4$	0·6862
2×10^8	10·28	$3{\cdot}92\times 10^3$	$1{\cdot}38\times 10^5$	0·5541
10^9	13·46	$6{\cdot}81\times 10^3$	$3{\cdot}54\times 10^5$	0·4310
10^{10}	19·7	$1{\cdot}50\times 10^4$	$1{\cdot}42\times 10^6$	0·2992
10^{11}	28·75	$3{\cdot}27\times 10^4$	$5{\cdot}83\times 10^6$	0·2069
10^{12}	41·7	$7{\cdot}18\times 10^4$	$2{\cdot}44\times 10^7$	0·1435

From these values, it can be estimated that

$$R = 1{\cdot}418\times 10^6 \text{ for } \sigma = 1{\cdot}4886\times 10^4. \quad (275)$$

By repeating such calculations for other a's, the minimum of R as a function of a may be determined. Thus, in the example considered, it was found that:

$$\left.\begin{aligned} &\text{for } a = 19{\cdot}6,\ \sigma = 1{\cdot}503\times 10^4,\ R = 1{\cdot}41765\times 10^6;\\ &\text{for } a = 19{\cdot}7,\ \sigma = 1{\cdot}496\times 10^4,\ R = 1{\cdot}4175\times 10^6;\\ &\text{for } a = 19{\cdot}8,\ \sigma = 1{\cdot}489\times 10^4,\ R = 1{\cdot}4177\times 10^6.\end{aligned}\right\} \quad (276)$$

Consequently, it may be concluded that for $T = 10^{10}$, the critical Rayleigh number for the onset of overstability is $1{\cdot}4175\times 10^6$ when $a = 19{\cdot}7$ and $\sigma = 1{\cdot}496\times 10^4$.

In Table XII the results of such calculations are summarized.

Similar, but less extensive, calculations were undertaken for the case when one of the bounding surfaces is free and the other is rigid. The results of these calculations are given in Table XIII.

The results on the critical Rayleigh number, for the onset of instability as stationary convection and as overstable oscillations for all three sets of boundary conditions, are all included in Figs. 21 and 22 (p. 96).

TABLE XIII

Critical Rayleigh numbers and related constants for the onset of overstability in case one bounding surface is free and the other is rigid and $\mathfrak{p} = 0{\cdot}025$

T	a_c	σ	R_c	p/Ω
10^7	5·85	$1{\cdot}44\times10^3$	$1{\cdot}71\times10^4$	0·9126
3×10^9	15·58	$1{\cdot}04\times10^4$	$4{\cdot}81\times10^5$	0·3801
10^{12}	40·5	$7{\cdot}46\times10^4$	$1{\cdot}87\times10^7$	0·1492

32. The case $\mathfrak{p} = 0$

The case when the Prandtl number $\mathfrak{p}$ tends to zero is singular with respect to the onset of overstable oscillations: thus, the asymptotic relations (234), (235), and (237) are valid for $\mathfrak{p}^2T \to \infty$, and they cannot be used when $\mathfrak{p} = 0$. The case $\mathfrak{p} = 0$ should be treated separately.

Returning then to equation (222) and setting $\mathfrak{p} = 0$, we have†

$$R^{(0)} = 2\pi^4\frac{(1+x)^3}{x}; \tag{277}$$

apart from the factor 2, this is the same formula which obtains in the absence of rotation (cf. equation II (193)). The minimum Rayleigh number occurs for $x = \frac{1}{2}$ when

$$R_c^{(0)} = 13{\cdot}5\pi^4 = 1315. \tag{278}$$

The frequency of the overstable oscillations, when instability occurs, can be obtained from equation (221) by putting $\mathfrak{p} = 0$ and $x = \frac{1}{2}$; we get

$$\sigma_1^2 = \tfrac{2}{3}T_1 - 2{\cdot}25. \tag{279}$$

Ignoring 2·25 in this equation and making use of the relation (236), we have

$$p = (\tfrac{2}{3})^{\frac{1}{2}}2\Omega, \tag{280}$$

where p is now the (circular) frequency in seconds.

The frequency given by equation (280) is different from the natural frequency of oscillation of the rotating fluid only by the factor $(2/3)^{\frac{1}{2}}$. It would, therefore, be physically correct to say that *in the limit* $\mathfrak{p} = 0$, *instability sets in by exciting the natural modes of oscillation of the rotating fluid*.

† We are restricting our present considerations to the case of two free boundaries.

33. Thermodynamic significance of the variational principles

We shall now show that as in the case of the simple Bénard problem (§ 14), the variational principles established in §§ 26 (*a*) and 30 have similar thermodynamic meanings.

Consider the average rate of viscous dissipation of energy in a unit column of the fluid and the average rate at which the buoyancy force $g|\delta\rho|$ $(= g\alpha\rho\theta)$ releases energy in the same unit column. These are given by (equations II (175) and (178))

$$\epsilon_\nu = -\frac{\rho\nu}{d^2}\int\limits_0^1 \{\langle w(D^2-a^2)w\rangle+\langle u(D^2-a^2)u\rangle+\langle v(D^2-a^2)v\rangle\}\,dz \tag{281}$$

and

$$\epsilon_g = \rho g\alpha \int\limits_0^1 \langle\theta w\rangle\,dz, \tag{282}$$

where angular brackets signify that the quantity enclosed is averaged over the horizontal plane.

Without loss of generality for our present purposes, we may suppose that the expressions for u, v, and w are those given in equations (186) and (187). According to these equations, we have, for example,

$$\int\limits_0^1 \langle u(D^2-a^2)u\rangle\,dz$$

$$= -\frac{1}{4a^4}\int\limits_0^1 \{a_x^2\,DW(D^2-a^2)\,DW+a_y^2\,d^2Z(D^2-a^2)Z\}\,dz. \tag{283}$$

It should be noted that the cross terms involving the products

$$DW(D^2-a^2)Z \quad \text{and} \quad Z(D^2-a^2)W$$

do not make any contributions to the foregoing expression; this is due to the quite general circumstance that the waves in the horizontal plane associated with DW and Z in the solution for $\mathbf{u}_\perp$ are exactly out of phase.

We have an expression similar to (283) for the term in v in equation (281); also,

$$\int\limits_0^1 \langle w(D^2-a^2)w\rangle\,dz = \tfrac{1}{4}\int\limits_0^1 W(D^2-a^2)W\,dz. \tag{284}$$

Combining these results, we have (cf. equation II (176))

$$\epsilon_\nu = -\frac{\rho\nu}{4d^2a^2}\int\limits_0^1 \{a^2W(D^2-a^2)W+DW(D^2-a^2)DW+d^2Z(D^2-a^2)Z\}\,dz. \tag{285}$$

After integrating by parts the second and the third terms on the right-hand side, we find

$$\epsilon_\nu = \frac{\rho\nu}{4a^2d^2}\int\limits_0^1 \{[(D^2-a^2)W]^2+d^2[(DZ)^2+a^2Z^2]\}\,dz. \tag{286}$$

In the further reduction of the expression for ϵ_g, we must distinguish the stationary and the oscillatory cases.

(a) The case when the marginal state is stationary

In this case, the equation relating θ and w is (cf. equation (86))

$$w = -\frac{\kappa}{\beta d^2}(D^2-a^2)\theta. \tag{287}$$

Accordingly,

$$\epsilon_g = -\frac{\rho g\alpha\kappa}{\beta d^2}\int\limits_0^1 \langle\theta(D^2-a^2)\theta\rangle\,dz$$

$$= -\frac{\rho g\alpha\kappa}{4\beta d^2}\int\limits_0^1 \Theta(D^2-a^2)\Theta\,dz. \tag{288}$$

After an integration by parts, we have

$$\epsilon_g = \frac{\rho g\alpha\kappa}{4\beta d^2}\int\limits_0^1 [(D\Theta)^2+a^2\Theta^2]\,dz. \tag{289}$$

On the other hand, according to equations (98) and (104),

$$\Theta = \frac{\nu}{g\alpha d^2a^2}F. \tag{290}$$

In terms of F the expression for ϵ_g becomes

$$\epsilon_g = \frac{\rho\kappa\nu^2}{4g\alpha\beta a^4d^6}\int\limits_0^1 [(DF)^2+a^2F^2]\,dz. \tag{291}$$

In a steady state, the kinetic energy dissipated by viscosity must be balanced by the internal energy released by the buoyancy force. We must, therefore, require

$$\epsilon_\nu = \epsilon_g. \tag{292}$$

Equations (286) and (291) now give

$$R = \frac{g\alpha\beta}{\kappa\nu}d^4 = \frac{\int\limits_0^1 [(DF)^2+a^2F^2]\,dz}{a^2\int\limits_0^1 \{[(D^2-a^2)W]^2+d^2[(DZ)^2+a^2Z^2]\}\,dz}. \tag{293}$$

But this is exactly the same expression for R which, according to the variational principle of § 26 (a), attains its absolute minimum for the marginal state.

(b) The case when the marginal state is oscillatory

The question now arises: how should one modify under circumstances of overstability, the principle which equates the rate of irreversible dissipation of energy with the rate of liberation of the thermodynamically available energy by the buoyancy force, as a criterion for marginal stability? It would appear natural that we generalize the principle along the following lines.

If the motion is periodic in time, then the kinetic energy of the fluid, as well as the release of the potential energy by the buoyancy force, will be subject to similar variations. Suppose that all quantities which describe the perturbation vary with a circular frequency p so that all the amplitudes have a time-dependent factor e^{ipt}. Then

$$u_i \frac{\partial u_i}{\partial t} = ipu_i^2. \tag{294}$$

We must allow for this change in the kinetic energy (per unit mass) in writing an equation of energy balance. Thus, it would appear that the quantity we must equate to ϵ_g is

$$\epsilon_\nu + i\rho p \int_0^d \langle u_i^2 \rangle \, dz. \tag{295}$$

Reverting to d as the unit of distance and defining σ ($= pd^2/\nu$) as in equation (92), we have to consider

$$\epsilon_\nu + i\sigma \frac{\rho\nu}{d^2} \int_0^1 \langle u_i^2 \rangle \, dz \tag{296}$$

in place of ϵ_ν for the stationary case. The criterion for the occurrence of an oscillatory marginal state should then be

$$\epsilon_\nu + i\sigma \frac{\rho\nu}{d^2} \int_0^1 \langle u_i^2 \rangle \, dz = \epsilon_g = g\alpha\rho \int_0^1 \langle \theta w \rangle \, dz, \tag{297}$$

where in evaluating ϵ_g we must allow for the fact that θ and w are no longer stationary.

At first sight, the appearance of the imaginary i, and complex numbers generally, in the equation for the energy balance, may strike one as very odd. Its origin must be traced to the fact that the oscillations in the velocity and in the acceleration are out of phase. Consequently, the excess (or defect) of energy dissipated during one phase of the cycle must be exactly compensated by a similar excess (or defect) of energy liberated in a synchronous manner. It is this need for synchronism which determines the period of oscillation, as well as the Rayleigh number, as characteristics of the marginal state. The mathematical counterpart of this is that the determination of R and σ is through the solution of a double characteristic value problem.

We shall now verify that the thermodynamic principle, as reformulated above, is in agreement with the variational principle of § 30. The expression (286) for ϵ_ν continues to be valid. And we must now evaluate the additional term in $\langle u_i^2\rangle$ in equation (297). Considering the contribution to the integral by the x-component of the velocity, we have (cf. equation (187))

$$\int_0^1 \langle u^2\rangle\, dz = \frac{1}{4a^4}\int_0^1 [a_x^2(DW)^2+a_y^2 d^2Z^2]\, dz. \tag{298}$$

We have a similar contribution from the y-component of the velocity; and altogether we have

$$i\sigma\frac{\rho\nu}{d^2}\int_0^1 \langle u_i^2\rangle\, dz = i\sigma\frac{\rho\nu}{4a^4d^2}\int_0^1 [a^4W^2+a^2(DW)^2+a^2d^2Z^2]\, dz. \tag{299}$$

This must be added to the expression (286) giving ϵ_ν.

Turning to the evaluation of ϵ_g, we must now use the relation (see equation (93))

$$(D^2-a^2-ip\sigma)\theta = -\left(\frac{\beta}{\kappa}d^2\right)w \tag{300}$$

for eliminating w. Accordingly, we now have

$$\epsilon_g = -\frac{g\rho\alpha\kappa}{4\beta d^2}\int_0^1 \Theta(D^2-a^2-ip\sigma)\Theta\, dz = \frac{g\rho\alpha\kappa}{4\beta d^2}\int_0^1 [(D\Theta)^2+(a^2+ip\sigma)\Theta^2]\, dz. \tag{301}$$

The relation between Θ and F is unaltered and is given, as before, by (290). Thus,

$$\epsilon_g = \frac{\rho\kappa\nu^2}{4g\alpha\beta a^4d^6}\int_0^1 [(DF)^2+(a^2+ip\sigma)F^2]\, dz. \tag{302}$$

Combining equations (286), (297), (299), and (302), we obtain

$$R = \frac{\int_0^1 [(DF)^2+(a^2+ip\sigma)F^2]\, dz}{a^2\int_0^1 \{[(D^2-a^2)W]^2+d^2[(DZ)^2+a^2Z^2]+i\sigma[(DW)^2+a^2W^2+d^2Z^2]\}\, dz}; \tag{303}$$

and this is, indeed, the expression for R which formed the basis of the variational principle of § 30.

We are thus enabled to formulate the following general principle.

Thermal instability as stationary convection will set in at the minimum (adverse) temperature gradient which is necessary to maintain a balance between the rate of dissipation of energy by viscosity and the rate of liberation of the thermodynamically available energy by the buoyancy force acting on the fluid. Likewise, the onset of thermal instability will be as overstable oscillations if it is possible (at a lower adverse temperature gradient) to balance in a synchronous manner the periodically varying amounts of kinetic energy with similarly varying amounts of dissipation and liberation of energy.

34. The case when $\mathbf{\Omega}$ and $\mathbf{g}$ act in different directions

We shall now briefly consider the case when $\mathbf{\Omega}$ and $\mathbf{g}$ act in different directions. For this purpose we must return to equations (84) and (85) in which the assumption that $\mathbf{\Omega}$ and $\mathbf{g}$ are parallel has not yet been made.

Let $\mathbf{\Omega}$ be inclined at an angle ϑ to the direction of the vertical. Also, let the direction of the x-axis be so chosen that $\mathbf{\Omega}$ lies in the xz-plane. Then

$$\boldsymbol{\lambda} = (0,0,1) \quad \text{and} \quad \mathbf{\Omega} = \Omega(\sin\vartheta, 0, \cos\vartheta), \tag{304}$$

and equations (84) and (85) become

$$\frac{\partial \zeta}{\partial t} = \nu\nabla^2\zeta + 2\Omega\left(\cos\vartheta\frac{\partial}{\partial z}+\sin\vartheta\frac{\partial}{\partial x}\right)w \tag{305}$$

and

$$\frac{\partial}{\partial t}\nabla^2 w = g\alpha\left(\frac{\partial^2}{\partial x^2}+\frac{\partial^2}{\partial y^2}\right)\theta+\nu\nabla^4 w-2\Omega\left(\cos\vartheta\frac{\partial}{\partial z}+\sin\vartheta\frac{\partial}{\partial x}\right)\zeta. \tag{306}$$

The remaining equation (86) is unaffected by this generalization.

If we seek solutions of equations (86), (305), and (306) which are independent of x and are of the forms

$$w = W(z)\cos\frac{ay}{d}, \quad \zeta = Z(z)\cos\frac{ay}{d}, \quad \text{and} \quad \Theta = \Theta(z)\cos\frac{ay}{d}, \tag{307}$$

then equations (305) and (306) reduce to equations (87) and (88) with the only difference that Ω is replaced by $\Omega\cos\vartheta$. Consequently, if we restrict ourselves to the onset of instability as rolls in the x-direction, then all of the discussions of the preceding sections, with the exception of those parts of § 28 which specifically deal with general cell patterns, will apply if we interpret Ω everywhere to mean the component of $\mathbf{\Omega}$ in the direction of $\mathbf{g}$. The question, whether lower Rayleigh numbers can be attained for the marginal states by considering more general patterns of motion than rolls, can be answered only by solving the necessary characteristic value problems in which a_x and a_y are explicitly retained; and minimizing the lowest characteristic values, for given a_x and a_y, as functions of these two parameters. Such an analysis has not been carried out for the present problem; but it has been carried out for the closely related problem of the inhibition of thermal convection by an impressed magnetic field (Chapter IV, § 47). If the conclusions reached in that context can be extended to the present problem, then it would appear that the minimum Rayleigh numbers do indeed occur for the onset of instability as rolls. This apparently paradoxical conclusion is clarified in Chapter IV, § 47.

35. Experiments on the onset of thermal instability in rotating fluids

We have seen that rotation influences the onset of thermal instability in many ways. The principal effects are these.

(1) Rotation inhibits the onset of instability; the extent of the inhibition depends on the Taylor number T $(= 4\Omega^2 d^4/\nu^2)$ and the Prandtl number $\mathfrak{p}$ $(= \nu/\kappa)$; and as $T \to \infty$, the Rayleigh number at which instability sets in, and the wave number a (in units of $1/d$) with which it is associated, have the asymptotic behaviours

$$R \to \text{constant } T^{\frac{2}{3}} \quad \text{and} \quad a \to \text{constant } T^{\frac{1}{6}}. \tag{308}$$

(2) The onset of instability will be as stationary convection so long as $\mathfrak{p}$ exceeds a certain critical value $\mathfrak{p}^*$. The precise value of $\mathfrak{p}^*$ depends on the nature of the bounding surfaces; but it can be inferred from the value $(= 0{\cdot}6766)$ determined for the case of two free boundaries that $\mathfrak{p}^* \sim 1$. When $\mathfrak{p} > \mathfrak{p}^*$, the (R_c, T)-relation is independent of $\mathfrak{p}$.

(3) The onset of instability will be as overstable oscillations if $\mathfrak{p} < \mathfrak{p}^*$ and the Taylor number exceeds a certain value $T^{(\mathfrak{p})}$ depending on $\mathfrak{p}$. For $T \leqslant T^{(\mathfrak{p})}$ the onset of instability will be as stationary convection. For a given $\mathfrak{p}$ $(< \mathfrak{p}^*)$ the asymptotic behaviours of R and a are again as in (1) above; but the constants of proportionality now depend on $\mathfrak{p}$. Also, the gyration frequency p of the overstable oscillations is essentially determined by Ω and the Taylor number; for $0 < \mathfrak{p} < \mathfrak{p}^*$,

$$p/\Omega \to \text{constant } T^{-\frac{1}{6}} \text{ as } T \to \infty, \tag{309}$$

where the constant depends on $\mathfrak{p}$.

From the foregoing summary of the principal effects to be expected, it is clear that in selecting the liquids for the experiments, it would be preferable to choose two liquids, one with a Prandtl number substantially larger than unity and the other with a Prandtl number substantially less than unity. Two such liquids are readily available: water with $\mathfrak{p} \sim 7 \cdot 5$ and mercury with $\mathfrak{p} = 0 \cdot 025$. Experiments with these two liquids have been carried out by Nakagawa and Frenzen, Fultz and Nakagawa, and Goroff. An account of these experiments will now be given.

(*a*) *Experiments with water*

In these experiments, water was contained in Pyrex glass cylinders of 12 inches outside diameter. To the bottoms of the cylinders were cemented electrically conducting Pyrex heating plates; they served as the heating elements to which measured amounts of electric power could be supplied. The cylinders containing the fluid and the necessary accessories were placed on a supporting steel ring mounted on a vertical shaft. The steel ring had provisions for levelling and for other adjustments. The fluid was set in rotation by driving the shaft with suitable belts and motors. In the experiments, layers of water of depths varying from 2 to 17 cm and speeds of rotation up to 50 rev/min were used. Taylor numbers up to 10^8 were reached in this manner.

The difference in the temperatures at two levels, one near the top and one near the bottom surface of the fluid, was measured by copper-constantan wire thermocouples. By using several junctions in series as a thermopile, it was possible to measure the difference in the average temperatures at the two levels with an accuracy of $\pm 0 \cdot 01^\circ$ C; this accuracy in the temperature measurements was found to be necessary and sufficient.

Observations with the foregoing experimental arrangement were of two kinds: visual and photographic observations made with the aid of a rotoscope which enabled the direct detection of the fluid motions relative to the rotating frame; and quantitative observations based on the records of the difference in temperature between the top and bottom surfaces made continuously during each experiment after varying amounts of electric power had been turned on to the heating elements.

By visual observations with the rotoscope, it was possible to observe the onset of cellular convection when the heating rate exceeded a certain amount. The motions could be made visible by scattering a small amount of aluminium powder on the top surface. The movements of the aluminium particles could be photographed by a time-exposure streak technique. An example of such a photograph is shown in Fig. 30. The internal motions in the cells could also be followed with black ink and

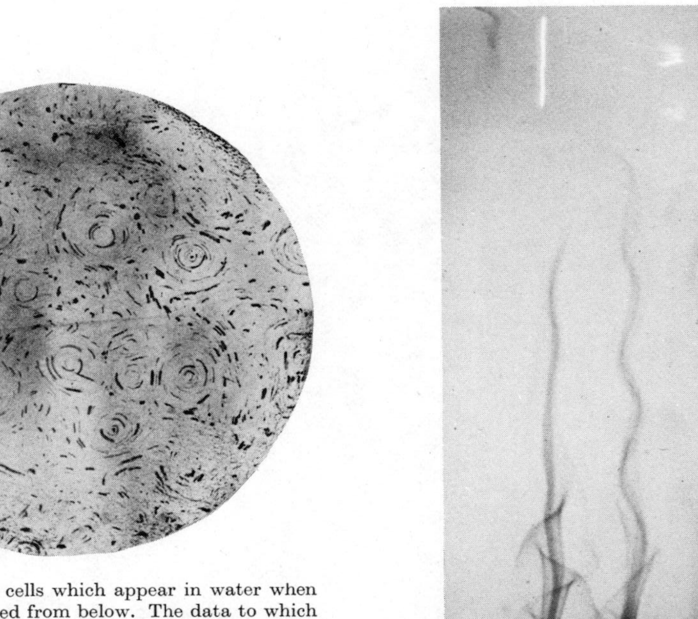

FIG. 30. Convection cells which appear in water when in rotation and heated from below. The data to which the illustration refers are: depth 18 cm, difference in temperature 0·7°; rate of rotation 5·0 rev/min; Taylor number $1{\cdot}2 \times 10^9$.

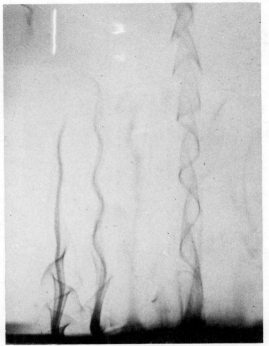

FIG. 31. Side view of ink rising from the bottom of a cylinder of water in the ascending cores of cells. The data to which the illustration refers are: depth 18 cm· difference in temperature 0·5°; rate of rotation 10,9 rev/min; Taylor number $5{\cdot}5 \times 10^9$.

photographed. Fig. 31 is an example of a photograph showing such internal motions.

For a quantitative determination of the critical Rayleigh number for the onset of instability, a careful study of the temperature records is necessary. In Fig. 32 examples of temperature records which were obtained for varying heating rates are shown. Fig. 32 *a* corresponds to

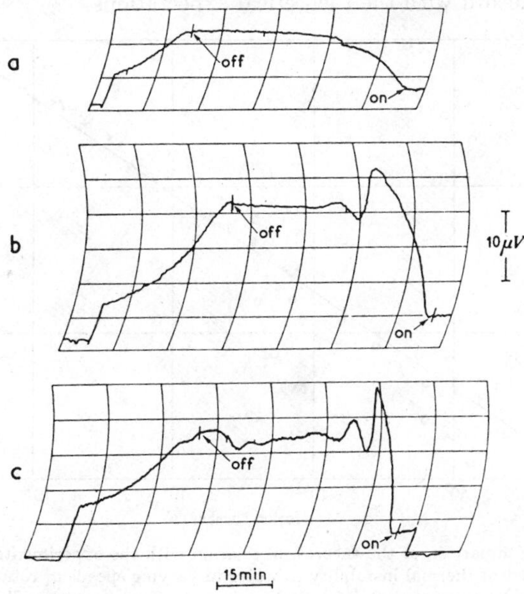

Fig. 32. Time records of the adverse temperature gradient for water for three different rates of heating (d = 3 cm, Ω = 10 rev/min) (from *Proc. Roy. Soc.* (*London*) A, **231**, 219 (1955)).

a case when the bottom plate was heated at a sufficiently low rate that the steady difference in temperature attained was insufficient to induce instability; Fig. 32 *b* is a typical record when the heating rate sufficed to surpass by some margin the temperature gradient necessary for the onset of instability; Fig. 32 *c* is an example when the heating rate was so much in excess of what is necessary to cause instability that the conditions were approaching turbulence. The difference between the records, when we are in the conductive régime and when we have passed it, is sufficiently striking that by examining them for various heating rates, the temperature gradient at which conditions are just past marginal could be determined with some precision. Simultaneous visual observations confirmed the criteria that were used. In this manner, Nakagawa and Frenzen determined the critical Rayleigh number as a function of the Taylor number. Their results are shown in Fig. 33. The theoretical (R_c, T)-relation for one rigid and one free surface derived in § 27 (c) is shown in the same plot; it will be observed that the theoretically predicted relation is fully confirmed.

Estimates of the cell dimensions based on photographic observations are also in accord with the theoretical expectations.

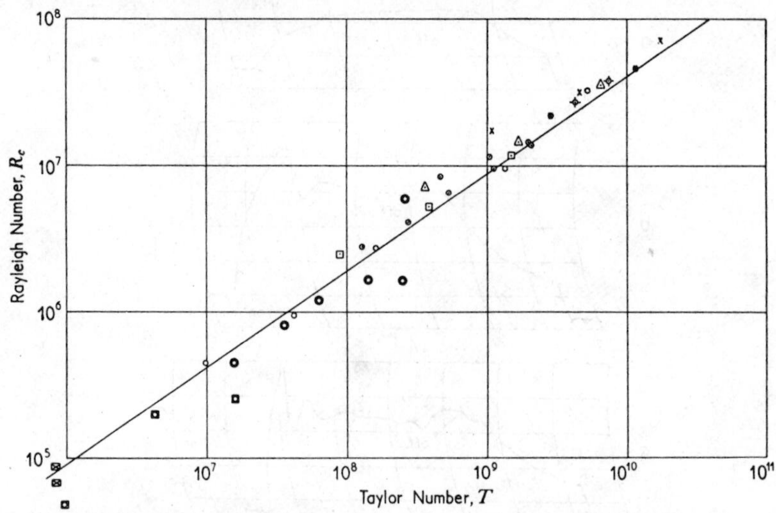

FIG. 33. Comparison of the theoretical relation with the experimental results on the onset of thermal instability in water in varying speeds of rotation: the critical Rayleigh number R_c is plotted against the Taylor number T. Experiments with different depths of water are distinguished: ⊠ 18 cm; -○- 14 cm; ⊙ 10 cm; ◎ 6 cm; × 17·3 cm; △ 13·3 cm; ⊡ 9·3 cm; o 5·3 cm; ⊡ 3 cm; ⊠ 2 cm.

(b) *Experiments with mercury*

The experiments of Fultz and Nakagawa with mercury were carried out with essentially the same experimental arrangement as described in § (a) above. But a number of precautions peculiar to mercury had to be observed. For example, Fultz and Nakagawa record that considerable trouble was experienced with the oxidation of the mercury surface. It was necessary on this latter account to replace the air above the heated surface by nitrogen. This was accomplished by circulating nitrogen; this also served the purpose of keeping the temperature of the top surface constant. Even with these precautions, it was found that some time after the heating current had been turned on, a thin contaminant film was formed on the surface. The film was rigid enough to prevent motions at the top surface. The layer of mercury was thus effectively confined between two rigid surfaces.

The temperature gradients at which instability occurred were again determined by an examination of the temperature records. This was supplemented by making plots of the temperature gradients against the heating rates and using the Schmidt–Milverton principle (Chapter II, § 18 (*b*)). Layers of mercury of depth up to 8 cm and speeds of rotation up to 30 rev/min were used. Taylor numbers in the range 10^8–10^{12} were thus accessible to the experiments.

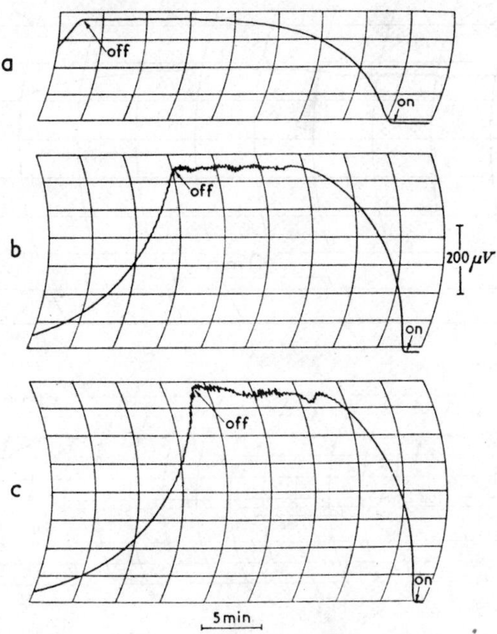

FIG. 34. Time records of the adverse temperature gradient for mercury for three different rates of heating ($d = 6$ cm, $\Omega = 15$ rev/min) (from *Proc. Roy. Soc.* (*London*) A, **231**, 220 (1955)).

In Fig. 34 a series of temperature records similar to those in Fig. 32 for water are shown. The difference in the two sets of temperature records, after instability has set in, is very striking; in the experiments with mercury, the temperature records exhibit very pronounced periodic fluctuations. Fig. 35, exhibiting records which show these periodic fluctuations with uniformity over durations as long as half an hour and more, leaves very little doubt that we are witnessing here the overstable oscillations predicted by the theory. Indeed, by counting the number of these fluctuations in known intervals of time, we can estimate the basic periods of the oscillations. The results of such estimates are shown in Fig. 36 together with the theoretically expected relation. It will be seen that the agreement is satisfactory.

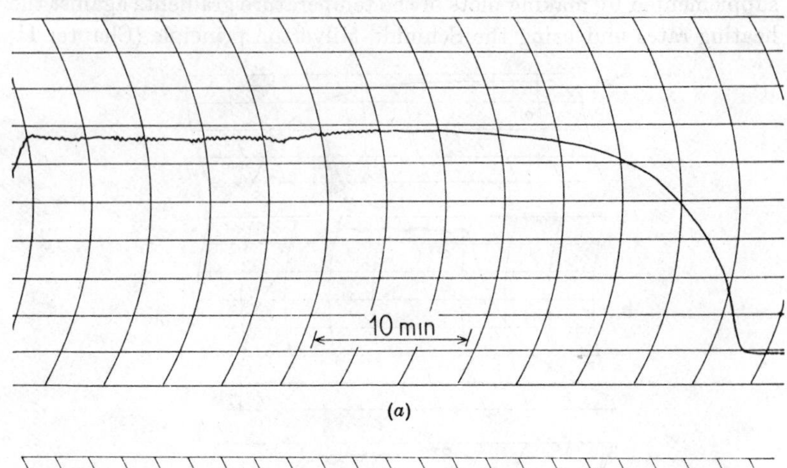

(*a*)

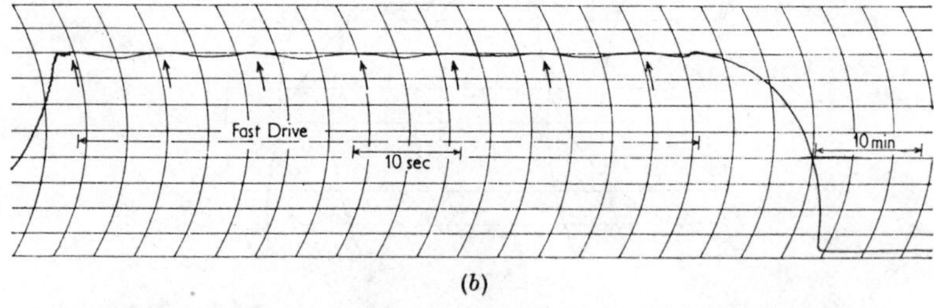

(*b*)

FIG. 35. Further time records of the adverse temperature gradient for mercury. The conditions to which they refer are given below:

	(*a*)	(*b*)
depth (cm)	6	6
Ω (sec^{-1})	1·11	3·08
temperature difference (degrees)	1·54	5·05
Taylor number (T)	5×10^9	$3{\cdot}9\times 10^{10}$
Rayleigh number (R)	$2{\cdot}04\times 10^6$	$6{\cdot}94\times 10^6$
period of oscillation (sec)	19½–20	9–10

In Fig. 37 the results on the critical Rayleigh numbers are shown together with the predicted relations for the three sets of boundary conditions. It is noteworthy that the agreement is best with the theoretical relation for two rigid boundaries: as we have already remarked, the experimental conditions do correspond to this case.

The experiments of Nakagawa confirm that the onset of thermal

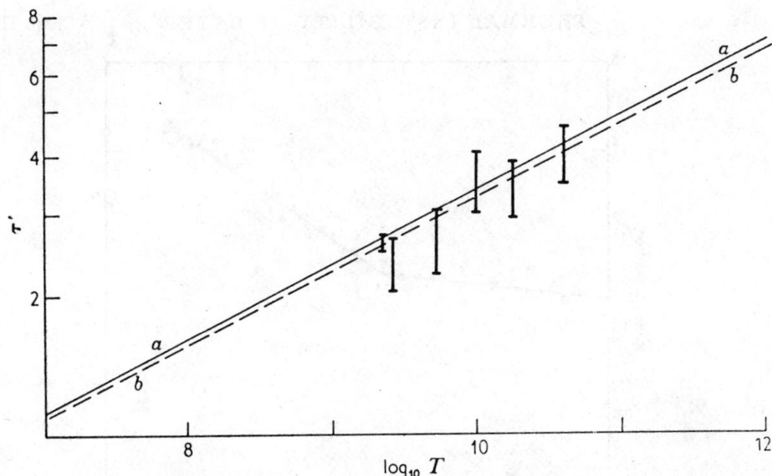

Fig. 36. A comparison of the observed periods of oscillation at marginal stability with the theoretical periods. The ordinate τ' gives the period in units of 2Ω. The curves *aa* and *bb* are the theoretical relations for $\mathfrak{p} = 0{\cdot}025$ and for the case of two bounding surfaces rigid (*aa*) and one bounding surface rigid and the other free (*bb*).

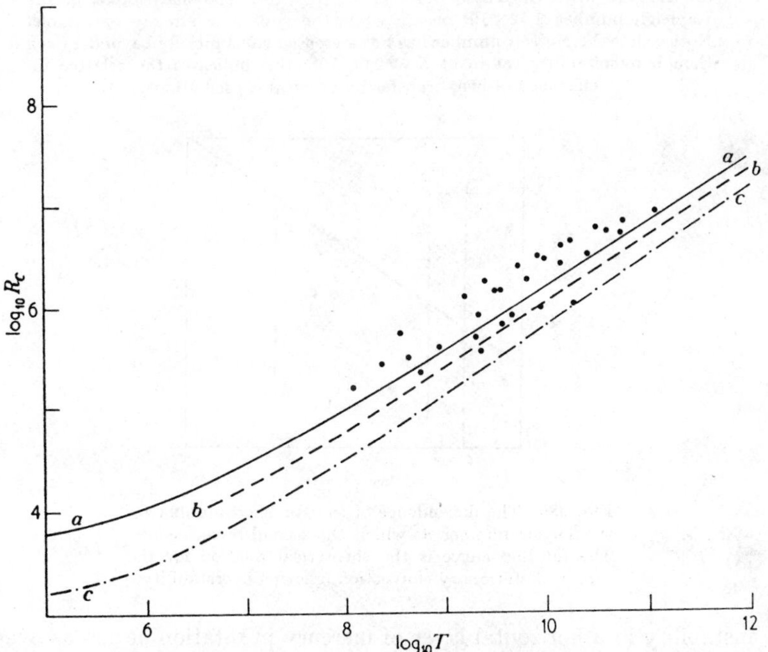

Fig. 37. A summary of the theoretical and experimental results on the critical Rayleigh numbers for the onset of instability in mercury. The curves *aa*, *bb*, and *cc* are the theoretical (R_c, T)-relations for the three cases (i) both bounding surfaces rigid, (ii) one bounding surface rigid and the other free, and (iii) both bounding surfaces free for $\mathfrak{p} = 0{\cdot}025$. The solid circles are the experimentally determined points for mercury ($\mathfrak{p} = 0{\cdot}025$).

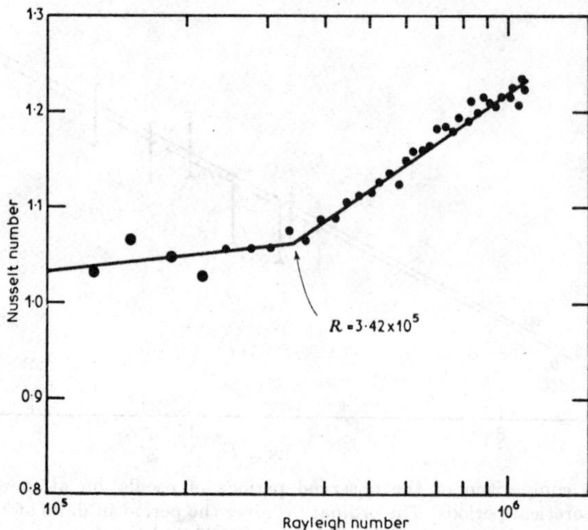

FIG. 38*a*. The variation of the Nusselt number for increasing Rayleigh numbers for the case when the Taylor number is $1{\cdot}12 \times 10^7$. The sharp break at the Rayleigh number $3{\cdot}42 \times 10^5$ occurs while the system is already overstable. Notice that the Nusselt number has not increased substantially beyond 1 even though overstability set in at $R \sim 2{\cdot}5 \times 10^4$; this indicates the relative inefficiency of heat-transfer by overstable oscillations.

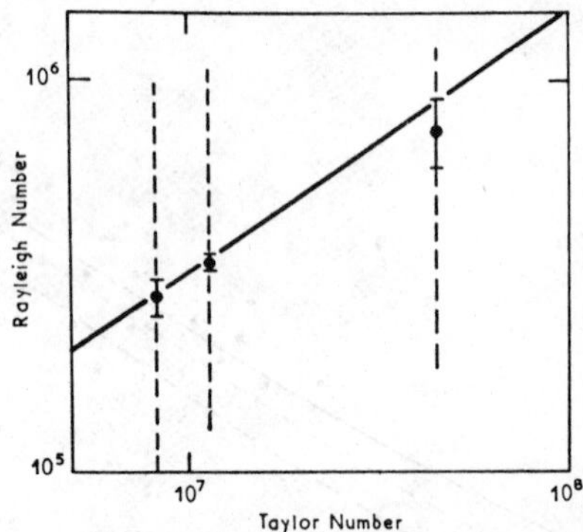

FIG. 38*b*. The dependence of the Rayleigh number on the Taylor number at which the second break occurs. The full-line curve is the theoretical relation for the onset of stationary convection ignoring overstability.

instability in a horizontal layer of mercury in rotation occurs as overstable oscillations at the predicted Rayleigh numbers $R_c^{(o)}$ and with the predicted characteristic frequencies. When the Rayleigh number exceeds $R_c^{(o)}$, the overstable oscillations will become of finite amplitude and non-linear effects, ignored in the linear stability theory, will become effective. At the same time, the mode with the real exponential time dependence will continue to be damped (on the assumption that the presence of the overstable oscillations does not seriously alter the static initial state to which the entire theory refers). When R becomes equal to $R_c^{(c)}$ ($> R_c^{(o)}$), this latter mode will begin to manifest itself. Since $R_c^{(c)}$ is about twenty-six times $R_c^{(o)}$ (for $T \to \infty$, when both bounding surfaces are rigid, and $\mathfrak{p} = 0{\cdot}025$), it is clear that the onset of stationary convection will occur superposed on overstable oscillations of quite finite amplitudes. However, on general grounds, one may expect that the overstable oscillations are not very efficient in transporting heat; and that on this account one may, nevertheless, be able to detect the onset of stationary convection in a Schmidt–Milverton plot. Some experiments of Dropkin and Globe seemed to indicate that a second break does occur in a Schmidt–Milverton plot after a first break† corresponding to the onset of overstability. But these experiments were not carried out with sufficient precautions to decide whether or not the second break occurs at the predicted Rayleigh number $R_c^{(c)}$. More recent experiments by Goroff carried out with the necessary care confirms that the second break does occur at the predicted place. In Fig. 38 *a* the results of his experiments for $T = 1{\cdot}12 \times 10^7$ are shown: in this plot of the Nusselt number against the Rayleigh number, a break clearly occurs at $R = 3{\cdot}45 \times 10^5$; and this is long after overstable oscillations had become a permanent feature of the temperature records. The value $R = 3{\cdot}45 \times 10^5$ should be compared with the theoretical value, $R_c^{(c)} = 3{\cdot}50 \times 10^5$ which the results of Table VIII predict for the onset of stationary convection for $T = 1{\cdot}12 \times 10^7$. In Fig. 38 *b* the values of $R_c^{(c)}$ obtained for three different values of T (including the one to which Fig. 38 *a* refers) are shown together with the theoretically expected relation. It would appear from these experiments that there can be little doubt that the onset of stationary convection occurs at the predicted Rayleigh number. They further establish the relative inefficiency of overstable oscillations for the transport of heat.

BIBLIOGRAPHICAL NOTES

For a general account of the subject of this chapter see:

1. S. Chandrasekhar, 'Thermal convection', *Daedalus*, **86**, 323–39 (1957).

† The first break was so weak in these experiments that one could not be certain of it in most cases.

§ 22. The fundamental theorem proved in this section is due to:

2. G. I. Taylor, 'Experiments with rotating fluids', *Proc. Roy. Soc.* (*London*) A, **100,** 114–21 (1921).

The same theorem, though not in as explicit a form, was independently stated by:

3. J. Proudman, 'On the motion of solids in a liquid possessing vorticity', *Proc. Roy. Soc.* (*London*) A, **92,** 408–24 (1916).

In Taylor's paper (reference 2), experiments are described which demonstrate very effectively the implications of the theorem for the character of the fluid motions which prevail under the circumstances. Taylor's experiments have been repeated by:

4. D. Fultz, 'A survey of certain thermally and mechanically driven systems of meteorological interest', *Proceedings of the First Symposium on the Use of Models in Geophysical Fluid Dynamics*, 27–63, edited by Robert R. Long, Johns Hopkins University, Baltimore, 1953.

The following references to related papers by Taylor may be noted:

5. G. I. Taylor, 'Motion of solids in fluids when the flow is not irrotational', *Proc. Roy. Soc.* (*London*), A, **93,** 99–113 (1917).

6. —— 'The motion of a sphere in a rotating liquid', ibid., **102,** 180–9 (1922).

7. —— 'Experiments on the motion of solid bodies in rotating fluids', ibid., **104,** 213–18 (1923).

The dynamics of rotating fluids have many fascinating aspects of which the Taylor–Proudman theorem is only one. For a general account of these other aspects see:

8. H. B. Squire, 'Rotating fluids', *Surveys in Mechanics*, 139–61, edited by G. K. Batchelor and R. M. Davies, Cambridge Monographs on Mechanics and Applied Mathematics, Cambridge, England, 1956.

See also:

9. G. W. Morgan, 'A study of motions in a rotating liquid', *Proc. Roy. Soc.* (*London*) A, **206,** 108–30 (1951.)

10. H. Görtler, 'Über eine Schwingungserscheinung in Flüssigkeiten mit stabiler Dichteschichtung', *Z. ang. Math. Mech.* **23,** 65–71 (1943).

11. —— 'Einige Bemerkungen über Strömungen in rotierenden Flüssigkeiten', ibid. **24,** 210–14 (1944).

§§ 24–27. The analyses in these sections are derived from:

12. S. Chandrasekhar, 'The instability of a layer of fluid heated below and subject to Coriolis forces', *Proc. Roy. Soc.* (*London*) A, **217,** 306–27 (1953).

13. —— and Donna D. Elbert, 'The instability of a layer of fluid heated below and subject to Coriolis forces. II', ibid., **231,** 198–210 (1955).

The method of solution described in § 27 (for cases (*b*) and (*c*)) is, however, different from that in reference 12; and the results included in Tables VIII and IX (due to Miss Donna Elbert) are based on the new formulae.

§ 28. The discussion of the cell patterns in this section follows:

14. G. Veronis, 'Cellular convection with finite amplitude in a rotating fluid', *J. Fluid Mech.*, **5,** 401–35 (1959).

§ 29. The principal results of this section are contained in references 12 and 13. But the arguments have been rearranged and made more explicit. The particular form of the discussion of the roots given in § (*a*) is due to Dr. John Sykes.

§§ 30–32. The analyses in these sections are again derived from references 12 and 13.

§ 33. See:

15. S. Chandrasekhar, 'The thermodynamics of thermal instability in liquids', *Max Planck Festschrift 1958*, 103–14, Veb Deutscher Verlag der Wissenschaften, Berlin, 1958.

§ 35. The references for the experimental work described in this section are:

16. D. Fultz, Y. Nakagawa, and P. Frenzen, 'An instance in thermal convection of Eddington's "overstability"', *Physical Rev.*, **94**, 1471–2 (1954).

17. —— —— 'Experiments on overstable thermal convection in mercury', *Proc. Roy. Soc.* (*London*) A, **231**, 211–25 (1955).

18. Y. Nakagawa and P. Frenzen, 'A theoretical and experimental study of cellular convection in rotating fluids', *Tellus*, **7**, 1–21 (1955).

19. D. Dropkin and S. Globe, 'Effect of spin on natural convection in mercury heated from below', *J. Appl. Phys.*, **30**, 84–89 (1959).

20. I. R. Goroff, 'An experiment on heat transfer by overstable and ordinary convection', *Proc. Roy. Soc.* (*London*) A, **254**, 537–41 (1960).

IV

THE THERMAL INSTABILITY OF A LAYER OF FLUID HEATED FROM BELOW

3. THE EFFECT OF A MAGNETIC FIELD

36. Hydromagnetics

IN this chapter we shall consider the effect of an externally impressed magnetic field on the onset of thermal instability in electrically conducting fluids. This brings us to the subject of *Hydromagnetics.* In broad terms, the subject of hydromagnetics is concerned with the ways in which magnetic fields can affect fluid behaviour. The subject has many ramifications; and the introduction of it in the context of a special problem is perhaps not the best way. But the problem of thermal instability in the presence of a magnetic field does illustrate the general principles of the subject. Moreover, the consideration of this problem in juxtaposition with a similar consideration of the effects of rotation brings out some striking similarities between the effects of rotation and magnetic field: they both impart to the fluid a certain rigidity; at the same time, they also impart to it certain properties of elasticity which enables it to transmit disturbances by new modes of wave propagation. To emphasize this similarity, we shall preface the discussion of thermal instability by a brief introduction to hydromagnetics and its basic theorems.

37. The basic equations of hydromagnetics

Consider a fluid which has the property of electrical conduction; and suppose also that magnetic fields are prevalent. The electrical conductivity of the fluid and the prevalence of magnetic fields contribute to effects of two kinds: *first,* by the motion of the electrically conducting fluid across the magnetic lines of force, electric currents are generated and the associated magnetic fields contribute to changes in the existing fields; and *second,* the fact that the fluid elements carrying currents traverse magnetic lines of force contributes to additional forces acting on the fluid elements. It is this twofold interaction between the motions and the fields that is responsible for patterns of behaviour which are often unexpected and striking.

We shall now write down the basic equations which express the interactions between the fluid motions and the magnetic fields. These are, of course, contained in Maxwell's equations and in the equations of hydrodynamics suitably modified. There is, however, one basic simplification which is possible. Since we shall not be concerned with effects which are related in any way to the propagation of electromagnetic waves, we can ignore the displacement currents in Maxwell's equations. Closely related to this approximation is the further possibility of avoiding any explicit reference to the charge density. The reason for this is not that it is small in itself, but rather that its variations affect the equation expressing the conservation of charge only by terms of order u^2/c^2; and terms of this order we can legitimately ignore.

With the displacement currents ignored, Maxwell's equations are

$$\operatorname{div} \mathbf{H} = 0, \tag{1}$$

$$\operatorname{curl} \mathbf{H} = 4\pi \mathbf{J}, \tag{2}$$

and

$$\operatorname{curl} \mathbf{E} = -\mu \frac{\partial \mathbf{H}}{\partial t}, \tag{3}$$

where in electromagnetic units, **E** and **H** are the intensities of the electric and the magnetic fields, **J** is the current density, and μ is the magnetic permeability. The magnetic permeability will be taken as unity in all applications; it is retained only to identify the units.

To complete the equations for the field, we need an equation for the current density. This requires some assumption concerning the nature of the fluid. In this book the assumption will be made that the fluid may be considered as continuous and that the macroscopic properties need be taken into account only indirectly through the effects of viscosity and the heat and electrical conductivities. And the coefficients expressing these latter effects will in turn be defined only phenomenologically.

Consider a fluid element. If it has a velocity **u**, the electric field it will experience is not **E**, as measured by a stationary observer, but $\mathbf{E}+\mu\mathbf{u}\times\mathbf{H}$. If in accordance with our assumptions we suppose that a coefficient of electrical conductivity σ can be defined, then the current density will be given by

$$\mathbf{J} = \sigma(\mathbf{E}+\mu\mathbf{u}\times\mathbf{H}). \tag{4}$$

Equations (1)–(4) are the basic equations of the field appropriate for hydromagnetics. Through the occurrence of the velocity **u** in the expression for **J**, the equations incorporate the effect of fluid motions on the electromagnetic field. The inverse effect of the field on the motions results from the force which the fluid elements experience in virtue of their carrying currents across magnetic lines of force. This is the *Lorentz force* given by

$$\mathscr{L} = \mu \mathbf{J} \times \mathbf{H}, \tag{5}$$

or, according to equation (2),

$$\mathscr{L} = \frac{\mu}{4\pi} \operatorname{curl} \mathbf{H} \times \mathbf{H}. \tag{6}$$

An alternative form for $\mathscr{L}$ is

$$\mathscr{L}_i = \frac{\mu}{4\pi} \epsilon_{ijk} \epsilon_{jlm} \frac{\partial H_m}{\partial x_l} H_k = \frac{\mu}{4\pi} (\delta_{im} \delta_{kl} - \delta_{il} \delta_{km}) \frac{\partial H_m}{\partial x_l} H_k$$

$$= \frac{\mu}{4\pi} H_k \left(\frac{\partial H_i}{\partial x_k} - \frac{\partial H_k}{\partial x_i} \right). \tag{7}$$

Since H_i is solenoidal, we can also write

$$\mathscr{L}_i = -\frac{\partial}{\partial x_i} \left(\mu \frac{|\mathbf{H}|^2}{8\pi} \right) + \frac{\partial}{\partial x_k} \left(\frac{\mu}{4\pi} H_i H_k \right). \tag{8}$$

This last form for the Lorentz force expresses it as the sum of a hydrostatic pressure, $\mu|\mathbf{H}|^2/8\pi$, and a tension, $\mu|\mathbf{H}|^2/4\pi$, along the lines of force; or, equivalently as the sum of a pressure, $\mu|\mathbf{H}|^2/8\pi$, transverse to the lines of force and a tension, $\mu|\mathbf{H}|^2/8\pi$, along the lines of force.

Including the Lorentz force among the other forces acting on the fluid, we have the equation of motion (cf. equation II (17))

$$\rho \frac{d\mathbf{u}}{dt} = \operatorname{div} \mathbf{P} + \rho \mathbf{X} + \mu \mathbf{J} \times \mathbf{H}, \tag{9}$$

where $\mathbf{P}$ is the total stress-tensor and $\mathbf{X}$ includes the external forces of non-electromagnetic origin.

For an incompressible fluid, the equation of motion takes the explicit form

$$\frac{\partial u_i}{\partial t} + u_j \frac{\partial u_i}{\partial x_j} - \frac{\mu H_j}{4\pi\rho} \frac{\partial H_i}{\partial x_j} = -\frac{\partial}{\partial x_i} \left(\frac{p}{\rho} + \mu \frac{|\mathbf{H}|^2}{8\pi\rho} \right) + \nu \nabla^2 u_i, \tag{10}$$

where the form (8) for the Lorentz force has been used.

38. The equation of motion governing the magnetic field and some of its consequences

We shall now obtain an equation of motion for the magnetic field. According to equation (4),

$$\mathbf{E} = \frac{1}{\sigma} \mathbf{J} - \mu \mathbf{u} \times \mathbf{H}, \tag{11}$$

or, making use of equation (2), we have

$$\mathbf{E} = \frac{1}{4\pi\sigma} \operatorname{curl} \mathbf{H} - \mu \mathbf{u} \times \mathbf{H}. \tag{12}$$

Inserting this expression for **E** in equation (3), we obtain

$$\frac{\partial \mathbf{H}}{\partial t}-\operatorname{curl}(\mathbf{u}\times\mathbf{H}) = -\operatorname{curl}(\eta\operatorname{curl}\mathbf{H}), \tag{13}$$

where

$$\eta = \frac{1}{4\pi\mu\sigma}. \tag{14}$$

We shall call η the *resistivity* though it differs from the usual definition by a factor $1/(4\pi)$. It may be further noted that η (like ν and κ) is of the dimension $\text{cm}^2\,\text{sec}^{-1}$.

Equation (13) is entirely general; in particular, it is not restricted to incompressible fluids.

If η is assumed to be a constant, equation (13), in Cartesian coordinates, takes the form (cf. equations III (13), (24), and (25))

$$\frac{\partial H_i}{\partial t}+\frac{\partial}{\partial x_j}(u_j H_i - u_i H_j) = \eta\nabla^2 H_i. \tag{15}$$

The elimination of **E** to obtain an equation for **H** represents an important simplification. For since the Lorentz force is also expressed in terms of **H** only, it follows that in the subsequent analysis we need not make any further reference to the electric field. Once **u** and **H** have been determined through a solution of equations (1), (9), and (15), we can, if we like, use equation (12) to evaluate **E**; but it plays no significant role in the solution of the problem itself. This minor role to which **E** is reduced is peculiar to hydromagnetics: it derives from the neglect of the displacement currents.

Several important consequences for the field follow from equation (13); we shall now consider some of them.

(a) *The decay of a magnetic field in the absence of fluid motions. The Joule dissipation*

We shall first consider the case when there are no fluid motions. In this case equation (13) becomes

$$\frac{\partial \mathbf{H}}{\partial t} = -\operatorname{curl}(\eta\operatorname{curl}\mathbf{H}). \tag{16}$$

From this equation it follows that the magnetic energy in any closed volume V must decrease monotonically with time so long as no energy flows into V across its boundary. To see this, multiply equation (16) scalarly by **H** and integrate over the volume. We obtain

$$\frac{1}{2}\frac{\partial}{\partial t}\int_V |\mathbf{H}|^2\,dV = -\int_V \mathbf{H}\,.\operatorname{curl}(\eta\operatorname{curl}\mathbf{H})\,dV. \tag{17}$$

Quite generally we have the identity

$$\int_V \boldsymbol{\varphi}\,.\,\mathrm{curl}\,\boldsymbol{\psi}\,dV = \int_V \boldsymbol{\psi}\,.\,\mathrm{curl}\,\boldsymbol{\varphi}\,dV - \int_S \boldsymbol{\varphi}\times\boldsymbol{\psi}\,.\,d\mathbf{S}, \tag{18}$$

where $\boldsymbol{\varphi}$ and $\boldsymbol{\psi}$ represent any two vector fields and S is the boundary of V.

Making use of the identity (18), we can rewrite equation (17) in the form

$$\frac{1}{2}\frac{\partial}{\partial t}\int_V |\mathbf{H}|^2\,dV = -\int_V \eta|\mathrm{curl}\,\mathbf{H}|^2\,dV + \int_S \eta(\mathbf{H}\times\mathrm{curl}\,\mathbf{H})\,.\,d\mathbf{S}. \tag{19}$$

The magnetic energy $\mathfrak{M}$ inside V is given by

$$\mathfrak{M} = \frac{\mu}{8\pi}\int_V |\mathbf{H}|^2\,dV. \tag{20}$$

Accordingly,

$$\frac{\partial\mathfrak{M}}{\partial t} = -\int_V \frac{|\mathbf{J}|^2}{\sigma}\,dV + \frac{1}{4\pi}\int_S \frac{1}{\sigma}\mathbf{H}\times\mathbf{J}\,.\,d\mathbf{S}, \tag{21}$$

where in reducing equation (19) to this form we have made use of equations (2) and (14). Since $\mathbf{J} = \sigma\mathbf{E}$ when there are no motions, a still further form of equation (21) is

$$\frac{\partial\mathfrak{M}}{\partial t} = -\int_V \frac{|\mathbf{J}|^2}{\sigma}\,dV + \frac{1}{4\pi}\int_S \mathbf{H}\times\mathbf{E}\,.\,d\mathbf{S}. \tag{22}$$

The change in the magnetic energy inside V, therefore, consists of a volume integral which represents a loss of energy resulting from the Joule heating by the currents flowing in the conductor, and a surface integral which represents the Poynting flux of energy flowing from the external field into the conductor. If there should be no inward directed Poynting flux at the boundary, the field must necessarily decay.

The precise manner in which a given initial field will decay can, in principle, be determined; but it will depend on the particular shape of volume enclosing the conductor and on the nature of the initial field. The general method, when η is constant, is the following. We first consider separable solutions of the equation

$$\frac{\partial\mathbf{H}}{\partial t} = -\eta\,\mathrm{curl}(\mathrm{curl}\,\mathbf{H}), \tag{23}$$

whose dependence on time is like $e^{-\lambda t}$. Then

$$\mathrm{curl}(\mathrm{curl}\,\mathbf{H}) = \frac{\lambda}{\eta}\mathbf{H}; \tag{24}$$

and we must solve this equation together with suitable boundary conditions on the surface S of the conductor and at infinity. The problem, in fact, reduces to one in characteristic values: the different characteristic values q_j (say) and the solutions $\mathbf{H}_j$ belonging to them form a *complete set* in the sense that any arbitrary field satisfying the same boundary conditions on S and at infinity can be expanded in terms of them. Thus, if $\mathbf{H}(0)$ is the field at time $t = 0$, we can expand it in the manner

$$\mathbf{H}(0) = \sum_j A_j \mathbf{H}_j, \tag{25}$$

where the coefficients A_j of the expansion are uniquely determined by $\mathbf{H}(0)$. In terms of such an expansion, the field at a later time is given by

$$\mathbf{H}(t) = \sum_j A_j e^{-q_j \eta t} \mathbf{H}_j. \tag{26}$$

This equation shows how the various modes of the field decay independently of each other.

While the procedure which we have described can be carried out explicitly for a number of special geometries, the essential physical content of the theory, for our present purposes, can be deduced quite simply by considering an infinite homogeneous medium and asking for the rate of decay of a periodic field of a given wavelength. Thus, considering the equation

$$\frac{\partial H_i}{\partial t} = \eta \nabla^2 H_i, \tag{27}$$

we seek a solution whose spatial dependence is $\exp(ik_j x_j)$. Equation (27) then gives

$$\frac{\partial H_i}{\partial t} = -k^2 \eta H_i. \tag{28}$$

Therefore,
$$\mathbf{H} = \mathbf{H}(0)\exp[i(k_j x_j - k^2 \eta t)], \tag{29}$$

where $\mathbf{H}(0)$ is the amplitude of the field at $t = 0$. It is seen that the field decays with a mean life

$$\tau = \frac{1}{k^2 \eta} = \frac{\lambda^2 \mu \sigma}{\pi}, \tag{30}$$

where λ $(= 2\pi/k)$ is the wavelength. Equation (30) establishes the general result: *a magnetic field of a linear scale L has a mean life of the order $L^2\sigma$.* An immediate consequence of this result is that magnetic fields in systems of large linear dimensions can endure for relatively long periods of time.

(b) The case when there are motions and the conductivity is infinite

The case when the electrical conductivity of the medium may be considered as infinite is of particular interest. The resistivity is then zero and the equation for **H** becomes

$$\frac{\partial \mathbf{H}}{\partial t} - \operatorname{curl}(\mathbf{u} \times \mathbf{H}) = 0. \tag{31}$$

Even if the conductivity should be finite, equation (31) will still suffice to give the variations in the field for times which are short compared to its decay time. Since this time can be very long for systems of large linear dimensions, it is clear that the variations of **H** in accordance with equation (31) are of particular interest for cosmical and geophysical problems.

When equation (31) is used, the corresponding equation for determining the electric field is

$$\mathbf{E} = -\mu \mathbf{u} \times \mathbf{H}. \tag{32}$$

(i) *Conservation theorems*

It will be observed that equation (31) for **H** is of exactly the same form as the equation for the vorticity derived in Chapter III (§ 20, equation (13)). The identity of the two equations is, in fact, complete since **H** is also solenoidal. All the theorems proved for the vorticity in § 20 have their counterparts for the magnetic field. In terms of **H**, the theorems are as follows.

First, we may observe that magnetic lines of force and magnetic tubes of force can be defined in the same way as vortex lines and vortex tubes were defined. And from $\operatorname{div} \mathbf{H} = 0$, it follows that *the normal magnetic flux across any section of a magnetic tube of force is the same*; *and the magnetic lines of force must either be closed or terminate on the boundaries.* But, of course, the most important theorem is the counterpart of the Helmholtz–Kelvin theorem; it states

$$\int_S \mathbf{H} . d\mathbf{S} = \text{constant}. \tag{33}$$

In words: *the integral of the normal component of* **H** *over any surface S bounded by a closed curve remains constant as we follow the surface S with the motion of the fluid elements constituting it.*

For an incompressible fluid, equation (31) has the alternative form (cf. equation (15))

$$\frac{dH_i}{dt} = \frac{\partial H_i}{\partial t} + u_j \frac{\partial H_i}{\partial x_j} = H_j \frac{\partial u_i}{\partial x_j}. \tag{34}$$

From this equation it follows (as in the corresponding discussion of $\boldsymbol{\omega}$ based on equation III (27)) that *a magnetic line of force consists of the same fluid elements: it moves with the fluid like a material substance and behaves as though it is permanently attached to the fluid.*

There is a simple generalization of the foregoing result for a compressible fluid which we may notice. The equation

$$\frac{\partial H_i}{\partial t}+\frac{\partial}{\partial x_j}(u_j H_i) = H_j\frac{\partial u_i}{\partial x_j} \tag{35}$$

can be combined with the equation of continuity,

$$\frac{\partial \rho}{\partial t}+\frac{\partial}{\partial x_j}(\rho u_j) = 0, \tag{36}$$

to give

$$\frac{d}{dt}\left(\frac{H_i}{\rho}\right) = \frac{\partial}{\partial t}\left(\frac{H_i}{\rho}\right)+u_j\frac{\partial}{\partial x_j}\left(\frac{H_i}{\rho}\right) = \frac{H_j}{\rho}\frac{\partial u_i}{\partial x_j}. \tag{37}$$

Thus H_i/ρ satisfies an equation of the same form as equation III (27); and the arguments following it can be repeated with H_i/ρ.

(ii) *The transformation of magnetic energy into kinetic energy and conversely*

Consider next what equation (31) implies for the change in the magnetic energy contained in a closed volume V of the fluid. We have

$$\frac{\partial \mathfrak{M}}{\partial t} = \frac{\mu}{4\pi}\int\limits_V \mathbf{H}\,.\,\mathrm{curl}(\mathbf{u}\times\mathbf{H})\,dV. \tag{38}$$

Making use of the identity (18), we now obtain

$$\frac{\partial \mathfrak{M}}{\partial t} = \frac{\mu}{4\pi}\int\limits_V (\mathbf{u}\times\mathbf{H})\,.\,\mathrm{curl}\,\mathbf{H}\,dV - \frac{\mu}{4\pi}\int\limits_S \mathbf{H}\times(\mathbf{u}\times\mathbf{H})\,.\,d\mathbf{S}. \tag{39}$$

If the fluid is confined inside V, the normal component of $\mathbf{u}$ on S must vanish and we can rewrite equation (39) in the form

$$\frac{\partial \mathfrak{M}}{\partial t} = \mu\int\limits_V \mathbf{u}\,.\,(\mathbf{H}\times\mathbf{J})\,dV + \frac{\mu}{4\pi}\int\limits_S (\mathbf{H}\,.\,\mathbf{u})\mathbf{H}\,.\,d\mathbf{S}. \tag{40}$$

We observe that the volume integral on the right-hand side of equation (40) is

$$-\int\limits_V \mathbf{u}\,.\,\mathcal{L}\,dV, \tag{41}$$

where $\mathcal{L}$ is the Lorentz force: it therefore represents the *loss* in the magnetic energy resulting from the work done by the field on the fluid.

For an incompressible fluid, we can obtain an alternative form of the energy equation by starting with the equation of motion in the form (35); thus,

$$\frac{\partial \mathfrak{M}}{\partial t} = -\frac{\mu}{4\pi}\int\limits_V H_i \frac{\partial}{\partial x_j}(u_j H_i - u_i H_j)\, dV$$

$$= -\frac{\mu}{4\pi}\int\limits_V \left\{\frac{1}{2}\frac{\partial}{\partial x_j}(u_j H_i^2) - H_i H_j \frac{\partial u_i}{\partial x_j}\right\} dV$$

$$= -\frac{\mu}{8\pi}\int\limits_S |\mathbf{H}|^2 u_j\, dS_j + \frac{\mu}{4\pi}\int\limits_V H_i \frac{\partial u_i}{\partial x_j} H_j\, dV. \tag{42}$$

The surface integral clearly vanishes and we are left with

$$\frac{\partial \mathfrak{M}}{\partial t} = \frac{\mu}{4\pi}\int\limits_V H_i \frac{\partial u_i}{\partial x_j} H_j\, dV. \tag{43}$$

This equation represents the change in the magnetic energy inside V as a *gain* resulting from the *stretching of the magnetic lines of force* by the fluid elements dragging them around with their motions.

(*c*) *The general energy equation*

In the general case when the dissipation of the magnetic energy by Joule heating and the gain in the energy by the mechanism of the stretching of the lines of force are both present, the corresponding energy equation can be obtained by combining equations (21) and (43); thus

$$\frac{\partial \mathfrak{M}}{\partial t} = -\int\limits_V \frac{|\mathbf{J}|^2}{\sigma}\, dV + \frac{\mu}{4\pi}\int\limits_V \mathbf{H}.(\nabla\mathbf{u}).\mathbf{H}\, dV + \frac{1}{4\pi}\int\limits_S \frac{1}{\sigma}\mathbf{H}\times\mathbf{J}.d\mathbf{S}, \tag{44}$$

where $\nabla\mathbf{u}$ stands for the dyadic $\partial u_i/\partial x_j$. Now substituting for $\mathbf{J}$ from equation (4), we have

$$\frac{\partial \mathfrak{M}}{\partial t} = -\int\limits_V \frac{|\mathbf{J}|^2}{\sigma}\, dV + \frac{\mu}{4\pi}\int\limits_V \mathbf{H}.(\nabla\mathbf{u}).\mathbf{H}\, dV +$$

$$+\frac{1}{4\pi}\int\limits_S \mathbf{H}\times\mathbf{E}.d\mathbf{S} + \frac{\mu}{4\pi}\int\limits_S \mathbf{H}\times(\mathbf{u}\times\mathbf{H}).d\mathbf{S}. \tag{45}$$

Expanding the vector triple product in the last term, we obtain

$$\frac{\partial \mathfrak{M}}{\partial t} = -\int\limits_V \frac{|\mathbf{J}|^2}{\sigma}\, dV + \frac{\mu}{4\pi}\int\limits_V \mathbf{H}.(\nabla\mathbf{u})\mathbf{H}\, dV +$$

$$+\frac{1}{4\pi}\int\limits_S \mathbf{H}\times\mathbf{E}.d\mathbf{S} - \frac{\mu}{4\pi}\int\limits_S (\mathbf{H}.\mathbf{u})(\mathbf{H}.d\mathbf{S}). \tag{46}$$

The physical meanings of the various terms in this equation have already been explained.

39. The Alfvén waves

We shall now consider solutions of the hydromagnetic equations which represent the propagation of waves.

Suppose that a uniform magnetic field **H** in the absence of motions prevails in an infinite homogeneous medium of an incompressible fluid of kinematic viscosity ν and electrical resistivity η; and consider small oscillations about the equilibrium state.

The linearized forms of equations (10) and (15) are

$$\frac{\partial u_i}{\partial t}-\frac{\mu H_j}{4\pi\rho}\frac{\partial h_i}{\partial x_j} = -\frac{\partial}{\partial x_i}\delta\varpi+\nu\nabla^2 u_i \tag{47}$$

and

$$\frac{\partial h_i}{\partial t}-H_j\frac{\partial u_i}{\partial x_j} = \eta\nabla^2 h_i, \tag{48}$$

where

$$\delta\varpi = \frac{\delta p}{\rho}+\mu\frac{\mathbf{H}\,.\,\mathbf{h}}{4\pi\rho}, \tag{49}$$

and **h** is the perturbation in the magnetic field. By taking the divergence of equation (47) and remembering that **u** and **h** are both solenoidal, we get

$$\nabla^2\delta\varpi = 0; \tag{50}$$

and the solution of this equation relevant to the problem on hand is

$$\delta\varpi \equiv 0. \tag{51}$$

Thus, equation (47) reduces to

$$\frac{\partial u_i}{\partial t} = \frac{\mu H_j}{4\pi\rho}\frac{\partial h_i}{\partial x_j}+\nu\nabla^2 u_i. \tag{52}$$

In addition to equations (48) and (52), we also have

$$\frac{\partial u_i}{\partial x_i} = 0 \quad \text{and} \quad \frac{\partial h_i}{\partial x_i} = 0. \tag{53}$$

We now seek solutions of equations (48), (52), and (53) whose dependence on space and time is given by

$$e^{ik_j x_j+i\omega t}; \tag{54}$$

the equations then give

$$(\omega-i\nu k^2)u_i = \frac{\mu}{4\pi\rho}(k_j H_j)h_i, \tag{55}$$

$$(\omega-i\eta k^2)h_i = (k_j H_j)u_i, \tag{56}$$

and

$$k_j h_j = 0, \qquad k_j u_j = 0. \tag{57}$$

From equations (55) and (56) we obtain

$$(\omega - i\nu k^2)(\omega - i\eta k^2)u_i = \frac{\mu}{4\pi\rho}(k_j H_j)^2 u_i, \tag{58}$$

and an exactly similar equation for h_i. The required 'dispersion relation' is, therefore,

$$(\omega - i\nu k^2)(\omega - i\eta k^2) = V_A^2 k^2, \tag{59}$$

where

$$V_A = \left(\frac{\mu}{4\pi\rho}\right)^{\frac{1}{2}} H \cos\vartheta, \tag{60}$$

and ϑ is the inclination of the direction of propagation to the direction of $\mathbf{H}$. The quantity V_A defines a velocity: it is called the *Alfvén velocity.*

According to equations (57) these waves are transverse; and when the viscosity and the resistivity are finite, the waves are damped.

(*a*) *The case when* $\nu = \eta = 0$

In this case the waves are undamped and

$$\frac{\omega}{k} = \pm V_A = \pm\left(\frac{\mu}{4\pi\rho}\right)^{\frac{1}{2}} H \cos\vartheta. \tag{61}$$

The velocity of propagation of these undamped waves is the Alfvén velocity. Also, in this case,

$$u_i = \frac{1}{\omega}\frac{\mu k}{4\pi\rho}(H\cos\vartheta)h_i, \tag{62}$$

or, making use of equation (61),

$$u_i = \pm\left(\frac{\mu}{4\pi\rho}\right)^{\frac{1}{2}} h_i \quad \text{and} \quad \tfrac{1}{2}\rho|\mathbf{u}|^2 = \frac{\mu}{8\pi}|\mathbf{h}|^2. \tag{63}$$

The mean energy of the wave in the magnetic field and in the kinetic motions is the same.

According to equation (61), the group velocity of the waves is given by

$$\frac{\partial\omega}{\partial k_i} = \pm\left(\frac{\mu}{4\pi\rho}\right)^{\frac{1}{2}} H_i. \tag{64}$$

A wave-packet must therefore follow the magnetic lines of force with a velocity $(\mu/4\pi\rho)^{\frac{1}{2}}\mathbf{H}$.

As Alfvén has pointed out, these hydromagnetic waves can be regarded as originating in the vibration of the magnetic lines of force as stretched strings. For as we have seen, the stresses caused by the magnetic field are equivalent to a hydrostatic pressure, $\mu|\mathbf{H}|^2/8\pi$, and a tension, $\mu|\mathbf{H}|^2/4\pi$, along the lines of force; if in addition to associating the magnetic lines of force with this tension we ascribe to them a line-density ρ, the formula,

$$\frac{\omega}{k} = \sqrt{\left(\frac{\text{tension}}{\text{line-density}}\right)}, \tag{65}$$

valid for the vibration of stretched strings, gives the correct Alfvén velocity.

(b) The effects of finite viscosity and resistivity

Returning to equation (59), we have the general solution

$$\omega = \pm k\sqrt{[V_A^2 - \tfrac{1}{4}(\nu-\eta)^2k^2]} + \tfrac{1}{2}i(\nu+\eta)k^2. \tag{66}$$

The effects of finite viscosity and resistivity are therefore: *first*, a damping of the waves in which ν and η occur additively; and *second*, an alteration of the Alfvén velocity in which ν and η occur differentially.

In the limit $\nu \to 0$ and $\eta \to 0$, we have the approximate formula

$$\omega = \pm V_A k\left[1 - \frac{1}{8}\frac{(\nu-\eta)^2}{V_A^2}k^2\right] + \tfrac{1}{2}i(\nu+\eta)k^2. \tag{67}$$

40. Some special solutions of the hydromagnetic equations. The analogue of the Taylor–Proudman theorem

For an incompressible, inviscid fluid of zero resistivity in a potential field, the basic hydrodynamic equations are:

$$\frac{\partial u_i}{\partial t} + u_j\frac{\partial u_i}{\partial x_j} - \frac{\mu}{4\pi\rho}H_j\frac{\partial H_i}{\partial x_j} = -\frac{\partial}{\partial x_i}\left(\frac{p}{\rho} + \mu\frac{|\mathbf{H}|^2}{8\pi\rho} + V\right), \tag{68}$$

$$\frac{\partial H_i}{\partial t} + u_j\frac{\partial H_i}{\partial x_j} = H_j\frac{\partial u_i}{\partial x_j}, \tag{69}$$

and the divergence conditions on $\mathbf{u}$ and $\mathbf{H}$. By making use of the identity given in equation III (9) and reverting to the original form of the Lorentz force given in equation (6), we can rewrite equation (68) in the form

$$\frac{\partial \mathbf{u}}{\partial t} - \mathbf{u}\times\operatorname{curl}\mathbf{u} + \frac{\mu}{4\pi\rho}\mathbf{H}\times\operatorname{curl}\mathbf{H} = -\operatorname{grad}\left(\frac{p}{\rho} + \tfrac{1}{2}|\mathbf{u}|^2 + V\right). \tag{70}$$

Under conditions of a steady state, the foregoing equations allow some special solutions which are worth noting.

(a) The equipartition solution

A particularly simple solution of equations (68) and (69) is given by

$$u_i = \pm\left(\frac{\mu}{4\pi\rho}\right)^{\frac{1}{2}} H_i \tag{71}$$

and

$$\frac{\partial}{\partial x_i}\left(\frac{p}{\rho} + \mu\frac{|\mathbf{H}|^2}{8\pi\rho} + V\right) = 0. \tag{72}$$

On this solution, the fluid velocity at every point is parallel to the direction of the magnetic field at that point; also, their relative magnitudes are such that the energies in the two forms are equal:

$$\tfrac{1}{2}\rho|\mathbf{u}|^2 = \mu\frac{|\mathbf{H}|^2}{8\pi}. \tag{73}$$

While this solution of the equations may appear artificial, it is important to note that except for the condition of *hydrostatic equilibrium* (72), there are no restrictions on the spatial dependence of the field (*or* the motions). Moreover, as we shall show in Chapter XII, § 113, these solutions are stable with respect to small perturbations.

(*b*) *Force-free fields*

A second class of special solutions arises when the Lorentz force vanishes, there are no fluid motions, and the condition of hydrostatic equilibrium is satisfied, i.e. when

$$4\pi\mathscr{L} = \operatorname{curl}\mathbf{H}\times\mathbf{H} = 0, \quad \mathbf{u} = 0, \quad \text{and} \quad \operatorname{grad}(V+p/\rho) = 0. \tag{74}$$

Fields for which $\mathscr{L} = 0$ are called *force-free*. Clearly, the vanishing of the Lorentz force requires that the field and the current density are everywhere parallel. One simple way in which this latter condition can be satisfied is to take

$$\operatorname{curl}\mathbf{H} = \alpha\mathbf{H}, \tag{75}$$

where α is a constant. Solutions of equation (75) for various simple geometries can be easily written down.

It is important to remark that force-free fields imply boundary conditions which it may not always be possible to satisfy. To see this, consider the volume integral

$$\int_V \mathbf{r}\,.\,(\operatorname{curl}\mathbf{H}\times\mathbf{H})\,dV = \int_V \operatorname{curl}\mathbf{H}\,.\,(\mathbf{H}\times\mathbf{r})\,dV. \tag{76}$$

Transforming this integral with the aid of the identity (18), we have

$$\int_V \mathbf{r}\,.\,(\operatorname{curl}\mathbf{H}\times\mathbf{H})\,dV = \int_V \mathbf{H}\,.\operatorname{curl}(\mathbf{H}\times\mathbf{r})\,dV - \int_S (\mathbf{H}\times\mathbf{r})\times\mathbf{H}\,.\,d\mathbf{S}. \tag{77}$$

On further reduction of the volume integral on the right-hand side, and expanding the vector triple product in the last term, we find

$$\int_V \mathbf{r}\,.\,(\operatorname{curl}\mathbf{H}\times\mathbf{H})\,dV = \tfrac{1}{2}\int_V |\mathbf{H}|^2\,dV - \tfrac{1}{2}\int_S |\mathbf{H}|^2 x_j\,dS_j + \\ + \int_S (\mathbf{H}\,.\,\mathbf{r})(\mathbf{H}\,.\,d\mathbf{S}). \tag{78}$$

Consequently, if the force-free condition is satisfied inside V,

$$\int_V |\mathbf{H}|^2\, dV = \int_S |\mathbf{H}|^2 \mathbf{r}\,.\,d\mathbf{S} - 2\int_S (\mathbf{H}\,.\,\mathbf{r})(\mathbf{H}\,.\,d\mathbf{S}). \tag{79}$$

Clearly, this condition cannot be met (except trivially) if $\mathbf{H}$ were to vanish on S; in other words, while we may be able to cancel the stresses inside a given region, we cannot arrange for its cancellation everywhere.

There is a simple generalization of the solution (75) allowing fluid motions. Since the equation of motion for $\mathbf{H}$ can be satisfied if $\mathbf{u}$ is everywhere parallel to $\mathbf{H}$, we take

$$\mathbf{u} = \beta\left(\frac{\mu}{4\pi\rho}\right)^{\frac{1}{2}}\mathbf{H}, \tag{80}$$

where β is a constant. If we suppose in addition that the field is force-free, equation (70) can be satisfied if

$$\operatorname{grad}(\tfrac{1}{2}|\mathbf{u}|^2 + V + p/\rho) = 0. \tag{81}$$

(c) *The analogue of the Taylor–Proudman theorem*

Suppose that a uniform magnetic field $\mathbf{H}$ is impressed on the fluid when there are no motions. Let the system be perturbed; and suppose that a steady state in which the departures from the initial state are small comes to prevail. Let $\mathbf{h}$ denote the perturbation in the magnetic field and $\mathbf{u}$ the velocity; according to equations (68) and (69) these quantities will be governed by

$$\mu H_j \frac{\partial h_i}{\partial x_j} = \frac{\partial}{\partial x_i}\left(\frac{\delta p}{\rho} + \mu\frac{\mathbf{H}\,.\,\mathbf{h}}{4\pi\rho} + \delta V\right) \tag{82}$$

and

$$H_j \frac{\partial u_i}{\partial x_j} = 0. \tag{83}$$

From equation (83) it follows that *the motions cannot vary in the direction of* $\mathbf{H}$. In other words: *all steady slow motions in the presence of a uniform magnetic field are necessarily two-dimensional.* This is exactly analogous to the Taylor–Proudman theorem for rotating fluids.

41. The problem of thermal instability in the presence of a magnetic field: general considerations

From the theorems proved in the preceding sections it is apparent that a strong magnetic field impressed on an electrically conducting fluid should have a profound effect on the onset of thermal instability. For convection implies that motions occur which have of necessity a three-dimensional character. Such motions are expressly forbidden for fluids of zero resistivity; as we have seen in § 40 (c), slow steady two-dimensional motions are the only ones that are allowed. A fluid of zero resistivity should, therefore, be thermally stable for all adverse temperature gradients. Moreover, it is clear that a magnetic field will generally have the effect of inhibiting the onset of convection. For, in addition to the dissipation of energy by viscosity, there will be the dissipation by Joule heating. In a steady state, the energy released by the buoyancy force acting on the fluid must balance the energy dissipated by both means (cf. § 43 (b) below); and this can be achieved only at higher adverse temperature gradients than are sufficient in the absence of Joule heating.

There is a further factor to remember: it is that in the presence of a magnetic field disturbances can be propagated as Alfvén waves. For this reason, the possibility of instability occurring as overstability should not be overlooked.

42. The perturbation equations

Consider then an infinite, horizontal layer of an electrically conducting fluid upon which is impressed a uniform magnetic field. As in Chapters II and III, it will suffice to consider the relevant equations of motion and heat conduction on the Boussinesq approximation. The only additional factor we must allow for now is the Lorentz force in the equation of motion. Thus, in place of equation II (43) we now have (cf. equation (10))

$$\frac{\partial u_i}{\partial t}+u_j\frac{\partial u_i}{\partial x_j}-\frac{\mu}{4\pi\rho}H_j\frac{\partial H_i}{\partial x_j} = -\frac{\partial}{\partial x_i}\left(\frac{p}{\rho}+\mu\frac{|\mathbf{H}|^2}{8\pi\rho}\right)+\left(1+\frac{\delta\rho}{\rho_0}\right)X_i+\nu\nabla^2 u_i. \tag{84}$$

Equations II (41), (44), and (46) are unaffected; and we also have the equations

$$\frac{\partial H_i}{\partial t}+u_j\frac{\partial H_i}{\partial x_j} = H_j\frac{\partial u_i}{\partial x_j}+\eta\nabla^2 H_i \tag{85}$$

and

$$\frac{\partial H_i}{\partial x_i} = 0. \tag{86}$$

We envisage now an initial state in which a steady adverse temperature gradient β is maintained and there are no motions. The characterization of this initial state differs in no respect from that described in Chapter II, § 9; and the equations governing small perturbations can be obtained similarly. Thus, in place of equation II (55) we now have

$$\frac{\partial u_i}{\partial t} = -\frac{\partial}{\partial x_i}(\delta\varpi)+g\alpha\theta\lambda_i+\nu\nabla^2 u_i+\frac{\mu H_j}{4\pi\rho}\frac{\partial h_i}{\partial x_j}, \tag{87}$$

where $\boldsymbol{\lambda} = (0, 0, 1)$ is a unit vector in the direction of the vertical, $\mathbf{H}$ is the impressed magnetic field, $\mathbf{h}$ is the perturbation in it, and

$$\delta\varpi = \frac{\delta p}{\rho_0} + \mu\frac{\mathbf{H}\,.\,\mathbf{h}}{4\pi\rho}. \tag{88}$$

The corresponding linearized forms of equations (85) and (86) are

$$\frac{\partial h_i}{\partial t} = H_j\frac{\partial u_i}{\partial x_j} + \eta\nabla^2 h_i \tag{89}$$

and

$$\frac{\partial h_i}{\partial x_i} = 0. \tag{90}$$

In addition to the foregoing equations, we have (cf. equations II (56) and (57))

$$\frac{\partial\theta}{\partial t} = \beta\lambda_j u_j + \kappa\nabla^2\theta \tag{91}$$

and

$$\frac{\partial u_i}{\partial x_i} = 0. \tag{92}$$

As in Chapters II and III, we eliminate the term in grad $\delta\varpi$ in equation (87) by applying the operator $\epsilon_{ijk}\,\partial/\partial x_j$ to the k-component of the equation. The result is (cf. equation II (67))

$$\frac{\partial\omega_i}{\partial t} = g\alpha\epsilon_{ijk}\frac{\partial\theta}{\partial x_j}\lambda_k + \nu\nabla^2\omega_i + \frac{\mu H_j}{4\pi\rho}\frac{\partial\upsilon_i}{\partial x_j}, \tag{93}$$

where in analogy with the vorticity $\boldsymbol{\omega}$ we have defined

$$\boldsymbol{\upsilon} = \text{curl}\,\mathbf{h}. \tag{94}$$

Apart from a factor $1/(4\pi)$, $\boldsymbol{\upsilon}$ is the current density induced by the perturbation. Taking the curl of equation (93) once again, we obtain (cf. equation II (72))

$$\frac{\partial}{\partial t}\nabla^2 u_i = g\alpha\left(\lambda_i\nabla^2\theta - \lambda_j\frac{\partial^2\theta}{\partial x_i\,\partial x_j}\right) + \nu\nabla^4 u_i + \frac{\mu H_j}{4\pi\rho}\frac{\partial}{\partial x_j}\nabla^2 h_i. \tag{95}$$

Similarly, taking the curl of equation (89) we have

$$\frac{\partial\upsilon_i}{\partial t} = H_j\frac{\partial\omega_i}{\partial x_j} + \eta\nabla^2\upsilon_i. \tag{96}$$

Now multiplying equations (89), (93), (95), and (96) by λ_i, we get the following set of equations:

$$\frac{\partial h_z}{\partial t} = \eta\nabla^2 h_z + H_j\frac{\partial w}{\partial x_j}, \tag{97}$$

$$\frac{\partial \xi}{\partial t} = \eta \nabla^2 \xi + H_j \frac{\partial \zeta}{\partial x_j}, \tag{98}$$

$$\frac{\partial \zeta}{\partial t} = \nu \nabla^2 \zeta + \frac{\mu H_j}{4\pi\rho} \frac{\partial \xi}{\partial x_j}, \tag{99}$$

and
$$\frac{\partial}{\partial t} \nabla^2 w = g\alpha \left(\frac{\partial^2 \theta}{\partial x^2} + \frac{\partial^2 \theta}{\partial y^2} \right) + \nu \nabla^4 w + \frac{\mu H_j}{4\pi\rho} \frac{\partial}{\partial x_j} \nabla^2 h_z, \tag{100}$$

where w, ζ, and $\xi/4\pi$ are the z-components of the velocity, the vorticity, and the current density, respectively.

Most of our discussions relating to the foregoing equations will be restricted to the case when **H** and **g** are parallel; the case when they are not will be considered in § 47.

When the direction of the impressed magnetic field coincides with the vertical, the relevant equations are:

$$\frac{\partial \theta}{\partial t} = \kappa \nabla^2 \theta + \beta w, \tag{101}$$

$$\frac{\partial h_z}{\partial t} = \eta \nabla^2 h_z + H \frac{\partial w}{\partial z}, \tag{102}$$

$$\frac{\partial \xi}{\partial t} = \eta \nabla^2 \xi + H \frac{\partial \zeta}{\partial z}, \tag{103}$$

$$\frac{\partial \zeta}{\partial t} = \nu \nabla^2 \zeta + \frac{\mu H}{4\pi\rho} \frac{\partial \xi}{\partial z}, \tag{104}$$

and
$$\frac{\partial}{\partial t} \nabla^2 w = g\alpha \left(\frac{\partial^2 \theta}{\partial x^2} + \frac{\partial^2 \theta}{\partial y^2} \right) + \nu \nabla^4 w + \frac{\mu H}{4\pi\rho} \frac{\partial}{\partial z} \nabla^2 h_z. \tag{105}$$

(a) The boundary conditions

We must seek solutions of equations (101)–(105) which satisfy various boundary conditions; those on w and ζ have already been described (Chapter II, § 9 (*a*)); it remains to consider the boundary conditions on h_z and ξ. These latter conditions depend upon the electrical properties of the medium adjoining the fluid. We shall consider two cases.

If the medium adjoining the fluid is electrically non-conducting, then no currents can cross the boundary and we must require that $J_z = 0$. Moreover, the field, $\mathbf{h}^{(ex)}$, in the electrically non-conducting medium must be that appropriate to a vacuum and be derivable from a potential. Thus,

$$\mathbf{h}^{(ex)} = \operatorname{grad} \psi, \quad \text{where } \nabla^2 \psi = 0; \tag{106}$$

and on the interface between the fluid and the medium the field must be continuous. We shall call this case A. Thus, the conditions on **h** and ξ $(= 4\pi J_z)$ are

$$\mathbf{h} = \mathbf{h}^{(ex)} \text{ and } \xi = 0 \text{ on the bounding surface in case A.} \tag{107}$$

If, on the other hand, the medium adjoining the fluid is a perfect conductor, then no magnetic field can cross the boundary and we must require that

$$h_z = 0 \text{ and } E_x = E_y = 0 \text{ on a plane boundary adjoining a perfect conductor.} \tag{108}$$

We shall associate the electromagnetic conditions (108) with a rigid boundary. In this case, there will be no motions at the boundary and the conditions (108) are equivalent to

$$h_z = 0 \quad \text{and} \quad J_x = J_y = 0. \tag{109}$$

We shall call this case B. From div $\mathbf{J} = 0$ and the fact that $J_x = J_y = 0$ on the plane boundary, we conclude that $\partial J_z/\partial z = 0$. Thus, the conditions on h_z and ξ are

$$h_z = 0 \text{ and } \frac{\partial \xi}{\partial z} = 0 \text{ on the bounding surface in case B.} \tag{110}$$

Since $\partial w/\partial z = 0$ on a rigid boundary, it follows from equation (102) that

$$\nabla^2 h_z = 0 \text{ on the bounding surface in case B.} \tag{111}$$

(*b*) *The analysis into normal modes*

Following our standard procedure, we analyse w, θ, ζ, h_z, and ξ into two-dimensional waves; and considering disturbances characterized by a particular wave number k, we suppose that w, θ, and ζ have the forms assumed in equation II (90); further, we now suppose that

$$\begin{aligned} \xi &= X(z)\exp[i(k_x x+k_y y)+pt], \\ h_z &= K(z)\exp[i(k_x x+k_y y)+pt]. \end{aligned} \tag{112}$$

Equations (101)–(105) become

$$p\Theta = \beta W+\kappa\left(\frac{d^2}{dz^2}-k^2\right)\Theta, \tag{113}$$

$$pK = H\frac{dW}{dz}+\eta\left(\frac{d^2}{dz^2}-k^2\right)K, \tag{114}$$

$$p\left(\frac{d^2}{dz^2}-k^2\right)W = -g\alpha k^2\Theta+\nu\left(\frac{d^2}{dz^2}-k^2\right)^2 W+\frac{\mu H}{4\pi\rho}\frac{d}{dz}\left(\frac{d^2}{dz^2}-k^2\right)K, \tag{115}$$

$$pX = H\frac{dZ}{dz}+\eta\left(\frac{d^2}{dz^2}-k^2\right)X, \tag{116}$$

and
$$pZ = \frac{\mu H}{4\pi\rho}\frac{dX}{dz}+\nu\left(\frac{d^2}{dz^2}-k^2\right)Z. \tag{117}$$

We observe that the equations for X and Z are formally independent of the others.

Measuring length in the unit d and letting

$$a = kd, \quad \sigma = pd^2/\nu, \quad \mathfrak{p}_1 = \nu/\kappa, \quad \text{and} \quad \mathfrak{p}_2 = \nu/\eta, \tag{118}$$

we can rewrite equations (113)–(117) in the forms

$$(D^2-a^2-\mathfrak{p}_1\sigma)\Theta = -\left(\frac{\beta d^2}{\kappa}\right)W, \tag{119}$$

$$(D^2-a^2-\mathfrak{p}_2\sigma)K = -\left(\frac{Hd}{\eta}\right)DW, \tag{120}$$

$$(D^2-a^2)(D^2-a^2-\sigma)W+\left(\frac{\mu Hd}{4\pi\rho\nu}\right)D(D^2-a^2)K = \left(\frac{g\alpha}{\nu}d^2\right)a^2\Theta, \tag{121}$$

$$(D^2-a^2-\mathfrak{p}_2\sigma)X = -\left(\frac{Hd}{\eta}\right)DZ, \tag{122}$$

and
$$(D^2-a^2-\sigma)Z = -\left(\frac{\mu Hd}{4\pi\rho\nu}\right)DX, \tag{123}$$

where $D = d/dz$ (z being measured in the unit d).

Solutions of equations (119)–(123) must be sought which satisfy the boundary conditions given in equations II (95) and (96) as well as the further conditions (see equations (107), (110), and (111)):

$X = 0$ and **h** is continuous with an external vacuum field on a non-conducting boundary, (124)

and

$DX = 0$ and $K = 0$ on a perfectly conducting boundary. (125)

(c) *The solutions for the horizontal components of the velocity and the magnetic field*

Equations (119)–(123) determine the z-components of the various quantities. To complete the solution we must determine the horizontal components of the velocity and the magnetic field. We have already shown in Chapter II, § 10 (*a*) how the horizontal components of the velocity can be determined from a knowledge of w and ζ. The horizontal components of the magnetic field can be determined, similarly, from a knowledge of h_z and ξ. Since h_x and h_y are related to the vertical components of **h** and **u** in exactly the same way as u and v are related to the vertical components of **u** and **ω**, it is clear that the required relations are

$$h_x = \frac{1}{a^2}\left(\frac{\partial^2 h_z}{\partial x \partial z} + d\frac{\partial \xi}{\partial y}\right) \tag{126}$$

and

$$h_y = \frac{1}{a^2}\left(\frac{\partial^2 h_z}{\partial y \partial z} - d\frac{\partial \xi}{\partial x}\right). \tag{127}$$

43. The case when instability sets in as stationary convection. A variational principle

As in Chapter III, we shall first consider the case when instability sets in as stationary convection. Whether instability can set in as overstability will be considered in § 46.

When instability sets in as ordinary convection, the marginal state will be characterized by $\sigma = 0$ and the basic equations reduce to (cf. equations (119)–(123))

$$(D^2 - a^2)\Theta = -\left(\frac{\beta d^2}{\kappa}\right)W, \tag{128}$$

$$(D^2 - a^2)K = -\left(\frac{Hd}{\eta}\right)DW, \tag{129}$$

$$(D^2 - a^2)^2 W + \left(\frac{\mu H d}{4\pi\rho\nu}\right)D(D^2 - a^2)K = \left(\frac{g\alpha d^2}{\nu}\right)a^2\Theta, \tag{130}$$

$$(D^2 - a^2)X = -\left(\frac{Hd}{\eta}\right)DZ, \tag{131}$$

and

$$(D^2 - a^2)Z = -\left(\frac{\mu H d}{4\pi\rho\nu}\right)DX. \tag{132}$$

We can eliminate K from equation (130) by making use of equation (129); thus,

$$(D^2 - a^2)^2 W - QD^2 W = \left(\frac{g\alpha d^2}{\nu}\right)a^2\Theta, \tag{133}$$

where (cf. equation (14))

$$Q = \frac{\mu H^2 d^2}{4\pi\rho\nu\eta} = \frac{\mu^2 H^2 \sigma}{\rho\nu}\, d^2 \tag{134}$$

is a non-dimensional 'number'. This number Q plays for problems involving magnetic fields the same role which the Taylor number T plays for problems involving rotation. (Notice that the elimination of K did not require any differentiation.)

We can eliminate Θ from equation (133) by applying the operator $(D^2 - a^2)$ and making use of equation (128). We obtain

$$(D^2 - a^2)[(D^2 - a^2)^2 - QD^2]W = -Ra^2 W, \tag{135}$$

where R is the Rayleigh number.

From equations (131) and (132), we obtain

$$[(D^2-a^2)^2-QD^2]X = [(D^2-a^2)^2-QD^2]Z = 0. \tag{136}$$

It is apparent that for the problem on hand

$$X = 0 \quad \text{and} \quad Z = 0. \tag{137}$$

In other words: *in this problem the z-components of the vorticity and the current density vanish identically.*

Returning to equations (129) and (135), we must seek solutions of these equations which satisfy the boundary conditions:

$$W = 0, \quad [(D^2-a^2)^2-QD^2]W = 0 \quad \text{for } z = 0 \text{ and } 1, \tag{138}$$

$$\left.\begin{aligned} &\textit{either} \quad DW = 0 \quad \text{(on a rigid surface)} \\ &\textit{or} \quad D^2W = 0 \quad \text{(on a free surface)} \end{aligned}\right\}, \tag{139}$$

and

$$\left.\begin{aligned} &\textit{either} \quad \mathbf{h} \text{ is continuous with an external field derived} \\ &\qquad \text{from a potential with a spatial dependence} \\ &\qquad e^{\pm az+i(a_x x+a_y y)} \text{ (on a non-conducting boundary)} \\ &\textit{or} \quad K = 0 \text{ (on a perfectly conducting boundary)} \end{aligned}\right\}. \tag{140}$$

Since neither equation (135) nor the boundary conditions (138) and (139) involve K, it is clear that *the solution for the underlying characteristic value problem can be carried out independently of the boundary conditions on the magnetic field.* However, the boundary conditions (140) are needed for a complete solution of the problem.

(a) A variational principle

As in the case of similar characteristic value problems considered in Chapters II and III, the problem presented by equation (135) and the boundary conditions (138) and (139) can also be formulated in terms of a variational principle.

Letting
$$F = (D^2-a^2)^2W-QD^2W, \tag{141}$$

we can rewrite the differential equation governing W in the form

$$(D^2-a^2)F = -Ra^2W. \tag{142}$$

The boundary conditions (138) require that

$$F = 0 \quad \text{and} \quad W = 0 \quad \text{for} \quad z = 0 \text{ and } 1. \tag{143}$$

Now multiply equation (142) by F and integrate over the range of z. After an integration by parts, we obtain

$$\int_0^1 [(DF)^2+a^2F^2]\,dz = Ra^2 \int_0^1 WF\,dz. \tag{144}$$

The integral on the right-hand side is

$$\int_0^1 WF\,dz = \int_0^1 W\{(D^2-a^2)^2W-QD^2W\}\,dz. \tag{145}$$

After further integrations by parts, this becomes (cf. similar reductions in equations II (120) and (121) and equation III (110))

$$\int_0^1 WF\,dz = \int_0^1 \{[(D^2-a^2)W]^2+Q(DW)^2\}\,dz. \tag{146}$$

The result of multiplying equation (142) by F and integrating over z is

$$R = \frac{\int_0^1 [(DF)^2+a^2F^2]\,dz}{a^2\int_0^1 \{[(D^2-a^2)W]^2+Q(DW)^2\}\,dz}. \tag{147}$$

This formula expresses R as the ratio of two positive definite integrals; and it can be shown, as in the other cases we have considered, that equation (147) provides the basis for a variational treatment of the problem. Indeed, *the critical Rayleigh number for the onset of instability as stationary convection is the absolute minimum of the quantity on the right-hand side of equation* (147).

(b) The thermodynamic significance of the variational principle

In Chapter II, § 14 we showed that for the simple Bénard problem, the quantity which is minimized in the variational treatment of the problem expresses, simply, the equality of the rate of dissipation of energy by viscosity and the rate of liberation of energy by the buoyancy force. In Chapter III, § 33 (*a*) we showed that the same thing was true when the system was subject to rotation. In both these cases viscosity provided the only source of irreversible dissipation of energy. But when a magnetic field is present, there is an additional source for irreversible dissipation of energy, namely, Joule heating. Under these circumstances, we should expect that the condition for marginal stability expresses the equality of the rate of dissipation of energy by both means with the rate of liberation of energy by the buoyancy force. We shall now show that this is indeed the case.

Since in the present problem, as in the simple Bénard problem, the z-component of vorticity vanishes, the expression obtained in Chapter II, § 14 for the average rate of viscous dissipation of energy in a unit column of the fluid is applicable. And since the equation relating θ and w is also the same, the expression for the average rate of liberation of energy by the buoyancy force is equally applicable. Therefore (cf. equations II (177) and (180))

$$\epsilon_\nu = \frac{\rho\nu}{4a^2d^2}\int_0^1 [(D^2-a^2)W]^2\,dz, \tag{148}$$

and

$$\epsilon_g = \frac{\rho g\alpha\kappa}{4\beta d^2}\int_0^1 [(D\Theta)^2+a^2\Theta^2]\,dz. \tag{149}$$

On the other hand, according to equations (133) and (141),

$$\Theta = \frac{\nu}{g\alpha d^2a^2}F. \tag{150}$$

In terms of F, the expression for ϵ_g is

$$\epsilon_g = \frac{\rho\kappa\nu^2}{4g\alpha\beta a^4d^6}\int_0^1 [(DF)^2+a^2F^2]\,dz. \tag{151}$$

It remains to evaluate the average rate of dissipation of energy ϵ_σ by Joule heating in a unit column of the fluid; this is given by

$$\epsilon_\sigma = \frac{1}{\sigma d^2}\int_0^1 \langle|\mathbf{J}|^2\rangle\,dz = \frac{\mu\eta}{4\pi d^2}\int_0^1 \langle|\mathrm{curl}\,\mathbf{h}|^2\rangle\,dz. \tag{152}$$

It should be noted that we are including under Joule heating only the volume integral over $|\mathbf{J}|^2/\sigma$ on the right-hand side of equation (46). Of the remaining integrals in equation (46), the second over the dyadic $\nabla\mathbf{u}$ represents no real loss (or gain) since this same term occurs with the opposite sign in the corresponding expression for the rate of change of the kinetic energy; and the surface integral over the Poynting flux clearly vanishes when averaged. However, the last surface integral will not in general vanish: it will vanish only when the boundary conditions are those appropriate to a rigid surface.

In the case under discussion $J_z \equiv 0$, and we can write the expression for ϵ_σ in the form

$$\epsilon_\sigma = \frac{\mu\eta}{4\pi d^2}\int_0^1 \left\langle\left(\frac{\partial h_y}{\partial z}-\frac{\partial h_z}{\partial y}\right)^2+\left(\frac{\partial h_x}{\partial z}-\frac{\partial h_z}{\partial x}\right)^2\right\rangle dz. \tag{153}$$

In the present connexion there is no loss of generality if we suppose that the expression for h_z (corresponding to the expression II (172) for W) is

$$h_z = K(z)\cos a_x x\cos a_y y. \tag{154}$$

Since the z-component of the current density is zero, the expressions for

h_x and h_y are:

$$\left.\begin{aligned} h_x &= \frac{1}{a^2}\frac{\partial^2 h_z}{\partial x \partial z} = -\frac{DK}{a^2} a_x \sin a_x x \cos a_y y \\ h_y &= \frac{1}{a^2}\frac{\partial^2 h_z}{\partial y \partial z} = -\frac{DK}{a^2} a_y \cos a_x x \sin a_y y \end{aligned}\right\}. \tag{155}$$

Inserting these expressions for the components of **h** in equation (153), we obtain

$$\epsilon_\sigma = \frac{\mu\eta}{4\pi d^2 a^4} \int_0^1 \{\langle [a_y \cos a_x x \sin a_y y (D^2-a^2)K]^2 + \\ +[a_x \sin a_x x \cos a_y y (D^2-a^2)K]^2 \rangle\} \, dz. \tag{156}$$

We thus obtain

$$\epsilon_\sigma = \frac{\mu\eta}{16\pi d^2 a^2} \int_0^1 [(D^2-a^2)K]^2 \, dz. \tag{157}$$

Now making use of equation (129) and recalling the definition of Q, we have

$$\epsilon_\sigma = \frac{\rho\nu}{4a^2 d^2} Q \int_0^1 (DW)^2 \, dz. \tag{158}$$

The average rate of dissipation of energy, by viscosity and by Joule heating, in a unit column of the fluid, is obtained by adding the contributions given by equations (148) and (158). Thus,

$$\epsilon_\nu + \epsilon_\sigma = \frac{\rho\nu}{4a^2 d^2} \int_0^1 \{[(D^2-a^2)W]^2 + Q(DW)^2\} \, dz. \tag{159}$$

By equating this expression for $\epsilon_\nu + \epsilon_\sigma$ with that for ϵ_g given in equation (151), we obtain

$$R = \frac{\int_0^1 [(DF)^2 + a^2 F^2] \, dz}{a^2 \int_0^1 \{[(D^2-a^2)W]^2 + Q(DW)^2\} \, dz}. \tag{160}$$

But this is exactly the same expression for R which is minimized in the variational method. The general principle formulated in Chapter III, § 33 is, therefore, valid if we include with the dissipation of energy by viscosity (which is explicitly mentioned) all manners of irreversible dissipation of energy which can take place.

44. Solutions for the case when instability sets in as stationary convection

We shall again obtain the solution for the following three cases: when both bounding surfaces are free; when both bounding surfaces are rigid; and when one bounding surface is rigid and the other is free.

(a) *The solution for two free boundaries*

In this case the boundary conditions (138) and (139) require

$$W = D^2W = D^4W = 0 \quad \text{for} \quad z = 0 \text{ and } 1. \tag{161}$$

From the equation satisfied by W (namely (135)), it follows that $D^6W = 0$ for $z = 0$ and 1. By differentiating equation (135) even

TABLE XIV

Critical Rayleigh numbers and the wave numbers of the unstable modes at marginal stability for the onset of stationary convection in the case when both bounding surfaces are free

Q	a_c	R_c	R_c/R_0	Q	a_c	R_c	R_c/R_0
0	2·233	657·511	1·0000	6,500	7·952	80343·6	122·19
5	2·432	796·573	1·2115	7,000	8·059	86034·0	130·85
10	2·590	923·070	1·4039	7,500	8·159	91705·7	139·47
20	2·826	1154·19	1·7554	8,000	8·253	97360·1	148·07
50	3·270	1762·04	2·6799	8,500	8·343	102998	156·65
100	3·702	2653·71	4·0360	9,000	8·429	108623	165·20
150	3·990	3475·67	5·2861	9,500	8·510	114234	173·74
200	4·210	4258·49	6·4767	10,000	8·588	119832	182·25
300	4·543	5752·65	8·7491	10,500	8·663	125419	190·75
400	4·794	7185·94	10·929	11,000	8·735	130995	199·23
500	4·998	8578·88	13·048	11,500	8·804	136560	207·69
600	5·171	9942·40	15·121	12,000	8·870	142116	216·14
700	5·321	11283·2	17·160	13,000	8·997	153202	233·00
800	5·455	12605·6	19·172	14,000	9·116	164255	249·81
1,000	5·684	15207·0	23·128	15,000	9·227	175279	266·58
1,500	6·123	21535·2	32·753	16,000	9·333	186276	283·30
2,000	6·453	27699·9	42·128	17,000	9·433	197249	299·99
2,500	6·720	33756·5	51·340	18,000	9·528	208199	316·65
3,000	6·945	39734·2	60·431	19,000	9·619	219129	333·27
3,500	7·140	45650·6	69·429	20,000	9·706	230038	349·86
4,000	7·313	51517·8	78·353	25,000	10·09	284341	432·45
4,500	7·442	57344·6	87·215	30,000	10·42	338308	514·53
5,000	7·585	63135·9	96·022	35,000	10·70	392013	596·21
5,500	7·717	68897·3	104·78	40,000	10·95	445507	677·56
6,000	7·839	74632·1	113·51				

numbers of times, we can successively conclude that all the even derivatives of W must vanish for $z = 0$ and 1. The proper solution for W appropriate for the lowest mode is therefore

$$W = A \sin \pi z, \tag{162}$$

where A is a constant. Substitution of the solution (162) in equation (135) leads to the characteristic equation

$$R = \frac{\pi^2+a^2}{a^2}[(\pi^2+a^2)^2+\pi^2 Q]. \tag{163}$$

Letting $$a^2 = \pi^2 x, \tag{164}$$

equation (163) becomes

$$R = \pi^4 \frac{1+x}{x}\left[(1+x)^2+\frac{Q}{\pi^2}\right]. \tag{165}$$

As a function of x, R given by equation (165) attains its minimum when

$$2x^3+3x^2 = 1+Q/\pi^2. \tag{166}$$

In some ways it is remarkable that this cubic equation is the same as the

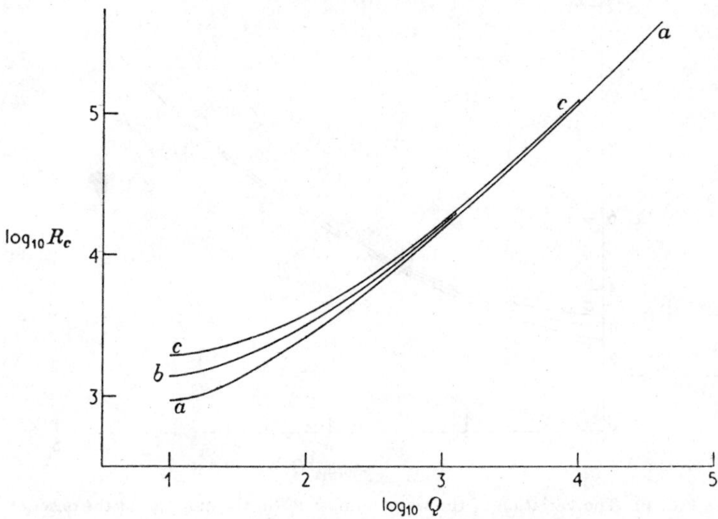

FIG. 39. The variation of the critical Rayleigh number R_c for the onset of instability as a function of Q for the three cases (i) both bounding surfaces free (curve labelled *aa*), (ii) one bounding surface free and the other rigid (curve labelled *b*), and (iii) both bounding surfaces rigid (curve labelled *cc*).

one which occurs in the corresponding rotational problem (see equation III (131)); the parameter Q/π^2 now plays the role of T/π^4.

With x determined as a solution of equation (166), equation (165) will give the required critical Rayleigh number R_c. Values of R_c determined in this fashion for various values of Q are given in Table XIV; the wave numbers characterizing the marginal states are also given. The results are further illustrated in Figs. 39 and 40. The inhibiting effect of the magnetic field on the onset of instability is apparent from these results.

For Q/π^2 sufficiently large, the required root of equation (166) tends to

$$x_{\min} \to (Q/2\pi^2)^{\frac{1}{3}} \quad (Q \to \infty). \tag{167}$$

The corresponding asymptotic behaviours of R_c and $a_{\min}$ are

$$R_c \to \pi^2 Q \quad \text{and} \quad a_{\min} \to (\tfrac{1}{2}\pi^4 Q)^{\frac{1}{6}}. \tag{168}$$

Substituting for R and Q in accordance with their definitions, we find that the formula which determines the critical temperature gradient for the onset of stationary convection, when $Q \to \infty$, is

$$g\alpha\beta_c \to \pi^2 \frac{\mu^2 H^2}{\rho} \sigma\kappa\, d^{-2} \quad (H \to \infty, \text{ or } \sigma \to \infty, \text{ or } \nu \to 0). \tag{169}$$

We observe that *in this limit, the critical temperature gradient has become*

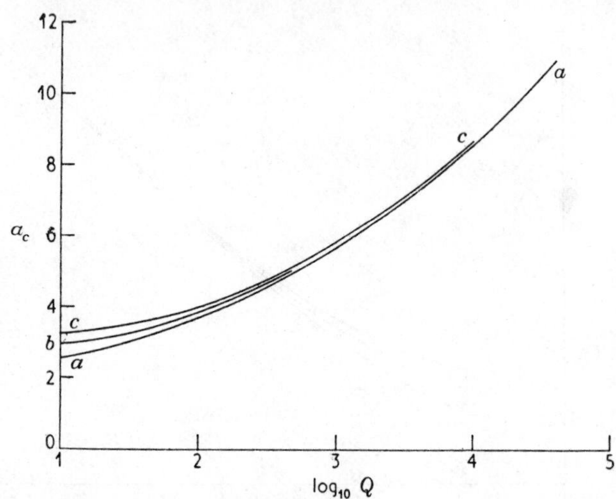

Fig. 40. The variation of the wave number a_c (in the unit $1/d$) at the onset of instability as a function of Q for the three cases (i) both bounding surfaces free (curve labelled *aa*), (ii) one bounding surface free and the other rigid (curve labelled *b*), and (iii) both bounding surfaces rigid (curve labelled *cc*).

independent of the kinematic viscosity. We shall return to the meaning of this result in § 45.

(*b*) *The solution for two rigid boundaries*

We shall obtain the solution for this case by the variational method.

In view of the symmetry of this problem with respect to the bounding planes, we shall find it convenient to translate the origin of z to be midway between the two planes. The fluid will then be confined between $z = \pm\frac{1}{2}$ and we shall have to seek solutions of equations (141) and (142) which satisfy the boundary conditions

$$F = W = DW = 0 \quad \text{for } z = \pm\tfrac{1}{2}. \tag{170}$$

It is apparent from the equations and the boundary conditions that the proper solutions of the problem fall into two non-combining groups of even and odd solutions, respectively; and it is also clear that the lowest characteristic values of R occur among the even solutions. We shall accordingly consider such solutions.

Since F is assumed to be even and is required to vanish at $z = \pm\frac{1}{2}$, we can expand it in a cosine series in the form

$$F = \sum_m A_m \cos[(2m+1)\pi z], \tag{171}$$

where the summation over m may be assumed to run from zero to infinity; but this is not necessary since we shall consider the coefficients A_m as variational parameters.

With the chosen form of F, the equation to be solved for W is

$$[(D^2-a^2)^2-QD^2]W = \sum_m A_m \cos[(2m+1)\pi z]. \tag{172}$$

Expressing W as a sum in the form

$$W = \sum_m A_m W_m, \tag{173}$$

we have to solve

$$[(D^2-a^2)^2-QD^2]W_m = \cos[(2m+1)\pi z]. \tag{174}$$

The general solution of this equation is

$$W_m = \gamma_{2m+1}\cos[(2m+1)\pi z]+\sum_{j=1}^{2} B_j^{(m)}\cosh q_j z, \tag{175}$$

where
$$\frac{1}{\gamma_{2m+1}} = [(2m+1)^2\pi^2+a^2]^2+(2m+1)^2\pi^2 Q; \tag{176}$$

the $B_j^{(m)}$'s $(j = 1, 2)$ are constants of integration; and the q_j^2's $(j = 1, 2)$ are the roots of the quadratic equation

$$(q^2-a^2)^2-Qq^2 = 0. \tag{177}$$

According to equation (177),

$$q_1 = \tfrac{1}{2}[\sqrt{(Q+4a^2)}+\sqrt{Q}] \quad \text{and} \quad q_2 = \tfrac{1}{2}[\sqrt{(Q+4a^2)}-\sqrt{Q}]; \tag{178}$$

an identity which follows from the same equation is

$$\frac{1}{\gamma_{2m+1}} = \prod_{j=1}^{2}[(2m+1)^2\pi^2+q_j^2]. \tag{179}$$

The constants $B_j^{(m)}$ in the solution (175) are determined by the boundary conditions on W_m which require that W_m and DW_m vanish for $z = \pm\frac{1}{2}$. These conditions give

$$\left.\begin{aligned} \sum_{j=1}^{2} B_j^{(m)}\cosh\tfrac{1}{2}q_j &= 0 \\ \sum_{j=1}^{2} B_j^{(m)} q_j \sinh\tfrac{1}{2}q_j &= (-1)^m(2m+1)\pi\gamma_{2m+1} \end{aligned}\right\}. \tag{180}$$

On solving these equations, we find

$$\left.\begin{aligned} B_1^{(m)} &= +(-1)^m(2m+1)\pi\gamma_{2m+1}\Delta\,\mathrm{sech}\,\tfrac{1}{2}q_1 \\ B_2^{(m)} &= -(-1)^m(2m+1)\pi\gamma_{2m+1}\Delta\,\mathrm{sech}\,\tfrac{1}{2}q_2 \end{aligned}\right\}, \tag{181}$$

where

$$\Delta = \frac{1}{q_1\tanh\tfrac{1}{2}q_1 - q_2\tanh\tfrac{1}{2}q_2}. \tag{182}$$

Substituting for F and W in accordance with equations (171), (173), and (175) in equation (142), we have

$$\sum_m A_m c_{2m+1}\cos[(2m+1)\pi z]$$
$$= Ra^2\sum_m A_m\Big\{\gamma_{2m+1}\cos[(2m+1)\pi z] + \sum_{j=1}^{2} B_j^{(m)}\cosh q_j z\Big\}, \tag{183}$$

where

$$c_{2m+1} = (2m+1)^2\pi^2 + a^2. \tag{184}$$

Multiplying equation (183) by $\cos[(2n+1)\pi z]$ and integrating over the range of z, we obtain

$$\tfrac{1}{2}c_{2n+1}A_n = Ra^2\Big\{\tfrac{1}{2}\gamma_{2n+1}A_n + \sum_m (n|m)A_m\Big\}, \tag{185}$$

where

$$\begin{aligned}(n|m) &= \sum_{j=1}^{2} B_j^{(m)}\int_{-\frac{1}{2}}^{+\frac{1}{2}}\cosh q_j z\cos[(2n+1)\pi z]\,dz \\ &= 2(2n+1)\pi(-1)^n\sum_{j=1}^{2}\frac{B_j^{(m)}\cosh\tfrac{1}{2}q_j}{(2n+1)^2\pi^2+q_j^2}.\end{aligned} \tag{186}$$

Equations (185) provide a set of linear homogeneous equations for the constants A_m; this same set of equations ensures that the expression (147) for R is a minimum for all variations of the A_m's which preserve the constancy of (146). In this latter formulation, Ra^2 is the Lagrangian undetermined multiplier.

The determinant of the system of equations represented by (185) must vanish; and this provides the characteristic equation for R; thus

$$\left\|\frac{1}{2}\left(\frac{c_{2n+1}}{Ra^2} - \gamma_{2n+1}\right)\delta_{nm} - (n|m)\right\| = 0. \tag{187}$$

Substituting for the $B_j^{(m)}$'s in accordance with equations (181), and making use of the identity (179), we find after some minor reductions that the expression for $(n|m)$ becomes

$$(n|m) = (-1)^{m+n}\,2(2n+1)(2m+1)\pi^2\gamma_{2n+1}\gamma_{2m+1}\Delta(q_2^2-q_1^2). \tag{188}$$

Inserting for Δ, q_1^2, and q_2^2 their values, we have

$$(n|m) = (-1)^{m+n+1}\,2(2n+1)(2m+1)\pi^2\gamma_{2n+1}\gamma_{2m+1}\times$$
$$\times\frac{\sqrt{[Q(Q+4a^2)]}}{q_1\tanh\tfrac{1}{2}q_1 - q_2\tanh\tfrac{1}{2}q_2}. \tag{189}$$

This expression for $(n|m)$ is manifestly symmetric in n and m.

By solving the determinantal equation (187) by including successively more rows and columns, we can evaluate the required characteristic values of R with increasing precision. The results of such calculations are summarized in Table XV and illustrated in Figs. 39 and 40. It will be seen that in no case is it really necessary to go to higher than the second approximation.

(c) *The solution for one rigid and one free boundary*

In this case the conditions to be satisfied on the two bounding surfaces are different. However, the solution for this case can be reduced to that

TABLE XV

Critical Rayleigh numbers and related constants when both bounding surfaces are rigid and the onset of instability is as stationary convection

		$F = \sum A_m \cos(2m+1)\pi z$			$F = \cos \pi z + A(1+\cos 2\pi z)$	
		R_c				
Q	a_c	First approximation	Second approximation	Third approximation	R_c	A
0	3·13	1715·1			1707·8	
10	3·25	1953·7	1946·0	1945·8	1945·9	0·09518
50	3·68	2811·4	2802·4		2802·1	0·09122
100	4·00	3767·6	3757·8	3757·3	3757·4	0·08556
200	4·45	5499·9	5489·3		5488·6	0·07746
500	5·16	10122	10111		10110	0·06283
1,000	5·80	17116	17105	17103	17103	0·05192
2,000	6·55	30139	30127		30125	0·04196
4,000	7·40	54712	54700		54697	0·03345
6,000	7·94	78405	78393		78391	0·02899
8,000	8·34	101622	101609		101606	0·02628
10,000	8·66	124523	124511	124509	124509	0·02471

of case (b) by considering odd (instead of even) solutions for W. For it is clear that an odd solution for W satisfying the boundary conditions appropriate to case (b) vanishes at $z = 0$; therefore, at $z = 0$, the boundary conditions on W appropriate to a free surface are satisfied. Consequently, a solution with W odd, suitable for case (b) and applicable to a cell depth d, provides a solution for case (c) applicable to a cell depth $\frac{1}{2}d$, a Rayleigh number sixteen times smaller, and a Q-number four times smaller.

By considering odd solutions for W appropriate to case (b), we can again obtain the solution by the variational method. Thus by expanding F in a sine series of the form

$$F = \sum_m A_m \sin 2m\pi z, \tag{190}$$

the solution can be found by following a procedure identical to that described in § (*b*) above. In this manner, we obtain the secular determinant

$$\left\| \frac{1}{2}\left(\frac{c_{2n}}{Ra^2}-\gamma_{2n}\right)\delta_{nm}-(n|m) \right\| = 0, \tag{191}$$

where c_{2n} and γ_{2n} are defined as in equations (176) and (184) (with $2m$ replacing $2m+1$) and $(n|m)$ is now given by (cf. equation (189))

$$(n|m) = (-1)^{n+m+1} 8nm\pi^2\gamma_{2n}\gamma_{2m}\frac{\sqrt{[Q(Q+4a^2)]}}{q_1\coth\frac{1}{2}q_1-q_2\coth\frac{1}{2}q_2}. \tag{192}$$

TABLE XVI

Critical Rayleigh numbers and related constants for the case when one bounding surface is rigid and the other is free and the onset of instability is as stationary convection

Q	a_c	R_c First approximation	R_c Second approximation†	A
0	2·68	1112·7	1100·75	0·2043
2·5	2·75	1179·4	1167·2	0·2016
12·5	2·97	1428·3	1415·5	0·1905
25	3·17	1712·7	1699·4	0·1790
50	3·45	2231·3	2217·6	0·1616
125	4·00	3600·2	3586·1	0·1326
250	4·50	5627·5	5613·3	0·1085
500	5·10	9318·7	9304·5	0·0865
1,000	5·75	16133	16119	0·0671
1,500	6·20	22606	22592	0·0578
2,000	6·50	28893	28879	0·0516
2,500	6·75	35058	35044	0·0472
5,000	7·65	64861	64847	0·0359
10,000	8·65	122155	122140	0·0270

† This approximation was obtained with the trial function
$$F = \sin 2\pi z + A(\sin \pi z + \sin 3\pi z).$$

The secular equation (191) has been solved in various approximations. The results of such calculations are summarized in Table XVI and illustrated in Figs. 39 and 40.

(*d*) *Cell patterns*

Since the z-component of the vorticity vanishes, the expressions for the horizontal components of the velocity in terms of $W(z)$ are formally the same as in the simple Bénard problem; and for the case of two free surfaces, even $W(z)$ is the same. The cell patterns which can emerge are, therefore, the same as those described in Chapter II, § 16. There is, however, one important quantitative difference: the actual wave numbers appropriate for the marginal state are different; they depend on Q, and as Q increases, the cells tend to become narrow and elongated; this is illustrated in Fig. 41.

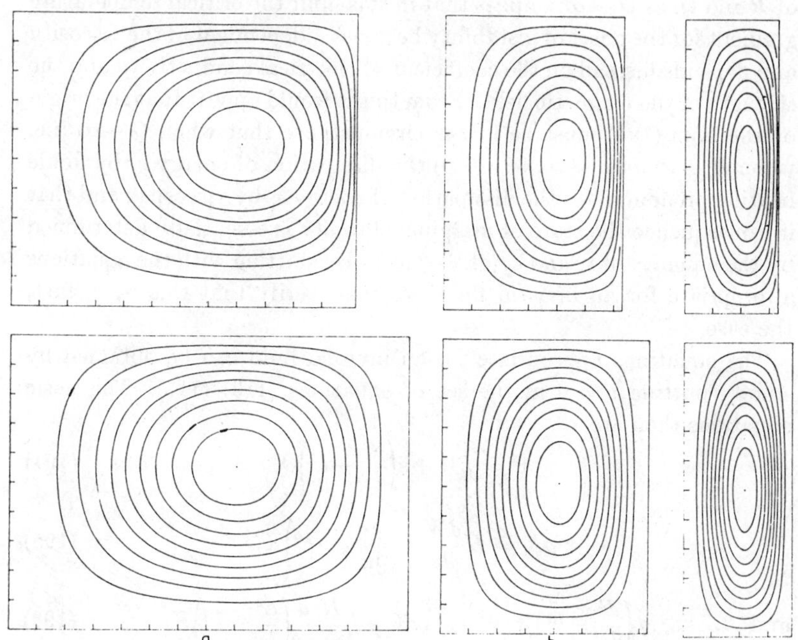

FIG. 41. The streamlines in a hexagonal cell in the planes of symmetry at the onset of instability for the first odd mode for various values of Q: (a) $Q = 12{\cdot}5$, $a_c = 2{\cdot}97$; (b) $Q = 1{,}000$, $a_c = 5{\cdot}75$; (c) $Q = 40{,}000$, $a_c = 10{\cdot}95$. Normalized to unity at the centre, the successive streamlines are for the values 0·9, 0·8,..., 0·1, respectively. The patterns on the top are in the plane of symmetry normal to a pair of opposite sides of the hexagon and the patterns on the bottom are in the plane of symmetry through two opposite vertices of the hexagon. Notice the progressive elongation of the cells with increasing Q (compare with Fig. 8).

45. The origin of the $\pi^2 Q$-law and an invariant

From the results for the three sets of boundary conditions illustrated in Figs. 39 and 40, it is apparent that all three cases exhibit the same general features. This is particularly true of the asymptotic dependence on Q of the critical Rayleigh number and the associated wave number. For the case of two free boundaries, the laws

$$R_c \to \pi^2 Q \quad \text{and} \quad a_c \to (\tfrac{1}{2}\pi^4 Q)^{\frac{1}{6}} \tag{193}$$

follow directly from the solution of the characteristic value problem. From the results of the calculations for the two other cases, it appears that the same power laws with the same constants of proportionality hold for them also. We shall now examine the physical origin of these laws.

We have seen in § 44 (a) (see equation (169)) that the proportionality of R and Q, as $Q \to \infty$, implies that in this limit the critical temperature gradient for the onset of instability becomes independent of the viscosity and depends instead on the coefficient of electrical conductivity and the strength of the magnetic field. From this it would appear that the origin of the laws (193) must lie in the circumstance that when $Q \to \infty$ (i.e. when $H \to \infty$, *or* $\sigma \to \infty$, *or* $\nu \to 0$) the dissipation of energy ϵ_σ by Joule heating predominates the dissipation of energy ϵ_ν by viscosity; and that in consequence the state of marginal stability is essentially determined by the equality of ϵ_σ and ϵ_g (cf. § 43 (b)). By starting with the equations appropriate for an inviscid fluid, we shall verify that this is, in fact, the case.

The equations appropriate for an inviscid fluid can be obtained by simply putting $\nu = 0$ in the set of equations (113)–(117). The basic equations then are

$$p\Theta = \beta W+\kappa\left(\frac{d^2}{dz^2}-k^2\right)\Theta, \tag{194}$$

$$pK = H\frac{dW}{dz}+\eta\left(\frac{d^2}{dz^2}-k^2\right)K, \tag{195}$$

and

$$p\left(\frac{d^2}{dz^2}-k^2\right)W = -g\alpha k^2\Theta+\frac{\mu H}{4\pi\rho}\frac{d}{dz}\left(\frac{d^2}{dz^2}-k^2\right)K. \tag{196}$$

For the onset of stationary convection, $p = 0$; and the elimination of K and Θ is immediate. We obtain

$$D^2(D^2-a^2)W = \frac{R}{Q}a^2W \tag{197}$$

together with the boundary conditions

$$W = 0 \quad \text{and} \quad D^2W = 0 \quad \text{for} \quad z = 0 \text{ and } 1. \tag{198}$$

(In passing from equations (194)–(196) to (197) we have reverted to d as the unit of distance.)

Observe that equation (197) is what we should have obtained from equation (135) by retaining only the term in Q.

Clearly the proper solution of equation (197) appropriate for the lowest mode is

$$W = A\sin\pi z, \tag{199}$$

where A is a constant. The corresponding characteristic equation is

$$R/Q = \frac{\pi^2(\pi^2+a^2)}{a^2}. \tag{200}$$

The minimum temperature gradient occurs for $a \to \infty$, when

$$R/Q = \pi^2 \quad (a \to \infty), \tag{201}$$

which is the correct asymptotic behaviour. *In this inviscid limit the cells which appear at marginal stability are infinitely narrow.*

According to equation (201), the critical temperature gradient at which instability sets in is given by (cf. equation (169))

$$g\alpha\beta_c = \pi^2\frac{\mu^2H^2}{\rho}\sigma\kappa\, d^{-2}. \tag{202}$$

Comparing this with the corresponding equation,

$$g\alpha\beta_c = \text{constant } \nu\kappa\, d^{-4}, \tag{203}$$

which obtains for a viscous liquid in the absence of a magnetic field, one may say *that the presence of the magnetic field imparts to the liquid an effective kinematic viscosity*

$$\nu_{\text{eff}} = \text{constant}\frac{\sigma}{\rho}(\mu H d)^2. \tag{204}$$

(We shall see in § 47 that only the component of the magnetic field parallel to $\mathbf{g}$ is effective.)

The foregoing discussion, applicable strictly for $\nu = 0$, can be extended to predict the correct dependence of the wave number a on Q, if in equation (135) we retain the term a^4W in addition to $-QD^2W$. We should then obtain

$$(D^2-a^2)(QD^2-a^4) = Ra^2W \tag{205}$$

together with the boundary conditions

$$W = 0 \quad \text{and} \quad (QD^2-a^4)W = 0 \quad \text{for } z = 0 \text{ and } 1. \tag{206}$$

The proper solution of equation (205) appropriate for the lowest mode is again given by (199); but in place of equation (200) we now have

$$R = \frac{\pi^2+a^2}{a^2}(a^4+Q\pi^2). \tag{207}$$

Comparison with equation (163) shows that equation (207) will suffice to predict the correct asymptotic behaviours of both R_c and a_{min}. The fact that we have to go to this one higher order to obtain a finite value for the wave number at marginal stability means that it is viscosity that prevents the cells from collapsing into lines.

(*a*) *An invariant*

Closely related to the remarks in the preceding paragraph on the role of viscosity in keeping the cross-sections of the cells finite is the existence of an invariant similar to the one which exists for the rotational problem (see Chapter III, § 28, equation (195)). It will be recalled that when the fluid is in rotation, the z-component of the vorticity does not vanish and contributes to the horizontal components of the velocity. The resulting increase in the mean kinetic energy in the horizontal motions compensates for the reduction in it caused by the increase in the wave number; and for the case of two free boundaries, the compensation is exact and $\langle|\mathbf{u}_\perp|^2\rangle$ becomes independent of rotation. In the present problem, there is no such additional contribution to $\mathbf{u}_\perp$; instead, a magnetic field in the horizontal directions comes to prevail; and a new invariant emerges which is associated with this magnetic energy in the horizontal directions.

The horizontal components of the magnetic field are given by equations (155); they depend on the scalar K which is to be determined from the equation (see equation (129))

$$(D^2-a^2)K = -\frac{Hd}{\eta}DW. \tag{208}$$

For the case of two free boundaries,

$$W = W_0 \sin \pi z, \tag{209}$$

and the required solution of equation (208) is

$$K = \frac{Hd}{\eta}\frac{\pi}{\pi^2+a^2}W_0 \cos \pi z + B\cosh az, \tag{210}$$

where B is a constant to be determined by the boundary conditions (140). The part of the magnetic field derived from $B\cosh az$ does not contribute to the Joule-dissipation ϵ_σ (cf. equation (157)); we shall ignore this part and consider only the part which contributes to ϵ_σ. The components of this latter field $\mathbf{h}_\perp^{(0)}$ are given by

$$h_x^{(0)} = W_0\frac{Hd}{\eta}\frac{\pi^2}{\pi^2+a^2}\frac{a_x}{a^2}\sin a_x x \cos a_y y \sin \pi z,$$

$$h_y^{(0)} = W_0\frac{Hd}{\eta}\frac{\pi^2}{\pi^2+a^2}\frac{a_y}{a^2}\cos a_x x \sin a_y y \sin \pi z. \tag{211}$$

The corresponding solutions for the horizontal components of the velocity are

$$u_x = -\pi\frac{a_x}{a^2}W_0 \sin a_x x \cos a_y y \cos \pi z,$$

$$u_y = -\pi\frac{a_y}{a^2}W_0 \cos a_x x \sin a_y y \cos \pi z. \tag{212}$$

According to these equations,

$$\langle\langle|\mathbf{h}_\perp^{(0)}|^2\rangle\rangle = \frac{1}{8}\frac{\pi^4}{a^2(\pi^2+a^2)^2}\frac{H^2 d^2}{\eta^2}W_0^2 \tag{213}$$

and

$$\langle\langle|\mathbf{u}_\perp|^2\rangle\rangle = \frac{1}{8}\frac{\pi^2}{a^2}W_0^2, \tag{214}$$

where the double angular brackets signify that the quantity enclosed is averaged both over the horizontal plane and the vertical direction.

Now consider the quantity

$$\Psi = \eta\frac{\mu}{8\pi}\langle\langle|\mathbf{h}_\perp^{(0)}|^2\rangle\rangle+\tfrac{1}{2}\rho\nu\langle\langle|\mathbf{u}_\perp|^2\rangle\rangle. \tag{215}$$

An equivalent expression for Ψ is

$\Psi = \eta\times$average energy in the magnetic field associated with $\mathbf{h}_\perp^{(0)}+$

$+\nu\times$average energy in the fluid motions associated with $\mathbf{u}_\perp$. (216)

Using equations (213) and (214), we obtain

$$\Psi = \tfrac{1}{16}\rho\nu\left[\frac{\pi^2}{a^2}+Q\frac{\pi^4}{a^2(\pi^2+a^2)^2}\right]W_0^2. \tag{217}$$

Letting $a^2 = \pi^2 x$ (as in § 44 (*a*), equation (164)), we have

$$\Psi = \tfrac{1}{16}\rho\nu\frac{(1+x)^2+Q/\pi^2}{x(1+x)^2}W_0^2. \tag{218}$$

On the other hand, the value of x at the onset of instability is related to Q by equation (166); using this equation to eliminate Q from equation (218), we find that we are left with

$$\Psi = \tfrac{1}{8}\rho\nu W_0^2. \tag{219}$$

Thus, *the energy associated with the horizontal components of the magnetic field and the fluid motions, weighted by their respective dissipative coefficients, is, for a constant amplitude of the vertical velocity, independent of the strength of the impressed magnetic field.*

46. On the onset of convection as overstability

We shall now take up the question set aside in § 43 of whether or not instability can arise as oscillations of increasing amplitude. This requires us to return to equations (119)–(123) which include the time constant σ. However, in discussing this question of overstability, we shall mostly restrict ourselves to the case when the fluid is confined between two free boundaries.

By applying the operator $(D^2-a^2-\mathfrak{p}_1\sigma)(D^2-a^2-\mathfrak{p}_2\sigma)$ to equation (121), we can eliminate K and Θ and obtain

$$(D^2-a^2)(D^2-a^2-\mathfrak{p}_1\sigma)[(D^2-a^2-\sigma)(D^2-a^2-\mathfrak{p}_2\sigma)-QD^2]W = -Ra^2(D^2-a^2-\mathfrak{p}_2\sigma)W. \quad (220)$$

As in § 44 (a), we can show that in this case also the proper solution for W belonging to the lowest mode is

$$W = \text{constant}\sin\pi z. \quad (221)$$

Substituting this solution for W in equation (220), we obtain the characteristic equation

$$(\pi^2+a^2)(\pi^2+a^2+\mathfrak{p}_1\sigma)[(\pi^2+a^2+\sigma)(\pi^2+a^2+\mathfrak{p}_2\sigma)+Q\pi^2] = Ra^2(\pi^2+a^2+\mathfrak{p}_2\sigma), \quad (222)$$

where it must be remembered that σ can be complex. Letting

$$x = \frac{a^2}{\pi^2}, \quad i\sigma_1 = \frac{\sigma}{\pi^2}, \quad R_1 = \frac{R}{\pi^4}, \quad \text{and} \quad Q_1 = \frac{Q}{\pi^2}, \quad (223)$$

we can rewrite equation (222) in the form

$$(1+x)(1+x+i\mathfrak{p}_1\sigma_1)[(1+x+i\sigma_1)(1+x+i\mathfrak{p}_2\sigma_1)+Q_1] = R_1 x(1+x+i\mathfrak{p}_2\sigma_1), \quad (224)$$

or

$$(1+x+i\sigma_1)(1+x+i\mathfrak{p}_1\sigma_1)(1+x+i\mathfrak{p}_2\sigma_1)+Q_1(1+x+i\mathfrak{p}_1\sigma_1) = R_1\frac{x}{1+x}(1+x+i\mathfrak{p}_2\sigma_1). \quad (225)$$

It is apparent from equation (225) that for an arbitrarily assigned σ_1, R_1 will be complex. But the physical meaning of R_1 requires it to be real. Consequently, the condition that R_1 be real implies a relation between the real and the imaginary parts of σ_1.

Since our present interest is to specify the critical Rayleigh number for the onset of instability via a state of purely oscillatory motions, it will suffice to seek the conditions that equation (225) will allow solutions for which σ_1 is real. Assuming, then, that σ_1 is real, and equating, separately, the real and the imaginary parts of equation (225), we obtain

$$R_1 x = (1+x)^3-\sigma_1^2(1+x)(\mathfrak{p}_1+\mathfrak{p}_2+\mathfrak{p}_1\mathfrak{p}_2)+Q_1(1+x) \quad (226)$$

and

$$R_1\frac{x}{1+x}\mathfrak{p}_2 = (1+x)^2(1+\mathfrak{p}_1+\mathfrak{p}_2)-\mathfrak{p}_1\mathfrak{p}_2\sigma_1^2+\mathfrak{p}_1 Q_1. \quad (227)$$

Alternatively, we can write

$$R_1 = \frac{1+x}{x}[(1+x)^2+Q_1-\sigma_1^2(\mathfrak{p}_1+\mathfrak{p}_2+\mathfrak{p}_1\mathfrak{p}_2)], \quad (228)$$

and $$\mathfrak{p}_1\mathfrak{p}_2\sigma_1^2 = (1+x)^2(1+\mathfrak{p}_1+\mathfrak{p}_2)+\mathfrak{p}_1 Q_1 - R_1\frac{x}{1+x}\mathfrak{p}_2. \tag{229}$$

Substituting for $R_1 x/(1+x)$ in (229) from (228), we obtain on simplification,

$$\sigma_1^2\mathfrak{p}_2^2 = \frac{\mathfrak{p}_2-\mathfrak{p}_1}{1+\mathfrak{p}_1}Q_1-(1+x)^2. \tag{230}$$

Using this expression for σ_1^2 in equation (228), we obtain the result

$$R_1 = \frac{(1+\mathfrak{p}_2)(\mathfrak{p}_1+\mathfrak{p}_2)}{\mathfrak{p}_2^2}\frac{1+x}{x}\left[(1+x)^2+Q_1\frac{\mathfrak{p}_1^2}{(1+\mathfrak{p}_1)(\mathfrak{p}_1+\mathfrak{p}_2)}\right]. \tag{231}$$

Equations (230) and (231) are the equations which must be satisfied if overstability is to occur for a wave number corresponding to x and a Q-value corresponding to Q_1.

One conclusion which equation (230) enables us to draw at once is that solutions describing overstability cannot occur if

$$\mathfrak{p}_2 < \mathfrak{p}_1; \tag{232}$$

for σ_1^2 would then be negative, contrary to hypothesis. Recalling the definitions of $\mathfrak{p}_1$ and $\mathfrak{p}_2$ (given in (118)), the foregoing condition is equivalent to

$$\kappa < \eta. \tag{233}$$

Thus, *for $\kappa < \eta$, overstability cannot occur and the principle of the exchange of stabilities is valid.*

The condition $\kappa < \eta$ will be met by a large margin under most terrestrial conditions. Thus, for mercury at room temperatures,

$$\eta = 7{\cdot}6\times10^3 \text{ cm}^2\text{ sec}^{-1} \quad\text{and}\quad \kappa = 4{\cdot}5\times10^{-2}\text{ cm}^2\text{ sec}^{-1}. \tag{234}$$

For overstability to be at all possible, $\mathfrak{p}_2 > \mathfrak{p}_1$; and even when this is the case, we shall obtain real frequencies σ_1 only if

$$Q_1 > (1+x)^2\frac{1+\mathfrak{p}_1}{\mathfrak{p}_2-\mathfrak{p}_1}. \tag{235}$$

For a given Q_1 overstable solutions are, therefore, possible only for $x < x_*$, where x_* is such that

$$(1+x_*)^2 = Q_1\frac{\mathfrak{p}_2-\mathfrak{p}_1}{1+\mathfrak{p}_1}. \tag{236}$$

When $x = x_*$, $\sigma_1^2 = 0$ and (cf. equation (228))

$$R_1 = \frac{1+x_*}{x_*}[(1+x_*^2)+Q_1]; \tag{237}$$

and this is, as one should expect, the value of R_1 at which stationary convection will occur for a wave number corresponding to x^*. For $x > x_*$, overstability is not possible for the given $\mathfrak{p}_1$, $\mathfrak{p}_2$, and Q_1; and the onset of instability as stationary convection remains the only possibility. For $x < x_*$, overstability is possible and the question of discriminating between the two manners of instability arises. As we have shown in detail in the discussion of the analogous problem in Chapter III (§§ 29 and 30), that manner of instability will occur that allows a solution for a lower Rayleigh number.

Consider the (x, R_1)-plane. In this plane we have, first, the curve,

$$R_1^{(c)} = \frac{1+x}{x}[(1+x)^2+Q_1], \tag{238}$$

which defines the locus of states which are marginal with respect to stationary convection. The overstable solutions branch off from this locus at the point x_* (see equation (237)), and for $x < x_*$ they are described by (cf. equations (228) and (231))

$$\begin{aligned} R_1^{(o)} &= R_1^{(c)} - (\mathfrak{p}_1+\mathfrak{p}_2+\mathfrak{p}_1\mathfrak{p}_2)\sigma_1^2\frac{1+x}{x} \\ &= \frac{(1+\mathfrak{p}_2)(\mathfrak{p}_1+\mathfrak{p}_2)}{\mathfrak{p}_2^2}\frac{1+x}{x}\left[(1+x)^2+Q_1\frac{\mathfrak{p}_1^2}{(1+\mathfrak{p}_1)(\mathfrak{p}_1+\mathfrak{p}_2)}\right], \end{aligned} \tag{239}$$

where the first form shows that $R_1^{(o)}$ (when this branch of the solution exists) is *always* less than $R_1^{(c)}$. Depending on $\mathfrak{p}_1$, $\mathfrak{p}_2$, and Q_1, the branch point x_* can occur either before or after the point $x_{\min}^{(c)}$ at which $R_1^{(c)}$ attains its minimum. If $x_* > x_{\min}^{(c)}$, then it is clear that for all $x < x_*$, overstability is the preferred manner of instability.

According to equations (168) and (236), the asymptotic behaviours of x_*, and $x_{\min}^{(c)}$, for $Q_1 \to \infty$, are

$$x_* \to \left(\frac{\mathfrak{p}_2-\mathfrak{p}_1}{1+\mathfrak{p}_1}Q_1\right)^{\frac{1}{2}} \quad \text{and} \quad x_{\min}^{(c)} \to (\tfrac{1}{2}Q_1)^{\frac{1}{3}} \quad (Q_1 \to \infty). \tag{240}$$

For a sufficiently large Q_1, $x_* > x_{\min}^{(c)}$; and from our earlier remarks it follows that for $\mathfrak{p}_2 > \mathfrak{p}_1$, overstability is the preferred manner of instability for Q_1 sufficiently large. Moreover, from the monotonic dependence of x_* and $x_{\min}^{(c)}$ on Q_1, we may conclude that *for* $\mathfrak{p}_2 > \mathfrak{p}_1$, *there exists a* $Q_1^{(\mathfrak{p}_1,\mathfrak{p}_2)}$, *such that for* $Q_1 \leqslant Q_1^{(\mathfrak{p}_1,\mathfrak{p}_2)}$ *the onset of instability will be as stationary convection, while for* $Q_1 > Q_1^{(\mathfrak{p}_1,\mathfrak{p}_2)}$ *it will be as overstability.*

There is no simple formula which gives $Q^{(\mathfrak{p}_1,\mathfrak{p}_2)}$ as a function of $\mathfrak{p}_1$ and $\mathfrak{p}_2$: it is simply determined by the condition that $R_{\min}^{(o)}$ and $R_{\min}^{(c)}$ are equal for $Q = Q^{(\mathfrak{p}_1,\mathfrak{p}_2)}$. However, in any given case, the complete (R_c, Q)-relation can be determined quite simply as follows.

The first part of the (R_c, Q)-relation, where instability sets in as stationary convection, is known from the results of Table XIV. This part of the relation terminates at $Q = Q^{(\mathfrak{p}_1,\mathfrak{p}_2)}$; and beyond this point the relation continues along the overstable branch. This overstable part of the relation can be deduced from the results given in Table XIV by a simple transformation of scales: for, according to equation (239), we have only to interpret Q in Table XIV as signifying $Q\mathfrak{p}_1^2/[(1+\mathfrak{p}_1)(\mathfrak{p}_1+\mathfrak{p}_2)]$ and multiply the values under R_c by $(1+\mathfrak{p}_2)(\mathfrak{p}_1+\mathfrak{p}_2)/\mathfrak{p}_2^2$. The intersection of the relation so deduced with the first part of the relation given directly by Table XIV determines $Q^{(\mathfrak{p}_1,\mathfrak{p}_2)}$ (see Fig. 42).

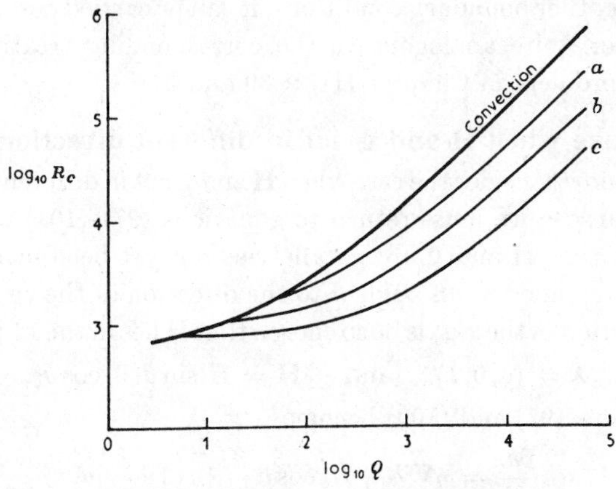

FIG. 42. The variation of the critical Rayleigh number R_c as a function of Q when the onset of instability can occur as overstability. The curves labelled a, b, and c are for values of the Prandtl numbers: $\mathfrak{p}_1 = 1$, $\mathfrak{p}_2 = 2$; $\mathfrak{p}_1 = 1$, $\mathfrak{p}_2 = 4$; $\mathfrak{p}_1 = 1$, $\mathfrak{p}_2 = 10$, respectively.

The asymptotic behaviours of the Rayleigh number and the related quantities, at the onset of overstability, and for $Q \to \infty$, may be noted; they are

$$\left.\begin{aligned} R_c^{(o)} &\to \pi^2 \frac{(1+\mathfrak{p}_2)\mathfrak{p}_1^2}{(1+\mathfrak{p}_1)\mathfrak{p}_2^2} Q \\ a_c^{(o)} &\to \left[\tfrac{1}{2}\pi^4 \frac{\mathfrak{p}_1^2}{(1+\mathfrak{p}_1)(\mathfrak{p}_1+\mathfrak{p}_2)} Q\right]^{\frac{1}{6}} \\ \sigma_1 &\to \frac{\sqrt{Q}}{\pi}\left(\frac{\mathfrak{p}_2-\mathfrak{p}_1}{1+\mathfrak{p}_1}\right)^{\frac{1}{2}} \frac{1}{\mathfrak{p}_2} \end{aligned}\right\} \quad (Q \to \infty). \tag{241}$$

In terms of the gyration frequency of the overstable oscillations, the last of the foregoing relations gives

$$|p| = \frac{\pi^2}{d^2}\nu|\sigma_1| \to \frac{\pi}{d^2}\nu\left(\frac{\mu}{4\pi\rho\nu\eta}\right)^{\frac{1}{2}} Hd \frac{1}{\mathfrak{p}_2}\left(\frac{\mathfrak{p}_2-\mathfrak{p}_1}{1+\mathfrak{p}_1}\right)^{\frac{1}{2}}, \tag{242}$$

or,

$$|p| \to \pi\left[\frac{\mathfrak{p}_2-\mathfrak{p}_1}{\mathfrak{p}_2(1+\mathfrak{p}_1)}\right]^{\frac{1}{2}} \frac{V_A}{d}, \tag{243}$$

where V_A is the Alfvén speed. According to equation (243), *the frequency of oscillation at marginal stability is essentially determined by the time required for the Alfvén wave to travel a distance equal to the depth of the layer.*

Since under terrestrial conditions the overstable case is not of much interest, extensive calculations such as those described for the rotational problem in Chapter III have not been undertaken for the present problem. For the same reason we shall omit the discussion of the problem for other boundary conditions; it can be carried out, if necessary, in a manner quite analogous to the corresponding treatment of the rotational problem in Chapter III, §§ 30 and 31.

47. The case when H and g act in different directions

We shall now consider the case when **H** and **g** act in different directions. For this purpose we must return to equations (97)–(101) in which the assumption that **H** and **g** are parallel has not yet been made.

Let **H** be inclined at an angle ϑ to the direction of the vertical. Also, let the direction of the x-axis be so chosen that **H** lies in the xz-plane. Then

$$\boldsymbol{\lambda} = (0,0,1) \quad \text{and} \quad \mathbf{H} = H(\sin\vartheta, 0, \cos\vartheta), \tag{244}$$

and equations (97) and (100) become

$$\frac{\partial h_z}{\partial t} = \eta\nabla^2 h_z + H\left(\cos\vartheta\frac{\partial}{\partial z}+\sin\vartheta\frac{\partial}{\partial x}\right)w \tag{245}$$

and

$$\frac{\partial}{\partial t}\nabla^2 w = g\alpha\left(\frac{\partial^2}{\partial x^2}+\frac{\partial^2}{\partial y^2}\right)\theta+\nu\nabla^4 w+\frac{\mu H}{4\pi\rho}\left(\cos\vartheta\frac{\partial}{\partial z}+\sin\vartheta\frac{\partial}{\partial x}\right)\nabla^2 h_z. \tag{246}$$

We shall not need equations (98) and (99), and equation (101) is unaffected.

If we seek solutions of equations (101), (245), and (246) which are independent of x, then equations (245) and (246) reduce to equations (102) and (105) with the only difference that H is replaced by $H\cos\vartheta$. Consequently, if we restrict ourselves to the onset of instability as rolls in the x-direction, then all of the discussions of the preceding sections will apply if we interpret H everywhere to mean the component of **H** in the direction of **g**. The question whether lower Rayleigh numbers can be reached by considering more general patterns of motion than rolls will now be considered; but we shall restrict our consideration to the case when instability sets in as stationary convection.

When instability sets in as stationary convection, the equations governing the marginal state are

$$\eta\nabla^2 h_z = -H\left(\cos\vartheta\frac{\partial}{\partial z}+\sin\vartheta\frac{\partial}{\partial x}\right)w \tag{247}$$

and $$\nu\nabla^4 w+\frac{\mu H}{4\pi\rho}\left(\cos\vartheta\frac{\partial}{\partial z}+\sin\vartheta\frac{\partial}{\partial x}\right)\nabla^2 h_z = -g\alpha\left(\frac{\partial^2}{\partial x^2}+\frac{\partial^2}{\partial y^2}\right)\theta. \tag{248}$$

The term in $\nabla^2 h_z$ in equation (248) can be directly eliminated by making use of equation (247); we obtain

$$\nabla^4 w-\frac{\mu H^2}{4\pi\rho\nu\eta}\left(\cos\vartheta\frac{\partial}{\partial z}+\sin\vartheta\frac{\partial}{\partial x}\right)^2 w = -\frac{g\alpha}{\nu}\left(\frac{\partial^2}{\partial x^2}+\frac{\partial^2}{\partial y^2}\right)\theta. \tag{249}$$

This equation must be considered together with the equation

$$\nabla^2\theta = -\frac{\beta}{\kappa}w. \tag{250}$$

If we measure all linear dimensions in units of the depth d of the layer, and let

$$Q = \frac{\mu H^2\cos^2\vartheta}{4\pi\rho\nu\eta}d^2 = \frac{\sigma\mu^2 H^2\cos^2\vartheta}{\rho\nu}d^2, \tag{251}$$

equations (249) and (250) become

$$\nabla^4 w-Q\left(\frac{\partial}{\partial z}+\tan\vartheta\frac{\partial}{\partial x}\right)^2 w = -\left(\frac{g\alpha d^2}{\nu}\right)\left(\frac{\partial^2}{\partial x^2}+\frac{\partial^2}{\partial y^2}\right)\theta \tag{252}$$

and $$\nabla^2\theta = -\left(\frac{\beta}{\kappa}d^2\right)w. \tag{253}$$

And eliminating θ between these equations, we have

$$\nabla^2\left[\nabla^4-Q\left(\frac{\partial}{\partial z}+\tan\vartheta\frac{\partial}{\partial x}\right)^2\right]w = R\left(\frac{\partial^2}{\partial x^2}+\frac{\partial^2}{\partial y^2}\right)w. \tag{254}$$

Following our standard procedure, we analyse the disturbances w and θ into two-dimensional waves of assigned wave numbers in the x- and the y-directions. Thus:

$$\left.\begin{aligned} w &= W(z)\exp[i(a_x x+a_y y)] \\ \theta &= \Theta(z)\exp[i(a_x x+a_y y)] \end{aligned}\right\}. \tag{255}$$

Equations (252) and (254) then become

$$[(D^2-a^2)^2-Q(D+ic)^2]W = \left(\frac{g\alpha d^2}{\nu}\right)a^2\Theta \tag{256}$$

and $$(D^2-a^2)[(D^2-a^2)^2-Q(D+ic)^2]W = -Ra^2 W, \tag{257}$$

where $$D = d/dz,\quad a^2 = a_x^2+a_y^2,\quad\text{and}\quad c = a_x\tan\vartheta. \tag{258}$$

The boundary conditions with respect to which equation (258) must be solved are: $$W = \Theta = 0 \quad\text{for } z = 0 \text{ and } 1; \tag{259}$$

and, *either* $DW = 0$ (on a rigid surface),

or $D^2W = 0$ (on a free surface). (260)

In view of equation (256), the boundary conditions (259) are equivalent to

$$W = 0 \quad \text{and} \quad [(D^2-a^2)^2-Q(D+ic)^2]W = 0 \quad \text{for } z = 0 \text{ and } 1. \tag{261}$$

Since W is complex, equation (257) together with the boundary conditions (260) and (261) constitutes a characteristic value problem in an equation effectively of order twelve. And the solution of the physical problem requires the minimum of the lowest characteristic value of equation (257) considered as a function of the two variables a and c.

It can be shown by methods with which we are now familiar that the characteristic value problem presented by equations (257), (260), and

Table XVII

Rayleigh numbers for $Q = 100$ and for various values of a and c for the case when both bounding surfaces are rigid

c	a	R	c	a	R
0·0	4·0	3,768	2·0	3·9	4,190
0·5	3·9	3,797		4·0	4,181
	4·0	3,793		4·1	4,178
	4·1	3,795	4·0	3·9	5,523
1·0	3·9	3,875		4·0	5,494
	4·0	3,878		4·1	5,474
	4·1	3,871			

(261) can be formulated in terms of a variational principle. The critical Rayleigh number R (for a given Q) is in fact the absolute minimum which can be attained by

$$\frac{\int\limits_0^1 [|DF|^2+a^2|F|^2]\,dz}{a^2\int\limits_0^1 \{|(D^2-a^2)W|^2+Q|DW+icW|^2\}\,dz}, \tag{262}$$

where

$$F = [(D^2-a^2)^2-Q(D+ic)^2]W. \tag{263}$$

By the variational method, R has been evaluated for a number of values of a and c for several different values of Q and for different boundary conditions. A sample set from such calculations is shown in Table XVII. In this table the entry for $c = 0$ is for the value of a for which it is known from the results given in Table XV that the minimum Rayleigh number occurs when the pattern of convection which appears at marginal stability is in the form of *longitudinal rolls* (i.e. rolls extended in directions parallel to the plane containing **H** and **g**). Thus, for a layer of liquid confined between two rigid planes, the lowest Rayleigh number is obtained for $a = 4{\cdot}00$ when $Q = 100$, and convection sets in as longitudinal rolls. From the results given in Table XVII, it follows that for c slightly different from zero, the minimum Rayleigh number still occurs for $a = 4{\cdot}00$; also that the new minimum slightly exceeds that for $c = 0$. For example, for $c = 0{\cdot}5$ and $Q = 100$, the minimum Rayleigh number is 3,793; this should be contrasted with $R_c = 3{,}768$ for $c = 0$. For larger values of c, the minimum appears to shift to somewhat higher values of a; but the Rayleigh number now exceeds R_c at $c = 0$ by a substantial margin. Thus, in the example considered, the minimum Rayleigh number for $c = 1{\cdot}0$ is 3,871 and this occurs for $a = 4{\cdot}1$. The results for other values of Q and other boundary conditions exhibit the same general behaviour and confirm that *when* **H** *and* **g** *act in different directions, convection when it first sets in appears as longitudinal rolls.*

In some ways the conclusion arrived at in the last paragraph is a paradoxical one. For when **H** and **g** are parallel, the Rayleigh number depends only on $a^2 = a_x^2 + a_y^2$ and we expect convection at marginal stability to have a cellular pattern. How then, one may ask, does this situation change discontinuously when **H** is inclined only very slightly to the vertical and when it is asserted convection at marginal stability must appear as longitudinal rolls? The resolution of this paradox is as follows.

When **H** and **g** act in different directions, differing Rayleigh numbers are required for the marginal appearance of convection of pre-assigned patterns. In particular, the most 'difficult' pattern to excite (other things being equal) is the system of the *transverse* rolls. Once the Rayleigh number is high enough to excite these transverse rolls, we may expect a proper cellular pattern of convection to emerge. The *difference* in the minimum Rayleigh numbers required to excite the longitudinal and the transverse rolls is, therefore, a measure of the extent to which a proper cellular pattern is suppressed at marginal stability when longitudinal rolls appear. But, and this is the main point of the argument, this difference in the two Rayleigh numbers tends to zero as c $(= a_x \tan\vartheta \leqslant a \tan\vartheta)$ tends to zero. In other words, when **H** is only very slightly inclined to the direction of gravity, the extent to which *transverse* rolls are suppressed at marginal stability (when *longitudinal* rolls appear) is also only very slight. Finally, when **H** is exactly parallel to **g** (or, when $|\mathbf{H}| \to 0$), this suppression of the cellular pattern ceases and longitudinal and transverse rolls appear simultaneously at marginal stability.

The picture we have thus arrived at as to what really 'happens' when **H** and **g** act in different directions constitutes an essential simplification for the general theory of thermal instability. For it effectively ensures that conclusions reached after an examination of special geometrical situations (such as **H** and **g** being parallel) are of wider scope and generality than the geometrical restrictions, under which they were arrived at, would appear to warrant. Thus in Chapter III, the principal results on the effect of rotation on thermal instability were derived for the case when **Ω** and **g** are parallel. One may now feel confident that if **Ω** and **g** are not parallel, convection at marginal stability will manifest itself as longitudinal rolls (cf. Chapter III, § 34).

48. Experiments on the inhibition of thermal convection by a magnetic field

Experiments on the onset of thermal instability in layers of mercury heated from below and subject to magnetic fields have been carried out by Nakagawa, Jirlow, and Lehnert and Little. A brief account of some of these experiments will be given.

In Nakagawa's experiments, the magnetic field was provided by a reconditioned electromagnet of a disused 36½-in. cyclotron of the Enrico Fermi Institute of the University of Chicago. The electromagnet provided a uniform magnetic field in a cylindrical volume, 78 cm in diameter and 22 cm in height; the strength of the field could be varied up to a maximum of 13,000 gauss.

The experimental arrangement was similar to that used in the experiments on overstable convection in rotating mercury described in Chapter III, § 35 (*b*). The critical temperature gradient at which instability set in was determined by the Schmidt–Milverton method. By working with layers of mercury 3, 4, 5, and 6 cm in depth and magnetic fields varying in strength from 250 to 8,000 gauss, Nakagawa was able to carry out experiments for Q-values in the range 10^2–10^6. The results of the experiments together with the appropriate theoretical relation is shown in Fig. 43. It will be seen that the agreement of the experimental results with the theoretical predictions is very satisfactory.

In another set of experiments, Nakagawa took a large number of streak photographs of the cell patterns which developed on the top surface at instability. An example is shown in Fig. 44. (The arrangements used in these experiments are described in Chapter V, § 54 (*b*).) The

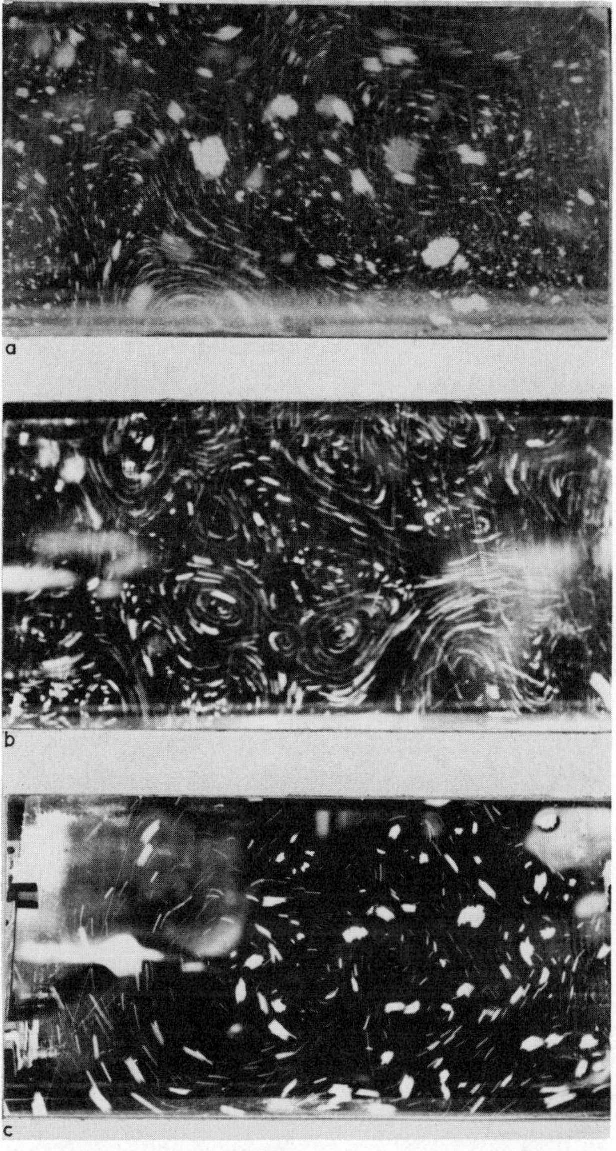

FIG. 44. Examples of streak photographs of the convective motion on the surface of mercury obtained for three different strengths of the magnetic field; (a) $H = 125$ G, $Q_1 = 9{\cdot}46$; (b) $H = 750$ G, $Q_1 = 3{\cdot}49\times10^2$; (c) $H = 3{,}000$ G, $Q_1 = 5{\cdot}76\times10^3$.

dimensions of the cells were estimated by measurements made on these photographs. The method was the following. As representative of the dimensions, the distances between the centres of adjacent cells were measured. It was found that in most cases they formed equilateral triangles. This is consistent with the supposition that the basic cell pattern is hexagonal.

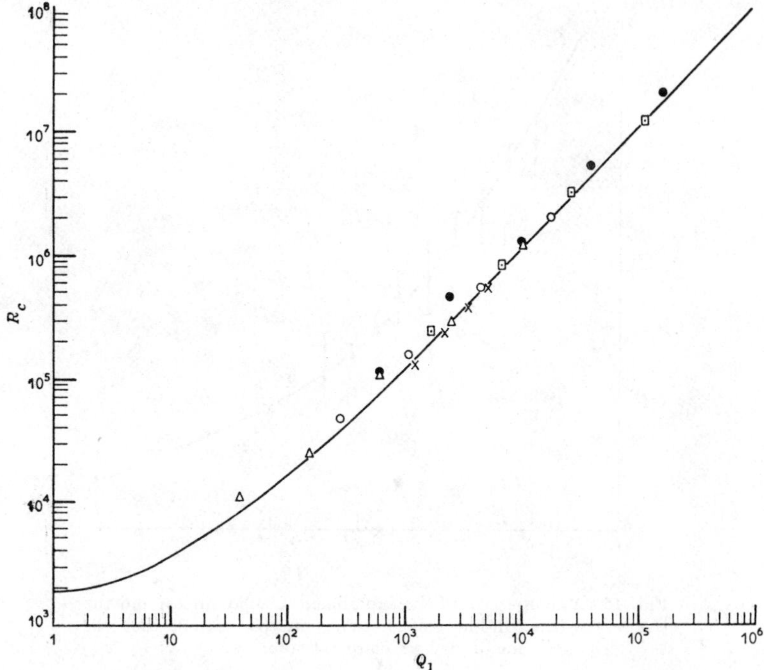

FIG. 43. A comparison of the experimental and theoretical results on the critical Rayleigh numbers for the onset of instability. The theoretical (R_c, Q_1)-relation is shown by the full-line curve. The solid circles (●), squares (⊡), open circles (○), and triangles (△) are the experimentally determined points with the $36\frac{1}{2}$-in. magnet for layers of depths $d = 6$, 5, 4, and 3 cm, respectively; the four crosses represent the result with a small magnet with $H = 1{,}500$ G and $d = 6$, 5, 4, and 3 cm, respectively.

The side L of the unit hexagon is related to the wave number a of the associated disturbance by (see § 16 (c), equation (252))

$$L = \frac{4\pi d}{3a}, \tag{264}$$

and the distance b between the centres of adjacent cells is $L\sqrt{3}$ (see Fig. 7 a, p. 49). Hence

$$b = 4\pi d/a\sqrt{3}. \tag{265}$$

Since a is known as a function of Q, the theory predicts a definite relation between b and Q; this relation together with the results of Nakagawa's measurements are shown in Fig. 45. It will be seen that the agreement between the measurements and the theory is again satisfactory.

Lehnert and Little, in their experiments, have verified another important aspect of the theoretical predictions. We have seen that when the

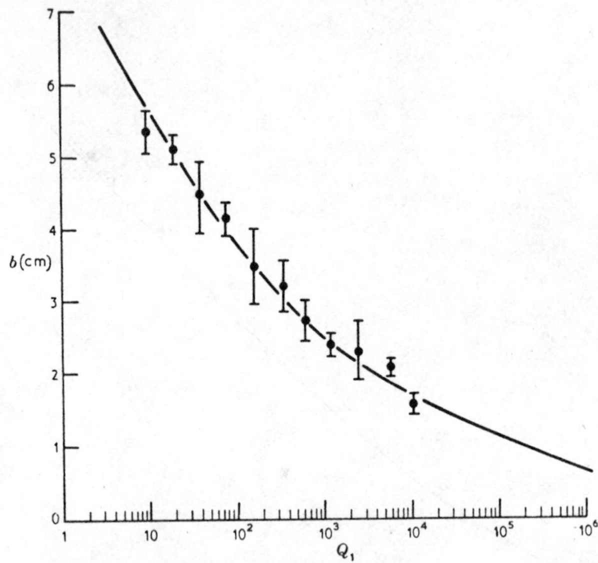

FIG. 45. A comparison of the experimental (solid circles) and the theoretical (the full-line curve) results on the sizes of the cells manifested at marginal stability.

direction of the impressed magnetic field is different from the vertical, only the component of the magnetic field in the direction of the vertical is effective; and further, that the cells which appear at onset must be longitudinal rolls in directions parallel to the plane containing $\mathbf{H}$ and $\mathbf{g}$. By using homogeneous oblique magnetic fields, Lehnert and Little were able to verify these predictions in a very striking way. They found, for example, that a magnetic field as strong as 4,500 gauss impressed in a horizontal direction had no discernible effect in inhibiting convection even though the field was five times that necessary to suppress convection, if acting in the vertical direction. Lehnert and Little further found that the pattern of convection which emerged under these conditions was, indeed, in the form of elongated cells extending across the entire vessel with the streamlines running, principally, parallel with the magnetic field. Fig. 46, which is a reproduction of one of their photographs, illustrates the phenomenon.

BIBLIOGRAPHICAL NOTES

The following general references on the subject of hydromagnetics may be noted:

1. T. G. COWLING, *Magnetohydrodynamics*, Interscience Tracts on Physics and Astronomy, No. 4, Interscience Publishers, Inc., New York, 1957.
2. W. M. ELSASSER, 'Hydromagnetism. I. A review', *American J. of Phys.* **23**, 590–609 (1955); 'Hydromagnetism. II. A review', ibid. **24**, 85–110 (1956).
3. S. LUNDQUIST, 'Studies in magneto-hydrodynamics', *Arkiv för Fysik*, **5**, 297–347 (1952).
4. G. H. A. COLE, 'Some aspects of magnetohydrodynamics', *Advances in Phys.* **5**, 452–97 (1956).

§ 38 (*a*). On the decay of magnetic fields in a conducting fluid sphere and the interaction of such fields with the prevailing motions, the fundamental papers are:

5. W. M. ELSASSER, 'Induction effects in terrestrial magnetism, Part I. Theory', *Physical Rev.* **69**, 106–16 (1946).
6. —— 'Induction effects in terrestrial magnetism, Part II. The secular variation', ibid. **70**, 202–12 (1946).
7. —— 'Induction effects in terrestrial magnetism, Part III. Electric modes', ibid. **72**, 821–33 (1947).

The decay of magnetic fields under astrophysical conditions is considered in:

8. T. G. COWLING, 'On the sun's general magnetic field', *Monthly Notices Roy. Astron. Soc. London*, **105**, 166–74 (1945).
9. —— 'The growth and decay of the sunspot magnetic field', ibid. **106**, 218–24 (1946).

On the related subject of dynamo action and the maintenance of magnetic fields against decay by fluid motions, see:

10. T. G. COWLING, 'The magnetic field of sunspots', *Monthly Notices Roy. Astron. Soc. London*, **94**, 39–48 (1933).
11. W. M. ELSASSER, 'The earth's interior and geomagnetism', *Rev. Modern Phys.* **22**, 1–35 (1950).
12. —— 'Hydromagnetic dynamo theory', ibid. **28**, 135–63 (1956).
13. E. C. BULLARD and H. GELLMAN, 'Homogeneous dynamos and terrestrial magnetism,' *Philos. Trans. Roy. Soc.* (*London*) A, **247**, 213–78 (1954).
14. G. E. BACKUS and S. CHANDRASEKHAR, 'On Cowling's theorem on the impossibility of self-maintained axisymmetric homogeneous dynamos', *Proc. Nat. Acad. Sci.* **42**, 105–9 (1956).
15. T. G. COWLING and A. HARE, 'Two-dimensional problems of the decay of magnetic fields in magnetohydrodynamics', *Quart. J. Mech. Appl. Math.* **10**, 385–405 (1957).

§ 38 (*b*). In his discussion of various physical problems, Alfvén has extensively used the fact that in a medium of infinite electrical conductivity, the magnetic lines of force behave as though they are permanently attached to the fluid elements. Many of Alfvén's ideas are described in:

16. H. ALFVÉN, *Cosmical Electrodynamics*, International Series of Monographs on Physics, Oxford, England, 1950.

See also:

17. T. G. COWLING, 'Solar electrodynamics', *The Sun*, chapter 8, edited by G. P. Kuiper, University of Chicago Press, Chicago, 1953.
18. J. W. DUNGEY, *Cosmic Electrodynamics*, Cambridge Monographs on Mechanics and Applied Mathematics, Cambridge, England, 1958.

§ 39. Alfvén's discovery of the waves known after him is announced in:

19. H. ALFVÉN, 'The existence of electromagnetic-hydrodynamic waves', *Nature*, **150**, 405–6 (1942).
20. —— 'On the existence of electromagnetic-hydrodynamic waves', *Arkiv f. mat. astr. o. Fysik*, **29**, 1–7, (1942).

An experimental demonstration of the Alfvén waves is described by:

21. S. LUNDQUIST, 'Experimental investigations of magneto-hydrodynamic waves', *Phys. Rev.* **76**, 1805–9 (1949).

Related experiments are described in:

22. B. LEHNERT, 'On the behaviour of an electrically conductive liquid in a magnetic field', *Arkiv för Fysik*, **5**, 69–90 (1952).
23. —— 'Experiments on non-laminar flow of mercury in presence of a magnetic field', *Tellus*, **4**, 63–67 (1952).
24. —— 'Magneto-hydrodynamic waves in liquid sodium', *Phys. Rev.* **94**, 815–24 (1954).

The propagation of hydromagnetic waves in a compressible medium is considered in:

25. H. C. VAN DE HULST, 'Interstellar polarization and magneto-hydrodynamic waves', *Problems of Cosmical Aerodynamics*, International Union of Theoretical and Applied Mechanics and International Astronomical Union, 45–56, Central Air Documents Office, Dayton, Ohio, 1951.
26. N. HERLOFSON, 'Magneto-hydrodynamic waves in a compressible fluid conductor', *Nature*, **165**, 1020–1 (1950).

§ 40. The stability of the equipartition solution considered in § 40 (a) is proved in:

27. S. CHANDRASEKHAR, 'On the stability of the simplest solution of the equations of hydromagnetics', *Proc. Nat. Acad. Sci.* **42**, 273–6 (1956).

Some references on force-free magnetic fields are:

28. S. LUNDQUIST, 'Magneto-hydrostatic fields', *Arkiv för Fysik*, **2**, 361–5 (1950).
29. R. LÜST and A. SCHLÜTER, 'Kraftfreie Magnetfelder', *Z. f. Astrophysik*, **34**, 263–82 (1954).
30. S. CHANDRASEKHAR, 'On force-free magnetic fields', *Proc. Nat. Acad. Sci.* **42**, 1–5, (1956).
31. —— and P. C. KENDALL, 'On force-free magnetic fields', *Astrophys. J.* **126**, 457–60 (1957).
32. L. WOLTJER, 'The Crab nebula', *Bull. Astr. Netherlands*, **14**, 39–80 (1958).

FIG. 46. Cellular convection in a layer of mercury in a magnetic field parallel to the free surface of the layer. This picture taken by Lehnert and Little shows the surface as seen from above. The magnetic field ($H = 4{,}500$ G) runs from the right to the left and the cells are seen to be elongated in the field direction and to extend across the whole vessel.

§ 41. The effect of an impressed magnetic field on the onset of thermal instability in electrically conducting fluids is considered in:

33. W. B. Thompson, 'Thermal convection in a magnetic field', *Phil. Mag.* Ser. 7, **42,** 1417–32 (1951).
34. S. Chandrasekhar, 'On the inhibition of convection by a magnetic field', ibid. **43,** 501–32 (1952).

§§ 42–46. The analyses in these sections are largely based on (34). The thermodynamic significance of the variational principle is discussed in:

35. S. Chandrasekhar, 'The thermodynamics of thermal instability in liquids', *Max Planck Festschrift 1958*, 103–14, Veb Deutscher Verlag der Wissenschaften, Berlin, 1958.

§ 47. This section is based on:

36. S. Chandrasekhar, 'On the inhibition of convection by a magnetic field. II', *Phil. Mag.* Ser. 7, **45,** 1177–91 (1954).

§ 48. The references for the experimental work described in this section are:

37. Y. Nakagawa, 'An experiment on the inhibition of thermal convection by a magnetic field,' *Nature,* **175,** 417–19 (1955).
38. Y. Nakagawa, 'Experiments on the inhibition of thermal convection by a magnetic field', *Proc. Roy. Soc.* (*London*) A, **240,** 108–13 (1957).
39. —— 'Experiments on the instability of a layer of mercury heated from below and subject to the simultaneous action of a magnetic field and rotation. II,' ibid. **249,** 138–45 (1959).
40. —— 'Apparatus for studying convection under the simultaneous action of a magnetic field and rotation,' *The Review of Scientific Instruments,* **28,** 603–9 (1957).
41. B. Lehnert and N. C. Little, 'Experiments on the effect of inhomogeneity and obliquity of a magnetic field in inhibiting convection', *Tellus,* **9,** 97–103 (1957).

See also:

42. K. Jirlow, 'Experimental investigation of the inhibition of convection by a magnetic field', ibid. **8,** 252–3 (1956).

V

THE THERMAL INSTABILITY OF A LAYER OF FLUID HEATED FROM BELOW

4. THE EFFECT OF ROTATION AND MAGNETIC FIELD

49. The like and the contrary effects of rotation and magnetic field on fluid behaviour

In the last two chapters we have studied the effect of rotation and magnetic field, acting separately, on the onset of thermal instability in layers of fluid heated from below. In some respects the effects are remarkably alike: they both inhibit the onset of instability; and they both elongate the cells which appear at marginal stability. These effects have likewise a common origin: in the Taylor–Proudman theorem in the case of rotation and in its exact analogue in the case of a magnetic field. On these accounts one must not suppose that acting together, rotation and magnetic field will reinforce each other; on the contrary, in other respects they will tend to oppose each other. Thus, we have seen that viscosity facilitates the onset of instability when rotation is present; and we have also seen that a magnetic field imparts to the fluid certain aspects of viscosity. Consequently, even though the two acting separately inhibit the onset of instability, they will have conflicting tendencies when acting together. Again, there is a fundamental difference in the character of the motions which prevail when rotation alone is present and when a magnetic field alone is present. Rotation induces a component of vorticity in the direction of $\mathbf{\Omega}$, and the effects arising from it are predominant; for large Taylor numbers it results in the streamlines becoming closely wound spirals with motions principally confined to planes transverse to $\mathbf{\Omega}$. But a magnetic field does not induce any similar component of vorticity and there are no comparable effects; instead, for large Q-numbers, the motions transverse to $\mathbf{H}$ are much reduced and the motions along the magnetic lines of force become predominant.

There is a further fact to remember: in liquid metals such as mercury, instability sets in mostly as overstability when rotation is present; but it sets in as stationary convection when a magnetic field is present. For all these reasons, the study of thermal instability in the presence of both rotation and magnetic field is an instructive one.

50. The propagation of hydromagnetic waves in a rotating fluid

In Chapters III and IV we saw that the propagation of waves in a rotating fluid and in the presence of a magnetic field bear on the respective problems in thermal instability. We shall see that the propagation of hydromagnetic waves in a rotating fluid has a similar bearing on the problem to be investigated in this chapter.

Consider an incompressible fluid in a state of uniform rotation with an angular velocity $\mathbf{\Omega}$. In a frame of reference rotating with the angular velocity $\mathbf{\Omega}$, the basic equations are (cf. equations III (44) and IV (10)):

$$\frac{\partial u_i}{\partial t}+u_j\frac{\partial u_i}{\partial x_j}-\frac{\mu H_j}{4\pi\rho}\frac{\partial H_i}{\partial x_j}$$
$$= 2\epsilon_{ijk}u_j\Omega_k+\nu\nabla^2 u_i-\frac{\partial}{\partial x_i}\left(\frac{p}{\rho}+\mu\frac{|\mathbf{H}|^2}{8\pi\rho}-\tfrac{1}{2}|\mathbf{\Omega}\times\mathbf{r}|^2+V\right), \quad (1)$$

$$\frac{\partial H_i}{\partial t}+u_j\frac{\partial H_i}{\partial x_j}-H_j\frac{\partial u_i}{\partial x_j} = \eta\nabla^2 H_i, \quad (2)$$

$$\frac{\partial u_i}{\partial x_i} = 0, \quad \text{and} \quad \frac{\partial H_i}{\partial x_i} = 0. \quad (3)$$

For an inviscid fluid of zero resistivity, the equations governing small departures from an initial state in which a uniform magnetic field H_j prevails are

$$\frac{\partial u_i}{\partial t}-\frac{\mu H_j}{4\pi\rho}\frac{\partial h_i}{\partial x_j} = -\frac{\partial}{\partial x_i}\delta\varpi+2\epsilon_{ijk}u_j\Omega_k, \quad (4)$$

$$\frac{\partial h_i}{\partial t} = H_j\frac{\partial u_i}{\partial x_j}, \qquad \frac{\partial u_i}{\partial x_i} = 0, \quad \text{and} \quad \frac{\partial h_i}{\partial x_i} = 0, \quad (5)$$

where h_i denotes the departure of the magnetic field from H_i.

We now seek solutions of equations (4) and (5) whose space-time dependence is given by

$$e^{i(\omega t+k_j x_j)}. \quad (6)$$

Equations (4) and (5) then give

$$\omega u_i-\frac{\mu(H_j k_j)}{4\pi\rho}h_i = -k_i\delta\varpi-2i\epsilon_{ijk}u_j\Omega_k, \quad (7)$$

$$\omega h_i = (H_j k_j)u_i, \quad (8)$$

$$u_j k_j = 0, \quad \text{and} \quad h_j k_j = 0. \quad (9)$$

Equation (9) shows that *the waves are transverse*.

Eliminating h_i between equations (7) and (8), we obtain

$$nu_i+k_i\delta\varpi+2i\epsilon_{ijk}u_j\Omega_k = 0, \quad (10)$$

where
$$n = \frac{1}{\omega}(\omega^2 - \omega_A^2) \tag{11}$$

and
$$\omega_A = \left(\frac{\mu}{4\pi\rho}\right)^{\frac{1}{2}}(H_j k_j). \tag{12}$$

It will be observed that equation (10) is identical with equation III (61); and the solution obtained there applies here. Thus, by choosing the orientation of the coordinate system such that the z-axis is the direction of $\mathbf{k}$ and the (x, z)-plane contains $\mathbf{\Omega}$, we have (cf. equations III (71) and (73)),

$$n = \pm 2\Omega \cos\vartheta, \tag{13}$$

$$u_x = \pm i u_y, \quad u_z = 0, \quad \text{and} \quad \delta\varpi = \mp\frac{2\Omega}{k} u_x \sin\vartheta, \tag{14}$$

where ϑ is the inclination of the direction of wave propagation to the direction of $\mathbf{\Omega}$. The corresponding amplitudes of the perturbations in the magnetic field follow from equation (8); we have

$$h_x = \frac{k_j H_j}{\omega} u_x, \quad h_x = \pm i h_y, \quad \text{and} \quad h_z = 0. \tag{15}$$

Now combining equations (11) and (13), we have

$$\omega^2 \mp (2\Omega\cos\vartheta)\omega - \omega_A^2 = 0. \tag{16}$$

Hence,
$$\omega = \pm[\Omega\cos\vartheta \pm \sqrt{(\Omega^2\cos^2\vartheta + \omega_A^2)}]. \tag{17}$$

If ω_1 and ω_2 are the two roots of equation (16), then

$$\omega_1 + \omega_2 = \pm 2\Omega\cos\vartheta, \qquad \omega_1\omega_2 = -\omega_A^2. \tag{18}$$

The velocity of propagation of the waves is given by

$$V = \frac{\omega}{k} = \pm\frac{1}{k}[\Omega\cos\vartheta \pm \sqrt{(\Omega^2\cos^2\vartheta + \omega_A^2)}]. \tag{19}$$

This is the phase velocity. The corresponding group velocity is given by (cf. equations III (75) and IV (64))

$$\frac{\partial\omega}{\partial\mathbf{k}} = \pm\left\{\frac{\mathbf{k}\times(\mathbf{\Omega}\times\mathbf{k})}{k^3} \pm \frac{\Omega\cos\vartheta[\mathbf{k}\times(\mathbf{\Omega}\times\mathbf{k})]k^{-3} + \omega_A \mathbf{H}(\mu/4\pi\rho)^{\frac{1}{2}}}{\sqrt{(\Omega^2\cos^2\vartheta + \omega_A^2)}}\right\}. \tag{20}$$

51. The perturbation equations

We consider an infinite horizontal layer of fluid in a state of uniform rotation with an angular velocity $\mathbf{\Omega}$ and subject to a uniform magnetic field $\mathbf{H}$. We envisage an initial state in which a steady adverse temperature gradient β is maintained and there are no motions. Examining the stability of this state, we find, by following the same procedures as in the last chapters, that the basic perturbation equations are (cf. equations III (78)–(80) and IV (87)–(92)):

$$\frac{\partial u_i}{\partial t} = -\frac{\partial}{\partial x_i}(\delta\varpi)+g\alpha\theta\lambda_i+\nu\nabla^2 u_i+2\epsilon_{ijk}u_j\Omega_k+\frac{\mu H_j}{4\pi\rho}\frac{\partial h_i}{\partial x_j}, \tag{21}$$

$$\frac{\partial h_i}{\partial t} = H_j\frac{\partial u_i}{\partial x_j}+\eta\nabla^2 h_i, \tag{22}$$

$$\frac{\partial\theta}{\partial t} = \beta w+\kappa\nabla^2\theta, \tag{23}$$

$$\frac{\partial u_i}{\partial x_i} = 0, \quad \text{and} \quad \frac{\partial h_i}{\partial x_i} = 0. \tag{24}$$

By taking the curl of equation (21), we can eliminate the term in $\delta\varpi$; we thus obtain (cf. equations III (82) and IV (93))

$$\frac{\partial\omega_i}{\partial t} = g\alpha\epsilon_{ijk}\frac{\partial\theta}{\partial x_j}\lambda_k+\nu\nabla^2\omega_i+2\Omega_j\frac{\partial u_i}{\partial x_j}+\frac{\mu H_j}{4\pi\rho}\frac{\partial \upsilon_i}{\partial x_j}, \tag{25}$$

where $$\boldsymbol{\omega} = \text{curl}\,\mathbf{u} \quad \text{and} \quad \boldsymbol{\upsilon} = \text{curl}\,\mathbf{h}. \tag{26}$$

Taking the curl of equation (25) once again, we obtain (cf. equations III (83) and IV (95))

$$\frac{\partial}{\partial t}\nabla^2 u_i = g\alpha\left(\lambda_i\nabla^2\theta-\lambda_j\frac{\partial^2\theta}{\partial x_j\,\partial x_i}\right)+\nu\nabla^4 u_i-2\Omega_j\frac{\partial\omega_i}{\partial x_j}+\frac{\mu H_j}{4\pi\rho}\frac{\partial}{\partial x_j}\nabla^2 h_i. \tag{27}$$

Now multiplying equations (25) and (27) by λ_i, we get

$$\frac{\partial\zeta}{\partial t} = \nu\nabla^2\zeta+2\Omega_j\frac{\partial w}{\partial x_j}+\frac{\mu H_j}{4\pi\rho}\frac{\partial\xi}{\partial x_j} \tag{28}$$

and $$\frac{\partial}{\partial t}\nabla^2 w = g\alpha\left(\frac{\partial^2\theta}{\partial x^2}+\frac{\partial^2\theta}{\partial y^2}\right)+\nu\nabla^4 w-2\Omega_j\frac{\partial\zeta}{\partial x_j}+\frac{\mu H_j}{4\pi\rho}\frac{\partial}{\partial x_j}\nabla^2 h_z, \tag{29}$$

where ζ and $\xi/4\pi$ are the z-components of the vorticity and the current density, respectively. Equations (28) and (29) replace equations IV (99) and (100); the remaining equations IV (96)–(98) are unaffected.

We shall restrict our discussion of this problem to the case when $\mathbf{g}$, $\mathbf{H}$, and $\boldsymbol{\Omega}$ act in the same direction. In this case the relevant equations are:

$$\frac{\partial\theta}{\partial t} = \kappa\nabla^2\theta+\beta w, \tag{30}$$

$$\frac{\partial h_z}{\partial t} = \eta\nabla^2 h_z+H\frac{\partial w}{\partial z}, \tag{31}$$

$$\frac{\partial\xi}{\partial t} = \eta\nabla^2\xi+H\frac{\partial\zeta}{\partial z}, \tag{32}$$

$$\frac{\partial\zeta}{\partial t} = \nu\nabla^2\zeta+2\Omega\frac{\partial w}{\partial z}+\frac{\mu H}{4\pi\rho}\frac{\partial\xi}{\partial z}, \tag{33}$$

and $$\frac{\partial}{\partial t}(\nabla^2 w) = g\alpha\left(\frac{\partial^2\theta}{\partial x^2}+\frac{\partial^2\theta}{\partial y^2}\right)+\nu\nabla^4 w-2\Omega\frac{\partial\zeta}{\partial z}+\frac{\mu H}{4\pi\rho}\frac{\partial}{\partial z}\nabla^2 h_z; \tag{34}$$

and we must seek solutions of these equations which satisfy the boundary conditions described in II, § 9 (*a*) and IV, § 42 (*a*).

Analysing the disturbances into two-dimensional waves, and considering disturbances characterized by a particular wave number, we find that equations (30)–(34) give

$$(D^2-a^2-\mathfrak{p}_1\sigma)\Theta = -\left(\frac{\beta d^2}{\kappa}\right)W, \tag{35}$$

$$(D^2-a^2-\mathfrak{p}_2\sigma)K = -\left(\frac{Hd}{\eta}\right)DW, \tag{36}$$

$$(D^2-a^2-\mathfrak{p}_2\sigma)X = -\left(\frac{Hd}{\eta}\right)DZ, \tag{37}$$

$$(D^2-a^2-\sigma)Z = -\left(\frac{2\Omega d}{\nu}\right)DW-\left(\frac{\mu Hd}{4\pi\rho\nu}\right)DX, \tag{38}$$

and

$$(D^2-a^2)(D^2-a^2-\sigma)W+\left(\frac{\mu Hd}{4\pi\rho\nu}\right)D(D^2-a^2)K-\left(\frac{2\Omega d^3}{\nu}\right)DZ = \left(\frac{g\alpha d^2}{\nu}\right)a^2\Theta, \tag{39}$$

where the notation is the same as in equations IV (118)–(123).

Eliminating X between equations (37) and (38), and similarly K between equations (36) and (39), we obtain

$$[(D^2-a^2-\sigma)(D^2-a^2-\mathfrak{p}_2\sigma)-QD^2]Z = -\left(\frac{2\Omega d}{\nu}\right)D(D^2-a^2-\mathfrak{p}_2\sigma)W$$

and

$$(D^2-a^2)[(D^2-a^2-\sigma)(D^2-a^2-\mathfrak{p}_2\sigma)+QD^2]W- \\ -\left(\frac{2\Omega d^3}{\nu}\right)D(D^2-a^2-\mathfrak{p}_2\sigma)Z = \left(\frac{g\alpha d^2}{\nu}\right)a^2(D^2-a^2-\mathfrak{p}_2\sigma)\Theta. \tag{41}$$

(40)

Eliminating Z between these last two equations, we have

$$\{(D^2-a^2)[(D^2-a^2-\sigma)(D^2-a^2-\mathfrak{p}_2\sigma)-QD^2]^2+TD^2(D^2-a^2-\mathfrak{p}_2\sigma)^2\}W \\ = \left(\frac{g\alpha d^2}{\nu}\right)a^2[(D^2-a^2-\sigma)(D^2-a^2-\mathfrak{p}_2\sigma)-QD^2](D^2-a^2-\mathfrak{p}_2\sigma)\Theta. \tag{42}$$

Finally, eliminating Θ between equations (35) and (42) we obtain

$$(D^2-a^2-\mathfrak{p}_1\sigma)\{(D^2-a^2)[(D^2-a^2-\sigma)(D^2-a^2-\mathfrak{p}_2\sigma)-QD^2]^2+ \\ +TD^2(D^2-a^2-\mathfrak{p}_2\sigma)^2\}W \\ = -Ra^2[(D^2-a^2-\sigma)(D^2-a^2-\mathfrak{p}_2\sigma)-QD^2](D^2-a^2-\mathfrak{p}_2\sigma)W. \tag{43}$$

Solutions of the foregoing equations must be sought which satisfy the boundary conditions given in equations II (95) and (96) and IV (124) and (125).

It may be recalled here that all equations following (31) apply only if

$\mathbf{g}$, H, and $\mathbf{\Omega}$ are parallel. The generalization to the case when $\mathbf{g}$, $\mathbf{H}$, and $\mathbf{\Omega}$ are coplanar is immediate. By restricting oneself to the onset of convection in the form of infinitely extended rolls (along directions parallel to the plane containing $\mathbf{g}$, $\mathbf{H}$, and $\mathbf{\Omega}$), instead of cells, one can readily verify that the same set of equations applies to this more general case if we interpret H and Ω in all the equations to mean the components of $\mathbf{H}$ and $\mathbf{\Omega}$ in the direction of the vertical. The restriction to the onset of convection in the form of rolls when $\mathbf{g}$, $\mathbf{H}$, and $\mathbf{\Omega}$ are coplanar can very likely be justified in the way in which the same restriction was justified in Chapter IV, § 47 for the case when $\mathbf{H}$ alone was acting. However, when $\mathbf{g}$, $\mathbf{H}$, and $\mathbf{\Omega}$ are not coplanar, the solution of the problem will require an explicit minimizing of the Rayleigh number as a function of the two wave numbers characterizing the disturbance in two directions at right angles to one another and in the horizontal plane. But it is unlikely that the solution will exhibit any essentially new feature not already disclosed by the solution for the case when $\mathbf{g}$, $\mathbf{H}$, and $\mathbf{\Omega}$ are parallel.

52. The case when instability sets in as stationary convection

We shall first consider the case when instability sets in as ordinary convection. The case when it sets in as overstability is considered in the following section.

When instability sets in as ordinary convection, the marginal state will be characterized by $\sigma = 0$; and the basic equations are:

$$(D^2-a^2)\Theta = -\left(\frac{\beta d^2}{\kappa}\right)W, \tag{44}$$

$$[(D^2-a^2)^2-QD^2]Z = -\left(\frac{2\Omega d}{\nu}\right)D(D^2-a^2)W, \tag{45}$$

$$(D^2-a^2)K = -\left(\frac{Hd}{\eta}\right)DW, \tag{46}$$

$$(D^2-a^2)X = -\left(\frac{Hd}{\eta}\right)DZ, \tag{47}$$

and
$$(D^2-a^2)\{[(D^2-a^2)^2-QD^2]^2+TD^2(D^2-a^2)\}W = -Ra^2[(D^2-a^2)^2-QD^2]W. \tag{48}$$

According to equations II (95) and (96) and IV (124) and (125), the boundary conditions with respect to which equations (44)–(48) must be solved are:

$$W = 0 \quad \text{and} \quad \Theta = 0 \quad \text{for } z = 0 \text{ and } 1, \tag{49}$$

$$\left.\begin{array}{lllll} \textit{either} & DW = 0 & \text{and} & Z = 0 & \text{(on a rigid boundary)} \\ \textit{or} & D^2W = 0 & \text{and} & DZ = 0 & \text{(on a free boundary)} \end{array}\right\}, \tag{50}$$

and *either* $DX = 0$ and $K = 0$ (on a perfectly conducting boundary),
or $X = 0$ (on a boundary adjoining a non-conducting medium). (51)

The foregoing equations and boundary conditions constitute a characteristic value problem in a differential equation of order ten. In spite of a variational formulation which is possible (see Appendix II), the problem is clearly of considerable complexity. For this reason, we shall restrict ourselves to the case when both the bounding surfaces are free and the medium adjoining the fluid is non-conducting. While this is admittedly an artificial case to consider, we do not expect to lose any of the essential features of the problem on this account; for, we have seen in Chapters III and IV that the solutions obtained for different sets of boundary conditions all exhibit the same features and show a like dependence on the parameters of the problem such as $\mathfrak{p}$, T, and Q.

(*a*) *The solution for the case of two free boundaries*

As in Chapter III, § 27 (*a*) and Chapter IV, § 44 (*a*), it can be shown that for this case, the proper solutions of equations (44)–(48) are given by

$$W = W_0 \sin n\pi z, \tag{52}$$

$$\Theta = \frac{\beta}{\kappa} \frac{d^2}{n^2\pi^2+a^2} W_0 \sin n\pi z, \tag{53}$$

$$Z = \frac{2\Omega d}{\nu} \frac{n\pi(n^2\pi^2+a^2)}{(n^2\pi^2+a^2)^2+Qn^2\pi^2} W_0 \cos n\pi z, \tag{54}$$

$$K = \frac{Hd}{\eta} \frac{n\pi}{n^2\pi^2+a^2} W_0 \cos n\pi z + \text{constant} \times \cosh az, \tag{55}$$

and
$$X = -\frac{2\Omega d}{\nu} \frac{Hd}{\eta} \frac{n^2\pi^2}{(n^2\pi^2+a^2)^2+Qn^2\pi^2} W_0 \sin n\pi z, \tag{56}$$

where n is an integer. The corresponding characteristic equation is obtained by substituting the solution (52) for W in equation (48). We find

$$R = \frac{(n^2\pi^2+a^2)\{[(n^2\pi^2+a^2)^2+Qn^2\pi^2]^2+Tn^2\pi^2(n^2\pi^2+a^2)\}}{a^2[(n^2\pi^2+a^2)^2+Qn^2\pi^2]}. \tag{57}$$

Letting $a^2 = \pi^2 x$, we can rewrite equation (57) in the form

$$R = \pi^4 \frac{n^2+x}{x}\left[(n^2+x)^2+\frac{Qn^2}{\pi^2}\right] + \frac{T}{x} \frac{1}{1/n^2+Q/\pi^2(n^2+x)^2}. \tag{58}$$

From this equation it follows that instability first sets in for the lowest mode $n = 1$. The corresponding expression for R is

$$R = \pi^4 \frac{(1+x)\{[(1+x)^2+Q_1]^2+T_1(1+x)\}}{x[(1+x)^2+Q_1]}, \tag{59}$$

where $$x = \frac{a^2}{\pi^2}, \quad Q_1 = \frac{Q}{\pi^2}, \quad \text{and} \quad T_1 = \frac{T}{\pi^4}. \tag{60}$$

As a function of x, R given by equation (59) attains its extremal value when
$$2x^3+3x^2-1 = Q_1+T_1\frac{(1+x)^4-Q_1(x^2-1)}{[(1+x)^2+Q_1]^2}. \tag{61}$$

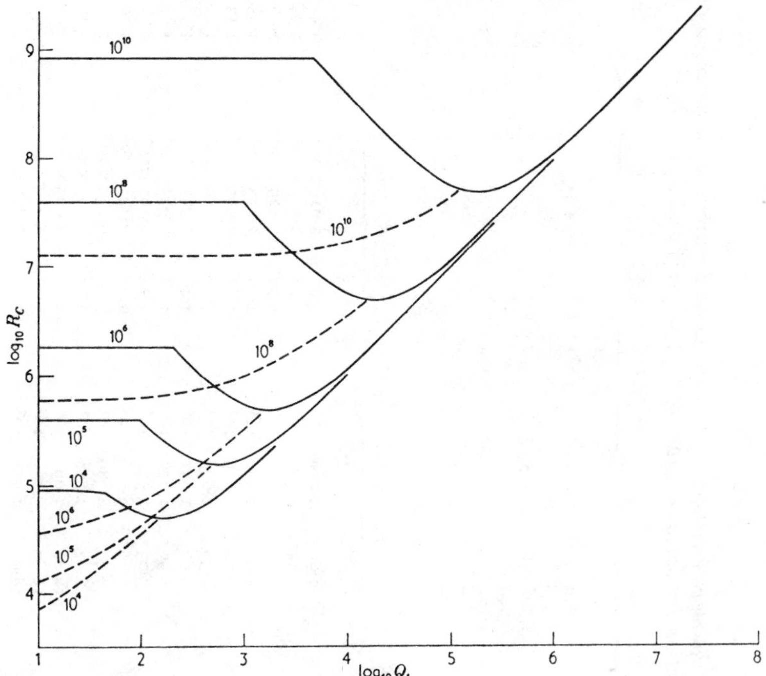

FIG. 47. The critical Rayleigh number R_c for the onset of ordinary cellular convection (solid line) and overstability (for $\mathfrak{p} = 0{\cdot}025$) (broken line) as a function of Q_1 $(= Q/\pi^2)$ for various assigned values of T_1 $(= T/\pi^4)$. The curves are labelled by the values of T_1 to which they refer. For a given value of T_1, instability will set in as overstability for all values of Q_1 less than that at the point of intersection of the corresponding full-line and dashed curves; for all larger values of Q_1, it will set in as ordinary convection.

However, this equation is not very useful for determining the critical Rayleigh numbers for assigned values of Q_1 and T_1. It is more convenient to evaluate R directly as a function of x (in accordance with equation (59)) and locate the minimum numerically. The critical numbers listed in Table XVIII and illustrated in Figs. 47 and 48 were determined in this fashion.

It is apparent from Table XVIII and Figs. 47 and 48 that the solution

TABLE XVIII

The critical Rayleigh numbers and the wave numbers of the associated disturbance for the onset of instability as stationary convection for various values of Q_1 ($= Q/\pi^2$) and T_1 ($= T/\pi^4$)

Q_1	$T_1 = 1$		$T_1 = 10$		$T_1 = 50$		$T_1 = 100$		$T_1 = 200$	
	a	R_c	a	R_c	a	R_c	a	R_c	a	R_c
10	3·70	$2{\cdot}657\times10^3$	3·76	$2{\cdot}886\times10^3$	4·00	$3{\cdot}881\times10^3$	4·31	$5{\cdot}079\times10^3$	4·98	$7{\cdot}325\times10^3$
40	—	—	—	—	—	—	—	—	4·69	$8{\cdot}923\times10^3$
100	5·67	$1{\cdot}504\times10^4$	5·66	$1{\cdot}508\times10^4$	5·65	$1{\cdot}527\times10^4$	5·62	$1{\cdot}550\times10^4$	5·58	$1{\cdot}595\times10^4$
500	7·59	$6{\cdot}238\times10^4$	7·59	$6{\cdot}240\times10^4$	7·59	$6{\cdot}245\times10^4$	7·58	$6{\cdot}252\times10^4$	7·57	$6{\cdot}267\times10^4$
1,000	8·59	$1{\cdot}184\times10^5$	8·59	$1{\cdot}184\times10^5$	8·58	$1{\cdot}184\times10^5$	8·58	$1{\cdot}185\times10^5$	8·57	$1{\cdot}185\times10^5$
10,000	12·80	$1{\cdot}065\times10^6$	—	—	—	—	12·80	$1{\cdot}065\times10^6$	12·80	$1{\cdot}065\times10^6$
50,000	16·84	$5{\cdot}129\times10^6$	—	—	—	—	16·83	$5{\cdot}129\times10^6$	16·83	$5{\cdot}129\times10^6$
100,000	18·94	$1{\cdot}015\times10^7$	—	—	—	—	18·94	$1{\cdot}015\times10^7$	18·94	$1{\cdot}015\times10^7$

Q_1	$T_1 = 500$		$T_1 = 1{,}000$		$T_1 = 1{,}500$		$T_1 = 2{,}000$	
	a	R_c	a	R_c	a	R_c	a	R_c
10	6·59	$1{\cdot}292\times10^4$	7·90	$2{\cdot}016\times10^4$	8·67	$2{\cdot}619\times10^4$	9·22	$3{\cdot}154\times10^4$
20	4·68	$1{\cdot}178\times10^4$	6·44	$1{\cdot}900\times10^4$	7·78	$2{\cdot}513\times10^4$	8·54	$3{\cdot}053\times10^4$
25	—	—	—	—	4·70	$2{\cdot}429\times10^4$	7·99	$2{\cdot}988\times10^4$
30	4·44	$1{\cdot}132\times10^4$	4·36	$1{\cdot}694\times10^4$	4·30	$2{\cdot}255\times10^4$	4·25	$2{\cdot}816\times10^4$
40	4·56	$1{\cdot}163\times10^4$	4·38	$1{\cdot}610\times10^4$	4·24	$2{\cdot}055\times10^4$	4·13	$2{\cdot}498\times10^4$
50	4·72	$1{\cdot}230\times10^4$	4·50	$1{\cdot}605\times10^4$	4·33	$1{\cdot}976\times10^4$	4·20	$2{\cdot}344\times10^4$
60	4·89	$1{\cdot}315\times10^4$	4·66	$1{\cdot}641\times10^4$	4·48	$1{\cdot}961\times10^4$	4·33	$2{\cdot}278\times10^4$
80	5·19	$1{\cdot}514\times10^4$	4·96	$1{\cdot}776\times10^4$	4·76	$2{\cdot}032\times10^4$	4·61	$2{\cdot}283\times10^4$
100	5·44	$1{\cdot}730\times10^4$	5·23	$1{\cdot}952\times10^4$	5·04	$2{\cdot}168\times10^4$	4·88	$2{\cdot}381\times10^4$
200	6·29	$2{\cdot}875\times10^4$	6·16	$3{\cdot}009\times10^4$	6·02	$3{\cdot}141\times10^4$	5·89	$3{\cdot}270\times10^4$
500	7·53	$6{\cdot}309\times10^4$	7·47	$6{\cdot}380\times10^4$	7·40	$6{\cdot}449\times10^4$	7·34	$6{\cdot}518\times10^4$
1,000	8·55	$1{\cdot}188\times10^5$	8·52	$1{\cdot}192\times10^5$	8·49	$1{\cdot}197\times10^5$	8·46	$1{\cdot}201\times10^5$
10,000	12·80	$1{\cdot}065\times10^6$	12·80	$1{\cdot}065\times10^6$	12·80	$1{\cdot}065\times10^6$		
50,000	16·83	$5{\cdot}129\times10^6$	16·83	$5{\cdot}129\times10^6$	16·83	$5{\cdot}129\times10^6$		
100,000	18·94	$1{\cdot}015\times10^7$	18·94	$1{\cdot}015\times10^7$	18·94	$1{\cdot}015\times10^7$		

Q_1	$T_1 = 3{,}000$: a		R_c		$T_1 = 10{,}000$: a		R_c	
10	10·02	—	$4{\cdot}102\times10^{4}$	—	12·59	—	$8{\cdot}979\times10^{4}$	—
20	9·52	—	$4{\cdot}004\times10^{4}$	—	12·36	—	$8{\cdot}885\times10^{4}$	—
30	8·75	4·18	$3{\cdot}881\times10^{4}$	$3{\cdot}937\times10^{4}$	12·09	—	$8{\cdot}784\times10^{4}$	—
40	—	4·00	—	$3{\cdot}380\times10^{4}$	11·78	3·74	$8{\cdot}673\times10^{4}$	$9{\cdot}522\times10^{4}$
50	—	4·04	—	$3{\cdot}073\times10^{4}$	11·40	3·68	$8{\cdot}550\times10^{4}$	$8{\cdot}118\times10^{4}$
60	—	4·12	—	$2{\cdot}903\times10^{4}$	—	3·68	—	$7{\cdot}192\times10^{4}$
80	—	4·36	—	$2{\cdot}777\times10^{4}$	—	3·77	—	$6{\cdot}104\times10^{4}$
100	—	4·63	—	$2{\cdot}796\times10^{4}$	—	3·91	—	$5{\cdot}544\times10^{4}$
200	—	5·66	—	$3{\cdot}522\times10^{4}$	—	4·73	—	$5{\cdot}134\times10^{4}$
500	—	7·22	—	$6{\cdot}654\times10^{4}$	—	6·51	—	$7{\cdot}545\times10^{4}$
1,000	—	8·39	—	$1{\cdot}210\times10^{5}$	—	7·98	—	$1{\cdot}267\times10^{5}$
2,000	—	—	—	—	—	9·39	—	$2{\cdot}324\times10^{5}$
4,000	—	—	—	—	—	10·79	—	$4{\cdot}432\times10^{5}$
10,000	—	12·80	—	$1{\cdot}065\times10^{6}$	—	12·80	—	$1{\cdot}067\times10^{6}$
50,000	—	16·83	—	$5{\cdot}129\times10^{6}$	—	16·80	—	$5{\cdot}130\times10^{6}$
100,000	—	18·94	—	$1{\cdot}015\times10^{7}$	—	18·94	—	$1{\cdot}015\times10^{7}$

Q_1	$T_1 = 30{,}000$: a		R_c		$T_1 = 100{,}000$: a		R_c	
10	15·3	—	$1{\cdot}843\times10^{5}$	—	18·9	—	$4{\cdot}067\times10^{5}$	—
20	15·2	—	$1{\cdot}833\times10^{5}$	—	18·8	—	$4{\cdot}057\times10^{5}$	—
30	15·1	—	$1{\cdot}824\times10^{5}$	—	18·7	—	$4{\cdot}048\times10^{5}$	—
40	14·9	3·64	$1{\cdot}814\times10^{5}$	$2{\cdot}703\times10^{5}$	18·7	3·61	$4{\cdot}038\times10^{5}$	$8{\cdot}830\times10^{5}$
50	14·8	3·54	$1{\cdot}803\times10^{5}$	$2{\cdot}246\times10^{5}$	18·6	3·48	$4{\cdot}028\times10^{5}$	$7{\cdot}261\times10^{5}$
60	14·6	3·50	$1{\cdot}792\times10^{5}$	$1{\cdot}933\times10^{5}$	18·5	3·43	$4{\cdot}018\times10^{5}$	$6{\cdot}175\times10^{5}$
80	14·2	3·50	$1{\cdot}769\times10^{5}$	$1{\cdot}540\times10^{5}$	18·3	3·38	$3{\cdot}998\times10^{5}$	$4{\cdot}781\times10^{5}$
100	—	3·54	—	$1{\cdot}311\times10^{5}$	18·2	3·37	$3{\cdot}977\times10^{5}$	$3{\cdot}933\times10^{5}$
200	—	3·97	—	$9{\cdot}244\times10^{4}$	—	3·50	—	$2{\cdot}283\times10^{5}$
500	—	5·43	—	$9{\cdot}722\times10^{4}$	—	4·33	—	$1{\cdot}604\times10^{5}$
1,000	—	7·09	—	$1{\cdot}414\times10^{5}$	—	5·67	—	$1{\cdot}829\times10^{5}$
2,000	—	8·85	—	$2{\cdot}423\times10^{5}$	—	7·57	—	$2{\cdot}716\times10^{5}$
4,000	—	10·5	—	$4{\cdot}496\times10^{5}$	—	9·67	—	$4{\cdot}700\times10^{5}$
10,000	—	12·6	—	$1{\cdot}070\times10^{6}$	—	12·3	—	$1{\cdot}082\times10^{6}$
50,000	—	16·8	—	$5{\cdot}131\times10^{6}$	—	16·7	—	$5{\cdot}135\times10^{6}$
100,000	—	18·9	—	$1{\cdot}015\times10^{7}$	—	18·9	—	$1{\cdot}015\times10^{7}$

Table XVIII (*cont.*)

$T_1 = 10^6$

Q_1	a		R_c	
10	27·88		$1·863\times10^{6}$	
20	27·9		$1·862\times10^{6}$	
40	27·8		$1·861\times10^{6}$	
60	27·8		$1·859\times10^{6}$	
80	27·7		$1·857\times10^{6}$	
100	27·7	3·29	$1·855\times10^{6}$	$3·760\times10^{6}$
200	27·4	3·23	$1·845\times10^{6}$	$1·948\times10^{6}$
225	27·4	3·23	$1·842\times10^{6}$	$1·744\times10^{6}$
250		3·25		$1·582\times10^{6}$
275		3·25		$1·449\times10^{6}$
300		3·25		$1·339\times10^{6}$
500		3·35		$8·678\times10^{5}$
1,000		3·76		$5·664\times10^{5}$
2,000		4·70		$5·103\times10^{5}$
4,000		6·35		$6·413\times10^{5}$
10,000		9·55		$1·200\times10^{6}$
50,000		16·0		$5·185\times10^{6}$
100,000		18·6		$1·018\times10^{7}$
1,000,000		27·9		$9·928\times10^{7}$

$T_1 = 10^7$

Q_1	a		R_c	
10	41·01		$8·594\times10^{6}$	
20	41·01		$8·593\times10^{6}$	
40	41·00		$8·591\times10^{6}$	
60	40·98		$8·589\times10^{6}$	
80	40·96		$8·587\times10^{6}$	
100	40·95		$8·585\times10^{6}$	
200	40·88		$8·576\times10^{6}$	
400	40·73		$8·556\times10^{6}$	
425	40·72	3·19	$8·553\times10^{6}$	$9·164\times10^{6}$
450	40·70	3·19	$8·551\times10^{6}$	$8·669\times10^{6}$
475	40·68	3·19	$8·548\times10^{6}$	$8·226\times10^{6}$
500		3·19		$7·827\times10^{6}$
1,000		3·23		$4·073\times10^{6}$
2,000		3·43		$2·318\times10^{6}$
4,000		4·00		$1·661\times10^{6}$
10,000		5·72		$1·816\times10^{6}$
50,000		12·1		$5·552\times10^{6}$
100,000		15·9		$1·046\times10^{7}$
1,000,000		27·6		$9·935\times10^{7}$

$T_1 = 10^8$

Q_1	a		R_c	
10	60·26		$3·977\times10^{7}$	
50	60·25		$3·976\times10^{7}$	
100	60·23		$3·976\times10^{7}$	
200	60·22		$3·975\times10^{7}$	
500	60·14		$3·972\times10^{7}$	
650	60·12	3·16	$3·971\times10^{7}$	$5·970\times10^{7}$
800	60·08	3·16	$3·969\times10^{7}$	$4·862\times10^{7}$
1,000	60·04	3·16	$3·967\times10^{7}$	$3·900\times10^{7}$
2,000		3·17		$1·983\times10^{7}$
4,000		3·26		$1·050\times10^{7}$
10,000		3·74		$5·676\times10^{6}$
50,000		7·09		$7·250\times10^{6}$
100,000		9·93		$1·191\times10^{7}$
1,000,000		25·4		$9·999\times10^{7}$

$T_1 = 10^9$

Q_1	a		R_c	
10	88·47		$1·843\times10^{8}$	
50	88·48		$1·843\times10^{8}$	
100	88·47		$1·843\times10^{8}$	
500	88·45		$1·843\times10^{8}$	
1,000	88·41		$1·842\times10^{8}$	
2,000	88·34	3·16	$1·841\times10^{8}$	$1·948\times10^{8}$
2,090	88·34	3·16	$1·841\times10^{8}$	$1·865\times10^{8}$
2,200	88·33	3·16	$1·841\times10^{8}$	$1·772\times10^{8}$
4,000		3·16		$9·809\times10^{7}$
10,000		3·22		$4·087\times10^{7}$
20,000		3·41		$2·321\times10^{7}$
50,000		4·30		$1·606\times10^{7}$
100,000		5·72		$1·815\times10^{7}$
200,000		7·95		$2·670\times10^{7}$
500,000		12·4		$5·529\times10^{7}$
1,000,000		17·4		$1·039\times10^{8}$
2,000,000		23·7		$2·014\times10^{8}$

$T_1 = 10^{10}$

Q_1	a		R_c	
10	129·9		$8·550\times10^{8}$	
50	129·9		$8·550\times10^{8}$	
100	129·9		$8·550\times10^{8}$	
500	129·9		$8·549\times10^{8}$	
1,000	129·9		$8·549\times10^{8}$	
2,000	129·8	3·14	$8·548\times10^{8}$	$1·945\times10^{9}$
4,000	129·8	3·14	$8·546\times10^{8}$	$9·739\times10^{8}$
4,600	129·8	3·14	$8·544\times10^{8}$	$8·472\times10^{8}$
4,800		3·14		$8·120\times10^{8}$
10,000		3·16		$3·914\times10^{8}$
20,000		3·17		$1·987\times10^{8}$
50,000		3·32		$8·739\times10^{7}$
100,000		3·73		$5·677\times10^{7}$
200,000		4·70		$5·100\times10^{7}$
500,000		7·10		$7·247\times10^{7}$
1,000,000		9·93		$1·189\times10^{8}$
3,000,000		17·2		$3·125\times10^{8}$
10,000,000		30·1		$9·947\times10^{8}$
30,000,000		44·6		$2·947\times10^{9}$

$T_1 = 10^{11}$

Q_1	a		R_c	
10	190·7		$3·967\times10^{9}$	
50	190·7		$3·967\times10^{9}$	
100	190·7		$3·967\times10^{9}$	
500	190·7		$3·967\times10^{9}$	
1,000	190·7		$3·967\times10^{9}$	
2,000	190·7		$3·967\times10^{9}$	
4,000	190·6		$3·967\times10^{9}$	
9,000	190·6	3·14	$3·966\times10^{9}$	$4·329\times10^{9}$
9,500	190·6	3·14	$3·966\times10^{9}$	$4·102\times10^{9}$
10,000	190·6	3·14	$3·966\times10^{9}$	$3·897\times10^{9}$
20,000		3·14		$1·952\times10^{9}$
50,000		3·16		$7·889\times10^{8}$
100,000		3·22		$4·089\times10^{8}$
200,000		3·41		$2·321\times10^{8}$
500,000		4·30		$1·606\times10^{8}$
1,000,000		5·72		$1·815\times10^{8}$
3,000,000		9·70		$3·607\times10^{8}$
10,000,000		17·6		$1·038\times10^{9}$
30,000,000		30·2		$2·985\times10^{9}$
100,000,000		50·4		$9·811\times10^{9}$
300,000,000		69·0		$2·932\times10^{10}$

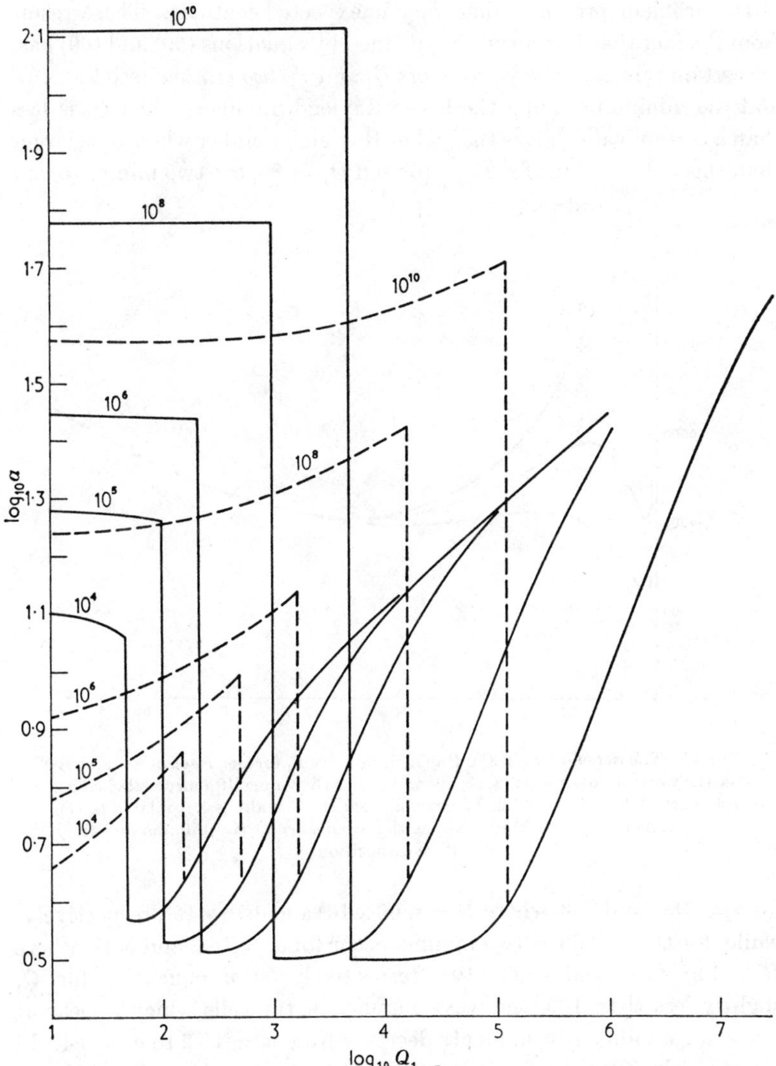

FIG. 48. The dependence on Q_1 (for various assigned values of T_1) of the wave number a (in the unit $1/d$) of the disturbance at which instability first sets in as convection (solid line) and as overstability (for $\mathfrak{p} = 0{\cdot}025$) (broken line). It will be observed that a discontinuous change in a occurs when (for increasing Q_1) the manner of instability changes from overstability to cellular convection.

of the problem presents some very unexpected features. These result from the fact that the curve $R(a)$ defined by equations (59) and (60) has, for certain ranges of the parameters Q_1 and T_1, *two minima* (see Fig. 49); and the minimum giving the lower Rayleigh number, when Q_1 is less than a certain value, gives the higher Rayleigh number when Q_1 is larger than this value. Thus, for $T_1 = 10^5$ and $Q_1 = 80$, the two minima occur

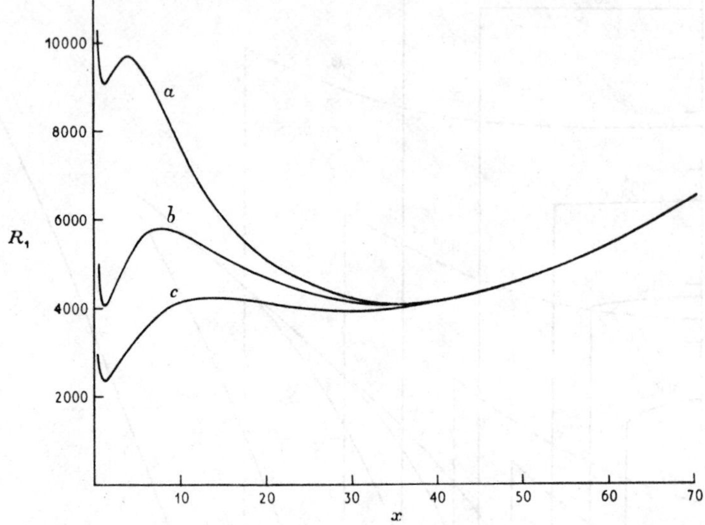

FIG. 49. The dependence of the Rayleigh number R_1 for the onset of instability on the wave number when $T_1 = 10^5$ and Q_1 has the values 40 (curve labelled *a*), 100 (curve labelled *b*), and 200 (curve labelled *c*). The abscissa is related to the wave number *a* by $x = a^2/\pi^2$. Notice the occurrence of two minima and their relative disposition.

for $a = 18{\cdot}3$ and $3{\cdot}38$ where $R = 4{\cdot}00 \times 10^5$ and $4{\cdot}78 \times 10^5$, respectively; while, for $Q_1 = 100$ the two minima occur for $a = 18{\cdot}2$ and $3{\cdot}37$, where $R = 3{\cdot}98 \times 10^5$ and $3{\cdot}93 \times 10^5$, respectively. Consequently, for Q_1 slightly less than 100, the wave number of the cells which appear at marginal stability will suddenly decrease from $a = 18{\cdot}2$ to $a = 3{\cdot}4$. In other words, if we start with an initial situation in which $T_1 = 10^5$ and no magnetic field is present, and gradually increase the strength of the magnetic field, then at first the cells which appear at marginal stability will be elongated; and when the magnetic field has increased to a value corresponding to $Q_1 = 100$, cells of two very different sizes will appear simultaneously: one set which will be highly elongated and another set which will, relatively, be highly flattened. As the magnetic field increases still further, the critical Rayleigh number will start *decreasing* and pass through a minimum; and eventually, the inhibition due to the magnetic field will predominate. The sequence of events we have just described occurs for all $T_1 > 2{,}500$. For T_1 somewhat less than this value, the wave number which will manifest itself at marginal stability varies extremely rapidly in a small interval of Q_1; thus for $T_1 = 1{,}500$, a varies from 7·29 to 4·7 as Q_1 varies between 23 and 25. However, for $T_1 < 200$, a appears to be a monotonic increasing function of Q_1.

It should be noted that for all $T_1 > 500$, the critical Rayleigh number *always* shows an initial decrease with Q_1; only, for $T_1 > 2{,}500$, this decrease becomes very pronounced when Q passes through the critical value when two sets of cells appear simultaneously. For $T_1 < 400$, R_c appears to be a monotonic increasing function of Q_1.

53. The case when instability sets in as overstability

We have seen that when liquid metals, such as mercury, are in rotation, thermal instability sets in mostly as overstability. But in the presence of a magnetic field, the principle of the exchange of stabilities is valid and instability sets in as stationary convection. When rotation and magnetic field are simultaneously present, the manner of the onset of instability must, in general, depend in an extremely complicated way on the relevant parameters T, Q, $\mathfrak{p}_1$ $(= \nu/\kappa)$, and $\mathfrak{p}_2$ $(= \nu/\eta)$. We shall seek to unravel the nature of this dependence in the particular case of mercury.

Returning to equation (43), and considering again the case of two free boundaries adjoining a non-conducting medium, we can obtain the required characteristic equation by inserting the solution $W = W_0 \sin \pi z$; thus

$$Ra^2(\pi^2+a^2+\mathfrak{p}_2\sigma)[(\pi^2+a^2+\sigma)(\pi^2+a^2+\mathfrak{p}_2\sigma)+Q\pi^2]$$
$$= (\pi^2+a^2+\mathfrak{p}_1\sigma)\{(\pi^2+a^2)[(\pi^2+a^2+\sigma)(\pi^2+a^2+\mathfrak{p}_2\sigma)+Q\pi^2]^2+ + T\pi^2(\pi^2+a^2+\mathfrak{p}_2\sigma)^2\}, \quad (62)$$

where it must be remembered that σ can be complex. Letting

$$x = \frac{a^2}{\pi^2}, \quad i\sigma_1 = \frac{\sigma}{\pi^2}, \quad R_1 = \frac{R}{\pi^4}, \quad T_1 = \frac{T}{\pi^4}, \quad \text{and} \quad Q_1 = \frac{Q}{\pi^2}, \quad (63)$$

we can rewrite equation (62) in the form

$$R_1 = \frac{1+x}{x}\left\{(1+x+i\mathfrak{p}_1\sigma_1)(1+x+i\sigma_1)+Q_1\frac{1+x+i\mathfrak{p}_1\sigma_1}{1+x+i\mathfrak{p}_2\sigma_1}+ + \frac{T_1}{1+x}\frac{(1+x+i\mathfrak{p}_1\sigma_1)(1+x+i\mathfrak{p}_2\sigma_1)}{(1+x+i\sigma_1)(1+x+i\mathfrak{p}_2\sigma_1)+Q_1}\right\}. \quad (64)$$

Since our present interest is to specify the critical Rayleigh number for the onset of instability via a state of purely oscillatory motions, it will suffice to seek the conditions that equation (64) will allow solutions for which σ_1 is real. Assuming then that this is the case, and equating the real and the imaginary parts of equation (64), we obtain the following pair of equations for determining R_1 and σ_1 for given x, Q_1, and T_1:

$$R_1 = \frac{1+x}{x}\left\{(1+x)^2 - \mathfrak{p}_1\sigma_1^2 + Q_1\frac{(1+x)^2+\mathfrak{p}_1\mathfrak{p}_2\sigma_1^2}{(1+x)^2+\mathfrak{p}_2^2\sigma_1^2} + \right.$$
$$\left. + \frac{T_1}{1+x}\frac{[(1+x)^2-\mathfrak{p}_2\sigma_1^2+Q_1][(1+x)^2-\mathfrak{p}_1\mathfrak{p}_2\sigma_1^2]+(1+x)^2(\mathfrak{p}_1+\mathfrak{p}_2)(1+\mathfrak{p}_2)\sigma_1^2}{[(1+x)^2-\mathfrak{p}_2\sigma_1^2+Q_1]^2+(1+x)^2(1+\mathfrak{p}_2)^2\sigma_1^2}\right\} \tag{65}$$

and

$$\frac{T_1}{1+x}\frac{(\mathfrak{p}_1-1)(1+x)^2+(\mathfrak{p}_1+\mathfrak{p}_2)Q_1+\mathfrak{p}_2^2(\mathfrak{p}_1-1)\sigma_1^2}{[(1+x)^2-\mathfrak{p}_2\sigma_1^2+Q_1]^2+(1+x)^2(1+\mathfrak{p}_2)^2\sigma_1^2} +$$
$$+Q_1\frac{\mathfrak{p}_1-\mathfrak{p}_2}{(1+x)^2+\mathfrak{p}_2^2\sigma_1^2}+1+\mathfrak{p}_1 = 0. \tag{66}$$

For assigned values of Q_1 and T_1, equations (65) and (66) define R_1 as a function of x; the minimum of this function determines the critical Rayleigh number for the onset of overstability (for the assigned Q_1 and T_1). This minimum value of the Rayleigh number for the onset of overstability should be compared with the corresponding value for the onset of convection as given by the formulae of § 52; the manner in which instability will first set in will depend on which of the two Rayleigh numbers is the smaller.

Of course, for certain ranges of the parameters, equation (66) will not allow positive solutions for σ_1^2; for these ranges of the parameters, overstability is not possible.

(a) *An approximate solution applicable to liquid metals*

The equations giving R_1 and σ_1 simplify considerably when applied to liquid metals such as mercury. These substances are characterized by values of $\mathfrak{p}_2$ which are exceedingly small; thus for mercury at room temperatures,

$$\mathfrak{p}_1 = \frac{1{\cdot}11}{4{\cdot}45}\times10^{-1} = 0{\cdot}025; \qquad \mathfrak{p}_2 = \frac{1{\cdot}11\times10^{-3}}{7{\cdot}6\times10^{3}} \simeq 1{\cdot}5\times10^{-7}. \tag{67}$$

Accordingly, in these cases, $\mathfrak{p}_2$ can be neglected in comparison with $\mathfrak{p}_1$ or unity. Moreover, it appears on examination that the terms in $\mathfrak{p}_2$ can be neglected even when they occur multiplied by a large factor such as

σ_1^2. Therefore, suppressing all terms in equations (65) and (66) which have $\mathfrak{p}_2$ as a factor, we obtain

$$R_1 = \frac{1+x}{x}\left\{(1+x)^2-\mathfrak{p}_1\sigma_1^2+Q_1+T_1\frac{(1+x)[(1+x)^2+Q_1+\mathfrak{p}_1\sigma_1^2]}{[(1+x)^2+Q_1]^2+(1+x)^2\sigma_1^2}\right\} \quad (68)$$

and

$$(1+x)(1+\mathfrak{p}_1)+Q_1\frac{\mathfrak{p}_1}{1+x} = T_1\frac{(1-\mathfrak{p}_1)(1+x)^2-\mathfrak{p}_1 Q_1}{[(1+x)^2+Q_1]^2+(1+x)^2\sigma_1^2}. \quad (69)$$

Equation (69) provides the following explicit formula for σ_1^2:

$$\sigma_1^2 = \frac{T_1}{1+x}\frac{(1+x)^2(1-\mathfrak{p}_1)-\mathfrak{p}_1 Q_1}{(1+x)^2(1+\mathfrak{p}_1)+\mathfrak{p}_1 Q_1}-\left[(1+x)+\frac{Q_1}{1+x}\right]^2. \quad (70)$$

Also, equations (68) and (69) can be combined to give

$$R_1 = 2\frac{1+x}{x}\frac{(1+x)^2+Q_1}{(1+x)^2(1-\mathfrak{p}_1)-\mathfrak{p}_1 Q_1}[(1+x)^2+\mathfrak{p}_1^2\sigma_1^2]. \quad (71)$$

(b) Numerical results for mercury

Equations (70) and (71) have been used to determine the critical Rayleigh numbers for the onset of overstability for various values of Q_1 and T for $\mathfrak{p}_1 = 0{\cdot}025$. The results of the calculations are summarized in Table XIX. The last column in this table gives the gyration frequency of the oscillation in the marginal state in units of Ω; according to equations (63) and the definition of T,

$$p/\Omega = 2\sigma_1/\sqrt{T_1}. \quad (72)$$

The critical Rayleigh numbers for the onset of overstability and stationary convection are exhibited in Fig. 47 as functions of Q_1 for various assigned values of T_1; the wave numbers of the associated disturbances are similarly exhibited in Fig. 48. We can directly read from Fig. 47 the manner of the onset of instability for the various values of T_1 for which the calculations have been made.

The phenomenon described in § 52 which gives rise to the peculiar, non-monotonic behaviour of the (R_c, Q)-curves when instability sets in as stationary convection does not occur when allowance is made for the onset of overstability for a value of $\mathfrak{p}_1$ as small as 0·025. However, the curves do show a discontinuous behaviour though of a much less pronounced character. From Fig. 47 it appears that for the cases illustrated, the transition from overstability to convection occurs at about the place where the convection curve passes through its minimum; and from Fig. 48 it is seen that this transition is accompanied by a substantial discontinuity in the wave number of the disturbance which manifests itself at marginal stability. This discontinuity is in the sense that the convection cells (for increasing Q_1) suddenly get very much widened.

TABLE XIX

Critical Rayleigh numbers and related constants for the onset of overstability for the case $\mathfrak{p}_1 = 0{\cdot}025$

Q_1	a	σ_1	R_c	p/Ω
		$T_1 = 10^4$		
10	4·56	53·52	$7{\cdot}0530 \times 10^3$	1·0704
30	5·35	45·55	$1{\cdot}3887 \times 10^4$	0·9110
100	6·63	30·12	$3{\cdot}5402 \times 10^4$	0·6024
150	7·20	18·44	$5{\cdot}0156 \times 10^4$	0·3688
		$T_1 = 10^5$		
10	6·03	140·8	$1{\cdot}284 \times 10^4$	0·8905
20	6·41	132·9	$1{\cdot}667 \times 10^4$	0·8405
30	6·66	127·8	$2{\cdot}019 \times 10^4$	0·8083
40	6·88	123·7	$2{\cdot}355 \times 10^4$	0·7823
50	7·05	120·4	$2{\cdot}679 \times 10^4$	0·7615
60	7·20	117·6	$2{\cdot}996 \times 10^4$	0·7438
100	7·66	109·0	$4{\cdot}217 \times 10^4$	0·6894
200	8·45	94·38	$7{\cdot}130 \times 10^4$	0·5969
400	9·48	72·53	$1{\cdot}278 \times 10^5$	0·4587
500	9·88	61·95	$1{\cdot}557 \times 10^5$	0·3918
		$T_1 = 10^6$		
10	8·35	342·1	$3{\cdot}650 \times 10^4$	0·6842
30	8·89	321·7	$4{\cdot}533 \times 10^4$	0·6434
100	9·78	291·0	$7{\cdot}072 \times 10^4$	0·5820
500	11·54	236·3	$1{\cdot}893 \times 10^5$	0·4726
1,000	12·64	202·5	$3{\cdot}271 \times 10^5$	0·4050
1,600	13·62	170·3	$4{\cdot}898 \times 10^5$	0·3406
		$T_1 = 10^8$		
10	17·4	1729	$6{\cdot}000 \times 10^5$	0·3458
100	18·0	1674	$6{\cdot}492 \times 10^5$	0·3348
1,000	20·3	1467	$1{\cdot}020 \times 10^6$	0·2934
10,000	25·0	1123	$3{\cdot}642 \times 10^6$	0·2246
16,000	26·4	1024	$5{\cdot}248 \times 10^6$	0·2048
		$T_1 = 10^{10}$		
100	37·7	8097	$1{\cdot}254 \times 10^7$	0·1619
1,000	38·4	7953	$1{\cdot}305 \times 10^7$	0·1591
10,000	41·9	7229	$1{\cdot}720 \times 10^7$	0·1446
40,000	46·3	6455	$2{\cdot}780 \times 10^7$	0·1291
100,000	50·1	5841	$4{\cdot}576 \times 10^7$	0·1168
130,000	51·3	5643	$5{\cdot}421 \times 10^7$	0·1129

54. Experiments on the onset of thermal instability in the presence of rotation and magnetic field

Experiments on the onset of thermal instability in the presence of rotation and magnetic field have been successfully carried out by Nakagawa.

In Nakagawa's experiments, a reconditioned electromagnet of a $32\frac{1}{2}$-in.

cyclotron of the University of Chicago was used to provide a uniform vertical magnetic field of variable strength. A Pyrex glass cylinder containing mercury mounted on levelled non-magnetic ball-bearings between the pole pieces of the electromagnet was rotated in the magnetic field by a constant speed electric motor (see Fig. 50 which shows the arrangement used in the experiments described in § (*b*) below). The angular velocity of the rotation was measured by an optical method using electrical timing devices; the accuracy attained in the time measurements was 10^{-5} sec.

A direct-current electric heater placed at the bottom of the container, below a stainless-steel plate, provided the means of heating. The temperature gradient which was established was measured by a nine-couple copper-constantan thermopile immersed at two fixed levels in the mercury. The average temperature of the mercury in any particular experiment was also measured by a thermocouple, one of whose junctions was placed at the middle of the mercury layer; the values of the various coefficients (like ν, κ, etc.) at this average temperature were used in evaluating the parameters such as R, T, and Q.

The e.m.f. of the thermopile and the thermocouple were amplified and recorded automatically and continuously. The accuracy in the thermopile measurements was $\pm 0{\cdot}001°$ C, while the average temperatures measured by the thermocouple were accurate to $\pm 0{\cdot}01°$ C.

The electrical connexions to the various components in the rotating part of the assembly were made through a system of mercury filled troughs with copper and constantan contacts.

A constant circulation of cooled nitrogen helped to maintain the conditions at the top surface of the mercury constant and uniform.

(*a*) *The results on the critical Rayleigh number and on the manner of the onset of instability*

All the experiments were carried out with a layer of mercury 3 cm in depth rotated with a constant angular velocity of 5 rev/min. The corresponding value of T_1 varied between $7{\cdot}5\times 10^5$ and $8{\cdot}5\times 10^5$ depending on the average temperature of the mercury. Experiments were performed for values of Q_1 in the range 10 to 2×10^4 by using magnetic fields of strength varying from 125 to 5,500 gauss.

The critical Rayleigh number for the onset of instability, for a constant speed of rotation and strength of magnetic field, was determined by measuring the steady temperature gradients which get established for various rates of heating, and using the Schmidt–Milverton principle;

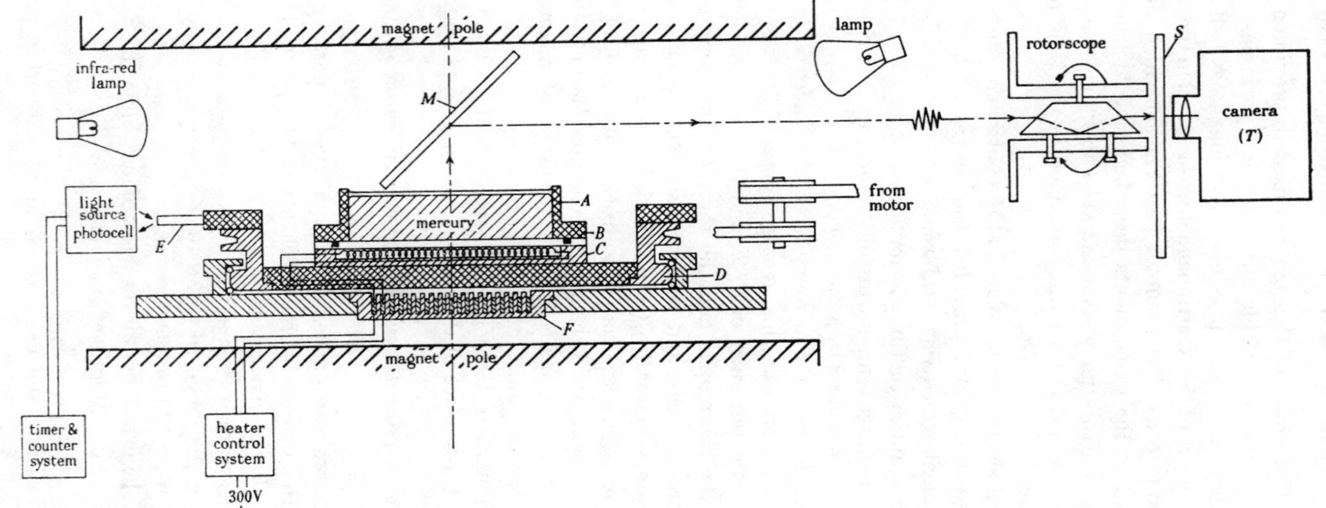

FIG. 50. A schematic diagram of the experimental arrangement: *A*, bakelite cylinder; *B*, stainless-steel plate; *C*, electric heater; *D*, non-magnetic ball-bearing; *E*, stainless-steel rod; *F*, mercury trough; *M*, front-surface mirror; *S*, rotary shutter; *T*, camera (*Proc. Roy. Soc.* (*London*) A, **249,** 140 (1958)).

and the manner of the onset of instability was ascertained in each case by inspecting the temperature records. The onset of overstability could always be distinguished by the nearly pure sinusoidal oscillations exhibited by the temperature records. A typical example of such a record is shown in Fig. 51.

In Fig. 52 the results of the experiments together with the theoretically expected relation for $T_1 = 10^6$, according to the calculations of §§ 52

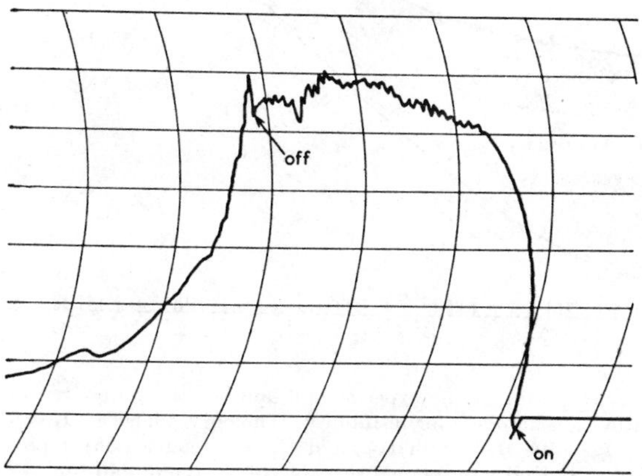

FIG. 51. A time record of the adverse temperature gradient for mercury: $d = 3$ cm; $\Omega = 5$ rev/min; $H = 125$ G; $Q_1 = 1{\cdot}01 \times 10^1$; $T = 7{\cdot}90 \times 10^5$.

and 53, are shown. Since the theoretical results have been derived for two free boundaries and refer, moreover, to a value of T_1 somewhat different from the average value to which the experiments correspond, a quantitative agreement between the experiments and the theory is not to be expected. All that can be expected is an agreement in the general features; and this is certainly present. In particular, the characteristic discontinuity in the dependence of the Rayleigh number on Q_1 with the accompanying change in the manner of the onset of instability is clearly demonstrated.

The periods $2\pi/p$ of the overstable oscillations were estimated by counting the number of oscillations in the temperature records which occur in known intervals of time. In Fig. 53 these results are shown together with the theoretically expected relations for $T_1 = 10^5$ and 10^6. Again the general agreement is satisfactory, though as we have explained, quantitative agreement is not to be expected.

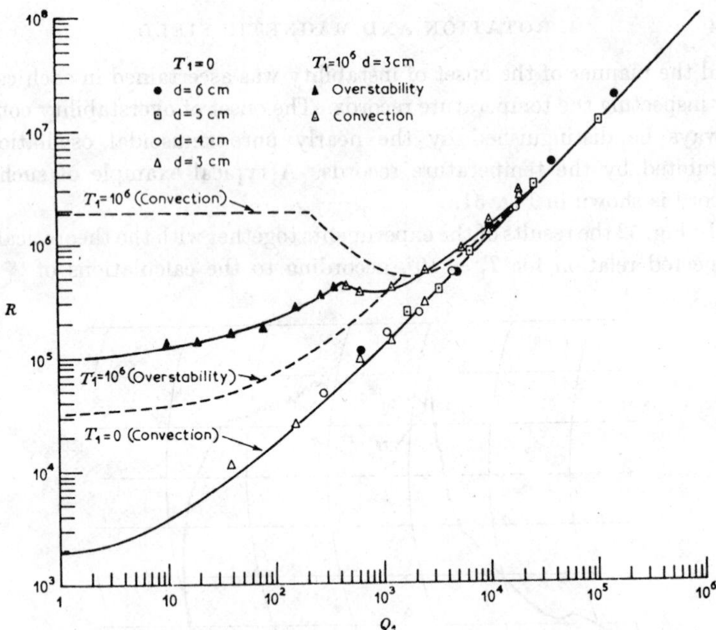

FIG. 52. A summary of the experimental and the theoretical results on the critical Rayleigh numbers for instability. The curves labelled $T_1 = 10^6$ (convection), $T_1 = 10^6$ (overstability), and $T_1 = 0$ (convection) represent the theoretically derived relations. The value of T_1 appropriate for the experimental points represented by the solid and shaded triangles is $7{\cdot}75 \times 10^5$.

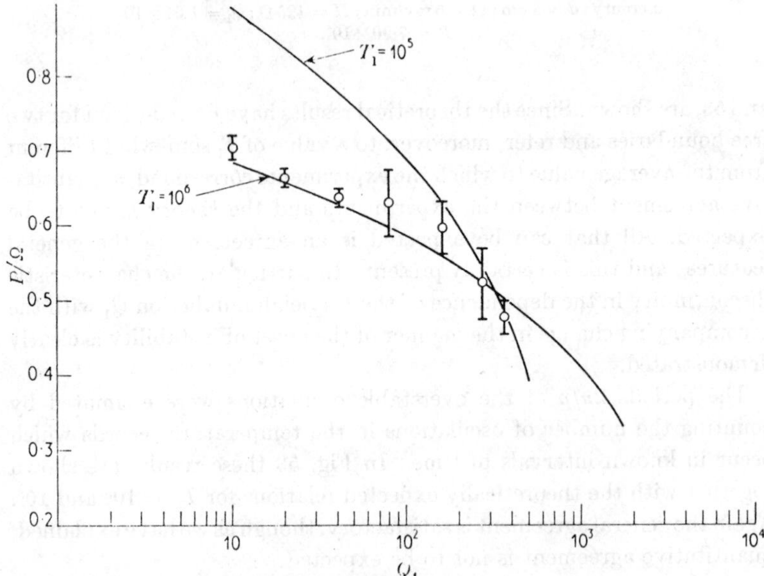

FIG. 53. A comparison of the observed periods of overstable oscillations with the theoretical values (the full line curves). The value of T_1 appropriate for the experimental results is $(8{\cdot}05 \pm 0{\cdot}07) \times 10^5$.

(*b*) *Optical observations and the discontinuous variation of the cell dimensions with the strength of the magnetic field*

For optical observations, some slight modifications in the arrangement described earlier were necessary. It is the arrangement used in these experiments that is illustrated in Fig. 50.

Mercury was contained in a Bakelite cylinder 24 cm in diameter and 4 cm in height. The bottom of the cylinder A was sealed by a stainless-steel plate B with an O-ring. The steel plate was carefully machined to eliminate surface irregularities which might disturb the ensuing convection patterns; freedom from such disturbances was found to be essential for successful optical observations.

Uniform heating from below was provided by a non-inductively wound electric heater C placed below the stainless-steel plate. The assembly consisting of the Bakelite cylinder, stainless-steel plate, and the electric heater was mounted, between the pole pieces of the electromagnet, on non-magnetic ball-bearing D. And this assembly was rotated in the magnetic field at 5 rev/min as in the previous experiments.

The optical part of the apparatus consisted of a mirror M mounted 45° to the vertical, the rotoscope, a rotary shutter S, and a single lens reflex camera T. The mirror was silvered on the front surface; and dimming of the mirror (by condensation of water-vapour during the experiments) was prevented by heat from an infra-red lamp.

A stationary image was secured at the camera by rotating the rotoscope with an angular velocity exactly one-half that of the apparatus. By a suitable sequence of openings and closings of the shutter, it was possible to obtain photographs every 15 sec.

As tracers of the motion, sand particles of diameter approximately $\frac{1}{2}$ mm were used. These particles floating freely on the surface, and following the motion of the mercury, enabled the emerging pattern of convection to be photographed. A thin layer of distilled water on the mercury surface eliminated much of the difficulty arising from surface oxidation.

The experiments were again carried out on a layer of mercury 3 cm in depth and an angular velocity of rotation of 5 rev/min. And as ascertained by the experiments in § (*a*), the rate of heating for each value of Q_1 was adjusted so that the conditions were just marginal. By repeating the experiments at least twice for each value of Q_1, Nakagawa took a total of 300 streak photographs. Measurements were made on prints of these photographs on 8×10-in. paper. Distances between the centres of adjacent cells were measured and in the manner described in Chapter

IV, § 48, the theoretical values to which these distances should correspond were inferred. In Fig. 54 the results of the measurements are compared with the expected theoretical relation for $T_1 = 10^6$. While, for reasons we have already explained, a quantitative agreement is not to be expected, the agreement in the general features of the relationship

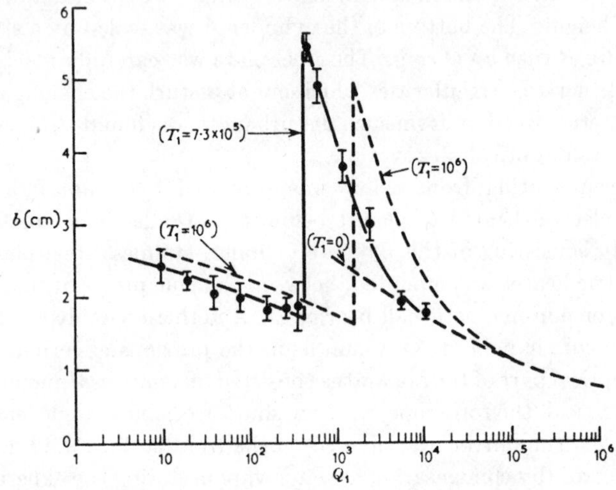

FIG. 54. A summary of the experimental and theoretical results on the sizes of the cells manifested at marginal stability. The theoretical relations for $T_1 = 10^6$ and $T_1 = 0$ are represented by the broken lines. Solid circles represent the experimental results for $T_1 = 7{\cdot}30 \times 10^5$; and the heavy solid lines are fitted to the experimental data. The lower branches of the broken line and the heavy solid line represent the case when the instability sets in as overstability. Both lines on the right-hand side of the figure represent the case when the instability sets in as steady convection.

is satisfactory. In particular, the predicted enlargement of the cell dimension at a critical field strength is strikingly confirmed.

In Fig. 55 some typical examples from Nakagawa's collection of streak photographs are illustrated. They illustrate the characteristic dependence of the size of the cells on the value of Q_1. The discontinuous change in the size of the cells which occurs for increasing field strength is clearly demonstrated by this sequence of photographs.

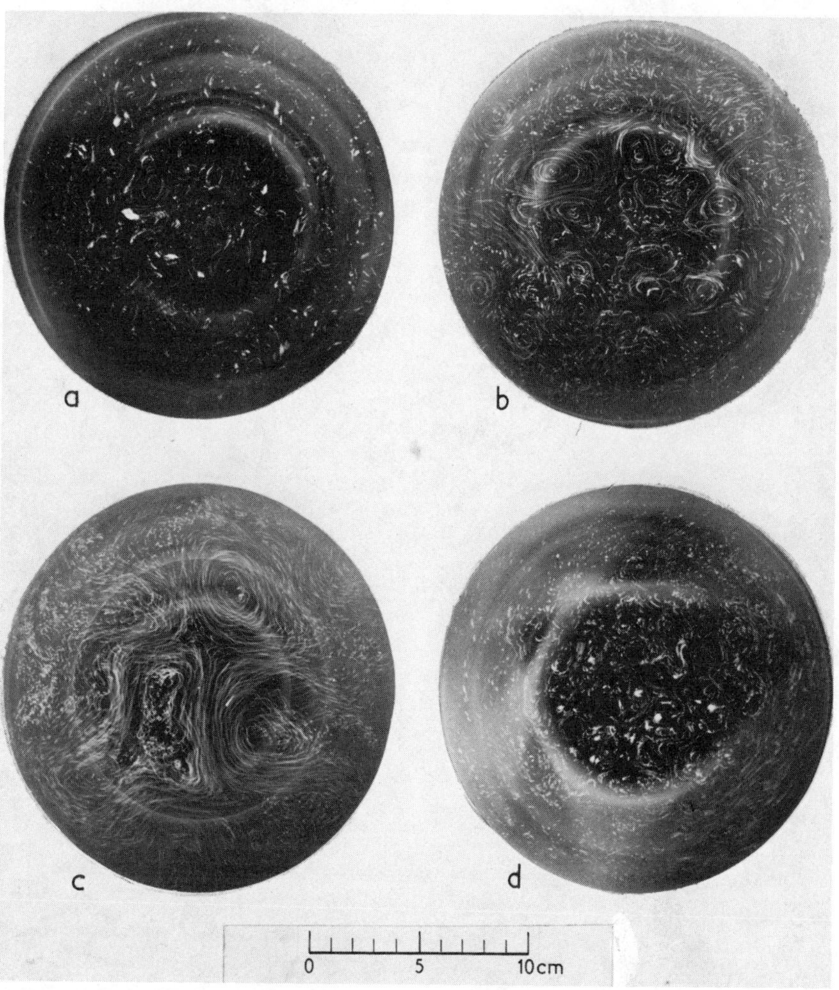

FIG. 55. Examples of streak photographs of the convective motion on the surface of a layer of mercury of depth 3 cm, rotating at 5 rev/min, and subject to impressed external magnetic fields of various strengths; (*a*) $H = 125$ G, $Q_1 = 9{\cdot}60$, and $T_1 = 6{\cdot}95 \times 10^5$; (*b*) $H = 750$ G, $Q_1 = 3{\cdot}52 \times 10^2$, and $T_1 = 7{\cdot}30 \times 10^5$; (*c*) $H = 1{,}000$ G, $Q_1 = 6{\cdot}25 \times 10^2$, and $T_1 = 7{\cdot}21 \times 10^5$; (*d*) $H = 3{,}000$ G, $Q_1 = 5{\cdot}61 \times 10^3$, and $T_1 = 7{\cdot}49 \times 10^5$. The sudden enlargement of the cell size as we pass from $H = 750$ G to $H = 1{,}000$ G is apparent.

BIBLIOGRAPHICAL NOTES

§ 50. Hydromagnetic waves in a rotating fluid have been considered by:

1. B. LEHNERT, 'Magnetohydrodynamic waves under the action of the Coriolis force, Part I', *Astrophys. J.* **119,** 647–54 (1954); Part II, ibid. **121,** 481–90 (1955).

See also:

2. S. CHANDRASEKHAR, 'The gravitational instability of an infinite homogeneous medium when Coriolis force is acting and a magnetic field is present', ibid. **119,** 7–9 (1954).
3. B. LEHNERT, 'The decay of magneto-turbulence in the presence of a magnetic field and Coriolis force', *Quart. Appl. Math.* **12,** 321–41 (1955).

§§ 51–53. The analyses in these sections are based on:

4. S. CHANDRASEKHAR, 'The instability of a layer of fluid heated below and subject to the simultaneous action of a magnetic field and rotation. I', *Proc. Roy. Soc.* (*London*) A, **225,** 173–84 (1954); II, ibid. **237,** 476–84 (1956).

The fact that the curve $R(a)$ defined by equations (59) and (60) has *two minima* for certain ranges of the parameters Q_1 and T_1 was first observed by Donna Elbert.

§ 54. The references for the experimental work described in this section are:

5. Y. NAKAGAWA, 'Experiments on the instability of a layer of mercury heated from below and subject to the simultaneous action of a magnetic field and rotation. I', *Proc. Roy. Soc.* (*London*) A, **242,** 81–88 (1957); II, ibid. **249,** 138–45 (1959).

Extending the experiments described in reference 5, Nakagawa has since demonstrated the occurrence of instability as stationary convection after the instability has already manifested itself as overstability; and he has, moreover, observed the predicted decrease of the Rayleigh number, with increasing strength of the magnetic field, over a part of the convective branch (see Fig. 52). [For the corresponding results for the case when rotation alone is present, see pp. 142 and 143.]

VI

THE ONSET OF THERMAL INSTABILITY IN FLUID SPHERES AND SPHERICAL SHELLS

55. Introduction

In this chapter we shall extend the considerations of the preceding chapters to a spherical geometry and to problems relating to the thermal instability of fluid spheres and spherical shells with internal heat sources. Apart from the general interest, from the point of view of applied mathematics, which attaches to such extensions of a basic physical theory, as the nature and origin of thermal instability, the extensions are also of interest to geophysics in several connexions; for example, with the pattern of the convective motions which might prevail in the fluid core of the earth and the relation of such motions to theories concerning the origin of the earth's magnetic field and its secular variations. There are other geophysical connexions in which the nature of convection in a spherical geometry might also be important: in connexion with the question whether the roughly equal division of the earth's surface into a land and an ocean hemisphere might be related to a convection pattern in an original fluid earth; and in connexion also with the question whether a slow, continuing, large scale convection in the earth's mantle might be related to the origin and the movements of the continents. The relevance of these latter questions in the particular geophysical and geological contexts are matters of controversy at the present time; but they do not affect the interest in the problems *per se*.

56. The perturbation equations

Consider a spherical shell of incompressible fluid subject to a spherically symmetric radial gravitational field $g(r)x_i$, where $g(r)$ is a function of r only. In any specific case, $g(r)$ will be a known function of r. Thus, if we should consider a homogeneous fluid sphere of density ρ,

$$g(r) = \tfrac{4}{3}\pi G\rho = \text{constant}, \tag{1}$$

where G is the constant of gravitation. Similarly, if we should consider a spherical shell of density ρ overlying a core of radius R_i and mass M_i,

$$g(r) = G\left\{(M_i - \tfrac{4}{3}\pi\rho R_i^3)\frac{1}{r^3} + \tfrac{4}{3}\pi\rho\right\}. \tag{2}$$

We shall suppose that there is a distribution of heat sources ϵ which maintains a radial temperature gradient in the fluid. The gradient will be determined by the equation of heat conduction,

$$\kappa\nabla^2 T = -\epsilon. \tag{3}$$

If κ and ϵ are assumed to be constants, equation (3) leads to the temperature distribution

$$T = \beta_0 - \beta_2 r^2 + \frac{\beta_1}{r}, \tag{4}$$

where
$$\beta_2 = \epsilon/6\kappa, \tag{5}$$

and β_0 and β_1 are constants. By considering the stability of this initial state, we readily find that the perturbation equations in the Boussinesq approximation are

$$\frac{\partial u_i}{\partial t} = -\frac{\partial}{\partial x_i}\left(\frac{\delta p}{\rho}\right) + \alpha g(r) x_i \theta + \nu\nabla^2 u_i \tag{6}$$

and
$$\frac{\partial \theta}{\partial t} = -u_i\frac{\partial T}{\partial x_i} + \kappa\nabla^2\theta, \tag{7}$$

where α is the coefficient of volume expansion, and δp and θ are the perturbations in the pressure and the temperature, respectively.

According to equation (4), we may write

$$\frac{\partial T}{\partial x_i} = -2\beta(r) x_i, \tag{8}$$

where
$$\beta(r) = \beta_2 + \frac{\beta_1}{2r^3}. \tag{9}$$

Equations (6) and (7) may now be rewritten in the forms:

$$\frac{\partial u_i}{\partial t} = -\frac{\partial}{\partial x_i}\left(\frac{\delta p}{\rho}\right) + \gamma(r)\theta x_i + \nu\nabla^2 u_i \tag{10}$$

and
$$\frac{\partial \theta}{\partial t} = 2\beta(r) u_i x_i + \kappa\nabla^2\theta, \tag{11}$$

where
$$\gamma(r) = \alpha g(r). \tag{12}$$

In addition to equations (10) and (11), we have the equation of continuity

$$\frac{\partial u_i}{\partial x_i} = 0. \tag{13}$$

Following our standard practice, we eliminate the term in $\text{grad}(\delta p/\rho)$ in equation (10) by taking its curl. We obtain

$$\frac{\partial \omega_i}{\partial t} = \gamma\epsilon_{ijk}\frac{\partial \theta}{\partial x_j} x_k + \nu\nabla^2\omega_i, \tag{14}$$

where $\boldsymbol{\omega}$ is the vorticity. Taking the curl of this equation once again, we obtain

$$\frac{\partial}{\partial t}\nabla^2 u_i = -O_i\,\theta+\nu\nabla^4 u_i, \tag{15}$$

where O_i stands for the differential operator,

$$O_i = -\epsilon_{ijk}\frac{\partial}{\partial x_j}\epsilon_{klm}\gamma x_l\frac{\partial}{\partial x_m} = \frac{\partial}{\partial x_j}\gamma\left(x_j\frac{\partial}{\partial x_i}-x_i\frac{\partial}{\partial x_j}\right). \tag{16}$$

On expanding, we find

$$O_i = \gamma\left(\frac{\partial}{\partial x_i}+\frac{\partial}{\partial x_i}x_j\frac{\partial}{\partial x_j}-x_i\nabla^2\right)+\frac{1}{r}\frac{\partial\gamma}{\partial r}\left(r^2\frac{\partial}{\partial x_i}-x_i x_j\frac{\partial}{\partial x_j}\right). \tag{17}$$

Now it can be directly verified that

$$x_i\nabla^2 f_i = \nabla^2 x_i f_i \quad \text{if } f_i \text{ is solenoidal;} \tag{18}$$

and since u_i and ω_i are both solenoidal, we obtain from equations (14) and (15)

$$\frac{\partial}{\partial t}(x_i\omega_i) = \nu\nabla^2(x_i\omega_i) \tag{19}$$

and

$$\nabla^2\left(\nu\nabla^2-\frac{\partial}{\partial t}\right)(u_i x_i) = \gamma L^2\theta, \tag{20}$$

where

$$\gamma L^2 = x_i O_i = \gamma\left(x_i\frac{\partial}{\partial x_i}+x_i\frac{\partial}{\partial x_i}x_j\frac{\partial}{\partial x_j}-r^2\nabla^2\right). \tag{21}$$

We can eliminate θ from equation (20) by observing that the operators ∇^2 and L^2 commute (see equation (25) below):

$$L^2\nabla^2 = \nabla^2 L^2. \tag{22}$$

Therefore,

$$\left(\kappa\nabla^2-\frac{\partial}{\partial t}\right)\gamma^{-1}\nabla^2\left(\nu\nabla^2-\frac{\partial}{\partial t}\right)(u_i x_i) = \left(\kappa\nabla^2-\frac{\partial}{\partial t}\right)L^2\theta = L^2\left(\kappa\nabla^2-\frac{\partial}{\partial t}\right)\theta. \tag{23}$$

Now making use of equation (11), we obtain

$$\left(\kappa\nabla^2-\frac{\partial}{\partial t}\right)\gamma^{-1}\nabla^2\left(\nu\nabla^2-\frac{\partial}{\partial t}\right)(u_i x_i) = -2\beta L^2(u_i x_i). \tag{24}$$

Equations (11), (19), (20), and (24) are the basic equations of the problem.

(a) The operator L^2

For future reference we may note here certain elementary properties of the operator L^2. In spherical polar coordinates r, ϑ, and φ,

$$L^2 = r\frac{\partial}{\partial r}+r\frac{\partial}{\partial r}r\frac{\partial}{\partial r}-r^2\nabla^2 = r^2\left(\frac{\partial^2}{\partial r^2}+\frac{2}{r}\frac{\partial}{\partial r}-\nabla^2\right)$$

$$= -\frac{1}{\sin\vartheta}\frac{\partial}{\partial\vartheta}\sin\vartheta\frac{\partial}{\partial\vartheta}-\frac{1}{\sin^2\vartheta}\frac{\partial^2}{\partial\varphi^2}; \tag{25}$$

L^2 is, indeed, the operator representing the square of the angular momentum. Its characteristic values are, therefore, $l(l+1)$; and the eigenfunctions belonging to them are the spherical harmonics $Y_l^m(\vartheta,\varphi)$:

$$L^2 Y_l^m(\vartheta,\varphi) = l(l+1)Y_l^m(\vartheta,\varphi), \tag{26}$$

where

$$Y_l^m(\vartheta,\varphi) = P_l^m(\cos\vartheta)e^{\pm im\varphi}, \tag{27}$$

and $P_l^m(\cos\vartheta)$ are the associated Legendre polynomials. The corresponding normalization integral is

$$\int_0^\pi \int_0^{2\pi} |Y_l^m(\vartheta,\varphi)|^2 \sin\vartheta\, d\vartheta d\varphi = N_l^m, \tag{28}$$

where

$$N_l^m = \frac{4\pi}{2l+1}\frac{(l+m)!}{(l-m)!}. \tag{29}$$

(b) The analysis into normal modes

Returning to equations (11), (19), (20), and (24), we must in accordance with our general procedure analyse the disturbance into normal modes. In the present instance, the analysis in terms of spherical harmonics is clearly indicated. Accordingly, we write

$$x_i\omega_i = r\omega_r = Z(r)Y_l^m(\vartheta,\varphi)e^{pt},$$
$$x_i u_i = ru_r = W(r)Y_l^m(\vartheta,\varphi)e^{pt}, \tag{30}$$

and

$$\theta = \Theta(r)Y_l^m(\vartheta,\varphi)e^{pt}.$$

Since

$$\nabla^2 = \frac{\partial^2}{\partial r^2}+\frac{2}{r}\frac{\partial}{\partial r}-\frac{L^2}{r^2}, \tag{31}$$

the effect of ∇^2 on a function which is a product of a function f of r only, and a spherical harmonic Y_l^m is given by

$$\nabla^2 Y_l^m(\vartheta,\varphi)f(r) = Y_l^m(\vartheta,\varphi)\mathscr{D}_l f(r), \tag{32}$$

where

$$\mathscr{D}_l = \frac{d^2}{dr^2}+\frac{2}{r}\frac{d}{dr}-\frac{l(l+1)}{r^2}. \tag{33}$$

In view of this identity, the insertion of the forms of the solution given in (30), in equations (11), (19), and (20), gives

$$(\mathscr{D}_l-\mathfrak{p}\sigma)\Theta = -\left(\frac{2\beta}{\kappa}R_1^2\right)W, \tag{34}$$

$$(\mathscr{D}_l-\sigma)Z = 0, \tag{35}$$

and

$$\mathscr{D}_l(\mathscr{D}_l-\sigma)W = \frac{\gamma}{\nu}R_1^4 l(l+1)\Theta, \tag{36}$$

where we have measured r in units of a suitable radius R_1 and let

$$\sigma = pR_1^2/\nu \quad \text{and} \quad \mathfrak{p} = \nu/\kappa. \tag{37}$$

Solutions of equations (34)–(36) must be sought which satisfy certain boundary conditions. These are described below.

(c) *Boundary conditions*

Let the spherical shell be confined between $r = 1$ and η. (The unit of length chosen is then the outer radius R_2 of the shell.)

In all cases we must require that

$$W = 0 \quad \text{and} \quad \Theta = 0 \quad \text{for } r = 1 \text{ and } r = \eta. \tag{38}$$

The remaining boundary conditions on W depend on the nature of the surfaces at $r = 1$ and η. We shall consider two cases: the case when the surface is rigid (as at the interface with the mantle of the liquid core of the earth); and the case when the surface is free (as at the boundary of an isolated fluid sphere in free space). On a rigid boundary, we must require that the transverse components of the velocity

$$u_\vartheta \quad \text{and} \quad u_\varphi \quad \text{also vanish;} \tag{39}$$

while on a free boundary, we must require that the tangential viscous stresses

$$p_{r\vartheta} \quad \text{and} \quad p_{r\varphi} \quad \text{vanish.} \tag{40}$$

Consider first the implications of imposing the conditions (39) on a spherical bounding surface. From the equation of continuity,

$$\frac{\partial u_r}{\partial r} + 2\frac{u_r}{r} + \frac{1}{r}\frac{\partial u_\vartheta}{\partial \vartheta} + \frac{u_\vartheta}{r}\cot\vartheta + \frac{1}{r\sin\vartheta}\frac{\partial u_\varphi}{\partial \varphi} = 0, \tag{41}$$

it follows from the vanishing of u_r, u_ϑ, and u_φ on a surface, $r =$ constant, for all values of ϑ and φ, that

$$\frac{\partial u_r}{\partial r} = 0 \quad \text{(on a rigid spherical boundary).} \tag{42}$$

An equivalent form of this condition is

$$\frac{dW}{dr} = 0 \quad \text{(on a rigid spherical boundary).} \tag{43}$$

Consider next the implications of imposing the conditions (40) on a spherical bounding surface. The expressions for the viscous stresses $p_{r\vartheta}$ and $p_{r\varphi}$ are

$$p_{r\vartheta} = \rho\nu\left(\frac{\partial u_r}{r\partial\vartheta} - \frac{u_\vartheta}{r} + \frac{\partial u_\vartheta}{\partial r}\right)$$

and

$$p_{r\varphi} = \rho\nu\left(\frac{1}{r\sin\vartheta}\frac{\partial u_r}{\partial\varphi} - \frac{u_\varphi}{r} + \frac{\partial u_\varphi}{\partial r}\right). \tag{44}$$

It follows from the vanishing of $p_{r\vartheta}$ and $p_{r\varphi}$ on a surface, $r =$ constant (on which u_r vanishes identically), that

$$\left(\frac{\partial}{\partial r}-\frac{1}{r}\right)u_\vartheta = r\frac{\partial}{\partial r}\left(\frac{u_\vartheta}{r}\right) = 0$$

and
$$\left(\frac{\partial}{\partial r}-\frac{1}{r}\right)u_\varphi = r\frac{\partial}{\partial r}\left(\frac{u_\varphi}{r}\right) = 0. \tag{45}$$

Therefore, applying the operator $r\partial/\partial r$ to the equation of continuity (41), we obtain

$$r\frac{\partial}{\partial r}\left(\frac{\partial u_r}{\partial r}+\frac{2}{r}u_r\right) = \frac{\partial^2}{\partial r^2}(ru_r)-\frac{2}{r}u_r = 0; \tag{46}$$

and since u_r vanishes on this surface, the condition (46) is equivalent to

$$\frac{d^2W}{dr^2} = 0 \quad \text{(on a free spherical boundary).} \tag{47}$$

If we should be considering a complete fluid sphere, certain boundary conditions must be satisfied at the centre which ensure that none of the quantities have singularities there. From the differential equations satisfied by W and Θ, it follows that

$$W \text{ and } \Theta \text{ must behave like } r^l \text{ as } r \to 0. \tag{48}$$

(d) The velocity field

After solving equations (34)–(36) we must complete the solution by specifying the entire velocity field; this requires the determination of the remaining components of the velocity u_ϑ and u_φ in terms of the radial components of the velocity and the vorticity. The solution can best be accomplished by making use of the following general representation of an arbitrary solenoidal vector field. (A more detailed account of this representation will be found in Appendix III.)

Any solenoidal vector field can be expressed as a linear combination of certain basic *toroidal* (**T**) and *poloidal* (**S**) fields given by

$$T_r = 0, \qquad T_\vartheta = \frac{T(r)}{r\sin\vartheta}\frac{\partial Y_l^m}{\partial\varphi}, \qquad T_\varphi = -\frac{T(r)}{r}\frac{\partial Y_l^m}{\partial\vartheta}, \tag{49}$$

and

$$S_r = \frac{l(l+1)}{r^2}S(r)Y_l^m, \qquad S_\vartheta = \frac{1}{r}\frac{\partial S}{\partial r}\frac{\partial Y_l^m}{\partial\vartheta}, \qquad S_\varphi = \frac{1}{r\sin\vartheta}\frac{\partial S}{\partial r}\frac{\partial Y_l^m}{\partial\varphi}. \tag{50}$$

The essential properties of the vector fields **S** and **T**, which are proved in Appendix III, may be recapitulated here.

(i) The fields **S** and **T** are both solenoidal.

(ii) For integrations over a surface of a sphere of radius r the fields **S** and **T** have the following orthogonality properties.

If $\mathbf{S}$ and $\mathbf{S}'$ are derived from different spherical harmonics,

$$\iint \mathbf{S}\,.\,\mathbf{S}'\,d\Sigma = 0, \tag{51}$$

where $d\Sigma = r^2 \sin\vartheta\, d\vartheta d\varphi$. Similarly, if $\mathbf{T}$ and $\mathbf{T}'$ are derived from different spherical harmonics,

$$\iint \mathbf{T}\,.\,\mathbf{T}'\,d\Sigma = 0. \tag{52}$$

(iii) If $\mathbf{S}$ and $\mathbf{S}'$ are derived from the same spherical harmonic,

$$\iint \mathbf{S}\,.\,\mathbf{S}'\,d\Sigma = l(l+1)N_l^m\left\{\frac{l(l+1)}{r^2}SS' + \frac{dS}{dr}\frac{dS'}{dr}\right\}, \tag{53}$$

where N_l^m has the same meaning as in equation (29). Similarly, if $\mathbf{T}$ and $\mathbf{T}'$ are derived from the same spherical harmonic,

$$\iint \mathbf{T}\,.\,\mathbf{T}'\,d\Sigma = l(l+1)N_l^m TT'. \tag{54}$$

(iv) Every poloidal field is orthogonal to every toroidal field:

$$\iint \mathbf{S}\,.\,\mathbf{T}\,d\Sigma = 0. \tag{55}$$

(v) curl $\mathbf{S}$ represents a toroidal field with the defining scalar

$$\tilde{S} = \frac{l(l+1)}{r^2}S - \frac{d^2S}{dr^2}; \tag{56}$$

and curl $\mathbf{T}$ represents a poloidal field with the defining scalar T.

(vi) curl$^2\mathbf{S}$ is again a poloidal field with $\tilde{S}$ as its defining scalar; similarly, curl$^2\mathbf{T}$ is a toroidal field with the defining scalar

$$\tilde{T} = \frac{l(l+1)}{r^2}T - \frac{d^2T}{dr^2}. \tag{57}$$

From the foregoing it follows that W and Z, as we have defined them in equations (30), are related to the scalars S and T by

$$S = \frac{rW}{l(l+1)} \quad \text{and} \quad T = \frac{rZ}{l(l+1)}. \tag{58}$$

57. The validity of the principle of the exchange of stabilities for the case β = constant and γ = constant

In this section the discussion will be restricted to the case when both β and γ are constants. Then (see equations (1), (5), (9), and (12))

$$\beta = \beta_2 = \epsilon/6\kappa \quad \text{and} \quad \gamma = g\alpha = \tfrac{4}{3}\pi G\rho\alpha. \tag{59}$$

Equations (34) and (36) can then be combined to give

$$\mathscr{D}_l(\mathscr{D}_l-\sigma)(\mathscr{D}_l-\mathfrak{p}\sigma)W = -l(l+1)C_l W, \tag{60}$$

where

$$C_l = \frac{2\beta\gamma}{\kappa\nu} R_1^6. \tag{61}$$

We shall now show that when β and γ are constants, the principle of the exchange of stabilities is valid. But first we shall prove certain properties of the operator $\mathscr{D}_l$ which we shall find useful in the subsequent reductions.

Consider the integral

$$\int\limits_b^a r^2\phi\mathscr{D}_l\psi\,dr = \int\limits_b^a r^2\phi\left\{\frac{d^2\psi}{dr^2}+\frac{2}{r}\frac{d\psi}{dr}-\frac{l(l+1)}{r^2}\psi\right\}dr$$

$$= \int\limits_b^a \left\{\phi\frac{d}{dr}\left(r^2\frac{d\psi}{dr}\right)-l(l+1)\phi\psi\right\}dr, \tag{62}$$

where $\phi(r)$ and $\psi(r)$ are any two functions which are bounded and continuous in the interval (a,b). By an integration by parts, we have

$$\int\limits_b^a r^2\phi\mathscr{D}_l\psi\,dr = r^2\phi\frac{d\psi}{dr}\bigg|_b^a - \int\limits_b^a \left\{r^2\frac{d\psi}{dr}\frac{d\phi}{dr}+l(l+1)\phi\psi\right\}dr. \tag{63}$$

And by a further integration by parts, we obtain

$$\int\limits_b^a r^2\phi\mathscr{D}_l\psi\,dr = r^2\left(\phi\frac{d\psi}{dr}-\psi\frac{d\phi}{dr}\right)_b^a + \int\limits_b^a r^2\psi\mathscr{D}_l\phi\,dr. \tag{64}$$

In particular, if ϕ and ψ vanish at both limits,

$$\int\limits_b^a r^2\phi^*\mathscr{D}_l\phi\,dr = -\int\limits_b^a \left\{r^2\left|\frac{d\phi}{dr}\right|^2+l(l+1)|\phi|^2\right\}dr \tag{65}$$

and

$$\int\limits_b^a r^2\phi\mathscr{D}_l\psi\,dr = \int\limits_b^a r^2\psi\mathscr{D}_l\phi\,dr. \tag{66}$$

This last equation means that *with respect to functions which vanish at the ends of an interval, the operator* $r^2\mathscr{D}_l$ *is Hermitian.*

Returning to equations (34) and (36), let

$$F = l(l+1)\frac{\gamma}{\nu}R_1^4\Theta; \tag{67}$$

the equations then become

$$\mathscr{D}_l(\mathscr{D}_l-\sigma)W = F \tag{68}$$

and

$$(\mathscr{D}_l-\mathfrak{p}\sigma)F = -l(l+1)C_l W. \tag{69}$$

The boundary conditions are

$$W = F = 0 \text{ and } \textit{either } \frac{dW}{dr} \textit{ or } \frac{d^2W}{dr^2} = 0 \text{ for } r = 1 \text{ and } \eta. \tag{70}$$

Now multiply equation (69) by r^2F^* and integrate over the range of r. Since F vanishes at both limits, we can use equation (65) and obtain

$$\int_\eta^1 \left\{r^2\left|\frac{dF}{dr}\right|^2 + l(l+1)|F|^2 + \mathfrak{p}\sigma r^2|F|^2\right\} dr = l(l+1)C_l \int_\eta^1 r^2WF^*\,dr. \tag{71}$$

By (68), the integral on the right-hand side of this equation is

$$\int_\eta^1 r^2WF^*\,dr = \int_\eta^1 r^2W\mathscr{D}_l^2\,W^*\,dr - \sigma^* \int_\eta^1 r^2W\mathscr{D}_l\,W^*\,dr. \tag{72}$$

Transforming the integrals on the right-hand side of equation (72) with the aid of equations (64) and (65), we obtain

$$\int_\eta^1 r^2WF^*\,dr = -\left(r^2\frac{dW}{dr}\mathscr{D}_l\,W^*\right)_\eta^1 + \int_\eta^1 r^2|\mathscr{D}_l\,W|^2\,dr +$$

$$+\sigma^* \int_\eta^1 \left\{r^2\left|\frac{dW}{dr}\right|^2 + l(l+1)|W|^2\right\} dr. \tag{73}$$

On a rigid boundary $dW/dr = 0$ and the integrated part in (73) vanishes; but if the boundary is free, then $d^2W/dr^2 = 0$ and

$$(\mathscr{D}_l\,W^*)_{r=1\text{ or }\eta} = \left[\frac{d^2W^*}{dr^2} + \frac{2}{r}\frac{dW^*}{dr} - l(l+1)\frac{W^*}{r^2}\right]_{r=1\text{ or }\eta}$$

$$= 2\left(\frac{1}{r}\frac{dW^*}{dr}\right)_{r=1\text{ or }\eta} \tag{74}$$

In either case, the integrated part can be written as

$$-2\left[r\left|\frac{dW}{dr}\right|^2\right]_\eta^1. \tag{75}$$

Thus equation (71) reduces to

$$\int_\eta^1 \left\{r^2\left|\frac{dF}{dr}\right|^2 + l(l+1)|F|^2 + \mathfrak{p}\sigma r^2|F|^2\right\} dr -$$

$$- l(l+1)C_l\left\{-2\left[r\left|\frac{dW}{dr}\right|^2\right]_\eta^1 + \int_\eta^1 r^2|\mathscr{D}_l\,W|^2\,dr +\right.$$

$$\left. +\sigma^* \int_\eta^1 \left[r^2\left|\frac{dW}{dr}\right|^2 + l(l+1)|W|^2\right] dr\right\} = 0. \tag{76}$$

The real and the imaginary parts of equation (76) must vanish separately; and the vanishing of the imaginary part gives

$$\operatorname{im}(\sigma)\left\{\mathrm{p}\int_{\eta}^{1} r^2|F|^2\,dr+l(l+1)C_l\int_{\eta}^{1}\left[r^2\left|\frac{dW}{dr}\right|^2+l(l+1)|W|^2\right]dr\right\}=0. \tag{77}$$

The factor of $\operatorname{im}(\sigma)$ in this equation is positive definite. Accordingly,

$$\operatorname{im}(\sigma)=0. \tag{78}$$

Hence σ is real and the onset of instability must be via a marginal state which is stationary. The principle of the exchange of stabilities is therefore valid.

58. A variational principle for the case when β and γ are constants

Since the principle of the exchange of stabilities is valid when β and γ are constants, the equations governing the marginal state in this instance are

$$\mathscr{D}_l Z=0, \tag{79}$$

$$\mathscr{D}_l^2 W=F, \tag{80}$$

and

$$\mathscr{D}_l F=-l(l+1)C_l W. \tag{81}$$

And the boundary conditions at $r=1$ and η are the same as those given in (70).

Equation (79) clearly requires that

$$Z\equiv 0. \tag{82}$$

The radial component of the vorticity, therefore, vanishes identically. According to the theorems stated in § 56 (*d*), *the velocity field is purely poloidal.*

Equations (80) and (81) together with the boundary conditions (70) constitute a characteristic value problem for C_l; and instability will set in with that value of l which leads to the lowest value for C_l.

The present characteristic value problem can also be formulated in terms of a variational principle. The basis for the variational principle follows from equation (76) which, when $\sigma=0$, gives

$$l(l+1)C_l=\frac{\int_{\eta}^{1}\{r^2(dF/dr)^2+l(l+1)F^2\}\,dr}{-2[r(dW/dr)^2]_{\eta}^{1}+\int_{\eta}^{1}r^2(\mathscr{D}_l W)^2\,dr}. \tag{83}$$

It can be readily proved by the methods described in Chapter II, § 13 that the lowest characteristic value for C_l represents the absolute minimum of the quantity on the right-hand side of equation (83). An equivalent formulation of the principle is that $l(l+1)C_l$ is the minimum value of

$$\int_\eta^1 \left\{r^2\left(\frac{dF}{dr}\right)^2+l(l+1)F^2\right\} dr, \tag{84}$$

for all variations of F which vanish at $r = 1$ and η and preserve the constancy of

$$\int_\eta^1 r^2 W \mathscr{D}_l^2 W\, dr, \tag{85}$$

where W is determined by F and the boundary conditions through the equation $\mathscr{D}_l^2 W = F$. In this latter formulation, $l(l+1)C_l$ appears as the Lagrangian undetermined multiplier.

(a) The thermodynamic significance of the variational principle

We shall now show that the physical content of equation (83) is the same as in the similar contexts in earlier chapters: it simply expresses the equality of the rates at which energy is dissipated by viscosity and is liberated by the buoyancy force as a sufficient means for characterizing the marginal state.

Consider then the rate at which energy is dissipated by viscosity. This is given by

$$\epsilon_\nu = \rho\nu \iiint \mathbf{u}\,.\,\mathrm{curl}^2\mathbf{u}\, dV, \tag{86}$$

where the integration is extended over the volume of the spherical shell. We have already seen that the velocity field is entirely poloidal. We can, therefore, write

$$\epsilon_\nu = \rho\nu \iiint \mathbf{S}\,.\,\mathrm{curl}^2\mathbf{S}\, dV. \tag{87}$$

By the theorems stated in § 56 (d), this is the same as

$$\begin{aligned}\epsilon_\nu &= \rho\nu \iiint \mathbf{S}\,.\,\tilde{\mathbf{S}}\, dV\\ &= \rho\nu l(l+1)N_l^m \int_{R_2}^{R_1} \left\{l(l+1)\frac{S\tilde{S}}{r^2}+\frac{dS}{dr}\frac{d\tilde{S}}{dr}\right\} dr.\end{aligned} \tag{88}$$

By an integration by parts, we obtain

$$\epsilon_\nu = \rho\nu l(l+1)N_l^m\left\{\int_{R_2}^{R_1}\left[l(l+1)\frac{S}{r^2}-\frac{d^2S}{dr^2}\right]\tilde{S}\, dr+\left(\tilde{S}\frac{dS}{dr}\right)_{R_2}^{R_1}\right\}. \tag{89}$$

Remembering the definition of $\tilde{S}$ and the requirement that S vanishes on the boundaries at $r = R_1$ and R_2, we have

$$\epsilon_\nu = \rho\nu l(l+1)N_l^m\left\{\int_{R_2}^{R_1}\tilde{S}^2\, dr-\left(\frac{d^2S}{dr^2}\frac{dS}{dr}\right)_{R_2}^{R_1}\right\}. \tag{90}$$

From the relation between W and the defining scalar S given in (58), it follows that

$$\tilde{S} = -\frac{1}{l(l+1)} r\mathscr{D}_l W; \tag{91}$$

and we can rewrite equation (90) in the form

$$\epsilon_\nu = \frac{\rho\nu N_l^m}{l(l+1)R_1}\left\{\int_\eta^1 r^2(\mathscr{D}_l W)^2\,dr - \left[\frac{d^2}{dr^2}(rW)\frac{d}{dr}(rW)\right]_\eta^1\right\}. \tag{92}$$

Since W and either dW/dr or d^2W/dr^2 vanish for $r = 1$ and $r = \eta$, we can also write

$$\epsilon_\nu = \frac{\rho\nu N_l^m}{l(l+1)R_1}\left\{\int_\eta^1 r^2(\mathscr{D}_l W)^2\,dr - 2\left[r\left(\frac{dW}{dr}\right)^2\right]_\eta^1\right\}. \tag{93}$$

Considering next the rate at which internal energy is released by the buoyancy force acting on the fluid, we must evaluate

$$\epsilon_g = \rho\gamma \iiint (u_i\, x_i)\theta\, dV, \tag{94}$$

or, according to equation (11),

$$\epsilon_g = -\rho\frac{\kappa\gamma}{2\beta}R_1 \iiint \theta\nabla^2\theta\, dV, \tag{95}$$

where the unit of length is again R_1. After performing the angle integrations, we are left with

$$\epsilon_g = -\rho\frac{\kappa\gamma}{2\beta}R_1 N_l^m \int_\eta^1 r^2\Theta\mathscr{D}_l\Theta\, dr. \tag{96}$$

Since Θ vanishes at both limits, we can use equation (65) to obtain

$$\epsilon_g = \rho\frac{\kappa\gamma}{2\beta}R_1 N_l^m \int_\eta^1 \left\{r^2\left(\frac{d\Theta}{dr}\right)^2 + l(l+1)\Theta^2\right\} dr. \tag{97}$$

Expressing Θ in terms of F in accordance with (67), we have

$$\epsilon_g = \rho\frac{\kappa\gamma}{2\beta}\,\frac{\nu^2}{\gamma^2[l(l+1)]^2}\,\frac{N_l^m}{R_1^7}\int_\eta^1 \left\{r^2\left(\frac{dF}{dr}\right)^2 + l(l+1)F^2\right\} dr. \tag{98}$$

Equating ϵ_ν and ϵ_g given by equations (93) and (98), we recover equation (83) which is the basis of the variational principle.

59. On the onset of thermal instability in a fluid sphere

For a homogeneous fluid sphere, β and γ are necessarily constants and the analysis of §§ 57 and 58 applies. We shall now obtain the solution of the associated characteristic value problem by the variational method.

Considering the onset of instability as a harmonic of order l, we expand F in a Fourier–Bessel series of the form

$$F = \frac{1}{\sqrt{r}} \sum_j A_j J_{l+\frac{1}{2}}(\alpha_{l,j} r), \tag{99}$$

where $J_{l+\frac{1}{2}}$ denotes the Bessel function of order $l+\frac{1}{2}$ and $\alpha_{l,j}$ is its jth zero. The summation over j in equation (99) may be considered as running from 1 to ∞; but this is not necessary as we shall consider the A_j's as variational parameters. With the foregoing choice, the boundary conditions on F at $r = 1$ and 0 are automatically satisfied.

The functions $J_{l+\frac{1}{2}}(\alpha_{l,j} r)$ for $j = 1, 2,...$ and given l form a complete set of orthogonal functions in the interval $(0, 1)$ and satisfy the orthogonality relation

$$\int_0^1 r J_{l+\frac{1}{2}}(\alpha_{l,j} r) J_{l+\frac{1}{2}}(\alpha_{l,k} r)\, dr = \tfrac{1}{2}\delta_{jk}[J'_{l+\frac{1}{2}}(\alpha_{l,j})]^2, \tag{100}$$

where a prime denotes differentiation with respect to the argument.

With the chosen form of F, the equation to be solved for W is

$$\mathscr{D}_l^2 W = \frac{1}{\sqrt{r}} \sum_j A_j J_{l+\frac{1}{2}}(\alpha_{l,j} r). \tag{101}$$

We may express the solution of this equation in the form

$$W = \sum_j A_j W_j, \tag{102}$$

where W_j is a solution of the equation

$$\mathscr{D}_l^2 W_j = \frac{1}{\sqrt{r}} J_{l+\frac{1}{2}}(\alpha_{l,j} r) \tag{103}$$

which satisfies the necessary boundary conditions. Since

$$\mathscr{D}_l \frac{J_{l+\frac{1}{2}}(\alpha r)}{\sqrt{r}} = -\frac{\alpha^2}{\sqrt{r}} J_{l+\frac{1}{2}}(\alpha r), \tag{104}$$

a particular integral of equation (103) which is free of singularity at the origin is

$$\frac{1}{\alpha_{l,j}^4} \frac{J_{l+\frac{1}{2}}(\alpha_{l,j} r)}{\sqrt{r}}. \tag{105}$$

Adding to (105) the complementary function $B^{(j)} r^l + C^{(j)} r^{l+2}$ (where $B^{(j)}$ and $C^{(j)}$ are constants), we have the general solution

$$W_j = \frac{1}{\alpha_{l,j}^4} \frac{J_{l+\frac{1}{2}}(\alpha_{l,j} r)}{\sqrt{r}} + B^{(j)} r^l + C^{(j)} r^{l+2}. \tag{106}$$

The condition $W_j = 0$ at $r = 1$ requires $B^{(j)} = -C^{(j)}$ and we have

$$W_j = \frac{1}{\alpha_{l,j}^4} \frac{J_{l+\frac{1}{2}}(\alpha_{l,j} r)}{\sqrt{r}} + B^{(j)}(r^l - r^{l+2}). \tag{107}$$

The constant $B^{(j)}$ is determined by the remaining boundary condition at $r = 1$, namely, that here either dW_j/dr or d^2W_j/dr^2 vanishes (depending on whether the bounding surface at $r = 1$ is rigid or free). We find

$$B^{(j)} = \tfrac{1}{4} q \frac{J'_{l+\frac{1}{2}}(\alpha_{l,j})}{\alpha_{l,j}^3}, \tag{108}$$

where

$$\left.\begin{aligned} q &= 2 && \text{for a rigid boundary at } r = 1 \\ &= -\frac{4}{2l+1} && \text{for a free boundary at } r = 1 \end{aligned}\right\}. \tag{109}$$

We now substitute for F and W according to equations (99), (102), and (107) in equation (81), and obtain

$$\sum_j A_j \alpha_{l,j}^2 \frac{J_{l+\frac{1}{2}}(\alpha_{l,j} r)}{\sqrt{r}} = l(l+1)C_l \sum_j A_j \left\{ \frac{1}{\alpha_{l,j}^4} \frac{J_{l+\frac{1}{2}}(\alpha_{l,j} r)}{\sqrt{r}} + B^{(j)}(r^l - r^{l+2}) \right\}. \tag{110}$$

Multiply this equation by $r^{\frac{3}{2}} J_{l+\frac{1}{2}}(\alpha_{l,k} r)$ and integrate over the range of r. Making use of the orthogonality relation (100), we obtain

$$\tfrac{1}{2}[J'_{l+\frac{1}{2}}(\alpha_{l,k})]^2 \alpha_{l,k}^2 A_k = l(l+1)C_l \left\{ \frac{1}{2} \frac{[J'_{l+\frac{1}{2}}(\alpha_{l,k})]^2}{\alpha_{l,k}^4} A_k + \sum_j (k|j) A_j \right\} \quad (k = 1, 2, \ldots), \tag{111}$$

where

$$(k|j) = B^{(j)} \int_0^1 (r^{l+\frac{3}{2}} - r^{l+\frac{7}{2}}) J_{l+\frac{1}{2}}(\alpha_{l,k} r)\, dr. \tag{112}$$

Equation (111) represents a set of linear homogeneous equations for the constants A_j. This same set of equations ensures that the expression (83) for C_l is a minimum for all variations of the A_j's which preserve the constancy of (85).

Before writing down the secular determinant which follows from equation (111), we shall first evaluate the matrix element $(k|j)$. We find that the integral defining $(k|j)$ can be evaluated explicitly if proper use is made of the various recurrence relations satisfied by the Bessel functions. We find

$$(k|j) = 2B^{(j)} \frac{J_{l+\frac{5}{2}}(\alpha_{l,k})}{\alpha_{l,k}^2}. \tag{113}$$

Making further use of the recurrence relations satisfied by the Bessel functions and remembering that $\alpha_{l,k}$ is a zero of $J_{l+\frac{1}{2}}(x)$, we find that

$$J_{l+\frac{5}{2}}(\alpha_{l,k}) = \frac{2(l+\frac{3}{2})}{\alpha_{l,k}} J_{l+\frac{3}{2}}(\alpha_{l,k}) = -\frac{2(l+\frac{3}{2})}{\alpha_{l,k}} J'_{l+\frac{1}{2}}(\alpha_{l,k}). \tag{114}$$

Hence,
$$(k|j) = -4\frac{l+\frac{3}{2}}{\alpha_{l,k}^3} J'_{l+\frac{1}{2}}(\alpha_{l,k})B^{(j)}, \tag{115}$$

or, substituting for $B^{(j)}$ its value (108), we have

$$(k|j) = -q(l+\tfrac{3}{2})\frac{J'_{l+\frac{1}{2}}(\alpha_{l,k})J'_{l+\frac{1}{2}}(\alpha_{l,j})}{\alpha_{l,k}^3\,\alpha_{l,j}^3}, \tag{116}$$

which is manifestly symmetric in k and j. This symmetry of $(k|j)$ reflects the self-adjoint character of the underlying characteristic value problem.

With $(k|j)$ given by equation (116), equation (111) becomes

$$\frac{J'_{l+\frac{1}{2}}(\alpha_{l,k})}{\alpha_{l,k}^3}\sum_j\left\{\tfrac{1}{2}[J'_{l+\frac{1}{2}}(\alpha_{l,k})]\left[\frac{\alpha_{l,k}^5}{l(l+1)C_l}-\frac{1}{\alpha_{l,k}}\right]\delta_{jk}+q(l+\tfrac{3}{2})\frac{J'_{l+\frac{1}{2}}(\alpha_{l,j})}{\alpha_{l,j}^3}\right\}A_j = 0. \tag{117}$$

Letting
$$\mathscr{A}_j = \frac{J'_{l+\frac{1}{2}}(\alpha_{l,j})}{\alpha_{l,j}^3}A_j, \tag{118}$$

we can rewrite equation (117) in the form

$$\sum_j\left\{\frac{\alpha_{l,k}^8}{q(2l+3)}\left[\frac{1}{l(l+1)C_l}-\frac{1}{\alpha_{l,k}^6}\right]\delta_{jk}+1\right\}\mathscr{A}_j = 0. \tag{119}$$

The required secular determinant is, therefore,

$$\left\|\frac{\alpha_{l,k}^8}{q(2l+3)}\left[\frac{1}{l(l+1)C_l}-\frac{1}{\alpha_{l,k}^6}\right]\delta_{jk}+1^{j+k}\right\| = 0. \tag{120}$$

A first approximation to the value of C_l is obtained by setting the (1, 1) element of the secular matrix equal to zero. Thus, we obtain

$$l(l+1)C_l = \frac{\alpha_{l,1}^8}{\alpha_{l,1}^2-q(2l+3)}. \tag{121}$$

Values in higher approximations can be obtained by retaining more rows and columns in the secular determinant. The values given in Tables XX and XXI were obtained in this manner; those listed under 'exact' in these tables are due to Backus.

The dependence of C_l on l and on the boundary conditions is exhibited in Fig. 56.

It will be observed that in both the cases considered, the easiest modes to excite are those belonging to $l = 1$.

(*a*) *The cell patterns*

The streamlines of the flow are governed by the equations

$$\frac{dr}{u_r} = \frac{r\,d\vartheta}{u_\vartheta} = \frac{r\sin\vartheta\,d\varphi}{u_\varphi}. \tag{122}$$

TABLE XX

The characteristic numbers C_l and related constants in case the bounding surface is free

l	First approximation	Second approximation	Third approximation	Exact	Second approximation A_2	Third approximation A_2	Third approximation A_3
1	3.0940×10^3	3.0914×10^3	3.0912×10^3	3.0912×10^3	-0.02202	-0.02203	0.00453
2	5.2274×10^3	5.2244×10^3	5.2242×10^3	5.2241×10^3	-0.01912	-0.01912	0.00461
3	8.7786×10^3	8.7750×10^3	8.7746×10^3	8.7745×10^3	-0.01664	-0.01664	0.00447
4	1.3986×10^4	1.3982×10^4	1.3981×10^4	1.3981×10^4	-0.01462	-0.01463	0.00426
5	2.1209×10^4	2.1204×10^4	2.1204×10^4	2.1204×10^4	-0.01298	-0.01298	0.00402
6	3.0853×10^4	3.0848×10^4	3.0847×10^4	3.0847×10^4	-0.01163	-0.01164	0.00378
7	4.3360×10^4	4.3354×10^4	4.3353×10^4		-0.01051	-0.01052	0.00355
8	5.9203×10^4	5.9196×10^4	5.9195×10^4		-0.00957	-0.00958	0.00334
9	7.8887×10^4	7.8880×10^4	7.8879×10^4		-0.00877	-0.00877	0.00315
10	1.0295×10^5	1.0294×10^5	1.0294×10^5		-0.00808	-0.00809	0.00297
11	1.3196×10^5	1.3195×10^5	1.3195×10^5		-0.00748	-0.00749	0.00280
12	1.6650×10^5	1.6649×10^5			-0.00696		
13	2.0721×10^5	2.0720×10^5			-0.00650		
14	2.5475×10^5	2.5474×10^5			-0.00609		
15	3.0978×10^5	3.0977×10^5			-0.00573		

TABLE XXI

The characteristic numbers C_l and related constants in case the bounding surface is rigid

l	First approximation	Second approximation	Third approximation	Exact	Second approximation A_2	Third approximation A_2	Third approximation A_3
1	8.1540×10^3	8.0471×10^3	8.0410×10^3	8.0401×10^3	0.08880	0.08866	-0.01747
2	1.0559×10^4	1.0403×10^4	1.0391×10^4	1.0389×10^4	0.09858	0.09830	-0.02251
3	1.5368×10^4	1.5132×10^4	1.5110×10^4	1.5105×10^4	0.10409	0.10365	-0.02634
4	2.2352×10^4	2.2006×10^4	2.1969×10^4	2.1959×10^4	0.10729	0.10668	-0.02929
5	3.1801×10^4	3.1315×10^4	3.1256×10^4	3.1238×10^4	0.10914	0.10837	-0.03158
6	4.4117×10^4	4.3459×10^4	4.3370×10^4	4.3342×10^4	0.11015	0.10922	-0.03337
7	5.9759×10^4	5.8892×10^4	5.8765×10^4		0.11062	0.10955	-0.03480
8	7.9221×10^4	7.8108×10^4	7.7934×10^4		0.11074	0.10954	-0.03593
9	1.0304×10^5	1.0164×10^5	1.0140×10^5		0.11061	0.10929	-0.03685
10	1.3176×10^5	1.3004×10^5	1.2973×10^5		0.11032	0.10888	-0.03759
11	1.6600×10^5	1.6390×10^5	1.6351×10^5		0.10991	0.10837	-0.03819
12	2.0637×10^5	2.0384×10^5			0.10942		
13	2.5357×10^5	2.5058×10^5			0.10888		
14	3.0814×10^5	3.0462×10^5			0.10830		
15	3.7094×10^5	3.6685×10^5			0.10770		

When the velocity field is purely poloidal (as in the present instance), the equations take the explicit forms

$$\frac{dr}{l(l+1)(S(r)/r^2)Y_l^m} = \frac{r\,d\vartheta}{(1/r)(dS/dr)(\partial Y_l^m/\partial\vartheta)} = \frac{r\sin\vartheta\,d\varphi}{(1/r\sin\vartheta)(dS/dr)(\partial Y_l^m/\partial\varphi)}. \tag{123}$$

These equations can be explicitly integrated for the lowest axisymmetric

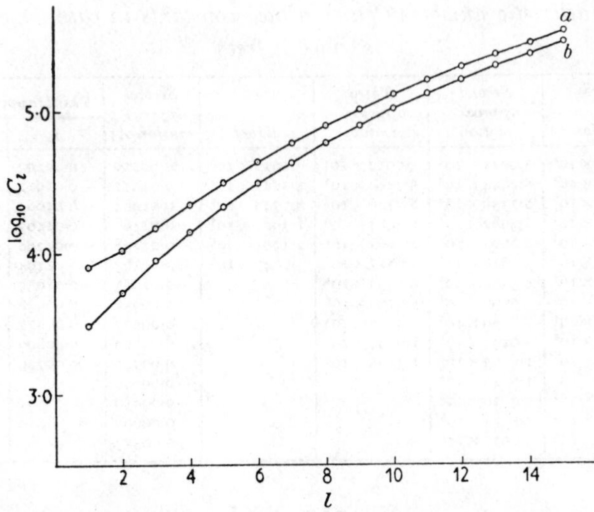

Fig. 56. The Rayleigh number C_l for the onset of thermal convection in a fluid sphere as a spherical harmonic disturbance of order l for the two cases: when the bounding surface is rigid (curve labelled a) and when the bounding surface is free (curve labelled b).

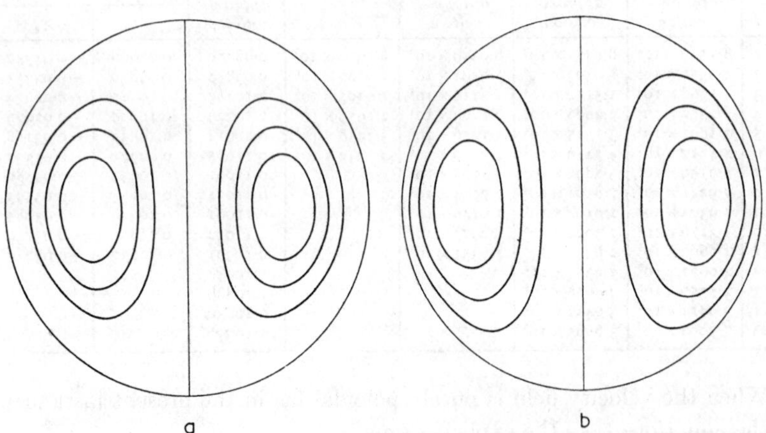

Fig. 57. The pattern of convection in a fluid sphere for the lowest symmetric mode of disturbance, $l = 1$, $m = 0$: (a) for a sphere with a rigid boundary, and (b) for a sphere with a free boundary.

mode $l = 1$ and $m = 0$: the streamlines are then confined to the meridional planes; and in these planes the equation governing them is

$$\frac{1}{2S}\frac{dS}{dr}\,dr = -\cot\vartheta\,d\vartheta. \tag{124}$$

The integral of this equation is

$$\sin\vartheta = \frac{\text{constant}}{\sqrt{S}}. \tag{125}$$

In terms of W the equation is

$$\sin\vartheta = \frac{\text{constant}}{\sqrt{(rW)}}. \tag{126}$$

The streamlines derived from this equation are illustrated in Fig. 57 *a*, *b*.

60. On the onset of thermal instability in spherical shells

On the assumption that the principle of the exchange of stabilities is valid,† the equations governing the marginal state are (cf. equations (34) and (36))

$$\mathscr{D}_l\,\Theta = -\frac{2\beta(r)}{\kappa}\,R_1^2\,W \tag{127}$$

and

$$\mathscr{D}_l^2\,W = \frac{\gamma(r)}{\nu}\,R_1^4\,l(l+1)\Theta, \tag{128}$$

where R_1 is the radius of the outer spherical boundary. Letting

$$\beta(r) = \beta_1\,b(r), \qquad \gamma(r) = \gamma_1\,c(r), \tag{129}$$

and

$$F = \frac{\gamma_1\,R_1^4}{\nu}\,l(l+1)c(r)\Theta, \tag{130}$$

where β_1 and γ_1 are the values of $\beta(r)$ and $\gamma(r)$ at $r = 1$, we can rewrite equations (127) and (128) in the forms

$$\mathscr{D}_l^2\,W = F \tag{131}$$

and

$$\frac{1}{b(r)}\,\mathscr{D}_l\left\{\frac{F}{c(r)}\right\} = -l(l+1)C_l\,W, \tag{132}$$

where

$$C_l = \frac{2\beta_1\gamma_1}{\kappa\nu}\,R_1^6. \tag{133}$$

Solutions of equations (131) and (132) must be sought which satisfy the boundary conditions (70).

The form of equation (131) which must be solved for W suggests that we expand F in a series of cylinder functions of order $l+\frac{1}{2}$ which vanish at $r = 1$ and η. Such functions can be constructed as follows. Let

$$\mathscr{C}_{l+\frac{1}{2},\nu}(z) = J_{-(l+\frac{1}{2})}(\alpha\eta)J_\nu(z) - J_{l+\frac{1}{2}}(\alpha\eta)J_{-\nu}(z), \tag{134}$$

† It should be noted that this principle has been proved only for the case β = constant and γ = constant.

where α is a constant unspecified for the present. Then,

$$\mathscr{C}_{l+\frac{1}{2},l+\frac{1}{2}}(\alpha r) = J_{-(l+\frac{1}{2})}(\alpha\eta)J_{l+\frac{1}{2}}(\alpha r)-J_{l+\frac{1}{2}}(\alpha\eta)J_{-(l+\frac{1}{2})}(\alpha r) \tag{135}$$

clearly vanishes at $r = \eta$; it will also vanish at $r = 1$ provided

$$J_{-(l+\frac{1}{2})}(\alpha\eta)J_{l+\frac{1}{2}}(\alpha)-J_{l+\frac{1}{2}}(\alpha\eta)J_{-(l+\frac{1}{2})}(\alpha) = 0. \tag{136}$$

It is known that equation (136) admits an infinity of roots, all of which are real and simple; and further, that if α_j $(j = 1, 2,...)$ are the distinct roots of the equation, the functions $\mathscr{C}_{l+\frac{1}{2},l+\frac{1}{2}}(\alpha_j r)$ form an orthogonal set of functions with the integral property

$$\int_\eta^1 r\mathscr{C}_{l+\frac{1}{2},l+\frac{1}{2}}(\alpha_j r)\mathscr{C}_{l+\frac{1}{2},l+\frac{1}{2}}(\alpha_k r)\,dr = N_{l+\frac{1}{2},j}\,\delta_{jk}, \tag{137}$$

where
$$N_{l+\frac{1}{2},j} = \frac{2}{\pi^2\alpha_j^2}\left\{\frac{J^2_{l+\frac{1}{2}}(\alpha_j\eta)}{J^2_{l+\frac{1}{2}}(\alpha_j)}-1\right\}. \tag{138}$$

For later use, we may note here that the derivatives of $\mathscr{C}_{l+\frac{1}{2},l+\frac{1}{2}}(\alpha_j r)$ at $r = 1$ and $r = \eta$ (which we shall denote by $\mathscr{C}'_{l+\frac{1}{2}}(\alpha_j)$ and $\mathscr{C}'_{l+\frac{1}{2}}(\alpha_j\eta)$, respectively) are given by

$$\mathscr{C}'_{l+\frac{1}{2}}(\alpha_j) = \left[\frac{1}{\alpha_j}\frac{d}{dr}\mathscr{C}_{l+\frac{1}{2},l+\frac{1}{2}}(\alpha_j r)\right]_{r=1} = (-1)^l\frac{2}{\pi\alpha_j}\frac{J_{l+\frac{1}{2}}(\alpha_j\eta)}{J_{l+\frac{1}{2}}(\alpha_j)} \tag{139}$$

and
$$\mathscr{C}'_{l+\frac{1}{2}}(\alpha_j\eta) = \left[\frac{1}{\alpha_j}\frac{d}{dr}\mathscr{C}_{l+\frac{1}{2},l+\frac{1}{2}}(\alpha_j r)\right]_{r=\eta} = (-1)^l\frac{2}{\pi\alpha_j\eta}. \tag{140}$$

Also, since $\mathscr{C}_{l+\frac{1}{2},\nu}(z)$ represents a solution of Bessel's equation of order ν, it must clearly satisfy the usual recurrence relations implied by the equations

$$\frac{d}{dz}\{z^{\nu+1}\mathscr{C}_{l+\frac{1}{2},\nu+1}(z)\} = +z^{\nu+1}\mathscr{C}_{l+\frac{1}{2},\nu}(z)$$

and
$$\frac{d}{dz}\{z^{-\nu+1}\mathscr{C}_{l+\frac{1}{2},\nu-1}(z)\} = -z^{-\nu+1}\mathscr{C}_{l+\frac{1}{2},\nu}(z). \tag{141}$$

Returning to the solution of equations (131) and (132), we express F as a series in the form

$$F = \frac{1}{\sqrt{r}}\sum_j A_j\mathscr{C}_{l+\frac{1}{2},l+\frac{1}{2}}(\alpha_j r), \tag{142}$$

where the A_j's are constants. With this choice for F, the solution for W can be written in the form

$$W = \sum_j A_j W_j, \tag{143}$$

where W_j is a solution of the equation

$$\mathscr{D}_l^2 W_j = \frac{1}{\sqrt{r}}\mathscr{C}_{l+\frac{1}{2},l+\frac{1}{2}}(\alpha_j r) \tag{144}$$

which satisfies the necessary boundary conditions at $r = 1$ and η. Since

$$\mathscr{D}_l \frac{\mathscr{C}_{l+\frac{1}{2},l+\frac{1}{2}}(\alpha_j r)}{\sqrt{r}} = -\alpha_j^2 \frac{\mathscr{C}_{l+\frac{1}{2},l+\frac{1}{2}}(\alpha_j r)}{\sqrt{r}}, \tag{145}$$

the general solution of equation (144) is given by

$$W_j = \frac{1}{\alpha_j^4 \sqrt{r}} \mathscr{C}_{l+\frac{1}{2},l+\frac{1}{2}}(\alpha_j r) + B_1^{(j)} r^l + B_2^{(j)} r^{l+2} + B_3^{(j)} r^{-(l+1)} + B_4^{(j)} r^{-(l-1)}, \tag{146}$$

where $B_1^{(j)},..., B_4^{(j)}$ are constants of integration to be determined by the boundary conditions. The conditions $W = 0$ at $r = 1$ and $r = \eta$ (which do not depend on the nature of the bounding surfaces) require

$$B_1^{(j)} + B_2^{(j)} + B_3^{(j)} + B_4^{(j)} = 0 \tag{147}$$

and

$$B_1^{(j)} \eta^l + B_2^{(j)} \eta^{l+2} + B_3^{(j)} \eta^{-(l+1)} + B_4^{(j)} \eta^{-(l-1)} = 0. \tag{148}$$

The remaining boundary conditions depend on the nature of the surfaces at $r = 1$ and $r = \eta$; and there are four cases to consider. We shall return to them presently.

We now substitute the solutions for F and W in accordance with equations (142), (143), and (146) into equation (132) and obtain

$$\sum_j A_j \frac{1}{b(r)} \mathscr{D}_l \left\{ \frac{\mathscr{C}_{l+\frac{1}{2},l+\frac{1}{2}}(\alpha_j r)}{c(r)\sqrt{r}} \right\} = -l(l+1)C_l \sum_j A_j \left\{ \frac{1}{\alpha_j^4} \frac{\mathscr{C}_{l+\frac{1}{2},l+\frac{1}{2}}(\alpha_j r)}{\sqrt{r}} + B_1^{(j)} r^l + B_2^{(j)} r^{l+2} + B_3^{(j)} r^{-(l+1)} + B_4^{(j)} r^{-(l-1)} \right\}. \tag{149}$$

Multiply this equation by $r^{\frac{3}{2}} \mathscr{C}_{l+\frac{1}{2},l+\frac{1}{2}}(\alpha_k r)$ and integrate over the range of r. We obtain

$$\sum_j A_j \int_\eta^1 r^2 \frac{\mathscr{C}_{l+\frac{1}{2},l+\frac{1}{2}}(\alpha_k r)}{b(r)\sqrt{r}} \mathscr{D}_l \left\{ \frac{\mathscr{C}_{l+\frac{1}{2},l+\frac{1}{2}}(\alpha_j r)}{c(r)\sqrt{r}} \right\} dr = -l(l+1)C_l \sum_j \left\{ \frac{N_{l+\frac{1}{2},j}}{\alpha_j^4} \delta_{jk} + Q_{kj} \right\} A_j, \tag{150}$$

where

$$Q_{kj} = \int_\eta^1 \mathscr{C}_{l+\frac{1}{2},l+\frac{1}{2}}(\alpha_k r) \{ B_1^{(j)} r^{l+\frac{3}{2}} + B_2^{(j)} r^{l+\frac{7}{2}} + B_3^{(j)} r^{-l+\frac{1}{2}} + B_4^{(j)} r^{-l+\frac{5}{2}} \} \, dr. \tag{151}$$

Letting

$$P_{k,j}^{b,c} = -\int_\eta^1 r^2 \frac{\mathscr{C}_{l+\frac{1}{2},l+\frac{1}{2}}(\alpha_k r)}{b(r)\sqrt{r}} \mathscr{D}_l \left\{ \frac{\mathscr{C}_{l+\frac{1}{2},l+\frac{1}{2}}(\alpha_j r)}{c(r)\sqrt{r}} \right\} dr, \tag{152}$$

we can rewrite equation (150) in the form

$$\sum_j \left\{ P_{k,j}^{b,c} - l(l+1)C_l \left[\frac{N_{l+\frac{1}{2},k}}{\alpha_k^4} \delta_{kj} + Q_{kj} \right] \right\} A_j = 0. \tag{153}$$

This leads to the secular equation

$$\left\|P_{k,j}^{b,c}-l(l+1)C_l\left[\frac{N_{l+\frac{1}{2},k}}{\alpha_k^4}\delta_{kj}+Q_{kj}\right]\right\| = 0. \tag{154}$$

From the Hermitian character of the operator $r^2\mathscr{D}_l$ with respect to functions which vanish at the ends of an interval, it follows that

$$P_{k,j}^{b,c} = -\int_\eta^1 r^2\frac{\mathscr{C}_{l+\frac{1}{2},l+\frac{1}{2}}(\alpha_j r)}{c(r)\sqrt{r}}\mathscr{D}_l\left\{\frac{\mathscr{C}_{l+\frac{1}{2},l+\frac{1}{2}}(\alpha_k r)}{b(r)\sqrt{r}}\right\}dr. \tag{155}$$

Hence,

$$P_{k,j}^{b,c} = P_{j,k}^{c,b}. \tag{156}$$

The matrix element Q_{kj} can be explicitly evaluated if proper use is made of the recurrence relations satisfied by the cylinder functions. We find after some lengthy but straightforward reductions that

$$Q_{kj} = -\frac{2}{\alpha_k^2}[\mathscr{C}_{l+\frac{1}{2},l+\frac{5}{2}}(\alpha_k)-\eta^{l+\frac{5}{2}}\mathscr{C}_{l+\frac{1}{2},l+\frac{5}{2}}(\alpha_k\eta)]B_2^{(j)}- \\ -\frac{2}{\alpha_k^2}[\mathscr{C}_{l+\frac{1}{2},l-\frac{3}{2}}(\alpha_k)-\eta^{-l+\frac{3}{2}}\mathscr{C}_{l+\frac{1}{2},l-\frac{3}{2}}(\alpha_k\eta)]B_4^{(j)}, \tag{157}$$

where the boundary conditions leading to equations (147) and (148) have contributed to the simplification.

From the recurrence relations satisfied by the functions $\mathscr{C}_{l+\frac{1}{2},\nu}(z)$ and the fact that $\mathscr{C}_{l+\frac{1}{2},l+\frac{1}{2}}(\alpha_k)$ and $\mathscr{C}_{l+\frac{1}{2},l+\frac{1}{2}}(\alpha_k\eta)$ are zero, it follows that

$$\mathscr{C}_{l+\frac{1}{2},l+\frac{5}{2}}(\alpha_k) = -\frac{2l+3}{\alpha_k}\mathscr{C}'_{l+\frac{1}{2}}(\alpha_k),$$

$$\mathscr{C}_{l+\frac{1}{2},l+\frac{5}{2}}(\alpha_k\eta) = -\frac{2l+3}{\alpha_k\eta}\mathscr{C}'_{l+\frac{1}{2}}(\alpha_k\eta),$$

$$\mathscr{C}_{l+\frac{1}{2},l-\frac{3}{2}}(\alpha_k) = +\frac{2l-1}{\alpha_k}\mathscr{C}'_{l+\frac{1}{2}}(\alpha_k),$$

and

$$\mathscr{C}_{l+\frac{1}{2},l-\frac{3}{2}}(\alpha_k\eta) = +\frac{2l-1}{\alpha_k\eta}\mathscr{C}'_{l+\frac{1}{2}}(\alpha_k\eta). \tag{158}$$

Further, let

$$\mathscr{G}_{l+\frac{1}{2},k} = \mathscr{C}'_{l+\frac{1}{2}}(\alpha_k)-\eta^{l+\frac{3}{2}}\mathscr{C}'_{l+\frac{1}{2}}(\alpha_k\eta), \tag{159}$$

and

$$\mathscr{H}_{l+\frac{1}{2},k} = \mathscr{C}'_{l+\frac{1}{2}}(\alpha_k)-\eta^{-l+\frac{1}{2}}\mathscr{C}'_{l+\frac{1}{2}}(\alpha_k\eta). \tag{160}$$

Making use of the foregoing relations and definitions, we find that we can express the matrix element Q_{kj} in the form

$$Q_{kj} = \frac{2}{\alpha_k^3}\{(2l+3)\mathscr{G}_{l+\frac{1}{2},k}B_2^{(j)}-(2l-1)\mathscr{H}_{l+\frac{1}{2},k}B_4^{(j)}\}. \tag{161}$$

The further reduction of Q_{kj} requires an explicit consideration of the nature of the surfaces at $r = 1$ and η; and we consider separately the four possible cases.

(i) *Free surfaces at $r = 1$ and $r = \eta$*

In this case, $d^2W_j/dr^2 = 0$ at $r = 1$ and $r = \eta$; and the conditions applied to the solution (146) give

$$l(l-1)[B_1^{(j)}+B_4^{(j)}]+(l+2)(l+1)[B_2^{(j)}+B_3^{(j)}] = 2\frac{\mathscr{C}'_{l+\frac{1}{2}}(\alpha_j)}{\alpha_j^3} \tag{162}$$

and

$$l(l-1)[B_1^{(j)}\eta^{l-2}+B_4^{(j)}\eta^{-(l+1)}]+(l+2)(l+1)[B_2^{(j)}\eta^{l}+B_3^{(j)}\eta^{-(l+3)}]$$
$$= 2\eta^{-\frac{3}{2}}\frac{\mathscr{C}'_{l+\frac{1}{2}}(\alpha_j\,\eta)}{\alpha_j^3}. \tag{163}$$

Solving these equations together with equations (147) and (148), we find

$$B_2^{(j)} = \frac{1}{(2l+1)(1-\eta^{2l+3})\alpha_j^3}\mathscr{G}_{l+\frac{1}{2},j} \tag{164}$$

and

$$B_4^{(j)} = \frac{1}{(2l+1)(\eta^{-2l+1}-1)\alpha_j^3}\mathscr{H}_{l+\frac{1}{2},j}. \tag{165}$$

Inserting the foregoing expressions for $B_2^{(j)}$ and $B_4^{(j)}$ in equation (161) for Q_{kj}, we obtain

$$Q_{kj} = \frac{2}{(2l+1)\alpha_k^3\,\alpha_j^3}\left\{\frac{2l+3}{1-\eta^{2l+3}}\mathscr{G}_{l+\frac{1}{2},k}\,\mathscr{G}_{l+\frac{1}{2},j}-\frac{2l-1}{\eta^{-2l+1}-1}\mathscr{H}_{l+\frac{1}{2},k}\,\mathscr{H}_{l+\frac{1}{2},j}\right\}. \tag{166}$$

We observe that Q_{kj} is symmetric in k and j.

(ii) *A rigid surface at $r = \eta$ and a free surface at $r = 1$*

In this case $dW_j/dr = 0$ at $r = \eta$ and $d^2W_j/dr^2 = 0$ at $r = 1$; and these conditions applied to the solution (146) give

$$l(l-1)[B_1^{(j)}+B_4^{(j)}]+(l+2)(l+1)[B_2^{(j)}+B_3^{(j)}] = 2\frac{\mathscr{C}'_{l+\frac{1}{2}}(\alpha_j)}{\alpha_j^3} \tag{167}$$

and

$$lB_1^{(j)}\eta^{l-1}+(l+2)B_2^{(j)}\eta^{l+1}-(l+1)B_3^{(j)}\eta^{-(l+2)}-(l-1)B_4^{(j)}\eta^{-l}$$
$$= -\eta^{-\frac{1}{2}}\frac{\mathscr{C}'_{l+\frac{1}{2}}(\alpha_j\,\eta)}{\alpha_j^3}. \tag{168}$$

Solving these equations together with equations (147) and (148), we find

$$B_j^{(2)} = -\frac{1}{\Delta_{rf}(l,\eta)\alpha_j^3}\left\{(2l+1)\left(\frac{1}{\eta^{2l+1}}-\frac{1}{\eta^2}\right)\mathscr{G}_{l+\frac{1}{2},j}-(2l-1)\left(\frac{1}{\eta^{2l+1}}-1\right)\mathscr{C}'_{l+\frac{1}{2}}(\alpha_j)\right\} \tag{169}$$

and

$$B_j^{(4)} = -\frac{1}{\Delta_{rf}(l,\eta)\alpha_j^3}\left\{(2l+1)\left(\frac{1}{\eta^2}-\eta^{2l+1}\right)\mathscr{H}_{l+\frac{1}{2},j}-(2l+3)(1-\eta^{2l+1})\mathscr{C}'_{l+\frac{1}{2}}(\alpha_j)\right\}, \tag{170}$$

where

$$\Delta_{rf}(l,\eta) = (2l+1)\left\{(2l+1)\left(\frac{1}{\eta^2}-\eta^2\right)-2\left(\frac{1}{\eta^{2l+1}}-\eta^{2l+1}\right)\right\}. \tag{171}$$

Inserting the foregoing expressions for $B_2^{(j)}$ and $B_4^{(j)}$ in equation (161) for Q_{kj}, we obtain

$$Q_{kj} = -\frac{2}{\Delta_{rf}(l,\eta)\alpha_k^3\alpha_j^3}\left\{(2l+1)(2l+3)\left(\frac{1}{\eta^{2l+1}}-\frac{1}{\eta^2}\right)\mathscr{G}_{l+\frac{1}{2},k}\,\mathscr{G}_{l+\frac{1}{2},j}-\right.$$

$$-(2l+1)(2l-1)\left(\frac{1}{\eta^2}-\eta^{2l+1}\right)\mathscr{H}_{l+\frac{1}{2},k}\,\mathscr{H}_{l+\frac{1}{2},j}+$$

$$\left.+(2l+3)(2l-1)\left(2-\eta^{2l+1}-\frac{1}{\eta^{2l+1}}\right)\mathscr{C}'_{l+\frac{1}{2}}(\alpha_k)\,\mathscr{C}'_{l+\frac{1}{2}}(\alpha_j)\right\}, \quad (172)$$

which is clearly symmetric in k and j.

(iii) *A free surface at $r = \eta$ and a rigid surface at $r = 1$*

In this case $d^2W_j/dr^2 = 0$ at $r = \eta$ and $dW_j/dr = 0$ at $r = 1$; and these conditions applied to the solution (146) give

$$lB_1^{(j)}+(l+2)B_2^{(j)}-(l+1)B_3^{(j)}-(l-1)B_4^{(j)} = -\frac{\mathscr{C}'_{l+\frac{1}{2}}(\alpha_j)}{\alpha_j^3} \quad (173)$$

and

$$l(l-1)[B_1^{(j)}\eta^{l-2}+B_4^{(j)}\eta^{-(l+1)}]+(l+2)(l+1)[B_2^{(j)}\eta^l+B_3^{(j)}\eta^{-(l+3)}]$$

$$= 2\eta^{-\frac{3}{2}}\frac{\mathscr{C}'_{l+\frac{1}{2}}(\alpha_j\,\eta)}{\alpha_j^3}. \quad (174)$$

Solving these equations together with equations (147) and (148), we find

$$B_2^{(j)} = \frac{1}{\Delta_{rf}(l,\eta)\alpha_j^3}\left\{(2l+1)\left(\frac{1}{\eta^{2l+1}}-\frac{1}{\eta^2}\right)\mathscr{G}_{l+\frac{1}{2},j}+(2l-1)\left(\frac{1}{\eta^{l+\frac{3}{2}}}-\eta^{l-\frac{1}{2}}\right)\mathscr{C}'_{l+\frac{1}{2}}(\alpha_j\,\eta)\right\}$$

and (175)

$$B_4^{(j)} = \frac{1}{\Delta_{rf}(l,\eta)\alpha_j^3}\left\{(2l+1)\left(\frac{1}{\eta^2}-\eta^{2l+1}\right)\mathscr{H}_{l+\frac{1}{2},j}+(2l+3)\left(\frac{1}{\eta^{l+\frac{3}{2}}}-\eta^{l-\frac{1}{2}}\right)\mathscr{C}'_{l+\frac{1}{2}}(\alpha_j\,\eta)\right\}, \quad (176)$$

where $\Delta_{rf}(l,\eta)$ has the same meaning as in equation (171). Inserting the foregoing expressions for $B_2^{(j)}$ and $B_4^{(j)}$ in equation (161) for Q_{kj}, we obtain (cf. equation (172))

$$Q_{kj} = \frac{2}{\Delta_{rf}(l,\eta)\alpha_k^3\alpha_j^3}\left\{(2l+1)(2l+3)\left(\frac{1}{\eta^{2l+1}}-\frac{1}{\eta^2}\right)\mathscr{G}_{l+\frac{1}{2},k}\,\mathscr{G}_{l+\frac{1}{2},j}-\right.$$

$$-(2l+1)(2l-1)\left(\frac{1}{\eta^2}-\eta^{2l+1}\right)\mathscr{H}_{l+\frac{1}{2},k}\,\mathscr{H}_{l+\frac{1}{2},j}-$$

$$\left.-(2l+3)(2l-1)\left(2-\eta^{2l+1}-\frac{1}{\eta^{2l+1}}\right)\mathscr{C}'_{l+\frac{1}{2}}(\alpha_k\,\eta)\mathscr{C}'_{l+\frac{1}{2}}(\alpha_j\,\eta)\right\}. \quad (177)$$

(iv) *Rigid surfaces at $r = 1$ and $r = \eta$*

In this case $dW_j/dr = 0$ at $r = 1$ and $r = \eta$; and these conditions applied to the solution (146) give

$$lB_1^{(j)}+(l+2)B_2^{(j)}-(l+1)B_3^{(j)}-(l-1)B_4^{(j)} = -\frac{\mathscr{C}'_{l+\frac{1}{2}}(\alpha_j)}{\alpha_j^3} \tag{178}$$

and

$$lB_1^{(j)}\eta^{l-1}+(l+2)B_2^{(j)}\eta^{l+1}-(l+1)B_3^{(j)}\eta^{-(l+2)}-(l-1)B_4^{(j)}\eta^{-l} = -\eta^{-\frac{1}{2}}\frac{\mathscr{C}'_{l+\frac{1}{2}}(\alpha_j\,\eta)}{\alpha_j^3}. \tag{179}$$

Solving these equations together with equations (147) and (148), we find

$$B_2^{(j)} = \frac{1}{\Delta_{rr}(l,\eta)\alpha_j^3}\left\{2\left(\frac{1}{\eta^{2l+1}}-\frac{1}{\eta^2}\right)\mathscr{G}_{l+\frac{1}{2},j}-(2l-1)\left(\frac{1}{\eta^2}-1\right)\mathscr{C}'_{l+\frac{1}{2}}(\alpha_j)+ +(2l-1)\left(\frac{1}{\eta^{l+\frac{3}{2}}}-\frac{1}{\eta^{l-\frac{1}{2}}}\right)\mathscr{C}'_{l+\frac{1}{2}}(\alpha_j\,\eta)\right\} \tag{180}$$

and

$$B_4^{(j)} = \frac{1}{\Delta_{rr}(l,\eta)\alpha_j^3}\left\{-2\left(\frac{1}{\eta^2}-\eta^{2l+1}\right)\mathscr{H}_{l+\frac{1}{2},j}+(2l+3)\left(\frac{1}{\eta^2}-1\right)\mathscr{C}'_{l+\frac{1}{2}}(\alpha_j)- -(2l+3)(\eta^{l-\frac{1}{2}}-\eta^{l+\frac{3}{2}})\mathscr{C}'_{l+\frac{1}{2}}(\alpha_j\,\eta)\right\}, \tag{181}$$

where

$$\Delta_{rr}(l,\eta) = (4l^2+4l+1)\left(\frac{1}{\eta^2}+\eta^2\right)-4\left(\frac{1}{\eta^{2l+1}}+\eta^{2l+1}\right)-(8l^2+8l-6). \tag{182}$$

Inserting the foregoing expressions for $B_2^{(j)}$ and $B_4^{(j)}$ in equation (161) for Q_{kj}, we obtain

$$Q_{kj} = \frac{2}{\Delta_{rr}(l,\eta)\alpha_k^3\,\alpha_j^3}\left\{2(2l+3)\left(\frac{1}{\eta^{2l+1}}-\frac{1}{\eta^2}\right)\mathscr{G}_{l+\frac{1}{2},k}\,\mathscr{G}_{l+\frac{1}{2},j}+ +2(2l-1)\left(\frac{1}{\eta^2}-\eta^{2l+1}\right)\mathscr{H}_{l+\frac{1}{2},k}\mathscr{H}_{l+\frac{1}{2},j}- -(2l-1)(2l+3)\left(\frac{1}{\eta^2}-1\right)(\mathscr{G}_{l+\frac{1}{2},k}\mathscr{H}_{l+\frac{1}{2},j}+\mathscr{G}_{l+\frac{1}{2},j}\mathscr{H}_{l+\frac{1}{2},k})\right\}, \tag{183}$$

which is again symmetric in k and j.

(*a*) *The case $b = c = 1$*

In this case we know that the underlying characteristic value problem is self-adjoint. Consequently, solving the secular equation (154) is equivalent to regarding the A_j's as variational parameters and minimizing the expression (83) for C_l.

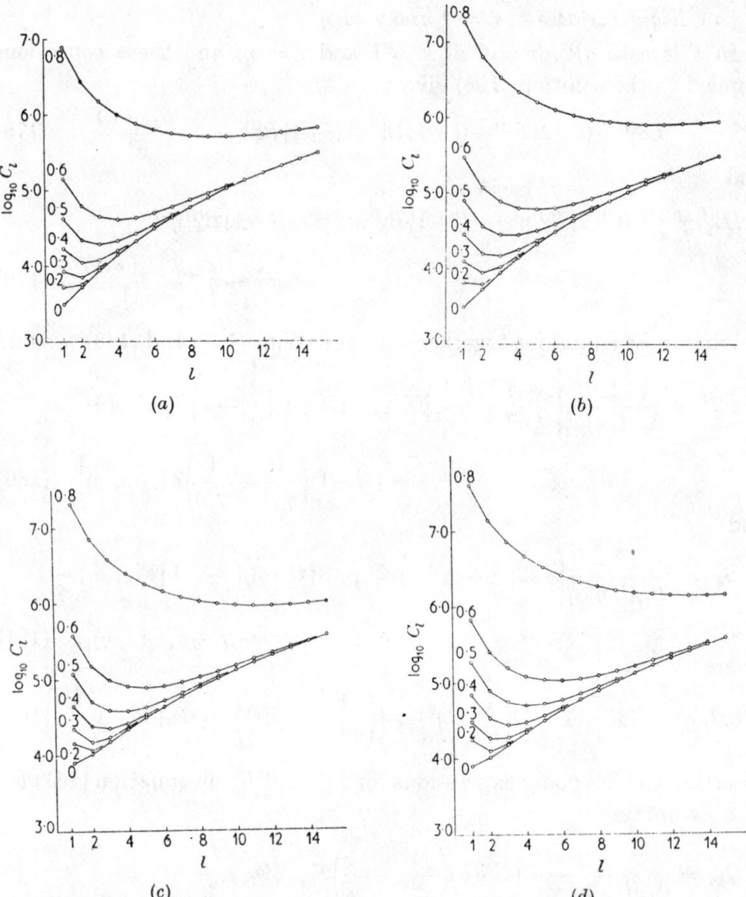

FIG. 58. The Rayleigh number C_l for the onset of convection in the mantle as a spherical harmonic disturbance of order l for various thicknesses of the mantle. The different curves are distinguished by the values of the fraction (η) of the radius of the sphere occupied by the core. The following cases are illustrated: (a) free surfaces at $r = 1$ and $r = \eta$; (b) a rigid surface at $r = \eta$ and a free surface at $r = 1$; (c) a free surface at $r = \eta$ and a rigid surface at $r = 1$; (d) rigid surfaces at $r = 1$ and $r = \eta$.

When $b = c = 1$,

$$P_{kj}^{1,1} = -\int\limits_{\eta}^{1} r^2 \frac{\mathscr{C}_{l+\frac{1}{2},l+\frac{1}{2}}(\alpha_k r)}{\sqrt{r}} \mathscr{D}_l\left\{\frac{\mathscr{C}_{l+\frac{1}{2},l+\frac{1}{2}}(\alpha_j r)}{\sqrt{r}}\right\} dr$$

$$= \alpha_j^2 \int\limits_{\eta}^{1} r\mathscr{C}_{l+\frac{1}{2},l+\frac{1}{2}}(\alpha_k r)\mathscr{C}_{l+\frac{1}{2}}(\alpha_j r)\, dr = \alpha_j^2 N_{l+\frac{1}{2},j}\, \delta_{kj}. \qquad (184)$$

The characteristic equation (154) can in this case be rewritten in the form

$$\left\|N_{l+\frac{1}{2},k}\left\{\frac{\alpha_k^2}{l(l+1)C_l}-\frac{1}{\alpha_k^4}\right\}\delta_{kj}-Q_{kj}\right\| = 0. \tag{185}$$

The symmetry of this matrix is a reflection of the self-adjoint character of the problem we are considering.

TABLE XXII

The characteristic numbers C_l for various values of l and η
(Free surfaces at $r = 1$ and $r = \eta$)

l	$\eta = 0{\cdot}2$	$\eta = 0{\cdot}3$	$\eta = 0{\cdot}4$	$\eta = 0{\cdot}5$	$\eta = 0{\cdot}6$	$\eta = 0{\cdot}8$
1	**$5{\cdot}211\times10^3$**	$8{\cdot}503\times10^3$	$1{\cdot}682\times10^4$	$4{\cdot}188\times10^4$	$1{\cdot}403\times10^5$	$7{\cdot}789\times10^6$
2	$5{\cdot}708\times10^3$	**$7{\cdot}113\times10^3$**	**$1{\cdot}091\times10^4$**	$2{\cdot}181\times10^4$	$6{\cdot}133\times10^4$	$2{\cdot}753\times10^6$
3	$8{\cdot}882\times10^3$	$9{\cdot}552\times10^3$	$1{\cdot}196\times10^4$	**$1{\cdot}924\times10^4$**	$4{\cdot}424\times10^4$	$1{\cdot}500\times10^6$
4	$1{\cdot}400\times10^4$	$1{\cdot}428\times10^4$	$1{\cdot}585\times10^4$	$2{\cdot}146\times10^4$	**$4{\cdot}076\times10^4$**	$1{\cdot}005\times10^6$
5	$2{\cdot}121\times10^4$	$2{\cdot}131\times10^4$	$2{\cdot}227\times10^4$	$2{\cdot}673\times10^4$	$4{\cdot}313\times10^4$	$7{\cdot}656\times10^5$
6		$3{\cdot}089\times10^4$	$3{\cdot}143\times10^4$	$3{\cdot}492\times10^4$	$4{\cdot}945\times10^4$	$6{\cdot}368\times10^5$
7			$4{\cdot}365\times10^4$	$4{\cdot}629\times10^4$	$5{\cdot}933\times10^4$	$5{\cdot}651\times10^5$
8				$6{\cdot}125\times10^4$	$7{\cdot}292\times10^4$	$5{\cdot}270\times10^5$
9				$8{\cdot}027\times10^4$	$9{\cdot}057\times10^4$	$5{\cdot}109\times10^5$
10				$1{\cdot}039\times10^5$	$1{\cdot}128\times10^5$	**$5{\cdot}104\times10^5$**
11				$1{\cdot}325\times10^5$	$1{\cdot}401\times10^5$	$5{\cdot}223\times10^5$
12				$1{\cdot}669\times10^5$	$1{\cdot}732\times10^5$	$5{\cdot}448\times10^5$
13				$2{\cdot}074\times10^5$	$2{\cdot}126\times10^5$	$5{\cdot}767\times10^5$
14				$2{\cdot}545\times10^5$	$2{\cdot}590\times10^5$	$6{\cdot}178\times10^5$
15				$3{\cdot}099\times10^5$	$3{\cdot}131\times10^5$	$6{\cdot}678\times10^5$

In the first approximation, equation (185) gives

$$l(l+1)C_l = \frac{N_{l+\frac{1}{2},1}\,\alpha_1^6}{N_{l+\frac{1}{2},1}+Q_{11}\,\alpha_1^4}. \tag{186}$$

This formula, together with the expressions for Q_{kj} given earlier, has been used to determine the characteristic numbers C_l for various values of l and η and for the sets of boundary conditions we have considered. The results are given in Tables XXII–XXV; they are further illustrated in Fig. 58*a–d*.

From Fig. 58*a–d* it is apparent that as the thickness of the shell decreases, the pattern of the convection which manifests itself at marginal stability shifts progressively to harmonics of the higher orders.

As we have stated, the results given in Tables XXII–XXV have been obtained only in the first approximation. Nevertheless, we may expect from past experience that the values tabulated are not in error by more than 2 or 3 per cent; this is confirmed by the comparison made in Table XXVI of the values obtained in the first and the second approximations for case (i) and $\eta = 0{\cdot}5$.

TABLE XXIII

The characteristic numbers C_l for various values of l and η
(A rigid surface at $r = \eta$ and a free surface at $r = 1$)

l	$\eta = 0.2$	$\eta = 0.3$	$\eta = 0.4$	$\eta = 0.5$	$\eta = 0.6$	$\eta = 0.8$
1	6.518×10^3	1.264×10^4	2.852×10^4	7.953×10^4	2.923×10^5	1.847×10^7
2	$\mathbf{6.213\times10^3}$	$\mathbf{8.809\times10^3}$	1.572×10^4	3.653×10^4	1.175×10^5	6.398×10^6
3	9.047×10^3	1.044×10^4	$\mathbf{1.486\times10^4}$	2.811×10^4	7.641×10^4	3.387×10^6
4	1.405×10^4	1.473×10^4	1.779×10^4	$\mathbf{2.783\times10^4}$	6.327×10^4	2.189×10^6
5	2.122×10^4	2.151×10^4	2.355×10^4	3.166×10^4	$\mathbf{6.070\times10^4}$	1.598×10^6
6		3.096×10^4	3.223×10^4	3.880×10^4	6.405×10^4	1.269×10^6
7			4.412×10^4	4.931×10^4	7.189×10^4	1.073×10^6
8				6.355×10^4	8.388×10^4	9.517×10^5
9				8.197×10^4	1.002×10^5	8.779×10^5
10				1.051×10^5	1.212×10^5	8.359×10^5
11				1.334×10^5	1.474×10^5	8.175×10^5
12				1.674×10^5	1.794×10^5	$\mathbf{8.159\times10^5}$
13				2.078×10^5	2.178×10^5	8.296×10^5
14				2.551×10^5	2.633×10^5	8.561×10^5
15				3.100×10^5	3.166×10^5	8.946×10^5

TABLE XXIV

The characteristic numbers C_l for various values of l and η
(A free surface at $r = \eta$ and a rigid surface at $r = 1$)

l	$\eta = 0.2$	$\eta = 0.3$	$\eta = 0.4$	$\eta = 0.5$	$\eta = 0.6$	$\eta = 0.8$
1	1.437×10^4	2.213×10^4	4.453×10^4	1.188×10^5	3.787×10^5	2.072×10^7
2	$\mathbf{1.173\times10^4}$	$\mathbf{1.504\times10^4}$	2.428×10^4	5.140×10^4	1.519×10^5	7.177×10^6
3	1.560×10^4	1.712×10^4	$\mathbf{2.258\times10^4}$	3.929×10^4	9.857×10^4	3.798×10^6
4	2.239×10^4	2.301×10^4	2.643×10^4	$\mathbf{3.854\times10^4}$	8.133×10^4	2.454×10^6
5	3.180×10^4	3.203×10^4	3.415×10^4	4.333×10^4	$\mathbf{7.769\times10^4}$	1.791×10^6
6		4.419×10^4	4.533×10^4	5.240×10^4	8.155×10^4	1.421×10^6
7			6.037×10^4	6.563×10^4	9.097×10^4	1.201×10^6
8				8.329×10^4	1.054×10^5	1.065×10^6
9				1.058×10^5	1.250×10^5	9.818×10^5
10				1.336×10^5	1.501×10^5	9.341×10^5
11				1.671×10^5	1.811×10^5	9.129×10^5
12				2.071×10^5	2.186×10^5	$\mathbf{9.103\times10^5}$
13				2.540×10^5	2.634×10^5	9.247×10^5
14				3.084×10^5	3.166×10^5	9.534×10^5
15				3.710×10^5	3.770×10^5	9.952×10^5

TABLE XXV

The characteristic numbers C_l for various values of l and η
(Rigid surfaces at $r = 1$ and $r = \eta$)

l	$\eta = 0{\cdot}2$	$\eta = 0{\cdot}3$	$\eta = 0{\cdot}4$	$\eta = 0{\cdot}5$	$\eta = 0{\cdot}6$	$\eta = 0{\cdot}8$
1	$1{\cdot}773\times10^{4}$	$3{\cdot}057\times10^{4}$	$6{\cdot}801\times10^{4}$	$1{\cdot}856\times10^{5}$	$6{\cdot}678\times10^{5}$	$4{\cdot}025\times10^{7}$
2	$\mathbf{1{\cdot}277\times10^{4}}$	$\mathbf{1{\cdot}837\times10^{4}}$	$3{\cdot}354\times10^{4}$	$7{\cdot}933\times10^{4}$	$2{\cdot}571\times10^{5}$	$1{\cdot}381\times10^{7}$
3	$1{\cdot}593\times10^{4}$	$1{\cdot}881\times10^{4}$	$\mathbf{2{\cdot}797\times10^{4}}$	$5{\cdot}560\times10^{4}$	$1{\cdot}575\times10^{5}$	$7{\cdot}214\times10^{6}$
4	$2{\cdot}247\times10^{4}$	$2{\cdot}381\times10^{4}$	$2{\cdot}990\times10^{4}$	$\mathbf{4{\cdot}986\times10^{4}}$	$1{\cdot}215\times10^{5}$	$4{\cdot}582\times10^{6}$
5	$3{\cdot}182\times10^{4}$	$3{\cdot}237\times10^{4}$	$3{\cdot}631\times10^{4}$	$5{\cdot}179\times10^{4}$	$1{\cdot}082\times10^{5}$	$3{\cdot}274\times10^{6}$
6		$4{\cdot}432\times10^{4}$	$4{\cdot}669\times10^{4}$	$5{\cdot}887\times10^{4}$	$\mathbf{1{\cdot}061\times10^{5}}$	$2{\cdot}536\times10^{6}$
7			$6{\cdot}114\times10^{4}$	$7{\cdot}055\times10^{4}$	$1{\cdot}115\times10^{5}$	$2{\cdot}086\times10^{6}$
8				$8{\cdot}695\times10^{4}$	$1{\cdot}229\times10^{5}$	$1{\cdot}796\times10^{6}$
9				$1{\cdot}084\times10^{5}$	$1{\cdot}400\times10^{5}$	$1{\cdot}605\times10^{6}$
10				$1{\cdot}354\times10^{5}$	$1{\cdot}628\times10^{5}$	$1{\cdot}479\times10^{6}$
11				$1{\cdot}684\times10^{5}$	$1{\cdot}919\times10^{5}$	$1{\cdot}401\times10^{6}$
12				$2{\cdot}079\times10^{5}$	$2{\cdot}277\times10^{5}$	$1{\cdot}351\times10^{6}$
13				$2{\cdot}545\times10^{5}$	$2{\cdot}709\times10^{5}$	$\mathbf{1{\cdot}329\times10^{6}}$
14				$3{\cdot}087\times10^{5}$	$3{\cdot}221\times10^{5}$	$1{\cdot}329\times10^{6}$
15				$3{\cdot}712\times10^{5}$	$3{\cdot}819\times10^{5}$	$1{\cdot}346\times10^{6}$

TABLE XXVI

The characteristic numbers C_l in the first and the second approximations for various values of l and $\eta = 0{\cdot}5$
(Free surfaces at $r = 1$ and $r = \eta$)

l	*First approximation*	*Second approximation*	A_2	l	*First approximation*	*Second approximation*	A_2
1	$4{\cdot}188\times10^{4}$	$4{\cdot}183\times10^{4}$	$-0{\cdot}03612$	9	$8{\cdot}027\times10^{4}$	$8{\cdot}026\times10^{4}$	$-0{\cdot}03490$
2	$2{\cdot}181\times10^{4}$	$2{\cdot}179\times10^{4}$	$-0{\cdot}03756$	10	$1{\cdot}0386\times10^{5}$	$1{\cdot}0385\times10^{5}$	$-0{\cdot}03446$
3	$1{\cdot}924\times10^{4}$	$1{\cdot}922\times10^{4}$	$-0{\cdot}03875$	11	$1{\cdot}3254\times10^{5}$	$1{\cdot}3253\times10^{5}$	$-0{\cdot}03447$
4	$2{\cdot}146\times10^{4}$	$2{\cdot}144\times10^{4}$	$-0{\cdot}03924$	12	$1{\cdot}6686\times10^{5}$	$1{\cdot}6685\times10^{5}$	$-0{\cdot}03470$
5	$2{\cdot}673\times10^{4}$	$2{\cdot}672\times10^{4}$	$-0{\cdot}03893$	13	$2{\cdot}0743\times10^{5}$	$2{\cdot}0742\times10^{5}$	$-0{\cdot}03567$
6	$3{\cdot}492\times10^{4}$	$3{\cdot}490\times10^{4}$	$-0{\cdot}03804$	14	$2{\cdot}5454\times10^{5}$	$2{\cdot}5453\times10^{5}$	$-0{\cdot}03868$
7	$4{\cdot}629\times10^{4}$	$4{\cdot}627\times10^{4}$	$-0{\cdot}03687$	15	$3{\cdot}0986\times10^{5}$	$3{\cdot}0985\times10^{5}$	$-0{\cdot}03793$
8	$6{\cdot}125\times10^{4}$	$6{\cdot}124\times10^{4}$	$-0{\cdot}03575$				

(*b*) *The case $b(r) = 1$*

This case corresponds to a uniform distribution of the heat sources; and the allowed non-constancy of $c(r)$ permits a variation of gravity in the mantle which is different from that in a homogeneous sphere.

When $b = 1$, the matrix element $P^{1,c}_{k,j}$ becomes

$$P^{1,c}_{k,j} = -\int\limits_{\eta}^{1} r^2 \frac{\mathscr{C}_{l+\frac{1}{2},l+\frac{1}{2}}(\alpha_k r)}{\sqrt{r}}\, \mathscr{D}_l\left\{\frac{\mathscr{C}_{l+\frac{1}{2},l+\frac{1}{2}}(\alpha_j r)}{c(r)\sqrt{r}}\right\} dr$$

$$= -\int\limits_{\eta}^{1} r^2 \frac{\mathscr{C}_{l+\frac{1}{2},l+\frac{1}{2}}(\alpha_j r)}{c(r)\sqrt{r}}\, \mathscr{D}_l\left\{\frac{\mathscr{C}_{l+\frac{1}{2},l+\frac{1}{2}}(\alpha_k r)}{\sqrt{r}}\right\} dr$$

$$= \alpha_k^2 \int\limits_{\eta}^{1} r\mathscr{C}_{l+\frac{1}{2},l+\frac{1}{2}}(\alpha_k r)\frac{1}{c(r)}\mathscr{C}_{l+\frac{1}{2},l+\frac{1}{2}}(\alpha_j r)\, dr. \tag{187}$$

We can, therefore, write

$$P^{1,c}_{k,j} = \alpha_k^2 (k|c^{-1}|j). \tag{188}$$

The characteristic equation (154) now becomes

$$\left\|\alpha_k^2(k|c^{-1}(r)|j) - l(l+1)C_l\left\{\frac{N_{l+\frac{1}{2},k}}{\alpha_k^4}\delta_{kj} + Q_{kj}\right\}\right\| = 0. \tag{189}$$

In the first approximation, equation (189) gives

$$l(l+1)C_l = \frac{(1|c^{-1}|1)\alpha_1^6}{N_{l+\frac{1}{2},1} + \alpha_1^4 Q_{11}}. \tag{190}$$

Comparing this formula with the corresponding formula (186) for the case $b = c = 1$, we find that in this approximation

$$C_{l;c;1} = C_{l;1;1}\frac{(1|c^{-1}|1)}{N_{l+\frac{1}{2},1}}. \tag{191}$$

It appears that in the mantle of the earth the value of gravity remains approximately constant. For this reason the model,

$$rg = \text{constant} \quad \text{and} \quad c(r) = r^{-1}, \tag{192}$$

has been considered in some detail by Lyttkens. The factors $(1|r|1)/N_{l+\frac{1}{2},1}$ which must be applied to the values listed in Tables XXII–XXV to obtain the corresponding numbers for the model (192) are given in Table XXVII. For case (i), the derived values of C_l are illustrated in Fig. 59.

A constant value of gravity is strictly inconsistent with the model of a mantle of constant density overlying a core: for the most general behaviour of $\gamma(r)$ compatible with this model is that given by equation (2). A variation of gravity compatible with this equation and which in the interval $(0, \frac{1}{2})$ does not depart, sensibly, from constancy is given by

$$rg = \text{constant}\left(\frac{6}{7}r + \frac{1}{7r^2}\right) \quad (\tfrac{1}{2} \leqslant r \leqslant 1). \tag{193}$$

TABLE XXVII

The factor $(1|r|1)/N_{l+\frac{1}{2},1}$

l	$\eta = 0$	$\eta = 0{\cdot}2$	$\eta = 0{\cdot}4$	$\eta = 0{\cdot}5$	$\eta = 0{\cdot}6$	$\eta = 0{\cdot}8$
1	0·590	0·621	0·704	0·751	0·800	0·900
2	0·648	0·654	0·711	0·754	0·801	0·900
3	0·686	0·687	0·721	0·758	0·803	0·900
4	0·715	0·715	0·733	0·763	0·805	0·900
5	0·738	0·738	0·747	0·770	0·807	0·900
6	0·756		0·760	0·777	0·809	0·900
7	0·772		0·773	0·784	0·812	0·901
8	0·785			0·792	0·816	0·901
9	0·796			0·800	0·819	0·901
10	0·806			0·808	0·823	0·901
11	0·814			0·816	0·827	0·901
12	0·822			0·823	0·831	0·902
13	0·829			0·829	0·836	0·902
14	0·835			0·835	0·840	0·902
15	0·841			0·841	0·844	0·902

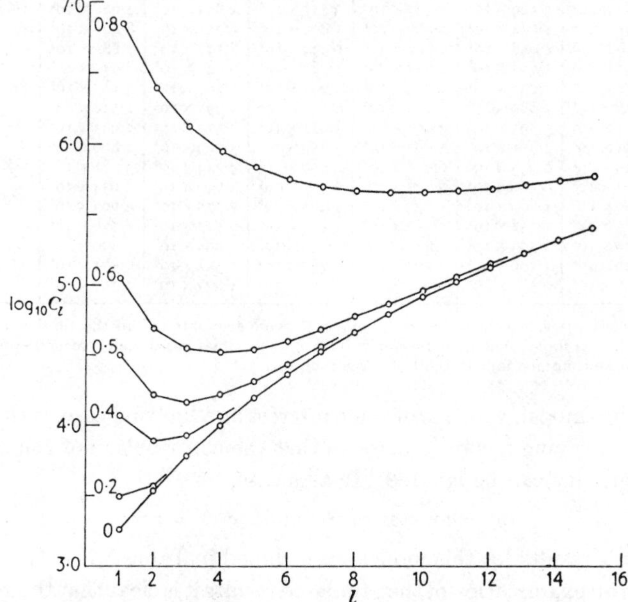

FIG. 59. The Rayleigh number C_l for the onset of convection in the mantle as a spherical harmonic disturbance of order l for various thicknesses of the mantle. For the case illustrated here, gravity has been assumed constant through the mantle (in contrast to Fig. 58 (*a*) which refers to the case when gravity varies as r). The different curves are distinguished by the value of the fraction (η) of the radius of the sphere occupied by the core.

The corresponding expression for $c(r)$ is

$$c(r) = \frac{6}{7}+\frac{1}{7r^3} \quad (\tfrac{1}{2} = \eta \leqslant r \leqslant 1). \tag{194}$$

Lyttkens has derived the characteristic numbers C_l for this model both on the first and the second approximations. His results are given in Table XXVIII. For comparison, the results for $\eta = \frac{1}{2}$ for the model $c(r) = r^{-1}$ are also included in this table.

TABLE XXVIII

The characteristic numbers C_l for $\eta = 0{\cdot}5$ and for different laws of variation of gravity

(*The surfaces at $r = 1$ and $r = \eta$ both free*)

	$c = r^{-1}$		$c = (6+r^{-3})/7$		$c = r^{-3}$		
l	*First approximation*	*Second approximation*	*First approximation*	*Second approximation*	*First approximation*	*Second approximation*	*Third approximation*
†1	$3{\cdot}147\times10^4$	$3{\cdot}129\times10^4$	$3{\cdot}459\times10^4$	$3{\cdot}447\times10^4$	$1{\cdot}854\times10^4$	$1{\cdot}680\times10^4$	$1{\cdot}672\times10^4$
2	$1{\cdot}645\times10^4$	$1{\cdot}634\times10^4$	$1{\cdot}807\times10^4$	$1{\cdot}800\times10^4$	$9{\cdot}75\ \times10^3$	$8{\cdot}82\ \times10^3$	
3	$\mathbf{1{\cdot}459\times10^4}$	$\mathbf{1{\cdot}448\times10^4}$	$\mathbf{1{\cdot}602\times10^4}$	$\mathbf{1{\cdot}594\times10^4}$	$\mathbf{8{\cdot}74\ \times10^3}$	$\mathbf{7{\cdot}88\ \times10^3}$	
4	$1{\cdot}638\times10^4$	$1{\cdot}625\times10^4$	$1{\cdot}797\times10^4$	$1{\cdot}787\times10^4$	$9{\cdot}94\ \times10^3$	$8{\cdot}94\ \times10^3$	
†5	$2{\cdot}058\times10^4$	$2{\cdot}039\times10^4$	$2{\cdot}255\times10^4$	$2{\cdot}241\times10^4$	$1{\cdot}267\times10^4$	$1{\cdot}135\times10^4$	$1{\cdot}130\times10^4$
6	$2{\cdot}712\times10^4$	$2{\cdot}686\times10^4$	$2{\cdot}969\times10^4$	$2{\cdot}949\times10^4$	$1{\cdot}698\times10^4$	$1{\cdot}518\times10^4$	
7	$3{\cdot}631\times10^4$	$3{\cdot}593\times10^4$	$3{\cdot}968\times10^4$	$3{\cdot}941\times10^4$	$2{\cdot}315\times10^4$	$2{\cdot}063\times10^4$	
8	$4{\cdot}854\times10^4$	$4{\cdot}801\times10^4$	$5{\cdot}295\times10^4$	$5{\cdot}259\times10^4$	$3{\cdot}149\times10^4$	$2{\cdot}806\times10^4$	
†9	$6{\cdot}425\times10^4$	$6{\cdot}355\times10^4$	$6{\cdot}998\times10^4$	$6{\cdot}950\times10^4$	$4{\cdot}244\times10^4$	$3{\cdot}782\times10^4$	$3{\cdot}765\times10^4$
10	$8{\cdot}394\times10^4$	$8{\cdot}302\times10^4$	$9{\cdot}125\times10^4$	$9{\cdot}065\times10^4$	$5{\cdot}640\times10^4$	$5{\cdot}034\times10^4$	
11	$1{\cdot}080\times10^5$	$1{\cdot}069\times10^5$	$1{\cdot}173\times10^5$	$1{\cdot}165\times10^5$	$7{\cdot}378\times10^4$	$6{\cdot}603\times10^4$	
12	$1{\cdot}373\times10^5$	$1{\cdot}358\times10^5$	$1{\cdot}487\times10^5$	$1{\cdot}480\times10^5$	$9{\cdot}517\times10^4$	$8{\cdot}546\times10^4$	
13	$1{\cdot}720\times10^5$	$1{\cdot}703\times10^5$	$1{\cdot}859\times10^5$	$1{\cdot}849\times10^5$	$1{\cdot}209\times10^5$	$1{\cdot}090\times10^5$	
14	$2{\cdot}126\times10^5$	$2{\cdot}106\times10^5$	$2{\cdot}294\times10^5$	$2{\cdot}283\times10^5$	$1{\cdot}514\times10^5$	$1{\cdot}371\times10^5$	
15	$2{\cdot}605\times10^5$	$2{\cdot}582\times10^5$	$2{\cdot}807\times10^5$	$2{\cdot}794\times10^5$	$1{\cdot}877\times10^5$	$1{\cdot}707\times10^5$	

† For these cases, the characteristic numbers C_l were also obtained in the third approximation; but it was found that they did not differ from the values listed under 'second approximation' by more than one unit in the last place retained.

Another model, which is of some interest as a limiting case, is that of a mantle overlying a core so massive that the contribution of the mantle to the gravity can be ignored. In this case,

$$rg = \text{constant } r^{-2} \quad \text{and} \quad c(r) = r^{-3}. \tag{195}$$

Lyttken's results for this model are included in Table XXVIII.

From an examination of the tabulated results it is clear that the general nature of the dependence of C_l on l and η is the same for all the models.

(c) *The case $c(r) = 1$*

In this case, the variation of gravity is the same as in a homogeneous sphere; but the allowed non-constancy of $b(r)$ permits a departure of the temperature gradient from that which obtains for a uniform distribution of heat sources in a fluid sphere. In this connexion, a case of particular interest is when all the heat sources are confined to the core and none exists in the mantle. Then, the general law (9) gives

$$\beta(r) = \frac{\beta_1}{2r^3}. \tag{196}$$

The corresponding expression for $b(r)$ is

$$b(r) = r^{-3}. \tag{197}$$

Now, in virtue of equations (156) and (188)

$$P^{b,1}_{k,j} = P^{1,b}_{j,k} = \alpha_j^2(j|b^{-1}|k); \tag{198}$$

and the characteristic equation (154) becomes

$$\left\|\alpha_j^2(j|b^{-1}|k) - l(l+1)C_l\left\{\frac{N_{l+\frac{1}{2},k}}{\alpha_j^4}\delta_{kj} + Q_{kj}\right\}\right\| = 0. \tag{199}$$

Since Q_{kj} is symmetric, the secular determinant obtained from (199) by transposing the rows and the columns is the same as the determinant (189) with c replaced by b. Hence

$$C_l\{\text{for } c(r) = f(r) \text{ and } b(r) = 1\}$$
$$= C_l\{\text{for } c(r) = 1 \text{ and } b(r) = f(r)\}. \tag{200}$$

The characteristic numbers C_l for the case (197) are therefore the same as for the case (195) considered in § (*b*) above.

61. On the effect of rotation on the onset of thermal instability in a fluid sphere. The formulation of the problem

The theory of thermal instability in a fluid sphere finds its most important applications to the liquid core of the earth. As the core partakes of the earth's rotation and is, moreover, the seat of the earth's magnetic field, it is clear that the solution of the complete problem must allow for the simultaneous presence of rotation and magnetic field. We have seen that an allowance for these effects is a complex matter even in a plane geometry. In a sphere in which the unperturbed temperature gradient is radial, the imposition of symmetry about an axis, which rotation introduces, makes for additional analytical difficulty. There is, indeed, no difficulty in writing down the equations of requisite generality; the difficulty is in obtaining solutions which are valid for the ranges of the parameters which the physical problems demand. Only when the perturbations are axisymmetric has some advance been made in these respects. For this reason, we shall restrict the discussion of the problem of the thermal instability in a rotating fluid sphere to the axisymmetric case.

(a) The representation of an axisymmetric solenoidal vector field

When a solenoidal vector field is axisymmetric, the restriction from the outset to a spherical harmonic analysis, which the representation in terms of the particular poloidal and toroidal fields described in § 56 (*d*) implies, can be relaxed; and this has on the formal side some advantages.

For definiteness suppose that **u** represents an axisymmetric solenoidal vector field. We shall express it as a superposition of a poloidal and a toroidal field in terms of two scalar functions U and V in the manner

$$\mathbf{u} = -\varpi\frac{\partial U}{\partial z}\mathbf{1}_\varpi + \varpi V\mathbf{1}_\varphi + \frac{1}{\varpi}\frac{\partial}{\partial\varpi}(\varpi^2 U)\mathbf{1}_z, \tag{201}$$

where ϖ, φ, and z define a system of cylindrical polar coordinates (with the axis of symmetry in the z-direction), $\mathbf{1}_\varpi$, $\mathbf{1}_\varphi$, and $\mathbf{1}_z$ are unit vectors along the three principal directions, and the defining scalars U and V are azimuth independent.

As is evident from equation (201), the field derived from the poloidal scalar U has non-vanishing components only in the meridional planes: U is in fact Stokes's stream function for motions in these planes. The field derived from the toroidal scalar V has non-vanishing components only in the transverse φ-direction. Thus U defines motions which are entirely meridional while V defines motions which are entirely rotational.

When we consider the vorticity of an axisymmetric solenoidal velocity field, we find a certain reciprocity: a poloidal velocity field gives rise to a toroidal vorticity and conversely. Specifically,

$$\text{curl}\,\mathbf{u} = -\varpi\frac{\partial V}{\partial z}\mathbf{1}_\varpi - \varpi\Delta_5 U\mathbf{1}_\varphi + \frac{1}{\varpi}\frac{\partial}{\partial\varpi}(\varpi^2 V)\mathbf{1}_z, \tag{202}$$

where

$$\Delta_5 = \frac{\partial^2}{\partial\varpi^2} + \frac{3}{\varpi}\frac{\partial}{\partial\varpi} + \frac{\partial^2}{\partial z^2} \tag{203}$$

is the Laplacian operator for axisymmetric functions in a five-dimensional Euclidean space.

An equivalent way of writing equations (201) and (202) is

$$\mathbf{u} = \mathbf{1}_z\times\mathbf{r}V + \text{curl}(\mathbf{1}_z\times\mathbf{r}U) \tag{204}$$

and

$$\text{curl}\,\mathbf{u} = -\mathbf{1}_z\times\mathbf{r}\Delta_5 U + \text{curl}(\mathbf{1}_z\times\mathbf{r}V). \tag{205}$$

The particular advantage of the representation (201) or (204) in the present connexion is that we can write down without any difficulty the result of the operation, any number of times, of curl on $\mathbf{u}$. Thus, by repeated applications of the formula (205), we obtain

$$\operatorname{curl}^2\mathbf{u} = -\mathbf{1}_z\times\mathbf{r}\Delta_5 V - \operatorname{curl}(\mathbf{1}_z\times\mathbf{r}\Delta_5 U), \tag{206}$$

$$\operatorname{curl}^3\mathbf{u} = +\mathbf{1}_z\times\mathbf{r}\Delta_5^2 U - \operatorname{curl}(\mathbf{1}_z\times\mathbf{r}\Delta_5 V), \tag{207}$$

and so on.

In spherical polar coordinates, the representation (201) takes the form

$$\mathbf{u} = -\frac{\partial}{\partial\mu}[(1-\mu^2)U]\mathbf{1}_r - \frac{(1-\mu^2)^{\frac{1}{2}}}{r}\frac{\partial}{\partial r}(r^2U)\mathbf{1}_\vartheta + r(1-\mu^2)^{\frac{1}{2}}V\mathbf{1}_\varphi, \tag{208}$$

where $\mathbf{1}_r$, $\mathbf{1}_\vartheta$, and $\mathbf{1}_\varphi$ are unit vectors along the arcs dr, $r\,d\vartheta$, and $r\sin\vartheta\,d\varphi$ in the three principal directions. Similarly, the expression for the Laplacian Δ_5 in terms of r and μ ($=\cos\vartheta$) is

$$\Delta_5 = \frac{\partial^2}{\partial r^2} + \frac{4}{r}\frac{\partial}{\partial r} + \frac{1-\mu^2}{r^2}\frac{\partial^2}{\partial\mu^2} - \frac{4\mu}{r^2}\frac{\partial}{\partial\mu}. \tag{209}$$

While the nature of the functions U and V are not specified except that they are azimuth independent, it is clear that on this theory the basic expansions will be in terms of the fundamental solutions of the equation

$$\Delta_5\psi = -\alpha^2\psi; \tag{210}$$

these are given by

$$\psi = \frac{J_{\pm(n+\frac{3}{2})}(\alpha r)}{r^{\frac{3}{2}}}C_n^{\frac{3}{2}}(\mu), \tag{211}$$

where $J_{\pm(n+\frac{3}{2})}(\alpha r)$ denote the Bessel functions of the order $n+\frac{3}{2}$, and $C_n^{\frac{3}{2}}(\mu)$ is the Gegenbauer polynomial defined as the coefficient of h^n in the expansion of $(1-2h\mu+h^2)^{-\frac{3}{2}}$ in ascending powers of h:

$$(1-2h\mu+h^2)^{-\frac{3}{2}} = \sum_{n=0}^{\infty} h^n C_n^{\frac{3}{2}}(\mu). \tag{212}$$

In this notation $C_n^{\frac{1}{2}}(\mu)$ will denote the usual Legendre polynomials.

The polynomial $C_n^{\frac{3}{2}}(\mu)$ is a solution of the differential equation

$$(1-\mu^2)\frac{d^2}{d\mu^2}C_n^{\frac{3}{2}}(\mu) - 4\mu\frac{d}{d\mu}C_n^{\frac{3}{2}}(\mu) = -n(n+3)C_n^{\frac{3}{2}}(\mu). \tag{213}$$

An alternative form of this equation is

$$\frac{d^2}{d\mu^2}[(1-\mu^2)C_n^{\frac{3}{2}}(\mu)] = -(n+1)(n+2)C_n^{\frac{3}{2}}(\mu). \tag{214}$$

And, finally, we may note here that the Gegenbauer polynomials satisfy the orthogonality relation

$$\int_{-1}^{+1} C_n^{\frac{3}{2}}(\mu)C_m^{\frac{3}{2}}(\mu)(1-\mu^2)\,d\mu = \frac{2(n+1)(n+2)}{2n+3}\delta_{mn}. \tag{215}$$

(*b*) *The perturbation equations*

Returning to the problem of thermal instability, we consider a homogeneous sphere of radius R rotating with an angular velocity Ω about the z-axis; and we shall suppose that there is a uniform distribution of heat sources ϵ such that the unperturbed temperature gradient is given by (cf. equations (8) and (9))

$$\frac{\partial T}{\partial x_i} = -2\beta x_i, \quad \text{where } \beta = \epsilon/6\kappa = \text{constant}. \tag{216}$$

In treating this problem, we shall ignore the rotational flattening of the sphere: this is justified in our present context since the principal effect we are seeking is that of the Coriolis acceleration term, $2\mathbf{u}\times\mathbf{\Omega}$, in the equation of motion; for this purpose the small departure of the equilibrium configuration from a sphere is not essential. Under these circumstances, the relevant perturbation equations are

$$\frac{\partial \theta}{\partial t} = \kappa\nabla^2\theta + 2\beta\mathbf{u}\,.\,\mathbf{r} \tag{217}$$

and

$$\frac{\partial \mathbf{u}}{\partial t} = -\operatorname{grad}\left(\frac{\delta p}{\rho}\right) + \gamma\theta\mathbf{r} - \nu\,\operatorname{curl}^2\mathbf{u} + 2\Omega\mathbf{u}\times\mathbf{1}_z, \tag{218}$$

where $\gamma = 4\pi G\alpha\rho/3$; and the velocity field is, of course, solenoidal.

As we have stated, we shall restrict our discussion to the case when all the quantities describing the perturbation are axisymmetric; in particular, we shall suppose that the velocity field is specified in terms of two scalar functions U and V in the manner (204).

We now eliminate the term $\operatorname{grad}(\delta p/\rho)$ in equation (218) by taking its curl. We then obtain

$$\frac{\partial}{\partial t}\operatorname{curl}\mathbf{u} = \gamma\operatorname{grad}\theta\times\mathbf{r} - \nu\,\operatorname{curl}^3\mathbf{u} + 2\Omega\,\operatorname{curl}(\mathbf{u}\times\mathbf{1}_z). \tag{219}$$

It can be readily verified that

$$\operatorname{grad}\theta\times\mathbf{r} = \mathbf{1}_z\times\mathbf{r}\left(\frac{\partial\theta}{\partial z} - \frac{z}{\varpi}\frac{\partial\theta}{\partial\varpi}\right). \tag{220}$$

Also,

$$\operatorname{curl}(\mathbf{u}\times\mathbf{1}_z) = \mathbf{1}_z\times\mathbf{r}\frac{\partial V}{\partial z} + \operatorname{curl}\left(\mathbf{1}_z\times\mathbf{r}\frac{\partial U}{\partial z}\right). \tag{221}$$

Making use of equations (205), (207), (220), and (221), we can rewrite equation (219) in the form

$$-\mathbf{1}_z\times\mathbf{r}\Delta_5\frac{\partial U}{\partial t}+\operatorname{curl}\left(\mathbf{1}_z\times\mathbf{r}\frac{\partial V}{\partial t}\right)=\gamma\mathbf{1}_z\times\mathbf{r}\left(\frac{\partial\theta}{\partial z}-\frac{z}{\varpi}\frac{\partial\theta}{\partial\varpi}\right)-$$
$$-\nu\{\mathbf{1}_z\times\mathbf{r}\Delta_5^2 U-\operatorname{curl}(\mathbf{1}_z\times\mathbf{r}\Delta_5 V)\}+$$
$$+2\Omega\left\{\mathbf{1}_z\times\mathbf{r}\frac{\partial V}{\partial z}+\operatorname{curl}\left(\mathbf{1}_z\times\mathbf{r}\frac{\partial U}{\partial z}\right)\right\}. \tag{222}$$

The poloidal and the toroidal fields represented in this equation must separately vanish. Thus, we must have

$$\Delta_5\frac{\partial U}{\partial t}=-\gamma\left(\frac{\partial}{\partial z}-\frac{z}{\varpi}\frac{\partial}{\partial\varpi}\right)\theta+\nu\Delta_5^2 U-2\Omega\frac{\partial V}{\partial z} \tag{223}$$

and
$$\frac{\partial V}{\partial t}=\nu\Delta_5 V+2\Omega\frac{\partial U}{\partial z}. \tag{224}$$

Making use of the relation

$$\mathbf{u}\cdot\mathbf{r}=-\left(\frac{\partial}{\partial z}-\frac{z}{\varpi}\frac{\partial}{\partial\varpi}\right)\varpi^2 U, \tag{225}$$

we can rewrite equation (217) in the form

$$\frac{\partial\theta}{\partial t}=\kappa\Delta_3\theta-2\beta\left(\frac{\partial}{\partial z}-\frac{z}{\varpi}\frac{\partial}{\partial\varpi}\right)\varpi^2 U, \tag{226}$$

where Δ_3 is the usual three-dimensional Laplacian for axisymmetric functions.

Seeking solutions of equations (223), (224), and (226) which have the time dependence e^{pt}, we obtain

$$\Delta_5\left(\Delta_5-\frac{p}{\nu}\right)U-\frac{2\Omega}{\nu}\frac{\partial V}{\partial z}=\frac{\gamma}{\nu}\left(\frac{\partial}{\partial z}-\frac{z}{\varpi}\frac{\partial}{\partial\varpi}\right)\theta, \tag{227}$$

$$\left(\Delta_5-\frac{p}{\nu}\right)V=-\frac{2\Omega}{\nu}\frac{\partial U}{\partial z}, \tag{228}$$

and
$$\left(\Delta_3-\frac{p}{\kappa}\right)\theta=\frac{2\beta}{\kappa}\left(\frac{\partial}{\partial z}-\frac{z}{\varpi}\frac{\partial}{\partial\varpi}\right)\varpi^2 U, \tag{229}$$

where it should be remembered that p can be complex.

By measuring distance in units of the radius R of the configuration, and making the transformations

$$\theta\to\left(\frac{2\beta}{\kappa}R^3\right)\theta \quad\text{and}\quad V\to\left(\frac{2\Omega}{\nu}R\right)V, \tag{230}$$

we can express equations (227)–(229) in the more convenient forms

$$(\Delta_3-\sigma\mathfrak{p})\theta = \left(\frac{\partial}{\partial z}-\frac{z}{\varpi}\frac{\partial}{\partial \varpi}\right)\varpi^2 U = \frac{1}{r}\frac{\partial}{\partial \mu}[(1-\mu^2)r^2 U], \tag{231}$$

$$(\Delta_5-\sigma)V = -\frac{\partial U}{\partial z}, \tag{232}$$

and
$$\Delta_5(\Delta_5-\sigma)U-T\frac{\partial V}{\partial z} = C\left(\frac{\partial}{\partial z}-\frac{z}{\varpi}\frac{\partial}{\partial \varpi}\right)\theta = C\,\frac{1}{r}\frac{\partial \theta}{\partial \mu}, \tag{233}$$

where

$$C = \frac{2\beta\gamma}{\kappa\nu}R^6, \qquad T = \frac{4\Omega^2}{\nu^2}R^4, \qquad \mathfrak{p} = \frac{\nu}{\kappa}, \quad \text{and} \quad \sigma = \frac{pR^2}{\nu}. \tag{234}$$

(c) *The boundary conditions*

Solutions of equations (231)–(233) must be sought which satisfy certain boundary conditions. These have been stated in § 56 (c) in terms of the components of **u** in spherical polar coordinates; and what they mean for the scalars U and V can be inferred from equation (208) which relates the components u_r, u_ϑ, and u_φ to the scalars U and V.

Since u_r must vanish on the boundary in all cases, we must require that

$$U = 0 \quad \text{for } r = 1. \tag{235}$$

If the bounding surface is rigid, then the remaining components, u_ϑ and u_φ, must also vanish on it; and from equation (208) it follows that

$$\frac{\partial U}{\partial r} = 0 \quad \text{and} \quad V = 0 \quad \text{if the surface } r = 1 \text{ is rigid.} \tag{236}$$

(Actually, the vanishing of u_ϑ requires only that $\partial(r^2U)/\partial r = 0$; but since $U = 0$ on the boundary, the condition stated follows.) On the other hand, if the bounding surface is free, then the vanishing of the tangential viscous stresses on this surface requires (cf. equation (45))

$$\frac{\partial}{\partial r}\left(\frac{u_\vartheta}{r}\right) = \frac{\partial}{\partial r}\left(\frac{u_\varphi}{r}\right) = 0. \tag{237}$$

According to equation (208), these conditions are equivalent to

$$\left.\begin{aligned} \frac{\partial}{\partial r}\left[\frac{1}{r^2}\frac{\partial}{\partial r}(r^2U)\right] &= \frac{\partial^2 U}{\partial r^2}+\frac{2}{r}\frac{\partial U}{\partial r}-2\frac{U}{r^2} = 0 \\ \text{and} \qquad \frac{\partial V}{\partial r} &= 0 \quad \text{for } r = 1 \end{aligned}\right\}. \tag{238}$$

Since $U = 0$ for $r = 1$, the first of these conditions can also be expressed as

$$\frac{\partial^2}{\partial r^2}(rU) = 0. \tag{239}$$

Collecting all these conditions, we have

$$U = 0, \qquad \frac{\partial U}{\partial r} = 0, \qquad V = 0, \text{ and } \quad \theta = 0 \quad \text{for } r = 1,$$
$$\text{if the bounding surface is rigid} \tag{240}$$

and

$$U = 0, \qquad \frac{\partial^2}{\partial r^2}(rU) = 0, \qquad \frac{\partial V}{\partial r} = 0, \quad \text{and} \quad \theta = 0 \quad \text{for } r = 1,$$
$$\text{if the bounding surface is free.} \tag{241}$$

(*d*) *The variational principle*

The solution of equations (231)–(233) together with the boundary conditions (240) or (241) constitutes a characteristic value problem. We shall now show that the problem can be formulated in terms of a variational principle.

Multiply equation (231) by θ and integrate over the volume of the unit three-dimensional sphere. Both sides of the equation can be reduced by integrations by parts; and we find

$$\iint \{|\text{grad}\,\theta|^2+\sigma\mathfrak{p}\theta^2\}r^2\,drd\mu = \iint \frac{1}{r}\frac{\partial\theta}{\partial\mu}Ur^4(1-\mu^2)\,drd\mu, \tag{242}$$

the integrated parts vanishing on account of the boundary conditions on θ. It will be observed that on the right-hand side the integration is over the unit five-dimensional sphere.

Making use of equation (233), we can rewrite equation (242) in the manner

$$C\iint \{|\text{grad}\,\theta|^2+\sigma\mathfrak{p}\theta^2\}r^2\,drd\mu$$
$$= \iint U\left(\Delta_5^2\,U-\sigma\Delta_5\,U-T\frac{\partial V}{\partial z}\right)\varpi^3\,d\varpi dz. \tag{243}$$

Considering first the term in V on the right-hand side of this equation, we can reduce it by making use of equation (232) and a sequence of integrations by parts. Thus,

$$-\iint U\frac{\partial V}{\partial z}\varpi^3\,d\varpi dz = \iint V\frac{\partial U}{\partial z}\varpi^3\,d\varpi dz$$
$$= -\iint V(\Delta_5-\sigma)Vr^4(1-\mu^2)\,drd\mu$$
$$= -\iint V\left\{(1-\mu^2)\frac{\partial}{\partial r}\left(r^4\frac{\partial V}{\partial r}\right)+r^2\frac{\partial}{\partial\mu}\left[(1-\mu^2)^2\frac{\partial V}{\partial\mu}\right]- \right.$$
$$\left. -\sigma V^2r^4(1-\mu^2)\right\}\,drd\mu$$
$$= \iint \left\{\left(\frac{\partial V}{\partial r}\right)^2+\frac{1-\mu^2}{r^2}\left(\frac{\partial V}{\partial\mu}\right)^2+\sigma V^2\right\}r^4(1-\mu^2)\,drd\mu$$
$$= \iint \{|\text{grad}\,V|^2+\sigma V^2\}r^4(1-\mu^2)\,drd\mu; \tag{244}$$

the integrated parts do not make any contributions in view of the conditions that U and either V or $\partial V/\partial r$ vanish on the boundary. Considering next the term in $\Delta_5 U$, we can reduce it by the same sequence of transformations which was used in the reduction of $V\Delta_5 V$ in (244). We find

$$-\iint U\Delta_5 U r^4(1-\mu^2)\,dr d\mu = \iint |\text{grad}\, U|^2 r^4(1-\mu^2)\,dr d\mu. \tag{245}$$

Finally, considering the term in $\Delta_5^2 U$ and letting

$$X = \Delta_5 U, \tag{246}$$

we have

$$\begin{aligned}\iint U\Delta_5 X r^4(1-\mu^2)\,dr d\mu &\\ = \iint U\Big\{(1-\mu^2)\frac{\partial}{\partial r}\Big(r^4\frac{\partial X}{\partial r}\Big)+r^2\frac{\partial}{\partial\mu}\Big[(1-\mu^2)^2\frac{\partial X}{\partial\mu}\Big]\Big\}\,dr d\mu &\\ = -\iint \Big\{r^4(1-\mu^2)\frac{\partial U}{\partial r}\frac{\partial X}{\partial r}+r^2(1-\mu^2)^2\frac{\partial U}{\partial\mu}\frac{\partial X}{\partial\mu}\Big\}\,dr d\mu &\\ = +\iint X\Delta_5 U r^4(1-\mu^2)\,dr d\mu - \int\Big(\frac{\partial U}{\partial r}\Big)_{r=1}(\Delta_5 U)_{r=1}(1-\mu^2)\,d\mu.&\end{aligned} \tag{247}$$

On a rigid boundary $\partial U/\partial r = 0$ and the surface integral in the last line of equation (247) will vanish. On the other hand, on a free boundary

$$\frac{\partial^2 U}{\partial r^2}+\frac{2}{r}\frac{\partial U}{\partial r} = 0. \tag{248}$$

Consequently,
$$(X)_{r=1} = (\Delta_5 U)_{r=1} = 2\Big(\frac{\partial U}{\partial r}\Big)_{r=1} \tag{249}$$

Hence, in all cases
$$\Big(\frac{\partial U}{\partial r}\Delta_5 U\Big)_{r=1} = 2\Big(\frac{\partial U}{\partial r}\Big)^2_{r=1} \tag{250}$$

and

$$\iint U\Delta_5^2 U r^4(1-\mu^2)\,dr d\mu = \iint (\Delta_5 U)^2 r^4(1-\mu^2)\,dr d\mu - 2\int_{-1}^{+1}\Big(\frac{\partial U}{\partial r}\Big)^2_{r=1}(1-\mu^2)\,d\mu. \tag{251}$$

Finally, combining equations (243), (244), (245), and (251), we obtain

$$\begin{aligned}C\int_0^1\int_{-1}^{+1}\{|\text{grad}\,\theta|^2+\sigma\mathfrak{p}\theta^2\}r^2\,dr d\mu &\\ = \int_0^1\int_{-1}^{+1}\{(\Delta_5 U)^2+\sigma|\text{grad}\,U|^2+T[|\text{grad}\,V|^2+\sigma V^2]\}r^4(1-\mu^2)\,dr d\mu &-\\ -2\int_{-1}^{+1}\Big(\frac{\partial U}{\partial r}\Big)^2_{r=1}(1-\mu^2)\,d\mu.&\end{aligned} \tag{252}$$

It can be readily shown by methods with which we are now familiar that formula (252) provides the basis for a variational treatment of the problem. In applying the variational method, it is, however, necessary to remember that equations (231) and (232) must be satisfied. Specifically, one makes some assumptions concerning U which are compatible with the boundary conditions on it; but whatever assumption one makes concerning U, one must determine θ and V as solutions of equations (231) and (232) which satisfy the necessary boundary conditions.

(*e*) *The thermodynamic significance of the variational principle*

The physical content of the variational principle which derives from equation (252) is the same as in the other cases we have considered. However, since we have retained the time dependence through σ and allowed for the possibility of overstability, we must, in accordance with the discussion in Chapter III, § 33 (*b*), expect that equation (252) expresses the equality

$$\epsilon_\nu+\rho p\iiint |\mathbf{u}|^2\,dV = \epsilon_g, \tag{253}$$

where ϵ_ν and ϵ_g are the rates at which energy is, respectively, dissipated by viscosity and liberated by the buoyancy force in the fluid sphere.

In the present instance, all the quantities occurring in (253) can be evaluated directly. Thus,

$$\epsilon_\nu = \rho\nu\int\limits_0^R\int\limits_{-1}^{+1} r^2\mathbf{u}\,.\,\text{curl}^2\mathbf{u}\,drd\mu.\dagger \tag{254}$$

Using the expressions (206) and (208) for $\text{curl}^2\mathbf{u}$ and $\mathbf{u}$, we have

$$\epsilon_\nu = \rho\nu\int\limits_0^R\int\limits_{-1}^{+1}\Big\{-\frac{\partial}{\partial\mu}[(1-\mu^2)U]\frac{\partial}{\partial\mu}[(1-\mu^2)\Delta_5\,U]-$$
$$-\frac{1-\mu^2}{r^2}\frac{\partial}{\partial r}(r^2\,U)\frac{\partial}{\partial r}(r^2\Delta_5\,U)-r^2(1-\mu^2)V\Delta_5\,V\Big\}r^2\,drd\mu. \tag{255}$$

After integrating by parts each of the three terms on the right-hand side, we obtain

$$\epsilon_\nu = \rho\nu\Big\{\int\limits_0^R\int\limits_{-1}^{+1}[(\Delta_5\,U)^2+|\text{grad}\,V|^2]r^4(1-\mu^2)\,drd\mu-$$
$$-2R^3\int\limits_{-1}^{+1}\Big(\frac{\partial U}{\partial r}\Big)^2_{r=R}(1-\mu^2)\,d\mu\Big\}. \tag{256}$$

† A factor 2π has been omitted.

Similarly, we find

$$\rho p \int_0^R \int_{-1}^{+1} |\mathbf{u}|^2 r^2 \, dr d\mu = \rho p \int_0^R \int_{-1}^{+1} [|\text{grad}\, U|^2 + V^2] r^4 (1-\mu^2) \, dr d\mu. \tag{257}$$

The rate ϵ_g at which energy is liberated by the buoyancy force acting on the fluid is given by

$$\epsilon_g = \rho\gamma \int_0^R \int_{-1}^{+1} \theta(\mathbf{u}\,.\,\mathbf{r}) r^2 \, dr d\mu. \tag{258}$$

Making use of equation (217), we can rewrite the foregoing as

$$\epsilon_g = -\rho \frac{\gamma}{2\beta} \int_0^R \int_{-1}^{+1} \theta(\kappa\nabla^2\theta - p\theta) r^2 \, dr d\mu; \tag{259}$$

or, after an integration by parts, we have

$$\epsilon_g = \rho \frac{\kappa\gamma}{2\beta} \int_0^R \int_{-1}^{+1} \left\{ |\text{grad}\, \theta|^2 + \frac{p}{\kappa} \theta^2 \right\} r^2 \, dr d\mu. \tag{260}$$

The energy principle (253) now gives

$$\int_0^R \int_{-1}^{+1} \left\{ (\Delta_5 U)^2 + |\text{grad}\, V|^2 + \frac{p}{\nu} [|\text{grad}\, U|^2 + V^2] \right\} r^4 (1-\mu^2) \, dr d\mu -$$

$$-2R^3 \int_{-1}^{+1} \left(\frac{\partial U}{\partial r}\right)^2_{r=R} (1-\mu^2) \, d\mu = \frac{\kappa\gamma}{2\beta\nu} \int_0^R \int_{-1}^{+1} \left\{ |\text{grad}\, \theta|^2 + \frac{p}{\kappa} \theta^2 \right\} r^2 \, dr d\mu. \tag{261}$$

Measuring distances in the unit R and making the transformations (230), we recover equation (252).

62. The effect of rotation on the onset of stationary convection in a fluid sphere

When instability sets in as ordinary convection, the marginal state will be characterized by $\sigma = 0$ and the relevant equations are (cf. equations (231–(233)):

$$\Delta_3 \theta = r \frac{\partial}{\partial \mu} [(1-\mu^2) U], \tag{262}$$

$$\Delta_5 V = -\frac{\partial U}{\partial z}, \tag{263}$$

and

$$\Delta_5^2 U = T \frac{\partial V}{\partial z} + C \frac{1}{r} \frac{\partial \theta}{\partial \mu}. \tag{264}$$

These equations must be solved together with the boundary conditions (240) or (241).

Making use of the identity

$$\frac{1}{r}\frac{\partial}{\partial \mu}\Delta_3 = \Delta_5 \frac{1}{r}\frac{\partial}{\partial \mu}, \tag{265}$$

we deduce from equation (262) that

$$\Delta_5\left(\frac{1}{r}\frac{\partial \theta}{\partial \mu}\right) = \frac{\partial^2}{\partial \mu^2}[(1-\mu^2)U]. \tag{266}$$

Now applying the operator Δ_5 to equation (264), we can eliminate V and θ by making use of equations (263) and (266). We thus obtain

$$\Delta_5^3 U + T\frac{\partial^2 U}{\partial z^2} = C\frac{\partial^2}{\partial \mu^2}[(1-\mu^2)U]. \tag{267}$$

The similarity of this equation with equation III (99) for the analogous problem in a plane geometry is to be noted.

As we have shown in § 61 (*d*), the characteristic value problem presented by equation (264), together with the subsidiary equations (262) and (263) and the boundary conditions (240) or (241), is a self-adjoint one. A variational solution is, therefore, possible; and this has been accomplished by Bisshopp, whose method we shall now describe.

Since equations (262) and (263) have to be solved for θ and V for some assumed form of U, it is necessary, if the method is to be practicable, that the functions in terms of which we expand U are such that explicit solutions are possible. At the same time the chosen functions must be such that the boundary conditions on U, namely,

$$U = 0 \quad \text{and} \quad \partial U/\partial r = 0 \quad \text{for } r = 1 \tag{268}$$

or

$$U = 0 \quad \text{and} \quad \partial^2(rU)/\partial r^2 = 0 \quad \text{for } r = 1, \tag{269}$$

are identically satisfied. And it would, of course, be preferable if the functions formed a complete orthogonal set. It will appear that all these conditions can be met by functions generated as solutions by the equation

$$\Delta_5^2 U = \alpha^4 U. \tag{270}$$

The general solution of equation (270) which is free of singularity at the origin is a linear combination of the fundamental solutions (cf. equations (210) and (211))

$$U_n = [A_n J_{n+\frac{3}{2}}(\alpha r) + B_n I_{n+\frac{3}{2}}(\alpha r)]\frac{C_n^{\frac{3}{2}}(\mu)}{r^{\frac{3}{2}}}, \tag{271}$$

where A_n and B_n are constants, $J_{n+\frac{3}{2}}$ and $I_{n+\frac{3}{2}}$ are the Bessel functions of order $n+\frac{3}{2}$ for real and imaginary arguments, and $C_n^{\frac{3}{2}}(\mu)$ is the

Gegenbauer polynomial defined as in equation (212). We now impose on U_n the boundary conditions required of U. Thus, if $r = 1$ is a rigid boundary, we require

$$\left.\begin{aligned} A_n J_{n+\frac{3}{2}}(\alpha)+B_n I_{n+\frac{3}{2}}(\alpha) &= 0 \\ \text{and} \qquad A_n J'_{n+\frac{3}{2}}(\alpha)+B_n I'_{n+\frac{3}{2}}(\alpha) &= 0 \end{aligned}\right\}, \tag{272}$$

while if $r = 1$ is a free boundary, we require

$$\left.\begin{aligned} A_n J_{n+\frac{3}{2}}(\alpha)+B_n I_{n+\frac{3}{2}}(\alpha) &= 0 \\ \text{and} \qquad A_n[2J'_{n+\frac{3}{2}}(\alpha)+\alpha J_{n+\frac{3}{2}}(\alpha)]+B_n[2I'_{n+\frac{3}{2}}(\alpha)-\alpha I_{n+\frac{3}{2}}(\alpha)] &= 0 \end{aligned}\right\}. \tag{273†}$$

In other words, we require α to be a root of

$$J_{n+\frac{3}{2}}(\alpha)I'_{n+\frac{3}{2}}(\alpha)-I_{n+\frac{3}{2}}(\alpha)J'_{n+\frac{3}{2}}(\alpha) = 0 \tag{274}$$

or

$$J_{n+\frac{3}{2}}(\alpha)I'_{n+\frac{3}{2}}(\alpha)-I_{n+\frac{3}{2}}(\alpha)J'_{n+\frac{3}{2}}(\alpha) = \alpha J_{n+\frac{3}{2}}(\alpha)I_{n+\frac{3}{2}}(\alpha), \tag{275}$$

depending on whether $r = 1$ is a rigid or a free surface.

Let α_{nj} $(j = 1, 2,...)$ denote the roots of equation (274) or (275) depending on the case we may be considering. The corresponding fundamental solutions are

$$U_{nj} = \left[\frac{J_{n+\frac{3}{2}}(\alpha_{nj} r)}{J_{n+\frac{3}{2}}(\alpha_{nj})}-\frac{I_{n+\frac{3}{2}}(\alpha_{nj} r)}{I_{n+\frac{3}{2}}(\alpha_{nj})}\right]\frac{C_n^{\frac{3}{2}}(\mu)}{r^{\frac{3}{2}}}. \tag{276}$$

It can be readily verified that the radial parts of the functions U_{nj} for a given n and different j's are orthogonal for integration over r after multiplication by r^4; and the functions belonging to different n's are orthogonal for integration over μ after multiplication by $1-\mu^2$; the latter orthogonality derives from that of the Gegenbauer polynomials.

The expansion for U which we assume is, then,

$$U = \sum_n \sum_j A_{nj} U_{nj}, \tag{277}$$

where the A_{nj}'s are constants. With the corresponding representation,

$$\theta = \sum_n \sum_j A_{nj}\,\theta_{nj} \quad \text{and} \quad V = \sum_n \sum_j A_{nj} V_{nj}, \tag{278}$$

the equations to be solved are

$$\Delta_5\left(\frac{1}{r}\frac{\partial \theta_{nj}}{\partial \mu}\right) = \frac{\partial^2}{\partial \mu^2}[(1-\mu^2)U_{nj}] \tag{279}$$

and

$$\Delta_5 V_{nj} = -\frac{\partial U_{nj}}{\partial z}. \tag{280}$$

† The second of these conditions follows most readily from the relation,

$$\Delta_5 U_n = -\alpha^2[A_n J_{n+\frac{3}{2}}(\alpha r)-B_n I_{n+\frac{3}{2}}(\alpha r)]C_n^{\frac{3}{2}}(\mu)/r^{\frac{3}{2}},$$

and the equivalent form of the boundary conditions given in equation (249).

Making use of equation (214) satisfied by the Gegenbauer polynomials, we can rewrite equation (279) in the form

$$\Delta_5\left(\frac{1}{r}\frac{\partial\theta_{nj}}{\partial\mu}\right) = -(n+1)(n+2)\left[\frac{J_{n+\frac{3}{2}}(\alpha_{nj}r)}{J_{n+\frac{3}{2}}(\alpha_{nj})}-\frac{I_{n+\frac{3}{2}}(\alpha_{nj}r)}{I_{n+\frac{3}{2}}(\alpha_{nj})}\right]\frac{C_n^{\frac{3}{2}}(\mu)}{r^{\frac{3}{2}}}. \quad (281)$$

The solution of this equation relevant for our purposes is

$$\frac{1}{r}\frac{\partial\theta_{nj}}{\partial\mu} = \frac{(n+1)(n+2)}{\alpha_{nj}^2}\left[\frac{J_{n+\frac{3}{2}}(\alpha_{nj}r)}{J_{n+\frac{3}{2}}(\alpha_{nj})}+\frac{I_{n+\frac{3}{2}}(\alpha_{nj}r)}{I_{n+\frac{3}{2}}(\alpha_{nj})}+B^{(nj)}r^{n+\frac{3}{2}}\right]\frac{C_n^{\frac{3}{2}}(\mu)}{r^{\frac{3}{2}}}, \quad (282)$$

where $B^{(nj)}$ is a constant. The condition $\theta_{nj} = 0$ for $r = 1$ can be satisfied by the choice $B^{(nj)} = -2$; and the corresponding solution is

$$\frac{1}{r}\frac{\partial\theta_{nj}}{\partial\mu} = \frac{(n+1)(n+2)}{\alpha_{nj}^2}\left[\frac{J_{n+\frac{3}{2}}(\alpha_{nj}r)}{J_{n+\frac{3}{2}}(\alpha_{nj})}+\frac{I_{n+\frac{3}{2}}(\alpha_{nj}r)}{I_{n+\frac{3}{2}}(\alpha_{nj})}-2r^{n+\frac{3}{2}}\right]\frac{C_n^{\frac{3}{2}}(\mu)}{r^{\frac{3}{2}}}. \quad (283)$$

The solution of equation (280) is somewhat more difficult. Following Bisshopp, we introduce the notations

$$f_{l|n}(\alpha_{nj}r) = \frac{J_{l+\frac{3}{2}}(\alpha_{nj}r)}{J_{n+\frac{3}{2}}(\alpha_{nj})}-\frac{I_{l+\frac{3}{2}}(\alpha_{nj}r)}{I_{n+\frac{3}{2}}(\alpha_{nj})} \quad (284)$$

and

$$g_{l|n}(\alpha_{nj}r) = \frac{J_{l+\frac{3}{2}}(\alpha_{nj}r)}{J_{n+\frac{3}{2}}(\alpha_{nj})}+\frac{I_{n+\frac{3}{2}}(\alpha_{nj}r)}{I_{n+\frac{3}{2}}(\alpha_{nj})}. \quad (285)$$

In this notation

$$U_{nj} = f_{n|n}(\alpha_{nj}r)\frac{C_n^{\frac{3}{2}}(\mu)}{r^{\frac{3}{2}}} \quad (286)$$

and

$$\frac{1}{r}\frac{\partial\theta_{nj}}{\partial\mu} = \frac{(n+1)(n+2)}{\alpha_{nj}^2}\left[g_{n|n}(\alpha_{nj}r)-2r^{n+\frac{3}{2}}\right]\frac{C_n^{\frac{3}{2}}(\mu)}{r^{\frac{3}{2}}}. \quad (287)$$

It is also apparent that

$$\Delta_5\left[f_{l|n}(\alpha_{nj}r)\frac{C_l^{\frac{3}{2}}(\mu)}{r^{\frac{3}{2}}}\right] = -\alpha_{nj}^2\, g_{l|n}(\alpha_{nj}r)\frac{C_l^{\frac{3}{2}}(\mu)}{r^{\frac{3}{2}}} \quad (288)$$

and

$$\Delta_5\left[g_{l|n}(\alpha_{nj}r)\frac{C_l^{\frac{3}{2}}(\mu)}{r^{\frac{3}{2}}}\right] = -\alpha_{nj}^2\, f_{l|n}(\alpha_{nj}r)\frac{C_l^{\frac{3}{2}}(\mu)}{r^{\frac{3}{2}}}. \quad (289)$$

Making use of the recurrence relations satisfied by the Bessel functions and the Gegenbauer polynomials, one can readily verify that

$$\frac{\partial U_{nj}}{\partial z} = \left(\mu\frac{\partial}{\partial r}+\frac{1-\mu^2}{r}\frac{\partial}{\partial\mu}\right)U_{nj}$$
$$= \frac{\alpha_{nj}}{r^{\frac{3}{2}}}\left[\frac{n+2}{2n+3}f_{n-1|n}(\alpha_{nj}r)C_{n-1}^{\frac{3}{2}}(\mu)-\frac{n+1}{2n+3}g_{n+1|n}(\alpha_{nj}r)C_{n+1}^{\frac{3}{2}}(\mu)\right]. \quad (290)$$

Inserting this last expression for $\partial U_{nj}/\partial z$ in equation (280), we find that the solution for V_{nj} relevant for our purposes is

$$V_{nj} = -\frac{1}{\alpha_{nj}}\left\{\frac{n+2}{2n+3}\left[g_{n-1/n}(\alpha_{nj}r)+D_1^{(nj)}r^{n+\frac{1}{2}}\right]\frac{C_{n-1}^{\frac{3}{2}}(\mu)}{r^{\frac{3}{2}}}-\right.$$
$$\left.-\frac{n+1}{2n+3}\left[f_{n+1/n}(\alpha_{nj}r)+D_2^{(nj)}r^{n+\frac{5}{2}}\right]\frac{C_{n+1}^{\frac{3}{2}}(\mu)}{r^{\frac{3}{2}}}\right\}, \qquad (291)$$

where $D_1^{(nj)}$ and $D_2^{(nj)}$ are constants to be determined by the boundary conditions on V. If the boundary at $r = 1$ is rigid, V_{nj} must vanish here; and this condition gives

$$D_1^{(nj)} = -g_{n-1/n}(\alpha_{nj}) \quad \text{and} \quad D_2^{(nj)} = -f_{n+1/n}(\alpha_{nj}). \qquad (292)$$

On the other hand, if the boundary at $r = 1$ is free, $\partial V_{nj}/\partial r$ must vanish here; and this condition gives

$$\left.\begin{aligned} (n-1)D_1^{(nj)} &= -\left\{\frac{d}{dr}\left(\frac{g_{n-1/n}(\alpha_{nj}r)}{r^{\frac{3}{2}}}\right)\right\}_{r=1} \\ \text{and} \qquad (n+1)D_2^{(nj)} &= -\left\{\frac{d}{dr}\left(\frac{f_{n+1/n}(\alpha_{nj}r)}{r^{\frac{3}{2}}}\right)\right\}_{r=1} \end{aligned}\right\}. \qquad (293)$$

With the solution for θ_{nj} and V_{nj} completed in this manner, we substitute for U, V, and θ in accordance with equations (277) and (278) in equation (264), and obtain

$$\sum_n\sum_j A_{nj}\,\alpha_{nj}^4\,U_{nj}-T\sum_n\sum_j A_{nj}\frac{\partial V_{nj}}{\partial z}$$
$$= C\sum_n\sum_j A_{nj}\frac{(n+1)(n+2)}{\alpha_{nj}^2}\left[g_{n/n}(\alpha_{nj}r)-2r^{n+\frac{3}{2}}\right]\frac{C_n^{\frac{3}{2}}(\mu)}{r^{\frac{3}{2}}}. \qquad (294)$$

By multiplying this equation by U_{mk}, integrating over the unit five-dimensional sphere, and setting the determinant of the resulting system of homogeneous linear equations for the A_{nj}'s equal to zero, we shall obtain the required characteristic equation for C. Bisshopp has shown how all the relevant matrix elements in the secular determinant can be explicitly evaluated. And he has solved for the least characteristic roots for various assigned values of T by retaining as many rows and columns as seemed necessary. His results are summarized in Tables XXIX and XXX; and the calculated dependence of C on T is further illustrated in Fig. 60.

The fact that a very large number of terms in the expansions for U have to be retained even for not too large values of T is noteworthy; and the shift in the relative importance of the various harmonics as T increases is also to be noted.

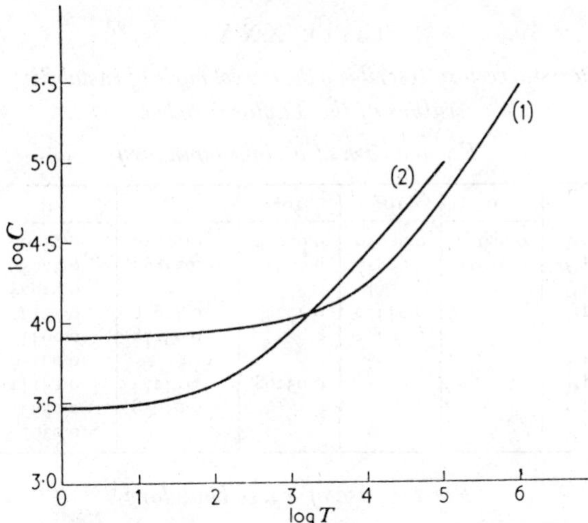

FIG. 60. The dependence of the critical Rayleigh number for the onset of the convective instability on the Taylor number in a rotating fluid sphere. The curves labelled (1) and (2) are for the cases of a rigid and a free bounding surface, respectively.

TABLE XXIX

The values of C for the onset of the lowest mode of instability for various values of the Taylor number

For the case of a rigid boundary

T	*First approximation*	*Second approximation*	*Fourth approximation*	*Sixth approximation*	*Ninth approximation*
0	8.0611×10^3	8.0421×10^3			
10^3	9.4882×10^3	9.3855×10^3	9.3383×10^3		
10^4		1.7306×10^4	1.6991×10^4		
10^5			5.6993×10^4	5.3344×10^4	
10^6				3.2122×10^5	2.9105×10^5

For the case of a free boundary

T	*First approximation*	*Second approximation*	*Fourth approximation*	*Sixth approximation*	*Ninth approximation*
0	3.0916×10^3	3.0912×10^3			
10^3	9.4715×10^3	9.2493×10^3	8.9870×10^3		
10^4			2.6395×10^4	2.5517×10^4	
10^5				9.6961×10^4	9.1229×10^4

TABLE XXX

The characteristic vectors describing the lowest mode of instability for various values of the Taylor number

For the case of a rigid boundary

$T =$	0	10^3	10^4	10^5	10^6
A_{01}	0·99982	0·99859	0·97204	0·82340	0·85096
A_{02}	0·01917	0·02847	0·00342	−0·33881	0·37755
A_{03}					0·02874
A_{21}		0·04470	0·23249	0·42816	0·33137
A_{22}		0·00362		0·03278	−0·00423
A_{23}					0·00159
A_{41}			0·03268	0·14274	0·13141
A_{42}				0·04958	0·05937
A_{43}					0·03938

For the case of a free boundary

$T =$	0	10^3	10^4	10^5
A_{01}	1·00000	0·99662	0·80118	0·61020
A_{02}	−0·00151	−0·04190	−0·36978	0·63219
A_{03}		−0·00388	−0·01094	−0·11656
A_{21}		0·07050	0·45896	−0·39936
A_{22}			−0·03995	0·16196
A_{23}				0·01481
A_{41}			0·09497	−0·16842
A_{42}				0·00975
A_{43}				0·00385

63. Some remarks on geophysical applications

As we have remarked earlier, the theory of thermal instability in fluid spheres and in spherical shells has bearings on a number of geophysical questions. While it is outside the scope of this book to go into them in any detail, some indications as to their nature may not be inappropriate.

First consider the results of § 59 on the manner of the onset of thermal instability in a homogeneous fluid sphere. We have seen that in this case, the pattern of convection at marginal stability belongs to the mode $l = 1$; in particular, the axisymmetric mode, $l = 1$ and $m = 0$, divides the sphere into two equal cells with oppositely directed current systems (see Fig. 57 *a*, *b*, p. 236). Further, the point on the surface of the sphere, where the material ascending from the centre emerges, is antipodal to the point where the material having traversed half a great circle starts descending towards the centre. In any actual case where such current systems are established, we may expect the two antipodal points to differ the most in temperature. More generally, of the two hemispheres centred on these points, the one in which the current system converges towards the centre will be cooler than the one in which it diverges from the centre.

It has been suggested that at an early stage in the history of its formation, the earth was a nearly homogeneous fluid sphere with convective motions of the type we have just described; and, further, that we can infer the existence, at one time, of such motions from the division of the earth's surface into a land and an ocean hemisphere. This division of the earth's surface reflects a higher deposition of sial in one hemisphere than in the other; and the advocates of the convection hypothesis see in this the systematic difference in temperature in two hemispheres which would accompany convective motions belonging to the pattern $l = 1$. Vening Meinesz, who has argued for this point of view, believes that convective motions which caused a P_1-distribution of sial over the globe were also responsible for the initiation of the formation of a heavy core at the centre.

We know that the present fluid core of the earth occupies about one-half of the radius of the earth; and that the basaltic mantle extends down to this level. According to Vening Meinesz: 'There are many strong arguments to support the theory that great current systems exist in the mantle' and that these current systems 'are chiefly the result of temperature gradients caused by the earth cooling at the surface'. If this view is accepted, then the results of § 60 become applicable; and we may infer from them that current systems in a mantle occupying half the radius of a sphere must belong principally to the modes $l = 3$, 4, and 5 (see Fig. 58*a–d*, p. 244). And Vening Meinesz believes that the present distribution of sialic matter over the surface of the earth supports this inference. Thus, using Prey's spherical harmonic analysis of the earth's topography, Vening Meinesz has derived the amplitudes which will result from a similar analysis of the thickness of the sialic matter over the globe; and he has combined the amplitudes to obtain the root-mean-square thicknesses associated with the spherical harmonics of the different orders l. The results of this analysis are shown in Fig. 61. The predominance of the amplitudes associated with $l = 3$, 4, and 5 is, indeed, striking; and in this predominance Vening Meinesz sees evidence for his hypothesis of 'great current systems' in the mantle.

The foregoing applications of the theory of thermal instability in fluid spheres and spherical shells are not universally subscribed. It cannot, however, be doubted that convective motions in the fluid core are relevant to all theories concerned with the origin of the earth's magnetic field and its secular variations. But the theory of thermal instability has not been worked out with sufficient generality for these purposes. Even the effect of rotation has been examined only in a very preliminary way; and the onset of instability as overstability—which, from the results described in Chapter III, should be expected to be the rule rather than the exception with liquid metals—requires investigation. And in addition to rotation, the effect of a magnetic field has also to be considered.

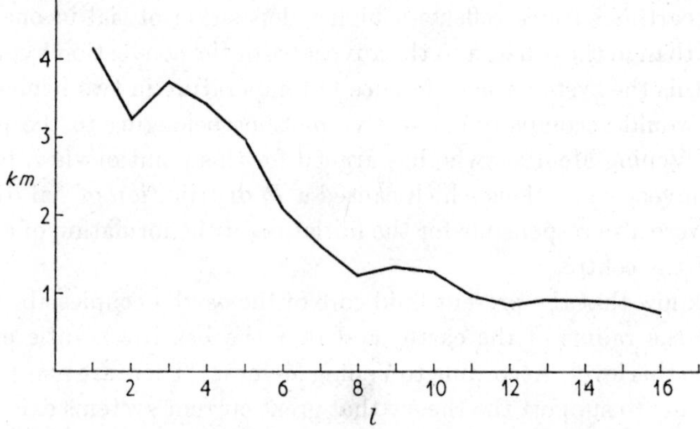

FIG. 61. Mean values (with respect to m) of the first sixteen terms in the spherical harmonic expansion of the thickness of the sialic crust over the earth's surface, according to Prey and Vening Meinesz.

The case of a uniform magnetic field presents no formal difficulty; but this is hardly appropriate for the problems in view. Without further knowledge, the choice of an initial field is so wide that the selection becomes almost arbitrary. It is, indeed, likely that the theory of the convective motions in the earth's core cannot be dissociated from the theory of the origin of the earth's magnetic field.

BIBLIOGRAPHICAL NOTES

The role of convection in geophysical problems is discussed in:

1. W. A. HEISKANEN and F. A. VENING MEINESZ, *The Earth and Its Gravity Field*, chapter 11, McGraw–Hill Series in the Geological Sciences, McGraw–Hill Book Company, Inc., New York, 1958.
2. C. L. PEKERIS, 'Thermal convection in the interior of the earth', *Monthly Notices Roy. Astron. Soc. London; Geophys. Suppl.* **3**, 343–67 (1935).
3. A. L. HALES, 'Convection currents in the earth', ibid. 372–9 (1936).
4. H. JEFFREYS, 'The earth's thermal history', ibid. **116**, 231–8 (1956).

§ 56. The problem of thermal instability in fluid spheres and spherical shells is considered in the following papers:

5. J. WASIUTYŃSKI, 'Studies in hydrodynamics and structure of stars and planets', *Astrophysica Norvegica,* **4,** 1–497 (1946); see particularly chapter 24.
6. H. JEFFREYS and M. E. M. BLAND, 'The instability of a fluid sphere heated within', *Monthly Notices Roy. Astron. Soc. London; Geophys. Suppl.* **6,** 148–58 (1951).
7. —— 'Problems of thermal instability in a sphere', ibid. 272–77 (1952).
8. S. CHANDRASEKHAR, 'The thermal instability of a fluid sphere heated within', *Phil. Mag.,* Ser. 7, **43,** 1317–29 (1952).
9. —— 'The onset of convection by thermal instability in spherical shells', ibid. **44,** 233–41 (1953); 'A correction', ibid. 1129–30 (1953).
10. G. E. BACKUS, 'On the application of eigenfunction expansions to the problem of the thermal instability of a fluid sphere heated within', ibid. **46,** 1310–27 (1955).
11. E. L. KOSCHMIEDER, 'Über Konvektionsströmungen auf einer Kugel, *Beiträge zur Physik der Atmosphäre,* **32,** 34–42 (1959).
12. E. LYTTKENS, 'The onset of convection in a mantle of a sphere with a heavy core' (unpublished).

For the representation of a solenoidal vector field in a sphere as a superposition of certain basic toroidal and poloidal fields, see:

13. J. A. STRATTON, *Electromagnetic Theory,* chapter 7, International Series in Pure and Applied Physics, McGraw-Hill Book Company, Inc., New York, 1941.
14. W. M. ELSASSER, 'Induction effects in terrestrial magnetism. I. Theory', *Physical Review,* **69,** 106–16 (1946).
15. E. C. BULLARD and H. GELLMAN, 'Homogeneous dynamos and terrestrial magnetism', *Phil. Trans. Roy. Soc. (London)* A, **247,** 213–78 (1954).

§ 57. The validity of the principle of the exchange of stabilities for this problem does not seem to have been examined before.

§ 58. The variational principle for this problem is formulated in references 6 and 8. The thermodynamic significance of the principle is considered in:

16. H. JEFFREYS, 'The thermodynamics of thermal instability in liquids', *Quart. J. Mech. Appl. Math.* **9,** 1–5 (1956).

However, the treatment followed in the text is different: the representation of the prevailing velocity field in terms of a single poloidal scalar simplifies the reductions very greatly.

§ 59. The analysis in this section is largely derived from reference 8.

§ 60. The method of treatment follows reference 9. The orthogonal functions, in terms of which the solution is carried out, are constructed in:

17. S. CHANDRASEKHAR and DONNA ELBERT, 'The roots of

$$J_{-(l+\frac{1}{2})}(\lambda\eta)J_{l+\frac{1}{2}}(\lambda)-J_{l+\frac{1}{2}}(\lambda\eta)J_{-(l+\frac{1}{2})}(\lambda) = 0\text{'},$$

Proc. Camb. Phil. Soc. **49,** 446–8 (1953).

In reference 9, only the case when the surfaces at $r = 1$ and η are both free is considered. The solutions for the other boundary conditions were obtained by Miss Ruby Ebisuzaki.

The results for the more general models considered in §§ (*b*) and (*c*) are due to Lyttkens (reference 12).

§ 61. The effect of rotation on the onset of thermal instability in fluid spheres is considered in the following papers:

18. H. TAKEUCHI and Y. SHIMAZU, 'Convective fluid motions in a rotating sphere', *J. Phys. of the Earth*, **2**, 13–26 (1954).
19. S. CHANDRASEKHAR, 'The thermal instability of a rotating fluid sphere heated within, Part I', *Phil. Mag.*, Ser. 8, **2**, 845–58 (1957); Part II, ibid. 1282–4 (1957).
20. F. E. BISSHOPP, 'On the thermal instability of a rotating fluid sphere', ibid. **3**, 1342–60 (1958).
21. T. NAMIKAWA, 'Fluid motions in a sphere. I. Thermal instability of a rotating fluid sphere heated within,' *J. Geomagnetism and Geoelectricity*, **9**, 182–92 (1957).

For the representation of axisymmetric solenoidal fields in terms of two scalars see:

22. R. LÜST and A. SCHLÜTER, 'Kraftfreie Magnetfelder', *Z. f. Astrophysik*, **34**, 263–82 (1954).
23. S. CHANDRASEKHAR, 'On force-free magnetic fields', *Proc. Nat. Acad. Sci.* **42**, 1–5 (1956).
24. —— 'Axisymmetric magnetic fields and fluid motions', *Astrophys. J.* **124**, 232–43 (1956).

The analysis in this section is largely based on reference 19. The variational principle formulated in § (*d*) is, however, more general than in this reference; also the analysis in § (*e*) is new.

§ 62. The analysis in this section is derived from reference 20.

The expansion of the velocity field in terms of orthogonal functions generated by suitable characteristic value problems in fourth-order differential equations has been discussed by:

25. S. CHANDRASEKHAR and W. H. REID, 'On the expansion of functions which satisfy four boundary conditions', *Proc. Nat. Acad. Sci.* **43**, 521–7 (1957).
26. D. L. HARRIS and W. H. REID, 'On orthogonal functions which satisfy four boundary conditions. I. Tables for use in Fourier-type expansions', *Astrophys. J. Supp., Ser.* **3**, 429–47 (1958).
27. S. CHANDRASEKHAR and DONNA D. ELBERT, 'On orthogonal functions which satisfy four boundary conditions. III. Tables for use in Fourier–Bessel-type expansions', ibid. 453–8 (1958).

§ 63. The principal source for the matters briefly discussed in this section is reference 1.

Namikawa has attempted the solution of the problem of thermal instability in a rotating fluid sphere with a uniform magnetic field:

28. T. NAMIKAWA, 'Fluid motions in a sphere II. Thermal instability of a

conducting fluid sphere heated within under a uniform magnetic field', *J. Geomagnetism and Geoelectricity*, **9**, 193–202 (1957).

29. T. Namikawa, 'Fluid motions in a sphere III. Thermal instability of a rotating fluid sphere heated within under a uniform magnetic field', ibid. 203–9 (1957).

Namikawa draws some general conclusions based on what is effectively a one-term approximation for the various fields. As Bisshopp's detailed calculations show, this is a dangerous procedure.

VII

THE STABILITY OF COUETTE FLOW

64. Introduction

IN the last five chapters we have considered the onset of instability under a variety of externally impressed conditions; but the origin of the instability was always the same: a potentially unstable arrangement of the fluid resulting from a prevailing adverse temperature gradient. In this and in the following two chapters we shall consider cases of instabilities which have a different origin: a potentially unstable arrangement of flow resulting from a prevailing adverse gradient of angular momentum. The simplest example of such instability occurs in *Couette flow*, i.e. in the steady circular flow of a liquid between two rotating coaxial cylinders.

In this chapter we shall consider the simple Couette flow of inviscid and viscid fluids. In Chapter VIII we shall consider some examples of more general curved flows; and in Chapter IX we shall consider the effect of an axial magnetic field on Couette flow.

65. The physical problem

We consider, then, the permissible stationary circular flows of an incompressible fluid between two rotating coaxial cylinders. As we shall presently see, in the absence of viscosity, the class of such permissible flows is very wide: indeed, if Ω denotes the angular velocity of rotation about the axis, then the equations of motion allow Ω to be an arbitrary function of the distance r from the axis, provided the velocities in the radial and the axial directions are zero. But if viscosity is present, the class becomes very restricted: in fact, in the absence of any transverse pressure gradient, the most general form of Ω which is allowed is

$$\Omega(r) = A+B/r^2, \tag{1}$$

where A and B are two constants which are related to the angular velocities Ω_1 and Ω_2 with which the inner and the outer cylinders are rotated. Thus, if R_1 and R_2 ($> R_1$) are the radii of the two cylinders, then

$$\Omega_1 = A+B/R_1^2 \quad\text{and}\quad \Omega_2 = A+B/R_2^2. \tag{2}$$

Solving for A and B in terms of Ω_1 and Ω_2, we have

$$A = -\Omega_1\eta^2\frac{1-\mu/\eta^2}{1-\eta^2} \quad\text{and}\quad B = \Omega_1\frac{R_1^2(1-\mu)}{1-\eta^2}, \tag{3}$$

where $$\mu = \Omega_2/\Omega_1 \quad \text{and} \quad \eta = R_1/R_2. \tag{4}$$

The questions to which we seek answers are the following. In the absence of viscosity when $\Omega(r)$ can be an arbitrary function of r, what is the necessary and sufficient condition for the stability of the flow? What does this condition imply for the distribution of Ω given by (1)? And, finally, how is the condition for inviscid fluids, as applied to (1), modified when account is taken of viscosity? These are the questions to which the present chapter is addressed.

66. Rayleigh's criterion

Rayleigh stated: *in the absence of viscosity, the necessary and sufficient condition for a distribution of angular velocity $\Omega(r)$ to be stable is*

$$\frac{d}{dr}(r^2\Omega)^2 > 0 \tag{5}$$

everywhere in the interval; and, further, that the distribution is unstable if $(r^2\Omega)^2$ should decrease anywhere inside the interval.

Since $|r^2\Omega|$ is the angular momentum, per unit mass, of a fluid element about the axis of rotation, an alternative way of stating Rayleigh's criterion is: *a stratification of angular momentum about an axis is stable if and only if it increases monotonically outward.*

Rayleigh did not establish his criterion by an analytical discussion of the relevant perturbation equations. Instead, he argued as follows.

The hydrodynamical equations governing an incompressible inviscid fluid in cylindrical polar coordinates (r, θ, z) are:†

$$\frac{\partial u_r}{\partial t}+(\mathbf{u}\,.\,\text{grad})u_r-\frac{u_\theta^2}{r} = -\frac{\partial}{\partial r}\left(\frac{p}{\rho}\right), \tag{6}$$

$$\frac{\partial u_\theta}{\partial t}+(\mathbf{u}\,.\,\text{grad})u_\theta+\frac{u_r u_\theta}{r} = -\frac{1}{r}\frac{\partial}{\partial \theta}\left(\frac{p}{\rho}\right), \tag{7}$$

$$\frac{\partial u_z}{\partial t}+(\mathbf{u}\,.\,\text{grad})u_z = -\frac{\partial}{\partial z}\left(\frac{p}{\rho}\right), \tag{8}$$

where $$(\mathbf{u}\,.\,\text{grad}) = u_r\frac{\partial}{\partial r}+\frac{u_\theta}{r}\frac{\partial}{\partial \theta}+u_z\frac{\partial}{\partial z}. \tag{9}$$

We also have the equation of continuity,

$$\frac{\partial u_r}{\partial r}+\frac{u_r}{r}+\frac{1}{r}\frac{\partial u_\theta}{\partial \theta}+\frac{\partial u_z}{\partial z} = 0. \tag{10}$$

† The considerations which follow are equally applicable if an external force derivable from an axisymmetric potential is present.

These equations clearly allow the stationary solution,

$$u_r = u_z = 0 \quad \text{and} \quad u_\theta = V(r), \tag{11}$$

where $V(r)$ is an arbitrary function of r. In terms of $V(r)$, the pressure distribution is determined by

$$p = \rho \int \frac{dr}{r} V^2(r). \tag{12}$$

For axisymmetric motions to which the rest of the discussion in this section will be restricted, equations (6)–(8) reduce to:

$$\frac{\partial u_r}{\partial t} + u_r \frac{\partial u_r}{\partial r} + u_z \frac{\partial u_r}{\partial z} = \frac{u_\theta^2}{r} - \frac{\partial}{\partial r}\left(\frac{p}{\rho}\right), \tag{13}$$

$$\frac{\partial u_\theta}{\partial t} + u_r \frac{\partial u_\theta}{\partial r} + u_z \frac{\partial u_\theta}{\partial z} + \frac{u_r u_\theta}{r} = 0, \tag{14}$$

$$\frac{\partial u_z}{\partial t} + u_r \frac{\partial u_z}{\partial r} + u_z \frac{\partial u_z}{\partial z} = -\frac{\partial}{\partial z}\left(\frac{p}{\rho}\right). \tag{15}$$

Equation (14) is equivalent to

$$\frac{d}{dt}(ru_\theta) = \frac{d}{dt}(r^2\Omega) = 0. \tag{16}$$

Therefore, the angular momentum L $(= r^2\Omega)$ of a fluid element, per unit mass, remains constant as we follow it with its motion. And the motions in the radial and the axial directions take place as though u_θ were absent and, instead, a force $u_\theta^2/r = L^2/r^3$ were acting in the radial direction. Since L is a constant of the motion, we can associate with the force L^2/r^3 a potential energy $\rho L^2/2r^2$. This potential energy is the same as the kinetic energy associated with the velocity u_θ in the transverse direction.

Suppose now that we interchange the fluid contained in two elementary rings, of equal heights and masses, at $r = r_1$ and $r = r_2$ $(> r_1)$. If dr_1 and dr_2 are the radial extents of the rings, the equality of their masses requires $2\pi r_1\,dr_1 = 2\pi r_2\,dr_2 = dS$ (say). In view of the constancy of L with the motion, the fluid at r_2, after the interchange, will have the same angular momentum (namely, L_1) which it had at r_1 before the interchange; similarly, the fluid at r_1, after the interchange, will have the same angular momentum (namely, L_2) which it had at r_2. As a result, the change in the kinetic energy (or, what is the same thing, the change in the centrifugal potential energy) is proportional to

$$\left\{\left(\frac{L_2^2}{r_1^2} + \frac{L_1^2}{r_2^2}\right) - \left(\frac{L_1^2}{r_1^2} + \frac{L_2^2}{r_2^2}\right)\right\} dS = (L_2^2 - L_1^2)\left(\frac{1}{r_1^2} - \frac{1}{r_2^2}\right) dS. \tag{17}$$

Remembering that $r_2 > r_1$, we observe that this is positive or negative according as L_2^2 is greater than L_1^2 or less than L_1^2. Consequently, if L^2

is a monotonic increasing function of r, no interchange of fluid rings such as we have imagined can occur without a source of energy; and this means stability. On the other hand, if L^2 should decrease anywhere, then an interchange of fluid rings in this region will result in a liberation of energy; and this means instability.

While the foregoing arguments of Rayleigh make his criterion a very likely one by drawing attention to its physical origin, one should still like to establish it directly from the relevant perturbation equations; this we shall do in § 67. Meantime, it is useful to see what Rayleigh's criterion implies for the particular distribution of Ω which is permissible for viscid fluids. As the *discriminant* for stability, we shall use

$$\Phi(r) = \frac{1}{r^3}\frac{d}{dr}(r^2\Omega)^2 = \frac{2}{r}\Omega\frac{d}{dr}(r^2\Omega). \tag{18}$$

Rayleigh's criterion requires

$$\Phi(r) > 0 \quad \text{for stability.} \tag{19}$$

When Ω has the form given by (1),

$$\Phi = 4A(A+B/r^2). \tag{20}$$

It is convenient to measure r in units of the radius R_2 of the outer cylinder. Then

$$\Phi = \frac{4AB}{R_2^2}\left(\frac{1}{r^2}+\frac{AR_2^2}{B}\right). \tag{21}$$

With the values of A and B given in (3), we have

$$\Phi = -4\Omega_1^2\,\eta^4\frac{(1-\mu)(1-\mu/\eta^2)}{(1-\eta^2)^2}\left(\frac{1}{r^2}-\kappa\right), \tag{22}$$

where
$$\kappa = -\frac{AR_2^2}{B} = \frac{1-\mu/\eta^2}{1-\mu}. \tag{23}$$

Clearly,
$$\Phi > 0 \quad \text{for} \quad \eta \leqslant r \leqslant 1, \tag{24}$$

so long as
$$\mu = \Omega_2/\Omega_1 > \eta^2. \tag{25}$$

Therefore, Rayleigh's criterion applied to the distribution (1) requires that *for stability, the outer cylinder must rotate with an angular speed greater than η^2-times that of the inner cylinder and in the same sense.* In the (Ω_2, Ω_1)-plane† (see Fig. 62), the regions of stability are delimited by the positive Ω_2-axis and the Rayleigh line $\Omega_2 = \Omega_1\,\eta^2$. If the effects of viscosity are ignored, we must have instability to the left of the Rayleigh line; and it is of interest to determine how Rayleigh's criterion

† Without loss of generality, we may suppose that the sense of rotation of the inner cylinder is positive.

is violated for flow represented in this part of the plane. With κ given by (23), we find

$$\frac{1}{r^2}-\kappa = \mu\frac{1-\eta^2}{\eta^2(1-\mu)} \quad \text{for } r = 1,$$

$$= \frac{(1-\eta^2)}{\eta^2(1-\mu)} \quad \text{for } r = \eta. \tag{26}$$

Accordingly, $$\Phi(r) < 0 \quad \text{for} \quad \eta \leqslant r \leqslant 1, \tag{27}$$

so long as $$0 < \mu < \eta^2. \tag{28}$$

Therefore, for $0 < \mu < \eta^2$, Rayleigh's criterion is violated everywhere

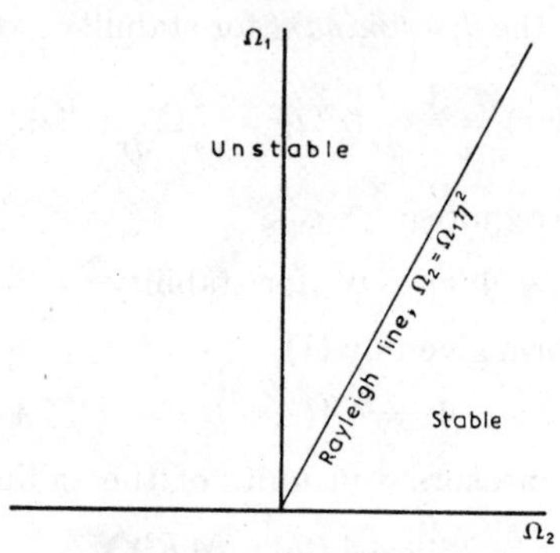

Fig. 62. The (Ω_1, Ω_2)-plane: in the limit of zero viscosity, the stable and the unstable regions in this plane are separated by the Rayleigh line, $\Omega_2 = \Omega_1(R_1/R_2)^2$; R_1 and R_2 are the radii and Ω_1 and Ω_2 are the angular velocities of the inner and the outer cylinders, respectively.

in the fluid; and we may say that instability occurs throughout the fluid. On the other hand, if the cylinders are rotating in opposite directions and $\mu < 0$, Φ is positive in a part of the fluid adjoining the outer cylinder; and it is negative in the remaining part of the fluid adjoining the inner cylinder. The two parts are separated by the *nodal surface* on which $\Omega = 0$. The radius η_0 of this surface in the unit R_2 is given by

$$\eta_0 = \frac{1}{\sqrt{\kappa}} = \eta\left(\frac{1+|\mu|}{\eta^2+|\mu|}\right)^{\frac{1}{2}}; \tag{29}$$

and

$$\left.\begin{aligned}\Phi(r) > 0 \quad \text{for } \eta_0 < r < 1\\ \Phi(r) < 0 \quad \text{for } \eta < r < \eta_0\end{aligned}\right\}. \tag{30}$$

Thus, when $\mu < 0$, Rayleigh's criterion is violated only interior to the nodal surface. The instability which is predicted for all negative μ in the absence of viscosity is, therefore, to be associated with this instability of the flow interior to η_0. On this account we may say that for $\mu < 0$, the instability is only partial. Also, as $\mu \to -\infty$, $\eta_0 \to \eta$ and the part of the fluid which is unstable according to Rayleigh's criterion becomes vanishingly small.

If we now consider the effect of viscosity on the stability of the flow prescribed by (1), we must on general grounds expect that it will postpone the onset of instability beyond the point predicted by Rayleigh's criterion; and that for a given κ (i.e. μ and η) the factor,

$$4\Omega_1^2\eta^4\frac{(1-\mu)(1-\mu/\eta^2)}{(1-\eta^2)^2} \quad (\mu < \eta^2), \tag{31}$$

which occurs in Rayleigh's discriminant must exceed a certain critical value, depending on ν and R_2, before instability can set in. The non-dimensional number in terms of which the criterion for stability will be expressed is, then,

$$T = \frac{4\Omega_1^2}{\nu^2}R_1^4\frac{(1-\mu)(1-\mu/\eta^2)}{(1-\eta^2)^2}. \tag{32}$$

This is the proper definition of the *Taylor number* for this problem. For a given η and μ, instability will set in for a certain critical Taylor number T_c; and the central problem of this subject is the determination of T_c as a function of η and μ.

67. Analytical discussion of the stability of inviscid Couette flow

We now return to an analytical discussion of the stability of the stationary flow,

$$u_r = u_z = 0 \quad \text{and} \quad u_\theta = V(r) = r\Omega(r), \tag{33}$$

which the equations of inviscid motion (6)–(9) allow. In (33), $V(r)$ is an arbitrary function of r.

Consider an infinitesimal perturbation of the flow represented by the solution (33). Let the perturbed state be described by

$$u_r, \quad V+u_\theta, \quad u_z, \quad \text{and} \quad \varpi \ (=\delta p/\rho). \tag{34}$$

The linear equations governing these perturbations are

$$\frac{\partial u_r}{\partial t}+\frac{V}{r}\frac{\partial u_r}{\partial \theta}-2\frac{V}{r}u_\theta = -\frac{\partial \varpi}{\partial r}, \tag{35}$$

$$\frac{\partial u_\theta}{\partial t}+\frac{V}{r}\frac{\partial u_\theta}{\partial \theta}+\left(\frac{V}{r}+\frac{dV}{dr}\right)u_r = -\frac{1}{r}\frac{\partial \varpi}{\partial \theta}, \tag{36}$$

$$\frac{\partial u_z}{\partial t}+\frac{V}{r}\frac{\partial u_z}{\partial \theta} = -\frac{\partial \varpi}{\partial z}, \tag{37}$$

and the equation of continuity,

$$\frac{\partial u_r}{\partial r}+\frac{u_r}{r}+\frac{1}{r}\frac{\partial u_\theta}{\partial \theta}+\frac{\partial u_z}{\partial z}=0. \tag{38}$$

In accordance with the general procedure of treating these problems, we analyse the disturbance into normal modes. In the present instance, it is natural to suppose that the various quantities describing the perturbation have a (t,θ,z)-dependence given by

$$e^{i(pt+m\theta+kz)}, \tag{39}$$

where p is a constant (which can be complex), m is an integer (which can be positive, zero, or negative), and k is the wave number of the disturbance in the z-direction.

Let $u_r(r)$, $u_\theta(r)$, $u_z(r)$, and $\varpi(r)$ now denote the amplitudes of the respective perturbations whose (t,θ,z)-dependence is given by (39). Equations (35)–(38) then give

$$i\sigma u_r-2\Omega u_\theta=-\frac{d\varpi}{dr}, \tag{40}$$

$$i\sigma u_\theta+\left(\Omega+\frac{d}{dr}r\Omega\right)u_r=-\frac{im}{r}\varpi, \tag{41}$$

$$i\sigma u_z=-ik\varpi, \tag{42}$$

and

$$\frac{du_r}{dr}+\frac{u_r}{r}+\frac{im}{r}u_\theta+iku_z=0, \tag{43}$$

where

$$\sigma=p+m\Omega. \tag{44}$$

(Note that through Ω, σ is a function of r.)

(a) The equations in terms of the Lagrangian displacement

Consider the variables ξ_r, ξ_θ, and ξ_z related to u_r, u_θ, and u_z by

$$u_r=i\sigma\xi_r,\quad u_\theta=i\sigma\xi_\theta-r\frac{d\Omega}{dr}\xi_r,\quad\text{and}\quad u_z=i\sigma\xi_z. \tag{45}$$

Defined in this manner, $\boldsymbol{\xi}$ is the *Lagrangian displacement* appropriate to the problem on hand.

It can be readily verified that consistent with the meaning of $\boldsymbol{\xi}$ as the Lagrangian displacement, the solenoidal character of $\mathbf{u}$ implies the solenoidal character of $\boldsymbol{\xi}$; thus

$$\frac{d\xi_r}{dr}+\frac{\xi_r}{r}+\frac{im}{r}\xi_\theta+ik\xi_z=0. \tag{46}$$

In terms of the variables ξ_r, ξ_θ, and ξ_z, equations (40)–(42) become

$$\left(\sigma^2-2r\Omega\frac{d\Omega}{dr}\right)\xi_r+2i\Omega\sigma\xi_\theta=\frac{d\varpi}{dr}, \tag{47}$$

$$\sigma^2\xi_\theta-2i\Omega\sigma\xi_r=\frac{im}{r}\varpi, \tag{48}$$

and

$$\sigma^2\xi_z=ik\varpi. \tag{49}$$

Multiplying equations (48) and (49) by im/r and ik, respectively, adding, and making use of equation (46), we obtain

$$\sigma^2\left(\frac{d\xi_r}{dr}+\frac{\xi_r}{r}\right)-\frac{2m\Omega\sigma}{r}\xi_r=\left(\frac{m^2}{r^2}+k^2\right)\varpi. \tag{50}$$

Next eliminating ξ_θ between equations (47) and (48), we have

$$\left(\sigma^2-2r\Omega\frac{d\Omega}{dr}\right)\xi_r+\frac{2i\Omega}{\sigma}\left(2i\Omega\sigma\xi_r+\frac{im}{r}\varpi\right)=\frac{d\varpi}{dr}. \tag{51}$$

Rearranging the terms in this equation, we obtain

$$[\sigma^2-\Phi(r)]\xi_r=\frac{d\varpi}{dr}+\frac{2m\Omega}{\sigma r}\varpi, \tag{52}$$

where

$$\Phi(r)=2r\Omega\frac{d\Omega}{dr}+4\Omega^2=\frac{2\Omega}{r}\frac{d}{dr}(r^2\Omega) \tag{53}$$

is the Rayleigh discriminant we have already introduced in § 66 (equation (18)).

Equation (52) must be considered in conjunction with equation (50), which we shall rewrite in the form

$$\frac{1}{r}\frac{d}{dr}(r\,\xi_r)-\frac{2m\Omega}{\sigma r}\xi_r=\frac{1}{\sigma^2}\left(\frac{m^2}{r^2}+k^2\right)\varpi. \tag{54}$$

If the fluid is confined between two coaxial cylinders of radii R_1 and R_2, we must require that the radial components of the velocity vanish for these values of r. Thus, equations (52) and (54) must be considered together with the boundary conditions

$$\xi_r=0 \quad \text{for } r=R_1 \text{ and } R_2. \tag{55}$$

We shall presently see that equations (52) and (54), for real σ, constitute a self-adjoint system with respect to the boundary conditions (55); and that in consequence their solution can be reduced to a variational problem; this was, indeed, the object of introducing the Lagrangian displacement as the variable.

(*b*) *The case $m = 0$*

When $m = 0$, $\sigma = p$ and equations (52) and (54) become

$$[p^2-\Phi(r)]\xi_r = \frac{d\varpi}{dr} \tag{56}$$

and

$$\frac{1}{r}\frac{d}{dr}(r\xi_r) = \frac{k^2}{p^2}\varpi. \tag{57}$$

Eliminating ϖ between these equations, we obtain

$$\frac{d}{dr}\left(\frac{1}{r}\frac{d}{dr}r\xi_r\right)-k^2\xi_r = -\frac{k^2}{p^2}\Phi(r)\xi_r. \tag{58}$$

The solution of this equation, together with the boundary conditions (55), constitutes a characteristic value problem of the classical Sturm–Liouville type; and by appealing to the standard theorems of the subject, we can reach the following conclusion.

The characteristic values of k^2/p^2 are all positive if $\Phi(r)$ is everywhere positive; and they are all negative if $\Phi(r)$ is everywhere negative. If $\Phi(r)$ should change sign anywhere in the interval (R_1, R_2), then there are two sets of real characteristic values which have the limit points $+\infty$ and $-\infty$.

Since a negative p^2 means instability, it is clear that the foregoing statements regarding the sign of the characteristic values of k^2/p^2 are equivalent to a restatement of Rayleigh's criterion.

An alternative way of establishing the same results is instructive. Multiply equation (58) by $r\xi_r$ and integrate over the range of r. The integral on the left-hand side can be transformed by an integration by parts. We obtain

$$\int \left\{\frac{1}{r}\left(\frac{d}{dr}r\xi_r\right)^2+k^2r\xi_r^2\right\}dr = \frac{k^2}{p^2}\int \Phi(r)r\,\xi_r^2\,dr \tag{59}$$

or

$$\frac{p^2}{k^2} = \frac{\int \Phi(r)r\xi_r^2\,dr}{\int [\{(d/dr)r\xi_r\}^2/r+k^2r\xi_r^2]\,dr}. \tag{60}$$

From equation (60) it is immediately apparent that p^2/k^2 is positive if $\Phi(r)$ is everywhere positive, and is negative if $\Phi(r)$ is everywhere negative. The further result that there exist unstable modes, if $\Phi(r)$ is anywhere negative, can be deduced from the fact that we may regard equation (58) as the condition that

$$I_1 = \int \Phi(r)r\xi_r^2\,dr \tag{61}$$

is a maximum, or a minimum, for a given

$$I_2 = \int \left\{\frac{1}{r}\left(\frac{d}{dr}r\xi_r\right)^2+k^2r\xi_r^2\right\}dr, \tag{62}$$

p^2/k^2 being, then, the value of I_1/I_2. If $\Phi(r)$ should anywhere be negative, I_1 admits a negative value and therefore a negative minimum, so that one value at least of p^2 is negative, and one mode of disturbance is unstable.

By comparing this alternative method of establishing Rayleigh's criterion with the corresponding method of establishing the stability of an incompressible fluid of variable density (Chapter X, § 92), we observe that $\Phi(r)$ in this problem, and the density gradient in the other problem, play entirely equivalent roles.

(c) *The case* $m \neq 0$

The treatment of the general case is not as straightforward as the case $m = 0$. It appears that when $m \neq 0$, it is more convenient to consider the problem of solving equations (52) and (54) subject to the boundary conditions (55) as a characteristic value problem for k^2 for a given p, rather than as a problem for p for a given k^2. This inversion of the usual procedure is advantageous in so far as the problem can then be formulated in terms of a variational principle. Moreover, we shall find that for real p the problem is Hermitian and the characteristic values of k^2 are real.

The question now arises whether the solution of the stability problem can be inferred from the solution of the characteristic value problem inverted in the manner we have proposed. That it should, in principle, be possible to infer the solution becomes clear when we realize that there is a unique dispersion relation which gives p as an analytic function of k^2 (with, possibly, more than one branch); and that the relationship is independent of how it is determined. And so long as the relation between p and k^2 is analytic, we should be able to answer the basic questions concerning stability by determining the relation between real p and k^2 (positive or negative) instead of, as is more usual, by determining the relation between positive k^2 and p (real or complex).

Returning to equations (52)–(55), we shall first show that for real p the characteristic values of k^2 are real. To show this, we shall multiply equation (52) by $r\xi_r^*$ and integrate over the range of r. We obtain

$$\int r[\sigma^2-\Phi(r)]|\xi_r|^2\,dr = \int \xi_r^*\left(r\frac{d\varpi}{dr}+\frac{2m\Omega}{\sigma}\varpi\right)dr. \tag{63}$$

After an integration by parts, the right-hand side becomes

$$-\int \varpi\left\{\frac{d}{dr}(r\xi_r^*)-\frac{2m\Omega}{\sigma}\xi_r^*\right\}dr; \tag{64}$$

and making use of equation (54), we can finally write

$$\int r[\sigma^2-\Phi(r)]|\xi_r|^2\,dr = -\int \frac{1}{\sigma^2}\left(\frac{m^2}{r^2}+k^{*2}\right)r|\varpi|^2\,dr. \tag{65}$$

By replacing every term in this equation by its complex conjugate, we infer the identity of k^2 and k^{*2}; k^2 is, therefore, real. When k^2 is real, the proper functions, ξ_r and ϖ, belonging to a particular k^2 (and a real p) are also real, and we can rewrite equation (65) in the manner

$$k^2 = \frac{\int r[(\Phi-\sigma^2)\xi_r^2-(m^2\varpi^2)/(r^2\sigma^2)]\,dr}{\int dr\; r\varpi^2/\sigma^2} = \frac{I_1}{I_2}\ \text{(say)}. \tag{66}$$

We shall now show that the expression (66) for k^2 provides the basis for a variational formulation of the problem.

First, it should be remarked that in evaluating k^2 in accordance with (66), we must suppose that ξ_r is determined in terms of ϖ by the equation

$$(\sigma^2-\Phi)\xi_r = \frac{d\varpi}{dr}+\frac{2m\Omega}{\sigma r}\varpi; \tag{67}$$

and that an allowed ϖ accordingly satisfies the boundary conditions

$$\frac{d\varpi}{dr}+\frac{2m\Omega}{\sigma r}\varpi = 0 \quad \text{for } r = R_1 \text{ and } R_2. \tag{68}$$

Consider now the effect on k^2 of an arbitrary variation $\delta\varpi$ in ϖ compatible only with the boundary conditions (68). If $\delta\xi_r$ denotes the corresponding variation in ξ_r, then, by (67)

$$(\sigma^2-\Phi)\delta\xi_r = \frac{d}{dr}\,\delta\varpi+\frac{2m\Omega}{\sigma r}\,\delta\varpi; \tag{69}$$

and

$$\delta\xi_r = 0 \quad \text{for } r = R_1 \text{ and } R_2. \tag{70}$$

From equation (66) it follows that

$$\delta k^2 = \frac{1}{I_2}\left(\delta I_1-\frac{I_1}{I_2}\,\delta I_2\right) = \frac{1}{I_2}(\delta I_1-k^2\delta I_2), \tag{71}$$

where

$$\delta I_1 = 2\int r\left[(\Phi-\sigma^2)\xi_r\,\delta\xi_r-\frac{m^2\varpi}{r^2\sigma^2}\,\delta\varpi\right]dr \tag{72}$$

and

$$\delta I_2 = 2\int r\varpi(\delta\varpi/\sigma^2)\,dr \tag{73}$$

are the variations in I_1 and I_2 resulting from the variation in ϖ.

The terms in $\delta\xi_r$ in the expression (72) for δI_1 are the same as

$$-2\int r\xi_r\left(\frac{d}{dr}\,\delta\varpi+\frac{2m\Omega}{\sigma r}\,\delta\varpi\right)dr. \tag{74}$$

After an integration by parts, (74) can be rewritten in the form

$$2\int \delta\varpi\left[\frac{d}{dr}(r\xi_r)-\frac{2m\Omega}{\sigma}\xi_r\right]dr. \tag{75}$$

The first-order change in k^2 resulting from the variation in ϖ is, therefore,

$$\delta k^2 = \frac{2}{I_2}\int \delta\varpi\left\{\frac{1}{r}\frac{d}{dr}(r\xi_r)-\frac{2m\Omega}{r\sigma}\xi_r-\frac{1}{\sigma^2}\left(\frac{m^2}{r^2}+k^2\right)\varpi\right\}r\,dr. \tag{76}$$

Consequently, $\delta k^2 = 0$ for an arbitrary variation $\delta\varpi$ compatible with the boundary conditions (68) if and only if

$$\frac{1}{r}\frac{d}{dr}(r\xi_r)-\frac{2m\Omega}{\sigma r}\xi_r = \frac{1}{\sigma^2}\left(\frac{m^2}{r^2}+k^2\right)\varpi. \tag{77}$$

Conversely, whenever equation (77) is satisfied, $\delta k^2 = 0$. But equation (77) is the same as equation (54). Therefore, solving the characteristic value problem presented by equations (52), (54), and (55) is equivalent to finding a maximum or a minimum of

$$I_1 = -\int\left\{\frac{1}{\sigma^2-\Phi}\left(\frac{d\varpi}{dr}+\frac{2m\Omega}{\sigma r}\varpi\right)^2+\frac{m^2\varpi^2}{r^2\sigma^2}\right\}r\,dr, \tag{78}$$

for given
$$I_2 = \int dr\, r\varpi^2/\sigma^2, \tag{79}$$

for arbitrary variations of ϖ subject only to the boundary conditions

$$\frac{d\varpi}{dr}+\frac{2m\Omega}{\sigma r}\varpi = 0 \quad \text{for } r = R_1 \text{ and } R_2; \tag{80}$$

and the ratio I_1/I_2 at such a maximum or a minimum is a characteristic value of k^2. [Note that the boundedness of ϖ/σ assumed here follows from equation (42).]

From equation (66) it is at once apparent that if $\Phi(r)$ is everywhere negative, k^2 cannot admit a positive characteristic value. For real p's, the characteristic values of k are necessarily imaginary. Therefore for real k's, p must necessarily be complex; and this means instability.

Consider next the case when $\Phi(r)$ changes sign in the interval (R_1, R_2) and it is negative over a part of the interval and positive over others. Under these circumstances, I_1 can assume a negative value for any arbitrarily assigned real p; and for a given I_2, I_1 can attain a negative minimum. Therefore, for any real p, there exists at least one negative characteristic value for k^2. At the same time, since $\Phi(r)$ is positive over some other parts of the interval I_1 can also assume positive values for suitably chosen real values of p. For a given I_2, there exist values of p for which I_1 can attain a positive maximum. Therefore, for some real values of p, there exist positive characteristic values for k^2. From our earlier arguments, these same values of p must also allow negative characteristic values of k^2. Thus, in this case the dispersion relation has two branches. One of them is such that for all real p's the characteristic values of k^2 are negative. The inverse of this latter branch must lead to complex p's for real k's; and this means instability.†

Consider finally the case when $\Phi(r)$ is everywhere positive. In this case it is convenient to rewrite equation (66) in the form

$$k^2 = \frac{\int r[\{(d\varpi/dr)+(2m\Omega/\sigma r)\varpi\}^2/(\Phi-\sigma^2)-(m^2\varpi^2/r^2\sigma^2)]\,dr}{\int dr\, r\varpi^2/\sigma^2}. \tag{81}$$

It is clear that for all real p such that $\sigma^2 = (p+m\Omega)^2 > \Phi(r)$, the characteristic values of k^2 are negative. For positive characteristic values it is clearly necessary that σ^2 be less than $\Phi(r)$ at least over a part of the interval (R_1, R_2). Indeed, it is clear that by requiring σ^2 to be less than $\Phi(r)$ over a part of the interval, we can obtain for k^2 arbitrary positive values; i.e. for all real k's there correspond real p's. The system therefore allows stable modes for all positive k^2. However, since negative values of k^2 are possible for some real p's, we cannot exclude instability in this case. Thus, while the condition that $\Phi(r) < 0$ for some range of r in (R_1, R_2) ensures instability, $\Phi(r) > 0$ for $R_1 \leqslant r \leqslant R_2$ does not allow us to exclude instability.

68. The periods of oscillation of a rotating column of liquid

In this section we shall consider the oscillations of a cylindrical column of rotating fluid. As Lord Kelvin has remarked in his first investigation of the subject, 'Crowds of extremely interesting cases present themselves'.

(a) *The case Ω = constant*

When Ω = constant,

$$\Phi(r) = \frac{1}{r^3}\frac{d}{dr}(r^2\Omega)^2 = 4\Omega^2, \tag{82}$$

and the relevant equations are (cf. equations (52) and (54)):

$$(\sigma^2-4\Omega^2)\xi_r = \frac{d\varpi}{dr}+\frac{2m\Omega}{\sigma r}\varpi \tag{83}$$

and

$$\frac{1}{r}\frac{d}{dr}(r\xi_r)-\frac{2m\Omega}{\sigma r}\xi_r = \frac{1}{\sigma^2}\left(\frac{m^2}{r^2}+k^2\right)\varpi, \tag{84}$$

† The assumption that is made here is that each branch of the dispersion relation provides a complex analytic function $k^2(p)$ whose behaviour for real p's can be deduced from the variational principle.

where
$$\sigma = p+m\Omega \tag{85}$$
is now a constant. Eliminating ξ_r between equations (83) and (84), we obtain
$$\frac{d^2\varpi}{dr^2}+\frac{1}{r}\frac{d\varpi}{dr}-\frac{m^2}{r^2}\varpi = -k^2\left(\frac{4\Omega^2}{\sigma^2}-1\right)\varpi. \tag{86}$$

It is convenient to measure r in units of the radius R_2 of the outer cylinder and let
$$a = kR_2 \quad \text{and} \quad R_1 = R_2\eta. \tag{87}$$
Equation (86) then becomes
$$\frac{d^2\varpi}{dr^2}+\frac{1}{r}\frac{d\varpi}{dr}+\left\{a^2\left(\frac{4\Omega^2}{\sigma^2}-1\right)-\frac{m^2}{r^2}\right\}\varpi = 0; \tag{88}$$
and the boundary conditions are (cf. equation (68))
$$\frac{d\varpi}{dr}+\frac{2m\Omega}{\sigma r}\varpi = 0 \quad \text{for } r = 1 \text{ and } \eta. \tag{89}$$

The general solution of equation (88) is
$$\varpi = AJ_m(\alpha r)+BY_m(\alpha r), \tag{90}$$
where
$$\alpha = a(4\Omega^2/\sigma^2-1)^{\frac{1}{2}}, \tag{91}$$
A and B are constants of integration, and J_m and Y_m are the Bessel functions of the two kinds of order m. The application of the boundary conditions (89) to the solution (90) leads to a transcendental equation relating α and σ ($= p+m\Omega$). We shall consider two special cases when the relation is fairly simple.

(i) *The case* $\eta = 0$

When there is no inner cylinder and $\eta = 0$, the freedom from singularity at $r = 0$ requires that the term in Y_m in the solution (90) be absent. We then have
$$\varpi = AJ_m(\alpha r); \tag{92}$$
and the boundary condition at $r = 1$ gives
$$\alpha J'_m(\alpha)+\frac{2m\Omega}{\sigma}J_m(\alpha) = 0, \tag{93}$$
where a prime denotes differentiation with respect to the argument. But according to equation (91),
$$\frac{\sigma}{2\Omega} = \frac{\pm 1}{\sqrt{(1+\alpha^2/a^2)}}. \tag{94}$$
Using this relation to eliminate Ω/σ in equation (93), we obtain
$$\alpha J'_m(\alpha)\pm m(1+\alpha^2/a^2)^{\frac{1}{2}}J_m(\alpha) = 0. \tag{95}$$

For any assigned value of α and m, equation (95) determines α/a; equation (94) then determines p; thus

$$\frac{p}{\Omega} = \frac{\sigma}{\Omega} - m = \pm\frac{2}{\sqrt{(1+\alpha^2/a^2)}} - m. \tag{96}$$

In this way the dispersion relation between a and p can be established.

The case $m = 0$ is particularly simple. In this case α must be a zero

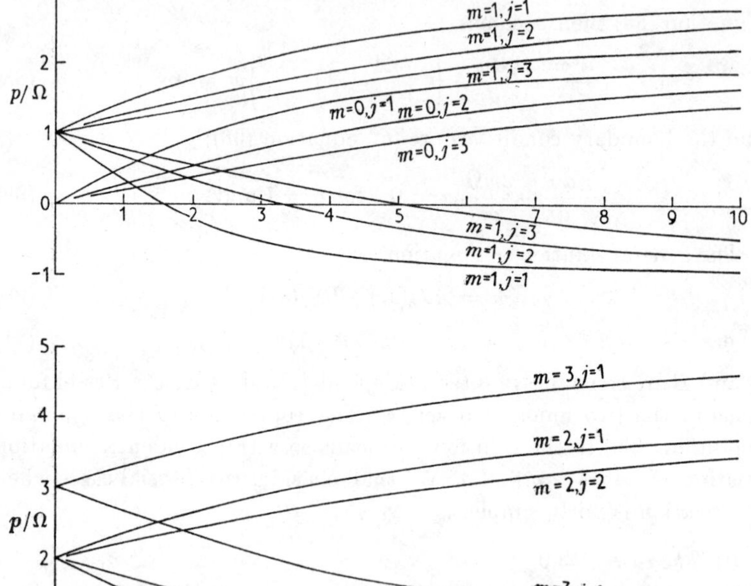

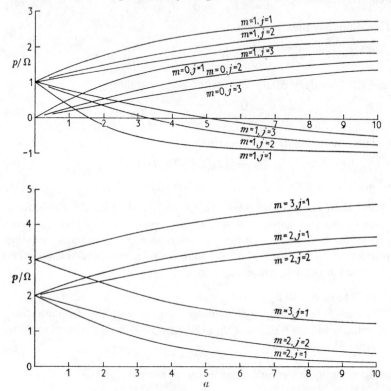

FIG. 63. The characteristic frequencies of oscillation of a rotating column of fluid of radius R_2 for various values of m and j; a measures the wave number in the unit $1/R_2$.

of $J_0'(x)$, i.e. of $J_1(x)$. If $\alpha_{1,j}$ denotes the jth zero of $J_1(x)$, the explicit form of the dispersion relation is

$$p = \pm\frac{2\Omega}{\sqrt{(1+\alpha_{1,j}^2/a^2)}}. \tag{97}$$

The dispersion relations for $m = 0, 1, 2,$ and 3 are illustrated in Fig. 63. If waves of the type we have been considering should occur as standing waves in a rotating column of liquid, then their half-wavelengths should divide the height H of the column an integral number of times; and we must have

$$a = n\pi R_2/H, \tag{98}$$

where n is an integer. By inserting this value of a in equations (96) and (97), we shall obtain the frequencies of the various natural modes of oscillation of the column.

In some recent experiments, Fultz has shown how these natural modes of oscillation of a rotating column of liquid can be excited and maintained. In these experiments the modes $m = 0$ in a rotating cylinder of water (of diameter 4·2 cm and height 15·3 cm) were excited by means of a small disk on the axis of the cylinder. This disk could be moved up and down the axis with a frequency which could be controlled. To excite a particular mode, the disk was so placed that its mean height was at the position of the topmost antinode of the mode (e.g. at the middle of the column to excite the fundamental mode). Fultz was able to detect with great sensitivity whether the disk was set at the resonant frequency by following the motions of the fluid by means of dye suitably introduced. In this way Fultz determined the frequencies of some of the lower modes of oscillation belonging to $m = 0$; and he found that the observed resonant frequencies agreed remarkably well with those given by equations (97) and (98). In Fig. 64 we reproduce some of Fultz's photographs which exhibit in a very striking manner these natural modes of oscillation of a rotating fluid.

(ii) *The case* $m = 0$

When $m = 0$, the equation governing ξ_r is (cf. equation (58))

$$\frac{d^2\xi_r}{dr^2}+\frac{1}{r}\frac{d\xi_r}{dr}+\left\{a^2\left(\frac{4\Omega^2}{p^2}-1\right)-\frac{1}{r^2}\right\}\xi_r = 0. \tag{99}$$

The solution of this equation which vanishes at $r = 1$ and η can be written in the form

$$\xi_r = \text{constant}\{Y_1(\alpha\eta)J_1(\alpha r)-J_1(\alpha\eta)Y_1(\alpha r)\}, \tag{100}$$

where α is a root of the equation

$$Y_1(\alpha\eta)J_1(\alpha)-J_1(\alpha\eta)Y_1(\alpha) = 0. \tag{101}$$

If α_j ($j = 1, 2,...$) denote the roots of this equation, then

$$a^2(4\Omega^2/p^2-1) = \alpha_j^2 \tag{102}$$

or

$$p = \pm\frac{2\Omega}{\sqrt{(1+\alpha_j^2/a^2)}}. \tag{103}$$

The values of α_1 for a few values of η are given in Table XXXI. For $\eta = 0$ and 0·5, the values of α_2 and α_3 are also given.

(*b*) *The case* $\Omega = A+B/r^2$ *and* $m = 0$

A further case of some interest is when the distribution of the angular velocity is that which obtains in a viscous liquid. In that case (cf. equation (20)),

$$\Phi = 4A(A+B/r^2); \tag{104}$$

TABLE XXXI

The roots α_j *of equation* (101) *for some values of* η

η	α_1	α_2	α_3
0·0	3·83171	7·015587	10·17347
0·2	4·23575		
0·3	4·70578		
0·4	5·39118		
0·5	6·39316	12·62470	18·88893
0·6	7·93009		
0·8	15·73755		

and equation (58) governing ξ_r (in case $m = 0$) can be written in the form

$$(DD_*-k^2)\xi_r = -4\frac{k^2}{p^2}A\left(A+\frac{B}{r^2}\right)\xi_r, \tag{105}$$

where $$D = \frac{d}{dr} \quad \text{and} \quad D_* = \frac{d}{dr}+\frac{1}{r}; \tag{106}$$

and the boundary conditions are

$$\xi_r = 0 \quad \text{for } r = R_1 \text{ and } R_2. \tag{107}$$

(i) *The solution for a narrow gap*

Reid has recently shown how equation (105) can be solved explicitly in terms of known, tabulated functions in the case when

$$d = (R_2-R_1) \ll \tfrac{1}{2}(R_2+R_1). \tag{108}$$

When (108) obtains, we need not distinguish between D and D_* and we can further replace $A+B/r^2$ which occurs on the right-hand side of equation (105) by

$$\Omega_1\left[1-(1-\mu)\frac{r-R_1}{R_2-R_1}\right]. \tag{109}$$

In rewriting equation (105) in the framework of the foregoing approximations, we shall find it convenient to measure radial distances from the surface of the inner cylinder in the unit $d = R_2-R_1$. Thus, letting

$$\zeta = \frac{r-R_1}{R_2-R_1} \quad \text{and} \quad a = k(R_2-R_1), \tag{110}$$

FIG. 64. Photographs illustrating the oscillations of a rotating column of fluid in the modes $m = 0$, $j = 1$ (photographs 1 and 2), $m = 0$, $j = 2$ (photographs 3, 4, 5, and 6), and $m = 0$, $j = 3$ (photographs 7 and 8). The data for the different photographs are as follows:

Photographs	*Rotation period* (sec)	Ω/p *observed*	Ω/p *theoretical*	*Intervals between photographs* (sec)
1, 2	2·295	1·316	1·318	1·25
3, 4	3·947	0·788	0·7886	1·50
5, 6	4·040	0·789	0·7886	2·00
7, 8	4·056	0·646	0·6444	1·25

(The height of the column is 8·25 cm in all cases.)

we have to solve

$$(D^2-a^2)\xi_r = -a^2\lambda[1-(1-\mu)\zeta]\xi_r, \tag{111}$$

where D now stands for $d/d\zeta$ and (cf. equations (3))

$$\lambda = \frac{4A\Omega_1}{p^2} = -\frac{4\Omega_1^2}{p^2}\eta^2\frac{1-\mu/\eta^2}{1-\eta^2}; \tag{112}$$

and the boundary conditions are

$$\xi_r = 0 \quad \text{for } \zeta = 0 \text{ and } 1. \tag{113}$$

In the case $\mu = 1$, equation (111) admits of elementary integration and the proper solutions which satisfy the boundary conditions (113) are given by

$$\xi_r = \text{constant} \sin n\pi\zeta, \tag{114}$$

where n is an integer. The corresponding characteristic equation is

$$\frac{1}{\lambda} = \frac{a^2}{a^2+n^2\pi^2}, \tag{115}$$

or, since $\mu = 1$,

$$p = \frac{2\Omega_1}{\sqrt{(1+n^2\pi^2/a^2)}}; \tag{116}$$

this is clearly the asymptotic form of (103) for $\eta \to 1$.

For $\mu < 1$, we can reduce equation (111) to the standard form

$$\frac{d^2\xi_r}{dx^2} = x\xi_r \tag{117}$$

by the substitution

$$x = \left[\frac{a}{\lambda(1-\mu)}\right]^{\frac{2}{3}}\{1-\lambda[1-(1-\mu)\zeta]\}. \tag{118}$$

The boundary conditions, then, are

$$\xi_r = 0 \quad \text{for } x = x_1 \text{ and } x = x_2, \tag{119}$$

where

$$x_1 = \left[\frac{a}{\lambda(1-\mu)}\right]^{\frac{2}{3}}(1-\lambda) \quad \text{and} \quad x_2 = \left[\frac{a}{\lambda(1-\mu)}\right]^{\frac{2}{3}}(1-\lambda\mu). \tag{120}$$

The general solution of equation (117) can be expressed in terms of Bessel's functions of order $\frac{1}{3}$, or more conveniently in terms of Airy's functions† Ai(x) and Bi(x); thus

$$\xi_r = A\,\text{Ai}(x)+B\,\text{Bi}(x), \tag{121}$$

† The definitions of these functions are

$$\text{Ai}(x) = \tfrac{1}{3}x^{\frac{1}{2}}\{I_{-\frac{1}{3}}(y)-I_{\frac{1}{3}}(y)\}$$

and

$$\text{Bi}(x) = (\tfrac{1}{3}x)^{\frac{1}{2}}\{I_{-\frac{1}{3}}(y)+I_{\frac{1}{3}}(y)\},$$

where $y = \frac{2}{3}x^{\frac{3}{2}}$.

where A and B are constants. The boundary conditions (119) lead to the characteristic equation

$$\frac{\mathrm{Ai}(x_1)}{\mathrm{Bi}(x_1)} = \frac{\mathrm{Ai}(x_2)}{\mathrm{Bi}(x_2)}. \tag{122}$$

Equation (122) determines x_2 as a function of x_1 for the different modes; the relationship for the lowest mode is illustrated in Fig. 65. With the

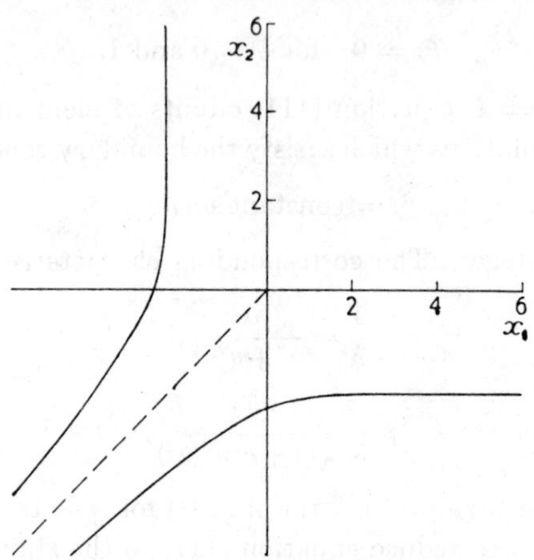

FIG. 65. The (x_1, x_2)-relation for the lowest mode as determined by equation (122).

(x_2, x_1)-relationship determined in this fashion, the required relation between a and $\sqrt{\lambda}$ is given, parametrically, by

$$a = (x_2 - x_1)\sqrt{\frac{x_2 - \mu x_1}{1-\mu}} \quad \text{and} \quad \frac{1}{\sqrt{\lambda}} = \sqrt{\frac{x_2 - \mu x_1}{x_2 - x_1}}. \tag{123}$$

The dispersion relations for $\mu = 1, 0$, and -1 are illustrated in Fig. 66.

In Fig. 67 *a* the limiting form of velocity profile for the lowest mode of instability when $\mu \to -\infty$ is illustrated; and the corresponding cell pattern is shown in Fig. 67 *b*.

(ii) *The formal solution for a wide gap*

When we wish to allow for the finite difference between R_2 and R_1, we must consider equation (105) without making any approximations.

Now measuring r in units of R_2 and letting $a = kR_2$, we have the equation (cf. equations (22) and (23))

$$(DD_* - a^2)\xi_r = a^2 \frac{4\Omega_1^2}{p^2} \eta^4 \frac{(1-\mu)(1-\mu/\eta^2)}{(1-\eta^2)^2}\left(\frac{1}{r^2} - \kappa\right)\xi_r. \tag{124}$$

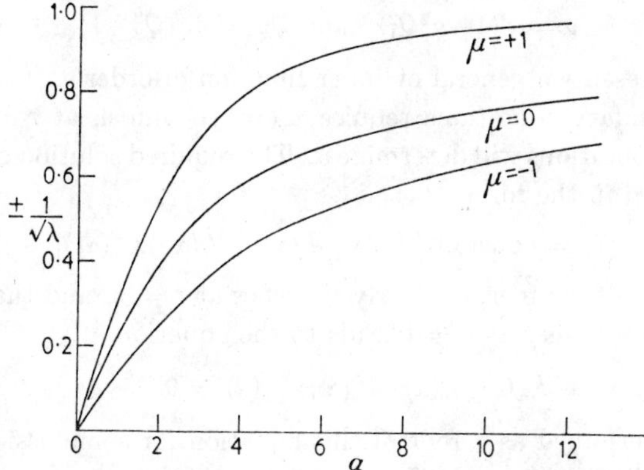

FIG. 66. The dispersion relations in the limit of zero viscosity for $\mu = 1$, 0, and -1 for the lowest modes of the flow between two cylinders with a narrow gap d. The abscissa gives the wave number in the unit $1/d$ and the ordinate gives the growth rate (or the circular frequency) in the unit $\sqrt{|4A\Omega_1|}$.

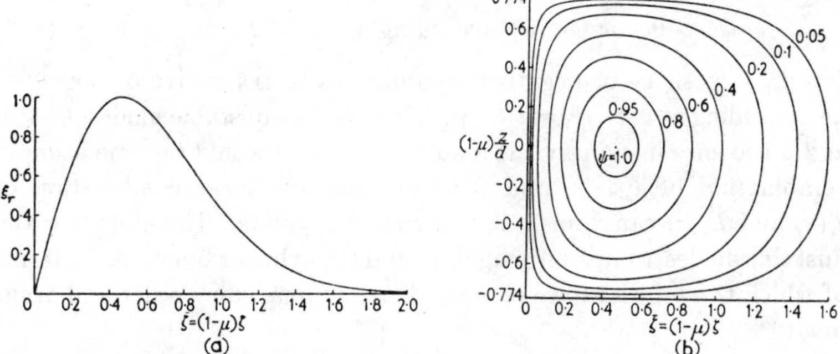

FIG. 67. The limiting form of the velocity profile (a) and the cell pattern (b) for lowest inviscid mode of instability as $\mu \to -\infty$ and $a/(1-\mu) = 2{\cdot}03$.

Letting
$$Q_1^2 = -\frac{4\Omega_1^2}{p^2}\eta^4\frac{(1-\mu)(1-\mu/\eta^2)}{(1-\eta^2)^2} \tag{125}$$
and
$$Q_2^2 = -Q_1^2\kappa = \frac{4\Omega_1^2}{p^2}\eta^4\frac{(1-\mu/\eta^2)^2}{(1-\eta^2)^2}, \tag{126}$$
we can rewrite equation (124) in the form
$$\frac{d^2\xi_r}{dr^2}+\frac{1}{r}\frac{d\xi_r}{dr}+\left\{a^2(Q_2^2-1)-\frac{1-a^2Q_1^2}{r^2}\right\}\xi_r = 0. \tag{127}$$
The general solution of this equation is
$$\xi_r = \mathscr{C}_\nu(\alpha r), \tag{128}$$

where $$\nu = \sqrt{(1-a^2Q_1^2)} \quad \text{and} \quad \alpha = a\sqrt{(Q_2^2-1)}, \tag{129}$$

and $\mathscr{C}_\nu$ represents a general cylinder function of order ν.

The boundary conditions require $\mathscr{C}_\nu(\alpha r)$ to vanish at $r = 1$ and η; and these conditions will determine α. The required solution can, in fact, be expressed in the form

$$\xi_r = \text{constant}\{J_{-\nu}(\alpha\eta)J_\nu(\alpha r) - J_\nu(\alpha\eta)J_{-\nu}(\alpha r)\}. \tag{130}$$

Defined in this manner, ξ_r clearly vanishes at $r = \eta$; and the condition that it also vanishes at $r = 1$ leads to the equation

$$J_{-\nu}(\alpha\eta)J_\nu(\alpha) - J_\nu(\alpha\eta)J_{-\nu}(\alpha) = 0. \tag{131}$$

With α determined as a root of this equation for some assigned ν, the required relation between a^2 and p^2 follows from equations (126) and (129).

From the general theory of §§ 66 and 67, we know that

$$p^2 > 0 \text{ for } \mu > \eta^2 \quad \text{and} \quad p^2 < 0 \text{ for } \mu < \eta^2. \tag{132}$$

Therefore,

$$Q_1^2 > 0 \quad \text{or} \quad < 0 \quad \text{according as} \quad \mu < 1 \quad \text{or} \quad > 1. \tag{133}$$

But Q_2^2 is positive or negative according as p^2 is positive or negative, i.e. according as $\mu > \eta^2$ or $\mu < \eta^2$. Thus, for the unstable modes, $Q_2^2 < 0$ and α becomes imaginary. If ν were real, $\mathscr{C}_\nu(\alpha r)$ would become a linear combination of $I_\nu(|\alpha|r)$ and $K_\nu(|\alpha|r)$. But no linear combination of $I_\nu(x)$ and $K_\nu(x)$ can vanish at two specified points. Therefore, for the unstable modes, ν must be imaginary and the cylinder functions, in terms of which the solutions are expressed, are of imaginary order and argument.

69. On viscous Couette flow

We now turn our attention to stationary viscous flow between two rotating coaxial cylinders.

In cylindrical polar coordinates, the Navier–Stokes equations for viscous incompressible fluids take the forms:

$$\frac{\partial u_r}{\partial t} + (\mathbf{u}\,.\,\text{grad})u_r - \frac{u_\theta^2}{r} = -\frac{\partial}{\partial r}\left(\frac{p}{\rho}\right) + \nu\left(\nabla^2 u_r - \frac{2}{r^2}\frac{\partial u_\theta}{\partial \theta} - \frac{u_r}{r^2}\right), \tag{134}$$

$$\frac{\partial u_\theta}{\partial t} + (\mathbf{u}\,.\,\text{grad})u_\theta + \frac{u_r u_\theta}{r} = -\frac{1}{r}\frac{\partial}{\partial \theta}\left(\frac{p}{\rho}\right) + \nu\left(\nabla^2 u_\theta + \frac{2}{r^2}\frac{\partial u_r}{\partial \theta} - \frac{u_\theta}{r^2}\right), \tag{135}$$

$$\frac{\partial u_z}{\partial t} + (\mathbf{u}\,.\,\text{grad})u_z = -\frac{\partial}{\partial z}\left(\frac{p}{\rho}\right) + \nu\nabla^2 u_z, \tag{136}$$

where
$$\mathbf{u}\,.\,\mathrm{grad} = u_r\frac{\partial}{\partial r}+\frac{u_\theta}{r}\frac{\partial}{\partial \theta}+u_z\frac{\partial}{\partial z} \tag{137}$$

and
$$\nabla^2 = \frac{\partial^2}{\partial r^2}+\frac{1}{r}\frac{\partial}{\partial r}+\frac{1}{r^2}\frac{\partial^2}{\partial \theta^2}+\frac{\partial^2}{\partial z^2}. \tag{138}$$

We have also the equation of continuity,
$$\frac{\partial u_r}{\partial r}+\frac{u_r}{r}+\frac{1}{r}\frac{\partial u_\theta}{\partial \theta}+\frac{\partial u_z}{\partial z} = 0. \tag{139}$$

These equations allow a stationary solution of the form
$$u_r = u_z = 0 \quad \text{and} \quad u_\theta = V(r) \tag{140}$$

provided
$$\frac{d}{dr}\left(\frac{p}{\rho}\right) = \frac{V^2}{r} \tag{141}$$

and
$$\nu\left(\nabla^2 V-\frac{V}{r^2}\right) = \nu\frac{d}{dr}\left(\frac{d}{dr}+\frac{1}{r}\right)V = 0. \tag{142}$$

From equation (142), it is apparent that the most general form of $V(r)$ which is compatible with $\nu \neq 0$ is
$$V = Ar+\frac{B}{r}, \tag{143}$$

where A and B are two arbitrary constants. The corresponding expression for the angular velocity is
$$\Omega = A+\frac{B}{r^2}. \tag{144}$$

This is the solution which was quoted in § 65. And as we remarked at that time, the appearance of two arbitrary constants in the solution (144) corresponds to the possibility of assigning to the two cylinders, confining the fluid, angular velocities at will. The constants A and B can, therefore, be related to the angular velocities Ω_1 and Ω_2 with which the two cylinders are rotated. We have (cf. equations (3) and (4))
$$A = -\Omega_1\eta^2\frac{1-\mu/\eta^2}{1-\eta^2} \quad \text{and} \quad B = \Omega_1\frac{R_1^2(1-\mu)}{1-\eta^2}, \tag{145}$$

where
$$\mu = \Omega_2/\Omega_1 \quad \text{and} \quad \eta = R_1/R_2. \tag{146}$$

As we have seen in § 66, Rayleigh's criterion for the stability of inviscid Couette flow applied to the distribution (144) yields
$$\mu > \eta^2 \quad \text{for stability}. \tag{147}$$

On general grounds, we should expect that the effect of viscosity is to postpone the onset of instability; and that the new criterion for stability must essentially take the form of specifying the extent to which the condition (147) can be violated before instability will manifest itself.

The precise form which the criterion takes was first investigated, both theoretically and experimentally, by G. I. Taylor. In the case when the gap $R_2 - R_1$ between the two cylinders can be considered as small compared to the mean radius $\frac{1}{2}(R_2 + R_1)$, Taylor found an explicit analytical expression for the criterion; and he was able to confirm by experiments that the marginal state is stationary and exhibits a break-up of the basic flow into a cellular pattern. Fig. 68, which is a reproduction of one of Taylor's photographs, clearly shows the manner of the onset of instability. In § 74 we return to a more detailed account of the experimental work which has been carried out since Taylor's early pioneering work.

70. The perturbation equations

We shall now investigate the stability of the flow described by equations (141) and (143). Let the perturbed state be characterized by

$$u_r, \quad V+u_\theta, \quad u_z, \quad \text{and} \quad \delta p/\rho = \varpi. \tag{148}$$

Assuming that the various perturbations are axisymmetric† and independent of θ, we obtain from (134)–(136) the linearized equations

$$\frac{\partial u_r}{\partial t} - 2\frac{V}{r}u_\theta = -\frac{\partial \varpi}{\partial r} + \nu\left(\nabla^2 u_r - \frac{u_r}{r^2}\right), \tag{149}$$

$$\frac{\partial u_\theta}{\partial t} + \left(\frac{dV}{dr} + \frac{V}{r}\right)u_r = \nu\left(\nabla^2 u_\theta - \frac{u_\theta}{r^2}\right), \tag{150}$$

and

$$\frac{\partial u_z}{\partial t} = -\frac{\partial \varpi}{\partial z} + \nu \nabla^2 u_z, \tag{151}$$

where ∇^2 has now the meaning

$$\nabla^2 = \frac{\partial^2}{\partial r^2} + \frac{1}{r}\frac{\partial}{\partial r} + \frac{\partial^2}{\partial z^2}. \tag{152}$$

Also, for axisymmetric motions, the equation of continuity reduces to

$$\frac{\partial u_r}{\partial r} + \frac{u_r}{r} + \frac{\partial u_z}{\partial z} = 0. \tag{153}$$

By analysing the disturbance into normal modes, we seek solutions of the foregoing equations which are of the forms

$$\left.\begin{aligned} u_r &= e^{pt}u(r)\cos kz; \qquad & u_z &= e^{pt}w(r)\sin kz \\ u_\theta &= e^{pt}v(r)\cos kz; \qquad & \varpi &= e^{pt}\varpi(r)\cos kz \end{aligned}\right\}, \tag{154}$$

where k is the wave number of the disturbance in the axial direction

† The general non-axisymmetric case has not been investigated for this problem.

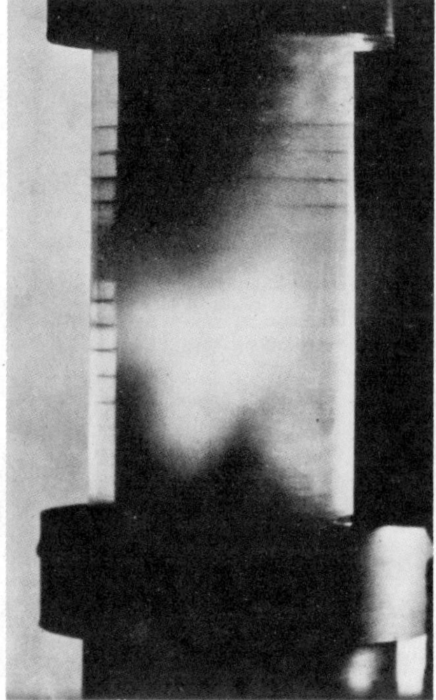

Fig. 68. One of G. I. Taylor's photographs illustrating the onset of instability in the flow between rotating cylinders. The radii of the two cylinders in this case are 4·035 cm and 3·25 cm respectively; the cylinders are rotating in the same direction.

and p is a constant which can be complex. For solutions of the form (154), equations (149)–(153) become

$$\nu\left(DD_*-k^2-\frac{p}{\nu}\right)u+2\frac{V}{r}v=\frac{d\varpi}{dr}, \tag{155}$$

$$\nu\left(DD_*-k^2-\frac{p}{\nu}\right)v-(D_*V)u=0, \tag{156}$$

$$\nu\left(D_*D-k^2-\frac{p}{\nu}\right)w=-k\varpi, \tag{157}$$

$$\nabla^2=\left(\frac{d}{dr}+\frac{1}{r}\right)\frac{d}{dr}-k^2=D_*D-k^2$$

$$=DD_*+\frac{1}{r^2}-k^2, \tag{158}$$

and
$$D_*u=-kw. \tag{159}$$

[In the foregoing equations, D and D_* have the same meanings as in § 68 (b), equation (106).]

Eliminating w between equations (157) and (159), we have

$$\frac{\nu}{k^2}\left(D_*D-k^2-\frac{p}{\nu}\right)D_*u=\varpi. \tag{160}$$

Inserting this expression for ϖ in equation (155), we find after some rearranging,

$$\frac{\nu}{k^2}\left(DD_*-k^2-\frac{p}{\nu}\right)(DD_*-k^2)u=2\frac{V}{r}v. \tag{161}$$

This equation must be considered together with

$$\nu\left(DD_*-k^2-\frac{p}{\nu}\right)v=(D_*V)u. \tag{162}$$

These equations are general and do not depend on any particular form of $V(r)$.

Measuring r in units of the radius R_2 of the outer cylinder and writing

$$k^2=a^2/R_2^2 \quad\text{and}\quad \sigma=pR_2^2/\nu, \tag{163}$$

equations (161) and (162) become (when $V(r)$ has the particular form (143))

$$(DD_*-a^2-\sigma)(DD_*-a^2)u=a^2\frac{2B}{\nu}\left(\frac{1}{r^2}+\frac{AR_2^2}{B}\right)v \tag{164}$$

and
$$(DD_*-a^2-\sigma)v=\frac{2A}{\nu}R_2^2u. \tag{165}$$

It is convenient to make the transformation

$$\frac{2AR_2^2}{\nu}u\rightarrow u; \tag{166}$$

the equations then take the more convenient forms

$$(DD_*-a^2-\sigma)(DD_*-a^2)u = -Ta^2\left(\frac{1}{r^2}-\kappa\right)v \tag{167}$$

and

$$(DD_*-a^2-\sigma)v = u, \tag{168}$$

where (cf. equations (21), (23), and (32))

$$T = -\frac{4AB}{\nu^2}R_2^2 = \frac{4\Omega_1^2 R_1^4}{\nu^2}\frac{(1-\mu)(1-\mu/\eta^2)}{(1-\eta^2)^2} \tag{169}$$

and

$$\kappa = -\frac{AR_2^2}{B} = \frac{1-\mu/\eta^2}{1-\mu}. \tag{170}$$

Solutions of equations (167) and (168) must be sought which satisfy the boundary conditions appropriate for no slip on the cylindrical walls at $r = 1$ and η. These conditions are that all three components of the velocity vanish on the walls; thus,

$$u = v = 0 \quad \text{and} \quad Du = 0 \text{ for } r = 1 \text{ and } \eta, \tag{171}$$

where the last of the three boundary conditions is equivalent to $w = 0$ (cf. equation (159)).

(a) The stability of the flow for $\mu > \eta^2$

We shall now show that when Rayleigh's criterion $\mu > \eta^2$ is satisfied, the flow is indeed stable.

First, we may notice certain elementary integral properties of the operator DD_*. If $f(r)$ and $g(r)$ are any two functions and if one of them, say $f(r)$, vanishes at the limits of integration,

$$\int rfDD_*g\,dr = -\int\left(r\frac{df}{dr}\frac{dg}{dr}+\frac{fg}{r}\right)dr; \tag{172}$$

and if the derivative of f also vanishes at the limits,

$$\int rfDD_*g\,dr = \int rg\,DD_*f\,dr. \tag{173}$$

These relations follow by successive integrations by parts. Thus, by writing

$$\int rf\,DD_*g\,dr = \int\left\{f\frac{d}{dr}\left(r\frac{dg}{dr}\right)-\frac{fg}{r}\right\}dr, \tag{174}$$

the truth of (172) becomes self-evident; and a further integration by parts leads to (173).

Now returning to equations (167) and (168), multiply (167) by ru^* and integrate over the range of r. We have

$$\int_\eta^1 ru^*\{(DD_*-a^2)^2u-\sigma(DD_*-a^2)u\}\,dr = -\mathscr{T}a^2\int_\eta^1 r\phi(r)vu^*\,dr, \tag{175}$$

where, for brevity, we have written

$$\mathscr{T} = \frac{T}{1-\mu} = \frac{4\Omega_1^2 R_1^4}{\nu^2}\frac{(1-\mu/\eta^2)}{(1-\eta^2)^2} \quad \text{and} \quad \phi = (1-\mu)\left(\frac{1}{r^2}-\kappa\right). \tag{176}$$

Since u and its derivative vanish at $r = 1$ and η, the integrals on the left-hand side of (175) can be transformed to positive definite forms by making use of (172) and (173). Thus,

$$\int_\eta^1 ru^*\{(DD_*-a^2)^2u-\sigma(DD_*-a^2)u\}\,dr$$
$$= \int_\eta^1 r|(DD_*-a^2)u|^2\,dr+\sigma\int_\eta^1 \left\{r\left|\frac{du}{dr}\right|^2+\left(\frac{1}{r}+a^2r\right)|u|^2\right\}dr. \tag{177}$$

Next, substituting for u^* (from equation (168)) in the integrand on the right-hand side of equation (175), we obtain

$$\int_\eta^1 r\phi(r)vu^*\,dr = \int_\eta^1 r\phi(r)v(DD_*-a^2-\sigma^*)v^*\,dr$$
$$= -(a^2+\sigma^*)\int_\eta^1 \phi(r)r|v|^2\,dr+\int_\eta^1 r\phi(r)vDD_*v^*\,dr. \tag{178}$$

Again, by making use of (172), we have

$$\int_\eta^1 r\phi(r)vDD_*v^*\,dr = -\int_\eta^1 \phi(r)\left(r\left|\frac{dv}{dr}\right|^2+\frac{|v|^2}{r}\right)dr+2(1-\mu)\int_\eta^1 \frac{v}{r^2}\frac{dv^*}{dr}\,dr. \tag{179}$$

Now combining equations (175), (177), (178), and (179), we obtain

$$\sigma I_1+I_2 = \mathscr{T}a^2\{(a^2+\sigma^*)I_3+I_4\}, \tag{180}$$

where

$$I_1 = \int_\eta^1 \left\{r\left|\frac{du}{dr}\right|^2+\left(\frac{1}{r}+a^2r\right)|u|^2\right\}dr, \tag{181}$$

$$I_2 = \int_\eta^1 |(DD_*-a^2)u|^2r\,dr, \tag{182}$$

$$I_3 = \int_\eta^1 \phi(r)r|v|^2\,dr, \tag{183}$$

and

$$I_4 = \int_\eta^1 \phi(r)\left(r\left|\frac{dv}{dr}\right|^2+\frac{|v|^2}{r}\right)dr-2(1-\mu)\int_\eta^1 \frac{v}{r^2}\frac{dv^*}{dr}\,dr. \tag{184}$$

The integrals I_1 and I_2 are clearly positive definite. For $\mu > 0$, $\phi(r) > 0$ (see equation (26)); so that in this case, I_3 is also positive definite. The first of the two integrals included in I_4 is positive definite for $\mu > 0$; but the second is complex. However, the real part of I_4 is positive definite for $\mu > 0$; in fact,

$$\text{re}(I_4) = \int_\eta^1 r\phi(r)\left|\frac{dv}{dr}-\frac{v}{r}\right|^2 dr. \tag{185}$$

For, expanding the integrand in (185), we have

$$\int_\eta^1 r\phi(r)\left|\frac{dv}{dr}-\frac{v}{r}\right|^2 dr = \int_\eta^1 \phi(r)\left(r\left|\frac{dv}{dr}\right|^2+\frac{|v|^2}{r}\right) dr - \int_\eta^1 \phi(r)\frac{d|v|^2}{dr}\, dr; \tag{186}$$

but

$$\int_\eta^1 \phi(r)\frac{d|v|^2}{dr}\, dr = (1-\mu)\int_\eta^1 \left(\frac{1}{r^2}-\kappa\right)\frac{d|v|^2}{dr}\, dr = (1-\mu)\int_\eta^1 \frac{1}{r^2}\frac{d|v|^2}{dr}\, dr. \tag{187}$$

Therefore, the right-hand side of (186) is, indeed, the real part of I_4.

Returning to equation (180) and equating the real parts of this equation, we obtain

$$\text{re}(\sigma)(I_1-\mathcal{T}a^2I_3)+I_2-\mathcal{T}a^2[a^2I_3+\text{re}(I_4)] = 0. \tag{188}$$

When $\mu > \eta^2$, $\mathcal{T} < 0$ and the coefficient of re(σ) in equation (188) is positive definite; and so also are the remaining terms in the equation. Therefore,

$$\text{re}(\sigma) < 0 \quad \text{for } \mu > \eta^2, \tag{189}$$

and the flow is stable; this result is entirely to be expected on physical grounds. Nevertheless, it appears to be the only one which can be established by general analytical arguments. In particular, it does not seem that one can deduce the general validity of the principle of the exchange of stabilities for this problem. For example, by equating the imaginary parts of equation (180), we obtain (cf. equation (176))

$$\text{im}(\sigma)(I_1+\mathcal{T}a^2I_3) = -2Ta^2\,\text{im}\int_\eta^1 \frac{v}{r^2}\frac{dv^*}{dr}\, dr, \tag{190}$$

and no general conclusions can be drawn from this equation; when $\mu < 0$, even I_3 is not positive definite!

71. The solution for the case of a narrow gap when the marginal state is stationary

If the gap R_2-R_1 between the two cylinders is small compared to their mean radius $\frac{1}{2}(R_2+R_1)$, we need not (as in § 68 (b)) distinguish between D and D_* in equations (161) and (162); and we can also replace $(A+B/r^2)$ which occurs on the right-hand side of equation (161) by

$$\Omega_1\left[1-(1-\mu)\frac{r-R_1}{R_2-R_1}\right]. \tag{191}$$

In rewriting equations (161) and (162) in the framework of these approximations, it will be convenient to measure radial distances from the surface of the inner cylinder in the unit $d = R_2-R_1$. Thus, letting

$$\zeta = (r-R_1)/d, \quad k = a/d, \quad \text{and} \quad \sigma = pd^2/\nu, \tag{192}$$

we have to consider the equations

$$(D^2-a^2-\sigma)(D^2-a^2)u = \frac{2\Omega_1 d^2}{\nu}a^2[1-(1-\mu)\zeta]v \tag{193}$$

and

$$(D^2-a^2-\sigma)v = \frac{2Ad^2}{\nu}u. \tag{194}$$

By the further transformation

$$u \to \frac{2\Omega_1 d^2a^2}{\nu}u, \tag{195}$$

the equations become

$$(D^2-a^2-\sigma)(D^2-a^2)u = (1+\alpha\zeta)v \tag{196}$$

and

$$(D^2-a^2-\sigma)v = -Ta^2u, \tag{197}$$

where, now,

$$T = -\frac{4A\Omega_1}{\nu^2}d^4 \tag{198}$$

and

$$\alpha = -(1-\mu). \tag{199}$$

Equations (196) and (197) must be considered together with the boundary conditions

$$u = Du = v = 0 \quad \text{for } \zeta = 0 \text{ and } 1. \tag{200}$$

We are primarily interested in the solutions of equations (196) and (197) (subject to the boundary conditions (200)) for various values of a for which the real part of σ is zero. The method described in Chapter III, § 31 in a different connexion is applicable to this problem. Thus, we must obtain solutions for two cases: when σ is zero and the marginal state is stationary and when σ is imaginary and the marginal state is oscillatory. In the latter case for each value of a, $i\sigma$ must be determined by the condition that T is real.† In either case, we must find the minimum of T as a function of a; and depending on which of the two minima is lower, we shall have the onset of instability as a stationary secondary flow or as overstability. Careful experiments on the onset of instability

† It is, of course, possible that under certain circumstances solutions with this property do not exist.

by Taylor and others have failed to reveal any suggestions of overstability. For this reason, the case $\sigma = 0$ is the only one which has been considered in the literature. However, as no general arguments for the validity of the principle of the exchange of stabilities have been found for this problem, the case of overstability requires investigation. We return to this question in § 72.

When the marginal state is stationary, the equations to be solved are

$$(D^2-a^2)^2u = (1+\alpha\zeta)v \tag{201}$$

and

$$(D^2-a^2)v = -Ta^2u, \tag{202}$$

together with the boundary conditions (200).

(a) The solution of the characteristic value problem for the case $\sigma = 0$

It can be readily verified that the characteristic value problem presented by equations (200)–(202) is not self-adjoint in the usual sense. For this reason, the method to be described below was patterned after the ones which have been found successful in cases where the problems are self-adjoint. However, Roberts has recently found a variational basis for the method; this is considered in Appendix IV.

The method of solution we shall adopt is the following.

Since v is required to vanish at $\zeta = 0$ and 1, we expand it in a sine series of the form

$$v = \sum_{m=1}^{\infty} C_m \sin m\pi\zeta. \tag{203}$$

Having chosen v in this manner, we next *solve* the equation,

$$(D^2-a^2)^2u = (1+\alpha\zeta)\sum_{m=1}^{\infty} C_m \sin m\pi\zeta, \tag{204}$$

obtained by inserting (203) in (201), and arrange that the solution satisfies the four remaining boundary conditions on u. With u determined in this fashion and v given by (203), equation (202) will lead, as we shall presently see, to a secular equation for T.

The solution of equation (204) is straightforward. The general solution can be written in the form

$$u = \sum_{m=1}^{\infty} \frac{C_m}{(m^2\pi^2+a^2)^2}\Big\{A_1^{(m)}\cosh a\zeta + B_1^{(m)}\sinh a\zeta + A_2^{(m)}\zeta\cosh a\zeta + \\ + B_2^{(m)}\zeta\sinh a\zeta + (1+\alpha\zeta)\sin m\pi\zeta + \frac{4\alpha m\pi}{m^2\pi^2+a^2}\cos m\pi\zeta\Big\}, \tag{205}$$

where the constants of integration $A_1^{(m)}$, $A_2^{(m)}$, $B_1^{(m)}$, and $B_2^{(m)}$ are to be determined by the boundary conditions $u = Du = 0$ at $\zeta = 0$ and 1. These latter conditions lead to the equations:

$$A_1^{(m)} = -\frac{4m\pi\alpha}{m^2\pi^2+a^2}, \qquad aB_1^{(m)}+A_2^{(m)} = -m\pi,$$

$$A_1^{(m)}\cosh a+B_1^{(m)}\sinh a+A_2^{(m)}\cosh a+B_2^{(m)}\sinh a = (-1)^{m+1}\frac{4m\pi\alpha}{m^2\pi^2+a^2},$$

$$A_1^{(m)}a\sinh a+B_1^{(m)}a\cosh a+A_2^{(m)}(\cosh a+a\sinh a)+ +B_2^{(m)}(\sinh a+a\cosh a) = (-1)^{m+1}(1+\alpha)m\pi. \quad (206)$$

On solving these equations, we find that

$$A_1^{(m)} = -\frac{4\alpha m\pi}{m^2\pi^2+a^2},$$

$$B_1^{(m)} = \frac{m\pi}{\Delta}\{a+\beta_m(\sinh a+a\cosh a)-\gamma_m\sinh a\},$$

$$A_2^{(m)} = -\frac{m\pi}{\Delta}\{\sinh^2 a+\beta_m a(\sinh a+a\cosh a)-\gamma_m a\sinh a\},$$

$$B_2^{(m)} = \frac{m\pi}{\Delta}\{(\sinh a\cosh a-a)+\beta_m a^2\sinh a-\gamma_m(a\cosh a-\sinh a)\}, \quad (207)$$

where
$$\Delta = \sinh^2 a-a^2,$$

$$\beta_m = \frac{4\alpha}{m^2\pi^2+a^2}[(-1)^{m+1}+\cosh a],$$

and
$$\gamma_m = (-1)^{m+1}(1+\alpha)+\frac{4\alpha}{m^2\pi^2+a^2}a\sinh a. \quad (208)$$

Now substituting for v and u from equations (203) and (205) in equation (202), we obtain

$$\sum_{n=1}^{\infty} C_n(n^2\pi^2+a^2)\sin n\pi\zeta = Ta^2\sum_{m=1}^{\infty}\frac{C_m}{(m^2\pi^2+a^2)^2}\Big\{A_1^{(m)}\cosh a\zeta+B_1^{(m)}\sinh a\zeta+A_2^{(m)}\zeta\cosh a\zeta+ +B_2^{(m)}\zeta\sinh a\zeta+(1+\alpha\zeta)\sin m\pi\zeta+\frac{4\alpha m\pi}{m^2\pi^2+a^2}\cos m\pi\zeta\Big\}. \quad (209)$$

Multiplying equation (209) by $\sin n\pi\zeta$ and integrating over the range of ζ, we obtain a system of linear homogeneous equations for the constants

$\mathscr{C}_m = C_m/(m^2\pi^2+a^2)^2$; and the requirement that these constants are not all zero leads to the secular equation

$$\left\| \frac{n\pi}{(n^2\pi^2+a^2)}\left\{[1+(-1)^{n+1}\cosh a]A_1^{(m)}+[(-1)^{n+1}\sinh a]B_1^{(m)}+\right.\right.$$
$$+(-1)^{n+1}\left[\cosh a-\frac{2a}{n^2\pi^2+a^2}\sinh a\right]A_2^{(m)}+$$
$$\left.+\left[(-1)^{n+1}\sinh a-\frac{2a}{n^2\pi^2+a^2}\{1+(-1)^{n+1}\cosh a\}\right]B_2^{(m)}\right\}+$$
$$\left.+\alpha X_{nm}+\tfrac{1}{2}\delta_{nm}-\tfrac{1}{2}(n^2\pi^2+a^2)^3\frac{\delta_{nm}}{a^2T}\right\| = 0, \quad (210)$$

$$\text{where } X_{nm} = \begin{cases} 0 \text{ if } m+n \text{ is even and } m \neq n, \\ \tfrac{1}{4} \text{ if } m = n, \\ \dfrac{4nm}{n^2-m^2}\left\{\dfrac{2}{m^2\pi^2+a^2}-\dfrac{1}{\pi^2(n^2-m^2)}\right\} \text{ if } m+n \text{ is odd.} \end{cases} \quad (211)$$

On using the first two equations of (206), equation (210) simplifies to the form

$$\left\| \frac{n\pi}{n^2\pi^2+a^2}\left\{\frac{4m\pi\alpha}{m^2\pi^2+a^2}[(-1)^{m+n}-1]-\right.\right.$$
$$\left.-\frac{2a}{n^2\pi^2+a^2}[(-1)^{n+1}\{A_2^{(m)}\sinh a+B_2^{(m)}\cosh a\}+B_2^{(m)}]\right\}+$$
$$\left.+\alpha X_{nm}+\tfrac{1}{2}\delta_{nm}-\tfrac{1}{2}(n^2\pi^2+a^2)^3\frac{\delta_{nm}}{a^2T}\right\| = 0; \quad (212)$$

and on substituting for the constants $A_2^{(m)}$ and $B_2^{(m)}$ their explicit solutions given in (207), we find that equation (212) simplifies greatly and we are left with

$$\left\| \frac{4mn\pi^2\alpha}{(n^2\pi^2+a^2)(m^2\pi^2+a^2)}[(-1)^{m+n}-1]-\right.$$
$$-\frac{2amn\pi^2}{(n^2\pi^2+a^2)^2(\sinh^2 a-a^2)}\left\{(\sinh a\cosh a-a)[1+(1+\alpha)(-1)^{m+n}]+\right.$$
$$+(\sinh a-a\cosh a)[(-1)^{n+1}+(1+\alpha)(-1)^{m+1}]-$$
$$\left.-\frac{4a\alpha\sinh a}{(m^2\pi^2+a^2)}[\sinh a+a(-1)^{m+1}][(-1)^{m+n}-1]\right\}+$$
$$\left.+\tfrac{1}{2}\delta_{nm}+\alpha X_{nm}-\tfrac{1}{2}(n^2\pi^2+a^2)^3\frac{\delta_{mn}}{a^2T}\right\| = 0. \quad (213)$$

A first approximation to the solution of equation (213) is obtained by setting the (1, 1)-element of the matrix equal to zero. We find

$$\tfrac{1}{2}(\pi^2+a^2)^3\frac{1}{Ta^2} = \tfrac{1}{4}\alpha+\tfrac{1}{2}-$$
$$-\frac{2a\pi^2(2+\alpha)}{(\pi^2+a^2)^2(\sinh^2 a-a^2)}[(\sinh a\cosh a-a)+(\sinh a-a\cosh a)]. \tag{214}$$

On further simplification, this gives

$$T = \frac{2}{2+\alpha}\,\frac{(\pi^2+a^2)^3}{a^2\{1-16a\pi^2\cosh^2\tfrac{1}{2}a/[(\pi^2+a^2)^2(\sinh a+a)]\}}. \tag{215}$$

We observe that, apart from the factor $2/(2+\alpha)$, this expression for T is identical with what was found in Chapter II (§ 17, equation (311)) for the Rayleigh number for the simple Bénard problem by the variational method in the first approximation for the case of two rigid boundaries. Consequently, in this approximation (cf. equation (199)),

$$T_c = \frac{2}{2+\alpha}\times 1715 = \frac{3430}{1+\mu} \quad\text{and}\quad a_{\min} = 3{\cdot}12. \tag{216}$$

We shall see below (§ (*b*)) that for $0 < \mu < 1$ equation (216) gives values for the critical Taylor number which do not differ from those obtained in the higher approximations by more than one per cent. The reason for this relatively high accuracy of the solution in the first approximation will be made apparent in § (*d*).

(*b*) *Numerical results*

A method of solving the infinite order characteristic equation which (213) provides for T would be to set the determinant formed by the first n rows and columns of the secular matrix equal to zero and let n take increasingly larger values. In practice, the usefulness of this method will depend largely on how rapidly the lowest positive root of the resulting equation of order n tends to its limit as $n\to\infty$. It appears that for the problem on hand, the process converges quite rapidly.

In Table XXXII the values of T obtained with the aid of equation (213) in the different approximations are listed for those values of a (for the assigned μ's) at which it was found (by trial and error) that T attained its minimum value. From an examination of this table, it would appear that for $\mu > -1{\cdot}0$, the third approximation provides T to well within one per cent of the true value. For $-3{\cdot}0 \leqslant \mu \leqslant -1{\cdot}0$, the calculations were carried out to as high approximations as seemed

Table XXXII

The critical Taylor numbers and the associated wave numbers for different values of μ

μ	a	*First approximation*	*Second approximation*	*Third approximation*	*Fourth approximation*	*Fifth approximation*	*Sixth approximation*
1·00	3·12	1715·1	1715·1	1708·0			
0·50	3·12	2286·7	2284·8	2275·3			
0·25	3·12	2744·1	2736·4	2725·3			
0·00	3·12	3429·9	3403·6	3390·3			
−0·25	3·13	4573·4	4477·9	4462·5			
−0·50	3·20	6866·2	6430·6	6417·1	6414·8		
−0·60	3·25	8595·8	7694·5	7687·7			
−0·70	3·34	11514	9422·7	9432·6			
−0·80	3·50		11774	11820			
−0·90	3·70		14846	14943			
−0·95	3·86		16647	16764			
−1·00	4·00		18615	18735	18677		
−1·25	4·60			30497	30458		
−1·50	5·05			46000	46192		
−1·75	5·60			67189	67592		
−2·00	6·10			97090	95610		95585
−2·50	7·10				176810	177107	177108
−3·00	8·09				320748	301116	302496
	8·14						302482

Table XXXIII

The critical Taylor numbers and related constants for various values of μ

μ	a	T_c	$\mathscr{C}_2/\mathscr{C}_1$	$\mathscr{C}_3/\mathscr{C}_1$	$\mathscr{C}_4/\mathscr{C}_1$	$\mathscr{C}_5/\mathscr{C}_1$	$\mathscr{C}_6/\mathscr{C}_1$
1·00	3·12	$1\cdot708\times10^{3}$	0	−0·001147			
0·50	3·12	$2\cdot275\times10^{3}$	0·001324	−0·001146			
0·25	3·12	$2\cdot725\times10^{3}$	0·002379	−0·001145			
0·00	3·12	$3\cdot390\times10^{3}$	0·003944	−0·001143			
−0·25	3·13	$4\cdot462\times10^{3}$	0·006537	−0·001144			
−0·50	3·20	$6\cdot417\times10^{3}$	0·01186	−0·001177			
−0·60	3·24	$7\cdot688\times10^{3}$	0·01569	−0·001194			
−0·70	3·34	$9\cdot433\times10^{3}$	0·02174	−0·001226			
−0·80	3·49	$1\cdot182\times10^{4}$	0·03195	−0·001262			
−0·90	3·70	$1\cdot494\times10^{4}$	0·04742	−0·001186			
−0·95	3·86	$1\cdot676\times10^{4}$	0·05904	−0·001083			
−1·00	4·00	$1\cdot868\times10^{4}$	0·07139	−0·000929	−0·00039		
−1·25	4·61	$3\cdot046\times10^{4}$	0·1472	+0·002421	−0·00087		
−1·50	5·06	$4\cdot619\times10^{4}$	0·2281	+0·0116	−0·0012		
−1·75	5·60	$6\cdot759\times10^{4}$	0·3205	+0·0312	−0·0008		
−2·00	6·10	$9\cdot558\times10^{4}$	0·4099	+0·0616	+0·00128	−0·000927	−0·000267
−2·50	7·10	$1\cdot771\times10^{5}$	0·5804	+0·1560	+0·01777	−0·000459	−0·000819
−3·00	8·14	$3\cdot025\times10^{5}$	0·7499	+0·2846	+0·0626	+0·006064	+0·001045

necessary. It would appear that for $\mu < -3\cdot 0$, one should go to approximations higher than the sixth to achieve comparable accuracy. Fortunately, the considerations to be set out in § (e) below make it unnecessary for us to obtain the solutions for $\mu < -3\cdot 0$ by the present method.

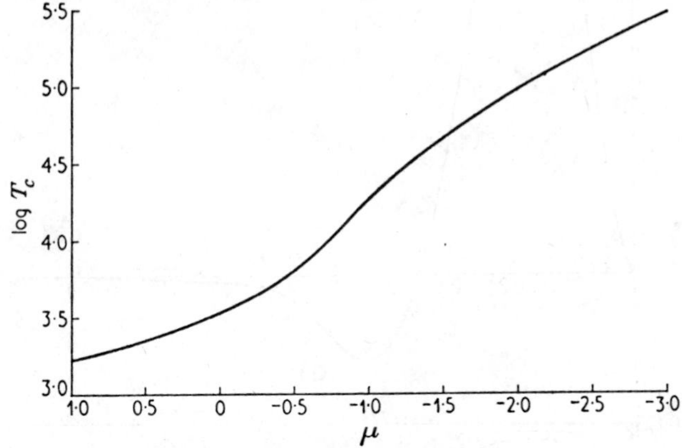

FIG. 69. The critical Taylor number T_c for the onset of instability as a function of $\mu = \Omega_2/\Omega_1$.

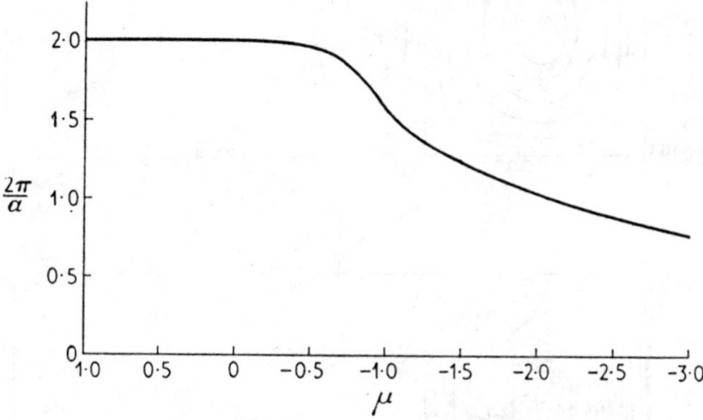

FIG. 70. The wavelength of the disturbance (in units of the gap width d) manifested at onset of instability as a function of $\mu = \Omega_2/\Omega_1$.

In Table XXXIII the coefficients $\mathscr{C}_m$ $(= C_m/(m^2\pi^2+a^2)^2)$ for the marginal state in the expansion (205) for u are given. This table also includes the critical Taylor numbers and the associated wave numbers. The (T_c, μ) and the $[a(T_c), \mu]$-relationships are further illustrated in Figs. 69 and 70.

Fig. 71 a shows the profile of the velocity u for $\mu = -3\cdot 0$; and in Fig. 71 b the corresponding cell pattern is exhibited. For $\mu = -3\cdot 0$, the

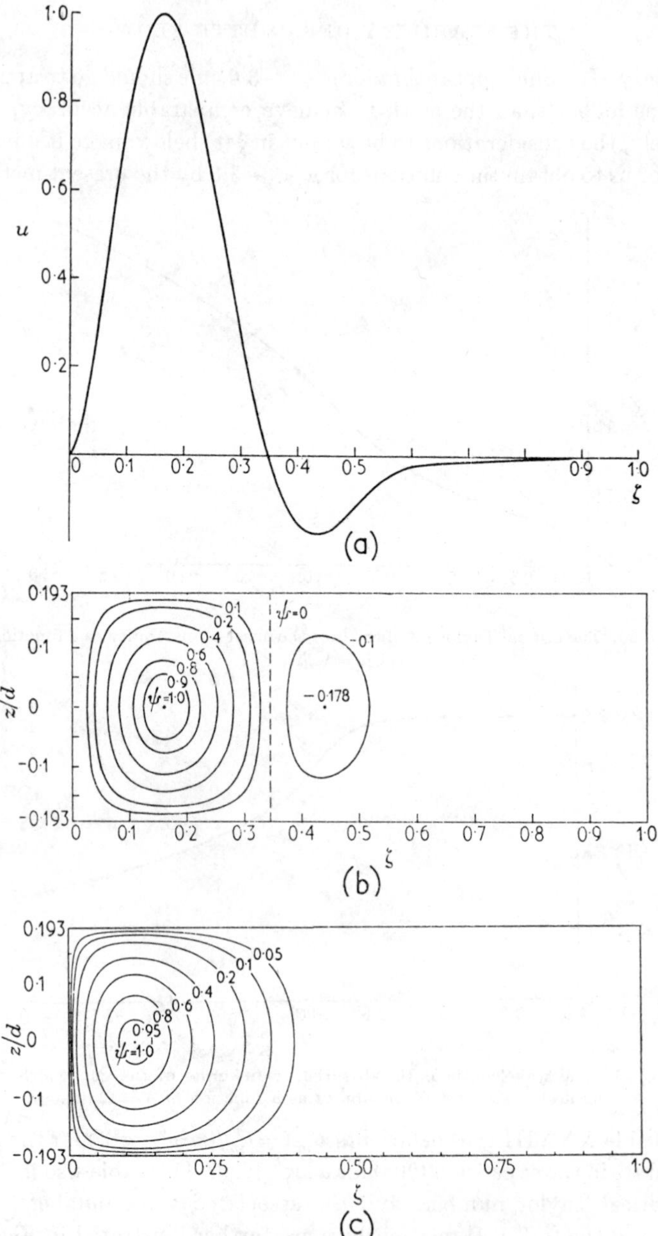

FIG. 71. The velocity profile (*a*) and the cell pattern (*b*) for $\mu = -3$ and $a = 8{\cdot}14$. In (*a*) the velocity is normalized to unit maximum amplitude; in (*b*) the line separating the positive and the negative values of the stream function ψ is indicated by the dashed line (note that this does not occur at the nodal surface which is at $\zeta = 0{\cdot}25$); (*c*) is the corresponding cell pattern for the lowest inviscid mode for the same value of a and μ.

nodal surface occurs at $\zeta = 0{\cdot}25$; and it is of interest to observe how rapidly the amplitude of u decreases as we go beyond this point. The velocity profile illustrated in Fig. 71 should be contrasted with that in Fig. 67 which corresponds to the lowest inviscid mode for the same value of a and μ.

(c) *An alternative method of solution*

Equations (200)–(202) present the first example of a characteristic value problem that we have encountered which is not self-adjoint. For this reason, it may be of interest to describe an alternative method of solution to determine the principles which might guide one in devising methods of solution which may be as equally convergent as the one we have followed in § (b).

Eliminating u between equations (201) and (202), we have

$$(D^2-a^2)^3 v = -Ta^2(1+\alpha\zeta)v \tag{217}$$

together with the boundary conditions

$$v = (D^2-a^2)v = D(D^2-a^2)v = 0 \quad \text{for } \zeta = 0 \text{ and } 1. \tag{218}$$

Rewriting equation (217) in the manner

$$(D^2-a^2)^3 v = (1+\alpha\zeta)\psi \tag{219}$$

and

$$\psi = -Ta^2 v, \tag{220}$$

we expand ψ and v in the forms

$$\psi = \sum_{m=1}^{\infty} C_m \sin m\pi\zeta \quad \text{and} \quad v = \sum_{m=1}^{\infty} C_m v_m, \tag{221}$$

where v_m is the solution of the equation

$$(D^2-a^2)^3 v_m = (1+\alpha\zeta)\sin m\pi\zeta \tag{222}$$

which satisfies the boundary conditions (218). We then insert the expansions (221) in equation (220) to obtain the secular equation.

The general solution of equation (222) is readily seen to be

$$v_m = -\frac{1}{(m^2\pi^2+a^2)^3}\Big\{A_1^{(m)}\cosh a\zeta + A_2^{(m)}\zeta\cosh a\zeta + A_3^{(m)}\zeta^2\cosh a\zeta + B_1^{(m)}\sinh a\zeta + B_2^{(m)}\zeta\sinh a\zeta + B_3^{(m)}\zeta^2\sinh a\zeta + (1+\alpha\zeta)\sin m\pi\zeta + \frac{6\alpha m\pi}{m^2\pi^2+a^2}\cos m\pi\zeta\Big\}, \tag{223}$$

where $A_1^{(m)},\ldots, B_3^{(m)}$ are constants of integration to be determined by the boundary conditions (218). These latter conditions lead to the equations

$$A_1^{(m)} = -\frac{6m\pi\alpha}{m^2\pi^2+a^2}, \qquad aA_2^{(m)}+3B_3^{(m)} = \frac{m\pi}{2a}(m^2\pi^2+a^2),$$

$$A_3^{(m)}+aB_2^{(m)} = 2m\pi\alpha,$$

$$(A_1^{(m)}+A_2^{(m)}+A_3^{(m)})\cosh a+(B_1^{(m)}+B_2^{(m)}+B_3^{(m)})\sinh a = (-1)^{m+1}\frac{6m\pi\alpha}{m^2\pi^2+a^2},$$

$$A_2^{(m)}a\sinh a+A_3^{(m)}(2a\sinh a+\cosh a)+ +B_2^{(m)}a\cosh a+B_3^{(m)}(2a\cosh a+\sinh a) = 2m\pi\alpha(-1)^m,$$

$$A_2^{(m)}a\cosh a+A_3^{(m)}(2a\cosh a+3\sinh a)+ +B_2^{(m)}a\sinh a+B_3^{(m)}(2a\sinh a+3\cosh a) = \frac{m\pi}{2a}(m^2\pi^2+a^2)(1+\alpha)(-1)^m. \quad (224)$$

Now substituting for ψ and v their expansions in equation (220), we obtain

$$\frac{1}{Ta^2}\sum_{m=1}^{\infty} C_m \sin m\pi\zeta = -\sum_{m=1}^{\infty} C_m v_m; \quad (225)$$

and this equation clearly leads to the secular equation

$$\left\|\frac{1}{Ta^2}\delta_{nm}+2(n|m)\right\| = 0, \quad (226)$$

where $(n|m)$ stands for the matrix element

$$(n|m) = \int_0^1 v_m \sin n\pi\zeta\, d\zeta. \quad (227)$$

On evaluating the matrix element $(n|m)$, we find after some lengthy but elementary reductions that the secular equation takes the explicit form

$$\left\|\frac{nm\pi^2}{n^2\pi^2+a^2}\left\{\left(\frac{6\alpha}{m^2\pi^2+a^2}+\frac{4\alpha}{n^2\pi^2+a^2}\right)[(-1)^{m+n}-1]+ +\frac{8a^2}{(n^2\pi^2+a^2)^2}\left[\frac{m^2\pi^2+a^2}{4a^2}\left(\frac{[1+(-1)^{m+1}\cosh a][(-1)^m+(-1)^n]}{\sinh a+(-1)^{m+1}a}+ +\alpha(-1)^m\frac{1+(-1)^{n+1}\cosh a}{\sinh a+(-1)^{n+1}a}\right)+\alpha[(-1)^{m+n}-1]\frac{\sinh a}{\sinh a+(-1)^m a}\right]\right\}+ +\tfrac{1}{2}\delta_{nm}+\alpha X_{nm}-\tfrac{1}{2}(n^2\pi^2+a^2)^3\frac{\delta_{nm}}{Ta^2}\right\| = 0, \quad (228)$$

$$\text{where } X_{nm} = \begin{cases} 0 \text{ if } m+n \text{ is even and } m \neq n, \\ \tfrac{1}{4} \text{ if } m = n, \\ \dfrac{4mn}{n^2-m^2}\left\{\dfrac{3}{m^2\pi^2+a^2}-\dfrac{1}{\pi^2(n^2-m^2)}\right\} \text{ if } m+n \text{ is odd.} \end{cases} \quad (229)$$

One might have thought that, since equation (228) is based on the solution of an equation of order six, the characteristic values derived from the present equation in the different approximations would show an even more rapid convergence than has been found by the method of § (*a*). However, this is not the case: the results given by equations (213) and (228) are practically indistinguishable. Thus:

$$T\begin{cases} = 6417{\cdot}8 \text{ (from equation (228) in 3rd approximation)} \\ = 6417{\cdot}1 \text{ (from equation (213) in 3rd approximation)} \end{cases}$$
$$\text{for } \mu = -0{\cdot}5 \text{ and } a = 3{\cdot}20$$

and

$$T\begin{cases} = 95624 \text{ (from equation (228) in 4th approximation)} \\ = 95625 \text{ (from equation (213) in 4th approximation)} \end{cases}$$
$$\text{for } \mu = -2{\cdot}0 \text{ and } a = 6{\cdot}05. \tag{230}$$

The explanation for this near efficacy of the two methods probably derives from the fact that the advantage gained in the present method by solving an equation of order six has been lost in the artificial treatment of the equation and the boundary conditions. In the method described in § (*a*), the variables used had direct physical meanings, and the boundary conditions relevant to each and the relations between them were satisfied identically; and this is perhaps the real key to the problem.

(*d*) *An approximate solution for* $\mu \to 1$

From the results given in Table XXXIII, it appears that the formula

$$T_c = \frac{3416}{1+\mu} \tag{231}$$

gives a very good approximation to the true values for $0 \leqslant \mu \leqslant 1$; and, moreover, the wave number ($\sim 3{\cdot}12$) at which instability sets in, hardly seems to depend on μ in the same range. Formula (231) was derived by Taylor by a method of solution very different from the ones we have described; and we have also seen how effectively the same formula is given by the *first* approximation to the solution of the secular equation (213).

We shall now show how one can derive (231) by a simple perturbation procedure applied to the solution of equation (217); at the same time, we shall obtain a correction term to (231) which extends its range of validity.

It is convenient for our present purposes to translate the origin of the coordinate system to be midway between the two cylinders. Thus, let

$$\zeta = x+\tfrac{1}{2}, \tag{232}$$

so that the limits of x are $\pm\frac{1}{2}$. In terms of x, equation (217) becomes

$$(D^2-a^2)^3v = -\lambda(1+\epsilon x)v, \tag{233}$$

where $$\lambda = \tfrac{1}{2}Ta^2(1+\mu) \quad\text{and}\quad \epsilon = -2\frac{1-\mu}{1+\mu}; \tag{234}$$

and the boundary conditions are

$$v = (D^2-a^2)v = D(D^2-a^2)v = 0 \quad \text{for } x = \pm\tfrac{1}{2}. \tag{235}$$

When $\epsilon = 0$, the characteristic value problem presented by equations (233) and (235) reduces to the one for the simple Bénard problem (for the case of two rigid boundaries) when formulated in terms of the amplitude Θ of the fluctuations in the temperature (see Chapter II, §§ 12 and 13 (*b*)). Let λ_j ($= R_j a^2$ in the notation of Chapter II) for $j = 0, 1,...,$ denote the different characteristic values of the equation

$$(D^2-a^2)^3\Theta = -\lambda\Theta, \tag{236}$$

for the boundary conditions

$$\Theta = (D^2-a^2)\Theta = D(D^2-a^2)\Theta = 0 \quad \text{for } x = \pm\tfrac{1}{2}, \tag{237}$$

and let the functions belonging to λ_j be distinguished by a subscript j. We know from the analysis in Chapter II, § 13 (*b*) that the functions†

$$\Theta_j \quad\text{and}\quad W_k = (D^2-a^2)\Theta_k \tag{238}$$

are mutually orthogonal‡ when $j \neq k$; without loss of generality, we may suppose that they are so normalized that

$$\int\limits_{-\frac{1}{2}}^{+\frac{1}{2}} \Theta_j W_k \, dx = \delta_{jk}. \tag{239}$$

Turning to the solution of (233), we first observe that according to the results given in Table III (Chapter II) the lowest value of λ/a^2, when $\epsilon = 0$, is 1708 and occurs for $a = 3{\cdot}117$. By (234) the corresponding value of T is $3416/(1+\mu)$; and this is exactly the same as (231). Thus, formula (231) appears as the zero-order term in a perturbation

† As defined here Θ and W are the same functions which describe the perturbations in the temperature and in the normal component of the velocity for the Bénard problem (see equations II (90)).

‡ Equation II (167) clearly implies that

$$\int\limits_{-\frac{1}{2}}^{+\frac{1}{2}} [(D\Theta_k)(D\Theta_j)+a^2\Theta_j\Theta_k]\, dx = 0 \quad \text{if } j \neq k.$$

Integrating by parts the first term in the integrand, we obtain

$$\int\limits_{-\frac{1}{2}}^{+\frac{1}{2}} \Theta_j(D^2-a^2)\Theta_k\, dx = \int\limits_{-\frac{1}{2}}^{+\frac{1}{2}} \Theta_j W_k\, dx = 0 \quad (j \neq k).$$

series. To obtain the higher-order terms, we expand the various quantities in powers of ϵ. Thus, we write

$$\left.\begin{aligned} v &= \Theta_0+\epsilon v^{(1)}+\epsilon^2 v^{(2)}+\dots \\ \text{and} \qquad \lambda &= \lambda_0+\epsilon\lambda^{(1)}+\epsilon^2\lambda^{(2)}+\dots \end{aligned}\right\}. \tag{240}$$

Substituting these expansions in equation (233) and equating the different powers of ϵ, we obtain:

$$(D^2-a^2)^3\Theta_0 = -\lambda_0\Theta_0, \tag{241}$$

$$(D^2-a^2)^3 v^{(1)} = -(\lambda_0 v^{(1)}+\lambda_0 x\Theta_0+\lambda^{(1)}\Theta_0), \tag{242}$$

$$(D^2-a^2)^3 v^{(2)} = -(\lambda_0 v^{(2)}+\lambda_0 x v^{(1)}+\lambda^{(1)}v^{(1)}+\lambda^{(1)}x\Theta_0+\lambda^{(2)}\Theta_0), \tag{243}$$

etc.

Equation (241) is clearly satisfied. To solve equation (242), we assume for $v^{(1)}$ a solution of the form

$$v^{(1)} = \sum_{j=0}^{\infty} A_j^{(1)}\Theta_j. \tag{244}$$

Inserting this solution in (242), we obtain

$$\sum_{j=0}^{\infty} A_j^{(1)}\lambda_j\Theta_j = \lambda_0\sum_{j=0}^{\infty} A_j^{(1)}\Theta_j+\lambda_0 x\Theta_0+\lambda^{(1)}\Theta_0. \tag{245}$$

Multiply this equation by W_k and integrate over the range of x. Making use of the orthogonality relation (239), we obtain

$$A_k^{(1)}\lambda_k = \lambda_0 A_k^{(1)}+\lambda_0(k|x|0)+\lambda^{(1)}\delta_{0k}, \tag{246}$$

where the notation
$$(k|x|j) = \int\limits_{-\frac{1}{2}}^{+\frac{1}{2}} W_k x\Theta_j\,dx \tag{247}$$

has been adopted. (Observe that the matrix $(k|x|j)$ is not Hermitian.)

For $k = 0$, equation (246) gives

$$\lambda^{(1)} = -\lambda_0(0|x|0) = -\lambda_0\int\limits_{-\frac{1}{2}}^{+\frac{1}{2}} W_0 x\Theta_0\,dx. \tag{248}$$

Since the proper functions W_0 and Θ_0 belonging to λ_0 are even, the matrix element $(0|x|0)$ is zero; and we conclude

$$\lambda^{(1)} = 0. \tag{249}$$

Therefore, formula (231) is in error by a term which is of the *second order* in $\epsilon = -2(1-\mu)/(1+\mu)$; and this explains its success in representing the results of more exact calculations as well as it does.

For $k \neq 0$, equation (246) gives

$$A_k^{(1)} = \frac{\lambda_0}{\lambda_k-\lambda_0}(k|x|0). \tag{250}$$

The coefficient $A_0^{(1)}$ is left unspecified; but as we shall presently see, this does not lead to any ambiguities in the solution for the physical problem.

Consider next equation (243). Since $\lambda^{(1)} = 0$, this equation reduces to

$$(D^2-a^2)^3 v^{(2)} = -(\lambda_0 v^{(2)}+\lambda_0 x v^{(1)}+\lambda^{(2)}\Theta_0). \tag{251}$$

In solving this equation, we again expand $v^{(2)}$ in terms of Θ_j. Thus, with the assumption

$$v^{(2)} = \sum_{j=0}^{\infty} A_j^{(2)}\Theta_j, \tag{252}$$

equation (251) gives

$$\sum_{j=0}^{\infty} A_j^{(2)}\lambda_j\Theta_j = \lambda_0 \sum_{j=0}^{\infty} A_j^{(2)}\Theta_j+\lambda_0 x v^{(1)}+\lambda^{(2)}\Theta_0. \tag{253}$$

Multiplying this equation by W_0 and integrating over the range of x, we obtain

$$A_0^{(2)}\lambda_0 = \lambda_0 A_0^{(2)}+\lambda_0 \int_{-\frac{1}{2}}^{+\frac{1}{2}} W_0 x v^{(1)}\,dx+\lambda^{(2)}. \tag{254}$$

Hence

$$\lambda^{(2)} = -\lambda_0 \int_{-\frac{1}{2}}^{+\frac{1}{2}} W_0 x v^{(1)}\,dx. \tag{255}$$

Substituting for $v^{(1)}$ from equation (244), we obtain

$$\lambda^{(2)} = -\lambda_0 \sum_{j=0}^{\infty} A_j^{(1)} \int_{-\frac{1}{2}}^{+\frac{1}{2}} W_0 x\Theta_j\,dx. \tag{256}$$

With the notation (247) for the matrix elements of x, we can write

$$\lambda^{(2)} = -\lambda_0 \sum_{j=0}^{\infty} A_j^{(1)}(0|x|j). \tag{257}$$

Since $(0|x|0) = 0$, the term $j = 0$ in (257) makes no contribution; and substituting for $A_j^{(1)}$ $(j \neq 0)$ from equation (250), we have

$$\lambda^{(2)} = -\lambda_0^2 \sum_{j=1}^{\infty} \frac{(0|x|j)(j|x|0)}{\lambda_j-\lambda_0}. \tag{258}$$

This expression for the second-order change in the characteristic value is very similar to the expression one has in the quantum theory for the second-order change in the energy levels of an atomic system caused by a perturbation. An important difference, however, is the non-Hermitian character of the matrix representing the perturbation.

In the particular problem to which equation (258) refers, the 'energy levels' are so widely spaced (e.g. $\lambda_1 \sim 15\lambda_0$) that it will be sufficient to retain only the first term in the infinite sum representing $\lambda^{(2)}$. Thus, we may write

$$\lambda^{(2)} \simeq -\frac{\lambda_0^2}{\lambda_1-\lambda_0}(0|x|1)(1|x|0). \tag{259}$$

To the second order in ϵ, the desired expression for the characteristic value is

$$\lambda = \lambda_0\left\{1-\epsilon^2\frac{\lambda_0}{\lambda_1-\lambda_0}(0|x|1)(1|x|0)\right\}. \tag{260}$$

Substituting for λ from equation (234) and remembering that $\lambda_j = R_j a^2$, we can rewrite equation (260) in the form

$$T = \frac{2R_0}{1+\mu}\left\{1-\epsilon^2\frac{R_0}{R_1-R_0}(0|x|1)(1|x|0)\right\}, \tag{261}$$

where R_0 and R_1 are the Rayleigh numbers for the lowest even and odd modes for the assigned a^2.

To obtain the critical Taylor number for the onset of instability, we must minimize the quantity on the right-hand side of equation (261) as a function of a. The value of a at which T given by (261) will attain its minimum value will differ from $a^{(0)}_{\min}$ $(= 3{\cdot}117)$ at which R_0 attains its minimum value, R_c $(= 1708)$, by an amount of order ϵ^2. At the displaced position of the new minimum, R_0 will differ from R_c only by an amount of order ϵ^4. Therefore, to order ϵ^2, we may write

$$T_c = \frac{2R_c}{1+\mu}\left\{1-4\left(\frac{1-\mu}{1+\mu}\right)^2\frac{R_c}{R_c^*-R_c}(0|x|1)(1|x|0)\right\}, \tag{262}$$

where R_c^* $(= 2{\cdot}4982\times 10^4)$ is the Rayleigh number for the first odd mode for $a^{(0)}_{\min}$ (see Table I, p. 38). [In (262) we have substituted for ϵ its value $-2(1-\mu)/(1+\mu)$.]

The solution for the first odd mode which is needed for the evaluation of the matrix elements in (262) can be found by the method described in Chapter II, § 15 (*b*). And it is found that the numerical form of equation (262) is

$$T_c = \frac{3416}{1+\mu}\left\{1-7{\cdot}61\times 10^{-3}\left(\frac{1-\mu}{1+\mu}\right)^2\right\}. \tag{263}$$

This formula gives $T_c = 3{\cdot}390\times 10^3$ and $6{\cdot}36\times 10^3$ for $\mu = 0$ and $-0{\cdot}5$, respectively; these values should be compared with $3{\cdot}390\times 10^3$ and $6{\cdot}42\times 10^3$ given by the 'exact' calculations.

(*e*) *The asymptotic behaviour for* $(1-\mu)\to\infty$

From an examination of the results given in Table XXXIII, it appears that for $(1-\mu)\to\infty$ the following asymptotic relations obtain:

$$T_c \to \tau(1-\mu)^4 \quad\text{and}\quad a(T_c)\to q(1-\mu) \text{ as } (1-\mu)\to\infty, \tag{264}$$

where τ and q are certain constants. Thus, from the tabulated values we find:

$$\frac{T_c}{(1-\mu)^4}\begin{cases} = 1180{\cdot}5 \\ = 1180{\cdot}2 \\ = 1181{\cdot}6 \end{cases} \quad\text{and}\quad \frac{a(T_c)}{1-\mu}\begin{cases} = 2{\cdot}03 \\ = 2{\cdot}03 \\ = 2{\cdot}035 \end{cases} \quad\text{for}\quad 1-\mu\begin{cases} = 3{\cdot}0 \\ = 3{\cdot}5 \\ = 4{\cdot}0. \end{cases} \tag{265}$$

The fact that asymptotic relations such as (264) should obtain can be readily understood. For, it will be recalled that the origin of the instability, in the case when the two cylinders are in counter rotation, lies in the violation of the Rayleigh criterion in the layers adjoining the inner cylinder and extending not much beyond the nodal surface. On these grounds, one may expect that as $(1-\mu) \to \infty$ the critical Taylor number for the onset of instability must be comparable to the critical Taylor number for the case $\mu = 0$, and for a gap width equal to the distance d_0 of the nodal surface to the inner cylinder. For the distribution of the angular velocity given by (191),

$$d_0 = d/(1-\mu). \tag{266}$$

Since the gap width occurs with the fourth power in the definition of T, the foregoing arguments will lead one to expect an asymptotic behaviour of the type

$$T_c \to \tau(1-\mu)^4, \tag{267}$$

where τ is comparable to the Taylor number for $\mu = 0$. For the same reasons, one may also expect that the wave number of the associated disturbance will show the behaviour

$$a(T_c) \to q(1-\mu), \tag{268}$$

where q is comparable to the wave number appropriate for $\mu = 0$.

The fact that the constants τ ($\sim$ 1182) and q ($\sim$ 2·035) indicated by the calculations are substantially different from the values 3416 and 3·12 appropriate for $\mu = 0$ should not cause much surprise; for, at $\zeta = d_0/d$, the conditions which prevail are far from those at $\zeta = 1$ for $\mu = 0$; and it is also not true, even asymptotically, that the disturbance does not extend beyond the nodal surface.

While the foregoing arguments give physical grounds for expecting asymptotic relations of the forms (267) and (268), we can equally trace their origin to the analytical structure of the underlying characteristic value problem. Thus, with the substitution

$$x = a\zeta, \tag{269}$$

equation (217) becomes

$$(D^2-1)^3 v = -\frac{T}{a^4}\left(1-\frac{1-\mu}{a}x\right)v; \tag{270}$$

and the corresponding boundary conditions are

$$v = (D^2-1)v = D(D^2-1)v = 0 \quad \text{for } x = 0 \text{ and } a. \tag{271}$$

Suppose now that

$$\frac{T}{a^4} \to \lambda \quad \text{and} \quad \frac{1-\mu}{a} \to \gamma \quad \text{as } (1-\mu) \to \infty \text{ and } a \to \infty. \tag{272}$$

In this limit equations (270) and (271) become

$$(D^2-1)^3 v = -\lambda(1-\gamma x)v \tag{273}$$

and

$$v = (D^2-1)v = D(D^2-1)v = 0 \quad \text{for } x = 0 \text{ and } \infty. \tag{274}$$

For an assigned value of γ, equations (273) and (274) will determine a sequence of possible values† of λ; and to determine the constants τ and q in the asymptotic relations (267) and (268), we must find the minimum of λ/γ^4 as a function of γ. Thus,

$$\tau = \min\left\{\frac{\lambda}{\gamma^4}(\gamma)\right\}; \tag{275}$$

and q is the reciprocal of the value of γ at which λ/γ^4 attains its minimum value.

The characteristic value problem presented by equations (273) and (274) has not been solved satisfactorily at the present time: for various reasons, the methods which have proved successful in other cases seem to fail when applied to the present problem.

72. On the principle of the exchange of stabilities

As we have remarked in § 71 and as we shall see in detail in § 74, experiments confirm that the instability of the Couette flow sets in as a secondary stationary flow. Without attempting a full theoretical justification of the principle of the exchange of stabilities, we shall give some reasons why we may expect its validity in this instance. For this purpose, we return to equations (196) and (197) in which the time constant σ is still retained. Translating the origin of the coordinate system to be midway between the two cylinders and replacing u by $u(1+\mu)/2$, we obtain (cf. equations (201)–(204))

$$(D^2-a^2)(D^2-a^2-\sigma)u = (1+\epsilon x)v \tag{276}$$

and

$$(D^2-a^2-\sigma)v = -\bar{T}a^2u, \tag{277}$$

where

$$\bar{T} = \tfrac{1}{2}(1+\mu)T \quad \text{and} \quad \epsilon = -2\frac{1-\mu}{1+\mu}. \tag{278}$$

And the boundary conditions are the same as hitherto.

Consider first the case $\mu > 0$. As we have seen in § 71 (*d*), in this case, when $\sigma = 0$, the lowest characteristic values of the problem hardly depend on the term linear in x on the right-hand side of equation (276). In fact, to order ϵ, the characteristic values are the same as when $\epsilon = 0$; and, moreover, on account of the wide spacing of the characteristic

† It is possible that real positive characteristic values for λ do not exist for all values of γ; they do not, for example, for $\gamma = 0$.

values of the 'unperturbed problem', the coefficient of ϵ^2 in a perturbation series for $\overline{T}$ is very small (see equation (263)). It would, therefore, appear that even when $\sigma \neq 0$, we can to the *first order* in ϵ ignore the term in ϵx on the right-hand side of equation (276). We shall, then, be left with

$$(D^2-a^2)(D^2-a^2-\sigma)u = v \tag{279}$$

and

$$(D^2-a^2-\sigma)v = -\overline{T}a^2u, \tag{280}$$

together with the boundary conditions

$$v = u = Du = 0 \quad \text{for } x = \pm\tfrac{1}{2}. \tag{281}$$

Now multiply equation (279) by u^* and integrate over the range of x. We obtain by an integration by parts

$$\int\limits_{-\frac{1}{2}}^{+\frac{1}{2}} vu^*\,dx = \int\limits_{-\frac{1}{2}}^{+\frac{1}{2}} u^*[(D^2-a^2)^2-\sigma(D^2-a^2)]u\,dx$$

$$= \int\limits_{-\frac{1}{2}}^{+\frac{1}{2}} |(D^2-a^2)u|^2\,dx+\sigma \int\limits_{-\frac{1}{2}}^{+\frac{1}{2}} (|Du|^2+a^2|u|^2)\,dx. \tag{282}$$

On the other hand, from equation (280) we find

$$-\overline{T}a^2 \int\limits_{-\frac{1}{2}}^{+\frac{1}{2}} vu^*\,dx = \int\limits_{-\frac{1}{2}}^{+\frac{1}{2}} v(D^2-a^2-\sigma^*)v^*\,dx$$

$$= -\int\limits_{-\frac{1}{2}}^{+\frac{1}{2}} [|Dv|^2+(a^2+\sigma^*)|v|^2]\,dx. \tag{283}$$

Combining equations (282) and (283), we obtain

$$\overline{T}a^2\left\{\int\limits_{-\frac{1}{2}}^{+\frac{1}{2}} |(D^2-a^2)u|^2\,dx+\sigma \int\limits_{-\frac{1}{2}}^{+\frac{1}{2}} (|Du|^2+a^2|u|^2)\,dx\right\}$$

$$= \int\limits_{-\frac{1}{2}}^{+\frac{1}{2}} (|Dv|^2+a^2|v|^2)\,dx+\sigma^* \int\limits_{-\frac{1}{2}}^{+\frac{1}{2}} |v|^2\,dx. \tag{284}$$

On equating the imaginary parts of equation (284), we obtain

$$\text{im}(\sigma)\left\{\overline{T}a^2 \int\limits_{-\frac{1}{2}}^{+\frac{1}{2}} (|Du|^2+a^2|u|^2)\,dx+ \int\limits_{-\frac{1}{2}}^{+\frac{1}{2}} |v|^2\,dx\right\} = 0. \tag{285}$$

From this equation it follows that

$$\text{im}(\sigma) = 0 \quad \text{when } \overline{T} > 0; \tag{286}$$

and we conclude that the principle of the exchange of stabilities is valid.

In view of the fact that for $\mu > 0$ the first-order formula (231) gives values which differ from the results of more exact calculations by less than one per cent, it would appear that the foregoing arguments effectively exclude the occurrence of overstability for $\mu > 0$.

As μ turns negative, the replacement of equations (276) and (277) by (279) and (280) will lead to rapidly increasing errors, and the possibility of overstability occurring cannot be excluded in the same way. On the other hand, from the discussion in § 71 (e), it would appear that for $(1-\mu) \to \infty$ the basic physical phenomenon is not very different from what happens when $\mu = 0$; the only difference is that it all happens in the layers immediately adjoining the inner cylinder; so that in this case also, overstability would seem unlikely.

While the foregoing arguments are not conclusive, it is not, in principle, difficult to settle the question by actual computation. Thus, the characteristic value problem presented by equations (196), (197), and (200) can be solved by the method described in § 71 (a). We find that equation (213) is replaced by

$$\Bigg\| \frac{\alpha nm\pi^2(2n^2\pi^2+a^2+b^2)(2m^2\pi^2+a^2+b^2)}{(n^2\pi^2+a^2)(n^2\pi^2+b^2)(m^2\pi^2+a^2)(m^2\pi^2+b^2)}[(-1)^{m+n}-1]+$$
$$+\frac{nm\pi^2(a^2-b^2)}{\Delta(n^2\pi^2+a^2)(n^2\pi^2+b^2)}\Big[\Big[(b\sinh a-a\sinh b)[(-1)^{n+1}+(1+\alpha)(-1)^{m+1}]+$$
$$+(b\cosh b\sinh a-a\sinh b\cosh a)[1+(1+\alpha)(-1)^{m+n}]+$$
$$+\frac{2\alpha(2m^2\pi^2+a^2+b^2)}{(m^2\pi^2+a^2)(m^2\pi^2+b^2)}\{ab(\cosh a-\cosh b)[(-1)^{n+1}-(-1)^{m+1}]+$$
$$+\tfrac{1}{2}(a^2-b^2)\sinh a\sinh b[(-1)^{m+n}-1]\}\Big]\Big]+$$
$$+\tfrac{1}{2}\delta_{nm}+\alpha X_{nm}-\tfrac{1}{2}(n^2\pi^2+a^2)(n^2\pi^2+b^2)^2\frac{\delta_{mn}}{Ta^2}\Bigg\| = 0, \qquad (287)$$

where
$$b^2 = a^2+i\sigma, \qquad (288)†$$
$$\Delta = 2ab(1-\cosh a\cosh b)+(a^2+b^2)\sinh a\sinh b, \qquad (289)$$
and
$$X_{nm} = \begin{cases} 0 \text{ if } m+n \text{ is even and } m \neq n, \\ \tfrac{1}{4} \text{ if } m = n, \\ \dfrac{4nm}{n^2-m^2}\left[\dfrac{2m^2\pi^2+a^2+b^2}{(m^2\pi^2+a^2)(m^2\pi^2+b^2)}-\dfrac{1}{\pi^2(n^2-m^2)}\right] \text{ if } m+n \text{ is odd.} \end{cases} \qquad (290)$$

And the question is: Is there a real σ for which (for some given a) the characteristic root of the secular equation (287) is real? If the answer is in the affirmative, then overstability is possible.

A detailed examination of the roots of equation (287) has not been undertaken. It would be particularly worthwhile to explore the case $\mu = -1$.

† We have replaced σ in equations (196) and (197) by $i\sigma$ to emphasize that we are now primarily interested in marginal states which are strictly oscillatory.

73. The solution for a wide gap when the marginal state is stationary

We now return to a consideration of the equations (167)–(170) without making any assumptions about the relative width of the gap R_2-R_1 between the cylinders. But we shall continue to restrict ourselves to the case when the onset of instability is as a stationary pattern of secondary flow. The relevant equations then are

$$(DD_*-a^2)^2u = -Ta^2\left(\frac{1}{r^2}-\kappa\right)v \tag{291}$$

and

$$(DD_*-a^2)v = u. \tag{292}$$

Eliminating u between equations (291) and (292), we obtain

$$(DD_*-a^2)^3v = -Ta^2\left(\frac{1}{r^2}-\kappa\right)v; \tag{293}$$

and the appropriate boundary conditions are

$$v = (DD_*-a^2)v = D(DD_*-a^2)v = 0 \quad \text{for } r = 1 \text{ and } \eta. \tag{294}$$

A solution of equation (293) patterned after the method described in § 71 (a) proves impracticable. An alternative method based on expansions in terms of orthogonal functions satisfying four boundary conditions appears well suited to the problem on hand. The method is the following.

Letting

$$G = (DD_*-a^2)v, \tag{295}$$

we observe that the boundary conditions (294) require that both G and its derivative vanish at $r = 1$ and η. Accordingly, we may expand G in terms of the set of orthogonal functions, $\mathscr{C}_1(\alpha_j r)$, determined by the characteristic value problem specified by the equation

$$(DD_*)^2y = \left(\frac{d^2}{dr^2}+\frac{1}{r}\frac{d}{dr}-\frac{1}{r^2}\right)^2 y = \alpha^4 y, \tag{296}$$

and the boundary conditions

$$y = 0 \quad \text{and} \quad dy/dr = 0 \quad \text{for } r = 1 \text{ and } \eta. \tag{297}$$

The required characteristic functions are expressible as linear combinations of the Bessel functions J_1, Y_1, I_1, and K_1 in the form

$$\mathscr{C}_1(\alpha_j r) = A_j J_1(\alpha_j r)+B_j Y_1(\alpha_j r)+C_j I_1(\alpha_j r)+D_j K_1(\alpha_j r), \tag{298}$$

where α_j is a root of a certain transcendental equation (given in Appendix V, equation (31)) and A_j, B_j, C_j, and D_j are constants determined apart from an arbitrary constant of proportionality.

The basic idea, then, is to expand G in terms of the functions $\mathscr{C}_1(\alpha_j r)$. Thus, we assume

$$G = (DD_*-a^2)v = \sum_{j=1}^{\infty} P_j \mathscr{C}_1(\alpha_j r), \tag{299}$$

where the coefficients P_j in the expansion are left unspecified at this stage. Having expressed G in this form, we next *solve* equation (299) as a differential equation for v and arrange that it satisfies the remaining boundary conditions, namely, that $v = 0$ for $r = 1$ and η. With v determined in this fashion, equation (291) will lead, as we shall presently see, to a secular equation for T.

(a) The characteristic equation

We shall now obtain the explicit form of the characteristic equation for T.

First we must solve equation (299) for v. For this purpose it is convenient to write the function $\mathscr{C}_1(\alpha_j r)$ in the manner

$$\mathscr{C}_1(\alpha_j r) = u_j(r)+v_j(r), \tag{300}$$

where

$$\left.\begin{aligned} u_j(r) &= A_j J_1(\alpha_j r)+B_j Y_1(\alpha_j r) \\ v_j(r) &= C_j I_1(\alpha_j r)+D_j K_1(\alpha_j r) \end{aligned}\right\}. \tag{301}$$

and

Clearly,

$$DD_* u_j = -\alpha_j^2 u_j \quad \text{and} \quad DD_* v_j = +\alpha_j^2 v_j. \tag{302}$$

Making use of the relations (302), we can readily verify that the general solution of equation (299) is

$$v = \sum_{j=1}^{\infty} P_j\left\{p_j I_1(ar)+q_j K_1(ar)-\frac{u_j(r)}{\alpha_j^2+a^2}+\frac{v_j(r)}{\alpha_j^2-a^2}\right\}, \tag{303}$$

where the constants of integration p_j and q_j are to be determined by the boundary conditions on v, namely, that v vanishes at $r = 1$ and η. These latter conditions lead to the equations

$$p_j I_1(a)+q_j K_1(a) = \frac{2\alpha_j^2}{\alpha_j^4-a^4}u_j(1)$$

and

$$p_j I_1(a\eta)+q_j K_1(a\eta) = \frac{2\alpha_j^2}{\alpha_j^4-a^4}u_j(\eta). \tag{304}$$

In obtaining these equations from (303), we have made use of the fact that

$$u_j(1) = -v_j(1) \quad \text{and} \quad u_j(\eta) = -v_j(\eta). \tag{305}$$

On solving equations (304), we find:

$$p_j = \frac{2\alpha_j^2}{\Delta(\alpha_j^4-a^4)}[+u_j(1)K_1(a\eta)-u_j(\eta)K_1(a)],$$

$$q_j = \frac{2\alpha_j^2}{\Delta(\alpha_j^4-a^4)}[-u_j(1)I_1(a\eta)+u_j(\eta)I_1(a)], \tag{306}$$

where

$$\Delta = I_1(a)K_1(a\eta)-I_1(a\eta)K_1(a). \tag{307}$$

Now substituting for v from equation (303) in equation (293), we obtain

$$\sum_{j=1}^{\infty} P_j\{(\alpha_j^2+a^2)^2 u_j+(\alpha_j^2-a^2)^2 v_j\}$$
$$= Ta^2\left(\kappa-\frac{1}{r^2}\right)\sum_{j=1}^{\infty} P_j\left\{p_j\, I_1(ar)+q_j\, K_1(ar)+\frac{a^2(u_j+v_j)-\alpha_j^2(u_j-v_j)}{\alpha_j^4-a^4}\right\}. \tag{308}$$

Finally, multiplying this equation by $r(u_k+v_k)$ and integrating over the range of r, we obtain

$$P_k(\alpha_k^4+a^4)N_k+2a^2\sum_{j=1}^{\infty} P_j\,\alpha_j^2\,\Delta_{jk}^{(1)}$$
$$= Ta^2\sum_{j=1}^{\infty} P_j\Big\{p_j[\kappa I_k^{(1)}(a)-I_k^{(-1)}(a)]+q_j[\kappa K_k^{(1)}(a)-K_k^{(-1)}(a)]+$$
$$+\frac{a^2}{\alpha_j^4-a^4}[\kappa N_k\,\delta_{kj}-M_{jk}]-\frac{\alpha_j^2}{\alpha_j^4-a^4}[\kappa\Delta_{jk}^{(1)}-\Delta_{jk}^{(-1)}]\Big\}, \tag{309}$$

where we have made use of the orthogonality relation

$$\int_\eta^1 (u_j+v_j)(u_k+v_k)r\,dr = N_j\,\delta_{jk}, \tag{310}$$

and introduced the abbreviations

$$\Delta_{jk}^{(\pm1)} = \int_\eta^1 (u_j-v_j)(u_k+v_k)r^{\pm1}\,dr,$$
$$I_k^{(\pm1)}(a) = \int_\eta^1 I_1(ar)(u_k+v_k)r^{\pm1}\,dr, \tag{311}$$
$$K_k^{(\pm1)}(a) = \int_\eta^1 K_1(ar)(u_k+v_k)r^{\pm1}\,dr,$$

and

$$M_{jk} = \int_\eta^1 (u_j+v_j)(u_k+v_k)\frac{dr}{r}. \tag{312}$$

Equation (309) leads to the secular equation

$$\Big\|N_j\left(\frac{\alpha_j^4+a^4}{a^2}-\kappa T\frac{a^2}{\alpha_j^4-a^4}\right)\delta_{jk}+2\alpha_j^2\,\Delta_{jk}^{(1)}-$$
$$-T\Big\{p_j[\kappa I_k^{(1)}(a)-I_k^{(-1)}(a)]+q_j[\kappa K_k^{(1)}(a)-K_k^{(-1)}(a)]-$$
$$-\frac{\alpha_j^2}{\alpha_j^4-a^4}[\kappa\Delta_{jk}^{(1)}-\Delta_{jk}^{(-1)}]-\frac{a^2}{\alpha_j^4-a^4}M_{jk}\Big\}\Big\| = 0. \tag{313}$$

Of the various matrix elements defined in equations (311) and (312), those with the superscript $(+1)$ in (311) can be evaluated explicitly. The remaining matrix elements must be evaluated numerically.

(b) *Numerical results for the case* $\eta = \frac{1}{2}$

First, we may note that when $\eta = \frac{1}{2}$ the definitions of T and κ are

$$T = \frac{64}{9}\left(\frac{\Omega_1 R_1^2}{\nu}\right)^2 (1-\mu)(1-4\mu) \tag{314}$$

and $$\kappa = (1-4\mu)/(1-\mu). \tag{315}$$

And Rayleigh's criterion for stability in this case is

$$\mu > \tfrac{1}{4} \quad \text{(for stability)}. \tag{316}$$

TABLE XXXIV

Taylor numbers for various assigned values of κ *and* a $(\eta = \frac{1}{2})$

κ	a	*First approximation*	*Second approximation*	*Third approximation*
0	6·0	$1{\cdot}5520\times10^4$	$1{\cdot}5470\times10^4$	$1{\cdot}5370\times10^4$
	6·2	$1{\cdot}5486\times10^4$	$1{\cdot}5434\times10^4$	$1{\cdot}5328\times10^4$
	6·4	$1{\cdot}5498\times10^4$	$1{\cdot}5444\times10^4$	$1{\cdot}5332\times10^4$
0·4	6·0	$1{\cdot}9832\times10^4$	$1{\cdot}9728\times10^4$	$1{\cdot}9594\times10^4$
	6·2	$1{\cdot}9788\times10^4$	$1{\cdot}9680\times10^4$	$1{\cdot}9539\times10^4$
	6·4	$1{\cdot}9804\times10^4$	$1{\cdot}9692\times10^4$	$1{\cdot}9542\times10^4$
0·6	6·2	$2{\cdot}2981\times10^4$	$2{\cdot}2811\times10^4$	$2{\cdot}2642\times10^4$
	6·4	$2{\cdot}3000\times10^4$	$2{\cdot}2823\times10^4$	$2{\cdot}2644\times10^4$
1·0	6·2	$3{\cdot}3929\times10^4$	$3{\cdot}3386\times10^4$	$3{\cdot}3110\times10^4$
	6·4	$3{\cdot}3958\times10^4$	$3{\cdot}3393\times10^4$	$3{\cdot}3100\times10^4$
	6·6	$3{\cdot}4081\times10^4$	$3{\cdot}3492\times10^4$	$3{\cdot}3182\times10^4$
1·333	6·2	$5{\cdot}6268\times10^4$	$5{\cdot}389\times10^4$	$5{\cdot}3352\times10^4$
	6·4	$5{\cdot}6318\times10^4$	$5{\cdot}386\times10^4$	$5{\cdot}3280\times10^4$
	6·6	$5{\cdot}6525\times10^4$	$5{\cdot}397\times10^4$	$5{\cdot}3354\times10^4$
1·6	6·4	$1{\cdot}1901\times10^5$	$1{\cdot}007\times10^5$	$9{\cdot}9072\times10^4$
	6·6	$1{\cdot}195\times10^5$	$1{\cdot}005\times10^5$	$9{\cdot}8831\times10^4$
	6·8	$1{\cdot}2023\times10^5$	$1{\cdot}006\times10^5$	$9{\cdot}8832\times10^4$
1·8	7·6	$7{\cdot}697\times10^5$	$2{\cdot}0840\times10^5$	$1{\cdot}9967\times10^5$
	7·8	$7{\cdot}843\times10^5$	$2{\cdot}0862\times10^5$	$1{\cdot}9954\times10^5$
	8·0	$8{\cdot}001\times10^5$	$2{\cdot}0917\times10^5$	$1{\cdot}9972\times10^5$
1·9	8·4	—	$2{\cdot}024\times10^5$	$2{\cdot}939\times10^5$
	8·6	—	$2{\cdot}042\times10^5$	$2{\cdot}9363\times10^5$
	8·8	—	$2{\cdot}065\times10^5$	$2{\cdot}9365\times10^5$
2·0	9·4	—	$5{\cdot}02\times10^5$	$4{\cdot}29\times10^5$
	9·6	—	$5{\cdot}046\times10^5$	$4{\cdot}2865\times10^5$
	9·8	—	$5{\cdot}092\times10^5$	$4{\cdot}305\times10^5$

In Table XXXIV the values of T obtained with the aid of the characteristic equation (313) in the different approximations (the 'order' of the approximation being the order of the determinant which is set equal to zero in the determination of T) are listed for various assigned values of a and κ. For each value of κ, values of a were chosen (by trial and error) in the range in which T (as a function of a) attains its minimum. From an examination of this table, it appears that for $\kappa < 1{\cdot}6$ the required values of T have been determined to within one per cent of the true values. For $\kappa = 1{\cdot}8$, $1{\cdot}9$, and $2{\cdot}0$, the results obtained in the second and the third approximations differ appreciably. Nevertheless, it is unlikely that even in these cases the results of the third approximation are in error by more than 3 per cent: this estimate of the error does not appear unreasonable when we observe that for $\kappa = 1{\cdot}6$, while the results of the first and the second approximations differ by 20 per cent, the

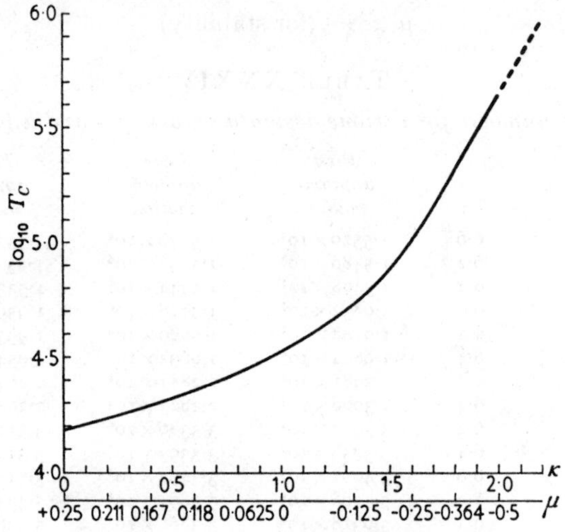

FIG. 72. The variation of the critical Taylor number T_c for the onset of instability as a function of κ $(= (1-4\mu)/(1-\mu))$. A scale of μ $(= \Omega_2/\Omega_1)$ is also shown.

results of the second and the third approximations do not differ by more than 2 per cent.

In Table XXXV the critical Taylor numbers (derived from the data of Table XXXIV) appropriate for the different values of κ are given; the values of a at which these minimum Taylor numbers are attained are also given. The derived (T_c, κ) and $[a(T_c), \kappa]$-relationships are further illustrated in Figs. 72 and 73. From Fig. 72 it appears that in a $(\log T_c, \kappa)$-plot, the relationship is very nearly a linear one in the neighbourhood of $\kappa = 2$.

Table XXXV also includes the values

$$\left.\begin{aligned} \omega_1 &= \frac{\Omega_1 R_1^2}{\nu} = 0{\cdot}375\sqrt{\left(\frac{T}{(1-\mu)(1-4\mu)}\right)} = \frac{0{\cdot}375}{1-\mu}\sqrt{\left(\frac{T}{\kappa}\right)} \\ \text{and} \quad \omega_2 &= \frac{\Omega_2 R_1^2}{\nu} = \mu\omega_1; \end{aligned}\right\} \quad (317)$$

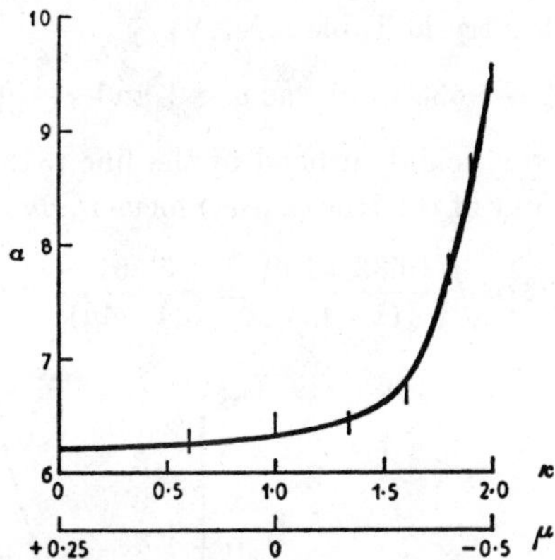

FIG. 73. The variation of the wave number a (in the unit $1/R_2$) of the disturbance at which instability first sets in as a function of κ and μ. The uncertainty in the determination of a is indicated by the height of the lines.

TABLE XXXV

Critical Taylor numbers and related constants for various values of κ ($\eta = \frac{1}{2}$)

κ	μ	a_c	T_c	ω_1	ω_2
0	+0·25	6·2	$1{\cdot}533\times10^4$	—	—
†0·1	+0·230769	6·2	$1{\cdot}621\times10^4$	196·3	+45·30
†0·2	+0·210526	6·2	$1{\cdot}719\times10^4$	139·3	+29·32
†0·3	+0·189189	6·2	$1{\cdot}829\times10^4$	114·2	+21·60
0·4	+0·166667	6·2	$1{\cdot}954\times10^4$	99·46	+16·58
0·6	+0·117647	6·2	$2{\cdot}264\times10^4$	82·56	+ 9·713
†0·8	+0·0625	—	$2{\cdot}693\times10^4$	73·39	+ 4·587
1·0	0	6·4	$3{\cdot}310\times10^4$	68·23	0
1·333	−0·125	6·4	$5{\cdot}328\times10^4$	66·63	− 8·329
1·6	−0·25	6·6	$9{\cdot}883\times10^4$	74·56	−18·64
†1·7	−0·304348	—	$1{\cdot}377\times10^5$	81·83	−24·90
1·8	−0·363636	7·8	$1{\cdot}995\times10^5$	91·56	−33·30
†1·85	−0·395349	—	$2{\cdot}419\times10^5$	97·19	−38·42
1·9	−0·428571	8·6	$2{\cdot}936\times10^5$	103·2	−44·22
†1·95	−0·463415	—	$3{\cdot}556\times10^5$	109·4	−50·71
2·0	−0·500	9·6	$4{\cdot}286\times10^5$	115·7	−57·87

† The entries for these values of κ were deduced by Lagrangian interpolation among the other computed (κ, log T)-values.

ω_1 and ω_2 are, therefore, the angular velocities Ω_1 and Ω_2 measured in the unit ν/R_1^2. In the (ω_2, ω_1)-plane, the locus determined by equations (317) separates the regions of stability from the regions of instability (see Fig. 74). In this plane, Rayleigh's criterion is represented by the straight line

$$\omega_2 = \tfrac{1}{4}\omega_1 \quad \text{(Rayleigh's criterion).} \tag{318}$$

Since (see the first entry in Table XXXV)

$$T_c \to 1{\cdot}533 \times 10^4 \quad \text{as } \mu \to \tfrac{1}{4} \text{ and } \kappa \to 0, \tag{319}$$

it is clear that in the neighbourhood of the line (318), as $\omega_2 \to \infty$, the asymptotic behaviour of the true (ω_2, ω_1)-locus is given by (cf. equation (317))

$$\omega_1 \to 0{\cdot}375 \sqrt{\left(\frac{1{\cdot}533 \times 10^4}{\frac{3}{4}(1-4\mu)}\right)} = \frac{53{\cdot}61}{\sqrt{(1-4\mu)}} \quad (\mu \to \tfrac{1}{4}). \tag{320}$$

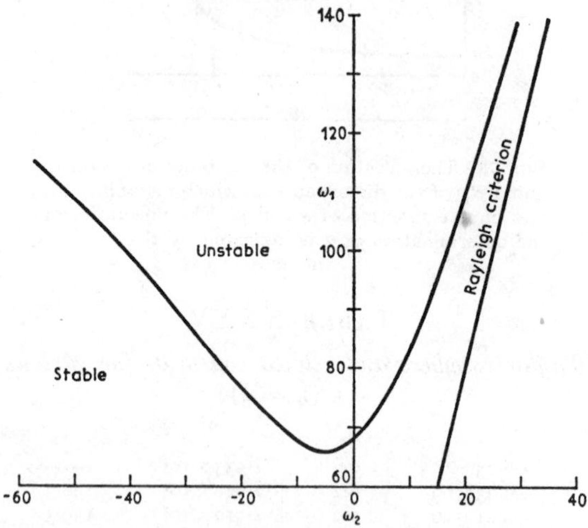

Fig. 74. The regions of stability and instability in the (ω_2, ω_1)-plane; ω_1 and ω_2 are the angular velocities of the inner and the outer cylinder in the units ν/R_1^2 and ν/R_2^2, respectively.

Finally, in Figs. 75 and 76, the profiles of the velocity in the radial direction and the corresponding cell patterns at marginal stability are shown for two typical cases ($\mu = 0$ and $\mu = -4/11$).

74. Experiments on the stability of viscous flow between rotating cylinders

The first experiments on the onset of instability in the viscous flow between rotating coaxial cylinders are those of Taylor. Experiments similar to Taylor's, and others related to the same phenomenon, have since been carried out by Lewis, Terada and Hattori, Wendt, Taylor himself, Schultz-Grunow and Hein, Donnelly, and Donnelly and Fultz. We shall restrict our account of the experiments to those of Donnelly, and Donnelly and Fultz as they are not only typical of the class, but they

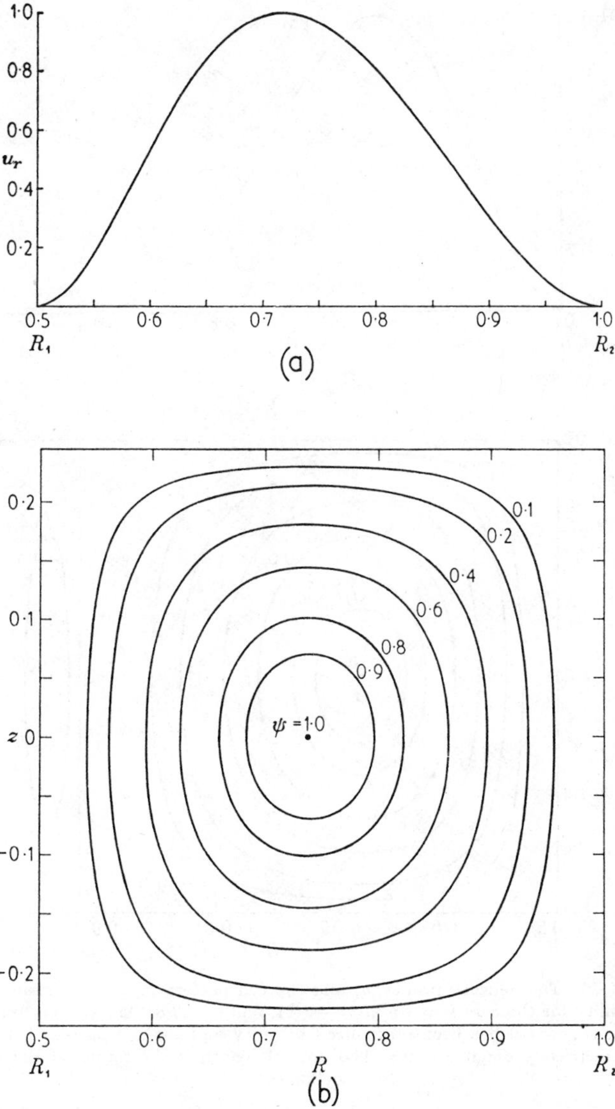

FIG. 75. The velocity profile (*a*) and the cell pattern (*b*) at the onset of instability for the case $\kappa = 1$, $\mu = 0$, and $a = 6{\cdot}4$. The stream function ψ ($\propto ru_r \cos az$) has been normalized to unity and the cell pattern is drawn symmetrically about $z = 0$. (The unit of length is the radius of the outer cylinder.)

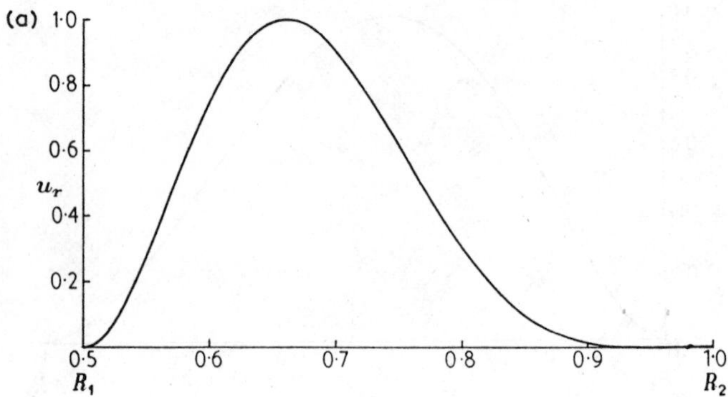

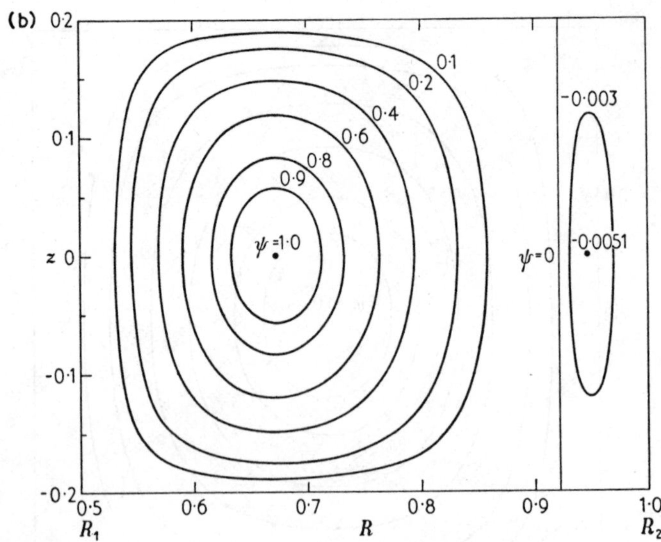

FIG. 76. The velocity profile (*a*) and the cell pattern (*b*) at the onset of instability for the case $\kappa = 1{\cdot}8$, $\mu = -4/11$, and $a = 7{\cdot}8$. The stream function ψ ($\propto ru_r \cos az$) has been normalized to unity and the cell pattern is drawn symmetrically about $z = 0$. (The unit of length is the radius of the outer cylinder.)

are also among the most precise and the most extensive; moreover, they bear directly on certain aspects of the theory considered in this chapter.

First, we may observe that the experiments one can perform in the present connexion are of two kinds: those which involve the measurement of the torque exerted on the outer cylinder (kept at rest) for different rates of rotation of the inner cylinder, and those in which the motions in the fluid are observed by suitable tracers.

In the experiments of the first kind, the onset of instability is detected by a principle not unlike the Schmidt–Milverton principle for the detection of the onset of thermal instability (cf. § 18 (*b*)). In the present instance, the torque exerted on the outer cylinder (which is suitably suspended) is measured as a function of the angular velocity Ω_1 of the inner cylinder. It is known that so long as the flow between the cylinders is laminar and conforms to the law (1), the relation is a linear one and is given by

$$\text{torque} = 4\pi\rho\nu\frac{R_1^2 R_2^2 H}{R_2^2-R_1^2}\Omega_1, \tag{321}$$

where H is the height of the suspended cylinder. At the onset of instability, the effective viscosity will abruptly start increasing, and this will be reflected in a sudden break in the relation (321).

In the experiments of the second kind, the onset of instability is detected by visual observations of sudden changes in the character of the fluid motion occurring as the relative angular velocity of the two cylinders is increased. While these experiments are not as precise as those on the torque measurements for determining the critical angular velocities at which instability arises, they are not restricted to any special value of μ $(= \Omega_2/\Omega_1)$ such as $\mu = 0$. Indeed, the important range of negative μ is readily accessible to observations only by this second method.

(*a*) *The determination of the critical Taylor numbers, for the case $\mu = 0$, by torque measurements*

The most precise experiments on the determination of the critical Taylor numbers for the onset of instability, for the case $\mu = 0$, are those of Donnelly.

In his experiments, Donnelly used a rotating cylinder viscometer (see Fig. 77) in which only a central section of the outer cylinder was suspended; the fixed end sections served as a guard cylinder for eliminating end effects. The viscometer was so designed that if the suspended cylinder departed by so much as 0·005 in. in any direction from exact concentricity with the guard cylinder, the suspension would not swing freely. The system was thus protected from external vibrations and lateral motions.

The suspension system for the outer cylinders consisted of the following parts. The straps by which the outer cylinder was carried were attached to a long quartz rod (0·018 in. in diameter and 11 cm long) at the end of which was a torsion fibre (0·005 in. in diameter and 10 cm long). The fibre was in turn attached to a shaft whose angular position could be read on a dial provided with a vernier. A small mirror affixed to the torsion fibre was directly in front of a stationary mirror. The torque exerted on the suspended cylinder was measured by a null method by turning the dial so as to bring into coincidence the image of a cross-hair

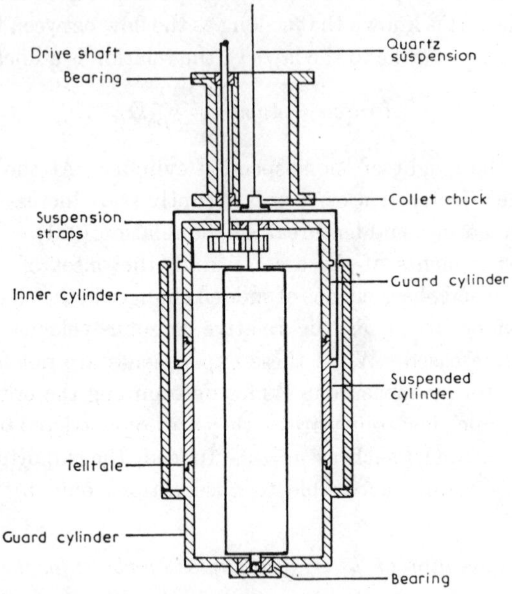

FIG. 77. Sketch of the viscometer used in Donnelly's experiments.

(observed through an auto-collimator) reflected in both mirrors. The angular deflexion necessary to bring about the coincidence is a measure of the torque.

The inner cylinder was driven by a shaft coupled to a motor and gear reducing system. The period of rotation was determined photo-electrically by using suitable scaling circuits. In the experiments, the speed of rotation was maintained constant to well within 0·1 per cent.

The viscometer provided two interchangeable inner cylinders: one of them left a gap width ($d/R_2 \sim 0{\cdot}05$) sufficiently small for the theory of § 71 to be applicable; and the other left a gap width equal to the radius of the inner cylinder so that the results of the experiments could be compared with the calculations of § 73 (*b*). The relevant dimensions of the viscometer are given in Table XXXVI.

To maintain the temperature of the liquid at a constant value during the experiments, the viscometer was placed in a beaker kept in a constant temperature bath.

The liquids used in the experiments and their relevant properties are listed in Table XXXVII.

TABLE XXXVI

The dimensions and constants of Donnelly's viscometer

Narrow gap	*Wide gap*
$R_1 = 1{\cdot}89936 \pm 0{\cdot}0001$ cm	$R_1 = 0{\cdot}99963 \pm 0{\cdot}0001$ cm
$R_2 = 2{\cdot}00023 \pm 0{\cdot}0001$ cm	$R_2 = 2{\cdot}00023 \pm 0{\cdot}0001$ cm
$H = 4{\cdot}9987 \pm 0{\cdot}0003$ cm	$H = 4{\cdot}9987 \pm 0{\cdot}0003$ cm
$C = 3{\cdot}63 \times 10^{-5}$	$C = 9{\cdot}67 \times 10^{-4}$

TABLE XXXVII

List of fluids and their physical properties

Fluid	*Temperature* (°C)	ρ (g)	ν (cm²/sec)
CCl_4	25·00	1·585	$(5{\cdot}796 \pm 0{\cdot}03) \times 10^{-3}$
CCl_4	20·95	1·592	$(6{\cdot}091 \pm 0{\cdot}03) \times 10^{-3}$
CS_2	25·00	1·255	$(2{\cdot}936 \pm 0{\cdot}02) \times 10^{-3}$
oil I, lot 11†	25·00	0·8404	$1{\cdot}226 \times 10^{-1}$
oil D, lot 11†	25·00	0·7813	$2{\cdot}347 \times 10^{-2}$

† These oils are provided by the National Bureau of Standards (Washington, D.C.) for calibrating viscometers. The National Bureau of Standards does not state the limits of error on their measurements.

If K is the torsion constant of the fibre and Φ is the deflexion in radians, then according to equation (321),

$$K\Phi = 4\pi\rho\nu \frac{R_1^2 R_2^2 H}{R_2^2 - R_1^2} \Omega_1. \tag{322}$$

In Donnelly's apparatus, the dial attached to the torsion fibre was divided into 1,000 parts so that if ϕ is the measured deflexion in dial divisions,

$$\rho\nu = \left(K \frac{R_2^2 - R_1^2}{4\pi R_1^2 R_2^2 H} \times 10^{-3}\right)\phi P = C\phi P \text{ (say)}, \tag{323}$$

where P $(= 2\pi/\Omega_1)$ is the period of rotation of the inner cylinder. Equation (323) applies when the flow is laminar; when it is not, we may still regard the quantity on the right-hand side as giving an *effective viscosity*.

The value of C was determined by calibration with liquids of known viscosity rather than by an absolute determination of the torsion constant K. The values of C given in Table XXXVI were obtained in this manner.

In carrying out the experiments, the speed of rotation of the inner cylinder was slowly increased from the lowest setting, keeping the outer cylinder near its null position all the time.

(i) *Results of the experiments with the narrow gap*

The definition of the Taylor number appropriate for this case is (see equation (198))

$$T = -\frac{4A\Omega_1}{\nu^2}d^4, \tag{324}$$

where A is given in equation (145). For $\mu = 0$, the explicit form of T is

$$T = \frac{4R_1^2 d^4}{R_2^2-R_1^2}\left(\frac{\Omega_1}{\nu}\right)^2. \tag{325}$$

According to the results given in Table XXXIII, the critical Taylor number for $\mu = 0$ is $3{\cdot}390\times 10^3$. The corresponding critical value of Ω_1 is given by

$$\Omega_1(\text{critical}) = \left(3{\cdot}390\times 10^3\frac{R_2^2-R_1^2}{4R_1^2 d^4}\right)^{\frac{1}{2}}\nu. \tag{326}$$

For the values of R_1 and R_2 given in Table XXXVI,

$$\Omega_1\,(\text{critical}) = 9{\cdot}448\times 10^2\nu. \tag{327}$$

For carbon tetrachloride at 25° C instability is predicted for a period of rotation

$$P\,(\text{critical}) = 1{\cdot}147\pm 0{\cdot}006\text{ s (calculated)}, \tag{328}$$

where the limits of error are due to the uncertainty in ν.

The results of Donnelly's experiments are shown in Fig. 78 *a*, *b*. In these figures, ϕP is plotted as a function of P. We observe that as the period of rotation of the inner cylinder is decreased, the viscosity as given by equation (323) begins to increase abruptly at a certain period. The break occurs with extreme sharpness; how sharply, can be seen from Fig. 78 *b* in which the measurements in the immediate vicinity of the critical period are exhibited.

The critical period of rotation determined, as intermediate between the last of the stable and the first of the unstable points of observation, is

$$P\,(\text{critical}) = 1{\cdot}1337\pm 0{\cdot}00095\text{ s (measured)}. \tag{329}$$

This agrees very well with the calculated value (328).

(ii) *Results of the experiments with the wide gap* ($\eta = \frac{1}{2}$)

The definition of T appropriate for this case is (cf. equation (314))

$$T = \frac{64}{9}\left(\frac{\Omega_1 R_1^2}{\nu}\right)^2. \tag{330}$$

According to the results given in Table XXXV, the critical Taylor number for this case ($\kappa = 1$ and $\mu = 0$) is $3{\cdot}310 \times 10^4$. The corresponding critical value of Ω_1 is given by

$$\Omega_1\,(\text{critical}) = \left(\frac{9 \times 3{\cdot}310 \times 10^4}{64 R_1^4}\right)^{\frac{1}{2}} \nu. \tag{331}$$

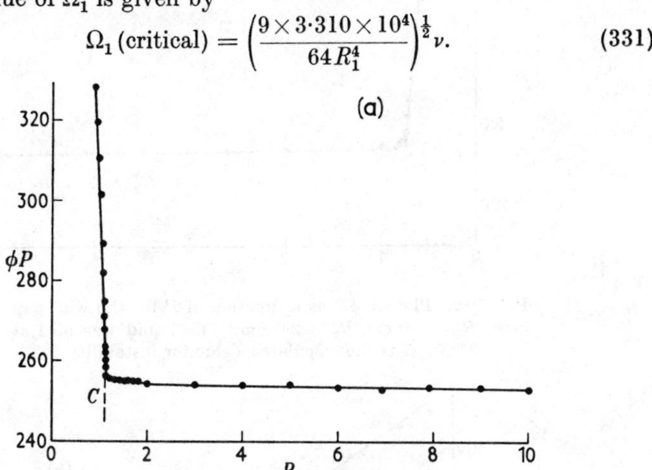

FIG. 78 *a*. Plot of ϕP (which is proportional to the effective viscosity) as a function of the period of rotation, P, of the inner cylinder for the narrow-gap case: $R_1 = 1{\cdot}9$ cm, $R_2 = 2{\cdot}0$ cm. The liquid was carbon tetrachloride at 25° C. C is the calculated value for instability.

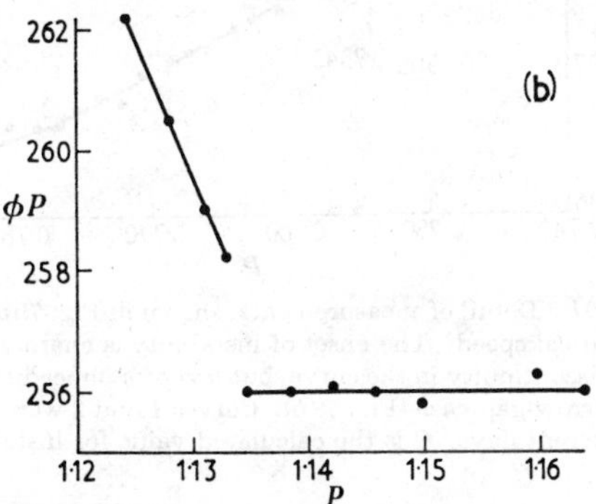

FIG. 78 *b*. Detail of measurements, shown in Fig. 78 *a*, about the critical speed. The onset of instability is characterized by a discontinuity both in the magnitude of ϕP and in the slope of the effective viscosity curve.

For the value of R_1 given in Table XXXVI,

$$\Omega_1\,(\text{critical}) = 68{\cdot}28\nu. \tag{332}$$

The experiments with this wide-gap viscometer were carried out with

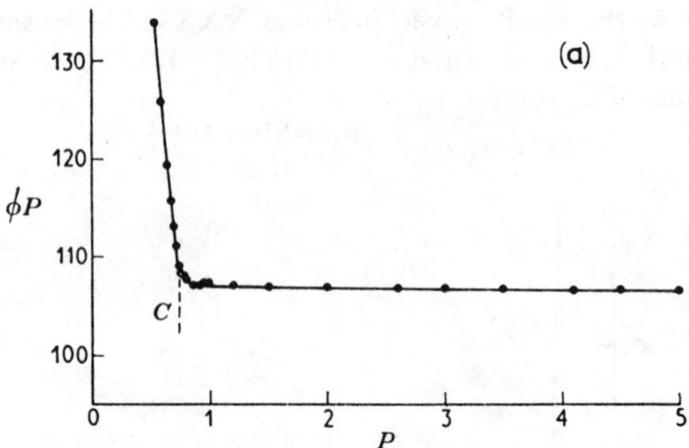

FIG. 79 *a*. Plot of ϕP as a function of P for the wide-gap case; $R_1 = 1{\cdot}0$ cm, $R_2 = 2{\cdot}0$ cm. The liquid was oil I at 25° C. C is the calculated value for instability.

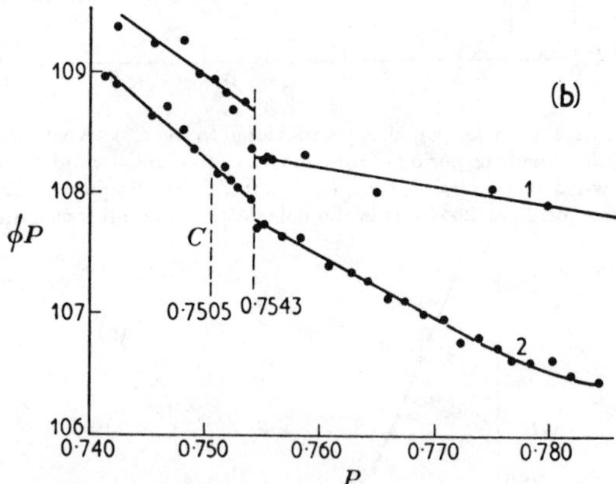

FIG. 79 *b*. Detail of measurements, shown in Fig. 79 *a*, near the critical speed. The onset of instability is characterized by a discontinuity in the curve, but less pronounced than in the narrow-gap case (Fig. 78 *b*). Curves 1 and 2 were taken on different days. C is the calculated value for instability.

an oil (oil I in Table XXXVII) supplied by the National Bureau of Standards for calibrating viscometers. For this oil, instability of flow is predicted for a period of rotation,

$$P\,(\text{critical}) = 0{\cdot}7506 \pm 0{\cdot}0004\,\text{s (calculated)}. \qquad (333)$$

The results of Donnelly's experiments in this case are shown in Fig. 79 *a*, *b*. We observe that, again, an increase in the viscosity as given by equation (323) begins abruptly at a certain critical period of rotation of the inner cylinder. The measurements in the immediate vicinity of the critical period exhibited in Fig. 79 *b* show how sharply it begins.

The two curves (1) and (2) in Fig. 79 *b* show that, while the magnitude of ϕP at the critical period varies from one sequence of measurements to another, the critical period itself is reproducible; the value of the critical period as experimentally determined is

$$P\,(\text{critical}) = 0{\cdot}7543 \pm 0{\cdot}0003\,\text{s}\,(\text{measured}). \tag{334}$$

Again, this agrees very well with the calculated value (333).

(*b*) *The dependence of the critical Taylor number on* Ω_2/Ω_1. *The results of visual and photographic observations*

While the experiments described in § (*a*) above verify the theoretical predictions regarding the critical Taylor numbers for the onset of instability for the particular case $\mu = 0$, they do not provide confirmations for other important aspects of the theory such as the dependence on μ $(= \Omega_2/\Omega_1)$ of the critical Taylor number and of the wave number of the associated disturbance. For these latter purposes, it is essential that experiments are carried out with an apparatus in which the two cylinders can be rotated independently of each other. The first experiments in this category are those of Taylor. Taylor's experiments were carried out with an apparatus with a narrow enough gap for the results of § 71 (*b*) to be applicable. They have been repeated by Donnelly and Fultz with an apparatus in which the gap was equal to the radius of the inner cylinder so that the results of § 73 (*b*) are similarly applicable.

Before presenting the results of Taylor's and Donnelly and Fultz's experiments, we shall give a brief description of the apparatus used by the latter authors (see Fig. 80). The outer cylinder was a precision bore Pyrex tube 94 cm long and inside diameter $R_2 = 6{\cdot}2846 \pm 0{\cdot}0006$ cm. The inner cylinder was a brass tube turned to a radius $R_1 = 3{\cdot}1432 \pm 0{\cdot}0001$ cm. The Pyrex cylinder rested on a groove cut in a plate which fitted on a brass base mounted on a rotating table. The brass cylinder was connected by means of a drive pin and locking collar to a shaft which could be rotated independently of the brass base. The cylinders were rotated by separate pulleys and by two motors with variable speed transmission. The two cylinders were aligned to be coaxial by suitable precision ball bearings.

The motion of the fluid was traced by ink emitted from three small holes at the middle of the inner cylinder. The ink was a solution of Nigrosine dye in the liquid used in the experiments and was of such

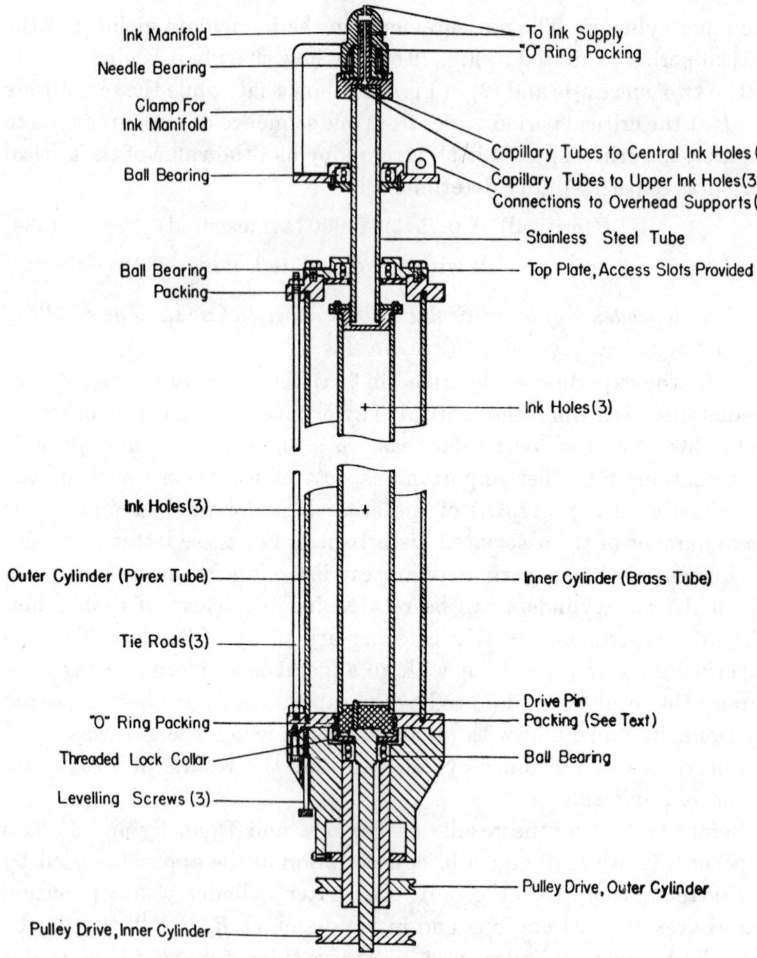

FIG. 80. Sketch of the apparatus used in the experiments of Donnelly and Fultz. The ratio of the radii of the two cylinders is one-half.

strength that drops of it neither sank nor rose in the liquid under the experimental conditions.

The liquids used in the experiments were distilled water and glycerine water mixtures of different strengths. Kinematic viscosities in the range 0·01–0·20 cm²/sec were thus available for the experiments.

In the experiments, the angular velocity of the outer cylinder was treated as an independent variable. The outer cylinder was first brought to a certain speed and was, thereafter, kept at that speed. After the conditions had become stationary, the inner cylinder was set in rotation and its speed was gradually increased until it was within a few per cent of the expected critical value (as determined by preliminary observations). A small amount of ink was then injected and allowed to spread over a small region of the inner cylinder. The speed of the inner cylinder was then increased, slowly and carefully, until a definite sustained three-dimensional motion was observed which indicated the onset of instability.

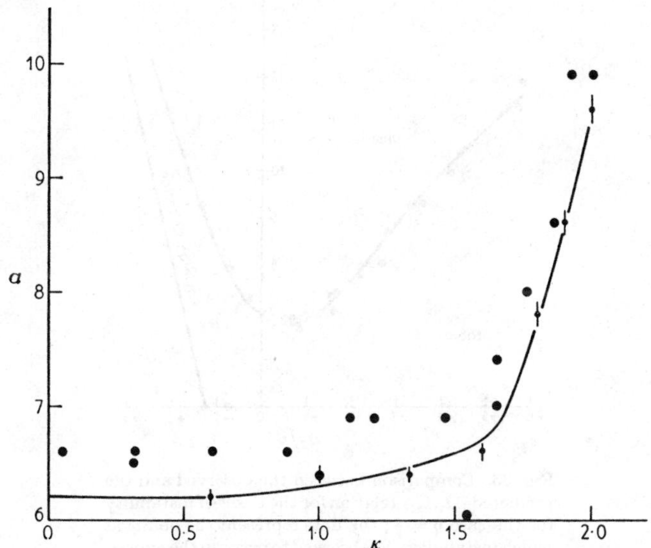

Fig. 81. Comparison of the calculated (full-line curve and the dots with vertical lines) and observed (dots) wave numbers at the onset of instability.

(i) *Observations on the wave numbers of the disturbance manifested at marginal stability*

In some ways, the most striking phenomenon observed in these experiments is the sudden appearance of regularly spaced toroidal vortices of rectangular cross-section. An example of such manifestation, as first photographed by Taylor, is shown in Fig. 68.

The height of the vortices, which appear at marginal stability, is related to the wave number a introduced in the theory by

$$\text{height of vortex} = R_2 \pi / a. \tag{335}$$

From the measured heights, the wave number a can be deduced. In Fig. 81 a comparison is made between the experimentally deduced and the theoretically predicted values of a.

Photographs of the vortices obtained by Donnelly and Fultz under various conditions are shown in Figs. 82–84.

The onset of instability at $\mu = 0$ is illustrated by the sequence of photographs a, b, and c in Fig. 82: (a) shows the liquid in laminar flow at a speed slightly below critical; (b) shows the beginnings of three-dimensional motion when the ink gathering together starts to move

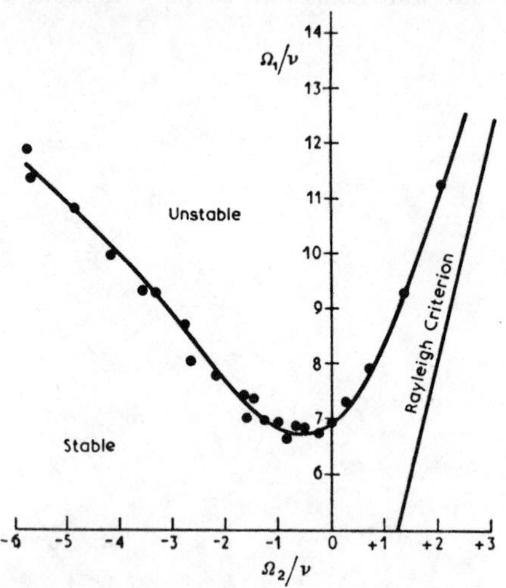

FIG. 85. Comparison between the observed and the predicted (Ω_2, Ω_1)-relation for the onset of instability for the case $\eta = \frac{1}{2}$; the dots represent the points at which instability was observed to occur in the experiments of Donnelly and Fultz and the full-line curve represents the theoretical relation.

radially outwards to form the first cell; and (c) shows the vortex fully developed from the initial state shown in (b) with no additional change in the speed.

When both cylinders are rotated in the same direction, the appearance of the cells is much the same as in the case $\mu = 0$. An example is shown in Fig. 82 d.

When the cylinders are rotated in opposite directions, the phenomena one observes are much more complex. Theoretically, we expect two cells in each compartment, with the motion in the outer cell relatively much weaker (see Fig. 71 a, b). Since the ink is ejected from holes in the inner cylinder, we should not, under the circumstances, expect to witness motions in the outer cell. Also, for reasons which we have explained in

FIG. 82. The onset of instability with the outer cylinder at rest, $\mu = 0$, $P_c = 4{\cdot}491$ sec: (a) laminar flow, $P = 4{\cdot}500$ sec; (b) beginning of radial motion at $P = 4{\cdot}483$ sec; (c) Appearance of cells with the outer cylinder at rest; cells at marginal stability, $P = P_c = 4{\cdot}466$ sec; (d) Appearance of cells with cylinders rotating in the same direction: $P = P_c = 3{\cdot}844$ sec, $\mu = 0{\cdot}1164$.

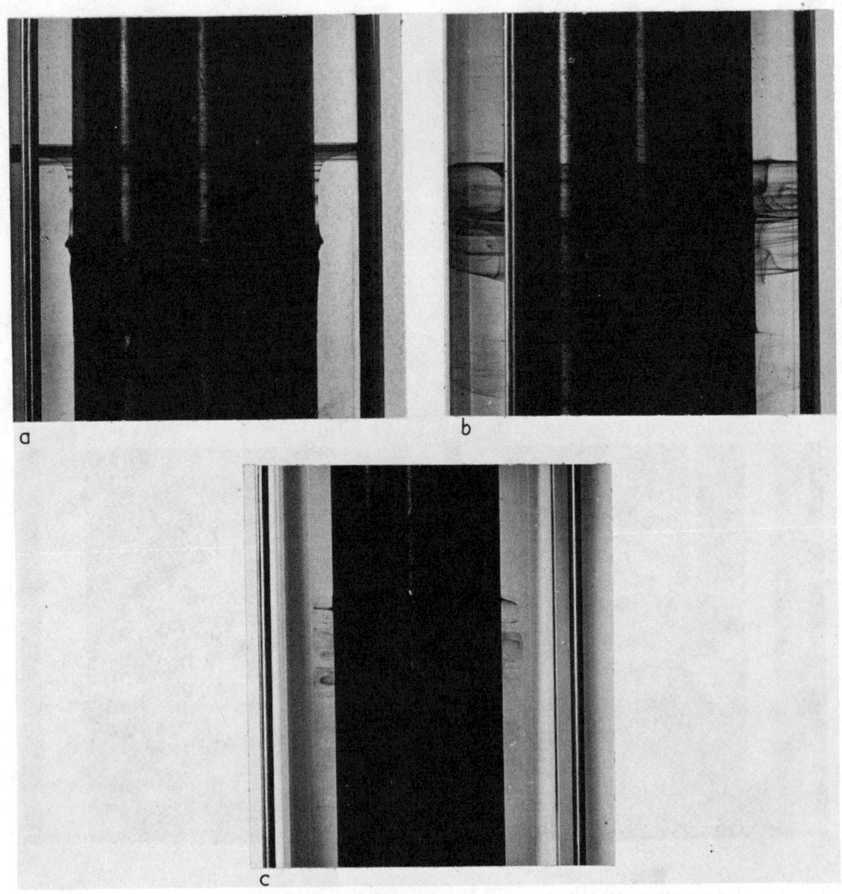

FIG. 83. Photographs of cells with cylinders ($\eta = \frac{1}{2}$) rotating in opposite directions: (*a*) $\mu = -0{\cdot}4116$; (*b*) $\mu = -0{\cdot}5$; and (*c*) $\mu = -1{\cdot}74$.

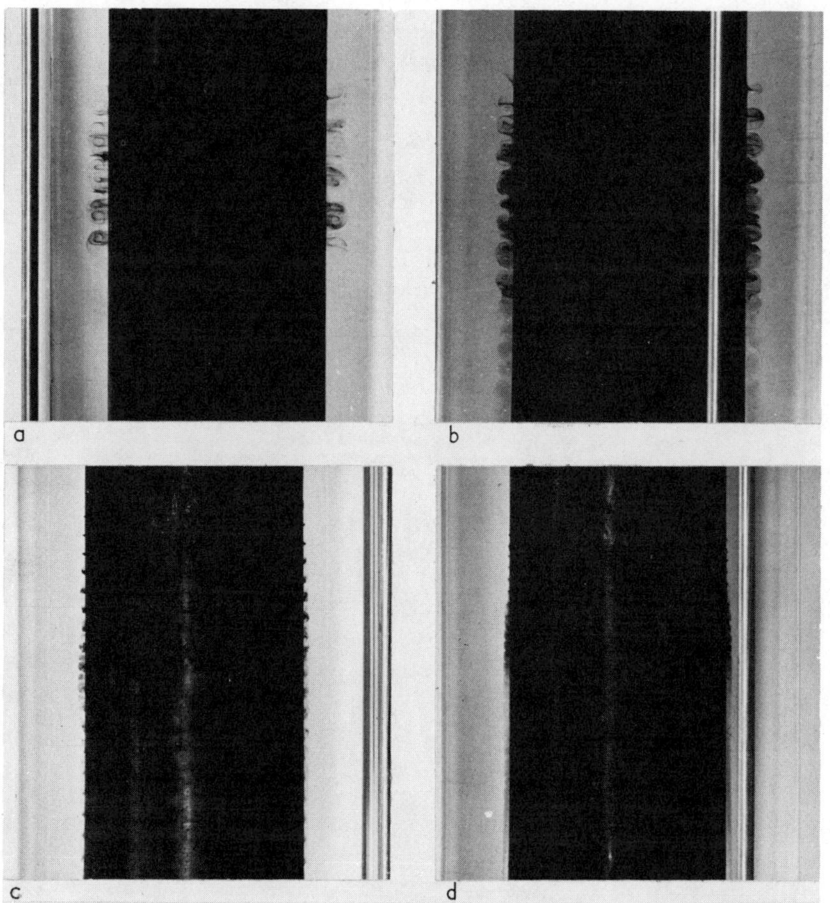

Fig. 84. Photographs of cells with cylinders ($\eta = \frac{1}{2}$) rotating in opposite directions: (a) $\mu = -2{\cdot}44$; (b) $\mu = -3{\cdot}02$; (c) $\mu = -5{\cdot}86$; and (d) $\mu = -6{\cdot}83$.

§§ 66 and 71 (*e*), we must expect that, as $\mu \to -\infty$, the cells at marginal stability appear very close to the inner cylinder and become increasingly tiny. All these expectations are strikingly confirmed by the sequence of photographs in Figs. 83 and 84.

(ii) *Comparison between the measured and the predicted* (T_c, μ)*-relations*

The results of Donnelly and Fultz, obtained with their apparatus on the critical angular velocities at which instability occurs, are exhibited

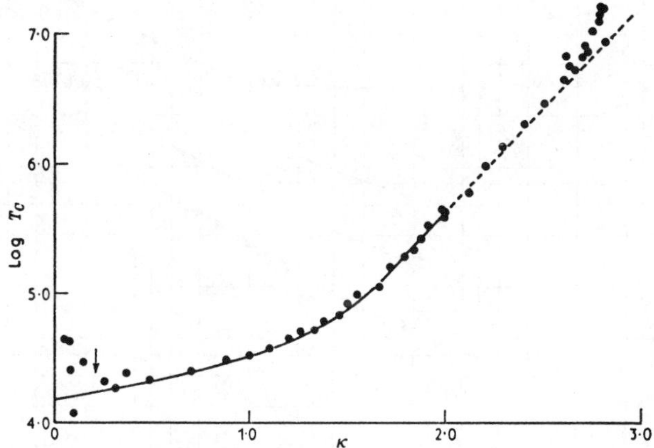

FIG. 86. The variation of the critical Taylor number (T_c) for the case $\eta = \frac{1}{2}$ as a function of κ ($= (1-4\mu)/(1-\mu)$). The solid line represents the theoretically calculated relation and the dashed line is a linear extrapolation of it. The dots represent the points at which instability was observed to occur. Data to the left of the arrow are unduly influenced by small errors in μ.

in Figs. 85 and 86 as plots analogous to those of Figs. 72 and 74. The agreement of the experimental results with the calculations of § 73 (*b*) is very satisfactory.

As we have already remarked, one of Taylor's original sets of measurements was obtained by an apparatus in which the gap width was a small fraction ($\sim$ 5 per cent) of the radius of the outer cylinder. These measurements can, therefore, be compared with the theoretical results of § 71 (*b*). This comparison is included in Fig. 87 (curve labelled $\eta = 0{\cdot}9418$). The agreement of the theoretical curve with the results of the measurements is good over the entire range.

Fig. 87 includes measurements by Taylor for other values of η; and also of Donnelly and Fultz for values of $-\mu$ beyond the range of the calculations of § 73 (*b*). From an examination of the data presented in

Fig. 87, Donnelly and Fultz have inferred that an asymptotic relation of the form,

$$(\Omega_1 R_1^2/\nu)^5 \to \text{constant}(|\Omega_2| R_2^2/\nu)^3 \quad \text{for } \Omega_2/\Omega_1 \to -\infty, \tag{336}$$

exists.

Donnelly and Fultz have pointed out that the asymptotic relation (336) has a simple physical interpretation. Arguing as in § 71 (*e*), they

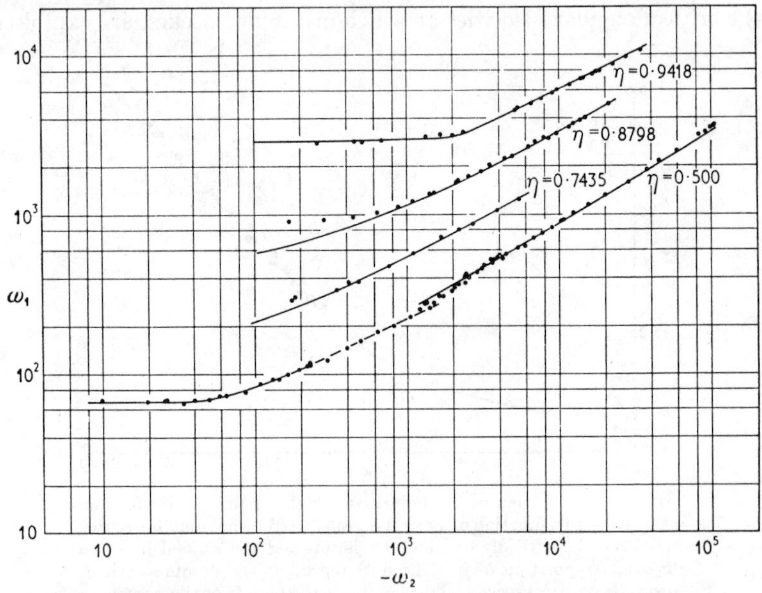

FIG. 87. Comparisons between the observed and the predicted (ω_2, ω_1)-relations for the onset of stability; $\omega_2 = \Omega_2 R_2^2/\nu^2$ and $\omega_1 = \Omega_1 R_1^2/\nu$. The uppermost solid curve and the one at the extreme left represent the theoretical relations derived in § 71 (*b*) and § 73 (*b*), respectively; the dashed-line continuation of the latter represents an extrapolation of the theoretical relation. The remaining solid lines are derived from the asymptotic relation (342). The experimental data for $\eta = 0{\cdot}9418$, $0{\cdot}8798$, and $0{\cdot}7435$ are those of G. I. Taylor; the data for $\eta = 0{\cdot}5$ are those of Donnelly and Fultz.

first observe that the onset of instability, in the case when $\mu < 0$, is primarily to be associated with the violation of Rayleigh's criterion in the layers immediately adjoining the inner cylinder and extending not much beyond the nodal surface (at $r = R_0$, say). Since the nodal surface approaches the inner cylinder as $\mu \to -\infty$, the fluid which becomes unstable is essentially confined to a narrow gap between R_1 and R_0. Accordingly, we may write (cf. equation (325))

$$\frac{4\Omega_1^2}{\nu^2} \frac{R_1^2 d_0^4}{R_0^2 - R_1^2} = T_0 \tag{337}$$

as the condition for the onset of instability, where $d_0 = R_0 - R_1$ and T_0 is a number of the order of 1000. Since the present arguments presuppose that $d_0 \ll R_1$, an equivalent form of (337) is

$$\left(\frac{\Omega_1 R_1^2}{\nu}\right)^2 = \tfrac{1}{2}T_0\left(\frac{R_1}{d_0}\right)^3. \tag{338}$$

But, according to equation (29),

$$d_0 = R_2\,\eta\left(\frac{1+|\mu|}{\eta^2+|\mu|}\right)^{\frac{1}{2}} - R_1 = R_1\left(\sqrt{\frac{1+|\mu|}{\eta^2+|\mu|}}-1\right). \tag{339}$$

For $\mu \to -\infty$,
$$\frac{d_0}{R_1} \to \frac{1-\eta^2}{2|\mu|} = \frac{1}{2}\left|\frac{\Omega_1}{\Omega_2}\right|(1-\eta^2). \tag{340}$$

Inserting this value of d_0/R_1 in equation (338), we obtain

$$\left(\frac{\Omega_1 R_1^2}{\nu}\right)^2 = \frac{4T_0}{(1-\eta^2)^3}\left|\frac{\Omega_2}{\Omega_1}\right|^3 \quad (\mu \to -\infty), \tag{341}$$

or, alternatively,

$$\left(\frac{\Omega_1 R_1^2}{\nu}\right)^5 = \frac{4\eta^6 T_0}{(1-\eta^2)^3}\left(\frac{|\Omega_2| R_2^2}{\nu}\right)^3 \quad (\mu \to -\infty), \tag{342}$$

in agreement with (336).

The values of T_0, empirically determined by Donnelly and Fultz for various values of η, are

$$T_0 = \begin{cases} 450 \\ 530 \\ 670 \\ 1{,}180 \end{cases} \text{for } \eta = \begin{cases} 0{\cdot}9418 \\ 0{\cdot}8798 \\ 0{\cdot}7435 \\ 0{\cdot}5. \end{cases} \tag{343}$$

BIBLIOGRAPHICAL NOTES

The fundamental papers on the stability of Couette flow are those of Lord Rayleigh and Sir Geoffrey Taylor:

1. Lord Rayleigh, 'On the dynamics of revolving fluids,' *Scientific Papers*, **6**, 447–53, Cambridge, England, 1920.
2. G. I. Taylor, 'Stability of a viscous liquid contained between two rotating cylinders', *Phil. Trans. Roy. Soc. (London)* A, **223**, 289–343 (1923).

In reference 1, Rayleigh treated the case of inviscid flow and derived the general criterion known after him. In reference 2, Taylor treated for the first time the corresponding problem in viscid flow, both theoretically and experimentally.

The following general reference, in which the problem of hydrodynamic stability is broadly surveyed with special references to Couette flow, may be noted:

3. J. L. Synge, 'Hydrodynamic stability', 227–69, *Semicentennial Addresses of the American Mathematical Society*, ii, New York, 1938.

We may also refer to the relevant sections in:

4. C. C. LIN, *The Theory of Hydrodynamic Stability*, Cambridge, England, 1955.

§ 66. As we have already remarked, Rayleigh developed his criterion for the stability of inviscid rotational flow in reference 1. In his discussion, Rayleigh argued largely in terms of the analogy between the present problem and the problem of the stability of an incompressible fluid of variable density. For his discussion of the latter problem, see:

5. LORD RAYLEIGH, 'Investigation of the character of the equilibrium of an incompressible heavy fluid of variable density', *Scientific Papers*, **2,** 200–7, Cambridge, England, 1900.

§ 67. The analytical derivation of Rayleigh's criterion (for the case $m = 0$) is due to:

6. J. L. SYNGE, 'The stability of heterogeneous liquids', *Trans. of the Royal Society of Canada,* **27,** 1–18 (1933).

The general case ($m \neq 0$) is considered in:

7. S. CHANDRASEKHAR, 'The stability of inviscid flow between rotating cylinders', *J. Indian Math. Soc.* (in press).

The discussion in this section is largely derived from reference 7.

A useful reference for the Sturmian theory and its later development is:

8. E. L. INCE, *Ordinary Differential Equations*, chapter x, Longmans, Green & Co. Ltd., London, 1927.

§ 68. The oscillations of a rotating column of liquid were first considered by Lord Kelvin:

9. LORD KELVIN, 'Vibrations of a columnar vortex', 152–65, *Mathematical and Physical Papers*, iv, *Hydrodynamics and General Dynamics*, Cambridge, England, 1910.

An extensive treatment of this subject with meteorological overtones will be found in:

10. V. BJERKNES, J. BJERKNES, H. SOLBERG, and T. BERGERON, *Physikalische Hydrodynamik,* chapter 11, Springer, Berlin, 1933.

§ 68 (*a*). Fultz first showed how the natural modes of oscillation of a rotating cylinder of liquid can be excited and maintained.

11. D. FULTZ, 'A note on overstability and the elastoid-inertia oscillations of Kelvin, Solberg, and Bjerknes', *J. Meteorology,* **16,** 199–208 (1959).

The solution of equation (96) for different values of m was carried out by Miss Donna Elbert; the results of her calculations are included in Fig. 63.

The roots listed in Table XXXI (with the exception of those for $\eta = 0$) are taken from:

12. S. CHANDRASEKHAR and DONNA ELBERT, 'The roots of

$$Y_n(\lambda\eta)J_n(\lambda)-J_n(\lambda\eta)Y_n(\lambda) = 0\text{'},$$

Proc. Camb. Phil. Soc. **50,** 266–8 (1954).

§ 68 (*b*). The solution for the narrow gap described in this section is due to Reid:

13. W. H. Reid, 'Inviscid modes of instability in Couette flow', *J. Math. Analysis and Applications*, **1**, 411–22 (1960).

I am grateful to Dr. Reid for providing me with copies for the illustrations (Figs. 65–67) included in this section.

A convenient reference for Airy's functions, in terms of which the solution for this case is found, is:

14. J. C. P. Miller, 'The Airy integral, giving tables of solutions of the differential equation $y'' = xy$', *British Assoc. Math. Tables*, Part-volume B, Cambridge, England, 1946.

§ 69. The illustration (Fig. 68) in this section is taken from Taylor's original paper (reference 2).

§ 70. The theorem proved in § (*a*) is due to Synge.

15. J. L. Synge, 'On the stability of a viscous liquid between rotating coaxial cylinders', *Proc. Roy. Soc.* (*London*) A, **167**, 250–6 (1938).

§ 71 (*a*). The analysis in this section is based on:

16. S. Chandrasekhar, 'The stability of viscous flow between rotating cylinders', *Mathematika*, **1**, 5–13 (1954).

Alternative treatments of the characteristic value problem presented by equations (200)–(202) have been given by:

17. D. Meksyn, 'Stability of viscous flow between rotating cylinders. I', *Proc. Roy. Soc.* (*London*) A, **187**, 115–28 (1946).

18. —— 'Stability of viscous flow between rotating cylinders. II. Cylinders rotating in opposite directions', ibid. 480–91 (1946).

19. —— 'Stability of viscous flow between rotating cylinders. III. Integration of a sixth order linear equation', ibid. 492–504 (1946.)

20. R. C. Di Prima, 'Application of the Galerkin method to problems in hydrodynamic stability', *Quart. Appl. Math.* **13**, 55–62 (1955).

The methods used by these authors (as well as by Taylor in reference 2) do not enable as systematic (or as accurate) a solution of the problem as the method described in the text.

An extension of the analysis in reference 16 to include a term in ζ^2 on the right-hand side of equation (201) has been given by:

21. H. Steinman, 'The stability of viscous flow between rotating cylinders,' *Quart. Appl. Math.* **14**, 27–33 (1956).

§ 71 (*b*). The numerical results given in reference 16 have been supplemented. These additional results (even as the original ones in reference 16) are due to Miss Donna Elbert.

§ 71 (*c*). The alternative method described in this section has not been published before.

§ 71 (*d*). The relation disclosed in this section between the present problem and the simple Bénard problem was first noticed by Low:

22. A. R. Low, 'Instability of viscous fluid motion', *Nature*, **115**, 299–300 (1925).

The analytical basis for this identity in the characteristic values of the two problems (in a certain approximation) was provided by:

23. H. Jeffreys, 'Some cases of instability in fluid motion', *Proc. Roy. Soc.* (*London*) A, **118**, 195–208 (1928).

The perturbation procedure described in this section is, however, new. I am indebted to Dr. P. Vandervoort for the evaluation of the matrix elements in the solution (262).

§ 71 (*e*). The essential arguments of this section can be found (in varying degrees of explicitness) in the writings of Meksyn (reference 18), Lin (reference 4), Di Prima (reference 20), and Donnelly and Fultz (reference 28 below).

§ 72. The discussion of overstability in this section has not been published before.

§ 73. The analysis in this section is based on:

24. S. Chandrasekhar, 'The stability of viscous flow between rotating cylinders', *Proc. Roy. Soc.* (*London*) A, **246**, 301–11 (1958).

The method of solution derives from:

25. S. Chandrasekhar and W. H. Reid, 'On the expansion of functions which satisfy four boundary conditions', *Proc. Nat. Acad. Sci.* **43**, 521–7 (1957).

The orthogonal functions, $\mathscr{C}_1(\alpha_j r)$, in terms of which the solution is found, are tabulated in:

26. S. Chandrasekhar and Donna D. Elbert, 'On orthogonal functions which satisfy four boundary conditions. III. Tables for use in Fourier–Bessel-type expansions', *Astrophys. J. Supp. Ser.* **3,** 453–8 (1958).

§ 74. The references to the experimental work described in this section are:

27. R. J. Donnelly, 'Experiments on the stability of viscous flow between rotating cylinders. I. Torque measurements', *Proc. Roy. Soc.* (*London*) A, **246,** 312–25 (1958).

28. —— and D. Fultz, 'Experiments on the stability of viscous flow between rotating cylinders. II. Visual observations', ibid. **258,** 101–23 (1960).

The results of similar and related experiments will be found in:

29. J. W. Lewis, 'An experimental study of the motion of a viscous liquid contained between two coaxial cylinders', *Proc. Roy. Soc.* (*London*) A, **117,** 388–407 (1928).

30. T. Terada and K. Hattori, 'Some experiments on motions of fluids, IV', *Rep. Aero. Res. Inst. Tokyo*, **2,** 287–326 (1926).

31. F. Wendt, 'Turbulente Strömungen zwischen zwei rotierenden konaxialen Zylindern', *Ingen. Arch.* **4,** 577–95 (1933).

32. G. I. Taylor, 'Fluid friction between rotating cylinders. I. Torque measurements', *Proc. Roy. Soc.* (*London*) A, **157,** 546–64 (1936).

33. —— 'Fluid friction between rotating cylinders. II. Distribution of velocity between concentric cylinders when outer one is rotating and inner one is at rest', ibid. 565–78 (1936).

34. F. Schultz-Grunow and H. Hein, 'Beitrag zur Couetteströmung,' *Z. f. Flugwiss.* **4,** 28–30 (1956).

VIII

THE STABILITY OF MORE GENERAL FLOWS BETWEEN COAXIAL CYLINDERS

75. Introduction

IN the last chapter we considered the stability of the simple Couette flow between rotating cylinders. In this chapter we shall consider some more general flows between coaxial cylinders when, in addition to rotation, a constant transverse or axial pressure gradient is present. In the former case, the streamlines continue to be circular and satisfy the requirement of a Couette flow; but the flow differs from the one considered in Chapter VII in that a Poiseuille flow in the transverse direction is superposed over the distribution of rotational velocities,

$$\Omega = A+B/r^2, \tag{1}$$

given in equation VII (144). On the other hand, when a constant pressure gradient $\partial p/\partial z$ along the common axis of the cylinders is present, a Poiseuille flow in the axial direction is superposed over the same distribution of rotational velocities. The streamlines are then no longer circular and the flow cannot be classed under Couette flow. Indeed, as we shall see, the superposition of an axial flow over a rotational flow introduces certain essentially new elements into the problem.

76. The stability of viscous flow in a curved channel

Consider equations VII (134)–(136) in the case when the conditions are stationary, a constant pressure gradient $(\partial p/\partial\theta)_0$ is acting in the transverse direction, the cylinders are at rest, and only transverse motions are present. Under these circumstances, the equations of motion allow the stationary solution (cf. equations VII (141) and (142))

$$u_r = u_z = 0 \quad \text{and} \quad u_\theta = V(r), \tag{2}$$

provided

$$\frac{1}{\rho}\frac{\partial p}{\partial r} = \frac{V^2}{r} \tag{3}$$

and

$$\nu\left(\nabla^2 V-\frac{V}{r^2}\right) = \nu\frac{d}{dr}\left(\frac{1}{r}\frac{d}{dr}rV\right) = \frac{1}{\rho r}\left(\frac{\partial p}{\partial\theta}\right)_0. \tag{4}$$

The general solution of equation (4) is

$$V = \frac{1}{2\rho\nu}\left(\frac{\partial p}{\partial\theta}\right)_0 r\log r+Cr+\frac{D}{r}, \tag{5}$$

where C and D are two constants of integration. If, as we have assumed, the flow takes place between two concentric cylinders (of radii R_1 and R_2 ($> R_1$)) at rest, we must require that V vanishes at $r = R_1$ and R_2. These latter conditions determine C and D; and we find

$$C = -\frac{1}{2\rho\nu}\left(\frac{\partial p}{\partial\theta}\right)_0 \frac{R_2^2\log R_2 - R_1^2\log R_1}{R_2^2 - R_1^2}$$

and
$$D = \frac{1}{2\rho\nu}\left(\frac{\partial p}{\partial\theta}\right)_0 \frac{R_1^2\,R_2^2}{R_2^2 - R_1^2}\log\frac{R_2}{R_1}. \tag{6}$$

When the difference d in the radii of the two cylinders is small compared with their mean radius $\frac{1}{2}(R_2+R_1)$, the velocity distribution given by equations (5) and (6) reduces to the familiar form of the Poiseuille flow between parallel planes. Thus, letting

$$\zeta = (r - R_1)/(R_2 - R_1), \tag{7}$$

we find that to order $(d/R_1)^2$ the velocity distribution approximates to

$$V(\zeta) = 6V_m\,\zeta(1-\zeta), \tag{8}$$

where
$$V_m = -\frac{d^2}{12\rho\nu R_1}\left(\frac{\partial p}{\partial\theta}\right)_0 \tag{9}$$

is the mean velocity across the channel.

The stability of the flow represented by the foregoing solution of the equations of motion was first considered by Dean. More recently, it has been reconsidered by Reid using methods similar to those described in other connexions in the earlier chapters.

Before we enter into a full discussion of the stability of the flow (8), we may note that Rayleigh's discriminant (defined in equation VII (18)) is positive for $0 \leqslant \zeta \leqslant \frac{1}{2}$ and negative for $\frac{1}{2} < \zeta \leqslant 1$. Accordingly, in the absence of viscosity the flow is unstable; and the instability is principally to be associated with the violation of Rayleigh's criterion beyond $\zeta = \frac{1}{2}$.

(a) *The perturbation equations*

Equations VII (161) and (162) are applicable to the problem on hand if we ascribe to V in these equations its present meaning.

When the gap $d = R_2 - R_1$ between the cylinders is small compared to their mean radius $\frac{1}{2}(R_2+R_1)$, the appropriate expression for V is that given in equation (8). In the same approximation, the perturbation equations are (cf. equations VII (193) and (194))

$$(D^2 - a^2 - \sigma)(D^2 - a^2)u = \frac{12V_m\,d^2}{R_1\,\nu}\,a^2\zeta(1-\zeta)v \tag{10}$$

and
$$(D^2-a^2-\sigma)v = \frac{6V_m d}{\nu}(1-2\zeta)u, \tag{11}$$

where a and σ have the same meanings as in equation VII (192). By the further transformation,

$$u \to \frac{12V_m d^2}{R_1 \nu} a^2 u, \tag{12}$$

the equations take the forms:

$$(D^2-a^2-\sigma)(D^2-a^2)u = \zeta(1-\zeta)v \tag{13}$$

and
$$(D^2-a^2-\sigma)v = \Lambda a^2(1-2\zeta)u, \tag{14}$$

where
$$\Lambda = \frac{72V_m^2 d^3}{R_1 \nu^2} = 72\,\mathrm{R}^2\frac{d}{R_1} \tag{15}$$

and
$$\mathrm{R} = V_m d/\nu \tag{16}$$

is the Reynolds number of the mean flow.

The boundary conditions with respect to which equations (13) and (14) must be solved are (cf. equation VII (200))

$$u = Du = v = 0 \quad \text{for } \zeta = 0 \text{ and } 1. \tag{17}$$

In view of the physical similarity of this problem with the problem considered in Chapter VII, § 71, it is likely that in this case also the onset of instability will be as a stationary secondary flow. No theoretical justification has been given for this supposition; and in spite of experimental evidence favouring it (cf. § 77 (c)), the matter deserves examination (cf. remarks on p. 317).

When the marginal state is stationary, the equations to be solved are

$$(D^2-a^2)^2 u = \zeta(1-\zeta)v \tag{18}$$

and
$$(D^2-a^2)v = \Lambda a^2(1-2\zeta)u, \tag{19}$$

together with the boundary conditions (17).

(*b*) *The solution of the characteristic value problem for the case* $\sigma = 0$

The characteristic value problem presented by equations (17)–(19) can be solved by the method described in § 71 (*a*).

Since v is required to vanish at $\zeta = 0$ and 1, we expand it in a sine series of the form

$$v = \sum_{m=1}^{\infty} C_m \sin m\pi\zeta. \tag{20}$$

With v chosen in this manner, we next solve the equation,

$$(D^2-a^2)^2 u = \zeta(1-\zeta)\sum_{m=1}^{\infty} C_m \sin m\pi\zeta, \tag{21}$$

obtained by inserting (20) in (18), and arrange that the solution satisfies the four remaining boundary conditions on u. With u determined in this fashion and v given by (20), equation (19) will lead to the required secular equation for Λ.

The solution of equation (21) is straightforward and can be written in the form (cf. equation VII (205))

$$u = \sum_{m=1}^{\infty} \frac{C_m}{(m^2\pi^2+a^2)^2}\Big\{A_1^{(m)}\cosh a\zeta+B_1^{(m)}\sinh a\zeta+A_2^{(m)}\zeta\cosh a\zeta+ +B_2^{(m)}\zeta\sinh a\zeta+\Big[\zeta(1-\zeta)+\frac{4(5m^2\pi^2-a^2)}{(m^2\pi^2+a^2)^2}\Big]\sin m\pi\zeta+ +\frac{4m\pi}{(m^2\pi^2+a^2)}(1-2\zeta)\cos m\pi\zeta\Big\}, \tag{22}$$

where the constants of integration $A_1^{(m)}$, $A_2^{(m)}$, $B_1^{(m)}$, and $B_2^{(m)}$ are to be determined by the boundary conditions $u = Du = 0$ at $\zeta = 0$ and 1. These latter conditions lead to the equations:

$$A_1^{(m)} = -\frac{4m\pi}{m^2\pi^2+a^2},$$

$$aB_1^{(m)}+A_2^{(m)} = -m\pi\Big[\frac{4(5m^2\pi^2-a^2)}{(m^2\pi^2+a^2)^2}-\frac{8}{m^2\pi^2+a^2}\Big] = -m\pi\frac{12(m^2\pi^2-a^2)}{(m^2\pi^2+a^2)^2},$$

$$A_1^{(m)}\cosh a+B_1^{(m)}\sinh a+A_2^{(m)}\cosh a+B_2^{(m)}\sinh a = (-1)^m\frac{4m\pi}{m^2\pi^2+a^2},$$

$$A_1^{(m)}a\sinh a+B_1^{(m)}a\cosh a+A_2^{(m)}(\cosh a+a\sinh a)+ +B_2^{(m)}(\sinh a+a\cosh a) = (-1)^{m+1}m\pi\frac{12(m^2\pi^2-a^2)}{(m^2\pi^2+a^2)^2}. \tag{23}$$

On solving these equations we find (cf. equations VII (207) and (208)):

$$A_1^{(m)} = -\frac{4m\pi}{m^2\pi^2+a^2},$$

$$B_1^{(m)} = +\frac{m\pi}{\Delta}\{\alpha_m a+\beta_m(\sinh a+a\cosh a)-\gamma_m\sinh a\},$$

$$A_2^{(m)} = -\frac{m\pi}{\Delta}\{\alpha_m\sinh^2 a+\beta_m a(\sinh a+a\cosh a)-\gamma_m a\sinh a\},$$

$$B_2^{(m)} = +\frac{m\pi}{\Delta}\{\alpha_m(\sinh a\cosh a-a)+\beta_m a^2\sinh a-\gamma_m(a\cosh a-\sinh a)\}, \tag{24}$$

where

$$\Delta = \sinh^2 a-a^2,$$

$$\alpha_m = \frac{12(m^2\pi^2-a^2)}{(m^2\pi^2+a^2)^2}, \qquad \beta_m = \frac{4}{m^2\pi^2+a^2}[(-1)^m+\cosh a],$$

and

$$\gamma_m = (-1)^{m+1}\alpha_m+\frac{4}{m^2\pi^2+a^2}a\sinh a. \tag{25}$$

Now substituting for v and u from equations (20) and (22) in equation (19), we obtain

$$\sum_{n=1}^{\infty} C_n(n^2\pi^2+a^2)\sin n\pi\zeta$$
$$= \Lambda a^2(2\zeta-1)\sum_{m=1}^{\infty}\frac{C_m}{(m^2\pi^2+a^2)^2}\Big\{A_1^{(m)}\cosh a\zeta+B_1^{(m)}\sinh a\zeta+$$
$$+A_2^{(m)}\zeta\cosh a\zeta+B_2^{(m)}\zeta\sinh a\zeta+\frac{4m\pi}{(m^2\pi^2+a^2)}(1-2\zeta)\cos m\pi\zeta+$$
$$+\left[\zeta(1-\zeta)+\frac{4(5m^2\pi^2-a^2)}{(m^2\pi^2+a^2)^2}\right]\sin m\pi\zeta\Big\}. \quad (26)$$

Multiplying equation (26) by $\sin n\pi\zeta$ and integrating over the range of ζ, we obtain a system of linear homogeneous equations for the constants $\mathscr{C}_m = C_m/(m^2\pi^2+a^2)^2$; and the requirement that these constants do not all vanish leads to the secular equation

$$\Big\|\langle(1-2\zeta)\cosh a\zeta\,|n\rangle A_1^{(m)}+\langle(1-2\zeta)\sinh a\zeta\,|n\rangle B_1^{(m)}+$$
$$+\langle(\zeta-2\zeta^2)\cosh a\zeta\,|n\rangle A_2^{(m)}+\langle(\zeta-2\zeta^2)\sinh a\zeta\,|n\rangle B_2^{(m)}+X_{mn}+$$
$$+\frac{4(5m^2\pi^2-a^2)}{(m^2\pi^2+a^2)^2}Z_{mn}+\frac{4m\pi}{m^2\pi^2+a^2}Y_{mn}+\tfrac{1}{2}(n^2\pi^2+a^2)^3\frac{\delta_{mn}}{a^2\Lambda}\Big\| = 0, \quad (27)$$

where

$$X_{mn} = \int_0^1 (1-2\zeta)\zeta(1-\zeta)\sin m\pi\zeta\sin n\pi\zeta\,d\zeta$$
$$= \frac{12}{\pi^4}\frac{(m+n)^4-(m-n)^4}{(m^2-n^2)^4}-\frac{4mn}{\pi^2(m^2-n^2)^2}$$
$$\text{if } n+m \text{ is odd and zero otherwise,} \quad (28)$$

$$Y_{mn} = \int_0^1 (1-2\zeta)^2\cos m\pi\zeta\sin n\pi\zeta\,d\zeta$$
$$= \frac{2n}{\pi(n^2-m^2)}-\frac{8}{\pi^3}\frac{(n+m)^3+(n-m)^3}{(n^2-m^2)^3}$$
$$\text{if } n+m \text{ is odd and zero otherwise,} \quad (29)$$

and $$Z_{mn} = \frac{8nm}{\pi^2(n^2-m^2)^2} \quad \text{if } n+m \text{ is odd and zero otherwise.} \quad (30)$$

The various matrix elements which occur in (27) are linear combinations of the elements:

$$\langle\cosh a\zeta\,|n\rangle = \frac{n\pi}{n^2\pi^2+a^2}[1+(-1)^{n+1}\cosh a],$$

$$\langle\sinh a\zeta\,|n\rangle = \frac{n\pi}{n^2\pi^2+a^2}(-1)^{n+1}\sinh a,$$

$$\langle \zeta \cosh a\zeta | n \rangle = (-1)^{n+1} \frac{n\pi}{n^2\pi^2+a^2}\left[\cosh a - \frac{2a}{n^2\pi^2+a^2}\sinh a\right],$$

$$\langle \zeta \sinh a\zeta | n \rangle = \langle \sinh a\zeta | n \rangle - \frac{2a}{n^2\pi^2+a^2}\langle \cosh a\zeta | n \rangle,$$

$$\langle \zeta^2 \cosh a\zeta | n \rangle = (-1)^{n+1} \frac{n\pi}{n^2\pi^2+a^2}\left[\cosh a - \frac{4a}{n^2\pi^2+a^2}\sinh a\right] - \\ - \frac{2(n^2\pi^2-3a^2)}{(n^2\pi^2+a^2)^2}\langle \cosh a\zeta | n \rangle,$$

$$\langle \zeta^2 \sinh a\zeta | n \rangle = (-1)^{n+1} \frac{n\pi}{n^2\pi^2+a^2}\left[\sinh a - \frac{4a}{n^2\pi^2+a^2}\cosh a\right] - \\ - \frac{2(n^2\pi^2-3a^2)}{(n^2\pi^2+a^2)^2}\langle \sinh a\zeta | n \rangle. \quad (31)$$

(c) *Numerical results*

Reid has solved the secular equation (27) in the second and in the fourth approximations for a number of values of a. The results of his

TABLE XXXVIII

Critical values of Λ and related constants

a	Λ *Second approximation*	Λ *Fourth approximation*	$\text{R}(d/R_1)^{\frac{1}{2}} = \sqrt{(\Lambda/72)}$ *Fourth approximation*
3·90	91,691		
3·96	91,650	92,975	35·935
4·00	91,665	93,001	35·940
4·10	91,789		

calculations for values of a in the range in which Λ attains its minimum are summarized in Table XXXVIII. From the results given in this table we infer that the Reynolds number R at which instability will set in and the wave number of the disturbance which will be manifested at onset are given by

$$\text{R}_c = 35{\cdot}94\sqrt{\frac{R_1}{d}} \quad \text{and} \quad a_c = 3{\cdot}96. \quad (32)$$

The corresponding velocity profiles and cell pattern are illustrated in Figs. 88 and 89.

A confirmation of the theoretical values (32) has recently been provided by the experiments of Brewster, Grosberg, and Nissan described in § 77 (c) (see Fig. 96 b, facing p. 359). They find

$$\mathrm{R}\sqrt{(d/R_1)} = 36{\cdot}5\pm1{\cdot}1 \quad \text{and} \quad a = 4{\cdot}9\pm0{\cdot}8 \text{ (measured)}, \tag{33}$$

in satisfactory agreement with the predicted values.

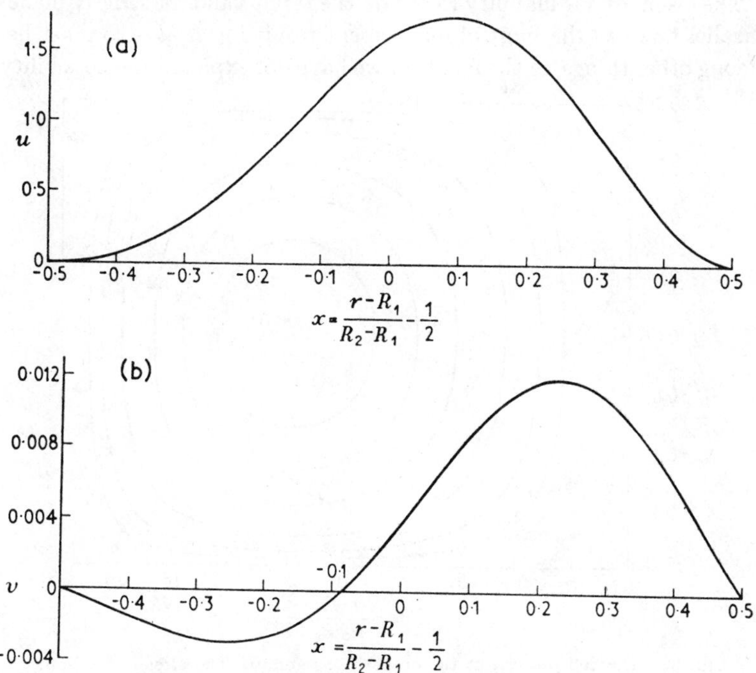

Fig. 88. The perturbation in the radial velocity (*a*) and the tangential velocity (*b*) at the onset of instability.

It is of interest to compare the Reynolds number for the onset of instability given by (32) with the Reynolds number at which instability sets in for strictly plane-parallel flows. It is known that in the latter case, the onset of instability occurs as overstable oscillations when†

$$\mathrm{R} = \tfrac{4}{3}\times5300 \simeq 7070. \tag{34}$$

Clearly, the criterion given by (32) must cease to be valid when

$$\mathrm{R} = 35{\cdot}94\sqrt{\frac{R_1}{d}} > 7070,$$

or

$$R_1 > 3{\cdot}85\times10^4 d. \tag{35}$$

† See, for example, C. C. Lin, *The Theory of Hydrodynamic Stability*, p. 29, Cambridge, England, 1955. The factor $\frac{4}{3}$ allows for the fact that Lin's definition of the Reynolds number is in terms of the velocity ($=\frac{3}{2}V_m$) at the centre of the channel and its half-width.

Conversely, for the instability of plane-parallel flows to be observable as theoretically predicted, the channel in which the experiments are carried out must be so accurately parallel that the radius of curvature at all points exceeds $4\times10^4 d$ by a sufficiently large margin.

The origin of our inability to obtain the result valid for strictly plane-parallel flows as the limit of our present result for $R_1/d\to\infty$ must lie, among other things, in the fact that we have not explored the possibility

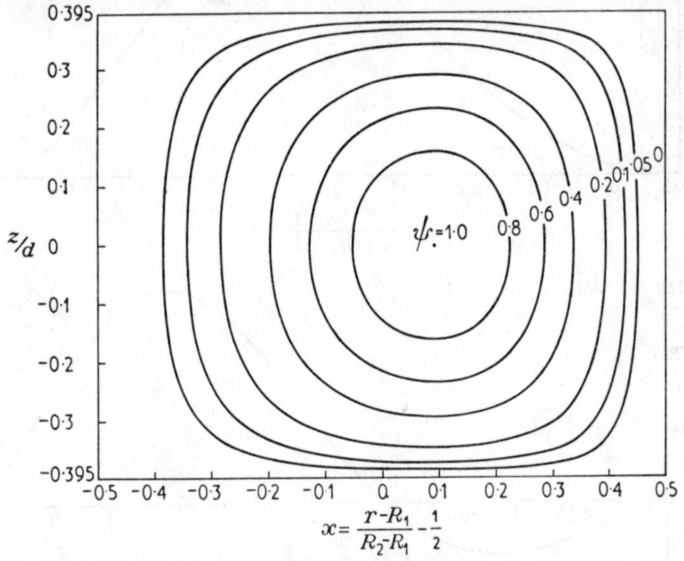

FIG. 89. The cell pattern at the onset of instability. The stream function, ψ ($\propto u_r \cos az$), has been normalized to unity and the cell pattern is drawn symmetrical about $z/d = 0$.

of overstable oscillations occurring. One may surmise that overstability is indeed possible; but that it normally occurs for a value of Λ in excess of that at which a stationary secondary flow manifests itself. However, the discovery of an overstable branch (if one such exists) will not, by itself, resolve the difficulty: the passage to the limit $R_1/d\to\infty$ clearly requires great care.

77. The stability of viscous flow between rotating cylinders when a transverse pressure gradient is present

We pass on now to a consideration of the stability of a general Couette flow which consists of a superposition of a Poiseuille flow in the transverse direction (maintained by a pressure gradient) and a distribution of angular velocities (maintained by the rotation of the two cylinders).

Such general Couette flows can be realized in an arrangement of the kind shown in Fig. 90: the cylinders are allowed to rotate independently of one another while a constant flow through the annulus is maintained by a suitable pumping circuit.

The stationary flow whose stability we wish to consider is represented by a superposition of the solutions given in §§ 69 and 76; thus,

$$V(r) = \left\{\frac{1}{2\rho\nu}\left(\frac{\partial p}{\partial \theta}\right)_0 r\log r+Cr+\frac{D}{r}\right\}+Ar+\frac{B}{r}, \tag{36}$$

where C and D have the values given in equation (6) and A and B have the values given in equation VII (145).

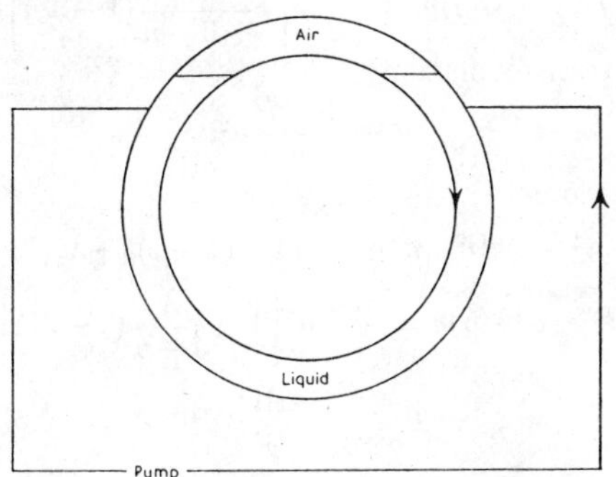

FIG. 90. Arrangement by which general Couette flows can be realized.

In the case when the gap width $d = R_2-R_1$ is small compared to the mean radius $\frac{1}{2}(R_2+R_1)$, we have the approximate relation (cf. equations VII (191) and equation (8))

$$\frac{V(r)}{r} = \Omega_1[1-(1-\mu)\zeta]+\frac{6V_m}{R_1}\zeta(1-\zeta). \tag{37}$$

We shall restrict our discussion to this case of narrow gap widths.

(a) The perturbation equations for the case $(R_2-R_1) \ll \frac{1}{2}(R_2+R_1)$

The relevant perturbation equations can be readily written down by combining the right-hand sides of equations VII (193) and (194) and equations § 76 (10) and (11). We thus obtain

$$(D^2-a^2-\sigma)(D^2-a^2)u = \frac{2d^2a^2}{\nu}\left\{\Omega_1[1-(1-\mu)\zeta]+\frac{6V_m}{R_1}\zeta(1-\zeta)\right\}v \tag{38}$$

and

$$(D^2-a^2-\sigma)v = \frac{2d^2}{\nu}\left\{A+\frac{3V_m}{d}(1-2\zeta)\right\}u. \tag{39}$$

Let

$$\lambda = \frac{6V_m}{R_1\Omega_1}. \tag{40}$$

With the value of A given in equation VII (145),

$$\frac{3V_m}{Ad} = -\frac{3V_m(1-\eta^2)}{\Omega_1\,\eta^2(1-\mu/\eta^2)\,d} = -\frac{3V_m(1+\eta)}{\Omega_1\,R_1\,\eta(1-\mu/\eta^2)}. \tag{41}$$

For the case of narrow gaps under consideration, $\eta \sim 1$ and we can write

$$\frac{3V_m}{Ad} = -\frac{6V_m}{R_1\Omega_1(1-\mu)} = -\frac{\lambda}{1-\mu}. \tag{42}$$

With these definitions we can rewrite equations (38) and (39) in the forms

$$(D^2-a^2-\sigma)(D^2-a^2)u = \frac{2\Omega_1\,d^2}{\nu}a^2[1-(1-\mu)\zeta+\lambda\zeta(1-\zeta)]v \tag{43}$$

and

$$(D^2-a^2-\sigma)v = \frac{2Ad^2}{\nu}\left[1-\frac{\lambda}{1-\mu}(1-2\zeta)\right]u. \tag{44}$$

By the further transformation

$$u \to \frac{2\Omega_1\,d^2}{\nu}a^2u, \tag{45}$$

the equations become

$$(D^2-a^2-\sigma)(D^2-a^2)u = [1-(1-\mu)\zeta+\lambda\zeta(1-\zeta)]v \tag{46}$$

and

$$(D^2-a^2-\sigma)v = -Ta^2\left[1-\frac{\lambda}{1-\mu}(1-2\zeta)\right]u, \tag{47}$$

where

$$T = -\frac{4A\Omega_1}{\nu^2}d^4 \tag{48}$$

is the Taylor number appropriate for narrow gaps (cf. equation VII (198)).

Solutions of equations (46) and (47) must be sought which satisfy the usual boundary conditions

$$u = Du = v = 0 \quad \text{for } \zeta = 0 \text{ and } 1. \tag{49}$$

(*b*) *The solution of the characteristic value problem for the case* $\sigma = 0$ *and* $\mu = 0$

For the problem under consideration, the principle of the exchange of stabilities has not been established. Nevertheless, if on the basis of experimental evidence we assume that the onset of instability is as a stationary secondary flow, the equations to be solved are

$$(D^2-a^2)^2u = [1-(1-\mu)\zeta+\lambda\zeta(1-\zeta)]v \tag{50}$$

and

$$(D^2-a^2)v = -Ta^2\left[1-\frac{\lambda}{1-\mu}(1-2\zeta)\right]u, \tag{51}$$

together with the boundary conditions (49).

The characteristic value problem presented by equations (49)–(51) can be solved by the same method which has proved successful for the separate problems. The details of the analysis need not be repeated since the relevant equations can be readily written down by suitably combining the corresponding equations of § 71 (*a*) and § 76 (*b*).

The solution for the case $\mu = 0$ has been explicitly carried out by

Table XXXIX

The critical Taylor numbers and the wave numbers of the associated disturbance for various values of λ

λ	a_c	T_c	λ	a_c	T_c
30	3·76	$8{\cdot}20\times 10^1$	−2·5†	5·00	$2{\cdot}37\times 10^4$
21	3·70	$1{\cdot}49\times 10^2$	−2·75	5·73	$3{\cdot}18\times 10^4$
15	3·60	$2{\cdot}56\times 10^2$	−3·00	6·35	$4{\cdot}10\times 10^4$
10	3·45	$4{\cdot}56\times 10^2$	−3·25	7·05	$5{\cdot}16\times 10^4$
6	3·30	$8{\cdot}32\times 10^2$	−3·50	7·40	$6{\cdot}38\times 10^4$
3	3·14	$1{\cdot}48\times 10^3$	−3·75	5·80	$6{\cdot}20\times 10^4$
1	3·13	$2{\cdot}43\times 10^3$	−4·00	5·50	$4{\cdot}60\times 10^4$
0·5	3·13	$2{\cdot}84\times 10^3$	−4·50	5·37	$2{\cdot}80\times 10^4$
0	3·12	$3{\cdot}39\times 10^3$	−5·00	5·20	$1{\cdot}83\times 10^4$
−0·5	3·17	$4{\cdot}18\times 10^3$	−6·00	5·00	$9{\cdot}61\times 10^3$
−1·0	3·24	$5{\cdot}42\times 10^3$	−8·00	4·70	$3{\cdot}84\times 10^3$
−1·5	3·40	$7{\cdot}66\times 10^3$	−10·00	4·55	$2{\cdot}02\times 10^3$
−2·0	3·80	$1{\cdot}26\times 10^4$	−15·00	4·35	$6{\cdot}92\times 10^2$

† For $-2{\cdot}5 \geqslant \lambda \geqslant -6{\cdot}0$, it was found necessary to go to the fourth approximation in the evaluation of T_c; for the remaining values of λ the third approximation sufficed.

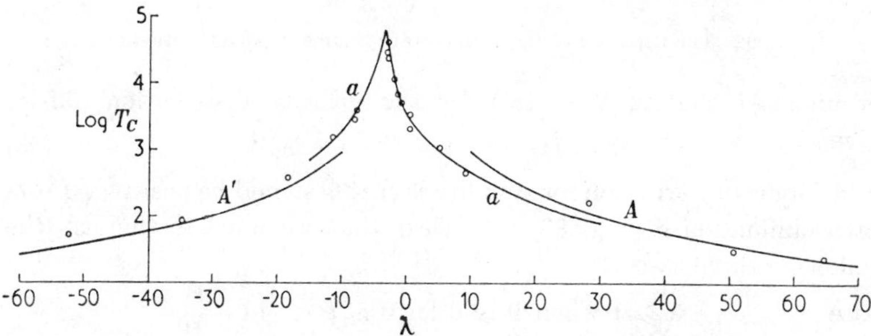

Fig. 91. The variation of the critical Taylor number T_c as a function of λ ($= 6V_m/R_1\Omega_1$). The calculated theoretical relation is the curve labelled *a* while the curves labelled *A* and *A'* are the theoretical asymptotic relations ($T_c \to 9{\cdot}30\lambda^{-2}$, $|\lambda| \to \infty$). The open circles are the experimental points as measured by Brewster, Grosberg, and Nissan.

Di Prima. His results for the critical Taylor numbers and the wave numbers of the disturbance which will be manifested at marginal stability are given in Table XXXIX. They are further illustrated in Figs. 91 and 92.

(c) *The physical interpretation of the results*

The very peculiar dependence of T_c and a_c on λ, exhibited by the calculations, can be understood by examining how Rayleigh's criterion for the stability of inviscid rotational flow is violated in the fluid.

The distribution of the transverse velocity in the fluid (for the case $\mu = 0$) is given by

$$1-\zeta+\lambda\zeta(1-\zeta) = (1-\zeta)(1+\lambda\zeta). \tag{52}$$

This distribution for a few values of λ is illustrated in Fig. 93. Apart from a constant of proportionality which we may ignore, Rayleigh's dis-

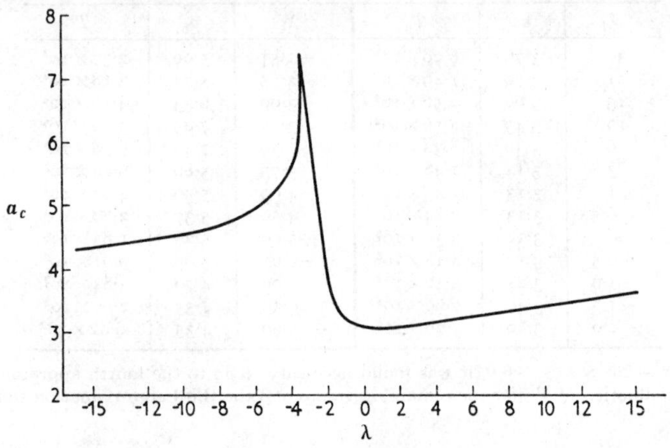

FIG. 92. The variation of the critical wave number a_c as a function of λ.

criminant (equation VII (18)) for the velocity distribution (52) is given by

$$\Phi = (1-\zeta)(1+\lambda\zeta)(\lambda-1-2\lambda\zeta); \tag{53}$$

and Rayleigh's criterion for stability is that Φ should be positive. From an examination of Fig. 93, it is clear that we must distinguish the following three cases:

case i: $\lambda \geqslant 1$ when $\Phi \geqslant 0$ for $0 \leqslant \zeta < \frac{1}{2}\left(1-\frac{1}{\lambda}\right)$,

and $\Phi \leqslant 0$ for $\frac{1}{2}\left(1-\frac{1}{\lambda}\right) \leqslant \zeta \leqslant 1$; (54)

case ii: $+1 > \lambda \geqslant -1$ when $\Phi \leqslant 0$ for $0 \leqslant \zeta \leqslant 1$; (55)

and

case iii: $\lambda < -1$ when $\Phi \geqslant 0$ for $\frac{1}{|\lambda|} \leqslant \zeta \leqslant \frac{1}{2}\left(1+\frac{1}{|\lambda|}\right)$,

and $\Phi \leqslant 0$ for $0 \leqslant \zeta \leqslant \frac{1}{|\lambda|}$

and $\frac{1}{2}\left(1+\frac{1}{|\lambda|}\right) \leqslant \zeta \leqslant 1$. (56)

The parts of the fluid which are stable or unstable according to Rayleigh's criterion are shown in Fig. 94. We observe that as $\lambda \to \pm\infty$, the parts of the fluid which are unstable become increasingly confined to the interval $\frac{1}{2} < \zeta < 1$. (As we shall presently explain, the instability in the zone $0 \leqslant \zeta \leqslant |\lambda|^{-1}$ plays no significant role for $\lambda \to -\infty$.) In the limit $\lambda \to \pm\infty$, the problem thus tends to the one which we have

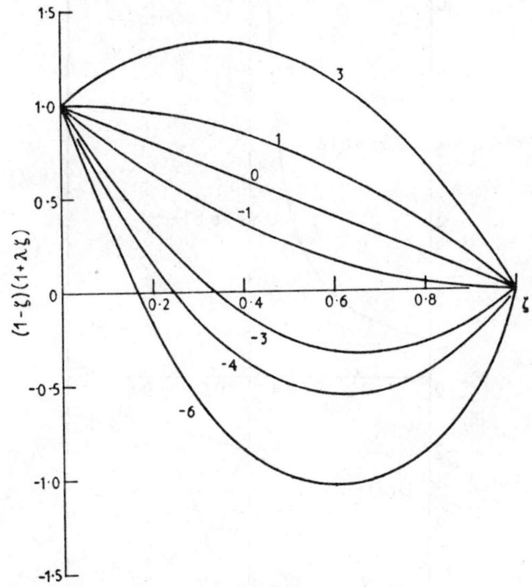

FIG. 93. The distribution of the transverse velocity $(1-\zeta)(1+\lambda\zeta)$ for various values of λ. The curves are labelled by the values of λ to which they refer.

considered in § 76. This is, indeed, evident from equations (50) and (51) which, for $|\lambda| \to \infty$, tend to

$$(D^2-a^2)^2 u = \lambda\zeta(1-\zeta)v \tag{57}$$

and

$$(D^2-a^2)v = \frac{T\lambda}{1-\mu}a^2(1-2\zeta)u. \tag{58}$$

By the further transformation $u \to \lambda u$, these equations become identical with equations (18) and (19) of § 76 with the only difference that Λ is now replaced by $T\lambda^2/(1-\mu)$. We conclude that for $|\lambda| \to \infty$, the critical Taylor number for the onset of instability tends asymptotically to the value given by (cf. Table XXXVIII)

$$\frac{T_c\lambda^2}{1-\mu} \to \Lambda_c = 9{\cdot}300\times 10^4. \tag{59}$$†

† Note that the definition of T includes, through A, a factor $(1-\mu)$.

Therefore, $$T_c \to 9{\cdot}300 \times 10^4 \frac{1-\mu}{\lambda^2} \quad \text{as } |\lambda| \to \infty. \tag{60}$$

The corresponding limiting value of a_c is given by

$$a_c \to 3{\cdot}96 \text{ as } |\lambda| \to \infty. \tag{61}$$

The asymptotic relations given by (60) for $\lambda \to \pm\infty$ are included in

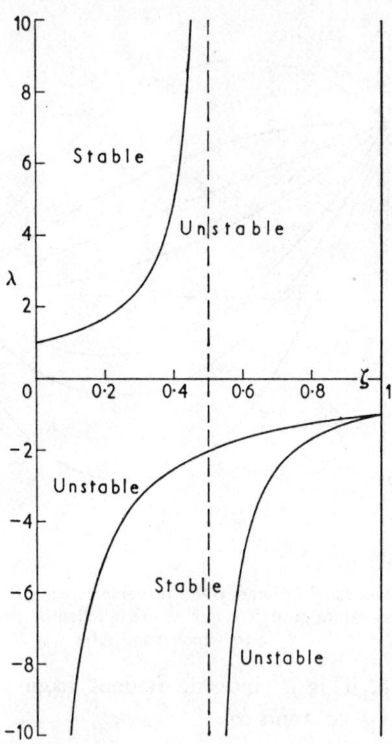

FIG. 94. The parts of the fluid which are stable or unstable according to Rayleigh's criterion for the flow $(1-\zeta)(1+\lambda\zeta)$ as a function of λ.

Fig. 91. The fact, that for $\lambda \to +\infty$ the asymptote is approached from below while it is approached from above for $\lambda \to -\infty$, can be understood when we observe that for large positive λ, the interval in which the flow is unstable is in *excess* of one-half, while for large negative λ, the wider of the two intervals in which the flow is unstable is *less* than one-half; and arguments similar to those used in § 71 (e) lead one to infer that the critical Taylor number varies approximately as the inverse fourth power of the extent of the widest zone of instability, provided the different zones are well separated. These same arguments also account for the fact that the values of a_c are less than the limiting value 3·96 for large positive λ, while they are greater than 3·96 for large negative λ.

Consider next the range $-1 \leqslant \lambda \leqslant +1$. In this range, Rayleigh's criterion is violated throughout the interval. Under these circumstances, we may, in a first approximation, replace the terms in equations (50) and (51), which allow for the variation of $V(\zeta)$ through the interval, by their average values. The arguments here are the same as in § 71 (*d*). With the replacements suggested, the equations become

$$(D^2-a^2)^2u = \tfrac{1}{2}(1+\tfrac{1}{3}\lambda)v \tag{62}$$

and

$$(D^2-a^2)v = -Ta^2u. \tag{63}$$

From these equations it is apparent that in the range $-1 \leqslant \lambda \leqslant +1$ we should expect the relations (cf. equation VII (231))

$$T_c = \frac{3416}{1+\lambda/3} \quad \text{and} \quad a_c = 3{\cdot}1 \tag{64}$$

to hold approximately. This expectation is confirmed by the results of Di Prima's exact calculations. Thus, for $\lambda = +1$ and -1, the approximate values of T_c we deduce from (64) are $2{\cdot}56\times10^3$ and $5{\cdot}12\times10^3$, respectively; and these values should be contrasted with the 'exact' values $2{\cdot}43\times10^3$ and $5{\cdot}42\times10^3$.

We now turn to the origin of the sharp maximum of T_c at $\lambda = -3{\cdot}5$. This must clearly be traced to the relative extents of the two zones of instability which occur for $\lambda < -1$. The argument here is that if there are different zones of instability, well separated, the onset of instability will be determined, principally, by the zone of widest extent. For $-3 < \lambda < -1$, the wider of the two zones of instability adjoins the inner cylinder, while for $\lambda < -3$, it adjoins the outer cylinder; for $\lambda = -3$, the two zones are of equal extent and they are separated by a stable zone also of the same extent. The minimum width of the largest extant zone of instability occurs, then, for $\lambda = -3$; and if our earlier argument (without the qualification 'well separated') should be taken literally, the maximum Taylor number should occur for this value of $\lambda = -3$. But the unqualified argument should not be taken literally in this instance: for, the two zones of instability are *not* well separated; and the amplitudes of the perturbations, which are expected to be large in the unstable zones, will not decrease sufficiently to evanescence in the intervening stable zone. We must attribute to this last circumstance the occurrence of the maximum Taylor number at the somewhat displaced value, $\lambda = -3{\cdot}5$, where the larger extent of the outer stable zone is compensated by its wider separation from the inner unstable zone. The fact that the maximum value of T_c should be associated with the maximum value of a_c is also clear.

(d) *Comparison with experimental results*

Experiments which verify the theoretical results of § (b) have been carried out by Brewster, Grosberg, and Nissan. Their experimental arrangement is shown in Fig. 95. The outer cylinder a (of inside diameter 13 in.) moulded in two semi-cylindrical sections from Perspex was transparent to allow visual observations. The inner cylinder b (of

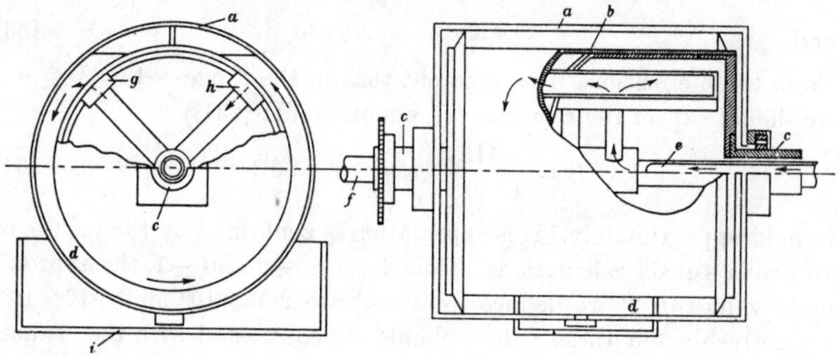

FIG. 95. A sketch of the experimental arrangement used by Brewster, Grosberg, and Nissan.

outside diameter 12 in.) made of copper was perforated to enable the flow of liquid through it. The inner cylinder was rotated by means of the half-shafts c; the outer cylinder was always at rest. A steady, controllable flow of liquid through the annulus d could be maintained by an external pump. The liquid flowed in and out of the apparatus through axial pipes e and f connected with stationary pressure and suction boxes g and h. A vertical meridional section of the annulus could be viewed through the tank i filled with the liquid. The onset of instability was viewed through this tank by transmitted light.

The experiments consisted in increasing either the rate of rotation or the rate of pumping until the onset of instability manifested itself by the formation of cells. The authors found that a small change in the speed of rotation (or the rate of pumping) usually resulted in the appearance or the disappearance of the cells.

The liquids used in the experiments were glycerine-water solutions. The flow through the annulus was visualized by one of three methods: by using the self-indicating properties of glycerine solutions; by introducing dyes; or, simply by observing the formation of bubbles at the onset of instability. The onset of instability visualized by two of these methods is shown in Fig. 96 *a* (by the first method) and Fig. 96 *b* (by the third method).

Fig. 96 *b* exhibits the onset of instability in the case when the cylinders are at rest: it occurs at the value of Λ predicted in § 76 (see equation (33)).

The parameter λ introduced in the theory is related to the net volume rate of flow $\mathcal{Q}$ per unit length of the cylinder. The relation in question is (cf. equations (37) and (40))

$$\mathcal{Q} = (\tfrac{1}{2}R_1\Omega_1+V_m)d = \tfrac{1}{2}R_1\Omega_1(1+\tfrac{1}{3}\lambda). \tag{65}$$

Since $\mathcal{Q}$ and Ω_1 are measured quantities, the value of λ can be deduced. [Note that $\mathcal{Q} = 0$ for $\lambda = -3$; for this reason Brewster, Grosberg, and Nissan call this the case of 'completely reversed flow'; it is also the case for which the extents of the two unstable zones and of the intermediate stable zone are all equal.]

The experimental results of Brewster, Grosberg, and Nissan on the critical Taylor numbers are included in Fig. 91. It will be observed that the agreement with the theoretical calculations is satisfactory.

78. The stability of inviscid flow between coaxial cylinders when an axial pressure gradient is present

The equations of inviscid flow (equations VII (13)–(15)) allow the stationary solution

$$u_r = 0, \quad u_\theta = V(r), \quad u_z = W(r), \tag{66}$$

and

$$p = \rho \int \frac{dr}{r} V^2(r), \tag{67}$$

where $V(r)$ and $W(r)$ are arbitrary functions of r.

In considering the stability of the flow represented by the foregoing solution of the equations of motion, we shall restrict ourselves to perturbations which are independent of θ. Let the perturbed state be described by

$$u_r, \quad V+u_\theta, \quad W+u_z, \quad \text{and} \quad \varpi = \delta p/\rho, \tag{68}$$

where, in accordance with our assumption, u_r, u_θ, u_z, and ϖ are functions only of r and z. The linear equations governing the perturbation are

$$\frac{\partial u_r}{\partial t}+W\frac{\partial u_r}{\partial z}-2\frac{V}{r}u_\theta = -\frac{\partial \varpi}{\partial r}, \tag{69}$$

$$\frac{\partial u_\theta}{\partial t}+W\frac{\partial u_\theta}{\partial z}+\left(\frac{dV}{dr}+\frac{V}{r}\right)u_r = 0, \tag{70}$$

$$\frac{\partial u_z}{\partial t}+W\frac{\partial u_z}{\partial z}+u_r\frac{dW}{dr} = -\frac{\partial \varpi}{\partial z}, \tag{71}$$

and the equation of continuity,

$$\frac{\partial u_r}{\partial r}+\frac{u_r}{r}+\frac{\partial u_z}{\partial z} = 0. \tag{72}$$

Following the standard procedure, we analyse the disturbance into normal modes and suppose that the various quantities describing the perturbation have a (z, t)-dependence given by

$$e^{i(pt+kz)}, \tag{73}$$

where p is a constant (which can be complex) and k is the wave number of the disturbance in the z-direction.

Let $u(r)$, $v(r)$, $w(r)$, and $\varpi(r)$ denote the amplitudes of the perturbations u_r, u_θ, u_z, and ϖ whose (z, t)-dependence is given by (73). Equations (69)–(72) then give

$$i(p+kW)u-2\Omega v = -D\varpi, \tag{74}$$

$$i(p+kW)v = -(D_* V)u, \tag{75}$$

$$i(p+kW)w+(DW)u = -ik\varpi, \tag{76}$$

and

$$D_* u = -ikw, \tag{77}$$

where

$$D = d/dr \quad \text{and} \quad D_* = D+1/r. \tag{78}$$

Eliminating w between equations (76) and (77), we find

$$(p+kW)D_* u-(kDW)u = ik^2\varpi. \tag{79}$$

Differentiating this equation with respect to r and making use of equation (74), we obtain

$$D[(p+kW)D_* u]-D[k(DW)u] = -ik^2[i(p+kW)u-2\Omega v]. \tag{80}$$

On simplification and rearrangement, equation (80) reduces to the form

$$(p+kW)(DD_*-k^2)u-k\Psi(r)u = 2ik^2\Omega v, \tag{81}$$

where

$$\Psi(r) = D^2W+(DW)(D-D_*) = D^2W-\frac{1}{r}DW = r\frac{d}{dr}\left(\frac{1}{r}\frac{dW}{dr}\right). \tag{82}$$

Finally, substituting for v from equation (75), we obtain

$$(p+kW)(DD_*-k^2)u-k\Psi(r)u = -k^2\Phi(r)\frac{u}{p+kW}, \tag{83}$$

where

$$\Phi(r) = 2\Omega D_* V = 2\frac{\Omega}{r}\frac{d}{dr}(rV) = 2\frac{\Omega}{r}\frac{d}{dr}(r^2\Omega) \tag{84}$$

is Rayleigh's discriminant as previously defined.

The boundary conditions are

$$u = 0 \quad \text{for } r = R_1 \text{ and } R_2, \tag{85}$$

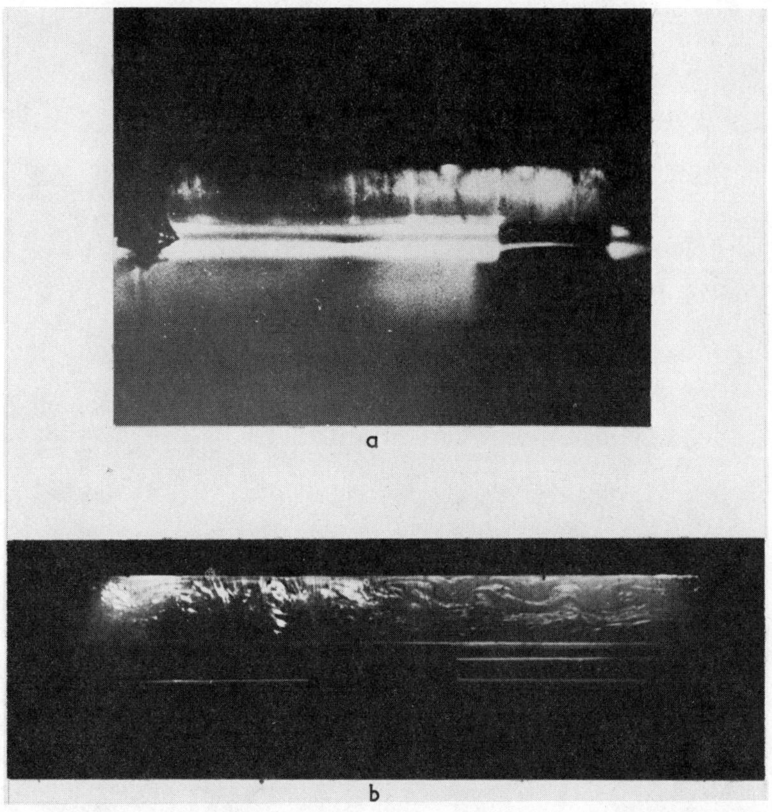

FIG. 96. The visualization of the onset of instability in general Couette flows by Brewster, Grosberg, and Nissan: (*a*) for the case of a pure pressure maintained flow; (*b*) for the case

$$\lambda\ (= 6V_m/R_1\Omega_1) = -3,$$

when the net flow is zero.

where R_1 and R_2 ($> R_1$) are the radii of the two cylinders confining the fluid.

An alternative form of equation (83) which we shall find useful is obtained by the substitution

$$i\chi = \frac{u}{p+kW}. \tag{86}$$

This substitution is generally permissible only if $D_* V \neq 0$; for, in this case (cf. equation (75))

$$v = -(D_* V)\chi, \tag{87}$$

and v is required to be a regular function in the interval (R_1, R_2) in the present framework of inviscid motion. [Note that for the form of V permissible under viscous flow, $D_* V = 2A =$ constant; in any event, the boundedness of v requires the boundedness of χ—a fact that is contrary to what holds in the absence of rotation (for real p's).] In terms of χ, equations (74) and (79) take the forms

$$(p+kW)^2\chi - \Phi(r)\chi = D\varpi \tag{88}$$

and

$$(p+kW)^2 D_* \chi = k^2\varpi. \tag{89}$$

Now eliminating ϖ between these equations, we obtain

$$D[(p+kW)^2 D_* \chi] - k^2(p+kW)^2\chi = -k^2\Phi(r)\chi; \tag{90}$$

and the appropriate boundary conditions are (cf. equation (86))

$$\chi = 0 \quad \text{for } r = R_1 \text{ and } R_2. \tag{91}$$

(a) *The case of a pure axial flow*

When the initial stationary flow is a purely axial one, the transformation (86) is, in general, not permissible: it is permissible if p is complex, or if $(p+kW)$ has no zero in the interval (R_1, R_2); and in no other case.

We begin our considerations, then, with the equation,

$$(p+kW)(DD_* - k^2)u - k\Psi u = 0, \tag{92}$$

which one obtains from equation (83) by setting $\Phi = 0$. Equation (92) has a singular point where $W = -p/k$. For stable oscillations (when p is real), this singular point can occur in the interval (R_1, R_2). However, if p should be complex (as it will be for damped and overstable oscillations), the equation is regular for all real values of r.

We shall now show that *a necessary condition for the occurrence of overstable oscillations is that* $\Psi(r)$ *changes sign in the interval* (R_1, R_2).

To prove the theorem, suppose that p ($= p_r + ip_i$) is complex. Then we can rewrite equation (92) in the form

$$(DD_* - k^2)u - \frac{k\Psi}{p+kW}u = 0. \tag{93}$$

Multiplying this equation by ru^* (where u^* is the complex conjugate of u) and integrating over the range of r, we obtain (after an integration by parts)

$$\int_{R_1}^{R_2} r\{|D_* u|^2+k^2|u|^2\}\,dr+k\int_{R_1}^{R_2}\frac{\Psi}{p+kW}r|u|^2\,dr=0. \tag{94}$$

(The integrated part vanishes on account of the boundary conditions.) The real and the imaginary parts of equation (94) must vanish separately. The vanishing of the imaginary part gives

$$p_i\int_{R_1}^{R_2}\frac{\Psi}{|p+kW|^2}r|u|^2\,dr=0. \tag{95}$$

If $p_i\neq 0$ (as we have supposed), the integral must vanish; and for this to happen, a necessary condition is clearly that $\Psi(r)$ vanishes for some value of r in the interval (R_1, R_2).

An alternative way of establishing the same result is instructive. Rewriting equation (93) in the form

$$\frac{d}{dr}\left(r\frac{du}{dr}\right)-\frac{u}{r}-k^2ru-\frac{k\Psi}{p+kW}ru=0, \tag{96}$$

we subtract from the equation

$$u^*\frac{d}{dr}\left(r\frac{du}{dr}\right)-\frac{|u|^2}{r}-k^2r|u|^2-\frac{k\Psi}{p+kW}r|u|^2=0 \tag{97}$$

its complex conjugate. We thus obtain

$$\frac{d}{dr}\left\{u^*\left(r\frac{du}{dr}\right)-u\left(r\frac{du^*}{dr}\right)\right\}=-2ip_i\frac{rk\Psi}{|p+kW|^2}|u|^2. \tag{98}$$

Letting

$$U=\tfrac{1}{2}ir\left(u^*\frac{du}{dr}-u\frac{du^*}{dr}\right) \tag{99}$$

stand for a real quantity, we can rewrite equation (99) in the form

$$\frac{dU}{dr}=p_i\frac{rk\Psi}{|p+kW|^2}|u|^2. \tag{100}$$

Since $U=0$ for $r=R_1$ and R_2 (in virtue of the boundary conditions on u), dU/dr must change its sign somewhere inside the interval (R_1, R_2); and this requires that $\Psi(r)$ do the same.

A further result which can be established when p is complex is that *the real part of* $-p$ *must lie between the maximum and the minimum values of* kW. This follows from the equation (cf. equation (90)),

$$D[(p+kW)^2D_*\chi]-k^2(p+kW)^2\chi=0 \tag{101}$$

which one obtains from equation (92) by the transformation (86). (The transformation is legitimate since we have supposed that p is complex.) By multiplying equation (101) by $r\chi^*$ and integrating over the range of r, we obtain (after an integration by parts)

$$\int_{R_1}^{R_2} (p+kW)^2 r(|D_*\chi|^2+k^2|\chi|^2)\,dr = 0. \tag{102}$$

The imaginary part of this equation gives

$$p_i \int_{R_1}^{R_2} (p_r+kW) r(|D_*\chi|^2+k^2|\chi|^2)\,dr = 0. \tag{103}$$

Since $p_i \neq 0$ (by hypothesis), the integral must vanish and the stated limits on p_r follow.

We have seen that the vanishing of Ψ somewhere inside the interval (R_1, R_2) is a necessary condition for the instability of the basic flow; we shall now prove that for a wide class of velocity distributions this condition is also a sufficient one. The proof consists in showing that under the conditions a wave number k_s exists for which the flow is stable and that it is unstable for the immediately neighbouring wave numbers.

Let

$$\Psi(r) = rDD_*(W/r) = 0 \quad \text{for } r = r_s \text{ where } W = W_s \text{ (say)}. \tag{104}$$

We shall suppose that

$$K(r) = -\frac{\Psi(r)}{r(W-W_s)} = -\frac{DD_*(W/r)}{W-W_s} > 0 \quad \text{throughout the interval.} \tag{105}$$

Further restrictions which we shall impose on W are:

$$W \geqslant 0 \quad \text{in the interval}$$

and

$$W = 0 \quad \text{for } r = R_1 \text{ and } R_2. \tag{106}$$

Consider, under these circumstances, the equation,

$$DD_* u + rKu = k^2 u, \tag{107}$$

together with the boundary conditions, $u = 0$ for $r = R_1$ and R_2. This is a characteristic value problem of the classical Sturm–Liouville type; and from the general theory it follows that under the conditions stated, the characteristic values k^2 can be arranged in a monotonic decreasing sequence. The characteristic value problem, moreover, allows a variational formulation: its solution is equivalent to finding extremal values of the expression,

$$k^2 = \frac{\int_{R_1}^{R_2} r\{rKu^2-(D_* u)^2\}\,dr}{\int_{R_1}^{R_2} ru^2\,dr}, \tag{108}$$

which are stationary with respect to arbitrary variations of u subject only to the boundary conditions. Therefore, the largest of the characteristic values of k^2 represents the absolute maximum which the quantity on the right-hand side of (108) can attain.

We shall now show that for a suitably chosen u, k^2 given by equation (108) can assume a positive value; in view of what has been said, the establishment of this fact will guarantee the existence of a positive characteristic value for k^2. The particular choice of u we make is W/r. With this choice, the numerator in the expression for k^2 becomes

$$\int\limits_{R_1}^{R_2} \{KW^2 - rD_*(W/r)[D_*(W/r)]\}\, dr. \tag{109}$$

After an integration by parts, we are left with

$$\int\limits_{R_1}^{R_2} W[KW + DD_*(W/r)]\, dr, \tag{110}$$

since the integrated part vanishes on account of the conditions (106) imposed on W. Now according to equation (105),

$$DD_*(W/r) = -K(W - W_s). \tag{111}$$

Inserting this expression in (110), we obtain (cf. equations (105) and (106))

$$W_s \int\limits_{R_1}^{R_2} WK\, dr > 0. \tag{112}$$

Thus, with the choice of u made, k^2 is indeed positive, and we infer that a positive characteristic value for k^2 exists. Let k_s^2 denote this value; and let u_s be the proper function belonging to it. Then

$$DD_* u_s - \frac{\Psi(r)}{W - W_s} u_s = k_s^2 u_s. \tag{113}$$

Now rewrite equation (93) in the form

$$DD_* u - \frac{\Psi(r)}{W - c} u = k^2 u, \tag{114}$$

where
$$c = -p/k = c_r + ic_i \text{ (say)} \tag{115}$$

is allowed to be complex. We shall suppose that k^2 is an analytic function of the complex variable c and consider its behaviour near k_s^2. From equations (113) and (114) we obtain

$$\frac{d}{dr}\left(ru_s \frac{du}{dr} - ru \frac{du_s}{dr}\right) - r\Psi uu_s\left(\frac{1}{W-c} - \frac{1}{W-W_s}\right) = (k^2 - k_s^2) ruu_s. \tag{116}$$

Integrating this equation over the range of r and remembering that both u and u_s vanish at the limits, we get

$$(k^2-k_s^2)\int\limits_{R_1}^{R_2} ruu_s\,dr = -(c-W_s)\int\limits_{R_1}^{R_2}\frac{r\Psi uu_s}{(W-c)(W-W_s)}\,dr. \tag{117}$$

We now let $k^2 \to k_s^2$, $c \to W_s$, and $u \to u_s$. In this limit, equation (117) becomes

$$\left(\frac{dk^2}{dc}\right)_{k^2=k_s^2}\int\limits_{R_1}^{R_2} ru_s^2\,dr = \lim_{c_r\to W_s;\,c_i\to 0}\int\limits_{R_1}^{R_2}\frac{Kr^2u_s^2}{W-(c_r+ic_i)}\,dr, \tag{118}$$

where we have reintroduced the function K defined in equation (105). Writing explicitly the real and the imaginary parts of the integral on the right-hand side of equation (118), we have

$$\int\limits_{R_1}^{R_2}\frac{Kr^2u_s^2}{W-c}\,dr = \int\limits_{R_1}^{R_2}\frac{K(W-c_r)}{(W-c_r)^2+c_i^2}r^2u_s^2\,dr + i\int\limits_{R_1}^{R_2}\frac{c_i K}{(W-c_r)^2+c_i^2}r^2u_s^2\,dr. \tag{119}$$

In the limit $c_i \to 0$ and $c_r \to W_s$, the real part in (119) tends to the principal value of the integral

$$\int\limits_{R_1}^{R_2}\frac{Kr^2u_s^2}{W-W_s}\,dr, \tag{120}$$

while the imaginary part tends to

$$\begin{aligned}
\lim_{c_i\to 0}\int\limits_{R_1}^{R_2}\frac{c_i K(r)r^2u_s^2}{(W-W_s)^2+c_i^2}\,dr &= [K(r)r^2u_s^2]_{r=r_s}\lim_{\substack{\epsilon\to 0\\ c_i\to 0}}\int\limits_{r_s-\epsilon}^{r_s+\epsilon}\frac{c_i\,dr}{(W-W_s)^2+c_i^2}\\
&= \left[\frac{K(r)r^2u_s^2}{dW/dr}\right]_{r=r_s}\lim_{\substack{\epsilon\to 0\\ c_i\to 0}}\int\limits_{W(r_s-\epsilon)}^{W(r_s+\epsilon)}\frac{c_i\,dW}{(W-W_s)^2+c_i^2}\\
&= \left[\frac{K(r)r^2u_s^2}{|dW/dr|}\right]_{r=r_s}\lim_{c_i\to 0}\int\limits_{-\infty}^{+\infty}\frac{c_i\,dx}{x^2+c_i^2}\\
&= \pm\pi\left[\frac{K(r)r^2u_s^2}{|dW/dr|}\right]_{r=r_s}\quad (c_i\to\pm 0).
\end{aligned} \tag{121}$$

Combining the foregoing results, we have an equation of the form

$$\left(\frac{dk^2}{dc}\right)_{k^2=k_s^2} = A\pm iB \quad \text{where } B>0. \tag{122}$$

Alternatively, we can write

$$\left(\frac{dc}{dk^2}\right)_{k^2=k_s^2} = \frac{A\mp iB}{A^2+B^2}\quad (B>0). \tag{123}$$

From equation (123) it follows that for values of k^2 (on the real axis) different from k_s^2 the imaginary part of c is finite. For these same neighbouring wave numbers, the imaginary part of p is finite; and this implies the instability of these modes.

We shall now return to consider the case when $\Psi(r)$ is of the same sign throughout. Then, from equation (95) it follows that $p_i = 0$; this means that p cannot be complex; and this implies that the flow is stable. In fact we shall verify directly that *any value of* $-Wk$ *is an admissible solution for* p. Suppose, for example, that for a particular p, $p+kW$ vanishes only once in (R_1, R_2). We may, then, start integrating equation (92) from either end with arbitrary initial values for du/dr. The integrations can be continued inward until we come to the point where $p+kW = 0$. At this point the two values of u and du/dr will presumably disagree. By a suitable choice of the relative initial values of du/dr, u can be made continuous. A discontinuity in du/dr will remain, but this is not inconsistent with inviscid flow.

Are other real values of p admissible? If so, $(p+kW)$ will be of one sign throughout. And this we can exclude. For, if $(p+kW)$ does not have a zero in the interval (R_1, R_2), the transformation (86) is a legitimate one and equation (101) is a valid equation to consider. From this latter equation we can deduce (cf. equation (102))

$$\int_{R_1}^{R_2} (p+kW)^2 r[(D_* \chi)^2 + r^2\chi^2]\, dr = 0, \tag{124}$$

an equation which cannot be satisfied by any real p. Thus, the assumption that the transformation (86) is a legitimate one has led to a contradiction. We conclude that *any value of* p *not included between* $-kW_{\max}$ *and* $-kW_{\min}$ *is not admissible as a solution.*

(*b*) *The general case when rotation is also present*

Returning to the general case, we first observe that the presence of rotation affects some aspects of the problem in a fundamental way. Thus, the existence of the relation (equation (75))

$$i(p+kW)v = -(D_* V)u \tag{125}$$

allows us to assume† that $(p+kW)$ has no zeros in the allowed range of r.

† For if $(p+kW)$ should vanish at, say, one point, $r = r_0$, then u must vanish at r_0 (we exclude the possibility that $D_* V$ may vanish here); and the problem considered becomes equivalent to two *independent* characteristic value problems in the intervals (R_1, r_0) and (r_0, R_2) both of which belong to the class considered in the text. And this argument clearly extends to the case when $p+kW$ has more than one zero in (R_1, R_2).

On this assumption, if p *is real, it cannot lie between* $-kW_{\max}$ *and* $-kW_{\min}$; and this result is exactly the opposite of that which obtains in the case of a pure axial flow. This circumstance already suggests that rotation alters the character of the problem in a *qualitative* way and that the results for the case of a pure axial flow cannot be obtained by a simple passage to the limit $\Omega = 0$. Indeed, the principal conclusions at which we shall arrive are:† *the instabilities associated with a pure axial flow cease to be relevant when rotation is present; and stability is determined exclusively by the distribution of the transverse velocities, i.e. by Rayleigh's criterion.*

The equation to be considered is (cf. equation (90))

$$D[(c-W)^2 D_* \chi]-k^2(c-W)^2\chi = -\Phi(r)\chi, \tag{126}$$

where $c = -p/k$; and the boundary conditions are

$$\chi = 0 \quad \text{for } r = R_1 \text{ and } R_2. \tag{127}$$

And the principal question concerns the reality, or otherwise, of c.

The natural and most direct way of looking at the problem presented by equation (126) and the boundary conditions (127) is to consider it as a characteristic value problem for c for an assigned k and given $W(r)$ and $\Phi(r)$. The problem then allows a variational formulation even though the characteristic value parameter c enters the problem non-linearly.

There is another useful way of looking at the problem. By writing

$$\Omega(r) = \lambda\omega(r), \tag{128}$$

where λ is some chosen unit of angular velocity (which might be taken to be the angular velocity at $r = R_1$ (say)), we can consider the problem as a characteristic value problem for λ^2. Thus, letting

$$\phi(r) = \frac{2\omega}{r}\frac{d}{dr}(r^2\omega), \tag{129}$$

we have
$$D[(c-W)^2 D_* \chi]-k^2(c-W)^2\chi = -\lambda^2\phi(r)\chi. \tag{130}$$

The advantage of this latter formulation is that the characteristic value parameter λ^2 enters the problem linearly; and, moreover, with some supplementary conditions, the known theorems of the classical Sturmian theory can be applied. However, it is important to remember that the solution of the physical problem requires that *every* positive $\lambda^2 > 0$ be derived as a characteristic value for a given $\phi(r)$ and an assigned k for suitably selected values of c (real or complex).

† We exclude here certain possible 'pathological' cases.

(i) *A variational principle for c*

Consider equations (126) and (127) as presenting a characteristic value problem for c. By multiplying equation (126) by $r\chi$ and integrating over the range of r, we obtain, after an integration by parts,

$$\int_{R_1}^{R_2} (c-W)^2[(D_*\chi)^2+k^2\chi^2]r\,dr = \int_{R_1}^{R_2} \Phi(r)r\chi^2\,dr. \tag{131}$$

Letting

$$I_1 = \int_{R_1}^{R_2} [(D_*\chi)^2+k^2\chi^2]r\,dr, \qquad I_3 = \int_{R_1}^{R_2} W^2[(D_*\chi)^2+k^2\chi^2]r\,dr,$$

$$I_2 = \int_{R_1}^{R_2} W[(D_*\chi)^2+k^2\chi^2]r\,dr, \quad \text{and} \quad I_4 = \int_{R_1}^{R_2} \Phi(r)\chi^2 r\,dr, \tag{132}$$

we can rewrite equation (131) in the form

$$c^2 I_1 - 2cI_2 + I_3 - I_4 = 0; \tag{133}$$

or solving for c, we have

$$c = \frac{1}{I_1}\{I_2 \pm \sqrt{(I_2^2 - I_1 I_3 + I_1 I_4)}\}. \tag{134}$$

Equation (133) provides the basis for a variational principle; to see this, consider the effect on c (determined in accordance with equation (134)) of an arbitrary variation $\delta\chi$ in χ compatible only with the boundary conditions on χ. To the first order in the variation, we have

$$2(cI_1 - I_2)\delta c = -(c^2\delta I_1 - 2c\delta I_2 + \delta I_3 - \delta I_4), \tag{135}$$

where δI_1, δI_2, δI_3, and δI_4 are the corresponding variations in the integrals I_1, I_2, I_3, and I_4. After an integration by parts (in each case), we find that the variations δI_1, δI_2, and δI_3 are given by:

$$\delta I_1 = -2\int_{R_1}^{R_2} \delta\chi[DD_*\chi - k^2\chi]r\,dr,$$

$$\delta I_2 = -2\int_{R_1}^{R_2} \delta\chi[D(WD_*\chi) - k^2W\chi]r\,dr,$$

$$\delta I_3 = -2\int_{R_1}^{R_2} \delta\chi[D(W^2D_*\chi) - k^2W^2\chi]r\,dr; \tag{136}$$

also,
$$\delta I_4 = 2\int_{R_1}^{R_2} \delta\chi\Phi(r)\chi r\,dr. \tag{137}$$

Inserting the foregoing expressions in (135), we obtain

$$(cI_1 - I_2)\delta c = \int_{R_1}^{R_2} \delta\chi\{D[(c-W)^2D_*\chi] - k^2(c-W)^2\chi + \Phi(r)\chi\}r\,dr. \tag{138}$$

We observe that the quantity which appears as a factor of $r\,\delta\chi$ under the integral sign on the right-hand side of equation (138) vanishes if equation (126) governing χ is satisfied.† Hence, a necessary and sufficient condition for δc to vanish identically to the first order, for all small variations in χ subject only to the boundary conditions, is that χ be a solution of the characteristic value problem. While this theorem provides the basis for a variational treatment of the problem, the principal conclusion we wish to draw from it and equation (134) is that *other things being equal, there are two, and only two, branches to the dispersion relation.*

(ii) *A variational principle for* λ^2

Now consider equations (127) and (130) as presenting a characteristic value problem for λ^2. If we restrict ourselves to real values of c not included between W_{max} and W_{min}, we can apply the standard theorems of the classical Sturmian theory (cf. § 67 (*b*)) and conclude: the characteristic values of λ^2 are all positive if $\phi(r)$ is everywhere positive and they are all negative if $\phi(r)$ is everywhere negative; but if $\phi(r)$ should change sign anywhere inside the interval (R_1, R_2), then there are two sets of real characteristic values which have the limit points $+\infty$ and $-\infty$. Moreover, solving equation (130) with the boundary conditions (127) is equivalent to finding extremal values of λ^2 given by

$$\lambda^2 = \frac{\int_{R_1}^{R_2} (c-W)^2[(D_*\chi)^2+k^2\chi^2]r\,dr}{\int_{R_1}^{R_2} \phi(r)r\chi^2\,dr}, \tag{139}$$

which are stationary with respect to arbitrary small variations of χ subject only to the boundary conditions (127).

(iii) *The criterion for stability*

The theorems enunciated in the preceding subsections enable us to derive a necessary and sufficient condition for stability.

Consider first the case when $\phi(r)$ is everywhere negative. Then, for every real k and for all real values of c (not included between W_{max} and W_{min}), the characteristic values of λ^2 are all negative. Consequently, if λ^2 is to be positive, c must be complex; and this means instability.

† Note that when c and χ are real,

$$cI_1 - I_2 = \int_{R_1}^{R_2} (c-W)[(D_*\chi)^2+k^2\chi^2]r\,dr$$

cannot vanish as long as $c > W_{max}$ or $c < W_{min}$; and these additional restrictions on c are necessary under the circumstances.

Consider next the case when $\phi(r)$ is everywhere positive. Then, for a given k and c (both assumed real and $c > W_{max}$ or $c < W_{min}$), the characteristic values, λ_n^2, of λ^2 are all positive. They can be arranged as an infinite, monotonic, increasing sequence; if λ_n^2 ($n = 1, 2,...$) represents this arrangement, the proper function χ_n belonging to λ_n^2 will have exactly $(n-1)$ nodes in the interval (R_1, R_2). To emphasize the dependence of λ_n^2 and χ_n on c (for assigned k) we shall write $\lambda_n^2(c)$ and $\chi_n(c)$.

Now by the variational principle of § (ii), the smallest of the characteristic values $\lambda_1^2(c)$ represents the *absolute minimum* which the quantity on the right-hand side of equation (139) can attain. Therefore, if $\lambda^2(c;\chi)$ denotes the value of λ^2 given by (139) for some assigned c and *chosen* χ, then

$$\lambda^2(c;\chi) \geqslant \lambda_1^2(c). \tag{140}$$

The equality sign in (140) can hold if and only if $\chi \equiv \chi_1(c)$, i.e.

$$\lambda^2(c;\chi_1(c)) = \lambda_1^2(c). \tag{141}$$

We shall now show that $\lambda_1^2(c)$ is a *monotonic increasing function for* $c > W_{max}$ *and a monotonic decreasing function for* $c < W_{min}$.

We first observe that according to equation (139)

$$\lambda^2(c_2;\chi_1(c_1)) < \lambda^2(c_1;\chi_1(c_1)) = \lambda_1^2(c_1) \quad \text{if } W_{max} < c_2 < c_1. \tag{142}$$

On the other hand, by (140)

$$\lambda^2(c_2;\chi_1(c_1)) > \lambda_1^2(c_2). \tag{143}$$

Hence, combining (142) and (143), we have

$$\lambda_1^2(c_2) < \lambda_1^2(c_1) \quad \text{if } W_{max} < c_2 < c_1. \tag{144}$$

In exactly the same way, we can show that

$$\lambda_1^2(c_2) > \lambda_1^2(c_1) \quad \text{if } c_2 < c_1 < W_{min}. \tag{145}$$

It is also clear from (139) that $\lambda^2(c)$ cannot be bounded as $c \to \pm\infty$. Hence,

$$\lambda_1^2(c) \to \infty \quad \text{as } c \to \pm\infty. \tag{146}$$

Further, by letting c approach W_{max} from above, or W_{min} from below, we can make λ^2 given by (139) as small as we please; consequently,

$$\lambda_1^2(c) \to 0 \quad \text{as } c \to W_{max}+0 \quad \text{or } W_{min}-0. \tag{147}$$

(Note that $c = W_{max}$ and $c = W_{min}$ are excluded.)

From the foregoing results, it follows that for *every* assigned k^2 and $\lambda^2 > 0$, there exist *two* real characteristic values for c, one larger than W_{max} and one less than W_{min}; and that the proper functions belonging to these two values have no nodes in (R_1, R_2). By similarly considering characteristic values λ_n^2 (whose proper functions have $n-1$ nodes), we can deduce that for every assigned k^2 and $\lambda^2 > 0$ there are two real characteristic values for c (one larger than $W_{\max}$ and one less than $W_{\min}$) whose proper functions have $n-1$ nodes. On the other hand, by the variational principle of § (i), there can be, in this case, no more than two branches to the dispersion relation. And since we have already accounted for two branches with real c's, there is no room for further complex values. The flow is, therefore, stable when $\phi(r)$ is everywhere positive.

When $\phi(r)$ changes sign in (R_1, R_2), then the possible characteristic values of λ^2 (for any real k and all real c's not included between $W_{\max}$ and $W_{\min}$) form two distinct sequences with limit points at $+\infty$ and $-\infty$. The existence of this latter sequence (with the limit point at $-\infty$) guarantees that for any given $\lambda^2 > 0$ there exist modes with complex characteristic values c which ensure instability.†

Thus, Rayleigh's criterion for the stability of a pure rotational flow continues to be valid in the presence of an arbitrary axial flow. This universality of Rayleigh's criterion is somewhat unexpected; but its origin is rooted in equation (75) which couples the radial and the transverse perturbations in the velocity. In this overpowering role of rotation, the present problem is, however, not unique: we have already encountered an example in Chapter III (§ 27 (*a*), p. 97).

79. The stability of viscous flow between rotating coaxial cylinders when an axial pressure gradient is present

Consider equations VII (134)–(136) in case the conditions are stationary, a constant pressure gradient $(\partial p/\partial z)_0$ is acting in the z-direction, the cylinders are in rotation, and no radial motions are present. Under these circumstances, the equations of motion allow the stationary solution

$$u_r = 0, \quad u_\theta = V(r), \quad \text{and} \quad u_z = W(r), \tag{148}$$

provided

$$\frac{1}{\rho}\frac{dp}{dr} = \frac{V^2}{r}, \tag{149}$$

$$\nu\left(\nabla^2 V - \frac{V}{r^2}\right) = \nu DD_* V = 0, \tag{150}$$

and

$$\nu\nabla^2 W = \nu D_* DW = \frac{1}{\rho}\left(\frac{\partial p}{\partial z}\right)_0. \tag{151}$$

The solution of equation (150) is given by

$$V = Ar + B/r, \tag{152}$$

where A and B are determined by the speeds of rotation of the inner and

† The argument is here similar to the one used on p. 284 (footnote).

the outer cylinders; they are given by equations VII (145) and (146). The corresponding solution of equation (151) is

$$W(r) = \frac{1}{4\rho\nu}\left(\frac{\partial p}{\partial z}\right)_0 (r^2+C\log r+D), \tag{153}$$

where C and D are constants. The requirement that $W(r)$ vanishes at $r = R_1$ and R_2 determines C and D and makes the solution determinate. We find

$$W(r) = -\frac{1}{4\rho\nu}\left(\frac{\partial p}{\partial z}\right)_0 \left\{R_1^2-r^2+\frac{2R_1 d+d^2}{\log(1+d/R_1)}\log\left(\frac{r}{R_1}\right)\right\}, \tag{154}$$

where d $(= R_2-R_1)$ is the gap width.

For small gap widths $(d \ll \frac{1}{2}(R_2+R_1))$, the velocity distribution given by equation (154) reduces to the familiar form of the Poiseuille flow between parallel planes. Thus, letting

$$\zeta = (r-R_1)/(R_2-R_1), \tag{155}$$

we find that, to order $(d/R_1)^2$, the velocity distribution becomes

$$W(r) = 6V_m \zeta(1-\zeta), \tag{156}$$

where

$$V_m = -\frac{d^2}{12\rho\nu}\left(\frac{\partial p}{\partial z}\right)_0 \tag{157}$$

is the mean axial flow.

(a) *The perturbation equations*

We shall now investigate the stability of the flow defined by equations (149), (152), and (153). Letting

$$u_r, \quad V+u_\theta, \quad W+u_z, \quad \text{and} \quad \varpi\,(= \delta p/\rho) \tag{158}$$

describe the perturbed state, and assuming, further, that the perturbations are axisymmetric, we readily obtain from equations VII (134)–(136) the linearized equations (cf. equations (69)–(71) and equations VII (149)–(153))

$$\frac{\partial u_r}{\partial t}+W\frac{\partial u_r}{\partial z}-2\Omega u_\theta = -\frac{\partial \varpi}{\partial r}+\nu\left(\nabla^2 u_r-\frac{u_r}{r^2}\right), \tag{159}$$

$$\frac{\partial u_\theta}{\partial t}+W\frac{\partial u_\theta}{\partial z}+\left(\frac{dV}{dr}+\frac{V}{r}\right)u_r = \nu\left(\nabla^2 u_\theta-\frac{u_\theta}{r^2}\right), \tag{160}$$

$$\frac{\partial u_z}{\partial t}+W\frac{\partial u_z}{\partial z}+u_r\frac{dW}{dr} = \nu\nabla^2 u_z-\frac{\partial \varpi}{\partial z}, \tag{161}$$

where ∇^2 has now the meaning

$$\nabla^2 = \frac{\partial^2}{\partial r^2}+\frac{1}{r}\frac{\partial}{\partial r}+\frac{\partial^2}{\partial z^2}; \tag{162}$$

and we also have the equation of continuity (72).

Seeking solutions of equations (159)–(161) which have the (z,t)-dependence given by (73), we obtain the equations

$$\nu\left(DD_*-k^2-i\frac{p}{\nu}-i\frac{k}{\nu}W\right)u+2\Omega v = D\varpi, \tag{163}$$

$$\nu\left(DD_*-k^2-i\frac{p}{\nu}-i\frac{k}{\nu}W\right)v = 2Au, \tag{164}$$

$$\nu\left(D_*D-k^2-i\frac{p}{\nu}-i\frac{k}{\nu}\right)w-uDW = ik\varpi, \tag{165}$$

and

$$D_*u = -ikw, \tag{166}$$

where D, D_*, u, v, and w have the same meanings as in equations (74)–(78).

Eliminating w between equations (165) and (166), we have

$$\frac{\nu}{k^2}\left(D_*D-k^2-i\frac{p}{\nu}-i\frac{k}{\nu}W\right)D_*u+\frac{i}{k}(DW)u = \varpi. \tag{167}$$

Inserting this expression for ϖ in equation (163), we obtain

$$\frac{\nu}{k^2}D\left(D_*D-k^2-i\frac{p}{\nu}-i\frac{k}{\nu}W\right)D_*u+\frac{i}{k}D(uDW)$$
$$= \nu\left(DD_*-k^2-i\frac{p}{\nu}-i\frac{k}{\nu}W\right)u+2\Omega v. \tag{168}$$

After some rearranging, equation (168) becomes

$$\left[DD_*-k^2-\frac{i}{\nu}(p+kW)\right](DD_*-k^2)u+i\frac{k}{\nu}ru\left[DD_*\left(\frac{W}{r}\right)\right] = \frac{2\Omega k^2}{\nu}v. \tag{169}$$

This equation must be considered together with the equation

$$\left[DD_*-k^2-\frac{i}{\nu}(p+kW)\right]v = \frac{2A}{\nu}u. \tag{170}$$

Solutions of equations (169) and (170) must be sought which satisfy the usual boundary conditions

$$u = v = 0 \quad\text{and}\quad Du = 0 \quad\text{for } r = R_1 \text{ and } R_2. \tag{171}$$

(b) The reduction to the case of a narrow gap

If the gap $d = R_2-R_1$ between the two cylinders is small compared to their mean radius $\frac{1}{2}(R_2+R_1)$, we need not, as in similar past instances, distinguish between D and D_* and use for Ω the linear representation

$$\Omega = \Omega_1[1-(1-\mu)\zeta]. \tag{172}$$

[In equation (172) ζ is the variable defined in equation (155) and Ω_1

and Ω_2 $(= \mu\Omega_1)$ are the angular velocities of the two cylinders.] In the same approximation (cf. equation (156)),

$$\frac{Wd}{\nu} = 6\frac{V_m d}{\nu}\zeta(1-\zeta) = 6\,\mathrm{R}\,\zeta(1-\zeta), \tag{173}$$

where R is the Reynolds number defined with respect to the mean axial flow. In the framework of these approximations which are appropriate for a narrow gap, equations (169) and (170) take the forms

$$\{(D^2-a^2)-i[\sigma+6\,\mathrm{R}\,a\zeta(1-\zeta)]\}(D^2-a^2)u-12i\,\mathrm{R}\,au$$
$$= 2\frac{\Omega_1 d^2}{\nu}a^2[1-(1-\mu)\zeta]v \tag{174}$$

and
$$\{(D^2-a^2)-i[\sigma+6\,\mathrm{R}\,a\zeta(1-\zeta)]\}v = \frac{2Ad^2}{\nu}u, \tag{175}$$

where
$$a = kd, \quad \sigma = pd^2/\nu, \tag{176}$$

and D stands for $d/d\zeta$. By translating the origin of ζ to be midway between the two cylinders and replacing

$$u \text{ by } \tfrac{1}{2}(1+\mu)\frac{2\Omega_1 d^2}{\nu}a^2u, \tag{177}$$

the equations become

$$\{(D^2-a^2)-i[\sigma+6\,\mathrm{R}\,a(\tfrac{1}{4}-\zeta^2)]\}(D^2-a^2)u-12i\,\mathrm{R}\,au = \left(1-2\frac{1-\mu}{1+\mu}\zeta\right)v \tag{178}$$

and
$$\{(D^2-a^2)-i[\sigma+6\,\mathrm{R}\,a(\tfrac{1}{4}-\zeta^2)]\}v = -\overline{T}a^2u, \tag{179}$$

where
$$\overline{T} = -\tfrac{1}{2}(1+\mu)\frac{4A\Omega_1}{\nu^2}d^4 \tag{180}$$

is the Taylor number as defined in equation VII (278).

Together with the boundary conditions,

$$u = Du = v = 0 \quad \text{for } \zeta = \pm\tfrac{1}{2}, \tag{181}$$

equations (178) and (179) constitute a characteristic value problem for $\overline{T}$ for assigned a, R, and σ. For an arbitrarily assigned σ, the characteristic values of $\overline{T}$ will in general be complex. The requirement that, for a given a and R, $\overline{T}$ be real determines σ. The critical Taylor number for the onset of instability (for a given R) is then given by the minimum (with respect to a) of the smallest real characteristic values, $\overline{T}$, so determined.

(c) *An approximate solution of the characteristic value problem for the case* $\mu > 0$

In Chapter VII, § 71 (*d*) we saw that the critical Taylor numbers for $\mu > 0$ (and R $= 0$) can be determined with fair precision by ignoring the variation of Ω across the gap and replacing $\Omega(\zeta)$ by its average value. In § 77 (c) we found the same thing with respect to the solution of equations (50) and (51) in case $-1 \leqslant \lambda \leqslant +1$. It would, therefore, appear that for $\mu > 0$ and for Reynolds numbers R, not too large, we can obtain the solutions of equations (178) and (179) with comparable precision by replacing the terms in these equations, which allow for the variation of $W(\zeta)$ and $\Omega(\zeta)$ across the gap, by their average values. Thus, we are led to consider the simpler equations

$$\{[(D^2-a^2)-i(\sigma+\mathrm{R}a)](D^2-a^2)-12i\,\mathrm{R}a\}u = v \tag{182}$$

and

$$[(D^2-a^2)-i(\sigma+\mathrm{R}a)]v = -\overline{T}a^2u \tag{183}$$

with the same boundary conditions (181).

The characteristic value problem presented by equations (181)–(183) can be solved quite readily by methods with which we are now familiar; in particular, the method which was used in § 44 (b) for solving equations IV (141) and (142) is applicable with only minor changes. Thus, we expand v in a cosine series in the form

$$v = \sum_{m=0}^{\infty} A_m \cos[(2m+1)\pi\zeta] \tag{184}$$

and express u in the manner

$$u = \sum_{m=0}^{\infty} A_m u_m, \tag{185}$$

where u_m is a solution of the equation

$$\{[(D^2-a^2)-i(\sigma+\mathrm{R}a)](D^2-a^2)-12i\,\mathrm{R}a\}u_m = \cos[(2m+1)\pi\zeta] \tag{186}$$

which satisfies the boundary conditions

$$u_m = Du_m = 0 \quad \text{for } \zeta = \pm\tfrac{1}{2}. \tag{187}$$

The required solution is (cf. equation IV (175))

$$u_m = \gamma_{2m+1}\cos[(2m+1)\pi\zeta]+\sum_{j=1}^{2} B_j^{(m)}\cosh q_j\,\zeta, \tag{188}$$

where

$$\frac{1}{\gamma_{2m+1}} = [(2m+1)^2\pi^2+a^2+i(\sigma+\mathrm{R}a)][(2m+1)^2\pi^2+a^2]-12i\,\mathrm{R}a, \tag{189}$$

the $B_j^{(m)}$'s $(j = 1,2)$ are constants of integration, and the q_j^2's $(j = 1,2)$ are the roots of the quadratic

$$(q^2-a^2)[(q^2-a^2)-i(\sigma+\mathrm{R}a)]-12i\,\mathrm{R}a = 0. \tag{190}$$

The constants $B_j^{(m)}$ in the solution (188) for u_m are determined by the boundary conditions (187); they are given by the same equations IV (181) and (182) if we ascribe to the γ's and the q's their present meanings.

Substituting for v and u in accordance with equations (184), (185), and (188) in equation (183), we obtain

$$\sum_{m=0}^{\infty} A_m c_{2m+1} \cos[(2m+1)\pi\zeta]$$
$$= \overline{T}a^2 \sum_{m=0}^{\infty} A_m\{\gamma_{2m+1} \cos[(2m+1)\pi\zeta] + \sum_{j=1}^{2} B_j^{(m)} \cosh q_j \zeta\}, \quad (191)$$

where
$$c_{2m+1} = (2m+1)^2\pi^2 + a^2 + i(\sigma + \text{R}a). \quad (192)$$

TABLE XL

The critical Taylor numbers and related constants for a few Reynolds numbers

R	a	$\overline{T}_c$	$-\sigma$
0	3·12	1,715	0
5	3·1	1,753	12·6
20	3·4	2,309	55·7
40	4·2	3,881	140·9
60	5·2	5,962	270·2
80	6·0	8,319	425
100	6·6	10,876	594

Equation (191) leads to the secular equation (cf. equations IV (182), (187), and (188))

$$\left\| \frac{1}{2}\left(\frac{c_{2n+1}}{\overline{T}a^2} - \gamma_{2n+1}\right)\delta_{nm} - (n|m) \right\| = 0, \quad (193)$$

where
$$(n|m) = (-1)^{m+n+1}2(2n+1)(2m+1)\pi^2\gamma_{2n+1}\gamma_{2m+1}\Delta(q_1^2 - q_2^2) \quad (194)$$

and
$$\Delta = (q_1 \tanh \tfrac{1}{2}q_1 - q_2 \tanh \tfrac{1}{2}q_2)^{-1}. \quad (195)$$

In the first approximation (in which we retain only the (0, 0)- element in the secular matrix), the solution for $\overline{T}$ is given by

$$\overline{T} = \frac{\pi^2 + a^2 + i(\sigma + \text{R}a)}{a^2\gamma_1[1 - 4\pi^2\gamma_1\Delta(q_1^2 - q_2^2)]}, \quad (196)$$

where
$$\frac{1}{\gamma_1} = [\pi^2 + a^2 + i(\sigma + \text{R}a)](\pi^2 + a^2) - 12i\,\text{R}a. \quad (197)$$

The critical Taylor numbers have been determined for a few values of the Reynolds numbers R by making use of the solution (196). The results are summarized in Table XL; they are further illustrated in Figs. 97 and 98.

(d) Comparison with experimental results

Several experiments reported in the literature bear on the effects of axial flow on the stability of transverse flow between rotating cylinders.

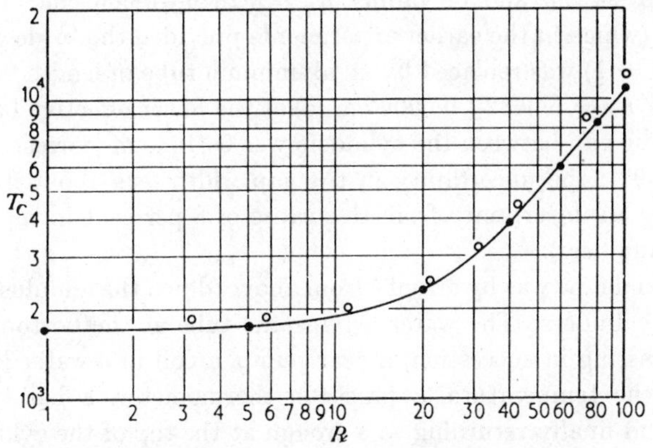

FIG. 97. A comparison of the observed and the theoretical dependence of the critical Taylor number T_c for the onset of instability as a function of the Reynolds number R of the axial flow. The full-line curve (passing through the solid circles) represents the theoretically derived relation; and the open circles represent the experimentally derived results of Donnelly and Fultz.

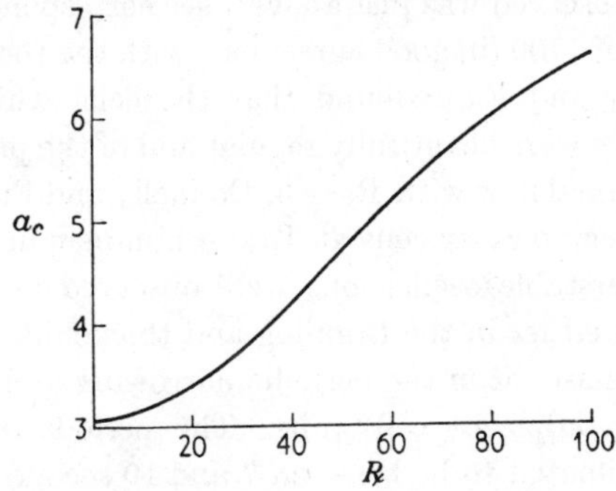

FIG. 98. The variation of the critical wave number a_c as a function of the Reynolds number R of the mean axial flow.

These experiments were not, however, especially designed for the study and verification of the basic hydrodynamic phenomenon; we shall, therefore, limit the present discussion to the experiments of Donnelly and Fultz which bear directly on the theoretical results of the preceding section.

The apparatus used in these experiments of Donnelly and Fultz was a modification of the one used in their earlier experiments described in § 74 (*b*). The outer cylinder was the same precision-bore Pyrex tube of length 94 cm and of radius $R_2 = 6{\cdot}2846 \pm 0{\cdot}006$ cm. The inner cylinder (which in the earlier experiments provided the 'wide gap' with a ratio $\eta = \frac{1}{2}$) was replaced by an aluminium tube of length 90 cm and of radius $R_1 = 5{\cdot}9682 \pm 0{\cdot}0006$ cm (over the lower effective half of the tube). The gap between the cylinders was 0·3164 cm (corresponding to $\eta = 0{\cdot}9497$); the uncertainty in the gap width was about 2 per cent (implying an uncertainty from this cause of 8 per cent in the deduced Taylor numbers).

The axial flow was by gravity from above, down the annulus between the two cylinders. The water leaving the tube at the bottom made a circuit passing in succession, a gear pump, a coil in a water bath controlling the temperature, a precision thermometer, a ball-type flow meter, and finally returning to a trough at the top of the cylinder providing the axial flow.

The apparatus could be successfully operated at Reynolds numbers of the mean axial flow up to 100 and Taylor numbers in the range 10^2 and $2{\cdot}5 \times 10^5$.

With zero axial flow, the period of rotation at which the onset of instability was observed was just above 7 sec corresponding to a critical Taylor number of 1700 (in good agreement with the theoretical value of 1708). Donnelly and Fultz found that the cells which appeared at marginal stability were beautifully regular and of the predicted spacing.

With a mean axial flow with $\mathrm{R} \sim 3$, Donnelly and Fultz observed the appearance of very regular cells at Taylor numbers in the range 1800–1900. Also, overstable oscillations were observed as predicted: they were manifested either in the thinning and thickening of alternate ink ribbons out of phase, or in the periodic narrowing and widening of the spacing between adjacent cell walls. The periods of oscillation for $\mathrm{R} \sim 3$ were estimated to be between 7 and 10 seconds (corresponding to σ in the range 9 to 5). For $\mathrm{R} \sim 95$, it was definitely established that $T_c \sim 10{,}000$ and that the periods of oscillation were considerably shorter: 2·8 sec at $\mathrm{R} \sim 95$ in contrast to the 7–10 sec for $\mathrm{R} \sim 3$.

It appears then that while the experiments are still of a preliminary character, they very definitely support the theoretical expectations.

BIBLIOGRAPHICAL NOTES

§ 76. The stability of viscous flow in a curved channel was first considered by Dean:

1. W. R. Dean, 'Fluid motion in a curved channel', *Proc. Roy. Soc.* (*London*) A, **121,** 402–20 (1928).

Later investigations of the same problem are those of:

2. W. H. Reid, 'On the stability of viscous flow in a curved channel', ibid. **244,** 186–98 (1958).
3. G. Hammerlin, 'Die Stabilität der Strömung in einem gekrümmten Kanal', *Arch. Rat. Mech. and Anal.* **1,** 212–24 (1958).

The treatment followed in the text is equivalent to one of two methods of solution described by Reid. Reid's other method is based on an expansion of u in terms of orthogonal functions which satisfy all four of the boundary conditions on u. Hammerlin's treatment of the problem is based on an analytical approach which is very different from the ones adopted in the book (which are, essentially, elementary). Hammerlin's treatment derives from his earlier paper:

4. G. Hammerlin, 'Über das Eigenwertproblem der dreidimensionalen Instabilität laminarer Grenzschichten an konkaven Wänden', *J. Rat. Mech. Analysis,* **4,** 279–321 (1955).

Applications of similar methods to Couette flow will be found in:

5. H. Witting, 'Über den Einfluß der Stromlinienkrümmung auf die Stabilität laminarer Strömungen', *Arch. Rat. Mech. and Anal.* **2,** 243–83 (1958).

The reference to the experimental work by Brewster, Grosberg, and Nissan, mentioned in § (*c*), is given below (reference 12). Some earlier experiments on the stability of the flow in a curved channel are those of:

6. M. Adler, 'Strömung in gekrümmten Rohren', *Z. angew. Math. Mech.* **14,** 257–75 (1934).

But Adler's experiments were not designed to determine the critical value of Λ with precision.

There are other important aspects of flow in a curved channel, and on curved surfaces generally, which we have not considered: they relate to boundary layer phenomena which are outside the scope of this book. But reference may be made to the basic papers of:

7. H. Görtler, 'Über den Einfluß der Wandkrümmung auf die Entstehung der Turbulenz', *Z. angew. Math. Mech.* **20,** 138–47 (1940).
8. —— 'Über eine dreidimensionale Instabilität laminarer Grenzschichten an konkaven Wänden,' *Nachr. Ges. Wiss. Göttingen, N.F.* **2,** 1–26 (1940).
9. —— 'Instabilität laminarer Grenzschichten an konkaven Wänden gegenüber gewissen dreidimensionalen Störungen', *Z. angew. Math. Mech.* **21,** 250–2 (1941).

A general account of these investigations will be found in:

10. H. Schlichting, *Boundary Layer Theory,* McGraw-Hill Book Company, Inc., New York, 1955; see particularly pp. 355–65.

§ 77. The theoretical treatment of the problem considered in this section is due to:

11. R. C. Di Prima, 'The stability of viscous flow between rotating concentric cylinders with a pressure gradient acting round the cylinders', *J. Fluid Mech.* **6,** 462–8 (1959).

The reference to the experimental work described in § (*d*) is:

12. D. B. Brewster, P. Grosberg, and A. H. Nissan, 'The stability of viscous flow between horizontal concentric cylinders,' *Proc. Roy. Soc.* (*London*) A, **251,** 76–91 (1959).

§ 78. The stability of inviscid flow between rotating coaxial cylinders when an axial pressure gradient is present does not seem to have attracted much attention. This section is largely an expansion of the results summarized in:

13. S. Chandrasekhar, 'The hydrodynamic stability of inviscid flow between coaxial cylinders', *Proc. Nat. Acad. Sci.* **46,** 137–41 (1960).

§ 78 (*a*). In considering the stability of the purely axial flow, we follow the classical discussions of the analogous problem in plane-parallel flows. The following papers by Kelvin and Rayleigh are particularly relevant:

14. Lord Kelvin, 'On a disturbing infinity in Lord Rayleigh's solution for waves in a plane vortex stratum', 186–7, *Mathematical and Physical Papers*, iv, *Hydrodynamics and General Dynamics*, Cambridge, England, 1910.

15. —— 'Rectilineal motion of viscous fluid between two parallel planes', 321–30, ibid.

16. Lord Rayleigh, 'On the stability, or instability, of certain fluid motions', *Scientific Papers*, i. 474–7, Cambridge, England, 1899.

17. —— 'On the stability, or instability, of certain fluid motions. II', ibid. iii, 17–23, Cambridge, England, 1902.

18. —— 'On the question of the stability of the flow of fluids', ibid. 575–84, Cambridge, England, 1902.

19. —— 'On the stability, or instability, of certain fluid motions. III', ibid. iv, 203–9, Cambridge, England, 1903.

20. —— 'On the stability of the laminar motion of an inviscid fluid', ibid. vi, 197–204, Cambridge, England, 1920.

The theorem, that a necessary condition for the existence of unstable modes is that $\Psi(r)$ $(= rDD_*(W/r))$ must vanish somewhere inside (R_1, R_2), is the exact analogue of the one proved in references 16 and 20 for plane-parallel flows. The remarks on the character of the stable solutions, in case $\Psi(r)$ is of the same sign throughout, parallel Rayleigh's in references 17 and 20.

In reference 18 Rayleigh briefly considers the case of pure axial flow between coaxial cylinders and shows that, for non-axisymmetric perturbations having an $e^{im\theta}$-dependence, the discriminant $\Psi(r)$ is replaced by

$$\frac{d^2W}{dr^2}-\frac{1}{r}\frac{dW}{dr}\frac{k^2r^2-m^2}{k^2r^2+m^2}.$$

The proof, that with certain restrictions on the velocity profile the condition that $\Psi(r)$ has a zero in (R_1, R_2) is both necessary and sufficient for instability, is similarly patterned after Lin's simplified version of the corresponding theorem of Tollmien in the theory of plane-parallel flows. The relevant references are:

21. W. Tollmien, 'Asymptotische Integration der Störungsdifferential-

gleichung ebener laminarer Strömungen bei hohen Reynoldsschen Zahlen', *Z. angew. Math. Mech.* **25/27**, 33–50 (1947).

22. —— Ibid. 70–83 (1947).
23. C. C. LIN, 'On the stability of two-dimensional parallel flows. I. General theory', *Quart. Appl. Math.* **3**, 117–42 (1945).
24. —— 'On the stability of two-dimensional parallel flows. II. Stability in an inviscid fluid', ibid. 218–34 (1945).
25. —— 'On the stability of two-dimensional parallel flows. III. Stability in a viscous fluid', ibid. 277–301 (1946).

For a concise account of these investigations see:

26. C. C. LIN, *The Theory of Hydrodynamic Stability*, Cambridge, England, 1955; see particularly §§ 4.3 and 8.2;

also, reference 10 (chapter XVI, pp. 314–23). I am grateful to Professor Lin for showing me a proof by Professor K. O. Friederichs on the existence of neutral modes (under certain circumstances) which ensure the instability of the immediately neighbouring modes. The discussion in the text following equation (107) is modelled after Lin's treatment in reference 26 (§ 8.2, pp. 122–3) with the amplification of Friederichs.

The stability of viscous flow along the axis of a cylinder (the so-called Hagen–Poiseuille flow for which $\Psi(r) = 0$) has been considered by:

27. J. PRETSCH, 'Über die Stabilität einer Laminarströmung in einem geraden Rohr mit kreisförmigem Querschnitt', *Z. angew. Math. Mech.* **21**, 204–17 (1941).

Pretsch finds that the flow (as in the inviscid case) is stable.

§ 78 (*b*). The discussion in this section is an expansion of reference 13.

§ 79. The first discussion of the problem treated in this section is due to:

28. S. GOLDSTEIN, 'The stability of viscous fluid flow between rotating cylinders', *Proc. Camb. Phil. Soc.* **33**, 41–61 (1937).

However, Goldstein's insufficient treatment of the characteristic value problem has led him into gross errors in his final results. He finds, for example, that after a small initial increase the critical Taylor number $\bar{T}_c$ decreases very sharply when the Reynolds number R exceeds 25; whereas the calculations summarized in § (*c*) would indicate that $\bar{T}_c$ is a monotonic increasing function of R.

The results summarized in Table XL are taken from:

29. S. CHANDRASEKHAR, 'The hydrodynamic stability of viscid flow between coaxial cylinders,' *Proc. Nat. Acad. Sci.* **46**, 141–3 (1960).

Very similar results were obtained independently and simultaneously by:

30. R. C. DI PRIMA (in press).

The references to the experimental work described in § (*d*) are:

31. R. J. CORNISH, 'Flow of water through fine clearances with relative motion of the boundaries', *Proc. Roy. Soc.* (*London*) A, **140**, 227–40 (1933).
32. C. GAZLEY, JR. 'Heat-transfer characteristics of the rotational and axial flow between concentric cylinders', *Trans. Amer. Soc. Mech. Eng.*, Paper No. 56-A-128, 1–16 (1956).
33. J. KAYE and E. C. ELGAR, 'Modes of adiabatic and diabatic fluid flow in an annulus with an inner rotating cylinder', ibid. 57-HT-14, 1–11 (1957).
34. R. J. DONNELLY and DAVE FULTZ, 'Experiments on the stability of spiral flow between rotating cylinders', *Proc. Nat. Acad. Sci.* **46**, 1150–54 (1960).

IX

THE STABILITY OF COUETTE FLOW IN HYDROMAGNETICS

80. The equations of hydromagnetics in cylindrical polar coordinates

In this chapter we shall extend to hydromagnetics the principal considerations of the last two chapters. In particular we shall be concerned with the stability of Couette flow when a uniform magnetic field is impressed in the direction of the axis and the fluid is an electrical conductor.

The equations governing the fluid under the general circumstances envisaged have been derived in Chapter IV, § 37 (equations IV (10) and (15)). However, for the treatment of problems in which the boundaries are cylinders and the basic flow is circular, it is convenient to have the fundamental equations written in cylindrical polar coordinates (r, θ, z). Denoting by u_r, u_θ, and u_z and H_r, H_θ, and H_z the components of the velocity and of the magnetic intensity in the radial r, the transverse θ, and the axial z, directions, respectively, we have the equations:

$$\frac{\partial u_r}{\partial t}+(\mathbf{u}\,.\,\mathrm{grad})u_r-\frac{u_\theta^2}{r}-\frac{\mu}{4\pi\rho}\left\{(\mathbf{H}\,.\,\mathrm{grad})H_r-\frac{H_\theta^2}{r}\right\}$$

$$= -\frac{\partial \Pi}{\partial r}+\nu\left(\nabla^2 u_r-\frac{2}{r^2}\frac{\partial u_\theta}{\partial \theta}-\frac{u_r}{r^2}\right), \qquad (1)$$

$$\frac{\partial u_\theta}{\partial t}+(\mathbf{u}\,.\,\mathrm{grad})u_\theta+\frac{u_\theta u_r}{r}-\frac{\mu}{4\pi\rho}\left\{(\mathbf{H}\,.\,\mathrm{grad})H_\theta+\frac{H_\theta H_r}{r}\right\}$$

$$= -\frac{1}{r}\frac{\partial \Pi}{\partial \theta}+\nu\left(\nabla^2 u_\theta+\frac{2}{r^2}\frac{\partial u_r}{\partial \theta}-\frac{u_\theta}{r^2}\right), \qquad (2)$$

$$\frac{\partial u_z}{\partial t}+(\mathbf{u}\,.\,\mathrm{grad})u_z-\frac{\mu}{4\pi\rho}(\mathbf{H}\,.\,\mathrm{grad})H_z = -\frac{\partial \Pi}{\partial z}+\nu\nabla^2 u_z, \qquad (3)$$

$$\frac{\partial H_r}{\partial t}+(\mathbf{u}\,.\,\mathrm{grad})H_r-(\mathbf{H}\,.\,\mathrm{grad})u_r = \eta\left(\nabla^2 H_r-\frac{2}{r^2}\frac{\partial H_\theta}{\partial \theta}-\frac{H_r}{r^2}\right), \qquad (4)$$

$$\frac{\partial H_\theta}{\partial t}+(\mathbf{u}\,.\,\mathrm{grad})H_\theta-(\mathbf{H}\,.\,\mathrm{grad})u_\theta+\frac{1}{r}(u_\theta H_r-u_r H_\theta)$$

$$= \eta\left(\nabla^2 H_\theta+\frac{2}{r^2}\frac{\partial H_r}{\partial \theta}-\frac{H_\theta}{r^2}\right), \qquad (5)$$

and $$\frac{\partial H_z}{\partial t}+(\mathbf{u}.\text{grad})H_z-(\mathbf{H}.\text{grad})u_z = \eta\nabla^2 H_z, \tag{6}$$

where $$\Pi = \frac{p}{\rho}+\frac{\mu|\mathbf{H}|^2}{8\pi\rho}+V, \tag{7}$$

and ($\mathbf{u}$.grad) and ($\mathbf{H}$.grad) have the meanings

$$\mathbf{u}.\text{grad} = u_r\frac{\partial}{\partial r}+\frac{u_\theta}{r}\frac{\partial}{\partial\theta}+u_z\frac{\partial}{\partial z} \tag{8}$$

and $$\mathbf{H}.\text{grad} = H_r\frac{\partial}{\partial r}+\frac{H_\theta}{r}\frac{\partial}{\partial\theta}+H_z\frac{\partial}{\partial z}. \tag{9}$$

Also, ∇^2 stands for the Laplacian

$$\nabla^2 = \frac{\partial^2}{\partial r^2}+\frac{1}{r}\frac{\partial}{\partial r}+\frac{1}{r^2}\frac{\partial^2}{\partial\theta^2}+\frac{\partial^2}{\partial z^2}. \tag{10}$$

Further, in cylindrical polar coordinates, the conditions

$$\text{div}\,\mathbf{u} = 0 \quad\text{and}\quad \text{div}\,\mathbf{H} = 0 \tag{11}$$

take the forms $$\frac{\partial u_r}{\partial r}+\frac{u_r}{r}+\frac{1}{r}\frac{\partial u_\theta}{\partial\theta}+\frac{\partial u_z}{\partial z} = 0 \tag{12}$$

and $$\frac{\partial H_r}{\partial r}+\frac{H_r}{r}+\frac{1}{r}\frac{\partial H_\theta}{\partial\theta}+\frac{\partial H_z}{\partial z} = 0. \tag{13}$$

It may be readily verified that the foregoing equations admit the stationary solution

$$\begin{gathered} u_r = u_z = 0, \quad u_\theta = V(r), \\ H_r = H_\theta = 0, \quad\text{and}\quad H_z = H = \text{constant}, \end{gathered} \tag{14}$$

provided $$\frac{d\Pi}{dr} = \frac{V^2}{r} \tag{15}$$

and $$\nu\left(\nabla^2 V-\frac{V}{r^2}\right) = \nu\frac{d}{dr}\left(\frac{dV}{dr}+\frac{V}{r}\right) = 0. \tag{16}$$

The general solution of equation (16) is

$$V(r) = Ar+B/r, \tag{17}$$

where A and B are two constants. Thus, the presence of a uniform magnetic field in the axial direction does not affect the distribution of the transverse velocity which is permissible in the absence of the field. The constants A and B in the solution (17) are related to the angular velocities Ω_1 and Ω_2 of the two cylinders confining the fluid; they are given by (cf. equations VII (145) and (146))

$$A = -\Omega_1\eta^2\frac{1-\mu/\eta^2}{1-\eta^2} \quad\text{and}\quad B = \Omega_1\frac{R_1^2(1-\mu)}{1-\eta^2}, \tag{18}$$

where $$\mu = \Omega_2/\Omega_1 \quad\text{and}\quad \eta = R_1/R_2. \tag{19}$$

If there should be a transverse pressure gradient (as in the problems considered in Chapter VIII, §§ 76 and 77), then a Poiseuille type of flow will be superposed on the distribution (17) exactly as when the field is absent; and the flow will be described by the same equations VIII (5) and (6), and (36).

81. The stability of non-dissipative Couette flow when a magnetic field parallel to the axis is present

Before we investigate the stability of the flow described by equations (14), (15), and (17), allowing fully for the effects of viscosity and resistivity, it will be useful to consider the simpler case when neither of the two dissipative mechanisms is operative. When this is the case, the terms in ν and η in equations (1)–(6) should be suppressed; and the resulting simpler set of equations allow the stationary solution

$$u_r = u_z = 0, \quad u_\theta = V(r) = r\Omega(r),$$
$$H_r = H_\theta = 0, \quad \text{and} \quad H_z = H \text{ constant}, \tag{20}$$

where $V(r)$ is now an *arbitrary function* of r.

Consider an infinitesimal perturbation of the flow represented by the solution (20). Let the perturbed state be described by

$$u_r, \quad V+u_\theta, \quad u_z, \quad \varpi\ (=\delta\Pi), \quad h_r, \quad h_\theta, \quad \text{and} \quad H+h_z. \tag{21}$$

The linear equations governing these perturbations are readily found; they are:

$$\frac{\partial u_r}{\partial t}+\Omega\frac{\partial u_r}{\partial \theta}-2\Omega u_\theta-\frac{\mu H}{4\pi\rho}\frac{\partial h_r}{\partial z} = -\frac{\partial \varpi}{\partial r}, \tag{22}$$

$$\frac{\partial u_\theta}{\partial t}+\Omega\frac{\partial u_\theta}{\partial \theta}+\left(\frac{dV}{dr}+\frac{V}{r}\right)u_r-\frac{\mu H}{4\pi\rho}\frac{\partial h_\theta}{\partial z} = -\frac{1}{r}\frac{\partial \varpi}{\partial \theta}, \tag{23}$$

$$\frac{\partial u_z}{\partial t}+\Omega\frac{\partial u_z}{\partial \theta}-\frac{\mu H}{4\pi\rho}\frac{\partial h_z}{\partial z} = -\frac{\partial \varpi}{\partial z}, \tag{24}$$

$$\frac{\partial h_r}{\partial t}+\Omega\frac{\partial h_r}{\partial \theta}-H\frac{\partial u_r}{\partial z} = 0, \tag{25}$$

$$\frac{\partial h_\theta}{\partial t}+\Omega\frac{\partial h_\theta}{\partial \theta}-H\frac{\partial u_\theta}{\partial z}-\left(\frac{dV}{dr}-\frac{V}{r}\right)h_r = 0, \tag{26}$$

and

$$\frac{\partial h_z}{\partial t}+\Omega\frac{\partial h_z}{\partial \theta}-H\frac{\partial u_z}{\partial z} = 0. \tag{27}$$

In addition to these equations, we have the conditions, (12) and (13), expressing the solenoidal character of **u** and **h**.

Analysing the disturbance into normal modes, we seek solutions of the foregoing equations whose dependence on t, θ, and z is given by

$$e^{i(pt+m\theta+kz)}, \tag{28}$$

where p is a constant (which can be complex), m is an integer (positive, zero, or negative), and k is the wave number of the disturbance in the z-direction.

Let $u_r(r)$, $u_\theta(r)$, etc., now denote the amplitudes of the various perturbations whose (t, θ, z)-dependence is given by (28). Equations (22)–(27) then give

$$i\sigma u_r - 2\Omega u_\theta - \frac{\mu H}{4\pi\rho} ikh_r = -\frac{d\varpi}{dr}, \tag{29}$$

$$i\sigma u_\theta + \left(\frac{dV}{dr} + \frac{V}{r}\right)u_r - \frac{\mu H}{4\pi\rho} ikh_\theta = -\frac{im}{r}\varpi, \tag{30}$$

$$i\sigma u_z - \frac{\mu H}{4\pi\rho} ikh_z = -ik\varpi, \tag{31}$$

$$i\sigma h_r = ikHu_r, \tag{32}$$

$$i\sigma h_\theta = ikHu_\theta + \left(\frac{dV}{dr} - \frac{V}{r}\right)h_r, \tag{33}$$

and

$$i\sigma h_z = ikHu_z, \tag{34}$$

where

$$\sigma = p + m\Omega. \tag{35}$$

Let $u_r = i\sigma\xi_r$, $u_\theta = i\sigma\xi_\theta - \left(\frac{dV}{dr} - \frac{V}{r}\right)\xi_r$, and $u_z = i\sigma\xi_z$ (36)

define the Lagrangian displacement $\boldsymbol{\xi}$ (cf. Chapter VII, § 67 (a)). From a comparison of this definition of $\boldsymbol{\xi}$ with equations (32)–(34) relating $\mathbf{h}$ and $\mathbf{u}$, it is clear that

$$\mathbf{h} = ikH\boldsymbol{\xi}. \tag{37}$$

This proportionality of $\mathbf{h}$ and $\boldsymbol{\xi}$ is an expression of the fact that in a medium of zero resistivity, the lines of force are dragged with the fluid.

In terms of $\boldsymbol{\xi}$, equations (29)–(31) become:

$$\left(\sigma^2 - 2r\Omega\frac{d\Omega}{dr} - \Omega_A^2\right)\xi_r + 2i\sigma\Omega\xi_\theta = \frac{d\varpi}{dr}, \tag{38}$$

$$(\sigma^2 - \Omega_A^2)\xi_\theta - 2i\sigma\Omega\xi_r = \frac{im}{r}\varpi, \tag{39}$$

$$(\sigma^2 - \Omega_A^2)\xi_z = ik\varpi, \tag{40}$$

where

$$\Omega_A^2 = \frac{\mu H^2}{4\pi\rho}k^2 \tag{41}$$

is the Alfvén frequency. In addition to these equations, we have the equation of continuity (cf. equation VII (46))

$$D_*\xi_r+\frac{im}{r}\xi_\theta+ik\xi_z = 0, \tag{42}$$

where

$$D_* = D+\frac{1}{r} = \frac{d}{dr}+\frac{1}{r}. \tag{43}$$

Multiplying equations (39) and (40) by $-im/r$ and $-ik$, respectively, adding and making use of equation (42), we obtain

$$(\sigma^2-\Omega_A^2)D_*\xi_r-\frac{2m\sigma\Omega}{r}\xi_r = \left(\frac{m^2}{r^2}+k^2\right)\varpi. \tag{44}$$

Next eliminating ξ_θ between equations (38) and (39), we have

$$\left(\sigma^2-\Omega_A^2-2r\Omega\frac{d\Omega}{dr}-\frac{4\Omega^2\sigma^2}{\sigma^2-\Omega_A^2}\right)\xi_r = D\varpi+\frac{2m\sigma\Omega}{\sigma^2-\Omega_A^2}\frac{\varpi}{r}. \tag{45}$$

On the other hand,

$$2r\Omega\frac{d\Omega}{dr}+\frac{4\Omega^2\sigma^2}{\sigma^2-\Omega_A^2} = 2r\Omega\frac{d\Omega}{dr}+4\Omega^2+\frac{4\Omega^2\Omega_A^2}{\sigma^2-\Omega_A^2} = \Phi(r)+\frac{4\Omega^2\Omega_A^2}{\sigma^2-\Omega_A^2}, \tag{46}$$

where $\Phi(r)$ is Rayleigh's discriminant. Equation (45) can now be rewritten in the form

$$\left(\sigma^2-\Omega_A^2-\Phi(r)-\frac{4\Omega^2\Omega_A^2}{\sigma^2-\Omega_A^2}\right)\xi_r = D\varpi+\frac{2m\sigma\Omega}{\sigma^2-\Omega_A^2}\frac{\varpi}{r}. \tag{47}$$

This equation must be considered together with equation (44).

If the fluid is confined between two rigid coaxial cylinders of radii R_1 and R_2, we must require that the radial component of the velocity vanishes at the walls. Thus, solutions of equations (44) and (46) must be sought which satisfy the boundary conditions

$$\xi_r = 0 \quad \text{for } r = R_1 \text{ and } R_2. \tag{48}$$

(*a*) *The case* $m = 0$

When $m = 0$, $\sigma = p$ and the basic equations are

$$\left(p^2-\Omega_A^2-\Phi(r)-\frac{4\Omega^2\Omega_A^2}{p^2-\Omega_A^2}\right)\xi_r = D\varpi \tag{49}$$

and

$$(p^2-\Omega_A^2)D_*\xi_r = k^2\varpi. \tag{50}$$

Eliminating ϖ between these equations, we obtain

$$\kappa(DD_*-k^2)\xi_r = -k^2\left[\Phi(r)+\frac{4\Omega^2\Omega_A^2}{\kappa}\right]\xi_r, \tag{51}$$

where

$$\kappa = p^2-\Omega_A^2. \tag{52}$$

Equation (51) together with the boundary conditions (48) constitute a characteristic value problem for κ. We shall first show that *the characteristic values of κ are all real.*

Multiply equation (51) by $r\xi_r^*$ and integrate over the range of r. We obtain, after an integration by parts,

$$\kappa \int_{R_1}^{R_2} r\{|D_* \xi_r|^2+k^2|\xi_r|^2\}\,dr = k^2 \int_{R_1}^{R_2} r\Phi(r)|\xi_r|^2\,dr+ + \frac{4\Omega_A^2\,k^2}{\kappa} \int_{R_1}^{R_2} r\Omega^2|\xi_r|^2\,dr. \quad (53)$$

The imaginary part of this equation gives

$$\text{im}(\kappa)\left[\int_{R_1}^{R_2} r\{|D_* \xi_r|^2+k^2|\xi_r|^2\}\,dr+\frac{4\Omega_A^2\,k^2}{|\kappa|^2}\int_{R_1}^{R_2} r\Omega^2|\xi_r|^2\,dr\right] = 0. \quad (54)$$

The factor of $\text{im}(\kappa)$ in this equation is positive definite. Hence

$$\text{im}(\kappa) = 0; \quad (55)$$

and this proves the reality of κ. The characteristic values of p^2 and the proper functions to which these belong must, therefore, also be real.

In view of the reality of the characteristic values of κ, we can rewrite equation (53) in the manner

$$\kappa^2 I_1-\kappa k^2 I_2-4k^2\Omega_A^2\,I_3 = 0, \quad (56)$$

where
$$I_1 = \int_{R_1}^{R_2} r\{(D_* \xi_r)^2+k^2\xi_r^2\}\,dr, \quad (57)$$

$$I_2 = \int_{R_1}^{R_2} r\Phi(r)\xi_r^2\,dr, \quad (58)$$

and
$$I_3 = \int_{R_1}^{R_2} r\Omega^2\xi_r^2\,dr. \quad (59)$$

Equation (56) provides the basis for a variational formulation of the problem. To see this, consider the effect on κ (determined as a root of equation (56)) of an arbitrary variation, $\delta\xi_r$, in ξ_r compatible only with the boundary conditions on ξ_r. To the first order in the variation,

$$-(2\kappa I_1-k^2 I_2)\delta\kappa = \kappa^2\delta I_1-\kappa k^2\delta I_2-4k^2\Omega_A^2\,\delta I_3, \quad (60)$$

where

$$\delta I_1 = 2\int_{R_1}^{R_2} r\{(D_* \xi_r)(D_* \delta\xi_r)+k^2\xi_r\,\delta\xi_r\}\,dr$$

$$= -2\int_{R_1}^{R_2} r\{(DD_*-k^2)\xi_r\}\,\delta\xi_r\,dr, \quad (61)$$

$$\delta I_2 = 2\int_{R_1}^{R_2} r\Phi(r)\xi_r\,\delta\xi_r\,dr, \tag{62}$$

and

$$\delta I_3 = 2\int_{R_1}^{R_2} r\Omega^2\xi_r\,\delta\xi_r\,dr \tag{63}$$

are the corresponding variations in the integrals I_1, I_2, and I_3. Inserting the foregoing expressions for δI_1, etc., in (60), we obtain

$$(\kappa I_1 - \tfrac{1}{2}k^2 I_2)\frac{\delta\kappa}{\kappa} = \int_{R_1}^{R_2} r\,\delta\xi_r\left\{\kappa(DD_* - k^2)\xi_r + k^2\left[\Phi(r) + \frac{4\Omega^2\Omega_A^2}{\kappa}\right]\xi_r\right\}dr. \tag{64}$$

From this last equation it follows that a *necessary and sufficient condition for $\delta\kappa$ to vanish to the first order, for all small variations in ξ_r which are compatible with the boundary conditions, is that ξ_r be a proper solution of the characteristic value problem.*

Now extracting the root of equation (56) (which is meaningful in view of the theorem we have just stated), we have

$$\kappa = p^2 - \Omega_A^2 = \frac{1}{2I_1}\{k^2 I_2 \pm \sqrt{(k^4 I_2^2 + 16k^2\Omega_A^2\, I_1\, I_3)}\}. \tag{65}$$

For stability it is necessary that

$$p^2 = \Omega_A^2 + \frac{1}{2I_1}\{k^2 I_2 - \sqrt{(k^4 I_2^2 + 16k^2\Omega_A^2\, I_1\, I_3)}\} > 0. \tag{66}$$

This condition is clearly equivalent to (since I_1 is positive definite)

$$2I_1\Omega_A^2 + k^2 I_2 > \sqrt{(k^4 I_2^2 + 16k^2\Omega_A^2\, I_1\, I_3)}. \tag{67}$$

On squaring this inequality (which is permissible since both sides of the inequality are positive), we have

$$4I_1^2\Omega_A^4 + 4k^2\Omega_A^2\, I_1\, I_2 > 16k^2\Omega_A^2\, I_1\, I_3, \tag{68}$$

or

$$I_1\Omega_A^2 > k^2(4I_3 - I_2). \tag{69}$$

Substituting for I_2 and I_3 from equations (58) and (59) in (69), we obtain

$$I_1\Omega_A^2 > k^2\int_{R_1}^{R_2}\{4\Omega^2 - \Phi(r)\}r\xi_r^2\,dr. \tag{70}$$

Making use of the definitions of the Alfvén frequency Ω_A and Rayleigh's discriminant $\Phi(r)$, we obtain

$$I_1\frac{\mu H^2}{4\pi\rho} > -\int_{R_1}^{R_2}\left(\frac{d\Omega^2}{dr}\right)r^2\xi_r^2\,dr \tag{71}$$

as the condition for stability.

From (71) it follows that *in the limit of zero magnetic field, a sufficient condition for stability is that the angular speed, $|\Omega|$, is a monotonic increasing function of r. At the same time, any adverse gradient of angular velocity can be stabilized by a magnetic field of sufficient strength.*†

In some ways it is remarkable that we do *not* recover Rayleigh's criterion, $\Phi(r) > 0$, in the limit of zero magnetic field. The origin of this must lie in the circumstance that in a fluid of zero resistivity the lines of magnetic force are permanently attached to the fluid (cf. Chapter IV, § 38 (*b*)); and the permanency of this attachment is in no way dependent on the strength of the magnetic field.

(*b*) *The case $m \neq 0$*

As in the corresponding hydrodynamic case (Chapter VII, § 67 (*c*)), equations (44), (47), and (48) can be considered as constituting a characteristic value problem for k^2; and considered as such, it is a self-adjoint one. Thus, we obtain from these equations

$$\int\limits_{R_1}^{R_2} r\left\{\sigma^2-\Omega_A^2-\Phi(r)-\frac{4\Omega^2\Omega_A^2}{\sigma^2-\Omega_A^2}\right\}\xi_r^2\,dr = \int\limits_{R_1}^{R_2}\left(r\xi_r\frac{d\varpi}{dr}+\frac{2m\sigma\Omega}{\sigma^2-\Omega_A^2}\varpi\xi_r\right)dr = -\int\limits_{R_1}^{R_2}\varpi\left\{\frac{d}{dr}(r\xi_r)-\frac{2m\sigma\Omega}{\sigma^2-\Omega_A^2}\xi_r\right\}dr = -\int\limits_{R_1}^{R_2}\frac{1}{\sigma^2-\Omega_A^2}\left(\frac{m^2}{r^2}+k^2\right)r\varpi^2\,dr. \tag{72}$$

Alternatively, we can write

$$k^2\int\limits_{R_1}^{R_2}\frac{r\varpi^2}{\sigma^2-\Omega_A^2}\,dr = \int\limits_{R_1}^{R_2}\left\{\frac{1}{\sigma^2-\Omega_A^2}\left(-\frac{m^2}{r^2}\varpi^2+4\Omega^2\Omega_A^2\,\xi_r^2\right)+(\Phi-\sigma^2+\Omega_A^2)\xi_r^2\right\}r.dr; \tag{73}$$

and it can be readily shown that this last equation provides the basis for a variational determination of k^2 for assigned values of the other parameters. However, the conditions for the stability of these modes, $m \neq 0$, have not been examined.

† This depends on the fact that I_1 is bounded, positive definite, and cannot take the value zero.

82. The periods of oscillation of a rotating column of liquid when a magnetic field is impressed in the direction of the axis

In this section we shall consider the solution of equations (44) and (47) in some special cases.

(*a*) *The case* Ω = *constant*

When Ω = constant, $\Phi = 4\Omega^2$ and

$$\Phi(r)+\frac{4\Omega^2\Omega_A^2}{\sigma^2-\Omega_A^2} = \frac{4\Omega^2\sigma^2}{\sigma^2-\Omega_A^2}, \tag{74}$$

where it will be recalled that

$$\sigma^2-\Omega_A^2 = (p+m\Omega)^2-\frac{\mu H^2}{4\pi\rho}k^2 = \kappa \text{ (say)}. \tag{75}$$

Equations (44) and (47) thus become, in this case,

$$\frac{d}{dr}(r\,\xi_r)-\frac{2m\sigma\Omega}{\kappa}\xi_r = \frac{1}{\kappa}\left(\frac{m^2}{r^2}+k^2\right)r\varpi \tag{76}$$

and

$$\left(\kappa-\frac{4\Omega^2\sigma^2}{\kappa}\right)\xi_r = \frac{d\varpi}{dr}+\frac{2m\sigma\Omega}{\kappa}\frac{\varpi}{r}. \tag{77}$$

Eliminating ξ_r between these equations and rearranging, we find

$$\frac{d^2\varpi}{dr^2}+\frac{1}{r}\frac{d\varpi}{dr}+\left\{k^2\left(\frac{4\sigma^2\Omega^2}{\kappa^2}-1\right)-\frac{m^2}{r^2}\right\}\varpi = 0; \tag{78}$$

and the boundary conditions require

$$\frac{d\varpi}{dr}+\frac{2m\sigma\Omega}{\kappa}\frac{\varpi}{r} = 0 \quad \text{for } r = R_1 \text{ and } R_2. \tag{79}$$

Comparing equations (78) and (79) with equations VII (88) and (89), we observe that the equations become identical if

$$\frac{\kappa}{\sigma} \text{ in equations (78) and (79) is replaced by } \sigma. \tag{80}$$

Hence, if σ_0 is the characteristic value (for a given k) of the hydrodynamic problem considered in § 68 (*a*), the characteristic value σ for the present hydromagnetic problem is given by

$$\sigma_0 = \frac{\kappa}{\sigma} = \sigma-\frac{\Omega_A^2}{\sigma}, \tag{81}$$

or

$$\sigma^2-\sigma_0\,\sigma-\Omega_A^2 = 0. \tag{82}$$

Thus, the presence of an axial magnetic field splits each characteristic value σ (in the absence of the field) into two, σ_1 and σ_2, such that

$$\sigma_1+\sigma_2 = \sigma_0 \quad\text{and}\quad \sigma_1\sigma_2 = -\Omega_A^2; \tag{83}$$

more particularly,

$$\sigma = \tfrac{1}{2}[\sigma_0\pm\sqrt{(\sigma_0^2+4\Omega_A^2)}]. \tag{84}$$

(b) The case $m = 0$ and $\Omega = A+B/r^2$; the stabilizing effect of a magnetic field in the case of narrow gaps

As a second example we shall consider equation (51) in case $\Omega(r)$ has the form, $A+B/r^2$, permissible under viscous flow. In this case

$$\Phi(r) = 4A\Omega(r), \tag{85}$$

and equation (51) has explicitly the form

$$(DD_*-k^2)\xi_r = -\frac{4k^2}{\kappa^2}\left(Ap^2+\Omega_A^2\frac{B}{r^2}\right)\Omega\xi_r. \tag{86}$$

If the gap $d = R_2-R_1$ between the cylinders is small compared with the mean radius $R_0 = \frac{1}{2}(R_1+R_2)$, then, we may, as in such past instances, ignore the difference between D and D_* and, further, replace $\Omega(r)$ by

$$\Omega = \Omega_1[1-(1-\mu)\zeta], \tag{87}$$

where

$$\zeta = (r-R_1)/d. \tag{88}$$

In the same approximation

$$A \simeq -\tfrac{1}{2}\Omega_1\frac{1-\mu}{1-\eta} \quad \text{and} \quad \frac{B}{r^2} \simeq \tfrac{1}{2}\Omega_1\frac{1-\mu}{1-\eta}. \tag{89}$$

Thus, in the framework of these approximations, equation (86) becomes

$$(D^2-a^2)\xi_r = a^2\frac{2\Omega_1^2(1-\mu)}{\kappa(1-\eta)}[1-(1-\mu)\zeta]\xi_r, \tag{90}$$

where

$$k = a/d \quad \text{and} \quad D = d/d\zeta. \tag{91}$$

If we let

$$\lambda = -\frac{2\Omega_1^2(1-\mu)}{\kappa(1-\eta)} = -\frac{2\Omega_1^2 R_0(1-\mu)}{d\kappa}, \tag{92}$$

equation (90) becomes identical with equation VII (111). Hence, if $\lambda(a;\mu)$ is the characteristic value determined by Reid for the problem considered in § 68 (*b*) (i), the required value of κ is given by

$$\kappa = -\frac{2\Omega_1^2 R_0(1-\mu)}{d\lambda(a;\mu)}. \tag{93}$$

Hence,

$$p^2 = \Omega_A^2-\frac{2\Omega_1^2 R_0(1-\mu)}{d\lambda(a;\mu)} \tag{94}$$

Inserting the expression (41) for Ω_A^2 in (94), we can write†

$$p^2 = k^2\left[\frac{\mu_* H^2}{4\pi\rho}-\frac{2\Omega_1^2 R_0 d(1-\mu)}{a^2\lambda(a;\mu)}\right]. \tag{95}$$

† To avoid using μ with two different meanings (once as the magnetic permeability and once as the ratio of the angular velocities), we are temporarily letting μ_* denote the magnetic permeability.

Therefore, disturbances with a wave number $k = a/d$ will be stabilized if

$$\frac{\mu_* H^2}{4\pi\rho} > 2\Omega_1^2 R_0 d[(1-\mu)\Lambda(a;\mu)], \tag{96}$$

where

$$\Lambda(a;\mu) = \frac{1}{a^2\lambda(a;\mu)}. \tag{97}$$

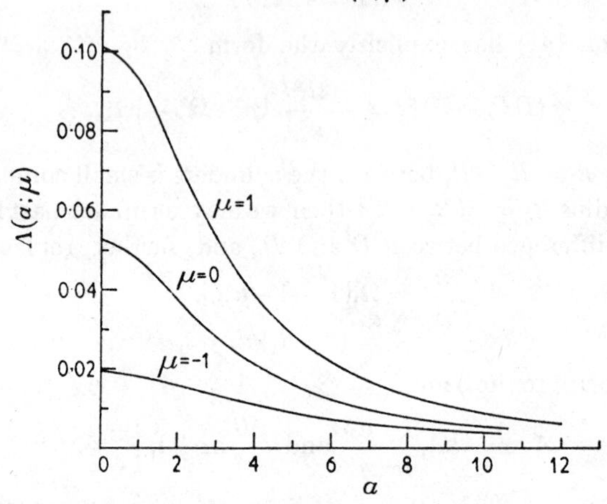

FIG. 99. The function $\Lambda(a;\mu)$ (see equation (97)) for various values of μ.

TABLE XLI

Values of $\Lambda_0(\mu)$ and $(1-\mu)\Lambda_0(\mu)$

$1-\mu$	$\Lambda_0(\mu)$	$(1-\mu)\Lambda_0(\mu)$	$1-\mu$	$\Lambda_0(\mu)$	$(1-\mu)\Lambda_0(\mu)$
0	0·10132	0	1·10	0·04831	0·05315
0·25	0·08843	0·02211	1·15	0·04621	0·05315
0·50	0·07634	0·03817	1·20	0·04418	0·05302
0·75	0·06435	0·04826	1·25	0·04208	0·05260
1·00	0·05277	0·05277	1·50	0·03269	0·04903
1·05	0·05053	0·05305	1·75	0·02513	0·04397

The function $\Lambda(a;\mu)$ is illustrated in Fig. 99 for $\mu = 1, 0,$ and -1. It will be observed that $\Lambda(a;\mu)$ is a monotonic decreasing function of a. Consequently, if

$$\Lambda_0(\mu) = \lim_{a\to 0}\Lambda(a;\mu), \tag{98}$$

the adverse flow represented by (87) (for $\mu < 1$) will be stabilized for all wave numbers (i.e. stabilized completely) if

$$\frac{\mu_* H^2}{4\pi\rho} > 2\Omega_1^2 R_0 d[(1-\mu)\Lambda_0(\mu)]. \tag{99}$$

Numerical values of $\Lambda_0(\mu)$ and $(1-\mu)\Lambda_0(\mu)$ are listed in Table XLI (see also Fig. 100).

It will be observed that $(1-\mu)\Lambda_0(\mu)$ exhibits a maximum at $\mu \simeq -0{\cdot}1275$, where it has a value $\sim 0{\cdot}0532$. Hence, *all adverse flows will be stabilized if*

$$\frac{\mu_* H^2}{4\pi\rho} > q_c(2\Omega_1^2 R_0 d), \text{ where } q_c \simeq 0{\cdot}0532. \tag{100}$$

For $(1-\mu) \to \infty$, it can be shown that

$$\Lambda_0(\mu) \to \frac{1}{(-a_1)^3(1-\mu)^2} = \frac{0{\cdot}07824}{(1-\mu)^2} \quad (1-\mu \to \infty), \tag{101}$$

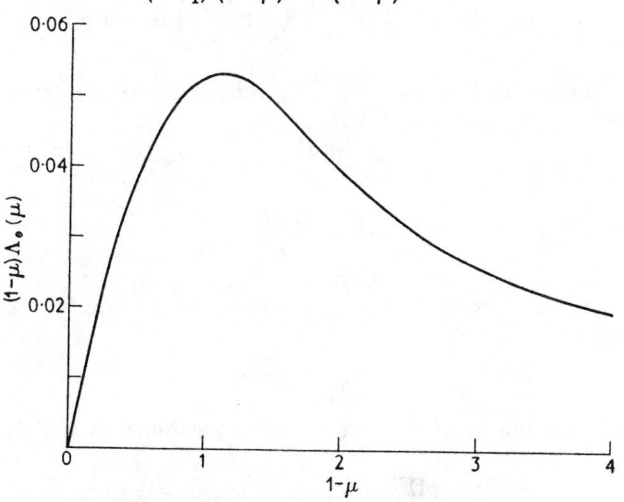

FIG. 100. The behaviour of $(1-\mu)\Lambda_0(\mu)$ (see equation (99)).

where a_1 is the first zero of Airy's function $\text{Ai}(x)$ (see p. 289); therefore, in this limit

$$\frac{\mu_* H^2}{4\pi\rho} > 0{\cdot}1565 \frac{\Omega_1^2 R_0 d}{1-\mu} \quad (1-\mu \to \infty). \tag{102}$$

83. The stability of non-dissipative Couette flow when a current flows parallel to the axis

Another interesting example of non-dissipative Couette flow occurs when a stationary distribution of current parallel to the axis is present. In this case, a magnetic field in the transverse θ-direction will prevail; and, as we shall see, its effect on the stability of the flow is quite different from that of a magnetic field in the axial z-direction.

Equations (1)–(6), with the terms in ν and η suppressed, allow the stationary solution

$$u_r = u_z = 0, \quad u_\theta = V(r), \quad H_r = H_z = 0, \quad H_\theta = H_\theta(r), \tag{103}$$

and
$$\frac{d\Pi}{dr} = -\frac{V^2}{r} + \frac{\mu}{4\pi\rho}\frac{H_\theta^2}{r}. \tag{104}$$

Let a perturbed state of this flow be described by

$$u_r,\quad u_z,\quad V(r)+u_\theta,\quad h_r,\quad h_z,\quad H_\theta(r)+h_\theta,\quad \text{and}\quad \varpi = \delta\Pi. \tag{105}$$

Restricting ourselves to a disturbance belonging to a particular mode by seeking solutions whose (t, θ, z)-dependence is given by (28), we readily find that the relevant equations are:

$$i\sigma u_r - 2\Omega u_\theta - \frac{\mu}{4\pi\rho}\left(\frac{imH_\theta}{r}h_r - 2\frac{H_\theta h_\theta}{r}\right) = -D\varpi, \tag{106}$$

$$i\sigma u_\theta + (D_* V)u_r - \frac{\mu}{4\pi\rho}\left\{\frac{imH_\theta}{r}h_\theta + (D_* H_\theta)h_r\right\} = -\frac{im}{r}\varpi, \tag{107}$$

$$i\sigma u_z - \frac{\mu}{4\pi\rho}\left(\frac{imH_\theta}{r}h_z\right) = -ik\varpi, \tag{108}$$

$$i\sigma h_r = \frac{imH_\theta}{r}u_r, \tag{109}$$

$$i\sigma h_\theta = -ru_r\frac{d}{dr}\left(\frac{H_\theta}{r}\right) + rh_r\frac{d\Omega}{dr} + \frac{imH_\theta}{r}u_\theta, \tag{110}$$

and
$$i\sigma h_z = \frac{imH_\theta}{r}u_z, \tag{111}$$

where the various symbols have the same meanings as in § 81; in particular,

$$\sigma = p + m\Omega \quad \text{and} \quad D_* = D + \frac{1}{r} = \frac{d}{dr} + \frac{1}{r}. \tag{112}$$

(Note that $h_r = h_z = 0$ for the $m = 0$ modes.)

Letting $\boldsymbol{\xi}$ denote the Lagrangian displacement defined as in equations (36) and inserting for $\mathbf{u}$ in terms of $\boldsymbol{\xi}$ in equations (109)–(111), we find that

$$h_r = \frac{imH_\theta}{r}\xi_r, \qquad h_z = \frac{imH_\theta}{r}\xi_z, \tag{113}$$

and
$$h_\theta = \frac{imH_\theta}{r}\xi_\theta - r\xi_r\frac{d}{dr}\left(\frac{H_\theta}{r}\right). \tag{114}$$

Now substituting for $\mathbf{u}$ and $\mathbf{h}$ in accordance with equations (36), (113), and (114) in equations (106)–(108), we find after some reductions

$$\left[\sigma^2 - m^2\Omega_H^2 - 2r\Omega\frac{d\Omega}{dr} + \frac{\mu r}{4\pi\rho}\frac{d}{dr}\left(\frac{H_\theta}{r}\right)^2\right]\xi_r + {} + 2i(\sigma\Omega - m\Omega_H^2)\xi_\theta = D\varpi, \tag{115}$$

$$(\sigma^2 - m^2\Omega_H^2)\xi_\theta - 2i(\sigma\Omega - m\Omega_H^2)\xi_r = \frac{im}{r}\varpi, \tag{116}$$

and $$(\sigma^2-m^2\Omega_H^2)\xi_z = ik\varpi, \tag{117}$$

where $$\Omega_H^2 = \frac{\mu}{4\pi\rho}\left(\frac{H_\theta}{r}\right)^2. \tag{118}$$

In addition to equations (115)–(117), we have the equation of continuity (42).

Multiplying equations (116) and (117) by $-im/r$ and $-ik$, respectively, adding and making use of the equation of continuity, we obtain

$$(\sigma^2-m^2\Omega_H^2)D_*\xi_r-\frac{2m}{r}(\sigma\Omega-m\Omega_H^2)\xi_r = \left(\frac{m^2}{r^2}+k^2\right)\varpi; \tag{119}$$

while eliminating ξ_θ between equations (115) and (116), we obtain

$$\left[\sigma^2-m^2\Omega_H^2-2r\Omega\frac{d\Omega}{dr}+\frac{\mu}{4\pi\rho}r\frac{d}{dr}\left(\frac{H_\theta}{r}\right)^2-4\frac{(\sigma\Omega-m\Omega_H^2)^2}{\sigma^2-m^2\Omega_H^2}\right]\xi_r$$
$$= D\varpi+\frac{2m}{r}\frac{\sigma\Omega-m\Omega_H^2}{\sigma^2-m^2\Omega_H^2}\varpi. \tag{120}$$

Solutions of equations (119) and (120) must be sought which satisfy the boundary conditions

$$\xi_r = 0 \quad \text{for } r = R_1 \text{ and } R_2. \tag{121}$$

(*a*) *The case* $m = 0$

When $m = 0$, equations (119) and (120) simplify considerably. And since, moreover, $\sigma = p$ in this case, we are left with

$$D_*\xi_r = \frac{k^2}{p^2}\varpi \tag{122}$$

and $$[p^2-\Psi(r)]\xi_r = D\varpi, \tag{123}$$

where $$\Psi(r) = 2r\Omega\frac{d\Omega}{dr}+4\Omega^2-\frac{\mu}{4\pi\rho}r\frac{d}{dr}\left(\frac{H_\theta}{r}\right)^2. \tag{124}$$

Recalling the definition of Rayleigh's discriminant $\Phi(r)$, we can write

$$\Psi(r) = \Phi(r)-\frac{\mu}{4\pi\rho}r\frac{d}{dr}\left(\frac{H_\theta}{r}\right)^2. \tag{125}$$

Now eliminating ϖ between equations (122) and (123), we obtain

$$(DD_*-k^2)\xi_r = -\frac{k^2}{p^2}\Psi(r)\xi_r. \tag{126}$$

Comparing equation (126) with equation VII (58) we observe that $\Psi(r)$ plays for this problem exactly the same role which Rayleigh's discriminant, $\Phi(r)$, does for the hydrodynamic problem. We may, therefore, conclude: *a necessary and sufficient condition, for the flow represented by* (103) *to be stable for axisymmetric perturbations, is that* $\Psi(r)$ *be positive throughout the interval* (R_1, R_2); *likewise, the flow is necessarily unstable if* $\Psi(r)$ *should change sign anywhere inside the interval.* This theorem was first established by Michael.

(*b*) *The case* $m \neq 0$

When $m \neq 0$, equations (119)–(121), considered as a characteristic value problem for k^2, can be formulated in variational form; and the basis for such a formulation is provided by the following integral relation which can be readily established:

$$\int\limits_{R_1}^{R_2} r\left\{\sigma^2-m^2\Omega_H^2-\Psi(r)+4\Omega^2-\frac{4(\sigma\Omega-m\Omega_H^2)^2}{\sigma^2-m^2\Omega_H^2}\right\}\xi_r^2\,dr$$

$$= -\int\limits_{R_1}^{R_2} \frac{1}{\sigma^2-m^2\Omega_H^2}\left(\frac{m^2}{r^2}+k^2\right)r\varpi^2\,dr. \tag{127}$$

However, the conditions for the stability of these modes, $m \neq 0$, have not been examined.

84. The stability of non-dissipative Couette flow when an axial and a transverse magnetic field are present

We shall briefly consider the case when superposed on a uniform axial magnetic field H_z there is a distribution of current density which produces an azimuthal magnetic field $H_\theta(r)$. In examining the stability of a non-dissipative Couette flow under these circumstances, we shall restrict ourselves to axisymmetric perturbations.

Proceeding as in §§ 81 and 83, we readily find that in place of equations (29)–(34) and (106)–(111) we now have

$$ipu_r-2\Omega u_\theta-\frac{\mu}{4\pi\rho}\left(ikH_z h_r-\frac{2H_\theta}{r}h_\theta\right) = -D\varpi, \tag{128}$$

$$ipu_\theta+(D_* V)u_r-\frac{\mu}{4\pi\rho}\{ikH_z h_\theta+h_r(D_* H_\theta)\} = 0, \tag{129}$$

$$ipu_z-\frac{\mu}{4\pi\rho}ikH_z h_z = -ik\varpi, \tag{130}$$

$$iph_r = ikH_z u_r, \tag{131}$$

$$iph_\theta = ikH_z u_\theta+rh_r\frac{d\Omega}{dr}-ru_r\frac{d}{dr}\left(\frac{H_\theta}{r}\right), \tag{132}$$

and

$$iph_z = ikH_z h_z. \tag{133}$$

We now define the Lagrangian displacement $\boldsymbol{\xi}$ by the equations

$$u_r = ip\xi_r, \quad u_\theta = ip\xi_\theta - r\frac{d\Omega}{dr}\xi_r, \quad \text{and} \quad u_z = ip\xi_z. \tag{134}$$

(These equations differ from (36) only by the replacement of σ by p as is appropriate for the case $m = 0$.) From a comparison of equations (131)–(133) with (134), it follows that

$$h_r = ikH_z\xi_r, \quad h_\theta = ikH_z\xi_\theta - r\xi_r\frac{d}{dr}\left(\frac{H_\theta}{r}\right), \quad \text{and} \quad h_z = ikH_z\xi_z. \tag{135}$$

With $\mathbf{u}$ and $\mathbf{h}$ expressed in this manner in terms of $\boldsymbol{\xi}$, equations (128) and (129) become

$$\left(p^2-\Omega_A^2-r\frac{d}{dr}\Omega^2+r\frac{d}{dr}\Omega_H^2\right)\xi_r+2i(p\Omega-\Omega_A\Omega_H)\xi_\theta = D\varpi \tag{136}$$

and

$$(p^2-\Omega_A^2)\xi_\theta-2i(p\Omega-\Omega_A\Omega_H)\xi_r = 0, \tag{137}$$

where

$$\Omega_A^2 = \frac{\mu H_z^2 k^2}{4\pi\rho} \quad \text{and} \quad \Omega_H^2 = \frac{\mu H_\theta^2}{4\pi\rho r^2} \tag{138}$$

have the same meanings as in §§ 81 and 83 (cf. equations (41) and (118)). Similarly, equation (130) becomes

$$(p^2-\Omega_A^2)\xi_z = ik\varpi; \tag{139}$$

but the equation of continuity in the present axisymmetric case requires

$$D_*\xi_r = -ik\xi_z; \tag{140}$$

hence,

$$(p^2-\Omega_A^2)D_*\xi_r = k^2\varpi. \tag{141}$$

By suitably combining equations (136), (137), and (141), we finally obtain

$$(p^2-\Omega_A^2)(DD_*-k^2)\xi_r = -k^2\left[r\frac{d}{dr}(\Omega^2-\Omega_H^2)+4\frac{(p\Omega-\Omega_A\Omega_H)^2}{p^2-\Omega_A^2}\right]\xi_r. \tag{142}$$

The full implications of this equation have not been explored; but the special case, $\Omega = 0$, offers no difficulty.

(*a*) *The case* $\Omega = 0$

In this case equation (142) reduces to

$$\kappa(DD_*-k^2)\xi_r = -k^2\left(-r\frac{d\Omega_H^2}{dr}+\frac{4\Omega_A^2\Omega_H^2}{\kappa}\right)\xi_r, \tag{143}$$

where

$$\kappa = p^2-\Omega_A^2. \tag{144}$$

On comparing equation (143) with equation (51), we observe that the two equations are of exactly the same form with Ω_H and $-r\,d\Omega_H^2/dr$

playing the roles of Ω and $\Phi(r)$, respectively. Accordingly, the theorems of § 81 (*a*) have now their exact counterparts. Thus, the characteristic values of κ (and, therefore, also of p^2) are real; and a necessary and sufficient condition for stability is (cf. equation (70))

$$I_1\Omega_A^2 > k^2\int\limits_{R_1}^{R_2}\left(4\Omega_H^2+r\frac{d\Omega_H^2}{dr}\right)r\xi_r^2\,dr. \tag{145}$$

Substituting for Ω_A^2 and Ω_H^2 from (138) in (145), we have

$$I_1 H_z^2 > \int\limits_{R_1}^{R_2}\left\{4\frac{H_\theta^2}{r^2}+r\frac{d}{dr}\left(\frac{H_\theta}{r}\right)^2\right\}r\xi_r^2\,dr. \tag{146}$$

An equivalent form of this inequality is

$$I_1 H_z^2 > \int\limits_{R_1}^{R_2}\frac{\xi_r^2}{r^2}\frac{d}{dr}(r^2H_\theta^2)\,dr. \tag{147}$$

From this it follows that *a sufficient condition for stability is that $H_\theta(r)$ decreases more rapidly than $1/r$.*

85. The stability of dissipative Couette flow in hydromagnetics. The perturbation equations

As we have seen in § 80, the equations of hydromagnetics allow, in the presence of an impressed, uniform, axial magnetic field, the same stationary distribution of $\Omega(r)$ as in the absence of a field. Nevertheless, in analogy with the Bénard problem considered in Chapter IV, we should expect that the magnetic field will have a very pronounced effect on the onset of instability. In this section we begin the investigation of this problem.

Let a perturbed state of the flow be described by

$$u_r,\quad V+u_\theta,\quad u_z,\quad h_r,\quad h_\theta,\quad H+h_z,\quad \text{and}\quad \varpi\ (=\delta\Pi). \tag{148}$$

We shall assume further that the various perturbations are axisymmetric and independent of θ. We then obtain from equations (1)–(6) the linearized equations

$$\frac{\partial u_r}{\partial t}-2\Omega u_\theta-\frac{\mu H}{4\pi\rho}\frac{\partial h_r}{\partial z}=-\frac{\partial\varpi}{\partial r}+\nu\left(\nabla^2u_r-\frac{u_r}{r^2}\right), \tag{149}$$

$$\frac{\partial u_\theta}{\partial t}+\left(\frac{dV}{dr}+\frac{V}{r}\right)u_r-\frac{\mu H}{4\pi\rho}\frac{\partial h_\theta}{\partial z}=\nu\left(\nabla^2u_\theta-\frac{u_\theta}{r^2}\right), \tag{150}$$

$$\frac{\partial u_z}{\partial t}-\frac{\mu H}{4\pi\rho}\frac{\partial h_z}{\partial z}=-\frac{\partial\varpi}{\partial z}+\nu\nabla^2u_z, \tag{151}$$

$$\frac{\partial h_r}{\partial t} - H\frac{\partial u_r}{\partial z} = \eta\left(\nabla^2 h_r - \frac{h_r}{r^2}\right), \tag{152}$$

$$\frac{\partial h_\theta}{\partial t} - H\frac{\partial u_\theta}{\partial z} - \left(\frac{dV}{dr} - \frac{V}{r}\right)h_r = \eta\left(\nabla^2 h_\theta - \frac{h_\theta}{r^2}\right), \tag{153}$$

and

$$\frac{\partial h_z}{\partial t} - H\frac{\partial u_z}{\partial z} = \eta\nabla^2 h_z, \tag{154}$$

where ∇^2 now has the meaning

$$\nabla^2 = \frac{\partial^2}{\partial r^2} + \frac{1}{r}\frac{\partial}{\partial r} + \frac{\partial^2}{\partial z^2}. \tag{155}$$

Also, for axisymmetric perturbations, the conditions requiring $\mathbf{u}$ and $\mathbf{h}$ to be solenoidal are

$$\frac{\partial u_r}{\partial r} + \frac{u_r}{r} + \frac{\partial u_z}{\partial z} = 0 \tag{156}$$

and

$$\frac{\partial h_r}{\partial r} + \frac{h_r}{r} + \frac{\partial h_z}{\partial z} = 0. \tag{157}$$

By analysing the disturbance into normal modes, we seek solutions of the foregoing equations which are of the forms

$$\begin{aligned} u_r &= e^{pt}u(r)\cos kz; \quad h_r = e^{pt}\phi(r)\sin kz, \\ u_\theta &= e^{pt}v(r)\cos kz; \quad h_\theta = e^{pt}\psi(r)\sin kz, \\ u_z &= e^{pt}w(r)\sin kz; \quad h_z = e^{pt}\chi(r)\cos kz, \end{aligned} \tag{158}$$

and

$$\varpi = \varpi(r)e^{pt}\cos kz,$$

where, as the notation implies, u, v, etc., are all functions of r only. For solutions of the form (158), equations (149)–(154) become

$$\nu\left(DD_* - k^2 - \frac{p}{\nu}\right)u + \frac{\mu Hk}{4\pi\rho}\phi + 2\Omega v = D\varpi, \tag{159}$$

$$\nu\left(DD_* - k^2 - \frac{p}{\nu}\right)v + \frac{\mu Hk}{4\pi\rho}\psi - (D_* V)u = 0, \tag{160}$$

$$\nu\left(D_* D - k^2 - \frac{p}{\nu}\right)w - \frac{\mu Hk}{4\pi\rho}\chi = -k\varpi, \tag{161}$$

$$\eta\left(DD_* - k^2 - \frac{p}{\eta}\right)\phi = Hku, \tag{162}$$

$$\eta\left(DD_* - k^2 - \frac{p}{\eta}\right)\psi = Hkv - \left(\frac{dV}{dr} - \frac{V}{r}\right)\phi, \tag{163}$$

$$\eta\left(D_* D - k^2 - \frac{p}{\eta}\right)\chi = -Hkw, \tag{164}$$

$$D_* u = -kw, \tag{165}$$

and

$$D_*\phi = +k\chi, \tag{166}$$

where D and D_* have the same meanings as in the preceding sections (cf. equation (112)).

Considering equations (159)–(166), we first observe that equations (162), (164), (165), and (166) are not all independent; for, by applying D_* to equation (162) and making use of equations (165) and (166), we recover equation (164).

Eliminating ϖ between equations (159) and (161), we obtain

$$\nu\left(DD_*-k^2-\frac{p}{\nu}\right)u+\frac{\mu Hk}{4\pi\rho}\phi+2\Omega v = -\frac{\nu}{k}\left(DD_*-k^2-\frac{p}{\nu}\right)Dw+\frac{\mu H}{4\pi\rho}D\chi. \quad (167)$$

Replacing w and χ in this equation by $-D_* u/k$ and $D_*\phi/k$ (in accordance with equations (165) and (166)), we obtain, after some rearrangement,

$$\left(DD_*-k^2-\frac{p}{\nu}\right)(DD_*-k^2)u+\frac{\mu Hk}{4\pi\rho\nu}(DD_*-k^2)\phi = 2\frac{\Omega}{\nu}k^2 v. \quad (168)$$

Equations (160), (162), (163), and (168) provide a system of equations of combined order ten for u, v, ϕ, and ψ.

(a) *The boundary conditions*

The boundary conditions on u, v, and w require that they vanish on the walls. According to equation (165) the condition $w = 0$ can be replaced by $D_* u = 0$, and since $u = 0$ on the walls, this last condition is also equivalent to $Du = 0$. Hence the conditions on u and v are

$$u = v = 0 \quad \text{and} \quad Du = 0 \quad \text{for } r = R_1 \text{ and } R_2. \quad (169)$$

The remaining boundary conditions refer to ϕ and ψ: and they depend on the nature of the walls confining the fluid. As we have seen in Chapter IV, § 42 (*a*), we might in this connexion consider two cases; when the walls are of non-conducting material and when they are of perfectly conducting material. (These are the cases distinguished as A and B in § 42 (*a*).) The boundary conditions appropriate to these two cases are (cf. equations IV (106) and (108))

$$J_r = 0 \quad \text{(on non-conducting walls)} \quad (170)$$

and
$$h_r = 0 \quad \text{and} \quad J_z = 0 \quad \text{(on conducting walls)}. \quad (171)$$

Equivalent forms of these conditions are

$$(\text{curl}\,\mathbf{h})_r = 0 \quad \text{(on non-conducting walls)} \quad (172)$$

and
$$h_r = 0 \quad \text{and} \quad (\text{curl}\,\mathbf{h})_z = 0 \quad \text{(on conducting walls)}. \quad (173)$$

Since we are considering axisymmetric perturbations, the first of the conditions in (172) requires $\partial h_\theta/\partial z = 0$; and according to the form of the solutions assumed (cf. equation (158)), this requires $\psi = 0$. Again, the condition $(\text{curl}\,\mathbf{h})_z = 0$, for the form of the solutions assumed, requires $D_* h_\theta = 0$, i.e. $D_* \psi = 0$. Thus, the conditions (172) and (173) require

$$\psi = 0 \quad \text{(on non-conducting walls)} \tag{174}$$

and

$$\phi = 0 \quad \text{and} \quad D_* \psi = 0 \quad \text{(on conducting walls)}. \tag{175}$$

(*b*) *The equations governing the marginal state when the onset of instability is as a stationary secondary flow*

So far no rigorous discussion has been undertaken to determine whether the principle of the exchange of stabilities is valid for the problem under consideration. However, on the basis of experimental evidence, we shall investigate only the case when instability sets in as a stationary secondary flow. In this case, the equations governing the marginal state can be obtained by setting $p = 0$ in the relevant equations, namely, (160), (162), (163), and (168). We have

$$(DD_* - k^2)^2 u + \frac{\mu H k}{4\pi\rho\nu}(DD_* - k^2)\phi = \frac{2\Omega}{\nu} k^2 v, \tag{176}$$

$$(DD_* - k^2) v + \frac{\mu H k}{4\pi\rho\nu}\psi = \frac{1}{\nu}(D_* V) u, \tag{177}$$

$$(DD_* - k^2)\phi = \frac{kH}{\eta} u, \tag{178}$$

and

$$(DD_* - k^2)\psi = \frac{kH}{\eta} v - \frac{1}{\eta}\left(\frac{dV}{dr} - \frac{V}{r}\right)\phi. \tag{179}$$

We can eliminate ϕ from equation (176) by directly substituting equation (178). We obtain

$$\left[(DD_* - k^2)^2 + \frac{\mu H^2}{4\pi\rho\nu\eta} k^2\right] u = \frac{2\Omega k^2}{\nu} v. \tag{180}$$

In deriving the foregoing equations, we have not made any use of the special distribution of the angular velocity given by (17); they are therefore of general applicability to *any* Couette flow. However, when $\Omega(r)$ has the form (17),

$$D_* V = 2A \quad \text{and} \quad \frac{dV}{dr} - \frac{V}{r} = -2\frac{B}{r^2}. \tag{181}$$

The boundary conditions with respect to which equations (177)–(180) must be solved are the same as those described in § (*a*) above.

(c) *The reduction to the case of a narrow gap*

We shall restrict our discussion of equations (177)–(180) to the case when the gap $d = R_2 - R_1$ between the cylinders is small compared to the mean radius $R_0 = \frac{1}{2}(R_1+R_2)$. Then in a scheme of approximation with which we are now familiar, equations (177)–(180) become

$$(D^2-a^2)v+\frac{\mu Hd}{4\pi\rho\nu}a\psi = \frac{2Ad^2}{\nu}u, \tag{182}$$

$$(D^2-a^2)\phi = \frac{Hd}{\eta}au, \tag{183}$$

$$(D^2-a^2)\psi = \frac{Hd}{\eta}av+\frac{2Bd^2}{R_0^2\eta}\phi, \tag{184}$$

and
$$[(D^2-a^2)^2+Qa^2]u = \frac{2\Omega_1 d^2}{\nu}a^2[1-(1-\mu)\zeta]v, \tag{185}$$

where
$$\zeta = (r-R_1)/d, \quad k = a/d, \quad \mu = \Omega_2/\Omega_1, \tag{186}$$

and†
$$Q = \frac{\mu H^2}{4\pi\rho\nu\eta}d^2 = \frac{\mu^2H^2\sigma}{\rho\nu}d^2 \tag{187}$$

is the same non-dimensional number which we introduced in Chapter IV (equation IV (134)).

By applying (D^2-a^2) to equation (184) and eliminating v and ϕ by making use of equations (182) and (183), we obtain

$$[(D^2-a^2)^2+Qa^2]\psi = 2d^2\left(\frac{A}{\nu}+\frac{B}{R_0^2\eta}\right)\frac{Hd}{\eta}au. \tag{188}$$

Under terrestrial conditions, the relative magnitude of ν and η enables an important simplification of the foregoing equations. Thus, considering mercury at room temperatures as typical of the fluids to which we may wish to apply the theory, we have

$$\eta = 7{\cdot}6\times10^3 \text{ cm}^2/\text{s while } \nu = 1{\cdot}1\times10^{-3} \text{ cm}^2/\text{s}; \tag{189}$$

whereas in the framework of the approximations underlying equation (188), the constants A and B/R_0^2 are of equal magnitude (though of opposite signs (cf. equation (89)). Therefore, in equation (188), $B/R_0^2\eta$ can be neglected, entirely, in comparison with A/ν; and we shall be left with

$$[(D^2-a^2)^2+Qa^2]\psi = \frac{2Ad^2}{\nu}\frac{Hd}{\eta}au. \tag{190}$$

† It is, again, unfortunate that μ occurs in these few equations with two meanings: as magnetic permeability and as Ω_2/Ω_1. Once the equations are reduced to their non-dimensional forms and the magnetic permeability has been 'absorbed' into Q, this ambiguity will cease; and no confusion is likely in the meantime.

For the same reason, we can neglect the term in ϕ on the right-hand side of equation (184) and write

$$(D^2-a^2)\psi = \frac{Hd}{\eta}av. \tag{191}$$

Substituting for v from this last equation in equation (185), we get

$$[(D^2-a^2)^2+Qa^2]u = \frac{2\Omega_1 d^2}{\nu}\frac{\eta}{Hda}a^2[1-(1-\mu)\zeta](D^2-a^2)\psi. \tag{192}$$

By the further transformation

$$\psi \to \frac{Hda}{\eta}\frac{2Ad^2}{\nu}\psi, \tag{193}$$

equations (190) and (192) become

$$[(D^2-a^2)^2+Qa^2]\psi = u \tag{194}$$

and
$$[(D^2-a^2)^2+Qa^2]u = -Ta^2[1-(1-\mu)\zeta](D^2-a^2)\psi, \tag{195}$$

where
$$T = -\frac{4A\Omega_1}{\nu^2}d^4 \tag{196}$$

is the Taylor number as we have usually defined it for narrow gaps.

The appropriate boundary conditions are

for $\zeta = 0$ and 1, $u = 0$, $Du = 0$, $(D^2-a^2)\psi = 0$, and *either* $\psi = 0$ *or* $D\psi = 0$ (197)

depending on whether the walls are of non-conducting or conducting material.

An important consequence of the circumstance, that under terrestrial conditions $\eta \sim 10^7\nu$, is that the order of the system of equations we have to solve has been reduced from 10 to 8.

Finally, we may note that equations (194) and (195) can be combined to give the single equation

$$[(D^2-a^2)^2+Qa^2]^2\psi = -Ta^2[1-(1-\mu)\zeta](D^2-a^2)\psi. \tag{198}$$

However, we shall not find this form very useful.

86. The solution of the characteristic value problem for the case $\mu > 0$

We have seen in Chapter VII, § 71 (*d*) (see also Chapter VIII, § 77 (*c*)) that, in the case when the cylinders are rotating in the same direction, we can replace the term which allows for the variation of Ω through the gap by its average value. With this replacement and the definitions

$$\bar{T} = \tfrac{1}{2}(1+\mu)T \tag{199}$$

and
$$G = (D^2-a^2)\psi, \tag{200}$$

the equations to be solved are

$$[(D^2-a^2)^2+Qa^2]\psi = u \tag{201}$$

and

$$[(D^2-a^2)^2+Qa^2]u = -\bar{T}a^2G. \tag{202}$$

In considering the equations in the forms (201) and (202), we shall find it convenient to translate the origin of ζ to be midway between the two cylinders. The boundary conditions, then, are

$$u = 0,\ Du = 0,\ G = 0, \text{ and } \textit{either } \psi \textit{ or } D\psi = 0 \text{ for } \zeta = \pm\tfrac{1}{2}. \tag{203}$$

From the symmetry of equations (201) and (202) and the boundary conditions (203), for reflection about $\zeta = 0$, it follows that the proper solutions of the characteristic value problem fall into two non-combining groups of even and odd solutions. And it is also clear that the lowest characteristic values must occur among the even solutions.

(a) *A variational principle*

As in the case of similar characteristic value problems considered in Chapters II, III, and IV, the problem presented by equations (201)–(203) allows a variational formulation; and the characteristic values of $\bar{T}$ represent extremal values which a certain ratio of two positive definite integrals can attain. To obtain this ratio, multiply equation (202) by u and integrate over the range of ζ. We have

$$\int_{-\frac{1}{2}}^{+\frac{1}{2}} u[(D^2-a^2)^2u+Qa^2u]\,d\zeta = -\bar{T}a^2\int_{-\frac{1}{2}}^{+\frac{1}{2}} u(D^2-a^2)\psi\,d\zeta$$

$$= -\bar{T}a^2\int_{-\frac{1}{2}}^{+\frac{1}{2}} [(D^2-a^2)\psi][(D^2-a^2)^2\psi+Qa^2\psi]\,d\zeta. \tag{204}$$

After one or more integrations by parts (in which the integrated parts vanish on account of one or other of the boundary conditions), we find that both sides of (204) can be brought to positive definite forms. The result is

$$\bar{T} = \frac{\int_{-\frac{1}{2}}^{+\frac{1}{2}} \{[(D^2-a^2)u]^2+Qa^2u^2\}\,d\zeta}{a^2\int_{-\frac{1}{2}}^{+\frac{1}{2}} \{(DG)^2+a^2G^2+Qa^2[(D\psi)^2+a^2\psi^2]\}\,d\zeta}. \tag{205}$$

And it can be shown, as in other similar instances, that the solution of the characteristic value problem is equivalent to finding extremal values of $\bar{T}$ given by the expression on the right-hand side of (205) for arbitrary variations of ψ subject to the boundary conditions and the additional 'constraint'

$$[(D^2-a^2)^2+Qa^2]\psi = u. \tag{206}$$

The critical Taylor number $\overline{T}$ for the onset of instability (for a given Q) represents, in fact, the *absolute minimum* which the quantity on the right-hand side of (205) can attain.

(*b*) *The solution of the characteristic value problem. The secular determinant*

In solving the characteristic value problem presented by equations (201)–(203), we shall follow a method similar to the ones we have used in other instances: we expand u in terms of a set of orthogonal functions each of which satisfies the boundary conditions on u; we then solve equation (201) for ψ and arrange that the boundary conditions on ψ are also satisfied. With all the boundary conditions thus met, we insert the assumed expansion for u and the derived expansion for ψ in equation (202); and this will lead us directly to the secular determinant in the usual manner. The underlying self-adjoint character of the problem must reflect itself in the symmetry of the deduced secular matrix.

Our first problem is then the selection of a suitable set of orthogonal functions in terms of which to expand u. The functions which have been constructed and studied in detail by Harris and Reid provide such a set. These functions are determined as the proper solutions of the characteristic value problem presented by the equation,

$$\frac{d^4y}{dx^4} = \alpha^4 y, \tag{207}$$

and the boundary conditions

$$y = y' = 0 \quad \text{for } x = \pm\tfrac{1}{2}. \tag{208}$$

The proper solutions of this problem fall into two non-combining groups of even and odd functions, respectively. The standard forms of these solutions are

$$C_m(x) = \frac{\cosh\lambda_m x}{\cosh\frac{1}{2}\lambda_m} - \frac{\cos\lambda_m x}{\cos\frac{1}{2}\lambda_m} \tag{209}$$

and

$$S_m(x) = \frac{\sinh\mu_m x}{\sinh\frac{1}{2}\mu_m} - \frac{\sin\mu_m x}{\sin\frac{1}{2}\mu_m}, \tag{210}$$

where λ_m and μ_m ($m = 1, 2, 3,...$) are the positive roots of the equations

$$\tanh\tfrac{1}{2}\lambda + \tan\tfrac{1}{2}\lambda = 0 \quad \text{and} \quad \coth\tfrac{1}{2}\mu - \cot\tfrac{1}{2}\mu = 0. \tag{211}$$

The functions $C_m(x)$ and $S_m(x)$ satisfy the orthogonality relations

$$\int_{-\frac{1}{2}}^{+\frac{1}{2}} C_m(x)C_n(x)\,dx = \int_{-\frac{1}{2}}^{+\frac{1}{2}} S_m(x)S_n(x)\,dx = \delta_{mn} \tag{212}$$

and

$$\int_{-\frac{1}{2}}^{+\frac{1}{2}} C_m(x)S_n(x)\,dx = 0. \tag{213}$$

Returning to equations (201)–(203) and restricting ourselves to the even solutions, we expand u in terms of the functions C_m; thus

$$u = \sum_m A_m C_m(\zeta), \tag{214}$$

where the summation over m may be considered as running from 1 to ∞. The corresponding expansion for ψ is given by

$$\psi = \sum_m A_m \psi_m(\zeta), \tag{215}$$

where $\psi_m(\zeta)$ is the solution of the equation

$$[(D^2-a^2)^2+Qa^2]\psi_m = C_m \tag{216}$$

which satisfies the boundary conditions on ψ.

The general solution of equation (216) which is even can be written in the form

$$\psi_m = \frac{1}{(\lambda_m^2-a^2)^2+Qa^2}\frac{\cosh\lambda_m\zeta}{\cosh\frac{1}{2}\lambda_m} - \frac{1}{(\lambda_m^2+a^2)^2+Qa^2}\frac{\cos\lambda_m\zeta}{\cos\frac{1}{2}\lambda_m} + B_1^{(m)}\cosh q_1\zeta + B_2^{(m)}\cosh q_2\zeta, \tag{217}$$

where $B_1^{(m)}$ and $B_2^{(m)}$ are constants of integration, and

$$q_1^2 = a^2+ia\sqrt{Q} \quad\text{and}\quad q_2^2 = a^2-ia\sqrt{Q} \tag{218}$$

are the roots of the equation

$$(q^2-a^2)^2+Qa^2 = 0. \tag{219}$$

Without loss of generality, we may write

$$q_1 = \alpha_1+i\alpha_2 \quad\text{and}\quad q_2 = \alpha_1-i\alpha_2, \tag{220}$$

where

$$\alpha_1 = \{\tfrac{1}{2}\sqrt{(a^4+Qa^2)}+\tfrac{1}{2}a^2\}^{\frac{1}{2}} \quad\text{and}\quad \alpha_2 = \{\tfrac{1}{2}\sqrt{(a^4+Qa^2)}-\tfrac{1}{2}a^2\}^{\frac{1}{2}}. \tag{221}$$

It is apparent that $B_1^{(m)}$ and $B_2^{(m)}$ are complex conjugates.

An identity which follows from equation (219) is

$$\begin{aligned}\Gamma_m &= \frac{1}{[(\lambda_m^2-a^2)^2+Qa^2][(\lambda_m^2+a^2)^2+Qa^2]} \\ &= \frac{1}{(\lambda_m^2-q_1^2)(\lambda_m^2-q_2^2)}\frac{1}{(\lambda_m^2+q_1^2)(\lambda_m^2+q_2^2)} \\ &= \frac{1}{(\lambda_m^4-q_1^4)(\lambda_m^4-q_2^4)} = \frac{1}{|\lambda_m^4-q_1^4|^2}.\end{aligned} \tag{222}$$

On the other hand, according to equations (218)

$$\lambda_m^4-q_{1,2}^4 = \lambda_m^4-a^4+a^2Q\mp 2ia^3\sqrt{Q}. \tag{223}$$

If we let

$$g_m = \frac{1}{a\sqrt{Q}}(\lambda_m^4-a^4+Qa^2), \tag{224}$$

then $$\lambda_m^4-q_{1,2}^4 = (g_m \mp 2ia^2)a\sqrt{Q}. \tag{225}$$

An alternative expression for Γ_m is, therefore,

$$\frac{1}{\Gamma_m} = (g_m^2+4a^4)a^2Q. \tag{226}$$

Making use of the foregoing relations and definitions, we can rewrite the solution (217) for ψ_m in the form

$$\psi_m = \Gamma_m\{(\lambda_m^4+a^4+Qa^2)C_m(\zeta)+2a^2C_m''(\zeta)\}+ \\ +B_1^{(m)}\cosh q_1\zeta+B_2^{(m)}\cosh q_2\zeta, \tag{227}$$

where $$C_m''(\zeta) = \frac{d^2C_m}{d\zeta^2} = \lambda_m^2\left(\frac{\cosh\lambda_m\zeta}{\cosh\frac{1}{2}\lambda_m}+\frac{\cos\lambda_m\zeta}{\cos\frac{1}{2}\lambda_m}\right). \tag{228}$$

From (227) we readily find

$$\begin{aligned} G_m &= (D^2-a^2)\psi_m \\ &= \Gamma_m\{a^2(\lambda_m^4-a^4-Qa^2)C_m(\zeta)+(\lambda_m^4-a^4+Qa^2)C_m''(\zeta)\}+ \\ &\qquad +ia\sqrt{Q}(B_1^{(m)}\cosh q_1\zeta-B_2^{(m)}\cosh q_2\zeta), \end{aligned} \tag{229}$$

where we have made use of equation (218).

The boundary conditions (197) require that G_m and either ψ_m or $D\psi_m$ vanish for $\zeta = \pm\frac{1}{2}$; and these conditions will determine the constants $B_1^{(m)}$ and $B_2^{(m)}$. Postponing for the present the explicit determination of these constants, we shall proceed with the setting up of the secular matrix.

Our next step then is the substitution of the expansions for u and ψ in equation (202); but first, we may observe that

$$[(D^2-a^2)^2+Qa^2]C_m = (\lambda_m^4+a^4+Qa^2)C_m-2a^2C_m''. \tag{230}$$

Therefore, the result of the substitution is

$$\frac{1}{\overline{T}a^2}\sum_m A_m\{(\lambda_m^4+a^4+Qa^2)C_m-2a^2C_m''\}+ \\ +\sum_m A_m\{\Gamma_m[a^2(\lambda_m^4-a^4-Qa^2)C_m+(\lambda_m^4-a^4+Qa^2)C_m'']+ \\ +ia\sqrt{Q}[B_1^{(m)}\cosh q_1\zeta-B_2^{(m)}\cosh q_2\zeta]\} = 0. \tag{231}$$

The required secular equation for $\overline{T}$ follows from this equation after multiplication by $C_n(\zeta)$ and integration over the range of ζ. Making use of the orthogonality property of the C-functions, and with the further definitions

$$X_{mn} = \int_{-\frac{1}{2}}^{+\frac{1}{2}} C_m''(\zeta)C_n(\zeta)\,d\zeta \tag{232}$$

and $$\langle\cosh q\zeta|C_n\rangle = \int_{-\frac{1}{2}}^{+\frac{1}{2}} C_n(\zeta)\cosh q\zeta\,d\zeta, \tag{233}$$

we can write

$$\left\| \frac{1}{Ta^2}\{(\lambda_m^4+a^4+Qa^2)\delta_{mn}-2a^2X_{mn}\}+ \right.$$
$$+\Gamma_m\{a^2(\lambda_m^4-a^4-Qa^2)\delta_{mn}+(\lambda_m^4-a^4+Qa^2)X_{mn}\}+$$
$$\left. +ia\sqrt{Q}\{B_1^{(m)}\langle\cosh q_1\zeta|C_n\rangle-B_2^{(m)}\langle\cosh q_2\zeta|C_n\rangle\} \right\| = 0. \quad (234)$$

By elementary calculations we find†

$$X_{mn} = \frac{2}{\lambda_m^4-\lambda_n^4}(C_m'''\,C_n''-C_n'''\,C_m'')_{\zeta=\frac{1}{2}} \quad (m \neq n)$$
$$= \frac{1}{\lambda_n^4}\{\tfrac{1}{2}C_n'''\,C_n''-\tfrac{1}{4}(C_n''')^2\}_{\zeta=\frac{1}{2}} \quad (m = n), \quad (235)$$

and
$$\langle\cosh q\zeta|C_n\rangle = \frac{2}{\lambda_n^4-q^4}\{C_n'''(\tfrac{1}{2})\cosh\tfrac{1}{2}q-C_n''(\tfrac{1}{2})q\sinh\tfrac{1}{2}q\}, \quad (236)$$

where it may be noted that

$$C_m''(\tfrac{1}{2}) = 2\lambda_m^2 \quad\text{and}\quad C_m'''(\tfrac{1}{2}) = 2\lambda_m^3\tanh\tfrac{1}{2}\lambda_m. \quad (237)$$

The last line in (234) can be simplified by making appropriate use of equations (222) and (224)–(226). Thus, we find successively

$$ia\sqrt{Q}\{B_1^{(m)}\langle\cosh q_1\zeta|C_n\rangle-B_2^{(m)}\langle\cosh q_2\zeta|C_n\rangle\}$$
$$= 4a\Gamma_n\sqrt{Q}\,\mathrm{re}\{i(\lambda_n^4-q_2^4)B_1^{(m)}[C_n'''(\tfrac{1}{2})\cosh\tfrac{1}{2}q_1-C_n''(\tfrac{1}{2})q_1\sinh\tfrac{1}{2}q_1]\}$$
$$= 4a^2Q\Gamma_n\,\mathrm{re}\{(-2a^2+ig_n)B_1^{(m)}[C_n'''(\tfrac{1}{2})\cosh\tfrac{1}{2}q_1-C_n''(\tfrac{1}{2})q_1\sinh\tfrac{1}{2}q_1]\}. \quad (238)$$

Thus, the secular equation finally takes the form

$$\left\| \frac{1}{Ta^2}\{(\lambda_m^4+a^4+Qa^2)\delta_{mn}-2a^2X_{mn}\}+ \right.$$
$$+\Gamma_m\{a^2(\lambda_m^4-a^4-Qa^2)\delta_{mn}+g_m X_{mn}\,a\sqrt{Q}\}+$$
$$\left. +4a^2Q\Gamma_n\,\mathrm{re}\{(-2a^2+ig_n)B_1^{(m)}[C_n'''(\tfrac{1}{2})\cosh\tfrac{1}{2}q_1-C_n''(\tfrac{1}{2})q_1\sinh\tfrac{1}{2}q_1]\} \right\| = 0. \quad (239)$$

The further reduction of the secular matrix requires an explicit evaluation of the constant $B_1^{(m)}$; and this depends on the choice of the boundary conditions. We consider the two cases separately.

(c) *The case of non-conducting walls*

In this case the boundary conditions are that G_m and ψ_m vanish at $\zeta = \pm\frac{1}{2}$. Applying these conditions to the solutions (227) and (229), we obtain

$$B_1^{(m)}\cosh\tfrac{1}{2}q_1+B_2^{(m)}\cosh\tfrac{1}{2}q_2 = -2a^2\Gamma_m\,C_m''(\tfrac{1}{2})$$

and

$$B_1^{(m)}\cosh\tfrac{1}{2}q_1-B_2^{(m)}\cosh\tfrac{1}{2}q_2 = ig_m\Gamma_m\,C_m''(\tfrac{1}{2}), \quad (240)$$

† The results quoted can be found in the very convenient tables of integrals which Reid and Harris have provided (see reference 10 in the Bibliographical Notes at the end of this chapter).

where g_m is defined in equation (224). Therefore,

$$B_1^{(m)} = \tfrac{1}{2}\Gamma_m C_m''(\tfrac{1}{2})(-2a^2+ig_m)\operatorname{sech}\tfrac{1}{2}q_1 \tag{241}$$

and $B_2^{(m)}$ is its complex conjugate. With $B_1^{(m)}$ given by equation (241), the last line of (239) becomes

$$\begin{aligned}2a^2Q\Gamma_m\Gamma_n C_m''(\tfrac{1}{2})\operatorname{re}\{(2a^2-ig_m)(2a^2-ig_n)[C_n'''(\tfrac{1}{2})-C_n''(\tfrac{1}{2})q_1\tanh\tfrac{1}{2}q_1]\}\\ = 2a^2Q\Gamma_m\Gamma_n C_m''(\tfrac{1}{2})C_n'''(\tfrac{1}{2})(4a^4-g_m g_n)-\\ -2a^2Q\Gamma_m\Gamma_n C_m''(\tfrac{1}{2})C_n''(\tfrac{1}{2})\{(4a^4-g_m g_n)\operatorname{re}(q_1\tanh\tfrac{1}{2}q_1)+\\ +2a^2(g_m+g_n)\operatorname{im}(q_1\tanh\tfrac{1}{2}q_1)\}.\end{aligned} \tag{242}$$

Therefore, defining

$$Z_{mn} = (4a^4-g_m g_n)C_m''(\tfrac{1}{2})C_n'''(\tfrac{1}{2})+\frac{g_m a\sqrt{Q}}{2a^2Q\Gamma_n}X_{mn} \tag{243}$$

and

$$\begin{aligned}\Sigma_{mn} = C_m''(\tfrac{1}{2})C_n''(\tfrac{1}{2})\{(g_m g_n-4a^4)\operatorname{re}(q_1\tanh\tfrac{1}{2}q_1)-\\ -2a^2(g_m+g_n)\operatorname{im}(q_1\tanh\tfrac{1}{2}q_1)\},\end{aligned} \tag{244}$$

we can rewrite the secular equation (239) in the form

$$\begin{Vmatrix}\frac{1}{Ta^2}\{(\lambda_m^4+a^4+Qa^2)\delta_{mn}-2a^2X_{mn}\}+\\ +a^2(\lambda_m^4-a^4-Qa^2)\Gamma_m\delta_{mn}+2a^2Q\Gamma_m\Gamma_n(Z_{mn}+\Sigma_{mn})\end{Vmatrix} = 0. \tag{245}$$

As we have remarked earlier, the self-adjoint character of the underlying problem requires that the secular matrix be symmetric. While the matrices X_{mn} and Σ_{mn} are manifestly so, Z_{mn} is not; but we can put Z_{mn} in a form in which its symmetry is equally apparent.

Making use of the relations (224) and (226), we can rewrite Z_{mn} in the form

$$Z_{mn} = (4a^4-g_m g_n)C_m''(\tfrac{1}{2})C_n'''(\tfrac{1}{2})+\tfrac{1}{2}(\lambda_m^4-a^4+Qa^2)(4a^4+g_n^2)X_{mn}. \tag{246}$$

Now substituting for X_{mn} in accordance with equation (235), we have

$$\begin{aligned}Z_{mn} = (4a^4-g_m g_n)C_m''(\tfrac{1}{2})C_n'''(\tfrac{1}{2})+\\ +(g_n^2+4a^4)\frac{g_m}{g_m-g_n}\{C_m'''(\tfrac{1}{2})C_n''(\tfrac{1}{2})-C_m''(\tfrac{1}{2})C_n'''(\tfrac{1}{2})\}\quad (m\neq n);\end{aligned} \tag{247}$$

on simplification this becomes

$$\begin{aligned}Z_{mn} = \frac{1}{g_m-g_n}[g_m(g_n^2+4a^4)C_m'''(\tfrac{1}{2})C_n''(\tfrac{1}{2})-\\ -g_n(g_m^2+4a^4)C_n'''(\tfrac{1}{2})C_m''(\tfrac{1}{2})]\quad (m\neq n).\end{aligned} \tag{248}$$

The symmetry of Z_{mn} is now manifest. When $m = n$ the corresponding expression for Z_{nn} is

$$Z_{nn} = (4a^4-g_n^2)C_n''(\tfrac{1}{2})C_n'''(\tfrac{1}{2})+ \\ +\frac{\lambda_n^4-a^4+Qa^2}{2\lambda_n^4}(4a^4+g_n^2)\{\tfrac{1}{2}C_n'''(\tfrac{1}{2})C_n''(\tfrac{1}{2})-\tfrac{1}{4}[C_n'''(\tfrac{1}{2})]^2\}. \quad (249)$$

Finally, we may note that the real and the imaginary parts of $q_1 \tanh \frac{1}{2}q_1$ which occur in the expression for Σ_{mn} are given by

$$\text{re}(q_1 \tanh \tfrac{1}{2}q_1) = \frac{\alpha_1 \sinh \alpha_1 - \alpha_2 \sin \alpha_2}{\cosh \alpha_1 + \cos \alpha_2} \quad (250)$$

and

$$\text{im}(q_1 \tanh \tfrac{1}{2}q_1) = \frac{\alpha_2 \sinh \alpha_1 + \alpha_1 \sin \alpha_2}{\cosh \alpha_1 + \cos \alpha_2}, \quad (251)$$

where α_1 and α_2 are defined in (221).

(d) The case of conducting walls

In this case the boundary conditions are that G_m and $D\psi_m$ vanish at $\zeta = \pm\frac{1}{2}$. Applying these conditions to the solutions (227) and (229), we obtain

$$B_1^{(m)}q_1 \sinh \tfrac{1}{2}q_1 + B_2^{(m)}q_2 \sinh \tfrac{1}{2}q_2 = -2a^2\Gamma_m C_m'''(\tfrac{1}{2})$$

and

$$B_1^{(m)}\cosh \tfrac{1}{2}q_1 - B_2^{(m)}\cosh \tfrac{1}{2}q_2 = +ig_m \Gamma_m C_m''(\tfrac{1}{2}). \quad (252)$$

On solving these equations, we find

$$B_1^{(m)} = \frac{\Gamma_m \operatorname{sech} \frac{1}{2}q_1}{2\,\text{re}(q_1 \tanh \frac{1}{2}q_1)}\{-2a^2C_m'''(\tfrac{1}{2})+ig_m C_m''(\tfrac{1}{2})q_2 \tanh \tfrac{1}{2}q_2\} \quad (253)$$

and $B_2^{(m)}$ is the complex conjugate of this. Inserting this value for $B_1^{(m)}$ in the last line of equations (239) and reducing the resulting expression in the manner of the preceding case (in § (*c*) above), we find that the secular matrix can be brought to exactly the same form (245) with, however, the definitions

$$Z_{mn} = (4a^4+g_m g_n)\times \\ \times\begin{cases} \dfrac{1}{g_m-g_n}\{g_n C_m'''(\tfrac{1}{2})C_n''(\tfrac{1}{2})-g_m C_n'''(\tfrac{1}{2})C_m''(\tfrac{1}{2})\} & (m \neq n), \\ \dfrac{\lambda_n^4-a^4+Qa^2}{2\lambda_n^4}\{\tfrac{1}{2}C_n''(\tfrac{1}{2})C_n'''(\tfrac{1}{2})-\tfrac{1}{4}[C_n'''(\tfrac{1}{2})]^2\}-C_n''(\tfrac{1}{2})C_n'''(\tfrac{1}{2}) & (m = n), \end{cases} \quad (254)$$

and

$$\Sigma_{mn} = \frac{1}{\text{re}(q_1 \tanh \frac{1}{2}q_1)}\{4a^4C_m'''(\tfrac{1}{2})C_n'''(\tfrac{1}{2})+ \\ +g_m g_n C_m''(\tfrac{1}{2})C_n''(\tfrac{1}{2})|q_1 \tanh \tfrac{1}{2}q_1|^2- \\ -2a^2[g_m C_m''(\tfrac{1}{2})C_n'''(\tfrac{1}{2})+g_n C_n''(\tfrac{1}{2})C_m'''(\tfrac{1}{2})]\text{im}(q_1 \tanh \tfrac{1}{2}q_1)\}, \quad (255)$$

where $$|q_1 \tanh \tfrac{1}{2}q_1|^2 = (\alpha_1^2+\alpha_2^2)\frac{\sinh^2\alpha_1+\sin^2\alpha_2}{(\cosh\alpha_1+\cos\alpha_2)^2}. \tag{256}$$

(e) *Numerical results*

The secular equation (245) has been solved for the cases considered in §§ (c) and (d) for a number of values of a and for several initially assigned

TABLE XLII

The critical Taylor numbers and related constants for different values of Q (The case of non-conducting walls and $\mu > 0$)

Q	a_c	$\overline{T}_c$ Second approximation	$\overline{T}_c$ Third approximation	A_2/A_1	A_3/A_1
30	2·68	$3{\cdot}9657\times10^3$		$+0{\cdot}01473$	
100	1·69	$1{\cdot}0821\times10^4$		$+0{\cdot}00887$	
300	0·904	$3{\cdot}2087\times10^4$		$+0{\cdot}00588$	
1,000	0·500	$1{\cdot}0722\times10^5$		$+0{\cdot}00499$	
3,000	0·272	$3{\cdot}2152\times10^5$	$3{\cdot}2152\times10^5$	$+0{\cdot}00477$	$+0{\cdot}00038$
10,000	0·150	$1{\cdot}0720\times10^6$	$1{\cdot}0720\times10^6$	$+0{\cdot}00448$	$+0{\cdot}00037$

TABLE XLIII

The critical Taylor numbers and related constants for different values of (The case of conducting walls and $\mu > 0$)

Q	a_c	$\overline{T}_c$ Second approximation	$\overline{T}_c$ Third approximation	$\overline{T}_c$ Fourth approximation	A_2/A_1	A_3/A_1	A_4/A_1
5	3·20	$2{\cdot}1853\times10^3$			$+0{\cdot}01759$		
10	3·30	$2{\cdot}6924\times10^3$			$+0{\cdot}01782$		
20	3·40	$3{\cdot}8093\times10^3$			$+0{\cdot}01733$		
50	3·45	$7{\cdot}9926\times10^3$			$+0{\cdot}01250$		
100	3·35	$1{\cdot}7573\times10^4$	$1{\cdot}7572\times10^4$		$-0{\cdot}00039$	$+0{\cdot}00096$	
200	2·90	$4{\cdot}4717\times10^4$			$-0{\cdot}03329$		
400	2·20	$1{\cdot}1795\times10^5$	$1{\cdot}1784\times10^5$		$-0{\cdot}08939$	$-0{\cdot}00509$	
1,000	1·45	$3{\cdot}7895\times10^5$	$3{\cdot}7808\times10^5$		$-0{\cdot}1739$	$-0{\cdot}00894$	
4,000	0·77	$1{\cdot}7384\times10^6$	$1{\cdot}7325\times10^6$		$-0{\cdot}2650$	$-0{\cdot}01254$	
10,000	0·50	$4{\cdot}4576\times10^6$	$4{\cdot}3589\times10^6$	$4{\cdot}3576\times10^6$	$-0{\cdot}2859$	$-0{\cdot}03353$	$-0{\cdot}00207$

values of Q. By determining the minimum of the characteristic roots $\overline{T}$ as a function of a (for a given Q), the critical Taylor numbers for the onset of instability were ascertained. The results of the calculations are summarized in Tables XLII and XLIII; they are further illustrated in Figs. 101 and 102.

It is remarkable that the results derived for the two cases differ as markedly as they do. In this respect the present problem differs from

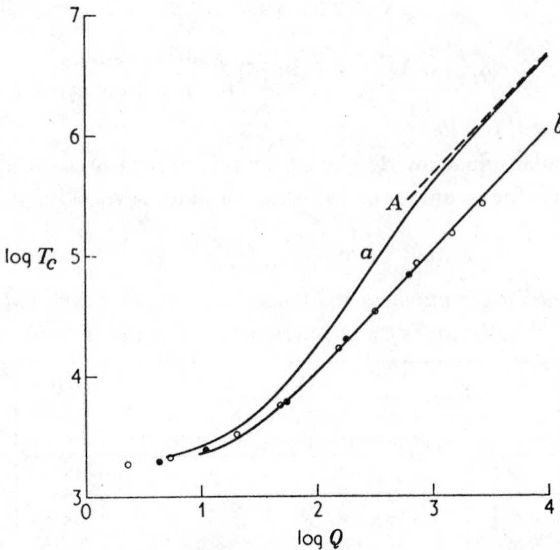

FIG. 101. A comparison of the observed and the theoretical dependence of the critical Taylor number T_c for the onset of instability as a function of Q. The curves labelled a and b are those derived from the theory for the cases of conducting and non-conducting walls, respectively; the dashed curve (labelled A) is the asymptote to which curve a (for the case of conducting walls) tends for $Q \to \infty$. The corresponding asymptote for the case of non-conducting walls cannot be distinguished (in the scale drawn) from the curve b for $Q > 1{,}000$. The experimental results of Donnelly and Ozima are represented by open circles (for the case when $R_1 = 1{\cdot}8$ cm and $R_2 = 2{\cdot}0$ cm) and solid circles (for the case $R_1 = 1{\cdot}9$ cm and $R_2 = 2{\cdot}0$ cm).

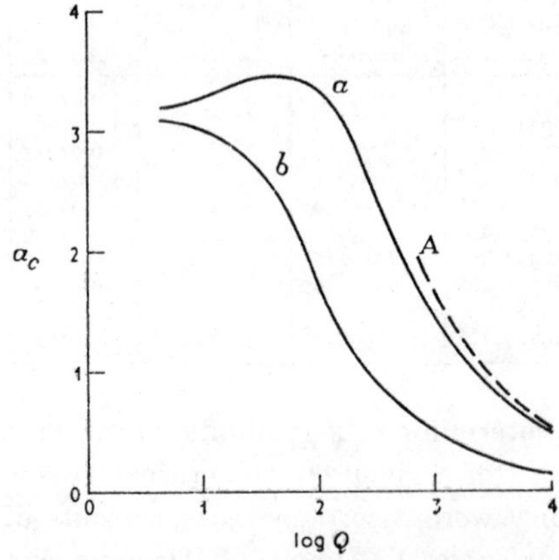

FIG. 102. The variation of the critical wave number a_c as a function of Q: the curves labelled a and b are for the cases of conducting and non-conducting walls, respectively; the dashed curve (labelled A) is the asymptote to which curve a (for the case of conducting walls) tends for $Q \to \infty$. The corresponding asymptote for the case of non-conducting walls cannot be distinguished (in the scale drawn) from the curve b for $Q > 1{,}000$.

the others considered in the earlier chapters. The origin of this difference will become apparent in § (*f*) below. There is, however, no difficulty in understanding why the presence of the magnetic field inhibits the onset of instability and why the wave number of the disturbance manifested at onset tends to zero as $Q \to \infty$. These latter effects are qualitatively the same as in the problem of thermal instability considered in Chapter IV: they clearly derive from the same cause.

(*f*) *The asymptotic behaviour for* $Q \to \infty$

The numerical results presented in Tables XLII and XLIII suggest that

$$a^2 \to 0 \text{ while } Qa^2 \to \text{a finite limit as } Q \to \infty; \tag{257}$$

and, further, that the limit to which Qa^2 tends depends significantly on the boundary conditions. On examining the original differential equations (201) and (202), we find that they are consistent with the following asymptotic behaviours:

$$Qa^2 \to Q_\infty \quad \text{and} \quad \bar{T}a^2 \to T_\infty \text{ as } Q \to \infty \text{ and } a \to 0. \tag{258}$$

On the assumption that these behaviours are valid, we find that the differential equations take the limiting forms

$$(D^4+Q_\infty)\psi = u \quad \text{and} \quad (D^4+Q_\infty)u = -T_\infty D^2\psi, \tag{259}$$

while the boundary conditions are unaffected and remain the same.

To determine then the correct asymptotic behaviours of the critical Taylor number and the associated wave number for $Q \to \infty$, we must solve equations (259) together with the proper boundary conditions, as a characteristic value problem for T_∞, for assigned values of Q_∞, and determine the minimum of the ratio T_∞/Q_∞ as a function of Q_∞. This problem can be solved exactly in the manner of § (*b*) above by setting $a^2 = 0$ in the various expressions except when it occurs in the combinations Qa^2 and Ta^2; they are then replaced by Q_∞ and T_∞, respectively. Thus equations (221), (222), and (224) defining the various quantities in the limit (258) become

$$\alpha_1 = \alpha_2 = \sqrt[4]{(Q_\infty/4)} \tag{260}$$

and

$$g_m \sqrt{Q_\infty} = \lambda_m^4+Q_\infty \quad \text{and} \quad \frac{1}{\Gamma_m} = g_m^2\, Q_\infty = (\lambda_m^4+Q_\infty)^2. \tag{261}$$

With these definitions, the limiting form of the secular equation (245) is

$$\left\| (\lambda_m^4+Q_\infty)\frac{\delta_{mn}}{T_\infty} + \frac{2}{g_m^2\, g_n^2\, Q_\infty}(Z_{mn}+\Sigma_{mn}) \right\| = 0. \tag{262}$$

Similarly passing to the limit appropriately in the expressions for Z_{mn} and Σ_{mn}, we find that

$$Z_{mn} = \frac{g_m g_n}{g_m - g_n}\{g_n C'''_m(\tfrac{1}{2})C''_n(\tfrac{1}{2}) - g_m C'''_n(\tfrac{1}{2})C''_m(\tfrac{1}{2})\} \quad (m \neq n)$$

$$= g_n^2 \frac{\lambda_n^4 + Q_\infty}{2\lambda_n^4}\{\tfrac{1}{2}C'''_n(\tfrac{1}{2})C''_n(\tfrac{1}{2}) - \tfrac{1}{4}[C'''_n(\tfrac{1}{2})]^2\} - g_n^2 C''_n(\tfrac{1}{2})C'''_n(\tfrac{1}{2}) \quad (m = n) \tag{263}$$

for *both* sets of boundary conditions, while

$$\Sigma_{mn} = g_m g_n C''_m(\tfrac{1}{2})C''_n(\tfrac{1}{2}) \times$$

$$\times \begin{cases} \mathrm{re}(q_1 \tanh \tfrac{1}{2}q_1) & \text{(for non-conducting walls)}, \\ \dfrac{|q_1 \tanh \tfrac{1}{2}q_1|^2}{\mathrm{re}(q_1 \tanh \tfrac{1}{2}q_1)} & \text{(for conducting walls)}. \end{cases} \tag{264}$$

In (264) the expressions involving q_1 must be evaluated in accordance with equations (250), (251), (256), and (260).

The secular determinant (262) has been solved in the second and in the third approximations for a number of values of Q_∞ and the minimum value attained by T_∞/Q_∞ has been determined for both sets of boundary conditions. The results of the calculations are

$$Q_\infty = 225, \quad T_\infty = 2{\cdot}4122 \times 10^4; \quad A_1 = 1, \quad A_2 = 0{\cdot}00465, \quad A_3 = 0{\cdot}00036$$

(for the case of non-conducting walls),

and (265)

$$Q_\infty = 2{,}700, \quad T_\infty = 1{\cdot}2184 \times 10^6; \quad A_1 = 1, \quad A_2 = -0{\cdot}32779,$$

$$A_3 = -0{\cdot}01436 \quad \text{(for the case of conducting walls)}.$$

The corresponding asymptotic behaviours are

$$\bar{T} \to 107{\cdot}2Q \quad \text{and} \quad a \to 15{\cdot}0/\sqrt{Q} \quad \text{as } Q \to \infty$$

(for the case of non-conducting walls)

and (266)

$$\bar{T} \to 451{\cdot}27Q \quad \text{and} \quad a \to 52{\cdot}0/\sqrt{Q} \quad \text{as } Q \to \infty$$

(for the case of conducting walls).

The fact that $A_3 \ll A_2$, for the solutions in both cases, may be taken to mean that the limiting values of T_∞/Q_∞ have been determined with adequate precision.

The role of the boundary conditions in determining the asymptotic behaviours for $Q \to \infty$ is apparent from the results given in (266). Moreover, on comparing the predicted asymptotic behaviours with the results of the exact calculations, we observe that while they are nearly attained for $Q = 1{,}000$ in the case of the non-conducting walls, they are not as well attained even for $Q = 10{,}000$ in the case of the conducting walls.

We may finally comment on why, for this problem, all the boundary conditions are relevant to determining the asymptotic behaviours for $Q \to \infty$; this has not been the case in any of the other problems we have considered thus far. Clearly, the reason must lie in the circumstance that, while the magnetic field has the usual effect of elongating the cells in its direction, it has the effect in this instance of making the flow adjoin the walls for longer distances so that the viscous dissipation remains comparable to the Joule dissipation for all strengths of the impressed magnetic field.

87. The solution of the characteristic value problem in the general case

With the origin of ζ midway between the two cylinders, the equations which have to be solved in the general case are

$$[(D^2-a^2)^2+Qa^2]\psi = u \tag{267}$$

and
$$[(D^2-a^2)^2+Qa^2]u = -Ta^2[\tfrac{1}{2}(1+\mu)-(1-\mu)\zeta](D^2-a^2)\psi \tag{268}$$

with the same boundary conditions (203). The occurrence of the term linear in ζ on the right-hand side of equation (268) destroys the symmetry of the problem with respect to reflection and the solutions no longer have a definite parity. On this account, we cannot expand u in terms of the C- or the S-functions alone: we must include both sets of functions. Apart from this necessity of including both the C- and the S-functions in the expansion of u, we can carry out the solution in exactly the same way as in § 86.

We assume, then, that we can expand u and ψ in the forms

$$u = \sum_m A_{C;m}\,C_m(\zeta)+\sum_m A_{S;m}\,S_m(\zeta) \tag{269}$$

and
$$\psi = \sum_m A_{C;m}\,\psi_{C;m}(\zeta)+\sum_m A_{S;m}\,\psi_{S;m}(\zeta), \tag{270}$$

where $\psi_{C;m}$ and $\psi_{S;m}$ are solutions of the equations

$$[(D^2-a^2)^2+Qa^2]\psi_{C;m} = C_m \tag{271}$$

and
$$[(D^2-a^2)^2+Qa^2]\psi_{S;m} = S_m \tag{272}$$

which satisfy the boundary conditions on ψ. The required solutions of equations (271) and (272) are readily found. They are (cf. equation (227))

$$\psi_{C;m} = \Gamma_{C;m}\{(\lambda_m^4+a^4+Qa^2)C_m+2a^2C''_m\}+B_{C;1}^{(m)}\cosh q_1\,\zeta+B_{C;2}^{(m)}\cosh q_2\,\zeta \tag{273}$$

and

$$\psi_{S;m} = \Gamma_{S;m}\{(\mu_m^4+a^4+Qa^2)S_m+2a^2S_m''\}+B_{S;1}^{(m)}\sinh q_1\zeta+B_{S;2}^{(m)}\sinh q_2\zeta, \tag{274}$$

where $B_{C;1}^{(m)}$, $B_{C;2}^{(m)}$, etc., are constants of integration, q_1 and q_2 have the same meanings as in § 86,

$$\Gamma_{C;m} = \frac{1}{|\lambda_m^4-q_1^4|^2}, \quad \text{and} \quad \Gamma_{S;m} = \frac{1}{|\mu_m^4-q_1^4|^2}. \tag{275}$$

The corresponding expressions for $G_{C;m}$ and $G_{S;m}$ are:

$$\begin{aligned} G_{C;m} &= (D^2-a^2)\psi_{C;m} \\ &= \Gamma_{C;m}\{a^2(\lambda_m^4-a^4-Qa^2)C_m+(\lambda_m^4-a^4+Qa^2)C_m''\}+ \\ &\qquad +ia\sqrt{Q}\{B_{C;1}^{(m)}\cosh q_1\zeta-B_{C;2}^{(m)}\cosh q_2\zeta\} \end{aligned} \tag{276}$$

and

$$\begin{aligned} G_{S;m} &= (D^2-a^2)\psi_{S;m} \\ &= \Gamma_{S;m}\{a^2(\mu_m^4-a^4-Qa^2)S_m+(\mu_m^4-a^4+Qa^2)S_m''\}+ \\ &\qquad +ia\sqrt{Q}\{B_{S;1}^{(m)}\sinh q_1\zeta-B_{S;2}^{(m)}\sinh q_2\zeta\}. \end{aligned} \tag{277}$$

The constants of integration follow from applying the boundary conditions to the foregoing solutions. We find (cf. equations (241) and (253))

$$\begin{aligned} B_{C;1}^{(m)} &= \tfrac{1}{2}\Gamma_{C;m}\,C_m''(\tfrac{1}{2})(-2a^2+ig_{C;m})\operatorname{sech}\tfrac{1}{2}q_1, \\ B_{S;1}^{(m)} &= \tfrac{1}{2}\Gamma_{S;m}\,S_m''(\tfrac{1}{2})(-2a^2+ig_{S;m})\operatorname{cosech}\tfrac{1}{2}q_1 \end{aligned} \tag{278}$$

in case of non-conducting walls; and

$$B_{C;1}^{(m)} = \frac{\Gamma_{C;m}\operatorname{sech}\frac{1}{2}q_1}{2\operatorname{re}(q_1\tanh\frac{1}{2}q_1)}\{-2a^2C_m'''(\tfrac{1}{2})+ig_{C;m}\,C_m''(\tfrac{1}{2})q_2\tanh\tfrac{1}{2}q_2\},$$

$$B_{S;1}^{(m)} = \frac{\Gamma_{S;m}\operatorname{cosech}\frac{1}{2}q_1}{2\operatorname{re}(q_1\coth\frac{1}{2}q_1)}\{-2a^2S_m'''(\tfrac{1}{2})+ig_{S;m}\,S_m''(\tfrac{1}{2})q_2\coth\tfrac{1}{2}q_2\} \tag{279}$$

in case of conducting walls. In equations (278) and (279)

$$g_{C;m} = \frac{1}{a\sqrt{Q}}(\lambda_m^4-a^4+Qa^2) \quad \text{and} \quad g_{S;m} = \frac{1}{a\sqrt{Q}}(\mu_m^4-a^4+Qa^2). \tag{280}$$

The constants $B_{C;2}^{(m)}$ and $B_{S;2}^{(m)}$ are the complex conjugates of $B_{C;1}^{(m)}$ and $B_{S;1}^{(m)}$.

The next step is to substitute the expansions for u and ψ in equation (268). We obtain (cf. equations (231))

$$\sum_m A_{C;m}\{(\lambda_m^4+a^4+Qa^2)C_m-2a^2C_m''\}+$$
$$+\sum_m A_{S;m}\{(\mu_m^4+a^4+Qa^2)S_m-2a^2S_m''\}$$
$$= -Ta^2[\tfrac{1}{2}(1+\mu)-(1-\mu)\zeta]\times$$
$$\times\Big[\Big[\sum_m A_{C;m}\{\Gamma_{C;m}[a^2(\lambda_m^4-a^4-Qa^2)C_m+(\lambda_m^4-a^4+Qa^2)C_m'']+$$
$$+ia\sqrt{Q}[B_{C;1}^{(m)}\cosh q_1\zeta-B_{C;2}^{(m)}\cosh q_2\zeta]\}+$$
$$+\sum_m A_{S;m}\{\Gamma_{S;m}[a^2(\mu_m^4-a^4-Qa^2)S_m+(\mu_m^4-a^4+Qa^2)S_m'']+$$
$$+ia\sqrt{Q}[B_{S;1}^{(m)}\sinh q_1\zeta-B_{S;2}^{(m)}\sinh q_2\zeta]\}\Big]\Big]. \quad (281)$$

The secular determinant for T now follows from equation (281) after multiplication by $C_n(\zeta)$, and $S_n(\zeta)$, and integration over the range of ζ. Because of the even and the odd characters of the functions C and S, not all matrix elements on the right-hand side will survive. Thus, no term in $A_{C;m}$ with the factor $(1-\mu)$ will survive on multiplication by C_n and integration; similarly, no term in $A_{S;m}$ with the factor $(1+\mu)$ will survive after the same multiplication and integration.

(*a*) *The case* $\mu = -1$

While there is no formal difficulty in carrying out all the necessary integrations for setting up the secular determinant for any value of μ, we shall restrict ourselves to the case $\mu = -1$ as, formally, the simplest and in some sense, physically, the most interesting. It is formally the simplest since only one-half of all the matrix elements on the right-hand side of (281) will survive in this case; and it is physically the most interesting as the portion of the fluid which is unstable by the criterion of § 81 occupies exactly one-half of the gap.

When $\mu = -1$, the set of equations we derive from (281) on multiplication by $C_n(\zeta)$, and $S_n(\zeta)$, and integration are, respectively,

$$\frac{1}{2Ta^2}\sum_m A_{C;m}\{(\lambda_m^4+a^4+Qa^2)\delta_{mn}-2a^2X_{m,n}^{(C)}\}-$$
$$-\sum_m A_{S;m}\{\Gamma_{S;m}[a^2(\mu_m^4-a^4-Qa^2)(S_m|\zeta|C_n)+(\mu_m^4-a^4+Qa^2)(S_m''|\zeta|C_n)]+$$
$$+2a\sqrt{Q}\,\mathrm{re}[iB_{S;1}^{(m)}\langle\zeta\sinh q_1\zeta|C_n\rangle\} = 0 \quad (282)$$

and

$$\frac{1}{2Ta^2}\sum_m A_{S;m}\{(\mu_m^4+a^4+Qa^2)\delta_{mn}-2a^2X_{m,n}^{(S)}\}-$$
$$-\sum_m A_{C;m}\{\Gamma_{C;m}[a^2(\lambda_m^4-a^4-Qa^2)(C_m|\zeta|S_n)+(\lambda_m^4-a^4+Qa^2)(C''_m|\zeta|S_n)]+$$
$$+2a\sqrt{Q}\,\mathrm{re}[iB_{C;1}^{(m)}\langle\zeta\cosh q_1\zeta|S_n\rangle]\}=0, \quad (283)$$

where

$$X_{m,n}^{(C)}=\int_{-\frac{1}{2}}^{+\frac{1}{2}}C''_m C_n\,d\zeta=\frac{2}{\lambda_m^4-\lambda_n^4}(C'''_m C''_n-C'''_n C''_m)_{\zeta=\frac{1}{2}} \quad (m\neq n)$$
$$=\frac{1}{\lambda_n^4}[\tfrac{1}{2}C'''_n C''_n-\tfrac{1}{4}(C'''_n)^2]_{\zeta=\frac{1}{2}} \quad (m=n); \quad (284)$$

$$X_{m,n}^{(S)}=\int_{-\frac{1}{2}}^{+\frac{1}{2}}S''_m S_n\,d\zeta=\frac{2}{\mu_m^4-\mu_n^4}(S'''_m S''_n-S'''_n S''_m)_{\zeta=\frac{1}{2}} \quad (m\neq n)$$
$$=\frac{1}{\mu_n^4}[\tfrac{1}{2}S'''_n S''_n-\tfrac{1}{4}(S'''_n)^2]_{\zeta=\frac{1}{2}} \quad (m=n); \quad (285)$$

$$(S_m|\zeta|C_n)=\int_{-\frac{1}{2}}^{+\frac{1}{2}}S_m\,\zeta C_n\,d\zeta=\frac{8}{(\lambda_n^4-\mu_m^4)^2}(C'''_n S'''_m)_{\zeta=\frac{1}{2}}; \quad (286)$$

$$(S''_m|\zeta|C_n)=\int_{-\frac{1}{2}}^{+\frac{1}{2}}S''_m\,\zeta C_n\,d\zeta$$
$$=\frac{1}{\lambda_n^4-\mu_m^4}\left[S''_m C'''_n-S'''_m C''_n-\frac{2(3\mu_m^4+\lambda_n^4)}{\lambda_n^4-\mu_m^4}C''_n S''_m\right]_{\zeta=\frac{1}{2}}, \quad (287)$$

$$(S_m|\zeta|C''_n)=\int_{-\frac{1}{2}}^{+\frac{1}{2}}S_m\,\zeta C''_n\,d\zeta$$
$$=\frac{1}{\lambda_n^4-\mu_m^4}\left[S''_m C'''_n-S'''_m C''_n-\frac{2(3\lambda_n^4+\mu_m^4)}{\lambda_n^4-\mu_m^4}C''_n S''_m\right]_{\zeta=\frac{1}{2}}; \quad (288)$$

$$\langle\cosh q\zeta|C_n\rangle=\frac{2}{\lambda_n^4-q^4}[C'''_n(\tfrac{1}{2})\cosh\tfrac{1}{2}q-C''_n(\tfrac{1}{2})q\sinh\tfrac{1}{2}q], \quad (289)$$

$$\langle\sinh q\zeta|S_n\rangle=\frac{2}{\mu_n^4-q^4}[S'''_n(\tfrac{1}{2})\sinh\tfrac{1}{2}q-S''_n(\tfrac{1}{2})q\cosh\tfrac{1}{2}q], \quad (290)$$

$$\langle\zeta\cosh q\zeta|S_n\rangle=\frac{1}{\mu_n^4-q^4}[4q^3\langle\sinh q\zeta|S_n\rangle+$$
$$+(S'''_n-2S''_n)_{\zeta=\frac{1}{2}}\cosh\tfrac{1}{2}q-S''_n(\tfrac{1}{2})q\sinh\tfrac{1}{2}q], \quad (291)$$

and

$$\langle\zeta\sinh q\zeta|C_n\rangle=\frac{1}{\lambda_n^4-q^4}[4q^3\langle\cosh q\zeta|C_n\rangle+$$
$$+(C'''_n-2C''_n)_{\zeta=\frac{1}{2}}\sinh\tfrac{1}{2}q-C''_n(\tfrac{1}{2})q\cosh\tfrac{1}{2}q]. \quad (292)$$

The secular equation derived from equations (282) and (283) has been solved for a number of different values of Q and a to determine the critical Taylor numbers for the onset of instability. The results are summarized in Table XLIV; they are further illustrated in Fig. 103. Fig. 104 shows the effect of the magnetic field on the velocity profile.

A comparison of the present results for the case of counter-rotation with those obtained in § 86 for the case $\mu > 0$ shows that the effect of the axial magnetic field on the onset of instability is very similar in the two cases. The similarity extends even to the occurrence of an early phase in which there is an increase in the wave number of the disturbance, a_c, which is manifested at marginal instability when the confining walls are perfect conductors. In this last respect the case when the confining walls are insulators is different: for the wave number a_c is, then, a monotonic decreasing function of Q. It is to this difference in the two cases that we must attribute the very large difference in the constants of proportionality in the asymptotic behaviours for $Q \to \infty$ (see equations (298) and (299) below).

TABLE XLIV

The critical Taylor numbers and related constants for the case $\mu = -1$

(a) The case of non-conducting walls ($A_{C;1} = 1$)

Q	a_c	T_c	$A_{S;1}$	$A_{C;2}$	$A_{S;2}$
0	4·00	$1{\cdot}870 \times 10^4$†	−0·7168	+0·1020	−0·0297
10	4·00	$2{\cdot}310 \times 10^4$	−0·6956	+0·1127	−0·0292
100	3·40	$7{\cdot}162 \times 10^4$	−0·6250	+0·1633	−0·0263
1,000	0·96	$7{\cdot}143 \times 10^5$	−0·4791	+0·1247	−0·0117
10,000	0·29	$7{\cdot}248 \times 10^6$	−0·4637	+0·1171	−0·0104

(b) The case of conducting walls ($A_{C;1} = 1$)

Q	a_c	T_c	$A_{S;1}$	$A_{C;2}$	$A_{S;2}$
0	4·0	$1{\cdot}870 \times 10^4$	−0·7168	+0·1020	−0·0297
10	3·9	$2{\cdot}457 \times 10^4$	−0·6721	+0·1086	−0·0256
30	4·3	$3{\cdot}693 \times 10^4$	−0·6613	+0·1433	−0·0255
100	4·7	$9{\cdot}200 \times 10^4$	−0·6925	+0·2294	−0·0276
1,000	4·9	$1{\cdot}795 \times 10^6$	−1·0318	+0·6477	−0·1699
3,000‡	4·0	$9{\cdot}064 \times 10^6$	−1·2285	+0·8612	−0·2832

† By the method described in § 71 (Table XXXIII, p. 304) the value of T for this same value of a is $1{\cdot}868 \times 10^4$.

‡ From the relatively high values of the coefficients $A_{C;2}$ and $A_{S;2}$, it would appear that for $Q \geqslant 3{,}000$ solutions in higher approximations should be found if we are to achieve the same accuracy as in the remaining parts of the table.

The asymptotic dependence of a_c and T_c on Q can be determined as in § 86 (*f*). Observing that the differential equations (267) and (268) are again consistent with the behaviours,

$$Qa^2 \to Q_\infty \quad \text{and} \quad Ta^2 \to T_\infty \quad \text{as } Q \to \infty \text{ and } a \to 0, \tag{293}$$

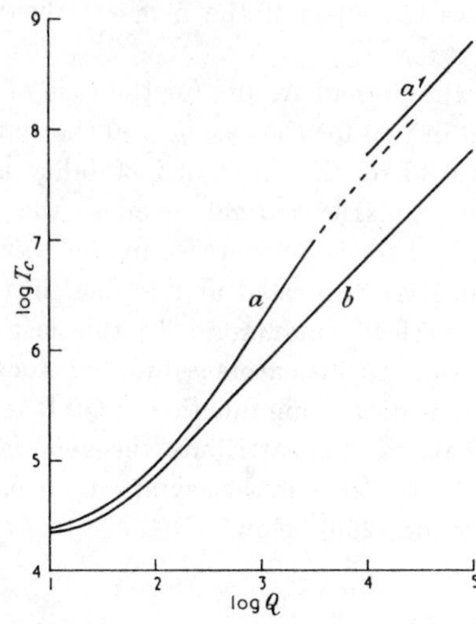

FIG. 103. The variation of the critical Taylor number (T_c) for the onset of instability as a function of Q for the case when the two cylinders are rotating in opposite directions and $\Omega_2/\Omega_1 = -1$. The curves labelled a and b are for the cases of conducting and non-conducting walls, respectively. (The dashed part of curve a is a free-hand extrapolation of the computed curve.) The curve labelled a' is the asymptote to which curve a (for the case of conducting walls) tends for $Q \to \infty$; the corresponding asymptote for the case of non-conducting walls cannot be distinguished (in the scale drawn) from the curve b for $Q > 1{,}000$. (Note the similarity with the corresponding curves in Fig. 102 and in particular the pronounced effect of the electromagnetic boundary conditions.)

where Q_∞ and T_∞ are certain constants, we conclude that in the limit $Q \to \infty$ the differential equations become

$$(D^4+Q_\infty)\psi = u \tag{294}$$

and
$$(D^4+Q_\infty)u = -T_\infty[\tfrac{1}{2}(1+\mu)-(1-\mu)\zeta]D^2\psi \tag{295}$$

with the same boundary conditions (203). To determine the correct asymptotic behaviours of the critical Taylor number and the associated wave number for $Q \to \infty$, we must solve equations (294) and (295) together with the proper boundary conditions, as a characteristic value problem for T_∞, for assigned values of Q_∞ and determine the minimum of the ratio T_∞/Q_∞ as a function of Q_∞. This problem can be solved in the same way as equations (267) and (268): the relevant formulae can in fact

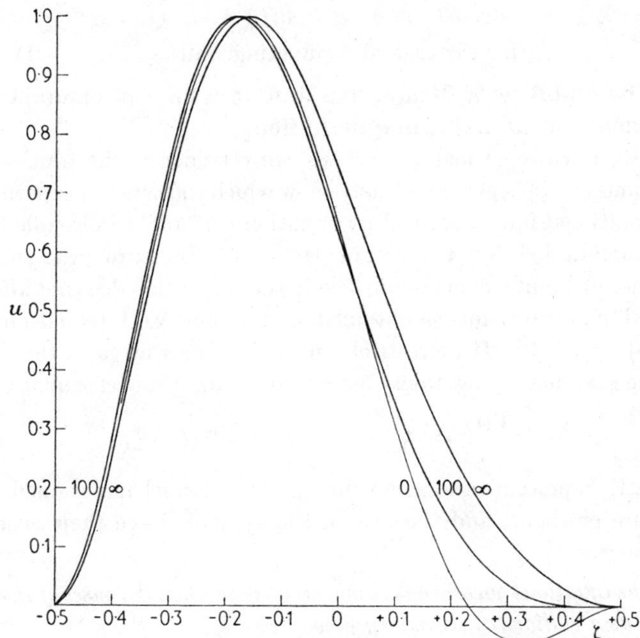

FIG. 104. The dependence of the velocity profile at the onset of instability on Q in case two insulating cylinders are in counter-rotation and Ω_2/Ω_1 is -1. The different curves are labelled by the value of Q to which they refer. The larger amplitudes in the inner half of the gap between the cylinders originate in the circumstance that in this half the distribution of velocities in the stationary flow is unstable according to the criterion valid for non-dissipative Couette flow.

be written down by setting $a^2 = 0$ in the various expressions except when it occurs in the combinations Qa^2 and Ta^2; they are then replaced by Q_∞ and T_∞. In this manner it is found that

$$Q_\infty = 817, \quad T_\infty = 5{\cdot}931\times 10^5; \quad A_{C;1} = 1, \quad A_{S;1} = -0{\cdot}4625,$$
$$A_{C;2} = 0{\cdot}1151, \quad \text{and} \quad A_{S;2} = -0{\cdot}0102$$
$$\text{(for the case of non-conducting walls and } \mu = -1) \qquad (296)$$

and

$$Q_\infty = 285{,}000, \quad T_\infty = 1{\cdot}768\times 10^9; \quad A_{C;1} = 1, \quad A_{S;1} = 4{\cdot}293,$$
$$A_{C;2} = -12{\cdot}311, \quad A_{S;2} = 15{\cdot}331, \quad A_{C;3} = -10{\cdot}709, \quad A_{S;3} = 3{\cdot}329$$
$$\text{(for the case of conducting walls and } \mu = -1). \qquad (297)$$

The corresponding asymptotic behaviours are

$$T \to 726Q \quad \text{and} \quad a \to 28{\cdot}6Q^{-\frac{1}{6}} \quad \text{as } Q \to \infty$$
$$\text{(for the case of non-conducting walls and } \mu = -1), \qquad (298)$$

and $$T \to 6203Q \quad \text{and} \quad a \to 534Q^{-\frac{1}{6}} \quad \text{as } Q \to \infty$$
$$\text{(for the case of conducting walls and } \mu = -1\text{).} \qquad (299)$$

88. The stability of dissipative flow in a curved channel in the presence of an axial magnetic field

As we have remarked in § 80, we can consider in the framework of equations (1)–(6) a general Couette flow which consists of a superposition of a rotational flow described by equation (17) and a Poiseuille type of flow maintained by a constant transverse pressure gradient. The presence of a uniform magnetic field in the z-direction does not affect the hydrodynamical solution obtained in Chapter VIII (equations VIII (5), (6), and (36)). If we restrict ourselves to narrow gaps, the solution for the stationary flow takes the approximate form (cf. equation VIII (37))

$$\frac{V(r)}{r} = \Omega_1[1-(1-\mu)\zeta]+\frac{6V_m}{R_1}\zeta(1-\zeta), \qquad (300)$$

where V_m represents the mean flow in the channel maintained by the pressure gradient, and the rest of the symbols have their customary meanings.

(*a*) *The equations governing the marginal state when the onset of instability is as a stationary secondary flow*

Equations (177)–(180) are applicable to any Couette flow; we have only to ascribe to V and Ω their current meanings. Accordingly, in the framework of the narrow gap approximation (and with the same neglect of the term in ϕ/η in equation (179) for the same reason), equations (190) and (192) are now replaced by

$$[(D^2-a^2)^2+Qa^2]\psi = \frac{2d^2}{\nu}\frac{Hda}{\eta}\left[A+\frac{3V_m}{d}(1-2\zeta)\right]u \qquad (301)$$

and

$$[(D^2-a^2)^2+Qa^2]u = \frac{2d^2a^2}{\nu}\frac{\eta}{Hda}\left\{\Omega_1[1-(1-\mu)\zeta]+\frac{6V_m}{R_1}\zeta(1-\zeta)\right\}(D^2-a^2)\psi; \qquad (302)$$

and the boundary conditions with respect to which these equations must be solved are the same as in § 85 (namely, those given in (197)).

(*b*) *The stability of a pure pressure maintained flow*

In view of the multiplicity of parameters which equations (301) and (302) contain, we shall restrict our consideration of these equations to the case of a pure pressure maintained flow. In this case $\Omega_1 = A = 0$, and equations (301) and (302) become

$$[(D^2-a^2)^2+Qa^2]\psi = \frac{6V_m d}{\nu}\frac{Hda}{\eta}(1-2\zeta)u \qquad (303)$$

and $$[(D^2-a^2)^2+Qa^2]u = \frac{12V_m d^2}{R_1 \nu}\frac{\eta}{Hda}a^2\zeta(1-\zeta)(D^2-a^2)\psi. \tag{304}$$

By the further transformation

$$\psi \to \frac{6V_m d}{\nu}\frac{H\,da}{\eta}\psi, \tag{305}$$

the equations take the more convenient forms

$$[(D^2-a^2)^2+Qa^2]\psi = (1-2\zeta)u \tag{306}$$

and $$[(D^2-a^2)^2+Qa^2]u = a^2\Lambda\zeta(1-\zeta)(D^2-a^2)\psi, \tag{307}$$

where $$\Lambda = \frac{72V_m^2 d^3}{R_1 \nu^2} = 72\mathrm{R}^2\frac{d}{R_1} \tag{308}$$

is the same non-dimensional number which we introduced in Chapter VIII, § 76 (*a*). And the boundary conditions are

$$u = 0,\quad Du = 0,\quad (D^2-a^2)\psi = 0,\quad \text{and } \textit{either } \psi \textit{ or } D\psi = 0$$
$$\text{for } \zeta = 0 \text{ and } 1. \tag{309}$$

TABLE XLV

The critical values of Λ and related constants for different values of Q

(*a*) *The case of non-conducting walls* ($A_{C;1} = 1$)

Q	a_c	Λ_c	$A_{S;1}$	$A_{C;2}$	$A_{S;2}$
0	3·96	9·2998 × 10⁴	0·2746	0·0276	+0·00123
50	3·65	2·1744 × 10⁵	0·4418	0·0477	+0·00057
100	3·185	3·6880 × 10⁵	0·5463	0·0657	−0·00029
1,000	0·866	3·7203 × 10⁶	0·6078	0·0584	−0·00396

(*b*) *The case of conducting walls* ($A_{C;1} = 1$)

Q	a_c	Λ_c	$A_{S;1}$	$A_{C;2}$	$A_{S;2}$
30	4·26	1·8635 × 10⁵	0·4351	0·0499	0·00086
100	4·57	4·7832 × 10⁵	0·7152	0·1338	0·00047
300	4·88	1·8297 × 10⁶	1·3174	0·5497	0·04357
1000	6·45(?)	1·252 × 10⁷			

The characteristic value problem presented by equations (306), (307), and (309) can be solved by a method similar to the one used in § 87: we expand u in terms of the C- and the S-functions; solve equation (306) for ψ and arrange that the boundary conditions on ψ are satisfied; substitute the assumed expansion for u and the derived expansion for ψ in equation (307) and derive the secular equation by multiplication by C_n and S_n and integration over ζ. The details of the analysis are somewhat long and complicated; and as no special novelty is involved, we

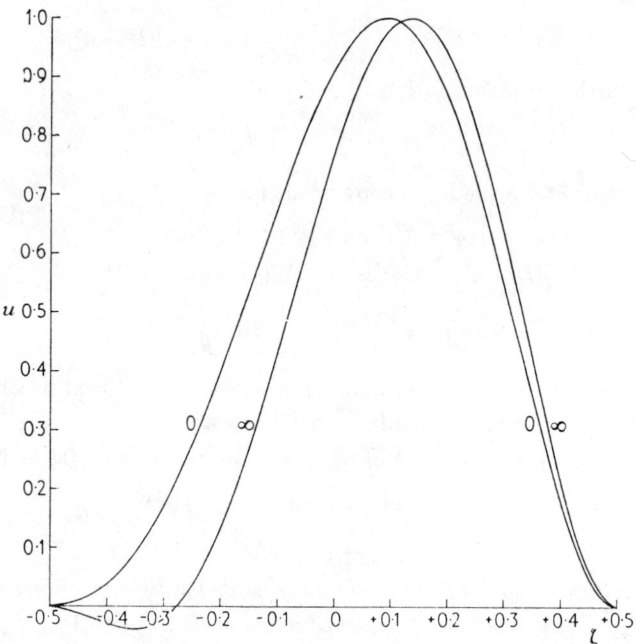

FIG. 105. The dependence of the velocity profile at the onset of instability on Q in the case of a pure pressure-maintained flow between two insulating cylinders. The curve labelled 0 refers to the hydrodynamic case when no magnetic field is present and the curve labelled ∞ refers to the limiting case when the axial magnetic field tends to infinity. The larger amplitudes in the outer half of the gap between the cylinders originate in the circumstance that in this half the distribution of velocities in the stationary flow is unstable according to the criterion valid for non-dissipative Couette flow (cf. Fig. 104, p. 421).

shall omit them. The results of the calculations which were undertaken to determine the dependence of Λ_c on Q are summarized in Table XLV. And the effect of the magnetic field on the velocity profile is shown in Fig. 105.

By arguments similar to those employed in § 86 (f), we can show that

$$\Lambda a^2 \to \Lambda_\infty, \quad Qa^2 \to Q_\infty \quad \text{as } Q \to \infty \text{ and } a \to 0, \tag{310}$$

where Λ_∞ and Q_∞ are certain determinate constants. The associated limiting form of the characteristic value problem can be solved in the same way as equations (299) and (300). We find

$$Q_\infty = 680, \quad \Lambda_\infty = 2{\cdot}5648\times 10^6; \quad A_{C;1} = 1, \quad A_{S;1} = 0{\cdot}5942,$$
$$A_{C;2} = 0{\cdot}0530, \quad \text{and} \quad A_{S;2} = 0{\cdot}00420$$
$$\text{(for the case of non-conducting walls).} \tag{311}$$

The corresponding asymptotic behaviour is

$$\Lambda_c \to 3{\cdot}7718 \times 10^3 Q \quad \text{and} \quad a_c \to 26{\cdot}1 Q^{-\frac{1}{6}} \quad \text{as } Q \to \infty$$

$$\text{(for the case of non-conducting walls).} \quad (312)$$

89. Experiments on the stability of viscous flow between rotating cylinders when an axial magnetic field is present

Experiments designed to ascertain the effects of an axial magnetic field on the stability of Couette flow have been undertaken by several authors; but the first really successful experiments were those of Donnelly and Ozima. The principles underlying the experiments of Donnelly and Ozima are the same as those described in § 74 (*a*): the inner cylinder of a viscometer is rotated at different speeds and the torque exerted on a freely suspended outer cylinder is measured by a null method; and the onset of instability is detected by the occurrence of a discontinuity in the measured viscosity as a function of the speed of rotation of the inner cylinder.

In the experiments of Donnelly and Ozima, mercury was used as the fluid; and the viscometer was of stainless steel. The torsion fibre used for suspension was of tungsten and of diameter 0·003 in. The cylinders were 11·8 cm long; the radius of the outer cylinder was 2 cm; and the gap width was 2 mm. The suspended part of the outer cylinder was 10 cm. The remaining fixed parts at the top and the bottom served as a guard against end effects.

The usual designs of a viscometer had to be modified to allow for the fact that steel floats in mercury. The outer cylinder was therefore attached at the bottom by a torsion fibre. Since the viscometer is used as a null-reading instrument and the outer cylinder is always brought back to the same position, the torsion constant of the suspension is not changed, though the sensitivity will be affected.

The overall height of the viscometer was $8\frac{3}{8}$ in. and could just be fitted between the pole pieces of a $38\frac{1}{2}$-in. cyclotron magnet of the University of Chicago. (The magnet was the same as that used in the experiments on thermal convection described in Chapters IV and V.) The apparatus could be run in and out of its location between the pole pieces on brass rails.

Adequate provisions for levelling and alignment were provided. The alignment was a particularly difficult matter: this was achieved in the experiments by having the guard cylinder and the suspended cylinder so machined that a misalignment of 0·004 in. would prevent a free turning of the outer cylinder. Also, by a suitable design of the suspension system, it could be removed and replaced in a reproducible manner.

The motor driving the viscometer had a transmission for close regulation of speed. Rotation speeds in the range 0·6 to 960 rev/min were available for the experiments. The period of rotation was measured by means of an electronic counter with a photoelectric cell and detector.

The experimental procedure was essentially the same as that described in § 76 (*b*). In the present series of experiments, magnetic fields in the range 250–10,000 gauss were used. The corresponding range of Q-values is 2–3,000.

It was found by Donnelly and Ozima that the onset of instability occurred almost as sharply as in the hydrodynamic experiments when the field strengths were not too high; thus, for field strengths of the order of a few hundred gauss, the period of rotation at which the onset of instability occurred could be determined with a precision of the order of one part in 300. The results of their experiments are included in Fig. 101, p. 412. It will be seen that the experiments amply confirm the broad aspects of the theoretical predictions; moreover, the experimental points fall rather closely on the theoretical curve for non-conducting walls.

BIBLIOGRAPHICAL NOTES

§ 81. The stability of non-dissipative Couette flow with an axial magnetic field is considered in:

1. E. P. Velikhov, 'Stability of an ideally conducting liquid flowing between cylinders rotating in a magnetic field', *J. Expl. Theoret. Phys.* (U.S.S.R.), **36,** 1398–1404 (1959).
2. S. Chandrasekhar, 'The stability of non-dissipative Couette flow in hydromagnetics', *Proc. Nat. Acad. Sci.* **46,** 253–7 (1960).

The treatment in the text is in large measure an expansion of reference 2.

§ 82 (*b*). The results given in Table XLI and illustrated in Figs. 99 and 100 are from:

3. W. H. Reid, 'The stability of non-dissipative Couette flow in the presence of an axial magnetic field', *Proc. Nat. Acad. Sci.* **46,** 370–3 (1960).

§ 83. The case treated in this section has been considered by:

4. D. H. Michael, 'The stability of an incompressible electrically conducting fluid rotating about an axis when current flows parallel to the axis', *Mathematika,* **1,** 45–50 (1954).

However, Michael's discussion was restricted, from the outset, to the case $m = 0$. The corresponding problem in dissipative flow has been considered by:

5. F. N. Edmonds, Jr., 'Hydromagnetic stability of a conducting fluid in a circular magnetic field', *The Physics of Fluids,* **1,** 30–41 (1958).

§ 85. The stability of viscous flow between rotating cylinders in the presence of an axial magnetic field has been considered by:

6. S. Chandrasekhar, 'The stability of viscous flow between rotating cylinders in the presence of a magnetic field', *Proc. Roy. Soc.* (*London*) A, **216,** 293–309 (1953).

In reference 6, attention was directed exclusively to the case $\mu > 0$. The treatment in the text is more general and allows for the customary linear variation of Ω through the gap.

§ 85 (*b*). In reference 6 only the case of perfectly conducting walls was considered. The importance of considering non-conducting walls was pointed out by:

7. E. R. Niblett, 'The stability of Couette flow in an axial magnetic field', *Canadian J. Phys.* **36,** 1509–25 (1958).

§ 86. In this section the problems considered in references 6 and 7 are treated afresh by a new method which enables the solutions to be obtained, systematically, in any desired order. The method is based on the expansion of u in terms of the proper functions C and S of the characteristic value problem

$$y^{\text{iv}} = \alpha^4 y \quad \text{and} \quad y = y' = 0 \quad \text{for} \quad x = \pm\tfrac{1}{2}.$$

The original suggestion of the usefulness of these functions for the solution of hydrodynamic problems is contained in:

8. S. Chandrasekhar and W. H. Reid, 'On the expansion of functions which satisfy four boundary conditions', *Proc. Nat. Acad. Sci.* **43,** 521–7 (1957).

However, the construction of these functions in their present useful and convenient forms is due to:

9. D. L. Harris and W. H. Reid, 'On orthogonal functions which satisfy four boundary conditions. I. Tables for use in Fourier-type expansions', *Astrophys. J. Supp., Ser.* **3,** 429–47 (1958).

Reid and Harris have further assembled very useful tables of integrals involving the C- and the S-functions:

10. W. H. Reid and D. L. Harris, 'On orthogonal functions which satisfy four boundary conditions. II. Integrals for use with Fourier-type expansions', *Astrophys. J. Supp., Ser.* **3,** 448–52 (1958).

The calculations summarized in Tables XLII and XLIII were carried out by Miss Donna Elbert expressly for inclusion in this book.

§ 87. The treatment follows:

11. S. Chandrasekhar and Donna D. Elbert, 'The stability of viscous flow between rotating cylinders in the presence of a magnetic field. II' *Proc. Roy. Soc.* (*London*) A, in press.

§ 88. The results quoted in this section are taken from:

12. S. Chandrasekhar, Donna D. Elbert, and N. Lebovitz, 'The stability of viscous flow in a curved channel in the presence of an axial magnetic field', in press.

§ 89. The first experiments on the stability of hydromagnetic Couette flow were carried out by Niblett (reference 7). Niblett's experiments were, however, inconclusive. The experiments described in this section are those of:

13. R. J. Donnelly and M. Ozima, 'Hydromagnetic stability of flow between rotating cylinders', *Phys. Rev. Letters*, **4,** 497–8 (1960).

X

THE STABILITY OF SUPERPOSED FLUIDS: THE RAYLEIGH–TAYLOR INSTABILITY

90. Introduction

IN the last three chapters we considered problems in which instabilities arose, principally, from adverse distributions of angular momentum. In the present chapter and in the next, we shall be concerned with instabilities which arise from different causes: they are of two kinds. The *first* derives from the character of the equilibrium of an incompressible heavy fluid of variable density (i.e. of a heterogeneous fluid). An important special case in this connexion is that of two fluids of different densities superposed one over the other (or accelerated towards each other); the instability of the plane interface between the two fluids, when it occurs (particularly in the second context), is called *Rayleigh–Taylor* instability. The *second* type of instability arises when the different layers of a stratified heterogeneous fluid are in relative horizontal motion. An important special case in this latter connexion is when two superposed fluids flow one over the other with a relative horizontal velocity; the instability of the plane interface between the two fluids, when it occurs in this instance, is called *Kelvin–Helmholtz instability*.

In this chapter we shall be concerned with Rayleigh–Taylor instability.

91. The character of the equilibrium of a stratified heterogeneous fluid. The perturbation equations

A static state, in which an incompressible fluid of variable density is arranged in horizontal strata and the pressure p and the density ρ are functions of the vertical coordinate z only, is clearly a kinematically realizable one. The character of the equilibrium of this initial static state can be determined, as usual, by supposing that the system is slightly disturbed and then by following its further evolution.

Let the actual density at any point (x, y, z) as a result of the disturbance be $\rho+\delta\rho$; and let δp denote the corresponding increment in the pressure. And finally, let u_i $(i = 1, 2, 3)$ denote the components of the velocity (considered small). The relevant equations of motion and continuity, then, are

$$\rho\frac{\partial u_i}{\partial t} = -\frac{\partial}{\partial x_i}\delta p + \frac{\partial}{\partial x_k}p_{ik} - g\,\delta\rho\lambda_i \tag{1}$$

and

$$\frac{\partial u_i}{\partial x_i} = 0, \tag{2}$$

where g is the acceleration due to gravity and $\boldsymbol{\lambda}$ is a unit vector in the direction of the vertical. In equation (1) p_{ik} is the viscous stress-tensor given by

$$p_{ik} = \mu\left(\frac{\partial u_k}{\partial x_i} + \frac{\partial u_i}{\partial x_k}\right), \tag{3}$$

where μ (which can be a function of z) denotes the coefficient of viscosity. In addition to equations (1) and (2), we have the equation,

$$\frac{\partial}{\partial t}\delta\rho + u_i\frac{\partial \rho}{\partial x_i} = 0, \tag{4}$$

which ensures that the density of every particle remains unchanged as we follow it with its motion.

Inserting equation (3) in equation (1) and making use of the solenoidal character of u_i, we find that the equation of motion reduces to the form

$$\rho\frac{\partial u_i}{\partial t} = -\frac{\partial}{\partial x_i}\delta p + \mu\nabla^2 u_i + \left(\frac{\partial u_k}{\partial x_i} + \frac{\partial u_i}{\partial x_k}\right)\frac{\partial \mu}{\partial x_k} - g\delta\rho\lambda_i, \tag{5}$$

or, since

$$\frac{\partial \mu}{\partial x_k} = \lambda_k\frac{d\mu}{dz}, \tag{6}$$

we have

$$\rho\frac{\partial u_i}{\partial t} = -\frac{\partial}{\partial x_i}\delta p + \mu\nabla^2 u_i + \left(\frac{\partial w}{\partial x_i} + \frac{\partial u_i}{\partial z}\right)\frac{d\mu}{dz} - g\delta\rho\lambda_i, \tag{7}$$

where w $(= \lambda_i u_i)$ is the z-component of the velocity.

It is now convenient to have the equations for the components of the velocity u, v, and w (in the x-, y-, and z-directions, respectively) written out explicitly. We have

$$\rho\frac{\partial u}{\partial t} = -\frac{\partial}{\partial x}\delta p + \mu\nabla^2 u + \left(\frac{\partial w}{\partial x} + \frac{\partial u}{\partial z}\right)\frac{d\mu}{dz}, \tag{8}$$

$$\rho\frac{\partial v}{\partial t} = -\frac{\partial}{\partial y}\delta p + \mu\nabla^2 v + \left(\frac{\partial w}{\partial y} + \frac{\partial v}{\partial z}\right)\frac{d\mu}{dz}, \tag{9}$$

$$\rho\frac{\partial w}{\partial t} = -\frac{\partial}{\partial z}\delta p + \mu\nabla^2 w + 2\frac{\partial w}{\partial z}\frac{d\mu}{dz} - g\delta\rho, \tag{10}$$

$$\frac{\partial u}{\partial x} + \frac{\partial v}{\partial y} + \frac{\partial w}{\partial z} = 0, \tag{11}$$

and

$$\frac{\partial}{\partial t}\delta\rho = -w\frac{d\rho}{dz}. \tag{12}$$

Analysing the disturbance into normal modes, we seek solutions whose dependence on x, y, and t is given by

$$\exp(ik_x x+ik_y y+nt), \tag{13}$$

where k_x, k_y, and n are constants. For solutions having this dependence on x, y, and t, equations (8)–(12) become

$$ik_x\,\delta p = -n\rho u+\mu(D^2-k^2)u+(D\mu)(ik_x w+Du), \tag{14}$$

$$ik_y\,\delta p = -n\rho v+\mu(D^2-k^2)v+(D\mu)(ik_y w+Dv), \tag{15}$$

$$D\delta p = -n\rho w+\mu(D^2-k^2)w+2(D\mu)(Dw)-g\delta\rho, \tag{16}$$

$$ik_x u+ik_y v = -Dw, \tag{17}$$

and

$$n\delta\rho = -wD\rho, \tag{18}$$

where

$$k^2 = k_x^2+k_y^2 \quad \text{and} \quad D = d/dz. \tag{19}$$

Multiplying equations (14) and (15) by $-ik_x$ and $-ik_y$, respectively, adding, and making use of equation (17), we obtain

$$k^2\delta p = [-n\rho+\mu(D^2-k^2)]Dw+(D\mu)(D^2+k^2)w; \tag{20}$$

whereas by combining equations (16) and (18) we have

$$D\delta p = -n\rho w+\mu(D^2-k^2)w+2(D\mu)(Dw)+\frac{g}{n}(D\rho)w. \tag{21}$$

Now eliminating δp between equations (20) and (21), we obtain

$$D\left\{\left[\rho-\frac{\mu}{n}(D^2-k^2)\right]Dw-\frac{1}{n}(D\mu)(D^2+k^2)w\right\}$$
$$= k^2\left\{-\frac{g}{n^2}(D\rho)w+\left[\rho-\frac{\mu}{n}(D^2-k^2)\right]w-\frac{2}{n}(D\mu)Dw\right\}. \tag{22}$$

This is the required equation.

If we suppose that the fluid is confined between two rigid planes, then the boundary conditions are

$$w = 0 \quad \text{and} \quad Dw = 0 \quad \text{on the bounding surfaces,} \tag{23}$$

where the latter condition ensures the vanishing of the horizontal components of the velocity.

(a) *Allowance for surface tension at interfaces between fluids*

In deriving equation (22) we have argued as though the density were continuously variable. The case when fluids of different densities are superposed one over the other can be formally included in the analysis by admitting discontinuities in the distribution of the density. This, however, is not entirely sufficient: for the interfaces between fluids will be subject to forces arising from surface tension; and we have not allowed for this possibility in deriving equation (22). We shall now consider the modifications which are necessary to allow for such forces coming into play at places where the density changes discontinuously.

Now the effect of surface tension is that, for equilibrium, the normal stresses acting on the two sides (+ and −) of an element of surface dS separating two fluids must differ by an amount depending on the *surface tension* T and given by

$$[(-p\delta_{ik}+p_{ik})_+-(-p\delta_{ik}+p_{ik})_-]N_k = -T\left(\frac{1}{R_1}+\frac{1}{R_2}\right), \tag{24}$$

where N_k is the unit outward normal to dS, and R_1 and R_2 are the principal radii of curvature of dS. The convention regarding signs is that R_1 and R_2 are to be considered positive if the corresponding centres of curvature lie on the side of dS to which the plus sign refers.

Returning to the problem we were considering, suppose that discontinuities in ρ occur at some preassigned levels, z_s ($s = 1, 2, 3,...$). In the perturbed state these plane surfaces will become slightly deformed. Let the surfaces of separation be then defined by

$$z_s+\delta z_s(x, y, t). \tag{25}$$

On account of (24), the discontinuity in the normal stresses required for equilibrium is

$$T_s\left(\frac{\partial^2}{\partial x^2}+\frac{\partial^2}{\partial y^2}\right)\delta z_s. \tag{26}$$

Therefore, to allow for the effects of surface tension, we must include on the right-hand side of equation (10) the additional term

$$\sum_s \left[T_s\left(\frac{\partial^2}{\partial x^2}+\frac{\partial^2}{\partial y^2}\right)\delta z_s\right]\delta(z-z_s), \tag{27}$$

where $\delta(z-z_s)$ denotes Dirac's δ-function. Thus, we replace equation (10) by

$$\rho\frac{\partial w}{\partial t} = -\frac{\partial}{\partial z}\delta p+\mu\nabla^2 w+2\frac{\partial w}{\partial z}\frac{d\mu}{dz}-g\,\delta\rho+\sum_s \left[T_s\left(\frac{\partial^2}{\partial x^2}+\frac{\partial^2}{\partial y^2}\right)\delta z_s\right]\delta(z-z_s). \tag{28}$$

Equations (8) and (9) remain unaffected.

Equation (21), appropriate for solutions having an (x, y, t)-dependence given by (13), is now replaced by

$$D\delta p = -n\rho w+\mu(D^2-k^2)w+2(D\mu)(Dw)+\frac{g}{n}(D\rho)w- -k^2\sum_s (T_s\,\delta z_s)\delta(z-z_s). \tag{29}$$

In this equation δz_s can be expressed in terms of the normal component of the velocity w_s at z_s by observing that

$$\frac{d}{dt}\delta z_s = w_s \tag{30}$$

and that, therefore,
$$\delta z_s = w_s/n. \tag{31}$$

The required generalization of equation (21) is thus:

$$D\delta p = -n\rho w + \mu(D^2 - k^2)w + 2(D\mu)(Dw) + \frac{g}{n}(D\rho)w - \\ - \frac{k^2}{n}\sum_s (T_s w_s)\delta(z - z_s). \tag{32}$$

To a certain extent, equation (32) has been obtained by a formal procedure. And the fact that the equation involves δ-functions means that to interpret the equation correctly at a point of discontinuity, we must integrate the equation, across the interface, over an infinitesimal element of z including the point of discontinuity. Before we effect such an integration, it is necessary to specify which quantities should be continuous and which bounded at an interface between two fluids. Clearly, all three components of the velocity, as well as the tangential viscous stresses, must be continuous. The continuity of Dw follows from equation (17) and the continuity of u and v. To obtain the condition which ensures the continuity of the tangential viscous stresses, we first observe that

$$p_{xz} = \mu\left(\frac{\partial u}{\partial z} + \frac{\partial w}{\partial x}\right) = \mu(Du + ik_x w)$$

and

$$p_{yz} = \mu\left(\frac{\partial v}{\partial z} + \frac{\partial w}{\partial y}\right) = \mu(Dv + ik_y w). \tag{33}$$

From these equations we obtain

$$i(k_x p_{xz} + k_y p_{yz}) = \mu[D(ik_x u + ik_y v) - k^2 w], \tag{34}$$

or, making use of equation (17), we have

$$i(k_x p_{xz} + k_y p_{yz}) = -\mu(D^2 + k^2)w. \tag{35}$$

Hence, $\mu(D^2+k^2)w$ must remain continuous across an interface. Altogether, the conditions at an interface are

w, Dw, and $\mu(D^2+k^2)w$ must be continuous across an interface between two fluids. (36)

The last of these conditions requires, in particular, that D^2w *must be bounded at an interface.*

In view of the continuity of w and Dw, and of the boundedness of

D^2w at an interface, the result of integrating equation (32) over an infinitesimal element of z including z_s (for example) is

$$\Delta_s(\delta p) = 2(Dw)_s \Delta_s(\mu) + \frac{g}{n} w_s \Delta_s(\rho) - \frac{k^2}{n} T_s w_s, \tag{37}$$

where
$$\Delta_s(f) = f(z_s+0) - f(z_s-0) \tag{38}$$

is the jump which a quantity f experiences at the interface $z = z_s$; and the subscript s distinguishes the value a quantity, known to be continuous at an interface, takes at $z = z_s$. On the other hand, according to equation (20) (the validity of which is unaffected by surface tension),

$$k^2\Delta_s(\delta p) = \Delta_s\{[-n\rho + \mu(D^2-k^2)]Dw + [(D^2+k^2)w]D\mu\}. \tag{39}$$

Combining equations (37) and (39), we have

$$\Delta_s\left\{\left[\rho - \frac{\mu}{n}(D^2-k^2)\right] Dw - \frac{1}{n}[(D^2+k^2)w]\, D\mu\right\}$$
$$= -\frac{k^2}{n^2}[g\Delta_s(\rho) - k^2 T_s] w_s - \frac{2k^2}{n}(Dw)_s \Delta_s(\mu). \tag{40}$$

This equation may be considered as a boundary condition (in addition to those stated in (36)) which must be satisfied at an interface between two fluids.

With the interpretation (37) of equation (32), we can eliminate δp between this equation and (20) to obtain

$$D\left\{\left[\rho - \frac{\mu}{n}(D^2-k^2)\right] Dw - \frac{1}{n}(D\mu)(D^2+k^2)w\right\}$$
$$= k^2\left\{\left[-\frac{g}{n^2}(D\rho) + \frac{k^2}{n^2}\sum_s T_s \delta(z-z_s)\right] w + \right.$$
$$\left. + \left[\rho - \frac{\mu}{n}(D^2-k^2)\right] w - \frac{2}{n}(D\mu)Dw\right\}. \tag{41}$$

This is the required generalization of equation (22). The correct interpretation of equation (41) is that for $z \neq z_s$ ($s = 1, 2,...$) the equation is identical with (22), while at $z = z_s$ it is equivalent to (40).

92. The inviscid case

We shall first consider the case when the fluid is inviscid and the stratification of density is moreover continuous. Then equation (22) becomes

$$D(\rho Dw) - \rho k^2 w = -\frac{k^2}{n^2} g(D\rho) w, \tag{42}$$

and the boundary conditions require only that

$$w = 0 \quad \text{on the boundaries.} \tag{43}$$

The characteristic value problem presented by equations (42) and (43) is of the standard type to which the classical Sturmian theory applies. We conclude: the characteristic values of n^2 are all positive if $D\rho$ is everywhere positive; they are all negative if $D\rho$ is everywhere negative; and if $D\rho$ should change sign anywhere inside the fluid, then both positive and negative characteristic values for n^2 exist. Thus, *the necessary and sufficient condition that a stratified heterogeneous fluid be stable is that $D\rho$ should be negative everywhere; moreover, if $D\rho$ should be positive anywhere inside the fluid, the stratification is unstable.*

It is also clear that the characteristic value problem is self-adjoint and allows a variational formulation. The basis for the latter is the integral relation,

$$\frac{n^2}{k^2} = \frac{g\int (D\rho)w^2\,dz}{\int \rho[(Dw)^2+k^2w^2]\,dz}, \tag{44}$$

which one readily obtains. We observe the complete similarity of equation (44) with equation VII (60); and the entirely equivalent roles of $D\rho$ and of $\Phi(r)$ for the two problems is apparent.

(a) *The case of two uniform fluids of constant density separated by a horizontal boundary*

Consider the case of two uniform fluids of densities ρ_1 and ρ_2 separated by a horizontal boundary at $z = 0$. For both regions of the fluid, the general equation (42) reduces to

$$(D^2-k^2)w = 0 \tag{45}$$

of which the general solution is

$$w = Ae^{+kz}+Be^{-kz}. \tag{46}$$

Since w must vanish both when $z \to -\infty$ (in the lower fluid) and $z \to +\infty$ (in the upper fluid), we must suppose that

$$w_1 = Ae^{+kz} \quad (z < 0) \tag{47}$$

and

$$w_2 = Ae^{-kz} \quad (z > 0), \tag{48}$$

where we have chosen the same constant A in the solutions for $z > 0$ and $z < 0$ to ensure the continuity of w across the interface at $z = 0$. The remaining condition (40), in the inviscid case, requires

$$\Delta_0(\rho Dw) = -\frac{k^2}{n^2}[g(\rho_2-\rho_1)-k^2T]w_0, \tag{49}$$

where w_0 is the common value of w at $z = 0$. Applying the condition (49) to the solution (47) and (48), we obtain

$$-k(\rho_2+\rho_1) = -\frac{k^2}{n^2}[g(\rho_2-\rho_1)-k^2T] \tag{50}$$

or

$$n^2 = gk\left\{\frac{\rho_2-\rho_1}{\rho_2+\rho_1}-\frac{k^2T}{g(\rho_2+\rho_1)}\right\}. \tag{51}$$

According to equation (51), *if* $\rho_2 < \rho_1$ *the arrangement is stable; while if* $\rho_2 > \rho_1$ *the arrangement is unstable for all wave numbers in the range* $0 < k < k_c$, *where*

$$k_c = [(\rho_2-\rho_1)g/T]^{\frac{1}{2}}. \tag{52}$$

However, in the latter case, the arrangement is stable for all disturbances with $k > k_c$. Thus, surface tension succeeds in stabilizing a potentially unstable arrangement for all sufficiently short wavelengths; but the arrangement remains unstable for all sufficiently long wavelengths. Moreover, unlike when surface tension is absent, there is a *mode of maximum instability* for which the amplitude of the disturbance grows most rapidly. The wave number k_* of this most unstable mode is given by

$$g(\rho_2-\rho_1) = 3k_*^2\,T, \tag{53}$$

or

$$k_* = k_c/\sqrt{3}. \tag{54}$$

The corresponding value of n is

$$n_* = \left[\frac{2}{3^{\frac{3}{2}}}\frac{(\rho_2-\rho_1)^{\frac{3}{2}}g^{\frac{3}{2}}}{(\rho_2+\rho_1)T^{\frac{1}{2}}}\right]^{\frac{1}{2}}. \tag{55}$$

(*b*) *The case of exponentially varying density*

A case of variable density for which a simple analytical solution can be found is

$$\rho = \rho_0\,e^{\beta z}, \tag{56}$$

where β is a constant. In this case equation (42) reduces to

$$D^2w+\beta Dw-k^2(1-g\beta/n^2)w = 0. \tag{57}$$

The general solution of this equation is

$$w = A_1\,e^{q_1 z}+A_2\,e^{q_2 z}, \tag{58}$$

where A_1 and A_2 are two arbitrary constants and q_1 and q_2 are the roots of the equation

$$q^2+q\beta-k^2(1-g\beta/n^2) = 0. \tag{59}$$

If we suppose that the fluid is confined between two rigid planes at $z = 0$ and $z = d$, then the vanishing of w at $z = 0$ is satisfied by the choice

$$w = A(e^{q_1 z}-e^{q_2 z}), \tag{60}$$

while the vanishing of w at $z = d$ requires

$$\exp[(q_1-q_2)d] = 1, \tag{61}$$

or
$$(q_1-q_2)d = 2im\pi, \tag{62}$$
where m is an integer.

Now writing the solution (60) in the manner
$$w = Ae^{\frac{1}{2}(q_1+q_2)z}\{e^{\frac{1}{2}(q_1-q_2)z}-e^{-\frac{1}{2}(q_1-q_2)z}\} \tag{63}$$
and making use of (62) and the relation
$$q_1+q_2 = -\beta \tag{64}$$
(which follows from (59)), we have the alternative form
$$w = \text{constant } e^{-\frac{1}{2}\beta z}\sin(m\pi z/d). \tag{65}$$

Returning to equation (62) and inserting for q_1 and q_2 their values
$$q_1 = \tfrac{1}{2}\{-\beta+\sqrt{[\beta^2+4k^2(1-g\beta/n^2)]}\}$$
and
$$q_2 = \tfrac{1}{2}\{-\beta-\sqrt{[\beta^2+4k^2(1-g\beta/n^2)]}\}, \tag{66}$$
we obtain
$$\beta^2+4k^2\left(1-\frac{g\beta}{n^2}\right) = -\frac{4m^2\pi^2}{d^2}, \tag{67}$$
or
$$\frac{g\beta}{n^2} = 1+\frac{\frac{1}{4}\beta^2d^2+m^2\pi^2}{k^2d^2}. \tag{68}$$

From equation (68) it follows that *the stratification is stable if β is negative, while it is unstable if β is positive.*

The smallest admissible value of m is 1, which leads to the greatest numerical value of n. If d, k, and m are given, n^2 is numerically greatest when β is such that
$$\tfrac{1}{4}\beta^2d^2 = k^2d^2+m^2\pi^2. \tag{69}$$
Rayleigh, who first worked out this example, has remarked: 'Contrary to what is met with in most vibrating systems, there is (in the case of stability) a limit on the side of rapidity of vibration; but none on the side of slowness.'

93. A general variational principle

In § 92 we saw that in the absence of viscosity and surface tension the underlying characteristic value problem allows a variational formulation. We shall now show that such a formulation is possible in the general case.

The basic equations are (cf. equations (20) and (32))
$$D\delta p = -n\rho w+\frac{g}{n}(D\rho)w-\frac{k^2}{n}\sum_s T_s w_s\,\delta(z-z_s)-\mu k^2w+\mu D^2w+2(D\mu)(Dw) \tag{70}$$
and
$$k^2\delta p = -n\rho Dw-k^2\mu Dw+k^2wD\mu+D(\mu D^2w); \tag{71}$$

and the boundary conditions with respect to which these equations must be solved are

$$w \text{ and } Dw \text{ vanish on a bounding surface.} \tag{72}$$

The requirement that a solution of equations (70) and (71) satisfy the necessary boundary conditions will lead to a determinate sequence of possible values for n. Let n_i and n_j denote two of these characteristic values; and let the solutions belonging to these characteristic values be distinguished by the subscripts i and j.

Now consider equation (70) for the characteristic value n_i and after multiplication by w_j (belonging to n_j) integrate over the range of z (which we shall assume to be $0 \leqslant z \leqslant d$). We have

$$\int_0^d w_j\, D\,\delta p_i\, dz = \int_0^d \left(-n_i\rho+\frac{g}{n_i}D\rho-k^2\mu\right)w_i\, w_j\, dz - \\ -\frac{k^2}{n_i}\sum_s T_s\, w_i(z_s)w_j(z_s)+\int_0^d w_j\{\mu D^2 w_i+2(D\mu)(Dw_i)\}\, dz. \tag{73}$$

On integrating by parts the left-hand side of this equation, we have

$$\int_0^d w_j\, D\delta p_i\, dz = -\int_0^d \delta p_i\, Dw_j\, dz, \tag{74}$$

the integrated part vanishing on account of the boundary conditions. On the right-hand side of equation (74) we now substitute for δp_i in accordance with (71). We thus obtain

$$-\int_0^d \delta p_i\, Dw_j\, dz = \int_0^d \left(\frac{n_i}{k^2}\rho+\mu\right)(Dw_i)(Dw_j)\, dz - \\ -\int_0^d (D\mu)w_i\, Dw_j\, dz-\frac{1}{k^2}\int_0^d (Dw_j)D(\mu D^2 w_i)\, dz. \tag{75}$$

The last integral on the right-hand side of this equation is

$$\int_0^d (Dw_j)D(\mu D^2 w_i)\, dz = -\int_0^d \mu(D^2 w_i)(D^2 w_j)\, dz, \tag{76}$$

the integrated part again vanishing. We can now rewrite equation (75) in the form

$$-\int_0^d \delta p_i\, Dw_j\, dz = \int_0^d \left(\frac{n_i}{k^2}\rho+\mu\right)(Dw_i)(Dw_j)\, dz + \\ +\frac{1}{k^2}\int_0^d \mu(D^2 w_i)(D^2 w_j)\, dz-\int_0^d (D\mu)w_i(Dw_j)\, dz. \tag{77}$$

Now combining equations (73), (74), and (77) we find after some further rearrangement

$$-n_i\int_0^d \rho\Big\{w_i w_j+\frac{1}{k^2}(Dw_i)(Dw_j)\Big\}\,dz+ \\ +\frac{g}{n_i}\int_0^d (D\rho)w_i w_j\,dz-\frac{k^2}{n_i}\sum_s T_s w_i(z_s)w_j(z_s) \\ =\int_0^d \mu\Big\{k^2 w_i w_j+(Dw_i)(Dw_j)+\frac{1}{k^2}(D^2w_i)(D^2w_j)\Big\}\,dz- \\ -\int_0^d \{w_j[\mu D^2 w_i+2(D\mu)(Dw_i)]+w_i(D\mu)(Dw_j)\}\,dz. \quad (78)$$

Considering the last integral on the right-hand side of this equation, we can rearrange it in the manner

$$-\int_0^d \{w_j D(\mu D w_i)+(D\mu)D(w_i w_j)\}\,dz, \quad (79)$$

or, after integrating by parts each of the two terms which occur under the integral sign, we are left with

$$\int_0^d \{\mu(Dw_i)(Dw_j)+(D^2\mu)w_i w_j\}\,dz. \quad (80)$$

Thus, we finally have

$$-n_i\int_0^d \rho\Big\{w_i w_j+\frac{1}{k^2}(Dw_i)(Dw_j)\Big\}\,dz+ \\ +\frac{g}{n_i}\int_0^d (D\rho)w_i w_j\,dz-\frac{k^2}{n_i}\sum_s T_s w_i(z_s)w_j(z_s) \\ =\int_0^d (D^2\mu)w_i w_j\,dz+ \\ +\int_0^d \mu\Big\{k^2 w_i w_j+2(Dw_i)(Dw_j)+\frac{1}{k^2}(D^2w_i)(D^2w_j)\Big\}\,dz. \quad (81)$$

Interchanging i and j in equation (81) and subtracting the resulting equation from (81), we obtain

$$(n_j-n_i)\Big[\int_0^d \rho\Big\{w_i w_j+\frac{1}{k^2}(Dw_i)(Dw_j)\Big\}\,dz+ \\ +\frac{1}{n_i n_j}\Big\{g\int_0^d (D\rho)w_i w_j\,dz-k^2\sum_s T_s w_i(z_s)w_j(z_s)\Big\}\Big]=0. \quad (82)$$

Hence, if $n_i \neq n_j$, we must have

$$\int_0^d \rho\left\{w_i w_j+\frac{1}{k^2}(Dw_i)(Dw_j)\right\} dz+$$
$$+\frac{1}{n_i n_j}\left\{g\int_0^d (D\rho)w_i w_j\, dz-k^2 \sum_s T_s w_i(z_s)w_j(z_s)\right\} = 0 \quad (i \neq j). \quad (83)$$

A further relation of this kind can be obtained by rewriting equation (81) in the manner,

$$g\int_0^d (D\rho)w_i w_j\, dz-k^2 \sum_s T_s w_i(z_s)w_j(z_s)$$
$$= n_i^2 \int_0^d \rho\left\{w_i w_j+\frac{1}{k^2}(Dw_i)(Dw_j)\right\} dz+n_i \int_0^d (D^2\mu)w_i w_j\, dz+$$
$$+n_i \int_0^d \mu\left\{k^2 w_i w_j+2(Dw_i)(Dw_j)+\frac{1}{k^2}(D^2w_i)(D^2w_j)\right\} dz, \quad (84)$$

and treating it in the same way. We then obtain

$$(n_i+n_j)\int_0^d \rho\left\{w_i w_j+\frac{1}{k^2}(Dw_i)(Dw_j)\right\} dz+\int_0^d (D^2\mu)w_i w_j\, dz+$$
$$+\int_0^d \mu\left\{k^2 w_i w_j+2(Dw_i)(Dw_j)+\frac{1}{k^2}(D^2w_i)(D^2w_j)\right\} dz = 0 \quad (i \neq j). \quad (85)$$

Equations (83) and (85) enable us to draw certain general conclusions. Thus, if n_i should be complex, we can suppose that n_i and n_j are complex conjugates and deduce from equation (83) that

$$\int_0^d \rho\left(|w|^2+\frac{1}{k^2}|Dw|^2\right) dz+\frac{1}{|n|^2}\left\{g\int_0^d (D\rho)|w|^2\, dz-k^2 \sum_s T_s|w(z_s)|^2\right\} = 0. \quad (86)$$

In the absence of surface tension, equation (86) cannot be true if $D\rho$ is everywhere positive so that in this case n must be real.

Again, if n should be complex we can deduce from equation (85) that

$$2\,\mathrm{re}(n)\int_0^d \rho\left(|w|^2+\frac{1}{k^2}|Dw|^2\right) dz$$
$$= -\int_0^d (D^2\mu)|w|^2\, dz-\int_0^d \mu\left(k^2|w|^2+2|Dw|^2+\frac{1}{k^2}|D^2w|^2\right) dz. \quad (87)$$

From this equation it follows that *if n is complex*, re$(n) < 0$ *if* $D^2\mu$ *is everywhere positive*. The restriction on $D^2\mu$ for the validity of this result is curious. Apart from this restriction, what the result implies is that if oscillatory modes exist they should be stable; conversely, overstability cannot occur.

(*a*) *The variational principle*

Returning to equation (81) and setting $i = j$, we get (on further suppressing the subscripts)

$$n\int_0^d \rho\left\{w^2+\frac{1}{k^2}(Dw)^2\right\}dz-\frac{g}{n}\int_0^d (D\rho)w^2\,dz+\frac{k^2}{n}\sum_s T_s w_s^2$$
$$= -\int_0^d \left\{\mu\left[k^2w^2+2(Dw)^2+\frac{1}{k^2}(D^2w)^2\right]+(D^2\mu)w^2\right\}dz. \quad (88)$$

This equation provides the basis for a variational formulation of the problem. To see this, consider the effect on n (determined in accordance with (88)) of an arbitrary variation δw in w compatible only with the boundary conditions on w. We have to the first order

$$-\left(I_1+\frac{1}{n^2}\left[gI_2-k^2\sum_s T_s w_s^2\right]\right)\delta n = n\delta I_1-\frac{1}{n}\left[g\delta I_2-2k^2\sum_s T_s w_s\,\delta w_s\right]+\delta I_3, \quad (89)$$

where

$$I_1 = \int_0^d \rho\left\{w^2+\frac{1}{k^2}(Dw)^2\right\}dz, \qquad I_2 = \int_0^d (D\rho)w^2\,dz,$$
$$I_3 = \int_0^d \left\{\mu\left[k^2w^2+2(Dw)^2+\frac{1}{k^2}(D^2w)^2\right]+(D^2\mu)w^2\right\}dz, \quad (90)$$

and δI_1, δI_2, and δI_3 are the corresponding variations in these integrals. After one or more integrations by parts, we find that these latter variations are given by

$$\tfrac{1}{2}\delta I_1 = \int_0^d \delta w\left\{\rho w-\frac{1}{k^2}D(\rho Dw)\right\}dz, \quad (91)$$

$$\tfrac{1}{2}\delta I_2 = \int_0^d \delta w\{(D\rho)w\}\,dz, \quad (92)$$

and $$\tfrac{1}{2}\delta I_3 = \int_0^d \delta w\left\{k^2\mu w-2D(\mu Dw)+(D^2\mu)w+\frac{1}{k^2}D^2(\mu D^2w)\right\}dz. \quad (93)$$

Combining the foregoing in accordance with (89), we find after some further reductions that

$$\tfrac{1}{2}k^2\Big(I_1+\frac{1}{n^2}\Big[gI_2-k^2\sum_s T_s w_s^2\Big]\Big)\frac{\delta n}{n}$$
$$=\int_0^d \delta w\Big\{D\Big[\rho Dw-\frac{\mu}{n}(D^2-k^2)Dw-\frac{1}{n}(D\mu)(D^2+k^2)w\Big]-$$
$$-k^2\Big[\rho w-\frac{\mu}{n}(D^2-k^2)w-\frac{g}{n^2}(D\rho)w-\frac{2}{n}(D\mu)(Dw)+$$
$$+\frac{k^2}{n^2}\sum_s T_s\delta(z-z_s)w\Big]\Big\}\,dz. \tag{94}$$

We observe that the quantity which appears as a factor of δw under the integral sign on the right-hand side of equation (94) vanishes if equation (41) governing w is satisfied. Hence, a necessary and sufficient condition for δw to be zero to the first order for all small arbitrary variations in w compatible with the boundary conditions is that w be a solution of the characteristic value problem. A variational procedure of solving for the characteristic values is therefore possible.

94. The case of two uniform viscous fluids separated by a horizontal boundary

Let two uniform fluids of densities ρ_1 and ρ_2 and viscosities μ_1 and μ_2 be separated by a horizontal boundary at $z = 0$. The subscripts 1 and 2 distinguish the lower and the upper fluids, respectively.

In each of the two regions of constant ρ and μ, equation (41) reduces to

$$D\Big[\rho-\frac{\mu}{n}(D^2-k^2)\Big]Dw = k^2\Big[\rho-\frac{\mu}{n}(D^2-k^2)\Big]w, \tag{95}$$

where, for the present, we are suppressing the subscripts distinguishing the two fluids. Since ρ and μ are constants, we can rewrite equation (95) in the form

$$\Big[1-\frac{\nu}{n}(D^2-k^2)\Big](D^2-k^2)w = 0, \tag{96}$$

where ν $(= \mu/\rho)$ is the coefficient of kinematic viscosity.

The general solution of equation (96) is a linear combination of the solutions

$$e^{\pm kz} \quad \text{and} \quad e^{\pm qz}, \tag{97}$$

where

$$q^2 = k^2+n/\nu. \tag{98}$$

Since w must vanish both when $z \to -\infty$ (in the lower fluid) and $z \to +\infty$

(in the upper fluid), we can write

$$w_1 = A_1 e^{+kz} + B_1 e^{q_1 z} \quad (z < 0) \tag{99}$$

and

$$w_2 = A_2 e^{-kz} + B_2 e^{-q_2 z} \quad (z > 0) \tag{100}$$

as the solutions appropriate for the two regions. In equations (99) and (100), A_1, B_1, A_2, and B_2 are constants of integration and

$$q_1 = \sqrt{(k^2+n/\nu_1)} \quad \text{and} \quad q_2 = \sqrt{(k^2+n/\nu_2)}. \tag{101}$$

It should be noted that in writing the solutions for w in the two regions $z < 0$ and $z > 0$ in the manner (99) and (100), we have assumed that q_1 and q_2 are so defined that their real parts are positive.

On the interface $z = 0$, certain conditions must be satisfied; these have been found in § 91 (a) and given by (36) and (40). The latter condition, under the circumstances of the present problem, gives

$$\left\{\left[\rho_2 - \frac{\mu_2}{n}(D^2-k^2)\right] Dw_2\right\}_{z=0} - \left\{\left[\rho_1 - \frac{\mu_1}{n}(D^2-k^2)\right] Dw_1\right\}_{z=0}$$
$$= -\frac{k^2}{n^2}\{g(\rho_2-\rho_1) - k^2 T\} w_0 - \frac{2k^2}{n}(\mu_2-\mu_1)(Dw)_0, \tag{102}$$

where w_0 and $(Dw)_0$ are the common values of w_1 and w_2 and, similarly, Dw_1 and Dw_2 at $z = 0$. Substituting for w_1 and w_2 in accordance with equations (99) and (100) in equation (102), we find

$$-k\rho_2 A_2 - k\rho_1 A_1 = \frac{gk^2}{n^2}\left\{(\rho_1-\rho_2) + \frac{k^2 T}{g}\right\} w_0 + \frac{2k^2}{n}(\mu_1-\mu_2)(Dw)_0. \tag{103}$$

Now applying the boundary conditions (36) and (103) to the solution given in (99) and (100) we obtain

$$A_1 + B_1 = A_2 + B_2 \quad (= w_0), \tag{104}$$

$$kA_1 + q_1 B_1 = -kA_2 - q_2 B_2 \quad (= Dw_0), \tag{105}$$

$$\mu_1\{2k^2 A_1 + (q_1^2+k^2)B_1\} = \mu_2\{2k^2 A_2 + (q_2^2+k^2)B_2\}, \tag{106}$$

and

$$-k\rho_2 A_2 - k\rho_1 A_1 = \frac{gk^2}{2n^2}\left\{(\rho_1-\rho_2) + \frac{k^2 T}{g}\right\}(A_1+B_1+A_2+B_2) +$$
$$+\frac{k^2}{n}(\mu_1-\mu_2)(kA_1+q_1 B_1 - kA_2 - q_2 B_2). \tag{107}$$

Letting

$$\alpha_1 = \frac{\rho_1}{\rho_1+\rho_2}, \qquad \alpha_2 = \frac{\rho_2}{\rho_1+\rho_2} \quad (\alpha_1+\alpha_2 = 1), \tag{108}$$

$$R = -\frac{gk}{n^2}\left\{\frac{\rho_1-\rho_2}{\rho_1+\rho_2} + \frac{k^2 T}{g(\rho_1+\rho_2)}\right\} = -\frac{gk}{n^2}\left\{(\alpha_1-\alpha_2) + \frac{k^2 T}{g(\rho_1+\rho_2)}\right\}, \tag{109}$$

and

$$C = \frac{k^2}{n}\frac{\mu_1-\mu_2}{\rho_1+\rho_2} = \frac{k^2}{n}(\alpha_1 \nu_1 - \alpha_2 \nu_2), \tag{110}$$

we can rewrite equations (104)–(107), in matrix notation, in the form of the single equation

$$\left|\begin{matrix} 1 & 1 & -1 & -1 \\ k & q_1 & k & q_2 \\ 2k^2\mu_1 & \mu_1(q_1^2+k^2) & -2k^2\mu_2 & -\mu_2(q_2^2+k^2) \\ \tfrac{1}{2}R-C-\alpha_1 & \tfrac{1}{2}R-q_1 C/k & \tfrac{1}{2}R+C-\alpha_2 & \tfrac{1}{2}R+q_2 C/k \end{matrix}\right|\left|\begin{matrix} A_1 \\ B_1 \\ A_2 \\ B_2 \end{matrix}\right| = 0. \tag{111}$$

The determinant of the linear system of equations which (111) represents must clearly vanish. The determinant can be reduced by subtracting the first column from the second and similarly the third column from the fourth and finally adding the first column to the third. By this procedure and by making further use of equations (98) and (108), we obtain

$$\left\|\begin{matrix} q_1-k & 2k & q_2-k \\ \alpha_1 n & 2k^2(\alpha_1\nu_1-\alpha_2\nu_2) & -\alpha_2 n \\ \alpha_1+C(1-q_1/k) & R-1 & \alpha_2-C(1-q_2/k) \end{matrix}\right\| = 0. \tag{112}$$

On expanding this determinant and simplifying, we find

$$-\left\{\frac{gk}{n^2}\left[(\alpha_1-\alpha_2)+\frac{k^2T}{g(\rho_1+\rho_2)}\right]+1\right\}(\alpha_2 q_1+\alpha_1 q_2-k)-4k\alpha_1\alpha_2+$$
$$+\frac{4k^2}{n}(\alpha_1\nu_1-\alpha_2\nu_2)\{(\alpha_2 q_1-\alpha_1 q_2)+k(\alpha_1-\alpha_2)\}+$$
$$+\frac{4k^3}{n^2}(\alpha_1\nu_1-\alpha_2\nu_2)^2(q_1-k)(q_2-k) = 0, \tag{113}$$

where it may be recalled that q_1 and q_2 are related to n by equations (101). Equation (113) is the required characteristic equation for n: it was first derived, essentially in this form, by Harrison in 1908.

(a) *The case* $\nu_1 = \nu_2$

In the further discussion of equation (113) we shall restrict ourselves to the case when the kinematic viscosities of the two fluids are the same, i.e. when $\nu_1 = \nu_2$. This assumption simplifies the characteristic equation (113) considerably; but one would not expect that any of the essential features of the problem would be obscured by this simplifying assumption. When $\nu_1 = \nu_2$,

$$q_1 = q_2 = k\sqrt{(1+n/k^2\nu)} = q \quad \text{(say)}, \tag{114}$$

and equation (113) reduces to

$$-\left\{\frac{gk}{n^2}\left[(\alpha_1-\alpha_2)+\frac{k^2T}{g(\rho_1+\rho_2)}\right]+1\right\}(q-k)-\frac{4k^2\nu}{n}(\alpha_1-\alpha_2)^2(q-k)+$$
$$+\frac{4k^3\nu^2}{n^2}(\alpha_1-\alpha_2)^2(q-k)^2-4k\alpha_1\alpha_2 = 0. \tag{115}$$

Letting
$$x = n/k^2\nu, \tag{116}$$
so that
$$q = k\sqrt{(1+x)}, \tag{117}$$
we can rewrite equation (115) in the form
$$\frac{g}{k^3\nu^2}\left[(\alpha_1-\alpha_2)+\frac{k^2T}{g(\rho_1+\rho_2)}\right]\frac{1}{x^2}+1+\frac{4}{x}(\alpha_1-\alpha_2)^2-$$
$$-\frac{4}{x^2}(\alpha_1-\alpha_2)^2[\sqrt{(1+x)}-1]+\frac{4\alpha_1\alpha_2}{\sqrt{(1+x)}-1} = 0. \tag{118}$$
With the further substitutions
$$y = q/k = \sqrt{(1+x)}, \quad x = y^2-1, \tag{119}$$
$$Q = \frac{g}{k^3\nu^2}, \qquad S = \frac{T}{(\rho_1+\rho_2)(g\nu^4)^{\frac{1}{3}}}, \quad \text{and} \quad Q^{-\frac{2}{3}}S = \frac{k^2T}{g(\rho_1+\rho_2)}, \tag{120}$$
we find that equation (118) reduces to the following quartic for y:
$$y^4+4\alpha_1\alpha_2y^3+2(1-6\alpha_1\alpha_2)y^2-4(1-3\alpha_1\alpha_2)y+$$
$$+(1-4\alpha_1\alpha_2)+Q(\alpha_1-\alpha_2)+Q^{\frac{1}{3}}S = 0. \tag{121}$$

Since in expressing the solutions for w_1 and w_2 in the manner (99) and (100) we have supposed that the real parts of q_1 and q_2 are positive, it is clear that in the present case (cf. equations (117) and (119)) only those roots of equation (121) which have positive real parts lead to physically valid solutions.

For assigned α_1, α_2 $(= 1-\alpha_1)$, Q, and S, a root of the quartic (121) whose real part is positive determines a pair of values of k and n which belong to each other; for, according to equations (116), (119), and (120),
$$k = \left(\frac{g}{\nu^2}\right)^{\frac{1}{3}}\frac{1}{Q^{\frac{1}{3}}} \quad \text{and} \quad n = k^2\nu(y^2-1) = \left(\frac{g^2}{\nu}\right)^{\frac{1}{3}}\frac{y^2-1}{Q^{\frac{2}{3}}}. \tag{122}$$
Hence if k and n are measured in the units
$$(g/\nu^2)^{\frac{1}{3}}\ \text{cm}^{-1} \quad \text{and} \quad (g^2/\nu)^{\frac{1}{3}}\ \text{sec}^{-1}, \quad \text{respectively,} \tag{123}$$
we can write
$$k = Q^{-\frac{1}{3}} \quad \text{and} \quad n = (y^2-1)Q^{-\frac{2}{3}}. \tag{124}$$

(b) *The modes of maximum instability for the case* $\nu_1 = \nu_2$, $\rho_2 > \rho_1$, *and* $S = 0$

When $S = 0$, the equation we have to consider is
$$y^4+4\alpha_1\alpha_2y^3+2(1-6\alpha_1\alpha_2)y^2-4(1-3\alpha_1\alpha_2)y+$$
$$+(1-4\alpha_1\alpha_2)+Q(\alpha_1-\alpha_2) = 0. \tag{125}$$
It can be readily shown that this equation, in the case when $\alpha_2 > \alpha_1$, allows only one root whose real part is positive. Indeed, this root is real and, moreover, $y > 1$. Accordingly, n is real and positive (cf. equation

(124)) and the amplitude of the disturbance will grow exponentially with time. *The arrangement is therefore unstable for disturbances of all wave numbers.*

The asymptotic behaviour of n for $k \to 0$ and $k \to \infty$ can be derived by considering equation (125) for $y \to \infty$ and $y \to 1$. In the former case

$$Q \to y^4/(\alpha_2-\alpha_1) \quad (y \to \infty), \tag{126}$$

and we deduce from (124) that simultaneously

$$k \to y^{-\frac{4}{3}}(\alpha_2-\alpha_1)^{\frac{1}{3}} \quad \text{and} \quad n \to y^{-\frac{2}{3}}(\alpha_2-\alpha_1)^{\frac{2}{3}} \tag{127}$$

so that

$$n^2 \to (\alpha_2-\alpha_1)k \quad (k \to 0). \tag{128}$$

This asymptotic relation for $k \to 0$ is exactly that which obtains in the absence of viscosity. This agreement is what one should expect on general grounds, namely, that viscosity plays no role among the very long wavelengths.

Considering next the limit $y \to 1$, we have (see equation (132) below)

$$Q \to 4(y-1)/(\alpha_2-\alpha_1) \quad (y \to 1), \tag{129}$$

and we deduce that simultaneously

$$k \to \left(\frac{\alpha_2-\alpha_1}{4}\right)^{\frac{1}{3}}(y-1)^{-\frac{1}{3}} \quad \text{and} \quad n \to 2\left(\frac{\alpha_2-\alpha_1}{4}\right)^{\frac{2}{3}}(y-1)^{\frac{1}{3}} \tag{130}$$

so that

$$n \to (\alpha_2-\alpha_1)/2k \quad (k \to \infty). \tag{131}$$

According to equations (128) and (131), $n \to 0$ both when $k \to 0$ and $k \to \infty$. There exists, therefore, *a mode of maximum instability*. To determine this most unstable mode it is necessary to establish first the complete (k, n)-relation. And this can be accomplished most easily in the following manner.

Rewriting equation (125) in the form

$$\begin{aligned} Q &= \frac{1}{\alpha_2-\alpha_1}\{y^4+4\alpha_1\alpha_2 y^3+2(1-6\alpha_1\alpha_2)y^2-4(1-3\alpha_1\alpha_2)+1-4\alpha_1\alpha_2\} \\ &= \frac{y-1}{\alpha_2-\alpha_1}\{y^3+(1+4\alpha_1\alpha_2)y^2+(3-8\alpha_1\alpha_2)y-(1-4\alpha_1\alpha_2)\} \end{aligned} \tag{132}$$

and evaluating Q for various assigned values of $y \geqslant 1$, we can first obtain corresponding pairs of values of Q and y; then, making use of equations (124), we can convert them into corresponding pairs of k and n. The (k, n)-relationships derived in this manner are illustrated in

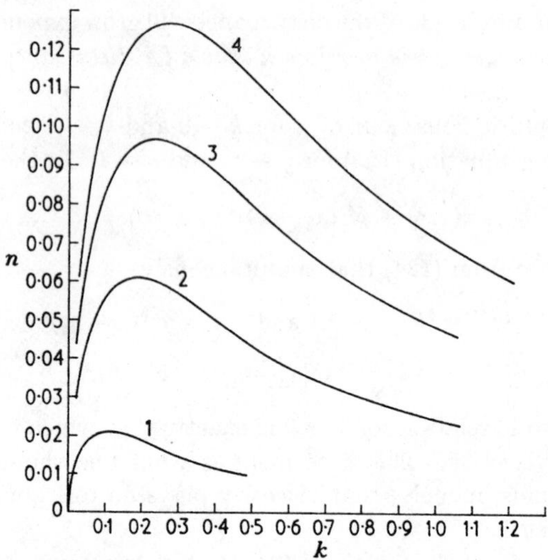

FIG. 106. The dependence of the rate of growth n (measured in the unit $(g^2/\nu)^{\frac{1}{3}}$) of a disturbance on its wave number k (measured in the unit $(g/\nu^2)^{\frac{1}{3}}$) in case the upper fluid is more dense and the kinematic viscosities are the same. The curves labelled 1, 2, 3, and 4 are for values of $(\rho_2-\rho_1)/(\rho_2+\rho_1)$ $(=(\alpha_2-\alpha_1))$ $= 0{\cdot}01$, $0{\cdot}05$, $0{\cdot}10$, and $0{\cdot}15$, respectively.

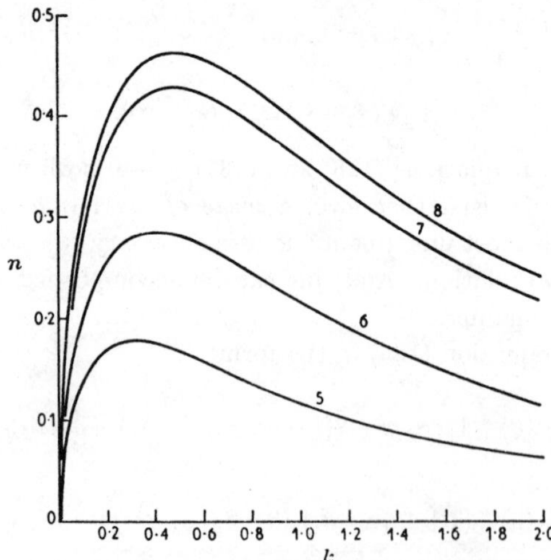

FIG. 107. Same as Fig. 106 but for values $\alpha_2-\alpha_1 = 0{\cdot}25$, $0{\cdot}5$, $0{\cdot}9$, and $1{\cdot}0$ (curves labelled 5, 6, 7, and 8, respectively).

Figs. 106 and 107; and the modes of maximum instability for various values of $\alpha_2-\alpha_1 = (\rho_2-\rho_1)/(\rho_1+\rho_2)$ are listed in Table XLVI.

TABLE XLVI

The modes of maximum instability for the case $\nu_2 = \nu_1$ and $\rho_2 > \rho_1$

$\dfrac{\rho_2-\rho_1}{\rho_2+\rho_1}$	$k(\nu^2/g)^{\frac{1}{3}}$	$n(\nu/g^2)^{\frac{1}{3}}$
0·01	0·1134	0·02081
0·05	0·1939	0·06086
0·10	0·2442	0·09663
0·15	0·2793	0·1267
0·25	0·3304	0·1782
0·50	0·4112	0·2842
0·90	0·4806	0·4265
1·00	0·4907	0·4599

(c) *The effect of surface tension on the unstable modes for $\nu_1 = \nu_2$ and $\rho_2 > \rho_1$*

When surface tension is taken into account, equation (132) is replaced by

$$Q^{\frac{1}{2}}\left(Q^{\frac{3}{2}}-\frac{S}{\alpha_2-\alpha_1}\right) = \frac{y-1}{\alpha_2-\alpha_1}\{y^3+(1+4\alpha_1\alpha_2)y^2+(3-8\alpha_1\alpha_2)y-(1-4\alpha_1\alpha_2)\}. \tag{133}$$

From this equation it follows that for a given S, as y goes from 1 to ∞, the left-hand side takes all values from 0 to ∞. The left-hand side is zero when

$$Q^{\frac{3}{2}} = S/(\alpha_2-\alpha_1) = Q_c^{\frac{3}{2}} \quad \text{(say)}; \tag{134}$$

so that $y = 1$ corresponds to $Q = Q_c$. This means that unstable modes occur only for $k < k_c$ where

$$k_c = Q_c^{-\frac{1}{2}} = \sqrt{[(\alpha_2-\alpha_1)/S]}. \tag{135}$$

It can be readily verified that this wave number k_c (now measured in the unit specified in (123)) is the same as (52) which we found in the absence of viscosity. *The wave numbers which are stabilized by surface tension are, therefore, independent of viscosity.* (This result does not, in fact, depend on the assumption $\nu_1 = \nu_2$ which underlies the present derivation.)

When $y \to \infty$, $Q \to \infty$ and the term in S on the left-hand side of (133) can be ignored. This means that for $k \to 0$ the same asymptotic behaviour (128) obtains as in the absence of surface tension.

For the particular case $\alpha_1 = 0$, the (n, k)-relationships, which have been derived by Reid for a number of values of S, are illustrated in Fig. 108. And in Table XLVII we present his results on the dependence of the modes of maximum instability on S.

In addition to the unstable modes we have considered, the problem also allows damped modes when $S \neq 0$. These damped modes exhibit a curious behaviour. For example, Reid has shown that in the case

$\alpha_1 = 0$, the damped modes are all aperiodic so long as $S < 1{\cdot}3165$; while for S greater than this critical value, damped oscillatory modes become possible for certain ranges of the wave number.

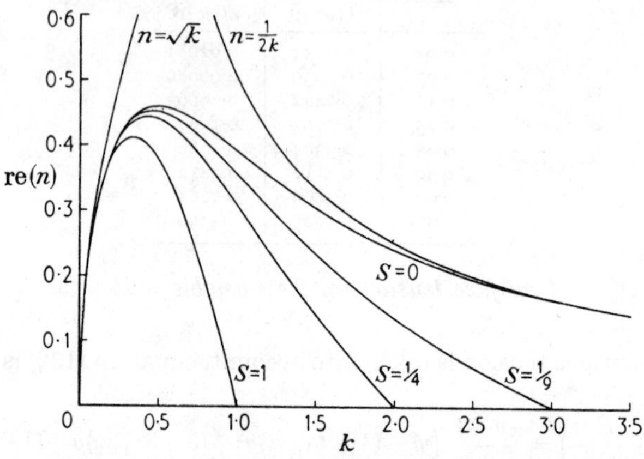

FIG. 108. The dependence of the rate of growth of a disturbance on its wave number and on the surface tension parameter S, in case the lower fluid is of zero density; n and k are measured in the same units as in Figs. 106 and 107.

TABLE XLVII

The modes of maximum instability for the case $\nu_1 = \nu_2$ *and* $\rho_1 = 0$ *and different values of* S

S	$k(\nu^2/g)^{\frac{1}{3}}$	$n(\nu/g^2)^{\frac{1}{3}}$
0	0·4907	0·4599
$\frac{1}{9}$	0·4591	0·4524
$\frac{1}{4}$	0·4227	0·4443
1	0·3400	0·4125

(*d*) *The manner of decay in the case,* $\nu_1 = \nu_2$, $\rho_2 < \rho_1$, *and* $S = 0$

When $\alpha_2 < \alpha_1$, equation (125) admits two roots whose real parts are positive. More precisely, the behaviour of these roots for assigned values of Q is as follows.

When $\alpha_2 < \alpha_1$, equation (132) gives positive values of Q for positive real values of y only in a range, $y_* \leqslant y < 1$, where $y_* > 0$ is a determinate constant depending only on α_1 and α_2 $(= 1-\alpha_1)$. Further, in the range, $y_* \leqslant y \leqslant 1$, Q attains a maximum (say Q_*) and vanishes at both ends. This means that there exists a value of k (say k_*) such that for $k > k_*$ $(= Q_c^{-\frac{1}{3}})$ there are two possible modes of decay: a '*viscous*' mode which decays very rapidly and a '*creeping*' mode which decays very slowly. The difference between these two modes becomes increasingly pronounced as $k \to \infty$. Both of these modes decay aperiodically. (We shall return, presently, to the difference in the proper functions belonging to the two modes.) For $Q > Q_*$ equation (125) admits two complex roots which are conjugates of one another and whose real parts

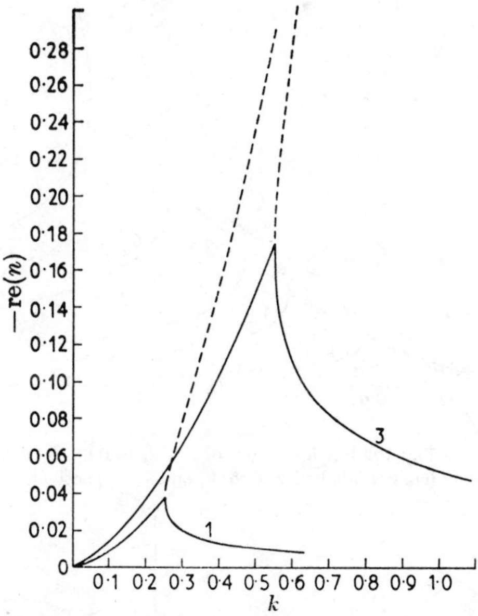

Fig. 109. The dependence of the rate of decay $-\text{re}(n)$ (measured in the unit $(g^2/\nu)^{\frac{1}{3}}$) of a disturbance on its wave number k (measured in the unit $(g/\nu^2)^{\frac{1}{3}}$) in case the lower fluid is more dense and the kinematic viscosities are the same. The curves labelled 1 and 3 are for values of $\alpha_1-\alpha_2 = 0{\cdot}01$ and $0{\cdot}10$, respectively. The dashed parts of each curve refer to the rapidly decaying 'viscosity' mode.

are positive. For $k < k_*$, the damping, therefore, takes place as oscillations of decreasing amplitude.

The establishment of the (k, n)-relationships for $k > k_*$ can be accomplished most simply by evaluating Q (according to equation (132)) for various values of y in the range, $y_* \leqslant y \leqslant 1$, and converting the resulting pairs of Q and y into corresponding pairs of k and n in accordance with (124). For $Q > Q_*$, there would appear to be no alternative to solving the quartic equation (125) explicitly.

The (k, n)-relationships derived in the manner we have described are illustrated in Figs. 109, 110, and 111. It will be observed that the imaginary part $\text{im}(n)$ of n has a maximum in the range $0 \leqslant k \leqslant k_*$;

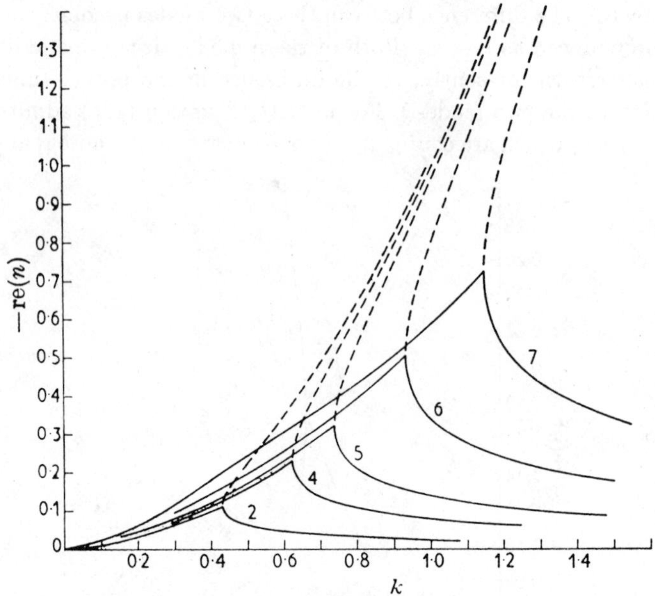

FIG. 110. Same as Fig. 109 but for values of $\alpha_1-\alpha_2 = 0{\cdot}05$, $0{\cdot}15$, $0{\cdot}25$, $0{\cdot}50$, and $0{\cdot}90$ (curves labelled 2, 4, 5, 6, and 7, respectively).

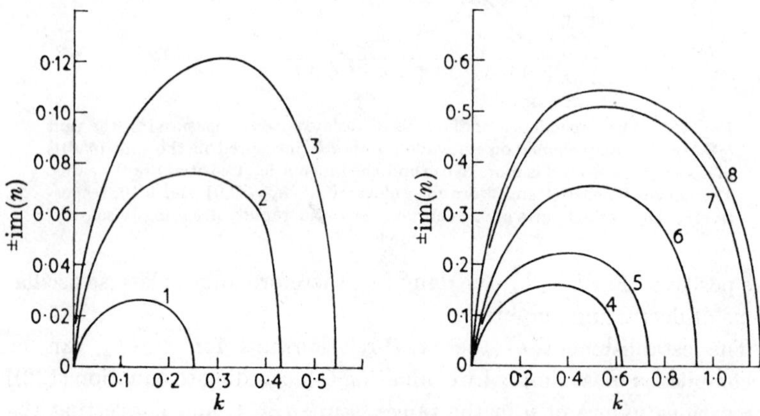

FIG. 111. The dependence of the imaginary part, im(n), of n on the wave number k in case the lower fluid is more dense. The curves labelled 1, 2, 3, 4, 5, 6, 7, and 8 are for values of $\alpha_1-\alpha_2 = 0{\cdot}01$, $0{\cdot}05$, $0{\cdot}10$, $0{\cdot}15$, $0{\cdot}25$, $0{\cdot}50$, $0{\cdot}90$, and $1{\cdot}0$, respectively.

this means that there is a maximum to the frequency of oscillations which can be set up or maintained.

In Table XLVIII the branch points (k_*, n) at which the manner of the decay changes its character from being aperiodic to being periodic are listed for various values of $\alpha_1 - \alpha_2$.

As we have already remarked, the two aperiodic modes of decay which are possible for $k > k_*$ are quite different in that one of them decays very slowly relative to the other. The physical origin for this difference in the two modes can be understood in terms of the proper solutions belonging to them. In Fig. 112 the solutions for the two modes (normalized to unit maximum amplitude) which occur for $k = 1{\cdot}1$ and

TABLE XLVIII

The branch point (k_*, n) at which the manner of decay (in the case $\nu_1 = \nu_2$ and $\rho_2 < \rho_1$) changes its character from being aperiodic to periodic

$\dfrac{\rho_1-\rho_2}{\rho_1+\rho_2}$	$k_*(\nu^2/g)^{\frac{1}{3}}$	$n(\nu/g^2)^{\frac{1}{3}}$
0·01	0·2527	−0·03767
0·05	0·4321	−0·1101
0·10	0·5446	−0·1747
0·15	0·6236	−0·2288
0·25	0·7400	−0·3206
0·50	0·9365	−0·5026
0·90	1·1527	−0·7215
1·00	1·1981	−0·7673

$\alpha_1 - \alpha_2 = 0{\cdot}5$ are illustrated. The two modes are characterized by $n = -0{\cdot}2757$ and $-1{\cdot}0439$. We observe that the solution belonging to the more rapidly decaying viscous mode decreases in its amplitude much more slowly than the solution belonging to the less rapidly decaying creeping mode: the origin of the greater damping of the viscous mode is thus made apparent. For comparison we have also included in Fig. 112 the proper solution for the oscillatory mode (characterized by $n = -0{\cdot}1947 \pm 0{\cdot}3523\,i$) which occurs for $k = 0{\cdot}5$.

(*e*) *Gravity waves*

An interesting by-product of the preceding analysis is the dispersion relation for the so-called gravity waves which occur in an infinitely deep ocean. We can obtain this relation by passing to the limit $\alpha_2 = 0$ and $\alpha_1 = 1$. In this limit, the characteristic equation becomes (cf. equation (121))

$$y^4 + 2y^2 - 4y + 1 + Q + Q^{\frac{1}{2}}S = 0. \tag{136}$$

The (k, n)-relationships derived from this equation (in the manner described in § (*c*) above) for several values of S are shown in Fig. 113.

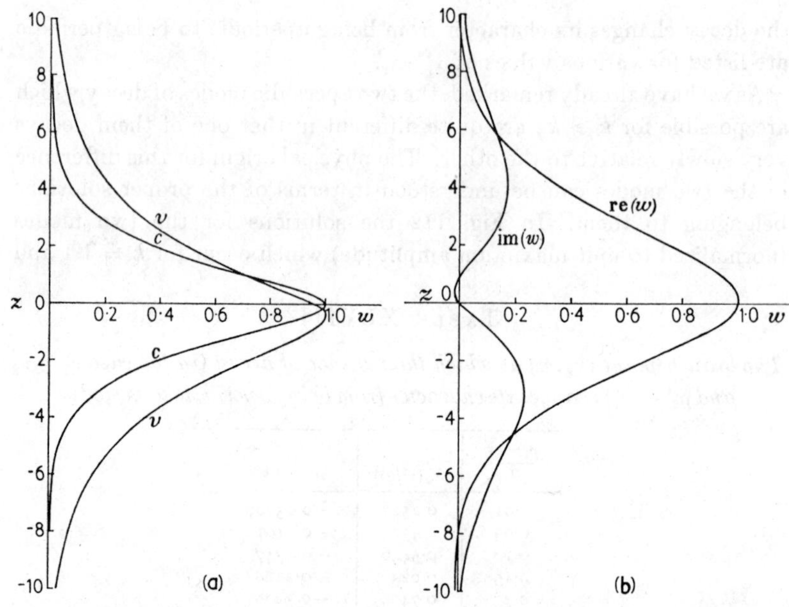

Fig. 112. (*a*) The proper solutions for the creeping *c* and the viscous *v* modes for the case $\alpha_1-\alpha_2 = 0{\cdot}5$ and $k = 1{\cdot}1$; the decay rates associated with these modes are respectively, $n = -0{\cdot}2757$ and $n = -1{\cdot}0439$. (*b*) The proper solutions for the oscillatory mode for the case $\alpha_1-\alpha_2 = 0{\cdot}5$ and $k = 0{\cdot}5$; the complex value of *n* associated with this mode is $n = -0{\cdot}1947 \pm 0{\cdot}3523i$.

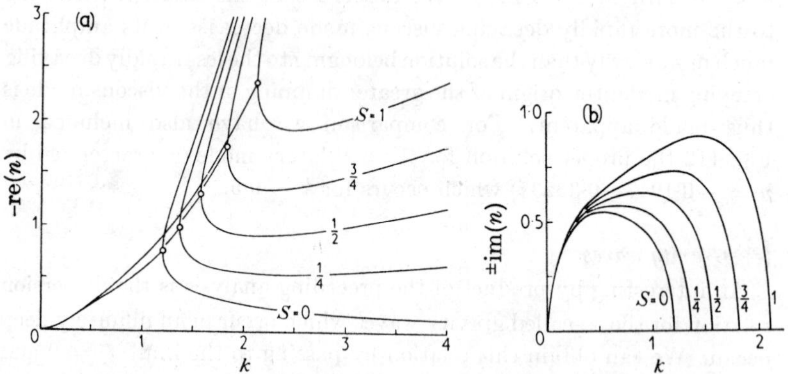

Fig. 113. The dependence of the real (*a*) and the imaginary (*b*) parts of *n* on the wave number *k* and the surface tension parameter *S* for gravity waves.

And in Table XLIX the coordinates of the branch point (where the manner of the damping changes) are listed. The proper solutions belonging to $k = 0{\cdot}8$ and $1{\cdot}4$ are shown in Fig. 114; the characteristic difference in the solutions belonging to the viscous and the creeping modes is clearly exhibited by the curves for $k = 1{\cdot}4$.

TABLE XLIX

The coordinates of the branch point at which the manner of decay changes from being aperiodic to periodic for gravity waves

S	$k_*(\nu^2/g)^{\frac{1}{3}}$	$n(\nu/g^2)^{\frac{1}{3}}$
0	1·1981	−0·7673
$\frac{1}{9}$	1·2653	−0·8557
$\frac{1}{4}$	1·3600	−0·9884
$\frac{1}{2}$	1·5635	−1·3065
$\frac{3}{4}$	1·8131	−1·7569
1	2·1073	−2·3732

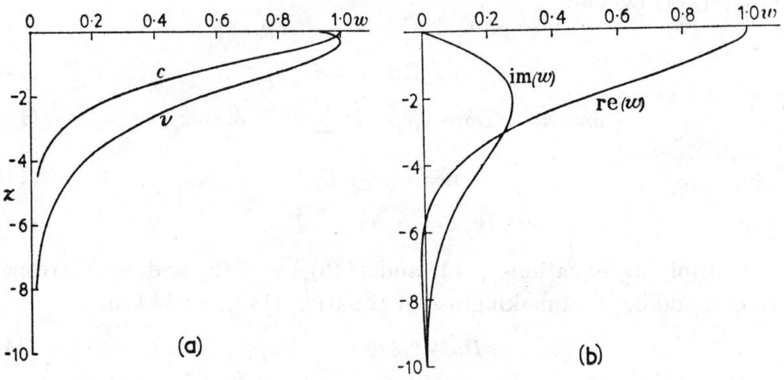

FIG. 114. (*a*) The proper solutions for the creeping c and the viscous v modes for gravity waves of wave number $k = 1{\cdot}4$. The damping constants associated with these modes are $n = -0{\cdot}4314$ and $n = -1{\cdot}5529$, respectively. (*b*) The proper solutions for gravity waves of wave number $k = 0{\cdot}8$; for this value of k the waves are periodically damped with $n = -0{\cdot}4404 \pm 0{\cdot}5061i$.

95. The effect of rotation

We shall consider the effect of rotation on the character of the equilibrium of a stratified heterogeneous inviscid fluid. The initial kinematic state is, then, the same as that postulated in § 91 except that now the fluid will be supposed to be inviscid but subject to a uniform rotation with an angular velocity Ω about the vertical. Under these circumstances the relevant perturbation equations are (cf. Chapter III, §§ 21 and 22):

$$\rho\frac{\partial u}{\partial t}-2\rho\Omega v = -\frac{\partial}{\partial x}\delta p, \tag{137}$$

$$\rho\frac{\partial v}{\partial t}+2\rho\Omega u = -\frac{\partial}{\partial y}\delta p, \tag{138}$$

$$\rho\frac{\partial w}{\partial t} = -\frac{\partial}{\partial z}\delta p-g\delta\rho+\sum_s T_s\left(\frac{\partial^2}{\partial x^2}+\frac{\partial^2}{\partial y^2}\right)\delta z_s\,\delta(z-z_s), \tag{139}$$

$$\frac{\partial}{\partial t}\delta\rho = -wD\rho, \tag{140}$$

and

$$\frac{\partial u}{\partial x}+\frac{\partial v}{\partial y}+\frac{\partial w}{\partial z} = 0, \tag{141}$$

where in equation (139) we have allowed for the possibility that at some preassigned levels z_s the density may change discontinuously and bring into play effects due to surface tension.

For solutions having the (x, y, t)-dependence given by (13), equations (137)–(141) become

$$\rho nu-2\rho\Omega v = -ik_x\,\delta p, \tag{142}$$

$$\rho nv+2\rho\Omega u = -ik_y\,\delta p, \tag{143}$$

$$\rho nw = -D\delta p-g\delta\rho-k^2\sum_s T_s\,\delta z_s\,\delta(z-z_s), \tag{144}$$

$$n\delta\rho = -wD\rho, \tag{145}$$

and

$$ik_x u+ik_y v = -Dw. \tag{146}$$

Multiplying equations (142) and (143) by $-ik_x$ and $-ik_y$, respectively, adding, and making use of equation (146), we obtain

$$\rho nDw+2\rho\Omega\zeta = -k^2\delta p, \tag{147}$$

where

$$\zeta = ik_x v-ik_y u \tag{148}$$

is the z-component of the vorticity. Similarly, by multiplying equations (142) and (143) by $-ik_y$ and $+ik_x$, respectively, adding, and making use of equations (146) and (148), we obtain

$$\rho n\zeta-2\rho\Omega Dw = 0 \tag{149}$$

or

$$\zeta = 2\Omega Dw/n. \tag{150}$$

Inserting this expression for ζ in equation (147), we find

$$\rho n\left(1+\frac{4\Omega^2}{n^2}\right)Dw = -k^2\delta p; \tag{151}$$

whereas by combining equations (144) and (145) and making use, also, of the relation (cf. equation (31))

$$\delta z_s = w_s/n \quad (w_s = w(z_s)), \tag{152}$$

we find
$$D\delta p = -\rho n w + \frac{g}{n}(D\rho)w - \frac{k^2}{n}\sum_s T_s w_s \delta(z-z_s). \tag{153}$$

Now eliminating δp between equations (151) and (153) we obtain

$$\left(1+\frac{4\Omega^2}{n^2}\right)D(\rho Dw) - k^2\rho w = -\frac{gk^2}{n^2}\left\{(D\rho) - \frac{k^2}{g}\sum_s T_s \delta(z-z_s)\right\}w. \tag{154}$$

Letting
$$\kappa^2 = \frac{k^2}{1+4\Omega^2/n^2}, \tag{155}$$

we can rewrite equation (154) in the form

$$D(\rho Dw) - \kappa^2\rho w = -\frac{g\kappa^2}{n^2}\left\{(D\rho) - \frac{k^2}{g}\sum_s T_s \delta(z-z_s)\right\}w. \tag{156}$$

(We shall presently see that either $n^2 > 0$ or $n^2 < -4\Omega^2$ so that κ^2 is always positive and the definition (155) a permissible one.)

At the interface z_s between two fluids, equation (156) leads to the condition (cf. equation (49))

$$\Delta_s(\rho Dw) = -\frac{\kappa^2}{n^2}\{g\Delta_s(\rho) - k^2 T_s\}w_s, \tag{157}$$

while
$$D(\rho Dw) - \kappa^2\rho w = -\frac{g\kappa^2}{n^2}(D\rho)w \quad (z \neq z_s). \tag{158}$$

We observe that equation (158) is formally identical with equation (42) which is valid in the absence of rotation; the difference in the two problems arises only from the circumstance that we now have κ^2 in place of k^2.

(a) The case of two uniform fluids separated by a horizontal boundary

In view of the formal identity of equations (42) and (158), we can write down directly the solution for the present problem from the solution for the same problem, in the absence of rotation, given in § 92 (*a*). Thus, in place of the solution (51), we now have

$$n^2 = g\kappa\left\{\frac{\rho_2-\rho_1}{\rho_2+\rho_1} - \frac{k^2 T}{g(\rho_1+\rho_2)}\right\}, \tag{159}$$

or, substituting for κ its value (155), we have

$$n^2\left(1+\frac{4\Omega^2}{n^2}\right)^{\frac{1}{2}} = gk\left\{\frac{\rho_2-\rho_1}{\rho_2+\rho_1} - \frac{k^2 T}{g(\rho_1+\rho_2)}\right\}. \tag{160}$$

In terms of n_0, the value of n appropriate for the wave number k in the absence of rotation, we can rewrite equation (160) in the manner,

$$n^2\sqrt{(1+4\Omega^2/n^2)} = n_0^2. \tag{161}$$

The solution of this equation is

$$n^2 = -2\Omega^2+\sqrt{(4\Omega^4+n_0^4)} \quad \text{if } n_0^2 > 0 \tag{162}$$

and

$$n^2 = -2\Omega^2-\sqrt{(4\Omega^4+n_0^4)} \quad \text{if } n_0^2 < 0. \tag{163}$$

From equations (162) and (163) it follows that in the present case rotation does not affect the instability or stability, as such, of a stratification: $n^2 > 0$ if $n_0^2 > 0$ and $n^2 < 0$ if $n_0^2 < 0$. On the other hand, according to equation (163)

$$n^2 \to -4\Omega^2 \quad \text{if } n_0^2 \to 0 \text{ through negative values.} \tag{164}$$

The frequency of the stable oscillations cannot, thus, be less than the natural frequency of wave propagation in a rotating fluid (cf. § 23, equation III (71)). It appears that this is a general result.

(*b*) *The case of exponentially varying density*

From the solution for the same problem in the absence of rotation given in § 92 (*b*) (equation (68)), we can now write down

$$\frac{g\beta}{n^2} = 1+\frac{\frac{1}{4}\beta^2d^2+m^2\pi^2}{\kappa^2d^2}, \tag{165}$$

or, substituting for κ its value (155), we have

$$\frac{g\beta}{n^2} = 1+\frac{\frac{1}{4}\beta^2d^2+m^2\pi^2}{k^2d^2}\left(1+\frac{4\Omega^2}{n^2}\right). \tag{166}$$

From this equation we obtain

$$n^2 = n_0^2\left(1-\frac{4\Omega^2}{g\beta}\,\frac{\frac{1}{4}\beta^2d^2+m^2\pi^2}{k^2d^2}\right), \tag{167}$$

where n_0^2 is the value of n^2 in the absence of rotation given by (68). Remembering that n_0^2 is positive or negative, according as β is positive or negative, we may conclude from (167) that if β is negative the stratification is stable for disturbances of all wave numbers, and moreover,

$$n^2 \to n_0^2 \text{ for } k \to \infty \quad \text{and} \quad n^2 \to -4\Omega^2 \text{ for } k \to 0; \tag{168}$$

whereas if β is positive, the stratification is unstable for disturbances of all wave numbers exceeding k_c given by

$$k_c^2\,d^2 = \frac{4\Omega^2}{g\beta}\,(\tfrac{1}{4}\beta^2d^2+m^2\pi^2), \tag{169}$$

while it is stable for disturbances with $k < k_c$. Thus, rotation stabilizes the potentially unstable arrangement for all wave numbers less than

$$k^2_{\text{min}} = \frac{4\Omega^2}{g\beta d^2}(\tfrac{1}{4}\beta^2 d^2+\pi^2). \tag{170}$$

We observe that as a function of β, k_{min} (for a given Ω) attains its minimum value for $\beta = 2\pi/d$: distributions with β less than, or greater than, $2\pi/d$ are stabilized by rotation for greater ranges of k.

96. The effect of a vertical magnetic field

We shall now consider the effect of a magnetic field in the direction of $\mathbf{g}$ on the character of the equilibrium of a heterogeneous fluid. We shall suppose that the fluid is inviscid and is of zero resistivity; and we shall not consider the effects of surface tension. It is not difficult to include the effects of finite viscosity, resistivity, and surface tension in the formal analysis; but the essential new elements which a magnetic field introduces can best be appreciated without the complications of many additional parameters which these other effects introduce.

Starting, then, from the equations of hydromagnetics in Chapter IV, § 38 (*b*), we readily find that under the circumstances postulated the relevant perturbation equations are:

$$\rho\frac{\partial u}{\partial t}-\frac{\mu H}{4\pi}\left(\frac{\partial h_x}{\partial z}-\frac{\partial h_z}{\partial x}\right) = -\frac{\partial}{\partial x}\delta p, \tag{171}$$

$$\rho\frac{\partial v}{\partial t}-\frac{\mu H}{4\pi}\left(\frac{\partial h_y}{\partial z}-\frac{\partial h_z}{\partial y}\right) = -\frac{\partial}{\partial y}\delta p, \tag{172}$$

$$\rho\frac{\partial w}{\partial t} = -\frac{\partial}{\partial z}\delta p-g\delta\rho, \tag{173}$$

$$\frac{\partial \mathbf{h}}{\partial t} = H\frac{\partial \mathbf{u}}{\partial z}, \quad\text{and}\quad \frac{\partial}{\partial t}\delta\rho = -w\frac{d\rho}{dz}, \tag{174}$$

where H denotes the strength of the uniform magnetic field impressed in the z-direction, $\mathbf{h} = (h_x, h_y, h_z)$ is the perturbation in H, and the remaining symbols have their usual meanings. In addition to (171)–(174) we have the equations expressing the solenoidal character of $\mathbf{u}$ and $\mathbf{h}$.

For solutions having the (x, y, t)-dependence given by (13), the perturbation equations become

$$\rho nu-\frac{\mu H}{4\pi}(Dh_x-ik_x h_z) = -ik_x\,\delta p, \tag{175}$$

$$\rho nv-\frac{\mu H}{4\pi}(Dh_y-ik_y h_z) = -ik_y\,\delta p, \tag{176}$$

$$\rho n w = -D\delta p + \frac{g}{n}(D\rho)w, \tag{177}$$

$$\mathbf{h} = \frac{H}{n}D\mathbf{u}, \tag{178}$$

$$ik_x u + ik_y v = -Dw, \quad \text{and} \quad ik_x h_x + ik_y h_y = -Dh_z. \tag{179}$$

Multiplying equations (175) and (176) by $-ik_x$ and $-ik_y$, respectively, adding, and making use of equations (179), we get

$$\rho n Dw - \frac{\mu H}{4\pi}(D^2 - k^2)h_z = -k^2\delta p; \tag{180}$$

but according to (178)
$$h_z = \frac{H}{n}Dw. \tag{181}$$

Hence
$$\rho Dw - \frac{\mu H^2}{4\pi n^2}(D^2 - k^2)Dw = -\frac{k^2}{n}\delta p. \tag{182}$$

Now eliminating δp between equations (177) and (182), we obtain

$$D(\rho Dw) - \frac{\mu H^2}{4\pi n^2}(D^2 - k^2)D^2 w = k^2 \rho w - \frac{gk^2}{n^2}(D\rho)w. \tag{183}$$

If the fluid is confined between two rigid planes at $z = z_1$ and z_2 (say), then the boundary conditions which have to be met at these planes are (see Chapter IV, § 42 (a))

$w = 0$ and either Dw or $D^2 w = 0$ depending on whether the confining walls are of perfectly conducting or of insulating material. (184)

Equation (183) together with the boundary conditions (184) leads to the integral formula

$$\int_{z_1}^{z_2} \rho\{|Dw|^2 + k^2|w|^2\}\,dz + \frac{\mu H^2}{4\pi n^2}\int_{z_1}^{z_2} \{|D^2 w|^2 + k^2|Dw|^2\}\,dz = \frac{gk^2}{n^2}\int_{z_1}^{z_2} (D\rho)|w|^2\,dz. \tag{185}$$

It can be readily shown that this formula provides the basis for a variational formulation of the problem; however, the principal conclusion which we wish to draw from (185) is that *the characteristic values of n^2 are necessarily real.*

If there should be discontinuities in ρ, as there would be if fluids of different densities are superposed one over the other, then the conditions which must be met at the interface are

$$w \text{ and } \mathbf{h} \text{ are both continuous.} \tag{186}$$

Normally, in a fluid of infinite electrical conductivity, the only condition on $\mathbf{h}$ we can impose is that the normal component of $\mathbf{h}$ (h_z in the present connexion) is continuous. However, since in the present instance the main field is normal to the boundary, the continuity of the *tangential stresses* requires that h_x and h_y are also continuous. The continuity of h_z requires, according to equation (181), the continuity of Dw; while the continuity of h_x and h_y, according to equations (179), requires the continuity of Dh_z and, therefore, of D^2w. Thus, altogether

w, Dw, and D^2w must be continuous at an interface between two fluids. (187)

A further condition follows from equation (183) by integrating it across the interface. We obtain

$$\Delta_s\left\{\rho Dw-\frac{\mu H^2}{4\pi n^2}(D^2-k^2)Dw\right\} = -\frac{gk^2}{n^2}\Delta_s(\rho)w_s. \tag{188}$$

(*a*) *Two uniform fluids separated by a horizontal boundary*: *the unstable case*

Consider the case of two uniform fluids of densities ρ_1 and ρ_2 separated by a horizontal boundary at $z = 0$. In each of the two regions of constant ρ, equation (183) reduces to

$$(D^2-k^2)w-\frac{\mu H^2}{4\pi\rho n^2}(D^2-k^2)D^2w = 0, \tag{189}$$

where, for the present, we are suppressing the subscripts distinguishing the two fluids.

The general solution of equation (189) is a linear combination of the integrals

$$e^{\pm kz} \quad\text{and}\quad e^{\pm qz}, \tag{190}$$

where

$$q^2 = \frac{4\pi\rho n^2}{\mu H^2}. \tag{191}$$

Now a result of general validity which we have derived is that n^2 must be real. Therefore, in (191) n^2 is either positive or negative. This means that the exponent q in the second of the two integrals in (190) is either real or purely imaginary. In the former case, we can choose a sign for q such that the solution vanishes at $\pm\infty$; in the latter case no such choice is possible. It is therefore clear that the stable ($n^2 < 0$) and the unstable ($n^2 > 0$) cases require separate treatments. Accordingly, we shall first carry through the analysis on the supposition that n^2 is positive. This will enable us to distinguish the stable from the unstable stratifications. The treatment (on the supposition $n^2 > 0$) will apply only to the unstable cases; it will not apply to the stable cases which will require a separate treatment.

We shall suppose, then, that

$$q = n\sqrt{\frac{4\pi\rho}{\mu H^2}} \quad \text{and} \quad n > 0. \tag{192}$$

Returning to the solution of equation (189), we first observe that since w must be bounded both when $z \to +\infty$ (in the upper fluid) and $z \to -\infty$ (in the lower fluid), we can write

$$\left.\begin{aligned} w_1 &= A_1 e^{+kz} + B_1 e^{+q_1 z} \quad (z < 0) \\ \text{and} \qquad w_2 &= A_2 e^{-kz} + B_2 e^{-q_2 z} \quad (z > 0) \end{aligned}\right\}, \tag{193}$$

where A_1, B_1, A_2, and B_2 are constants of integration,

$$q_1 = n\sqrt{\frac{4\pi\rho_1}{\mu H^2}}, \quad \text{and} \quad q_2 = n\sqrt{\frac{4\pi\rho_2}{\mu H^2}}. \tag{194}$$

In accordance with (192) we suppose that $n > 0$; only then is the foregoing manner of writing the solutions in the two regions ($z > 0$ and $z < 0$) valid.

On the common boundary ($z = 0$) between the two fluids, w, Dw, and D^2w must be continuous. Applying these conditions to the solution (193), we find:

$$A_1 + B_1 = +A_2 + B_2 \quad (= w_0), \tag{195}$$

$$kA_1 + q_1 B_1 = -kA_2 - q_2 B_2 \quad (= (Dw)_0), \tag{196}$$

and

$$k^2A_1 + q_1^2 B_1 = +k^2A_2 + q_2^2 B_2 \quad (= (D^2w)_0). \tag{197}$$

The remaining condition (188) now requires

$$\rho_2\left\{Dw_2 - \frac{1}{q_2^2}(D^2-k^2)Dw_2\right\}_{z=0} - \rho_1\left\{Dw_1 - \frac{1}{q_1^2}(D^2-k^2)Dw_1\right\}_{z=0} = -\frac{gk^2}{n^2}(\rho_2-\rho_1)w_0. \tag{198}$$

Substituting for w_1 and w_2 from (193) and simplifying, we are left with

$$-\rho_2\left(A_2 k + \frac{k^2}{q_2} B_2\right) - \rho_1\left(A_1 k + \frac{k^2}{q_1} B_1\right) = -\frac{gk^2}{2n^2}(\rho_2-\rho_1)(A_1+B_1+A_2+B_2). \tag{199}$$

Equations (195)–(197) and (199) lead to the secular equation

$$\Delta(q_1, q_2) = \left\| \begin{matrix} 1 & 1 & -1 & -1 \\ k & q_1 & k & q_2 \\ k^2 & q_1^2 & -k^2 & -q_2^2 \\ \frac{1}{2}R-\alpha_1 & \frac{1}{2}R-\alpha_1 k/q_1 & \frac{1}{2}R-\alpha_2 & \frac{1}{2}R-\alpha_2 k/q_2 \end{matrix} \right\| = 0, \tag{200}$$

where (cf. equations (108) and (109))

$$\alpha_1 = \frac{\rho_1}{\rho_1+\rho_2}, \quad \alpha_2 = \frac{\rho_2}{\rho_1+\rho_2}, \quad \text{and} \quad R = \frac{gk}{n^2}(\alpha_2-\alpha_1). \tag{201}$$

The determinant, $\Delta(q_1, q_2)$, can be reduced to give

$$\Delta(q_1,q_2) = \left\| \begin{matrix} q_1-k & 2k & q_2-k \\ q_1^2-k^2 & 0 & -(q_2^2-k^2) \\ \alpha_1(q_1-k)/q_1 & R-1 & \alpha_2(q_2-k)/q_2 \end{matrix} \right\|$$

$$= (q_1-k)(q_2-k) \left\| \begin{matrix} 1 & 2k & 1 \\ q_1+k & 0 & -(q_2+k) \\ \alpha_1/q_1 & R-1 & \alpha_2/q_2 \end{matrix} \right\| = 0. \tag{202}$$

Accordingly, $q_1 = k$ and $q_2 = k$ are two characteristic roots; but these two roots lead to trivial solutions: for, one may verify that the proper functions w_1 and w_2 belonging to these two roots are identically zero. Therefore, removing the factors q_1-k and q_2-k and expanding the remaining determinant, we find

$$(R-1)(q_1+q_2+2k) = 2k\left\{\frac{\alpha_2}{q_2}(q_1+k)+\frac{\alpha_1}{q_1}(q_2+k)\right\}. \tag{203}$$

Define the Alfvén velocity

$$V_A = \sqrt{\frac{\mu H^2}{4\pi(\rho_1+\rho_2)}}. \tag{204}$$

In terms of V_A,

$$q_1 = \frac{n}{V_A}\sqrt{\alpha_1} \quad \text{and} \quad q_2 = \frac{n}{V_A}\sqrt{\alpha_2}. \tag{205}$$

Inserting these values in (203) and substituting also for R in accordance with its definition, we obtain after some minor rearrangements

$$\frac{gk}{n^2}(\alpha_2-\alpha_1)\left(\frac{n}{kV_A}+\frac{2}{\sqrt{\alpha_1}+\sqrt{\alpha_2}}\right) = \frac{n}{kV_A}+2(\sqrt{\alpha_1}+\sqrt{\alpha_2})+\frac{2kV_A}{n}. \tag{206}$$

(It may be noted here for future reference that, in passing from equation (203) to (206), a factor $k/(\sqrt{\alpha_2}+\sqrt{\alpha_1})$ has been removed.)

By measuring n and k in the units

$$(g/V_A)\text{sec}^{-1} \quad \text{and} \quad (g/V_A^2)\text{cm}^{-1}, \text{ respectively}, \tag{207}$$

we can rewrite equation (206) in the non-dimensional form

$$\frac{k}{n^2}(\alpha_2-\alpha_1)\left(\frac{n}{k}+\frac{2}{\sqrt{\alpha_1}+\sqrt{\alpha_2}}\right) = \frac{n}{k}+2(\sqrt{\alpha_1}+\sqrt{\alpha_2})+2\frac{k}{n}. \tag{208}$$

On further reduction, equation (208) becomes

$$n^3+2k(\sqrt{\alpha_2}+\sqrt{\alpha_1})n^2+k(2k+\alpha_1-\alpha_2)n-2k^2(\sqrt{\alpha_2}-\sqrt{\alpha_1}) = 0. \tag{209}$$

If $\alpha_2 > \alpha_1$, the constant term in (209) is negative. The equation must, therefore, allow at least one positive real root; and it can be readily shown that the two remaining roots are either real and negative, or complex conjugates with negative real parts. Since the analysis has presupposed that n is positive, it is clear that the only admissible root of (209), in case $\alpha_2 > \alpha_1$, is the single positive root which it allows. The root is

TABLE L

The dispersion relations for Rayleigh–Taylor instability when the impressed magnetic field is parallel to **g**

	α_2				α_2		
k	1·0	0·8	0·6	k	1·0	0·8	0·6
0	0	0	0				
0·02	—	0·09796	0·05079	1·0	0·69562	0·35893	0·13020
0·03	0·16932	—	—	1·2	0·72788	0·36946	0·13193
0·05	0·21583	0·14489	0·07095	1·6	0·77544	0·38433	0·13422
0·10	0·29674	0·18999	0·08775	2·0	0·80890	0·39437	0·13566
0·20	0·40000	0·24164	0·10399	2·5	0·83899	0·40313	0·13687
0·40	0·52298	0·29544	0·11772	3·0	0·86096	0·40939	0·13769
0·60	0·60000	0·32536	0·12407	5·0	0·91024	0·42306	0·13941
0·80	0·65452	0·34495	0·12777	∞	1·00000	0·44721	0·14214

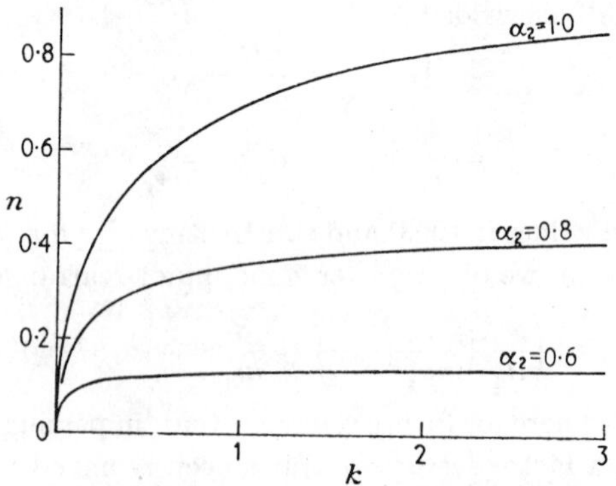

FIG. 115. The dispersion relations for Rayleigh–Taylor instability when the impressed magnetic field is parallel to **g**. The abscissa measures the wave number in the unit $(g/V_A^2)\text{cm}^{-1}$ and the ordinate measures the growth rate in the unit $(g/V_A)\text{sec}^{-1}$, where V_A is the Alfvén speed.

listed in Table L as a function of k for some values of α_2; the derived dispersion relations are further illustrated in Fig. 115.

From equation (209) we derive the following asymptotic relations:

$$\left.\begin{aligned} n^2 &\to k(\alpha_2-\alpha_1) \qquad (k\to 0) \\ \text{and} \qquad n &\to (\sqrt{\alpha_2}-\sqrt{\alpha_1}) \quad (k\to\infty) \end{aligned}\right\}. \tag{210}$$

For $k \to 0$ the relation between n and k, therefore, approaches the corresponding hydrodynamic relation (cf. equation (128)): *the instability of the very long waves is unaffected by the presence of the magnetic field.* However, in contrast to the hydrodynamic case, *n does not increase indefinitely with k: it tends to a definite limit.*

(*b*) *Two uniform fluids separated by a horizontal boundary: the stable case*

If $\alpha_2 < \alpha_1$, the constant term in equation (209) is positive and the equation must allow at least one real negative root. Indeed, the real parts of all three roots must be negative. This follows from the theorem that a necessary and sufficient condition for the real parts of all three roots of the cubic,

$$n^3+Bn^2+Cn+D = 0 \tag{211}$$

to be negative is that $BC > D > 0$. For the cubic (209), $D > 0$ (when $\alpha_1 > \alpha_2$) and

$$\begin{aligned} BC-D &= 2k^2(\sqrt{\alpha_1}+\sqrt{\alpha_2})(2k+\alpha_1-\alpha_2)-2k^2(\sqrt{\alpha_1}-\sqrt{\alpha_2}) \\ &= \frac{2k^2}{\sqrt{\alpha_1}+\sqrt{\alpha_2}}\{[1+2\sqrt{(\alpha_1\alpha_2)}](2k+\alpha_1-\alpha_2)-(\alpha_1-\alpha_2)\} \\ &= \frac{4k^2}{\sqrt{\alpha_1}+\sqrt{\alpha_2}}\{[1+2\sqrt{(\alpha_1\alpha_2)}]k+(\alpha_1-\alpha_2)\sqrt{(\alpha_1\alpha_2)}\} > 0. \end{aligned} \tag{212}$$

Since the characteristic equation (209) allows no positive root when $\alpha_1 > \alpha_2$, we conclude that these cases must be stable and that n must be imaginary. Under these circumstances, the solution for w cannot be written as in (193) by discarding one of the two integrals e^{+qz} and e^{-qz}. Letting $n = i\sigma$ (to emphasize that we are now dealing with an imaginary n) we should, in fact, write

$$w_1 = A_1 e^{+kz}+B_1^{(\text{out};-\infty)}e^{i\sigma z/V_1}+B_1^{(\text{inc};-\infty)}e^{-i\sigma z/V_1} \quad (z<0) \tag{213}$$

and

$$w_2 = A_2 e^{-kz}+B_2^{(\text{out};+\infty)}e^{-i\sigma z/V_2}+B_2^{(\text{inc};+\infty)}e^{+i\sigma z/V_2} \quad (z>0), \tag{214}$$

where V_1 and V_2 denote the Alfvén speeds in the two media.

In equation (213), we may regard the terms in $\exp(-i\sigma z/V_1)$ and $\exp(+i\sigma z/V_1)$ as representing incoming and outgoing Alfvén waves in the first medium. Similarly, in equation (214) we may regard the terms in $\exp(+i\sigma z/V_2)$ and $\exp(-i\sigma z/V_2)$ as representing incoming and outgoing Alfvén waves in the second medium.

By applying the boundary conditions (187) and (188) to the solutions (213) and (214), we will obtain a set of four linear equations for the six constants in the solutions. Consequently, the equations can be solved for arbitrary k and σ; and, moreover, four of the constants A_1, $B_1^{(\text{out};-\infty)}$, A_2, and $B_2^{(\text{out};+\infty)}$ (say), can be expressed linearly in terms of the remaining two constants, $B_1^{(\text{inc};-\infty)}$ and $B_2^{(\text{inc};+\infty)}$ (say). This means that *in this case there is no dispersion relation between k and σ*; and, moreover, the solution for w involves two complex constants $B^{(\text{inc};-\infty)}$ and $B_1^{(\text{inc};+\infty)}$. The possibility of being able to specify the two latter constants, arbitrarily, ensures our ability to specify, at will, the amplitudes and the phases of the incoming Alfvén waves, from $-\infty$ and $+\infty$, in the two media; and the lack of a dispersion relation, similarly, ensures our ability to specify, at will, the frequency of the incoming Alfvén waves. Given, then, the amplitudes, the phases, and the frequency of the incoming Alfvén waves and given, also, the wave number of the deformed interface, we must expect a unique state of perturbation to emerge; and this is exactly the content of the theory.

In some ways it is remarkable that, while in the absence of a magnetic field we have a valid dispersion relation in all cases, it ceases to exist in the presence of a vertical magnetic field when the circumstances are such as to lead to stability.

97. The effect of a horizontal magnetic field

We shall now consider the effect of a magnetic field transverse to the direction of $\mathbf{g}$ on the character of the equilibrium of a heterogeneous inviscid fluid of zero resistivity. Except for the change in the direction of the impressed magnetic field relative to $\mathbf{g}$, the physical circumstances of the problem are the same as in § 96.

Let the direction of the impressed, uniform magnetic field be along the x-axis. Equations (175)–(178) are then replaced by

$$\rho n u = -ik_x\,\delta p, \tag{215}$$

$$\rho n v - \frac{\mu H}{4\pi}(ik_x h_y - ik_y h_x) = -ik_y\,\delta p, \tag{216}$$

$$\rho n w - \frac{\mu H}{4\pi}(ik_x h_z - Dh_x) = -D\delta p + \frac{g}{n}(D\rho)w, \tag{217}$$

and
$$\mathbf{h} = \frac{ik_x}{n}H\mathbf{u}. \tag{218}$$

Inserting for the components of $\mathbf{h}$ in equations (216) and (217) in accordance with equation (218), we obtain

$$\rho n v - \frac{ik_x}{n}\frac{\mu H^2}{4\pi}\zeta = -ik_y\,\delta p \tag{219}$$

and
$$\rho n w - \frac{ik_x}{n}\frac{\mu H^2}{4\pi}(ik_x w - Du) = -D\delta p + \frac{g}{n}(D\rho)w, \tag{220}$$

where
$$\zeta = ik_x v - ik_y u \tag{221}$$

is the z-component of the vorticity. Now multiplying equations (215) and (219) by $-ik_y$ and ik_x, respectively, and adding, we obtain

$$\left(\rho n+\frac{k_x^2}{n}\frac{\mu H^2}{4\pi}\right)\zeta = 0 \tag{222}$$

from which we conclude that

$$\zeta = ik_x v - ik_y u = 0. \tag{223}$$

Hence, equation (219) becomes

$$\rho nv = -ik_y \delta p. \tag{224}$$

Equations (215) and (224) can be combined to give (on addition, after multiplication by $-ik_x$ and $-ik_y$, respectively)

$$\rho nDw = -k^2\delta p. \tag{225}$$

From equation (223) and the equation,

$$ik_x u + ik_y v = -Dw, \tag{226}$$

expressing continuity, we find

$$u = i\frac{k_x}{k^2}Dw. \tag{227}$$

Inserting this expression for u in equation (220), we obtain

$$\rho nw - \frac{\mu H^2 k_x^2}{4\pi k^2 n}(D^2-k^2)w = -D\delta p + \frac{g}{n}(D\rho)w. \tag{228}$$

From equations (225) and (228), δp can be eliminated to give

$$D(\rho Dw) + \frac{\mu H^2 k_x^2}{4\pi n^2}(D^2-k^2)w - k^2\rho w = -\frac{gk^2}{n^2}(D\rho)w. \tag{229}$$

In case we consider solutions for which $k_x = 0$, equation (229) reduces to

$$D(\rho Dw) - k^2\rho w = -\frac{gk^2}{n^2}(D\rho)w; \tag{230}$$

this is clearly the same as the equation which obtains in the absence of any field. Hence for disturbances which are independent of the coordinate along the direction of $\mathbf{H}$ (assumed perpendicular to $\mathbf{g}$), the presence of the magnetic field does not in any way affect the development of the Rayleigh–Taylor instability. However, for disturbances with $k_x \neq 0$, there is no longer this identity between the hydrodynamic and the hydromagnetic problems. To see what effect the additional term in k_x^2 in equation (229) has on the problem, consider once again the case of two uniform fluids separated by a horizontal boundary at $z = 0$. We must require that

$$w \text{ and } h_z \text{ are continuous at } z = 0. \tag{231}$$

According to equation (218), the continuity of w ensures the continuity of h_z. Also, we can verify that the continuity of the tangential stresses leads to no additional conditions in the first order. Consequently, apart from the condition

$$w \text{ is continuous at } z = 0, \tag{232}$$

we have only to satisfy the condition

$$\Delta_0(\rho Dw)+\frac{\mu H^2 k_x^2}{4\pi n^2}\Delta_0(Dw) = -\frac{gk^2}{n^2}(\rho_2-\rho_1)w_0 \tag{233}$$

which follows from integrating equation (229) across the interface. Applying the condition (233) to the solutions given in (47) and (48) (which continue to be valid for this problem), we obtain

$$n^2 = gk\left\{\frac{\rho_2-\rho_1}{\rho_2+\rho_1}-\frac{\mu H^2 k_x^2}{2\pi(\rho_2+\rho_1)gk}\right\}. \tag{234}$$

Comparing this equation with equation (51), we observe that the additional term in k_x^2 in equation (230) has, for this problem, an effect which is the same as a surface tension. We may, indeed, define an equivalent surface tension by the formula

$$T_{\text{eff}} = \frac{\mu H^2}{2\pi k}\cos^2\vartheta, \tag{235}$$

where ϑ is the inclination of the wave vector (k_x, k_y) to the direction of **H**; this surface tension has its origin, clearly, in the tension $\mu H^2/4\pi$ which exists along the lines of force (cf. Chapter IV, § 37).

98. The oscillations of a viscous liquid globe

The problems considered in the preceding sections can naturally be extended to a spherical geometry. In view of possible applications, such extensions have more than a formal interest; but in most instances, the problems become of such complexity and involve so many parameters that 'elementary' methods of solution seem impracticable. Nevertheless, it will be useful to consider at least one problem for which an analytical solution can be found and which will illustrate the type of difficulties one must confront in the other problems. For this illustrative purpose, we shall consider the oscillations of a viscous liquid globe; this problem is the analogue, in a sphere, of the problem of the gravity waves considered in § 94 (*e*).

(a) The perturbation equations and their solution

Quite generally, the equations governing the departures from an equilibrium state are

$$\frac{\partial \mathbf{u}}{\partial t} = -\operatorname{grad}\varpi - \nu \operatorname{curl}^2\mathbf{u}, \tag{236}$$

$$\operatorname{div}\mathbf{u} = 0, \quad \nabla^2 \delta V = 0, \tag{237}$$

and

$$\varpi = -\delta V + \delta p/\rho, \tag{238}$$

where $\mathbf{u}$ denotes the velocity and δp and δV are the changes in the pressure and the internal gravitational potential consequent to the perturbation.

Since we are considering departures from an equilibrium spherical state of an incompressible fluid, a normal mode can be specified uniquely in terms of the shape of the deformed surface of the fluid. Suppose, then, that the deformed surface of the configuration is described by the equation

$$r = R + \epsilon Y_l^m(\vartheta, \varphi), \tag{239}$$

where R is the radius of the unperturbed sphere, ϵ is some function of the time (which we shall presently specify to be of the form $\epsilon_0 e^{-\sigma t}$), and Y_l^m is a spherical harmonic.

For deformations of the surface described by (239), the change in the gravitational potential δV can be readily found from the corresponding linearized forms of Poisson's and Laplace's equations and the conditions that the potential and its derivative are continuous on (239). We find†

$$\delta V = \epsilon \frac{3}{2l+1} \frac{GM}{R^{l+2}} r^l Y_l^m. \tag{240}$$

To determine $\delta p/\rho$, we first observe that by taking the divergence of equation (237) and remembering that $\nabla^2 \delta V = 0$ we have

$$\nabla^2(\delta p/\rho) = 0. \tag{241}$$

We may, therefore, write

$$\frac{\delta p}{\rho} = \epsilon P_0 \frac{r^l}{R^l} Y_l^m, \tag{242}$$

where P_0 is some constant.

† The solutions of Poisson's and Laplace's equations which have to be matched are

$$V_{\text{internal}} = \tfrac{2}{3}\pi G\rho(3R^2 - r^2) + \epsilon B r^l Y_l^m$$

and

$$V_{\text{external}} = GM/r + \epsilon A r^{-(l+1)} Y_l^m.$$

The conditions that V and its radial derivative are continuous on (239) give

$$A/R^{l+1} = BR^l \quad \text{and} \quad lBR^{l-1} + (l+1)A/R^{l+2} = 3GM/R^3.$$

From these equations we find

$$A = 3GMR^{l-1}/(2l+1) \quad \text{and} \quad B = 3GM/(2l+1)R^{l+2},$$

and the result quoted follows.

Combining equations (240) and (242), we have

$$\varpi = -\epsilon \frac{3}{2l+1} \frac{GM}{R^{l+2}} r^l Y_l^m + \frac{\delta p}{\rho} = \epsilon(l+1)\Pi_0 r^l Y_l^m, \tag{243}$$

where Π_0 is a constant undetermined for the present. The problem is now reduced to solving equation (236) as an inhomogeneous equation for $\mathbf{u}$ with ϖ given by (243).

We now consider $\mathbf{u}$ as expressed in terms of the toroidal and the poloidal functions described in Chapter VI, § 56 (*d*). In the terminology of these functions, grad ϖ is a solenoidal vector which is poloidal; and its defining scalar function is

$$\epsilon \Pi_0 r^{l+1}. \tag{244}$$

In view of this purely poloidal character of the inhomogeneous term in equation (236), it follows that the velocity field $\mathbf{u}$ must also be purely poloidal and specifically of the type given in equation VI (50).

We now suppose that the dependence on time of the various quantities is given by

$$\epsilon_0 e^{-\sigma t}, \tag{245}$$

where σ is a constant to be determined. (Note the minus sign in this definition.) Then $\mathbf{u}$ can be expressed in terms of a single scalar function $U(r)$ (now only a function of r) in the manner

$$u_r = e^{-\sigma t} \frac{l(l+1)}{r^2} U(r) Y_l^m, \quad u_\theta = e^{-\sigma t} \frac{1}{r} \frac{dU}{dr} \frac{\partial Y_l^m}{\partial \theta},$$

and

$$u_\varphi = e^{-\sigma t} \frac{1}{r \sin\theta} \frac{dU}{dr} \frac{\partial Y_l^m}{\partial \varphi}; \tag{246}$$

grad ϖ is similarly defined in terms of the scalar function

$$e^{-\sigma t} \epsilon_0 \Pi_0 r^{l+1}, \tag{247}$$

where we have written $\epsilon_0 e^{-\sigma t}$ in place of ϵ.

(i) *The solution for the inviscid case: the Kelvin modes*

Before we proceed to a general discussion of equation (236), we shall consider the case $\nu = 0$. Equation (236) then becomes

$$\frac{\partial \mathbf{u}}{\partial t} = -\text{grad}\,\varpi. \tag{248}$$

Replacing each side of this equation by its defining scalar function, we have

$$\sigma U(r) = \epsilon_0 \Pi_0 r^{l+1}, \tag{249}$$

so that in this case the radial component of the velocity is given by

$$u_r = \frac{\epsilon_0 \Pi_0}{\sigma} l(l+1) r^{l-1} Y_l^m e^{-\sigma t}. \tag{250}$$

The boundary conditions which must be satisfied in this inviscid case are: (1) the radial component of the velocity given by (250) must be compatible with the assumed form of the deformed boundary, namely,

$$r = R + \epsilon_0 e^{-\sigma t} Y_l^m; \tag{251}$$

and (2) the pressure must vanish on the same boundary. The first of these conditions clearly requires

$$\frac{\epsilon_0 \Pi_0}{\sigma} l(l+1) R^{l-1} = -\sigma \epsilon_0 \tag{252}$$

or

$$\sigma^2 = -l(l+1)\Pi_0 R^{l-1}. \tag{253}$$

Now the pressure p_0 in the unperturbed configuration is given by

$$p_0 = \tfrac{2}{3}\pi G \rho^2 (R^2 - r^2). \tag{254}$$

The vanishing of the pressure $p_0 + \delta p$ in the perturbed configuration on the perturbed boundary (251) requires

$$-(\tfrac{4}{3}\pi G \rho R)\epsilon Y_l^m + \left(\frac{\delta p}{\rho}\right)_{r=R} = 0. \tag{255}$$

On the other hand, according to equation (243),

$$-\epsilon \frac{3}{2l+1} \frac{GM}{R^2} Y_l^m + \left(\frac{\delta p}{\rho}\right)_{r=R} = \epsilon (l+1) \Pi_0 R^l Y_l^m. \tag{256}$$

Eliminating $(\delta p/\rho)_{r=R}$ between these equations, we have

$$(l+1)\Pi_0 R^l = \frac{2(l-1)}{2l+1} \frac{GM}{R^2}. \tag{257}$$

Finally, combining equations (253) and (257), we obtain

$$\sigma^2 = -\frac{2l(l-1)}{2l+1} \frac{GM}{R^3} = -\tfrac{8}{3}\pi G \rho \frac{l(l-1)}{2l+1}. \tag{258}$$

This formula giving the frequencies of oscillation of an inviscid liquid globe is due to Kelvin.

(ii) *The solution for the general case*

Returning to equation (236) and replacing each term by its defining scalar function, we have

$$-\sigma U = -\epsilon_0 \Pi_0 r^{l+1} + \nu \left\{ \frac{d^2 U}{dr^2} - \frac{l(l+1)}{r^2} U \right\}, \tag{259}$$

where we have made use of the result quoted in § 56 (*d*) that the defining scalar of $\text{curl}^2 S$ is $\tilde{S}$ (defined in equation VI (56)). We rewrite equation (259) in the form

$$\frac{d^2 U}{dr^2} - \frac{l(l+1)}{r^2} U + \frac{\sigma}{\nu} U = \frac{\epsilon_0}{\nu} \Pi_0 r^{l+1}. \tag{260}$$

The general solution of this equation which is free of singularity at the origin is

$$U = Ar^{\frac{1}{2}}J_{l+\frac{1}{2}}(qr)+\frac{\epsilon_0}{\sigma}\Pi_0\, r^{l+1}, \tag{261}$$

where A is a constant, $J_{l+\frac{1}{2}}(x)$ is the Bessel function of order $l+\frac{1}{2}$, and

$$q = \sqrt{(\sigma/\nu)}. \tag{262}$$

According to equations (246) and (261), the radial component of the velocity is given by

$$u_r = l(l+1)\left\{A\frac{J_{l+\frac{1}{2}}(qr)}{r^{\frac{3}{2}}}+\frac{\epsilon_0}{\sigma}\Pi_0\, r^{l-1}\right\}Y_l^m\, e^{-\sigma t}. \tag{263}$$

(b) The boundary conditions and the characteristic equation

The solution represented by equation (261) must satisfy certain boundary conditions. These are: (1) the radial component of the velocity given by (263) must be compatible with the assumed form of the deformed boundary, namely, (251); (2) the tangential viscous stresses must vanish at $r = R$; and (3) the (r,r)-component of the total stress must vanish on the deformed boundary.

The first of the foregoing conditions clearly requires (cf. equation (252))

$$l(l+1)\left\{A\frac{J_{l+\frac{1}{2}}(x)}{R^{\frac{3}{2}}}+\frac{\epsilon_0}{\sigma}\Pi_0\, R^{l-1}\right\} = -\epsilon_0\sigma, \tag{264}$$

where

$$x = qR = \sqrt{(\sigma R^2/\nu)}. \tag{265}$$

The general expressions for the tangential viscous stresses in spherical polar coordinates have been given in Chapter VI (equation (44)). The boundary conditions require that these stresses vanish at $r = R$. Substituting for the components of $\mathbf{u}$ from (246) in equations VI (44), we verify that the conditions require, only,

$$\frac{l(l+1)}{r^2}U-\frac{2}{r}\frac{dU}{dr}+\frac{d^2U}{dr^2} = 0 \quad \text{for } r = R. \tag{266}$$

Applying this condition to U given by (261), we find after some reductions (in which use is made of the various recurrence relations satisfied by the Bessel functions)

$$\frac{A}{R^{\frac{3}{2}}}[2xJ_{l+\frac{3}{2}}(x)-x^2J_{l+\frac{1}{2}}(x)]+2(l^2-1)\left[A\frac{J_{l+\frac{1}{2}}(x)}{R^{\frac{3}{2}}}+\frac{\epsilon_0}{\sigma}\Pi_0\, R^{l-1}\right] = 0. \tag{267}$$

From equations (264) and (267) it follows that

$$A = \frac{2(l-1)\epsilon_0\sigma R^{\frac{3}{2}}}{l[2xJ_{l+\frac{3}{2}}(x)-x^2J_{l+\frac{1}{2}}(x)]}. \tag{268}$$

Inserting this value of A in equation (264) and rearranging, we obtain

$$-l(l+1)\Pi_0 R^{l-1} = \sigma^2\left[\frac{2(l^2-1)}{2xQ_{l+\frac{1}{2}}(x)-x^2}+1\right], \tag{269}$$

where we have introduced the function

$$Q_\nu(x) = J_{\nu+1}(x)/J_\nu(x). \tag{270}$$

It remains to satisfy the last of the boundary conditions which requires the (r,r)-component of the total stress tensor to vanish on the deformed boundary of the configuration. The required component of the stress tensor is given by

$$p_{rr} = p_0+\delta p-2\nu\rho\frac{\partial u_r}{\partial r}, \tag{271}$$

where p_0 is the pressure in the unperturbed configuration given by (254). Retaining only terms of order ϵ, we find that the vanishing of p_{rr} on $R+\epsilon Y_l^m$ requires

$$-(\tfrac{4}{3}\pi G\rho R)\epsilon Y_l^m+\left(\frac{\delta p}{\rho}\right)_{r=R}-2\nu\left(\frac{\partial u_r}{\partial r}\right)_{r=R} = 0, \tag{272}$$

where (cf. equation (256))

$$\left(\frac{\delta p}{\rho}\right)_{r=R} = \left\{(l+1)\Pi_0 R^l+\frac{3}{2l+1}\frac{GM}{R^2}\right\}\epsilon Y_l^m. \tag{273}$$

Inserting this latter expression for $(\delta p/\rho)_{r=R}$ in equation (272), we find

$$\left\{-\frac{2(l-1)}{2l+1}\frac{GM}{R^2}+(l+1)\Pi_0 R^l\right\}\epsilon Y_l^m-2\nu\left(\frac{\partial u_r}{\partial r}\right)_{r=R} = 0. \tag{274}$$

This equation can be written more conveniently in the form

$$\{-\sigma_{l;0}^2+l(l+1)\Pi_0 R^{l-1}\}\epsilon Y_l^m-\frac{2\nu l}{R}\left(\frac{\partial u_r}{\partial r}\right)_{r=R} = 0, \tag{275}$$

where $\sigma_{l;0}$ is the Kelvin frequency (cf. equation (258)).

Now evaluating $\partial u_r/\partial r$ in accordance with equations (263) and (268), we find

$$\left(\frac{\partial u_r}{\partial r}\right)_{r=R} = l(l^2-1)\left\{\frac{2\sigma}{lR}\frac{(l-1)-xQ_{l+\frac{1}{2}}(x)}{2xQ_{l+\frac{1}{2}}(x)-x^2}+\frac{\Pi_0}{\sigma}R^{l-2}\right\}\epsilon Y_l^m. \tag{276}$$

Using this result in equation (275), we have

$$l(l+1)\Pi_0 R^{l-1}\left[1-2l(l-1)\frac{\nu}{\sigma R^2}\right]+\frac{4\nu\sigma}{R^2}l(l^2-1)\frac{(l-1)-xQ_{l+\frac{1}{2}}(x)}{x^2-2xQ_{l+\frac{1}{2}}(x)} = \sigma_{l;0}^2. \tag{277}$$

Remembering that $x = \sqrt{(\sigma R^2/\nu)}$, we can rewrite the foregoing equation in the form

$$l(l+1)\Pi_0 R^{l-1}\left[1-\frac{2l(l-1)}{x^2}\right]+4\sigma^2\frac{l(l^2-1)}{x^2}\frac{(l-1)-xQ_{l+\frac{1}{2}}(x)}{x^2-2xQ_{l+\frac{1}{2}}(x)} = \sigma_{l;0}^2. \tag{278}$$

Now eliminating Π_0 between equations (269) and (278), we obtain

$$\sigma^2\left[1-\frac{2l(l-1)}{x^2}\right]\left[\frac{2(l^2-1)}{x^2-2xQ_{l+\frac{1}{2}}(x)}-1\right]+4\sigma^2\frac{l(l^2-1)}{x^2}\frac{(l-1)-xQ_{l+\frac{1}{2}}(x)}{x^2-2xQ_{l+\frac{1}{2}}(x)}=\sigma_{l;0}^2. \tag{279}$$

After some simplifications, equation (279) can be brought to the form

$$\frac{\sigma_{l;0}^2}{\sigma^2}=\frac{2(l^2-1)}{x^2-2xQ_{l+\frac{1}{2}}(x)}-1+\frac{2l(l-1)}{x^2}\left[1-\frac{(l+1)Q_{l+\frac{1}{2}}(x)}{\frac{1}{2}x-Q_{l+\frac{1}{2}}(x)}\right]. \tag{280}$$

This is the required characteristic equation for σ; it will determine how viscosity damps the Kelvin modes.

(c) *The manner of decay of the Kelvin modes. Numerical results*

Let

$$\alpha^2=\sigma_{l;0}R^2/\nu. \tag{281}$$

The parameter α^2 is determined by the constitution of the globe; its density ρ (which determines $\sigma_{l;0}$), its viscosity ν, and its radius R. Indeed, α^{-2} is a measure of the kinematic viscosity in the unit $(R^2\sigma_{l;0})^{-1}$.

In terms of α,

$$x=\sqrt{(\sigma R^2/\nu)}=\alpha\sqrt{(\sigma/\sigma_{l;0})}; \tag{282}$$

and we may rewrite equation (280) in the form

$$\alpha^4=x^4\left[\frac{2(l^2-1)}{x^2-2xQ_{l+\frac{1}{2}}(x)}-1\right]+2l(l-1)x^2\left[1-\frac{(l+1)Q_{l+\frac{1}{2}}(x)}{\frac{1}{2}x-Q_{l+\frac{1}{2}}(x)}\right]$$

$$=2(l-1)x^2\left\{l+(l+1)\frac{x-2lQ_{l+\frac{1}{2}}(x)}{x-2Q_{l+\frac{1}{2}}(x)}\right\}-x^4=\Psi(x)\text{ (say)}. \tag{283}$$

Considering first the aperiodic modes of decay (for which x must be real), we observe that when $x\to 0$, the function $\Psi(x)$ is positive and has the behaviour

$$\Psi(x)=2\frac{(l-1)(2l^2+4l+3)}{(2l+1)}x^2+O(x^4)\quad(x\to 0). \tag{284}$$

Hence

$$\alpha^2\sim x\sqrt{\frac{2(l-1)(2l^2+4l+3)}{2l+1}},\qquad\frac{\sigma}{\sigma_{l;0}}\sim x\sqrt{\frac{2l+1}{2(l-1)(2l^2+4l+3)}}, \tag{285}$$

and

$$\frac{\sigma}{\sigma_{l;0}}\sim\alpha^2\frac{2l+1}{2(l-1)(2l^2+4l+3)}\quad(x\to 0). \tag{286}$$

Substituting for α^2 from equation (281) in (286), we obtain

$$\sigma\to\sigma_{l;0}^2\frac{2l+1}{2(l-1)(2l^2+4l+3)}\frac{R^2}{\nu}\quad\text{as }\nu\to\infty. \tag{287}$$

The infinitely slow damping which the foregoing relation predicts for $\nu\to\infty$ means only that in this problem we have the analogues of the 'creeping' modes which we have encountered in § 94 (*e*).

As x increases from zero, $\Psi(x)$ starts increasing, attains a maximum (at which point α^2 also attains a maximum $\alpha^2_{\max}$, say), and then decreases to zero at a finite x. This is the first of a succession of zeros and there are in reality an infinity of intervals in which $\Psi(x)$ is positive. Postponing the consideration of these other intervals and restricting ourselves for the present to the first interval (including the origin) in which $\Psi(x)$ is

TABLE LI

The aperiodic modes of decay in the oscillations of a viscous liquid globe

$l = 2$		$l = 3$		$l = 4$	
$\sigma_{2;0} R^2/\nu$	$\sigma/\sigma_{2;0}$	$\sigma_{2;0} R^2/\nu$	$\sigma/\sigma_{2;0}$	$\sigma_{2;0} R^2/\nu$	$\sigma/\sigma_{2;0}$
0	0	0	0	0	0
0·2755	0·03630	0·5926	0·06750	2·2167	0·22105
1·8630	0·26301	1·1809	0·13548	4·0288	0·41948
2·5583	0·39089	2·0461	0·23948	5·1504	0·56112
2·9588	0·48669	3·1435	0·38492	6·1714	0·71459
3·2917	0·59543	4·1405	0·54341	7·0597	0·88531
3·5385	0·72347	4·9908	0·72334	7·4420	0·97957
3·6743	0·88180	5·4974	0·88041	8·0530	1·1933
3·6902†	**0·96799**	5·7554	1·0008	8·3459	1·3851
3·6848	1·0214	5·9914	1·2168	8·4432	1·5350
3·4459	1·4046	**6·0260†**	**1·3574**	**8·4574†**	**1·6275**
3·0802	1·7929	6·0213	1·4064	8·4458	1·7097
2·6827	2·2375	5·9252	1·6219	8·3348	1·9197
2·4118	2·5914	5·6603	1·9239	8·0837	2·1822
2·2157	2·8889	5·1645	2·3720	7·6532	2·5297
2·0663	3·1469	4·3085	3·1774	6·5163	3·3900
1·8047	3·6884	3·6506	3·9555	5·5913	4·2069
1·5955	4·2369	3·0239	4·9529	4·5353	5·4464
1·3434	5·1098	2·5594	5·9733	3·9523	6·3762
0·7976	8·8048	0·3831	41·764	2·6573	9·7880

† For a coefficient of kinematic viscosity less than that which corresponds to this entry, oscillations with decreasing amplitude occur.

positive, we observe that for any $\alpha^2 < \alpha^2_{\max}$ there are two real roots of equation (283) which lead to two aperiodic modes of decay. In this respect, also, the solution for the present problem resembles the solution for the plane problem. These facts are apparent from the numerical results given in Table LI and illustrated in Fig. 116.

In addition to the lowest modes of aperiodic decay which we have described and illustrated, there are an infinity of other (higher order) modes (with larger damping constants) which can be derived from equation (283) by considering intervals, beyond the first, in which $\Psi(x)$ is positive. The existence of these other intervals is associated with the poles of $Q_{l+\frac{1}{2}}(x)$ which occur at the zeros, $x_{j;l+\frac{1}{2}}$, of $J_{l+\frac{1}{2}}(x)$. In the neighbourhood of each of these poles of $Q_{l+\frac{1}{2}}(x)$ the expression $x^2-2xQ_{l+\frac{1}{2}}(x)$ has a zero; and at these zeros $\Psi(x)$ has poles. A little consideration shows that in these intervals (beyond the first) in which $\Psi(x)$ is positive, it takes all values from zero to infinity. The intervals themselves become increasingly narrow: a consequence largely of the stringent requirements that not only must $Q_{l+\frac{1}{2}}(x)$ approach $\frac{1}{2}x$, but the expression on the right-hand side of (283) must also be positive. To a good approximation, these higher modes of decay are, therefore, characterized by the damping constants

$$\sigma \sim x^2_{j;l+\frac{1}{2}}\nu/R^2 \quad (j \geqslant 2). \tag{288}$$

We have seen that for $\alpha^2 < \alpha^2_{\max}$ there are two principal modes of

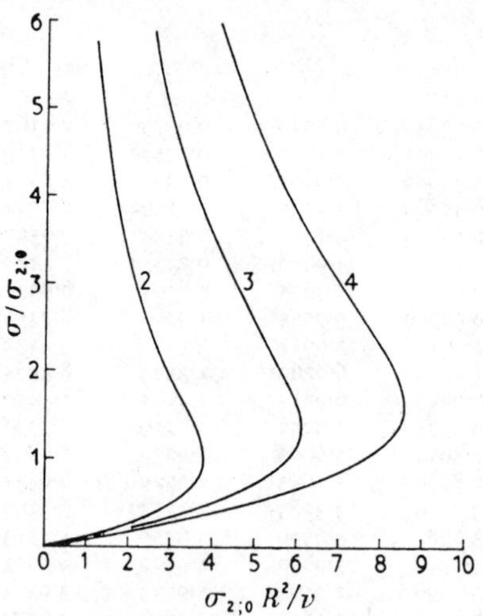

FIG. 116. The dependence of the damping constant σ for the aperiodic modes on the constitution of the globe R, ρ, and ν. The ordinate measures σ in units of the circular frequency $\sigma_{2;0}$ of the undamped $l = 2$ Kelvin mode. The abscissa measures the reciprocal of the kinematic viscosity in the unit $(R^2\sigma_{2;0})^{-1}$. The curves are labelled by the values of l to which they refer.

aperiodic decay. For $\alpha^2 > \alpha^2_{\max}$, the lowest modes of decay must be characterized by complex σ's with negative real parts. A complete numerical analysis of the corresponding characteristic values is difficult because of the lack of tables of spherical Bessel functions with complex arguments. However, the behaviour of σ for $\nu \to 0$ can be readily established: for, as $\nu \to 0$, $|x| \to \infty$ and the asymptotic behaviour of σ is determined by (cf. equation (280))

$$\frac{\sigma^2_{l;0}}{\sigma^2} = \frac{2(l-1)(2l+1)}{x^2} - 1. \tag{289}$$

Combining this relation with the definition of x (equation (282)), we find

$$\sigma = \pm i\sigma_{l;0} + (l-1)(2l+1)\nu/R^2 \quad \text{as } \nu \to 0. \tag{290}$$

Therefore, in the limit of zero viscosity, the Kelvin modes are damped with mean lives given by

$$\tau_l = \frac{R^2}{\nu(l-1)(2l+1)} \quad (\nu \to 0). \tag{291}$$

99. The oscillations of a viscous liquid drop

In this section, we shall show that in spite of appearances the solution of the problem of the oscillations of a viscous liquid drop can be reduced to that of the viscous liquid globe considered in § 98. The restoring forces maintaining equilibrium in the two cases are very different: in the case of the liquid globe it is gravitation, while in the case of the liquid drop it is surface tension. Nevertheless, it will appear that if we express the time constant σ in each case in terms of the appropriate proper frequency $\sigma_{l;0}$, in the absence of viscosity, the characteristic equations determining σ become identical. This identity of the two problems was first established by Reid, though in the limit $\nu \to 0$ the theorem was stated by Lamb.

Consider, then, the perturbation of a viscous liquid drop held together by surface tension at the free boundary. Equations (236) and (237) governing departures from equilibrium are, of course, of general validity; but in place of equation (238) we now have

$$\varpi = \delta p/\rho, \tag{292}$$

since in this case there is no additional contribution derived from a potential.

A normal mode of perturbation of the drop can again be specified by defining the shape of the deformed boundary as by equation (239). And it is clear that the arguments leading to the solution (242) continue to be valid. Accordingly, we can write

$$\varpi = \delta p/\rho = \epsilon(l+1)\Pi_0 r^l Y_l^m, \tag{293}$$

where Π_0 is a constant. The solution for $\mathbf{u}$ now follows exactly as before; in particular, it can be expressed as in equations (246).

(a) The solution for the inviscid case

The arguments leading to equation (253) are unaffected, and we have

$$\sigma^2 = -l(l+1)\Pi_0 R^{l-1}. \tag{294}$$

The analysis following equation (253), which seeks to determine Π_0, is not valid now; for, in the unperturbed state, the pressure p_0 inside the drop is constant and is given by

$$p_0 = 2T/R, \tag{295}$$

where T denotes the surface tension. (We are restricting our discussion to the case when the drop is in free space and the external pressure is zero.) According to equation (24), the condition which we must satisfy in the perturbed state is that on the deformed boundary,

$$p_0+\delta p = T\left(\frac{1}{R_1}+\frac{1}{R_2}\right), \tag{296}$$

where R_1 and R_2 are the principal radii of curvature of the surface

$$r = R+\epsilon Y_l^m. \tag{297}$$

It can be shown† that for the surface defined by (297)

$$\frac{1}{R_1}+\frac{1}{R_2} = \frac{2}{R}+\frac{\epsilon}{R^2}(l-1)(l+2)Y_l^m. \tag{298}$$

Hence, in this case

$$\left(\frac{\delta p}{\rho}\right)_{r=R} = \epsilon(l-1)(l+2)\frac{T}{\rho R^2}Y_l^m. \tag{299}$$

By comparison with equation (293) it follows that

$$(l+1)\Pi_0 R^{l-1} = (l-1)(l+2)\frac{T}{\rho R^3}. \tag{300}$$

Using this result in equation (294), we obtain

$$\sigma^2 = -l(l-1)(l+2)\frac{T}{\rho R^3}. \tag{301}$$

(*b*) *The solution for the general case*

Returning to the general case, we first observe that the arguments leading to equation (269) continue to be valid; but we must now replace the condition expressed in (271) by the requirement that

$$p_0+\delta p-2\nu\rho\frac{\partial u_r}{\partial r} = T\left(\frac{1}{R_1}+\frac{1}{R_2}\right) \tag{302}$$

on the deformed boundary. By making use of equations (293), (295), and (298), we find that equation (302) gives

$$\left\{(l+1)\Pi_0 R^l-(l-1)(l+2)\frac{T}{\rho R^2}\right\}\epsilon Y_l^m-2\nu\left(\frac{\partial u_r}{\partial r}\right)_{r=R} = 0. \tag{303}$$

We can rewrite this equation in the form

$$\{-\sigma_{l;0}^2+l(l+1)\Pi_0 R^{l-1}\}\epsilon Y_l^m-\frac{2\nu l}{R}\left(\frac{\partial u_r}{\partial r}\right)_{r=R} = 0, \tag{304}$$

where (cf. equation (301))

$$\sigma_{l;0}^2 = -l(l-1)(l+2)\frac{T}{\rho R^3} \tag{305}$$

† Cf. Lamb, *Hydrodynamics*, Cambridge, England, 1932; p. 475.

defines the frequencies of oscillation of the normal modes in the absence of viscosity.

Equations (275) and (304) are seen to be of identical forms; and since equation (269) is also true, it is apparent that we shall be led to the same characteristic equation (280) for $\sigma/\sigma_{l;0}$. The formal identity of the two problems is thus established.

In the examples considered in § 98 and in the present section, the forces maintaining equilibrium in the unperturbed spherical forms seem to be relevant only for determining the frequencies $\sigma_{l;0}$ of the normal inviscid modes; they do not seem to be significant for determining the manner of their decay by viscosity.

We have seen that the characteristic equation (280) (or (283)) allows aperiodic modes so long as α^2 $(= \sigma_{l;0}\, R^2/\nu)$ is less than a certain maximum value $\alpha^2_{\max}$; for α^2 in excess of this value, damped oscillations occur. For the principal mode $l = 2$, the critical point is (see Table LI)

$$\sigma_{2;0}\, R^2/\nu = 3{\cdot}69 \quad \text{and} \quad \sigma_{2;\nu}/\sigma_{2;0} = 0{\cdot}968. \tag{306}$$

For a drop of water surrounded by air ($T = 74$ dynes/cm), this gives $R = 2{\cdot}3\times 10^{-6}$ cm; drops of radii larger than this value will oscillate with decreasing amplitude while drops of smaller radii will be aperiodically damped.

BIBLIOGRAPHICAL NOTES

The following general references may be noted:

1. Sir Horace Lamb, *Hydrodynamics*, Cambridge, England, 1932.
2. J. Proudman, *Dynamical Oceanography*, John Wiley & Sons Inc., New York, 1953.

Several of the topics treated in this chapter are considered by Lamb: superposed fluids in § 231 (pp. 370–2); the effects of surface tension in §§ 266 and 267 (pp. 456–61); gravity waves in § 349 (pp. 625–8); and the oscillations of a liquid drop in § 275 (pp. 473–5) and § 355 (pp. 639–41). Some of these same topics (with interesting oceanographical overtones) are considered by Proudman in chapters XV and XVI of his book (reference 2).

§ 91. The fundamental paper in the subject is that of Rayleigh:

3. Lord Rayleigh, 'Investigation of the character of the equilibrium of an incompressible heavy fluid of variable density', *Scientific Papers*, ii, 200–7, Cambridge, England, 1900.

It is clear that the stability of heterogeneous fluid accelerated in a direction perpendicular to the plane of stratification can be treated by the same formalism: we have merely to replace the acceleration g due to gravity by the actual acceleration g_* (say) to which the fluid is subject. The special case of the stability of the interface between two fluids of differing densities in this latter context was treated by:

4. Sir Geoffrey Taylor, 'The instability of liquid surfaces when accelerated

in a direction perpendicular to their planes. I', *Proc. Roy. Soc.* (*London*) **A, 201,** 192–6 (1950).

An experimental demonstration of the development of the 'Rayleigh–Taylor' instability (in the case a heavier fluid, overlying a lighter one, is accelerated towards it) is described by:

5. D. J. Lewis, 'The instability of liquid surfaces when accelerated in a direction perpendicular to their planes. II', *Proc. Roy. Soc.* (*London*) A, **202,** 81–96 (1950).

The analytical treatment in the text follows largely the presentation in:

6. S. Chandrasekhar, 'The character of the equilibrium of an incompressible heavy viscous fluid of variable density', *Proc. Camb. Phil. Soc.* **51,** 162–78 (1955).

However, the effects arising from surface tension are not considered in reference 6.

§ 92. The results described in this section are largely derived from Rayleigh (reference 3). See also the appropriate sections in Lamb (reference 1) and Proudman (reference 2).

§ 93. The variational formulation given in this section is based on reference 6. Applications of the variational principle to the solution of particular problems will be found in:

7. R. Hide, 'The character of the equilibrium of an incompressible heavy viscous fluid of variable density: an approximate theory', *Proc. Camb. Phil. Soc.* **51,** 179–201 (1955).

Unfortunately, Hide's particular comparisons with some exact results in reference 6 are vitiated by an oversight to which Reid has drawn attention.

8. W. H. Reid, 'The effects of surface tension and viscosity on the stability of two superposed fluids', *Proc. Camb. Phil. Soc.* (in press).

§ 94. The problem considered in this section was first treated by:

9. W. J. Harrison, 'The influence of viscosity on the oscillations of superposed fluids', *Proc. London Math. Soc.* **6,** 396–405 (1908).

Harrison's discussion of the problem was quite complete from an analytical standpoint: he derived the basic characteristic equation (equation (113) in the text); he obtained series expansions for the characteristic values valid in the limit of zero viscosity; and he considered also the effects arising from surface tension. A later discussion of the problem (covering much the same analytical ground) is due to:

10. R. Bellman and R. H. Pennington, 'Effects of surface tension and viscosity on Taylor instability', *Quart. Appl. Math.* **12,** 151–62 (1954).

Detailed numerical results illustrating the effects of viscosity will be found in reference 6. Similarly, the most detailed consideration of the effects arising from surface tension is due to Reid (reference 8). The results quoted and illustrated in this section are taken from these two references.

Discussions on these same and related matters will be found in:

11. C.-M. Tchen, 'Stability of oscillations of superposed fluids', *J. Appl. Phys.* **27,** 760–7 (1956).

12. G. F. Carrier and C. T. Chang, 'On an initial value problem concerning Taylor instability of incompressible fluids', *Quart. Appl. Math.* **16,** 436–9 (1959).

13. S. Feldman, 'On the instability theory of the melted surface of an ablating body when entering the atmosphere', *J. Fluid Mech.* **6,** 131–55 (1959).

An example of a continuously stratified heterogeneous viscous fluid, in which the density varies exponentially (as in Rayleigh's example considered in § 92 (*b*)) and the kinematic viscosity remains constant, has been investigated in considerable detail by:

14. T. Y. Teng Fan, 'The character of the instability of an incompressible fluid of constant kinematic viscosity and exponentially varying density', *Astrophys. J.* **121,** 508–20 (1955).

A problem which includes features both of heterogeneity and of variation in density due to thermal expansion has recently been considered by:

15. B. R. Morton, 'On the equilibrium of a stratified layer of fluid', *Quart. J. Mech. Appl. Math.* **10,** 433–47 (1957).

A related investigation is that of:

16. A. Skumanich, 'On thermal convection in a polytropic atmosphere', *Astrophys. J.* **121,** 408–17 (1955).

§ 95. The effect of rotation has been considered by:

17. R. Hide, 'The character of the equilibrium of a heavy, viscous, incompressible, rotating fluid of variable density. I. General theory', *Quart. J. Mech. Appl. Math.* **9,** 22–34 (1956).

18. —— 'The character of the equilibrium of a heavy, viscous, incompressible, rotating fluid of variable density. II. Two special cases', ibid. 35–50 (1956).

Hide's treatment by including the effects of viscosity and of inclination between the directions of $\mathbf{\Omega}$ and $\mathbf{g}$ tends to obscure the essential elements of the problem. For this reason, the inviscid case has been treated *de novo*.

§ 96. The effect of a vertical magnetic field has been considered by:

19. R. Hide, 'Waves in a heavy, viscous, incompressible, electrically conducting fluid of variable density, in the presence of a magnetic field', *Proc. Roy. Soc.* (*London*) A, **233,** 376–96 (1955).

In his treatment of this problem, Hide includes the effects of finite viscosity and resistivity; and the analysis becomes encumbered by many parameters. For the sake of simplicity, the treatment in the text has been restricted to the non-dissipative case.

§ 97. Kruskal and Schwarzschild first showed that a horizontal magnetic field has no effect on the development of the Rayleigh–Taylor instability:

20. M. Kruskal and M. Schwarzschild, 'Some instabilities of a completely ionized plasma', *Proc. Roy. Soc.* (*London*) A, **223,** 348–60 (1954).

§ 98. The oscillations of a viscous liquid globe were first treated by:

21. H. Lamb, 'On the oscillations of a viscous spheroid', *Proc. London Math. Soc.* **13,** 51–66 (1881).

The treatment in the text, however, follows:

22. S. Chandrasekhar, 'The oscillations of a viscous liquid globe', ibid. (3) **9,** 141–9 (1959).

The numerical results given in Table LI are from this reference. Related problems are considered in:

23. S. Chandrasekhar, 'The character of the equilibrium of an incompressible fluid sphere of variable density and viscosity subject to radial acceleration', *Quart. J. Mech. Appl. Math.* **8,** 1–21 (1955).

24. W. H. Reid, 'The oscillations of a viscous liquid globe with a core', *Proc. London Math. Soc.* (3) **9,** 388–96 (1959).

§ 99. The principal result established in this section is due to:

25. W. H. Reid, 'The oscillations of a viscous liquid drop', *Quart. Appl. Math.* **18,** 86–89 (1960).

XI

THE STABILITY OF SUPERPOSED FLUIDS: THE KELVIN–HELMHOLTZ INSTABILITY

100. The perturbation equations

As we have already remarked in § 90, the Kelvin–Helmholtz instability arises when we consider the character of the equilibrium of a stratified heterogeneous fluid when the different layers are in relative motion. The initial stationary state whose stability we wish to examine, then, is that of an incompressible, inviscid fluid in which there is horizontal streaming. To be specific, we shall suppose that the streaming takes place in the x-direction with a velocity U. The assumption that the fluid is inviscid allows us to consider U as an arbitrary function of the height z.

Let the actual density at any point (x, y, z), as the result of a disturbance, be $\rho+\delta\rho$. Let the corresponding change in the pressure be δp, and finally, let the components of the velocity in the perturbed state be $U+u$, v, and w. The equations governing the perturbation are (cf. equations X (8), (9), and (28)):

$$\rho\frac{\partial u}{\partial t}+\rho U\frac{\partial u}{\partial x}+\rho w\frac{dU}{dz} = -\frac{\partial}{\partial x}\delta p, \tag{1}$$

$$\rho\frac{\partial v}{\partial t}+\rho U\frac{\partial v}{\partial x} = -\frac{\partial}{\partial y}\delta p, \tag{2}$$

$$\rho\frac{\partial w}{\partial t}+\rho U\frac{\partial w}{\partial x} = -\frac{\partial}{\partial z}\delta p-g\delta\rho+\sum_s T_s\left[\left(\frac{\partial^2}{\partial x^2}+\frac{\partial^2}{\partial y^2}\right)\delta z_s\right]\delta(z-z_s), \tag{3}$$

$$\frac{\partial}{\partial t}\delta\rho+U\frac{\partial}{\partial x}\delta\rho = -w\frac{d\rho}{dz}, \tag{4}$$

$$\frac{\partial}{\partial t}\delta z_s+U_s\frac{\partial}{\partial x}\delta z_s = w(z_s), \tag{5}$$

and

$$\frac{\partial u}{\partial x}+\frac{\partial v}{\partial y}+\frac{\partial w}{\partial z} = 0. \tag{6}$$

In writing equation (3) we have allowed for the possibility that at some prescribed levels z_s the density may change discontinuously and bring into play effects due to surface tension. Further, in equation (5) a subscript s distinguishes the value of the quantity at $z = z_s$.

Analysing the disturbance into normal modes, we seek solutions whose dependence on x, y, and t is given by

$$\exp i(k_x x+k_y y+nt). \tag{7}$$†

For solutions having this dependence on x, y, and t, equations (1)–(6) become:

$$i\rho(n+k_x U)u+\rho(DU)w = -ik_x \delta p, \tag{8}$$

$$i\rho(n+k_x U)v = -ik_y \delta p, \tag{9}$$

$$i\rho(n+k_x U)w = -D\delta p-g\delta\rho-k^2 \sum_s T_s \delta z_s \delta(z-z_s), \tag{10}$$

$$i(n+k_x U)\delta\rho = -wD\rho, \tag{11}$$

$$i(n+k_x U_s)\delta z_s = w_s, \tag{12}$$

and

$$i(k_x u+k_y v) = -Dw, \tag{13}$$

where D denotes d/dz.

Multiplying equations (8) and (9) by $-ik_x$ and $-ik_y$, respectively, adding, and making use of equation (13), we obtain

$$i\rho(n+k_x U)Dw-i\rho k_x(DU)w = -k^2\delta p; \tag{14}$$

whereas, by combining equations (10), (11), and (12), we have

$$i\rho(n+k_x U)w = -D\delta p-ig(D\rho)\frac{w}{n+k_x U}+ik^2\sum_s T_s\left(\frac{w}{n+k_x U}\right)_s \delta(z-z_s). \tag{15}$$

Now eliminating δp between equations (14) and (15), we obtain

$$D\{\rho(n+k_x U)Dw-\rho k_x(DU)w\}-k^2\rho(n+k_x U)w$$
$$= gk^2\left\{(D\rho)-\frac{k^2}{g}\sum_s T_s \delta(z-z_s)\right\}\frac{w}{n+k_x U}. \tag{16}$$

If we suppose that the fluid is confined between two rigid planes at $z = 0$ and $z = d$ (say), then we must require that the solutions of equation (16) must satisfy the boundary conditions

$$w = 0 \text{ at } z = 0 \quad\text{and}\quad z = d. \tag{17}$$

The fact that equation (16) involves a δ-function means that to interpret the equation correctly at the interface $z = z_s$ (for example) we must integrate the equation over an infinitesimal element $(z_s-\epsilon, z_s+\epsilon)$ and pass to the limit $\epsilon = 0$. Before we effect such an integration and carry out the stated limiting process, it is necessary to observe that

$$w/(n+k_x U) \quad \text{must be continuous at an interface.} \tag{18}$$

If U is continuous at an interface, (18) simply reduces to a condition

† Notice that we are now assuming a time dependence e^{int}, in contrast to Chapter X where it was e^{nt}.

requiring the continuity of w; but if U should be discontinuous at $z = z_s$, then we must require, instead, the uniqueness of the normal displacement of any point on the interface; and according to equation (12) this latter condition is equivalent to (18).

Now integrating equation (16) between $z_s-\epsilon$ and $z_s+\epsilon$ and passing to the limit $\epsilon = 0$, we obtain, in view of (18),

$$\Delta_s\{\rho(n+k_x U)Dw-\rho k_x(DU)w\} = gk^2\left\{\Delta_s(\rho)-\frac{k^2}{g}T_s\right\}\left(\frac{w}{n+k_x U}\right)_s, \quad (19)$$

where
$$\Delta_s(f) = f_{z=z_s+0}-f_{z=z_s-0} \quad (20)$$
is the jump a quantity f experiences at $z = z_s$, and the subscript s denotes the value which a quantity, known to be continuous at an interface, takes at the interface $z = z_s$. We may consider equation (19) as a boundary condition which must be satisfied at a surface of discontinuity, while the equation

$$D\{\rho(n+k_x U)Dw-\rho k_x(DU)w\}-k^2\rho(n+k_x U)w = gk^2(D\rho)\frac{w}{n+k_x U} \quad (21)$$

is valid everywhere else.

On expanding equation (21) and rearranging, we have

$$(n+k_x U)(D^2-k^2)w-k_x(D^2U)w-gk^2\frac{D\rho}{\rho}\frac{w}{n+k_x U}+$$
$$+\frac{D\rho}{\rho}[(n+k_x U)Dw-k_x(DU)w] = 0. \quad (22)$$

It may be noted here that the last term in equation (22), appearing with the factor $D\rho/\rho$, represents the effect of the heterogeneity of the fluid on the inertia; in most cases of interest, the neglect of this effect, in comparison with the effect on the potential energy, is justified.

101. The case of two uniform fluids in relative horizontal motion separated by a horizontal boundary

Let two uniform fluids of densities ρ_1 and ρ_2 be separated by a horizontal boundary at $z = 0$. Let the density ρ_2 of the upper fluid be less than the density ρ_1 of the lower fluid so that, in the absence of streaming, the arrangement is a stable one. We shall further suppose that the two fluids are streaming with the constant velocities U_1 and U_2.

In each of the two regions of constant ρ and U, equation (21) reduces to

$$(D^2-k^2)w = 0. \quad (23)$$

(A factor $n+k_x U$ has been discarded; this is permissible under the circumstances.) The general solution of equation (23) is a linear combination of the integrals e^{+kz} and e^{-kz}. Since w cannot increase exponentially on either side of the interface at $z = 0$ and since $w/(n+k_x U)$ must be continuous on this surface, we can write

$$w_1 = A(n+k_x U_1)e^{+kz} \quad (z < 0) \tag{24}$$

and

$$w_2 = A(n+k_x U_2)e^{-kz} \quad (z > 0) \tag{25}$$

as the solutions appropriate for the two regions. Applying the condition (19) to the solution given by these equations, we obtain the characteristic equation

$$\rho_2(n+k_x U_2)^2+\rho_1(n+k_x U_1)^2 = gk\left\{(\rho_1-\rho_2)+\frac{k^2T}{g}\right\}. \tag{26}$$

Letting

$$\alpha_1 = \frac{\rho_1}{\rho_1+\rho_2} \quad \text{and} \quad \alpha_2 = \frac{\rho_2}{\rho_1+\rho_2}, \tag{27}$$

we can rewrite equation (26) in the form

$$\alpha_2(n+k_x U_2)^2+\alpha_1(n+k_x U_1)^2 = gk\left\{(\alpha_1-\alpha_2)+\frac{k^2T}{g(\rho_1+\rho_2)}\right\}. \tag{28}$$

Expanding equation (28), we have

$$n^2+2k_x(\alpha_1 U_1+\alpha_2 U_2)n+k_x^2(\alpha_1 U_1^2+\alpha_2 U_2^2)- -gk\left\{(\alpha_1-\alpha_2)+\frac{k^2T}{g(\rho_1+\rho_2)}\right\} = 0. \tag{29}$$

The roots of this equation are given by

$$n = -k_x(\alpha_1 U_1+\alpha_2 U_2)\pm\left[gk\left\{(\alpha_1-\alpha_2)+\frac{k^2T}{g(\rho_1+\rho_2)}\right\}-k_x^2\alpha_1\alpha_2(U_1-U_2)^2\right]^{\frac{1}{2}}. \tag{30}$$

(*a*) *The case when surface tension is absent*

In the absence of surface tension, equation (30) gives

$$n = -k_x(\alpha_1 U_1+\alpha_2 U_2)\pm[gk(\alpha_1-\alpha_2)-k_x^2\alpha_1\alpha_2(U_1-U_2)^2]^{\frac{1}{2}}. \tag{31}$$

Several conclusions of interest can be drawn from this equation. (1) When $k_x = 0$,

$$n = \pm\sqrt{[gk(\alpha_1-\alpha_2)]}; \tag{32}$$

i.e. perturbations transverse to the direction of streaming are unaffected by its presence. (2) In every other direction, instability occurs when

$$k_x^2\alpha_1\alpha_2(U_1-U_2)^2 > gk(\alpha_1-\alpha_2); \tag{33}$$

i.e. for a given difference in velocity U_1-U_2 and for a given direction of the wave vector **k**, instability occurs for all wave numbers

$$k > \frac{g(\alpha_1-\alpha_2)}{\alpha_1\alpha_2(U_1-U_2)^2\cos^2\vartheta}, \tag{34}$$

where ϑ is the angle between the directions of **k** and **U**. According to

(34), for a given (U_1-U_2), instability occurs for the least wave number when $\mathbf{k}$ is in the direction of $\mathbf{U}$; and this minimum wave number, k_{min}, is given by

$$k_{\text{min}} = \frac{g(\alpha_1-\alpha_2)}{\alpha_1\alpha_2(U_1-U_2)^2}. \tag{35}$$

We have instability for $k > k_{\text{min}}$.

The striking aspect of the instability predicted by (34) and (35) is that it occurs no matter how small U_1-U_2 may be. The stability of the static arrangement, in the absence of streaming, is unable to inhibit the instability, in the presence of streaming, for disturbances of sufficiently small wavelengths. This is the Kelvin–Helmholtz instability. Helmholtz stated this result in the following form: '*Every perfect geometrically sharp edge by which a fluid flows must tear it asunder and establish a surface of separation, however slowly the rest of the fluid may move.*'

One may say that the Kelvin–Helmholtz instability arises by the crinkling of the interface by the shear which is present; and this crinkling occurs even for the smallest differences in the velocities of the two fluids. The question occurs whether this latter circumstance may not, in part, be due to the discontinuous distributions of ρ and U which underlie the arrangement which has been considered; and may not be present if the distributions of ρ and U are continuous and the gradient of U is sufficiently small compared to the gradient of ρ. We return to this question in §§ 102–4.

(*b*) *The stabilizing effect of surface tension*

When surface tension is present, equation (30) will predict stability if

$$k^2\alpha_1\alpha_2(U_1-U_2)^2 < gk\left\{(\alpha_1-\alpha_2)+\frac{k^2T}{g(\rho_1+\rho_2)}\right\}. \tag{36}$$

(We have set $k_x = k$ in equation (30), since these are the disturbances which are most sensitive to the Kelvin–Helmholtz instability.) An alternative form of (36) is

$$\alpha_1\alpha_2(U_1-U_2)^2 < g\left\{\frac{\alpha_1-\alpha_2}{k}+\frac{kT}{g(\rho_1+\rho_2)}\right\}. \tag{37}$$

The right-hand side of this inequality has a minimum when

$$\frac{\alpha_1-\alpha_2}{k^2} = \frac{T}{g(\rho_1+\rho_2)}. \tag{38}$$

If k_* denotes the value of k given by (38), we shall have stability if

$$(U_1-U_2)^2 < \frac{2g}{k_*}\frac{\alpha_1-\alpha_2}{\alpha_1\alpha_2}. \tag{39}$$

Inserting for k_* in accordance with (38), we conclude that *surface tension will suppress the Kelvin–Helmholtz instability if*

$$(U_1-U_2)^2 < \frac{2}{\alpha_1\alpha_2}\sqrt{\frac{Tg(\alpha_1-\alpha_2)}{\rho_1+\rho_2}}. \tag{40}$$

This result is due to Kelvin.

For air over sea-water ($\alpha_2 = 0{\cdot}00126$; $\rho_1 = 1{\cdot}02$ g/cm^3; $T = 74$ dynes/cm; and $g = 981$ cm/sec^2) (40) predicts that we will have stability for

$$|U_1-U_2| < 650 \text{ cm/sec}; \tag{41}$$

and when $|U_1-U_2|$ has this maximum value compatible with stability, we find:

$$k_* = 3{\cdot}68 \text{ cm}^{-1}, \quad \lambda_* = 2\pi/k_* = 1{\cdot}71 \text{ cm},$$

$$n_* = \alpha_2 k_* |U_1-U_2| = 3{\cdot}02 \text{ sec}^{-1},$$

and

$$n/k_* = 0{\cdot}82 \text{ cm/sec}. \tag{42}$$

Therefore, when $|U_1-U_2|$ exceeds 650 cm/sec (12·5 nautical miles per hour) instability will manifest itself as surface waves with wavelength 1·71 cm and wave velocity 0·82 cm/sec. Commenting on this result, Kelvin stated in his original paper: 'Observation shows the sea to be ruffled by a wind of much smaller velocity than this. Such ruffling is, therefore, due to the viscosity of the air.' The first part of this statement of Kelvin has generally been repeated though other causes have been advanced for the 'discrepancy'. However, Munk has recently pointed out that a wind velocity of 650 cm/sec is associated with several phenomena observed on the surface of the seas. The number of breaking waves ('white caps') suddenly increases; the convection processes, as indicated by the soaring of sea gulls, seem to change; and the rate of evaporation also increases suddenly. Munk ascribes all these to a sudden change in the pattern of air-flow over the waves consequent to the onset of the Kelvin–Helmholtz instability at the predicted wind velocity of 650 cm/sec.

In this connexion we may also refer to some experiments of Francis in which air was blown over lubricating oil in a wind-tunnel. Francis observed that at a sharply defined critical wind-speed small ripples appeared which were unstable and grew rapidly (see Fig. 117). The critical wind-speeds observed in this manner are well in accord with those predicted by (40). It would thus appear that these experiments demonstrate, at any rate, the occurrence of the Kelvin–Helmholtz instability under controlled laboratory conditions.

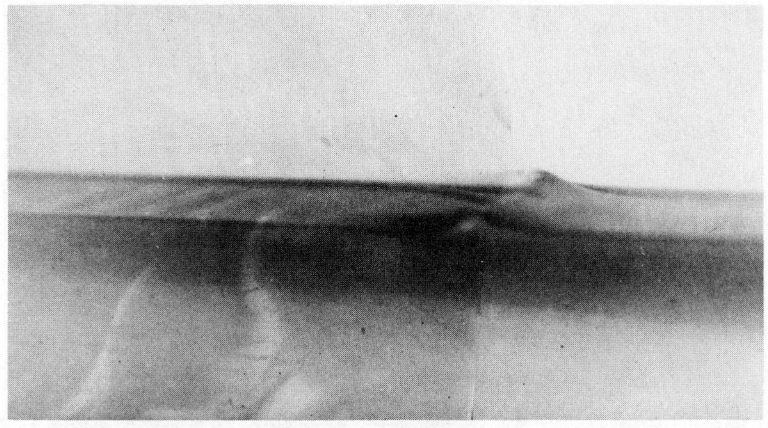

Fig. 117. Photograph by J. R. D. Francis of an unstable wave produced by air blown over a viscous oil in a wind tunnel; in the illustration, the wave has just grown from small ripples. Wind from left to right; camera looking slightly upwind and downward on to oil surface. Ripples show as transverse bars of light.

102. The effect of a continuous variation of U on the development of the Kelvin–Helmholtz instability

As we have remarked, the striking aspect of the *occurrence* of the Kelvin–Helmholtz instability at the interface between two fluids, in the absence of surface tension, is its independence on the magnitude of $|U_1-U_2|$. It is a matter of some interest to determine whether this aspect of the Kelvin–Helmholtz instability is due to the sharp discontinuities in ρ and U which has been assumed in its derivation. Before we consider this matter generally in §§ 103 and 104, we shall examine the case of two superposed fluids of different densities, separated by a transition layer of intermediate density in which the velocity of streaming varies continuously from that of the lower fluid to that of the upper fluid. Specifically, the distribution of ρ and U which we shall investigate is the following (see Fig. 118):

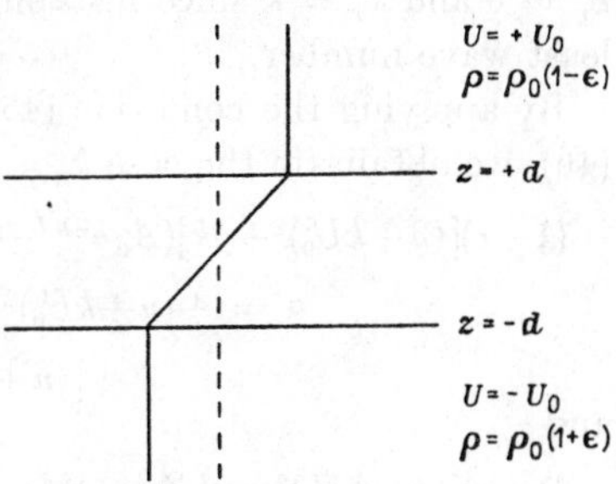

Fig. 118. The distributions of ρ and U for which the Kelvin–Helmholtz instability is investigated.

$$\begin{aligned} z &> +d, & \rho &= \rho_0(1-\epsilon), & U &= U_0, \\ +d > z &> -d, & \rho &= \rho_0, & U &= U_0 z/d, \\ z &< -d, & \rho &= \rho_0(1+\epsilon), & U &= -U_0. \end{aligned} \tag{43}$$

Since U is continuous at the interfaces at $z = +d$ and $z = -d$, the boundary conditions which must be satisfied at these surfaces are (cf. equations (18) and (19)):

$$w \text{ is continuous at } z = +d \text{ and } -d \tag{44}$$

and

$$\Delta_{\pm d}\{\rho(n+k_x U)^2 Dw - \rho k_x(n+k_x U)(DU)w\} - gk^2\Delta_{\pm d}(\rho)w_{\pm d} = 0. \tag{45}$$

The equation satisfied by w in each of the regions of constant ρ is the same equation (23). We may accordingly write for the solutions in the three regions:

$$\begin{aligned} w_+ &= A_+ e^{-kz} & &(z > d), \\ w_0 &= A_0 e^{-kz} + B_0 e^{+kz} & &(d > z > -d), \\ w_- &= B_- e^{+kz} & &(z < -d). \end{aligned} \tag{46}$$

The solutions in the regions $z > d$ and $z < d$ have been so chosen that they vanish at $+\infty$ and $-\infty$, respectively. The continuity of w at $z = +d$ and $-d$ leads to the conditions

$$A_+ e^{-kd} = A_0 e^{-kd} + B_0 e^{+kd} = w_{+d},$$
$$B_- e^{-kd} = A_0 e^{+kd} + B_0 e^{-kd} = w_{-d}. \tag{47}$$

We shall restrict our further discussion to perturbations for which $k_y = 0$ and $k_x = k$ since instability will set in the x-direction for the least wave number.

By applying the condition (45) at $z = \pm d$ to the solutions given in (46), we obtain (in the case $k_x = k$):

$$(1-\epsilon)[(n+kU_0)^2+gk](A_0 e^{-kd}+B_0 e^{+kd})$$
$$= +(n+kU_0)^2(A_0 e^{-kd}-B_0 e^{+kd})+$$
$$+[(n+kU_0)U_0/d+gk](A_0 e^{-kd}+B_0 e^{+kd}) \tag{48}$$

and

$$(1+\epsilon)[(n-kU_0)^2-gk](A_0 e^{+kd}+B_0 e^{-kd})$$
$$= -(n-kU_0)^2(A_0 e^{+kd}-B_0 e^{-kd})-$$
$$-[(n-kU_0)U_0/d+gk](A_0 e^{+kd}+B_0 e^{-kd}). \tag{49}$$

These equations can be reduced to give

$$-\frac{A_0}{B_0} e^{-2kd} = \frac{\epsilon[(n+kU_0)^2+gk]+(n+kU_0)U_0/d-2(n+kU_0)^2}{\epsilon[(n+kU_0)^2+gk]+(n+kU_0)U_0/d} \tag{50}$$

and

$$-\frac{B_0}{A_0} e^{-2kd} = \frac{\epsilon[(n-kU_0)^2-gk]+(n-kU_0)U_0/d+2(n-kU_0)^2}{\epsilon[(n-kU_0)^2-gk]+(n-kU_0)U_0/d}. \tag{51}$$

Let $$\nu = n/kU_0 \quad \text{and} \quad \kappa = 2kd \tag{52}$$

so that ν and κ measure n and k in the units $U_0 k$ and $1/(2d)$, respectively. Further, let

$$J = \frac{\epsilon g d}{U_0^2} = \frac{1}{2}\frac{\epsilon g \kappa}{U_0^2 k}; \tag{53}$$

this non-dimensional 'number' has the meaning

$$J = -\frac{g}{\rho_0}\frac{\Delta\rho/2d}{(dU/dz)^2}, \tag{54}$$

where $\Delta\rho$ $(= -2\epsilon\rho_0)$ is the difference in the density of the two fluids between which, in the transition layer, the velocity changes continuously from $-U$ (in the lower fluid) to $+U$ (in the upper fluid); it is the *Richardson number* for the problem on hand. (For the general definition of the Richardson number, its meaning, and its significance, see §§ 103 and 104 below.)

In terms of ν, κ, and J, equations (50) and (51) take the more convenient forms

$$-\frac{A_0}{B_0}e^{-\kappa} = 1-\frac{\kappa(\nu+1)^2}{J+(\nu+1)+\frac{1}{2}\epsilon\kappa(\nu+1)^2} \tag{55}$$

and

$$-\frac{B_0}{A_0}e^{-\kappa} = 1-\frac{\kappa(\nu-1)^2}{J-(\nu-1)-\frac{1}{2}\epsilon\kappa(\nu-1)^2}. \tag{56}$$

Eliminating A_0/B_0, we obtain the characteristic equation

$$e^{-2\kappa} = \left[1-\frac{\kappa(\nu+1)^2}{J+(\nu+1)+\frac{1}{2}\epsilon\kappa(\nu+1)^2}\right]\left[1-\frac{\kappa(\nu-1)^2}{J-(\nu-1)-\frac{1}{2}\epsilon\kappa(\nu-1)^2}\right]. \tag{57}$$

While there is no formal difficulty in discussing equation (57) quite generally, the main interest in the present problem lies in cases in which the change in density is so small that we may neglect it in all terms except in the combination J. This neglect is equivalent to omitting the last term in equation (22) which occurs with the factor $D\rho/\rho$; as we have already remarked, its omission is equivalent to ignoring the effect of the heterogeneity of the fluid on the inertia while retaining its effect on the potential energy. On this approximation, equation (57) gives

$$[(J+1)^2-\nu^2]e^{-2\kappa} = [(J+\nu+1)-\kappa(\nu+1)^2][(J-\nu+1)-\kappa(\nu-1)^2]. \tag{58}$$

On expansion and simplification, equation (58) becomes

$$\kappa^2\nu^4-(2J\kappa+2\kappa^2-2\kappa+1-e^{-2\kappa})\nu^2+$$
$$+[(J+1)(1-e^{-\kappa})-\kappa][(J+1)(1+e^{-\kappa})-\kappa] = 0. \tag{59}$$

For stability, ν^2 should clearly be real and positive. First, we may verify that the roots ν^2 of equation (59) are always real. By evaluating the discriminant Δ of the equation, we find

$$\Delta = 4\kappa^2(J+1)^2e^{-2\kappa}+f(4\kappa J+1-e^{-2\kappa}), \tag{60}$$

where

$$f = (2\kappa-1)^2-e^{-2\kappa} = (2\kappa-1-e^{-\kappa})(2\kappa-1+e^{-\kappa}). \tag{61}$$

The expression for Δ can be rewritten in the form

$$\Delta = 4\kappa^2(J+1)^2e^{-2\kappa}+4\kappa(J+1)f+(f-4\kappa^2)f \tag{62}$$

or

$$\Delta = [2\kappa(J+1)e^{-\kappa}+e^{\kappa}f]^2-e^{2\kappa}f^2+(f-4\kappa^2)f. \tag{63}$$

The last two terms in equation (63) can be combined to give

$$\Delta = [2\kappa(J+1)e^{-\kappa}+e^{\kappa}f]^2-f[(2\kappa-1)e^{\kappa}+e^{-\kappa}]^2. \tag{64}$$

Now let κ_0 be such that

$$2\kappa_0-1 = e^{-\kappa_0} \quad (\kappa_0 = 0{\cdot}7388). \tag{65}$$

Then (cf. equation (61))

$$f > 0 \text{ for } \kappa > \kappa_0 \quad\text{and}\quad f < 0 \text{ for } \kappa < \kappa_0, \tag{66}$$

since $2\kappa > (1-e^{-\kappa})$ for all $\kappa > 0$. Equation (60) shows that $\Delta > 0$ for

$\kappa > \kappa_0$; and equation (64) shows that $\Delta > 0$ for $\kappa < \kappa_0$. Hence, Δ is always positive; and this proves the reality of the roots, ν^2, of equation (59). Turning next to a consideration of their signs, we first observe that their sum is positive, while their product is given by

$$[(J+1)(1-e^{-\kappa})-\kappa][(J+1)(1+e^{-\kappa})-\kappa]. \tag{67}$$

Hence, one of the two roots must be negative if

$$\frac{\kappa}{1+e^{-\kappa}} < J+1 < \frac{\kappa}{1-e^{-\kappa}}. \tag{68}$$

For values of the Richardson number in the range specified in (68), the motion is unstable; and outside this range, it is stable.† The region of instability in the (J, κ)-plane is shown in Fig. 119. It is seen that for any assigned value of the Richardson number, there is a band of wavelengths for which the motion is unstable. For large J the critical wave number, in the immediate vicinity of which there is instability, approaches $\kappa = J+1$; the corresponding wave velocity approaches the shear velocity $(= 2U_0)$ at the transition layer.

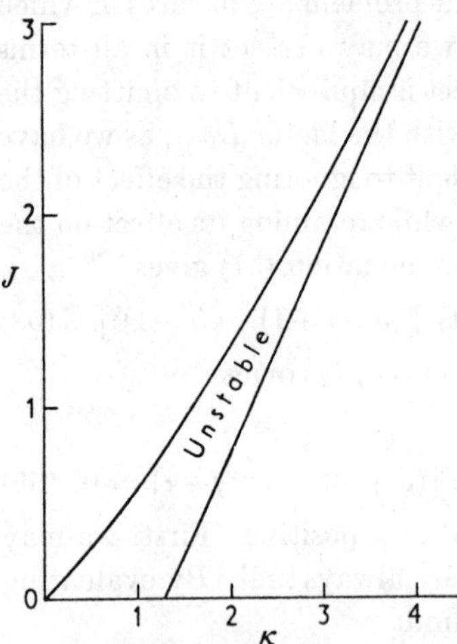

FIG. 119. The region of instability in the (J, κ)-plane for the distributions shown in Fig. 118; κ is the wave number in the unit $1/(2d)$ and J $(= \epsilon g d/U_0^2)$ is the Richardson number.

Goldstein has extended these considerations to the case when the transition layer is further subdivided so that the density changes from $\rho_0+\epsilon$ to $\rho_0-\epsilon$ in five steps

$(\rho_0+\epsilon \to \rho_0+\frac{1}{2}\epsilon \to \rho_0 \to \rho_0-\frac{1}{2}\epsilon \to \rho_0-\epsilon)$

instead of three, while the velocity increases linearly through the three intermediate layers. Goldstein found that in this case, also, there are bands of wavelengths for which the motion is unstable for any assigned value of the Richardson number. Moreover, it appears that by further subdivision of the transition layer we do not gain anything: there are always bands of wavelengths for which the motion is unstable and we cannot avoid the Kelvin–Helmholtz instability for any value of the Richardson number.

† Notice that when $J = 0$ the motion is stable for $\kappa > 1{\cdot}2785$ ($1{\cdot}2785$ is the root of the equation $x = 1+e^{-x}$); this result is due to Rayleigh.

103. The Kelvin–Helmholtz instability in a fluid in which both ρ and U are continuously variable

We shall preface the consideration of the mathematical aspects of the problem by seeking the origin of the Kelvin–Helmholtz instability in the manner we sought the origin of Rayleigh's criterion for the stability of revolving fluids in § 66.

The source of the Kelvin–Helmholtz instability clearly lies in the energy stored in the kinetic energy of relative motion of the different layers: the tendency towards mixing and instability will be greater, the larger the prevailing shear as measured by dU/dz. The only counteracting force damping this tendency is derived from inertia; and so long as inertia can maintain a sufficient pressure gradient and prevent the mixing, instability will not occur. To obtain a quantitative expression of this condition, suppose that two equal neighbouring volumes, at heights z and $z+\delta z$, are interchanged. The work that must be done to effect this interchange against the acceleration of gravity, per unit volume, is given by

$$\delta W = -g\delta\rho\delta z, \tag{69}$$

where $\delta\rho$ is the difference in the density at the two heights. The kinetic energy which is available to do this work (again, per unit volume) is given by

$$\tfrac{1}{2}\rho[U^2+(U+\delta U)^2-\tfrac{1}{2}(U+U+\delta U)^2] = \tfrac{1}{4}\rho(\delta U)^2. \tag{70}$$

A sufficient condition for stability is, therefore,

$$\tfrac{1}{4}\rho(\delta U)^2 < -g\delta\rho\delta z. \tag{71}$$

An equivalent form of this condition is

$$\left(\frac{dU}{dz}\right)^2 < -4\frac{g}{\rho}\frac{d\rho}{dz}. \tag{72}$$

The number (cf. equation (54))

$$J = -\frac{g}{\rho}\frac{d\rho/dz}{(dU/dz)^2} \tag{73}$$

which measures the ratio of the buoyancy force to the inertia is the general definition of the Richardson number. The inequality (72), therefore, suggests that *a necessary condition for stability is that the Richardson number be everywhere greater than* $\frac{1}{4}$:

$$J > \tfrac{1}{4} \quad \textit{for stability}. \tag{74}$$

The arguments leading to (74) do not, of course, justify one's inferring that its violation must lead to instability: for, it can very well happen that although stored energy is available for initiating instability, no mechanism exists for transforming the energy into possible hydrodynamic modes. The mathematical difficulty of establishing a general condition such as (74), under suitable restrictions, appears to stem from this cause.

(a) Analytical results for the case $\rho = \rho_0 e^{-\beta z}$ and $U = U_0 z/d$

Mathematically, the simplest distributions for which one may seek to establish a condition such as (74) are

$$\rho = \rho_0 e^{-\beta z} \quad \text{and} \quad U = U_0 z/d \quad (z > 0), \tag{75}$$

where β and d are constants. The Richardson number is then a constant throughout the fluid and has the value

$$J = g\beta d^2/U_0^2. \tag{76}$$

For the distributions of ρ and U given in (75), equation (22) becomes

$$(n+kU_0 z/d)^2\left\{D^2w-\beta Dw+\left[-k^2+\frac{\beta kU_0/d}{n+kU_0 z/d}+\frac{k^2 g\beta}{(n+kU_0 z/d)^2}\right]w\right\} = 0. \tag{77}$$

By measuring z in the unit d and letting

$$n = U_0 \nu/d, \qquad k = \kappa/d, \quad \text{and} \quad \beta = \lambda/d, \tag{78}$$

we can rewrite equation (77) in the form

$$(\nu+\kappa z)^2\left\{D^2w-\lambda Dw+\left[-\kappa^2+\frac{\lambda\kappa}{\nu+\kappa z}+\frac{\kappa^2 J}{(\nu+\kappa z)^2}\right]w\right\} = 0. \tag{79}$$

Solutions of this equation must be sought which satisfy the boundary conditions

$$w = 0 \quad \text{for } z = 0 \text{ and } z \to \infty. \tag{80}$$

By the change of variable

$$w = e^{\lambda z/2}W, \tag{81}$$

equation (79) reduces to

$$(\nu+\kappa z)^2\left\{\frac{d^2W}{dz^2}+\left[-(\kappa^2+\tfrac{1}{4}\lambda^2)+\frac{\lambda}{z+\nu/\kappa}+\frac{J}{(z+\nu/\kappa)^2}\right]W\right\} = 0. \tag{82}$$

Now let

$$\zeta = (z+\nu/\kappa)\sqrt{(4\kappa^2+\lambda^2)}. \tag{83}$$

Equation (82) then simplifies to the form

$$\zeta^2\left\{\frac{d^2W}{d\zeta^2}+\left[-\frac{1}{4}+\frac{j}{\zeta}+\frac{\frac{1}{4}-m^2}{\zeta^2}\right]W\right\} = 0, \tag{84}$$

where

$$j = \frac{\lambda}{\sqrt{(4\kappa^2+\lambda^2)}} \quad \text{and} \quad m^2 = \tfrac{1}{4}-J. \tag{85}$$

If we were to discard the factor ζ^2 in equation (84), then

$$\mathscr{L}W = \frac{d^2W}{d\zeta^2}+\left[-\frac{1}{4}+\frac{j}{\zeta}+\frac{\frac{1}{4}-m^2}{\zeta^2}\right]W = 0. \tag{86}$$

We recognize in equation (86) Whittaker's standard form of the equation for the confluent hypergeometric function. Indeed, the condition at infinity requires that the solution of equation (86), appropriate to the problem on hand, be, in fact, Whittaker's function,† $W_{j,m}(\zeta)$. Thus

$$W \equiv W_{j,m}(\zeta). \tag{87}$$

This solution has the asymptotic behaviour

$$W_{j,m}(\zeta) \sim \zeta^j e^{-\zeta/2} \quad (\zeta \to \infty). \tag{88}$$

The corresponding form of the solution for w is

$$w(z) = e^{\lambda z/2} W_{j,m}[(z+\nu/\kappa)\sqrt{(4\kappa^2+\lambda^2)}]. \tag{89}$$

The boundary condition at $z = 0$ now requires

$$W_{j,m}\left[\frac{\nu}{\kappa}\sqrt{(4\kappa^2+\lambda^2)}\right] = 0. \tag{90}$$

Hence, $\quad \dfrac{\nu}{\kappa}\sqrt{(4\kappa^2+\lambda^2)}$ must be a zero of $W_{j,m}(x)$. (91)

Now Dyson has investigated the zeros of the Whittaker function $W_{j,m}(x)$ and has shown that it has no complex zeros; that it has an infinite sequence of real zeros when $m^2 < 0$; and that it has at most one real zero for $m^2 > 0$. From these results it follows that when $J > \frac{1}{4}$ and $m^2 < 0$, there exists, for each wave number κ, an infinite set of distinct, real characteristic values for ν; and, when $J < \frac{1}{4}$ and $m^2 > 0$, there exists, again, for each wave number κ, either one real characteristic value for ν or none. Thus, when $J > \frac{1}{4}$, the flow would appear to be stable; and when $J < \frac{1}{4}$, while the flow does not appear unstable, the solution of the characteristic value problem, as formulated, does not seem to lead to a complete set of proper values and proper functions. This paradoxical situation was clarified by Case who pointed out that the solutions of equation (86) do not, by any means, exhaust the possible solutions: for, the equation

$$x^2 f(x) = 0 \tag{92}$$

has in addition to the solution $f(x) = 0$, the solutions

$$f(x) = \delta(x) \quad \text{and} \quad f(x) = \delta'(x). \tag{93}$$

Therefore, we must supplement the solution (87) by the solutions of the equations

$$\mathscr{L}W = \delta(\zeta) \quad \text{and} \quad \mathscr{L}W = \delta'(\zeta). \tag{94}$$

The solutions of these latter equations can be expressed in terms of suitably defined Green's functions. It is found that the corresponding

† For the definition and properties of Whittaker's function see E. T. Whittaker and G. N. Watson, *Modern Analysis* (Cambridge: at the University Press, 1927), chapter XVI, particularly pp. 339–42.

characteristic values form real continua. The existence of these continua of characteristic values means that in the expansion of an arbitrary perturbation into normal modes, we must include the proper functions derived as solutions of equations (94) along with those of equation (86). And when these are included, the result, as Case has shown, is the following: for no positive J can an arbitrary perturbation grow indefinitely with time; for $J > \frac{1}{4}$, an initial perturbation becomes a sum of oscillatory terms plus a term (derived from the continua) which tends to zero as $t^{-\frac{1}{2}}$; for $J < \frac{1}{4}$, the corresponding asymptotic behaviour is $t^{-0\cdot5+\sqrt{(0\cdot25-J)}}$. It would, therefore, be strictly correct to say that *for the distributions* (75) *extending over a semi-infinite region, the flow is stable for all Richardson numbers.*

104. An example of the instability of a shear layer in an unbounded heterogeneous inviscid fluid

The analysis of Dyson and Case, summarized in § 103 (a), has shown that a linear variation of the stream velocity in a semi-infinite heterogeneous fluid with an exponentially decreasing density is, strictly, stable for all values of the Richardson number. While one cannot, of course, be certain of this, one *is* left with the suspicion that the result *may* in part derive from the assumed linear variation of the stream velocity U which, on the one hand, eliminates a principal term from equation (22) (namely, the one in D^2U) and, on the other, makes U unbounded. On these grounds, one may wonder if the result would not have been different had D^2U been different from zero and U been bounded.

A remarkable example to which the suspicions we have referred do not apply, and which at the same time allows an elementary and an explicit solution, has recently been discovered by Drazin. Drazin's example is

$$U = U_0 \tanh(z/d), \tag{95}$$

where U_0 and d are constants. On this distribution

$$U \to +U_0 \text{ as } z \to +\infty \quad \text{and} \quad U \to -U_0 \text{ as } z \to -\infty. \tag{96}$$

Setting $k_x = k$ in equation (22) and ignoring the effect of the heterogeneity on the inertia (by omitting the last term in $D\rho/\rho$), we have

$$(n/k+U)(D^2-k^2)w-(D^2U)w-g\frac{D\rho}{\rho}\frac{w}{U+n/k} = 0. \tag{97}$$

We shall again suppose that

$$\rho = \rho_0 e^{-\beta z} \quad \text{and} \quad D\rho/\rho = -\beta = \text{constant}. \tag{98}$$

For this distribution of the density, equation (97) takes the form

$$(U-c)(D^2-k^2)w-(D^2U)w+J\frac{w}{U-c} = 0, \tag{99}$$

if we measure length and velocity in the units d and U_0, respectively, and put

$$c = -n/kU_0 \quad \text{and} \quad J = g\beta d^2/U_0^2. \tag{100}$$

In the chosen units,

$$U = \tanh z; \tag{101}$$

and we shall seek solutions of equation (99) which satisfy the boundary conditions

$$w \to 0 \quad \text{as} \quad z \to \pm\infty. \tag{102}$$

(The infinity in ρ, as $z \to -\infty$, would not seem to be relevant since the phenomenon we are concerned with is essentially limited to the neighbourhood $|z| \leqslant 1$.)

Returning to equation (99), we may first observe that if c is a characteristic value belonging to a proper solution w and assigned k and J, then c^* is also a characteristic value for the same k and J and belongs to w^*. Moreover, if U is an odd function of z (as is the case with (101)) then $-c$ is also a characteristic value and belongs to $w(-z)$. On these symmetry grounds, it would appear that the principle of the exchange of stabilities is valid and that the state of marginal stability is characterized by $c = 0$. Assuming, then, that this is the case, we consider the equation

$$U(D^2-k^2)w-(D^2U)w+J\frac{w}{U} = 0. \tag{103}$$

We shall first transform equation (103) by using U as the independent variable. Making use of the relations

$$\frac{dU}{dz} = \operatorname{sech}^2 z = 1-U^2 \quad \text{and} \quad \frac{d^2U}{dz^2} = -2U(1-U^2), \tag{104}$$

we find that equation (103) becomes

$$\frac{d}{dU}\left[(1-U^2)\frac{dw}{dU}\right]+\left[2-\frac{k^2}{1-U^2}+\frac{J}{U^2(1-U^2)}\right]w = 0; \tag{105}$$

and the corresponding boundary conditions are

$$w = 0 \quad \text{for} \quad U = \pm 1. \tag{106}$$

The singular points of equation (105) are

$$U = \pm 1 \quad \text{and} \quad U = 0. \tag{107}$$

The behaviour of the solutions of equation (105) at these singular points is readily determined. Thus, by putting $U = 1 \pm y$ and treating y as small, we obtain the equation

$$y\frac{d^2w}{dy^2}+\frac{dw}{dy}+\frac{1}{4y}(J-k^2)w = 0. \tag{108}$$

This equation allows a solution of the form

$$w = y^\nu, \tag{109}$$

where
$$\nu^2 = \tfrac{1}{4}(k^2-J). \tag{110}$$

Therefore, a solution of equation (105), regular at $U = +1$ and -1, must have the behaviour

$$w \to \text{constant}(1-U^2)^\nu \quad (U \to \pm 1), \tag{111}$$

where
$$\nu = \tfrac{1}{2}\sqrt{(k^2-J)}; \tag{112}$$

and it is necessary that the real part of ν is non-negative. In a similar manner, we find that the indicial equation, which determines the behaviour of w at $U = 0$, is

$$\mu^2-\mu+J = 0. \tag{113}$$

Therefore, a solution of equation (105), regular at $U = 0$, must have the behaviour
$$w \to \text{constant}\, U^\mu, \tag{114}$$

where
$$\mu = \tfrac{1}{2}+\tfrac{1}{2}\sqrt{(1-4J)}. \tag{115}$$

The required behaviours (111) and (114) of w suggest that we seek a solution of the form

$$w = U^\mu(1-U^2)^\nu\chi, \tag{116}$$

where χ is regular at $U = 0, +1$, and -1. For the more general substitution
$$w = f\chi \tag{117}$$

equation (105) becomes

$$\frac{d^2\chi}{dU^2}+2\left(\frac{1}{f}\frac{df}{dU}-\frac{U}{1-U^2}\right)\frac{d\chi}{dU}+$$
$$+\left\{\frac{1}{1-U^2}\left[2-\frac{k^2}{1-U^2}+\frac{J}{U^2(1-U^2)}\right]-\frac{2U}{1-U^2}\frac{1}{f}\frac{df}{dU}+\frac{1}{f}\frac{d^2f}{dU^2}\right\}\chi = 0. \tag{118}$$

For the particular substitution (116),

$$f = U^\mu(1-U^2)^\nu, \tag{119}$$

and we readily verify that

$$\frac{1}{f}\frac{df}{dU} = -\frac{2\nu U}{1-U^2}+\frac{\mu}{U} \tag{120}$$

and
$$\frac{1}{f}\frac{d^2f}{dU^2} = 4\nu(\nu-1)\frac{U^2}{(1-U^2)^2}-\frac{2\nu(2\mu+1)}{1-U^2}+\mu(\mu-1)\frac{1}{U^2}. \tag{121}$$

Inserting these expressions in equation (118) and making use of the relations (110) and (113) in the reductions, we find that the equation takes the simple form

$$\frac{d^2\chi}{dU^2}+\left\{\frac{\mu}{U}-\frac{2(\nu+1)U}{1-U^2}\right\}\frac{d\chi}{dU}-\frac{(2\nu+\mu+2)(2\nu+\mu-1)}{1-U^2}\chi = 0. \quad (122)$$

It is a remarkable fact noticed by Drazin that the special solution,

$$\chi = \text{constant} \quad (123)$$

and

$$(2\nu+\mu+2)(2\nu+\mu-1) = 0, \quad (124)$$

which we can write down by simply looking at equation (122), provides the solution for the problem on hand.

Of the two factors in (124), the first of them

$$2\nu+\mu+2 = \tfrac{5}{2}+\sqrt{(k^2-J)}+\tfrac{1}{2}\sqrt{(1-4J)} \neq 0; \quad (125)$$

hence, it is the second factor which must vanish. Thus,

$$2\nu+\mu-1 = \sqrt{(k^2-J)}-\tfrac{1}{2}+\tfrac{1}{2}\sqrt{(1-4J)} = 0, \quad (126)$$

or

$$\sqrt{(k^2-J)} = \tfrac{1}{2}[1-\sqrt{(1-4J)}]. \quad (127)$$

On squaring relation (127), we obtain

$$\sqrt{(1-4J)} = 1-2k^2; \quad (128)$$

and on squaring this relation, we obtain

$$J = k^2(1-k^2). \quad (129)$$

The corresponding expressions for ν and μ are

$$\nu = \tfrac{1}{2}k^2 \quad \text{and} \quad \mu = 1-k^2, \quad (130)$$

so that the proper function which belongs to the characteristic value (129) is

$$w = \text{constant } (1-U^2)^{k^2/2}U^{(1-k^2)}; \quad (131)$$

or, reverting to the variable z, we have

$$w = \text{constant } (\operatorname{sech} z)^{k^2}(\tanh z)^{1-k^2}. \quad (132)$$

[Notice that when $J = 0$ equation (105) becomes

$$\frac{d}{dU}\left[(1-U^2)\frac{dw}{dU}\right]+\left(2-\frac{k^2}{1-U^2}\right)w = 0. \quad (133)$$

The solutions of this equation are the associated Legendre functions of degree 1 and order k. Only for $k^2 = 0$ and 1 does equation (133) allow solutions which satisfy the boundary conditions (106); and, in these special cases, the solutions are

$$w = \text{constant } U \quad (k^2 = 0)$$

and

$$w = \text{constant}\sqrt{(1-U^2)} \quad (k^2 = 1). \quad (134)$$

It will be observed that these solutions are included in (131).]

The curve of marginal stability defined by equation (129) is shown in Fig. 120. The possibility of closed unstable regions above the curve, and similar stable regions below the curve, cannot be excluded; but it does not seem likely on energetic considerations such as those which led to the condition (74). It would appear, then, that the Richardson

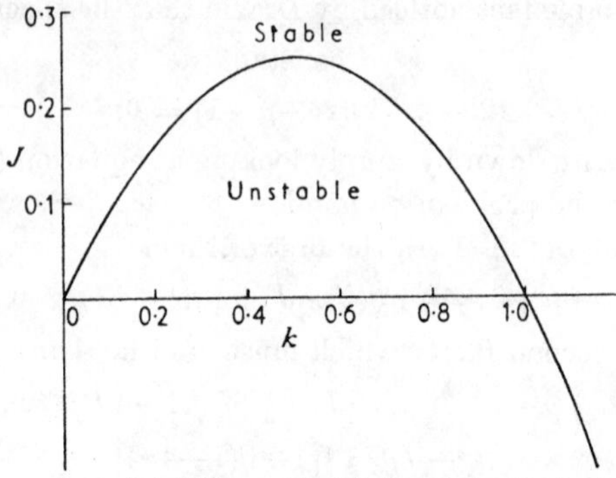

FIG. 120. The regions of stability and instability in the (J, k)-plane for a shear layer with the velocity distribution $U = U_0 \tanh(z/d)$; k is the wave number measured in the unit $1/d$ and J is the Richardson number.

number defining the limit of instability in the (J, k)-plane is determined by equation (129); it is given by

$$J_{\max} = \tfrac{1}{4} \quad \text{for} \quad k^2 = \tfrac{1}{2}. \tag{135}$$

This result is in exact agreement with (74) and lends support to the physical arguments which were advanced in § 103.

105. The effect of rotation

We shall now consider the effect of rotation on the development of the Kelvin–Helmholtz instability. The initial stationary state is, then, the same as that postulated in § 100, except that the fluid will now be supposed to be in uniform rotation about the z-axis with an angular velocity Ω. Under these circumstances, equations (1) and (2) will be replaced by

$$\rho\frac{\partial u}{\partial t}+\rho U\frac{\partial u}{\partial x}+\rho w\frac{dU}{dz}-2\rho\Omega v = -\frac{\partial}{\partial x}\delta p, \tag{136}$$

$$\rho\frac{\partial v}{\partial t}+\rho U\frac{\partial v}{\partial x}+2\rho\Omega u = -\frac{\partial}{\partial y}\delta p, \tag{137}$$

while equations (3)–(6) remain unaltered. For solutions having the (x, y, t)-dependence given by (7), equations (136) and (137) give

$$i\rho(n+k_x U)u+\rho(DU)w-2\rho\Omega v = -ik_x\,\delta p \tag{138}$$

and $$i\rho(n+k_x U)v+2\rho\Omega u = -ik_y\,\delta p. \tag{139}$$

From equations (138) and (139) we obtain (in place of equation (14))

$$i\rho(n+k_x U)Dw-i\rho k_x(DU)w+2\rho\Omega\zeta = -k^2\delta p, \tag{140}$$

where $$\zeta = ik_x v-ik_y u \tag{141}$$

is the z-component of the vorticity. By multiplying equations (138) and (139) by $-ik_y$ and $+ik_x$, respectively, adding, and making use of the equation of continuity (13), we also obtain

$$i\rho(n+k_x U)\zeta = ik_y\rho(DU)w+2\rho\Omega Dw. \tag{142}$$

Substituting for ζ from this equation in equation (140), we obtain after some rearrangement

$$i\rho(n+k_x U)\left[1-\frac{4\Omega^2}{(n+k_x U)^2}\right]Dw- \\ -i\rho\left(k_x+\frac{2ik_y\Omega}{n+k_x U}\right)(DU)w = -k^2\delta p. \tag{143}$$

Now eliminating δp from equation (15) (which continues to be valid) and equation (143), we obtain

$$D\left\{\rho(n+k_x U)\left[1-\frac{4\Omega^2}{(n+k_x U)^2}\right]Dw-\rho\left(k_x+\frac{2ik_y\Omega}{n+k_x U}\right)(DU)w\right\}- \\ -k^2\rho(n+k_x U)w = gk^2\left\{(D\rho)-\frac{k^2}{g}\sum_s T_s\,\delta(z-z_s)\right\}\frac{w}{n+k_x U}. \tag{144}$$

As in § 100, we must require that at the interface between two fluids $w/(n+k_x U)$ is continuous. By integrating equation (144) across such an interface, we obtain the further condition

$$\Delta_s\left\{\rho(n+k_x U)\left[1-\frac{4\Omega^2}{(n+k_x U)^2}\right]Dw-\rho\left(k_x+\frac{2ik_y\Omega}{n+k_x U}\right)(DU)w\right\} \\ = gk^2\left\{\Delta_s(\rho)-\frac{k^2 T_s}{g}\right\}\left(\frac{w}{n+k_x U}\right)_s, \tag{145}$$

where $\Delta_s(f)$ denotes the jump which a quantity f experiences at the interface $z = z_s$; and the subscript s signifies the value which a quantity, known to be continuous at an interface, takes at $z = z_s$.

(a) *The case of two uniform fluids in relative horizontal motion separated by a horizontal boundary*

Consider the case of two uniform fluids of densities ρ_1 and ρ_2 separated by a horizontal boundary at $z = 0$; and let the velocities of streaming in the two fluids be U_1 and U_2. In the two regions of constant ρ and U, equation (144) becomes

$$\left[1-\frac{4\Omega^2}{(n+k_x U)^2}\right]D^2w = k^2w, \tag{146}$$

where, for the present, we are suppressing the subscripts distinguishing the two fluids. The general solution of equation (146) is a linear combination of the integrals

$$e^{+\kappa z} \quad \text{and} \quad e^{-\kappa z}, \tag{147}$$

where

$$\kappa = \frac{k}{\sqrt{[1-4\Omega^2/(n+k_x U)^2]}}. \tag{148}$$

Since w must vanish when $z \to \pm\infty$ and $w/(n+k_x U)$ must be continuous at $z = 0$, we may write

$$w_1 = A(n+k_x U_1)e^{+\kappa_1 z} \quad (z < 0) \tag{149}$$

and

$$w_2 = A(n+k_x U_2)e^{-\kappa_2 z} \quad (z > 0) \tag{150}$$

as the solutions appropriate in the two regions provided, in the definitions of κ_1 and κ_2, the signs of the square roots are so chosen that the real parts of κ_1 and κ_2 are positive.

When $DU = 0$ on either side of the interface (as in the present instance), the condition (145) which must obtain generally reduces to

$$\Delta_s\left\{\rho(n+k_x U)\frac{Dw}{\kappa^2}\right\} = g\left\{\Delta_s(\rho)-\frac{k^2T_s}{g}\right\}\left(\frac{w}{n+k_x U}\right)_s. \tag{151}$$

Applying this condition to the solutions given by (149) and (150), we find

$$\frac{\rho_2}{\kappa_2}(n+k_x U_2)^2+\frac{\rho_1}{\kappa_1}(n+k_x U_1)^2 = g\left\{\rho_1-\rho_2+\frac{k^2T}{g}\right\}. \tag{152}$$

Substituting for κ_1 and κ_2 in accordance with (148) and letting α_1 and α_2 have the meanings assigned in (27), we obtain the characteristic equation

$$\alpha_1(n+k_x U_1)^2\left[1-\frac{4\Omega^2}{(n+k_x U_1)^2}\right]^{\frac{1}{2}}+\alpha_2(n+k_x U_2)^2\left[1-\frac{4\Omega^2}{(n+k_x U_2)^2}\right]^{\frac{1}{2}}$$
$$= gk\left\{(\alpha_1-\alpha_2)+\frac{k^2T}{g(\rho_1+\rho_2)}\right\}. \tag{153}$$

When $k_x = 0$, equation (153) reduces to equation X (160) appropriate for Rayleigh–Taylor instability in the presence of rotation.† Clearly, as in the non-rotating case, the Kelvin–Helmholtz instability is least

† It may be recalled here that what was denoted by n in Chapter X is now denoted by in.

uninhibited for perturbations in the direction of streaming. Accordingly, replacing k_x by k in equation (153), we have

$$\alpha_1(n+kU_1)^2\left[1-\frac{4\Omega^2}{(n+kU_1)^2}\right]^{\frac{1}{2}}+\alpha_2(n+kU_2)^2\left[1-\frac{4\Omega^2}{(n+kU_2)^2}\right]^{\frac{1}{2}}$$
$$= gk\left\{(\alpha_1-\alpha_2)+\frac{k^2T}{g(\rho_1+\rho_2)}\right\}. \quad (154)$$

(b) The discussion of the characteristic equation

We shall restrict our discussion of equation (154) to the case $T = 0$; and consider, first, the stable modes of oscillation which are possible by a graphical method used by Taylor in a different connexion.

Ignoring the term in the surface tension, we can rewrite equation (154) in the form

$$\alpha_1(U_1-c)^2\left[1-\frac{4\Omega^2}{k^2(U_1-c)^2}\right]^{\frac{1}{2}}+\alpha_2(U_2-c)^2\left[1-\frac{4\Omega^2}{k^2(U_2-c)^2}\right]^{\frac{1}{2}} = \frac{g}{k}(\alpha_1-\alpha_2), \quad (155)$$

where we have put $$c = -n/k. \quad (156)$$

By measuring velocities in the unit $\sqrt{(g/k)}$ and letting

$$U_1-c = \xi\sqrt{\frac{g}{k}} \quad \text{and} \quad U_2-c = \eta\sqrt{\frac{g}{k}}, \quad (157)$$

we have $$\frac{\alpha_1}{\alpha_1-\alpha_2}\xi^2\sqrt{\left(1-\frac{\omega^2}{\xi^2}\right)}+\frac{\alpha_2}{\alpha_1-\alpha_2}\eta^2\sqrt{\left(1-\frac{\omega^2}{\eta^2}\right)} = 1, \quad (158)$$

where $$\omega^2 = \frac{4\Omega^2}{gk}, \quad (159)$$

and the signs of the square roots in equation (158) must be so chosen that their real parts are positive.

So long as ξ and η are assumed to be real, equation (158) can be satisfied only when ξ^2 and η^2 are each greater than ω^2; and when this is the case, equation (158) defines a real locus in the (ξ, η)-plane. This locus consists of four disjointed curves (the parts designated by P in Fig. 121 a, b) symmetrically disposed with respect to the axes. The part of the locus in the first quadrant connects the points

$$(\omega, \eta_0) \quad \text{and} \quad (\xi_0, \omega), \quad (160)$$

where ξ_0 and η_0 are determined by the equations

$$\alpha_1\xi^2\sqrt{\left(1-\frac{\omega^2}{\xi^2}\right)} = \alpha_1-\alpha_2 \quad \text{and} \quad \alpha_2\eta^2\sqrt{\left(1-\frac{\omega^2}{\eta^2}\right)} = \alpha_1-\alpha_2. \quad (161)$$

The explicit expressions for ξ_0 and η_0 are

$$\xi_0 = \left\{\tfrac{1}{2}\omega^2+\tfrac{1}{2}\left[\omega^4+4\left(\frac{\alpha_1-\alpha_2}{\alpha_1}\right)^2\right]^{\frac{1}{2}}\right\}^{\frac{1}{2}}$$

and
$$\eta_0 = \left\{\tfrac{1}{2}\omega^2+\tfrac{1}{2}\left[\omega^4+4\left(\frac{\alpha_1-\alpha_2}{\alpha_2}\right)^2\right]^{\frac{1}{2}}\right\}^{\frac{1}{2}}. \tag{162}$$

Since $\alpha_1 > \alpha_2$,
$$\eta_0 > \xi_0 > \omega. \tag{163}$$

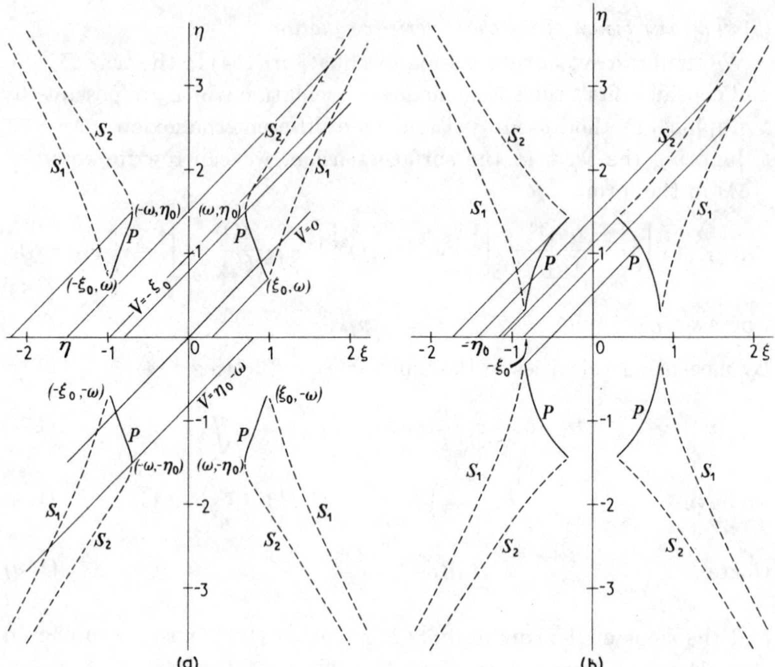

FIG. 121. The loci defined by the equations
$$\pm\alpha_1\,\xi^2\sqrt{(1-\omega^2/\xi^2)}\pm\alpha_2\,\eta^2\sqrt{(1-\omega^2/\eta^2)} = \alpha_1-\alpha_2.$$
When the signs are dissimilar they define the dashed line curves; the branch designated S_1 corresponds to the first sign in the equation being positive and the second sign negative; and the branch designated S_2 corresponds to the first sign in the equation being negative and the second sign positive. The branches S_1 and S_2 are connected by the branch P which corresponds to both signs in the equation being positive. The cases illustrated are for $\alpha_1-\alpha_2 = 0{\cdot}5$ and $\omega^2 = 0{\cdot}5$ (the figure on the right) and $\omega^2 = 0{\cdot}1$ (the figure on the left).

It is also clear that at (ω, η_0) the locus has a vertical tangent, while at (ξ_0, ω) it has a horizontal tangent.

In Table LII, ξ_0 and η_0 are listed for a few values of ω for the case $\alpha_1-\alpha_2 = 0{\cdot}5$.

For a given difference V in the streaming velocities U_1 and U_2, ξ and η must lie on the line

$$\xi-\eta = V\sqrt{(k/g)}. \tag{164}$$

In order to find whether stable modes are possible for a given $V\sqrt{(k/g)}$, we have only to determine whether the corresponding line (164) has any intersections with the real part of the (ξ, η)-locus. Similarly, the effect of varying V (keeping k fixed) can be found by drawing a series of parallel

TABLE LII

The coordinates of the end points ξ_0, ω and ω, η_0 of the (ξ, η)-locus $(\alpha_1-\alpha_2 = 0\cdot5)$

ω	ξ_0	η_0	ω	ξ_0	η_0
0	0·8164	1·4142	0·8	1·0293	1·5315
0·1	0·8196	1·4160	1·0	1·1548	1·6005
0·2	0·8289	1·4213	1·2	1·3043	1·6869
0·3162	0·8476	1·4320	1·4	1·4715	1·7909
0·4	0·8669	1·4428	1·6	1·6502	1·9117
0·6	0·9331	1·4792	1·8	1·8362	2·0479
0·7071	0·9808	1·5052	2·0	2·0269	2·1974

lines inclined to the ξ-axis by 45° and determining when intersections with the (ξ, η)-locus occur.

While the allowed real modes can be enumerated and studied in the manner we have described, it is more important to be able to determine when equations (158) and (164) permit a complex root and entail instability. In this latter connexion, particular interest attaches to the maximum value of $|V|$ (for a given k) such that for all $|V|$ less than this maximum, we can be assured of stability.

The problem of ascertaining the number and the character of the roots of equations (158) and (164) is not a straightforward one: the difficulty arises from the curious restriction on the signs of the square roots in equation (158), namely, that the real parts must be positive. This latter restriction makes the problem non-algebraic and precludes the application of the fundamental theorem of algebra. However, we can restore to the problem its algebraic character by considering together with equation (164) the following set of four equations:

$$+\alpha_1\xi^2\sqrt{(1-\omega^2/\xi^2)}+\alpha_2\eta^2\sqrt{(1-\omega^2/\eta^2)} = \alpha_1-\alpha_2, \tag{165}$$

$$+\alpha_1\xi^2\sqrt{(1-\omega^2/\xi^2)}-\alpha_2\eta^2\sqrt{(1-\omega^2/\eta^2)} = \alpha_1-\alpha_2, \tag{166}$$

$$-\alpha_1\xi^2\sqrt{(1-\omega^2/\xi^2)}+\alpha_2\eta^2\sqrt{(1-\omega^2/\eta^2)} = \alpha_1-\alpha_2, \tag{167}$$

$$-\alpha_1\xi^2\sqrt{(1-\omega^2/\xi^2)}-\alpha_2\eta^2\sqrt{(1-\omega^2/\eta^2)} = \alpha_1-\alpha_2 \tag{168}$$

(with the same restriction, as before, on the choice of the signs of the square roots). The four equations (together with (164)) constitute a pure algebraic problem; for, they are together equivalent to the single eighth degree polynomial equation which can be derived from any one of the four equations (165)–(168) by clearing the square roots by two successive squarings. The fundamental theorem of algebra, therefore, applies to the equations (164)–(169) together; and we can conclude that *they must have a total of exactly eight roots under all circumstances.* In addition to this result, there are a number of properties of the equations which we shall now enumerate. (1) If any of the four equations allows a complex root, it must also allow its complex conjugate as a root, i.e. each of the equations can allow complex roots only in conjugate pairs. (2) The roots which follow from any of the four equations (165)–(168) are continuous functions of the parameters of the equation except at the singular points $(\pm\xi_0, \pm\omega)$, $(\pm\omega, \pm\eta_0)$, $(0, \pm\eta_0)$, and $(\pm\xi_0, 0)$. (3) Only at the singular points can the number of roots, which any of them allows, change.

We shall now show how the properties of the system (164)–(168) which we have enumerated enable us to ascertain in a systematic way the number and the character of the roots of each of the four equations for any assigned value of $|V|$. For this purpose, it is convenient to introduce the following terminology.

We have already seen how equation (165) defines in the (ξ, η)-plane a locus consisting of four disjointed pieces. We shall call this locus the principal or the P-branch of the equations (165)–(168). Similarly we shall call the loci defined by equations (166) and (167) the subsidiary branches S_1 and S_2, respectively. Equation (168) cannot be satisfied (with the imposed restriction on the signs of the square roots) by any real pair of values of ξ and η; the equation accordingly defines a purely complex branch which we shall designate by C.

In Fig. 121 we illustrate the disposition of the loci of the P- and S-branches for $\alpha_1-\alpha_2 = 0\cdot5$ and $\omega^2 = 0\cdot5$ and $0\cdot1$.

Now consider the equations (164)–(168) for the special case

$$V = 0 \quad\text{and}\quad \xi = \eta. \tag{169}$$

The equations become

$$\xi^2\surd(1-\omega^2/\xi^2) = \alpha_1-\alpha_2 \quad (P), \tag{170}$$

$$\xi^2\surd(1-\omega^2/\xi^2) = 1 \quad (S_1), \tag{171}$$

$$-\xi^2\surd(1-\omega^2/\xi^2) = 1 \quad (S_2), \tag{172}$$

and

$$-\xi^2\surd(1-\omega^2/\xi^2) = \alpha_1-\alpha_2 \quad (C). \tag{173}$$

By squaring these equations, we obtain

$$\xi^4-\omega^2\xi^2-(\alpha_1-\alpha_2)^2 = 0 \quad (P \text{ and } C) \tag{174}$$

and

$$\xi^4-\omega^2\xi^2-1 = 0 \quad (S_1 \text{ and } S_2). \tag{175}$$

The roots of these equations are

$$\xi^2 = \tfrac{1}{2}\omega^2 \pm \tfrac{1}{2}\sqrt{[\omega^4+4(\alpha_1-\alpha_2)^2]} \quad (P \text{ and } C) \tag{176}$$

and

$$\xi^2 = \tfrac{1}{2}\omega^2 \pm \tfrac{1}{2}\sqrt{(\omega^4+4)} \qquad (S_1 \text{ and } S_2). \tag{177}$$

By substituting these roots back into the original equations (170)–(173) we directly verify that the positive sign in equation (176) goes with equation (170) while the negative sign goes with equation (173). Similarly, the positive sign in equation (177) goes with the S_1-branch while the negative sign goes with the S_2-branch. Therefore, for $V = 0$, the P- and the S_1-branches each allow two real roots, while the S_2- and the C-branches each allow a conjugate pair of complex roots; this is in agreement with what we observe in Fig. 121.

The roots of equations (158), for $\alpha_1-\alpha_2 = 0{\cdot}5$ and $\omega^2 = 0{\cdot}1$ and $0{\cdot}5$, are listed in Table LIIA; and their dependence on $|\xi-\eta|$ is further illustrated in Fig. 122. Since there is no essential loss of generality in supposing that U_1 (or U_2) is zero, we may interpret the tabulated roots ξ as giving the characteristic frequencies of oscillation measured in the unit $\sqrt{(gk)}$.

An examination of the results for the two cases considered shows that the allowed proper solutions, of the prescribed form (7), can be classified as follows:

$$
\begin{array}{lll}
\text{(I)} & 0 \leqslant |\xi-\eta| \leqslant \xi_0-\omega: & \text{a real branch,} \\
\text{(II)} & 0 \leqslant |\xi-\eta| \leqslant \eta_0-\omega: & \text{a real branch,} \\
\text{(III)} & \xi_0 < |\xi-\eta| < \xi_0+\omega: & \text{a complex branch,}\dagger \\
\text{(IV)} & \xi_0+\omega \leqslant |\xi-\eta| \leqslant \eta_0+\omega: & \text{a real branch,} \\
\text{(V)} & \eta_0 < |\xi-\eta| < \infty: & \text{a complex branch.}
\end{array} \tag{178}
$$

It would appear then that instability, via a mode of the prescribed form, will first manifest itself when $|U_1-U_2|$ surpasses the value given by

$$|U_1-U_2|\sqrt{\frac{k}{g}} = \xi_0 = \left\{\tfrac{1}{2}\omega^2+\tfrac{1}{2}\left[\omega^4+4\left(\frac{\alpha_1-\alpha_2}{\alpha_1}\right)^2\right]^{\frac{1}{2}}\right\}^{\frac{1}{2}}. \tag{179}$$

† In view of the P-branch having a horizontal tangent at $(\pm\xi_0, \omega)$, real roots can appear 'slightly' before $\xi_0+\omega$, as they in fact do for $\omega^2 = 0{\cdot}5$.

If $\xi_0 > (\eta_0-\omega)$ (as it is for the case $\omega^2 = 0{\cdot}5$ and for all sufficiently large values of ω^2) then in the interval $\xi_0 > |\xi-\eta| > (\eta_0-\omega)$ the characteristic equation allows no solution, real or complex. From this

Table LIIa

Characteristic roots of equation (158) *for* $\alpha_1-\alpha_2 = 0{\cdot}5$, *and* $\omega^2 = 0{\cdot}1$ *and* $0{\cdot}5$

($\omega^2 = 0{\cdot}1$)

$\lvert\xi-\eta\rvert$	ξ(I)	ξ(II)	ξ(III)	$\lvert\xi-\eta\rvert$	ξ(IV)	ξ(V)
0	0·74330	0·74330		$\xi_0+\omega$	0·84767	
0·2	0·78838	0·68834		1·166	0·84374	
0·4	0·82388	0·62326		1·17	0·84206	
0·52	0·84337	0·57914		1·2	0·83442	
$\xi_0-\omega$	0·84767	0·57473		1·3	0·81116	
0·6		0·54752		1·4	0·78026	
0·7		0·50543		η_0		0
0·8		0·46041		1·44	0·76464	0·00721±0·00238i
ξ_0			0·84767±0i	1·5	0·73642	0·06160±0·02005i
1·0		0·36278	0·84802±0·00827i	1·6	0·66971	0·15787±0·04716i
1·1		0·31870	0·84745±0·00946i	1·74	0·31751	0·41604±0·11444i
$\eta_0-\omega$		0·31623	0·84726±0·00886i	$\eta_0+\omega$	0·31623	0·42134±0·13873i
1·15			0·84670±0·00546i	1·76		0·42722±0·16657i
1·16			0·84651±0·00283i			
$\xi_0+\omega$			0·84518±0i			

$\xi_0-\omega = 0{\cdot}53144$; $\xi_0 = 0{\cdot}84767$; $\eta_0-\omega = 1{\cdot}11577$; $\xi_0+\omega = 1{\cdot}16389$; $\eta_0 = 1{\cdot}43200$; $\eta_0+\omega = 1{\cdot}74823$

($\omega^2 = 0{\cdot}5$)

$\lvert\xi-\eta\rvert$	ξ(I)	ξ(II)	$\lvert\xi-\eta\rvert$	ξ(III)	ξ(IV)	ξ(V)
0	0·89945	0·89945	ξ_0	0·98082±0i		
0·15	0·93699	0·86107	1·2	0·98322±0·02323i		
0·20	0·95050	0·84782	1·4	0·98292±0·03816i		
0·26	0·97168	0·83164	η_0			0
$\xi_0-\omega$	0·98082	0·82790	1·6	0·97684±0·03433i		0·06406±0·04400i
0·28		0·82618	1·6878	0·96158±0i	0·98069	0·12462±0·08454i
0·4		0·79297	$\xi_0+\omega$	0·96143±0i	0·98082	0·12471±0·08459i
0·5		0·76533	1·688		0·96134	0·12476±0·08463i
0·6		0·73912	1·9		0·84532	0·28115±0·19012i
$\eta_0-\omega$		0·70711	$\eta_0+\omega$		0·70711	0·50266±0·42942i
			2·4			0·57548±0·58286i

$\xi_0-\omega = 0{\cdot}27371$; $\eta_0-\omega = 0{\cdot}79807$; $\xi_0 = 0{\cdot}98082$; $\eta_0 = 1{\cdot}50518$; $\xi_0+\omega = 1{\cdot}68792$; $\eta_0+\omega = 2{\cdot}21229$

fact and, as well as, from the variable number of allowed solutions of the prescribed form, as $|\xi-\eta|$ increases, it would follow that a complete solution to the stability problem can be achieved only in terms of a properly formulated initial value problem.

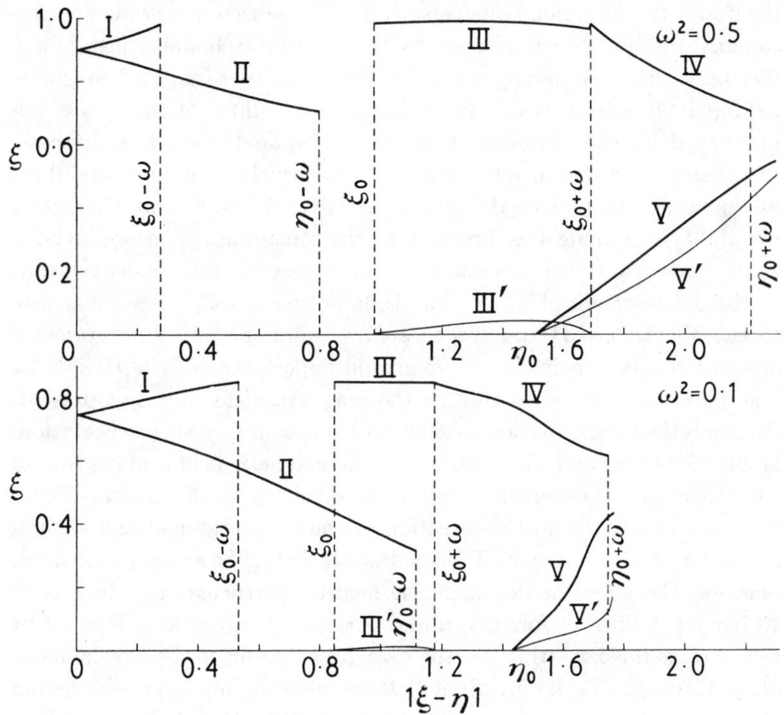

FIG. 122. The solutions of the characteristic equation (158) for $\alpha_1-\alpha_2 = 0\cdot5$ and $\omega^2 = 0\cdot1$ and $0\cdot5$. The various branches of the solution (indicated by roman numerals) follow the classification scheme (178) (see also Table LIIA). For the complex branches III and V, the real and the imaginary parts of ξ are indicated by III and III′ (and V and V′) respectively.

106. The effect of a horizontal magnetic field

We shall now consider the effect of a uniform horizontal magnetic field on the development of the Kelvin–Helmholtz instability. While the direction of the impressed magnetic field, in the horizontal plane, can be specified arbitrarily with respect to the direction of the streaming, two directions are clearly distinguished: the direction of the streaming itself and the direction transverse to it. These two directions are profoundly different with respect to the Kelvin–Helmholtz instability: for, perturbations transverse to the direction of a uniform magnetic field (which merely result in exchanging the lines of force) are uninfluenced by the presence of the magnetic field—this was the case, for example, in the development of the Rayleigh–Taylor instability; at the same time perturbations most sensitive to the Kelvin–Helmholtz instability are along the direction of the streaming. Consequently, if **H** is transverse to the direction of **U**, the perturbations most relevant to the development of the Kelvin–Helmholtz instability are transverse to the direction of **H** and are, therefore, uninfluenced by its presence; accordingly, in this instance, we should expect the Kelvin–Helmholtz instability to develop as though the magnetic field were not present. An analytical examination of this problem confirms this expectation. If, on the other hand, the direction of the magnetic field is along that of the streaming, the perturbations most sensitive to the Kelvin–Helmholtz instability are in the direction of the magnetic field and should, therefore, be affected by it. Indeed, the effect must be a very pronounced one: for, the effect of the magnetic field on perturbations which tend to bend the lines of force is, under certain circumstances, equivalent to a surface tension—this was the case, for example, in the development of the Rayleigh–Taylor instability. Remembering this and remembering also that surface tension can suppress the Kelvin–Helmholtz instability for sufficiently small differences in the stream velocities of two superposed fluids, one can expect that a magnetic field in the direction of the streaming will have a similar inhibiting effect. A detailed analysis confirms this expectation also.

(*a*) *The effect of a magnetic field in the direction of streaming*

When the direction of the impressed magnetic field coincides with that of the streaming, the relevant equations governing perturbations which have the (x, y, t)-dependence given by (7) are (cf. equations (8)–(13) and equations X (215)–(218)):

$$i\rho(n+k_x U)u+\rho(DU)w = -ik_x\,\delta p, \tag{180}$$

$$i\rho(n+k_x U)v-\frac{\mu H}{4\pi}(ik_x h_y-ik_y h_x) = -ik_y\,\delta p, \tag{181}$$

$$i\rho(n+k_x U)w-\frac{\mu H}{4\pi}(ik_x h_z-Dh_x) = -D\delta p-g\delta\rho, \tag{182}$$

$$i(n+k_x U)\mathbf{h} = ik_x H\mathbf{u}+h_z D\mathbf{U}, \tag{183}$$

and
$$i(n+k_x U)\delta\rho = -(D\rho)w; \tag{184}$$

and besides, we have the equations which express the solenoidal characters of **u** and **h**. (We are not including effects arising from surface tension.)

Writing equation (183) for the components separately, we find

$$h_x = \frac{k_x H}{n+k_x U}\left(u-\frac{iDU}{n+k_x U}w\right), \tag{185}$$

$$h_y = \frac{k_x H}{n+k_x U}v, \quad \text{and} \quad h_z = \frac{k_x H}{n+k_x U}w. \tag{186}$$

Inserting these expressions for the components of **h** in equations (181) and (182), we obtain

$$i\rho(n+k_x U)v-\frac{k_x}{n+k_x U}\frac{\mu H^2}{4\pi}\left(\zeta-\frac{k_y DU}{n+k_x U}w\right) = -ik_y\,\delta p \tag{187}$$

and

$$i\rho(n+k_x U)w+\frac{\mu H^2 k_x}{4\pi}D\left\{\frac{1}{n+k_x U}\left(u-\frac{iDU}{n+k_x U}w\right)\right\}-$$
$$-\frac{ik_x^2}{n+k_x U}\frac{\mu H^2}{4\pi}w = -D\delta p-ig\frac{D\rho}{n+k_x U}w, \tag{188}$$

where
$$\zeta = ik_x v-ik_y u \tag{189}$$

is the z-component of the vorticity; also, in rewriting equation (182) in the form (188) we have made use of equation (184).

Multiplying equations (180) and (187) by $-ik_y$ and $+ik_x$, respectively, and adding, we obtain

$$\zeta = \frac{k_y DU}{n+k_x U}w, \tag{190}$$

so that equation (187) reduces to

$$i\rho(n+k_x U)v = -ik_y\,\delta p. \tag{191}$$

Now multiplying equations (180) and (191) by $-k_x$ and $-k_y$, respectively, and adding, we obtain

$$\rho(n+k_x U)Dw-\rho k_x(DU)w = ik^2\delta p. \tag{192}$$

Next, rewriting equation (188) in the form

$$ik^2 D\delta p = \rho k^2(n+k_x U)w-\frac{k^2 k_x^2}{n+k_x U}\frac{\mu H^2}{4\pi}w-$$
$$-k_x k^2\frac{\mu H^2}{4\pi}D\left\{\frac{1}{n+k_x U}\left(iu+\frac{DU}{n+k_x U}w\right)\right\}+gk^2\frac{D\rho}{n+k_x U}w, \tag{193}$$

we eliminate u by making use of the relation,

$$ik^2u = -(k_x Dw+k_y \zeta) = -\left(k_x Dw+\frac{k_y^2 DU}{n+k_x U}w\right), \tag{194}$$

which follows from equations (13), (189), and (190). We thus find

$$ik^2D\delta p = \rho k^2(n+k_x U)w+k_x^2\frac{\mu H^2}{4\pi}\left\{D\left(\frac{Dw}{n+k_x U}\right)-\frac{k^2w}{n+k_x U}\right\}-$$
$$-k_x^3\frac{\mu H^2}{4\pi}D\left\{\frac{DU}{(n+k_x U)^2}w\right\}+gk^2\frac{D\rho}{n+k_x U}w. \tag{195}$$

Finally eliminating δp between equations (192) and (195) we obtain

$$D\{\rho(n+k_x U)Dw-\rho k_x(DU)w\}$$
$$= k^2\rho(n+k_x U)w+k_x^2\frac{\mu H^2}{4\pi}\left\{D\left(\frac{Dw}{n+k_x U}\right)-\frac{k^2w}{n+k_x U}\right\}-$$
$$-k_x^3\frac{\mu H^2}{4\pi}D\left\{\frac{DU}{(n+k_x U)^2}w\right\}+gk^2\frac{D\rho}{n+k_x U}w. \tag{196}$$

(i) *The case of two uniform fluids in relative horizontal motion separated by a horizontal boundary*

We shall apply the foregoing equations to the case of two uniform fluids of densities ρ_1 and ρ_2 separated by a horizontal boundary at $z = 0$. Let the velocities of streaming of the two fluids be U_1 and U_2.

In the two regions of constant ρ and U, equations (190) and (196) give

$$\zeta = 0 \tag{197}$$

and

$$\left[\rho(n+k_x U)-\frac{k_x^2}{n+k_x U}\frac{\mu H^2}{4\pi}\right](D^2-k^2)w = 0, \tag{198}$$

where we have suppressed the subscripts distinguishing the two fluids. The solutions in the two regions $z < 0$ and $z > 0$ can be written in the forms (cf. equations (24) and (25))

$$w_1 = A(n+k_x U_1)e^{+kz} \quad (z < 0) \tag{199}$$

and

$$w_2 = A(n+k_x U_2)e^{-kz} \quad (z > 0). \tag{200}$$

By this choice of the solutions, the condition requiring the continuity of $w/(n+k_x U)$, as well as the conditions at infinity, have been met. It remains to satisfy the further condition which follows from equation (196) by integrating it across the interface at $z = 0$. This condition is, clearly,

$$\Delta_0\{\rho(n+k_x U)Dw\} = k_x^2\frac{\mu H^2}{4\pi}\Delta_0\left(\frac{Dw}{n+k_x U}\right)+gk^2\Delta_0(\rho)\left(\frac{w}{n+k_x U}\right)_0, \tag{201}$$

where $\Delta_0(f)$ denotes the jump a quantity f experiences at $z = 0$ and $[w/(n+k_x U)]_0$ is the unique value which this quantity has at $z = 0$.

Now applying the condition (201) to the solutions given by equations (199) and (200), we obtain the characteristic equation

$$\rho_2(n+k_x U_2)^2+\rho_1(n+k_x U_1)^2 = gk(\rho_1-\rho_2)+k_x^2\frac{\mu H^2}{2\pi}. \tag{202}$$

On comparing this equation with equation (26), we observe that in this instance the effect of the magnetic field is equivalent to that of a surface tension of amount

$$T_{\text{eff}} = \frac{\mu H^2}{2\pi}\frac{k_x^2}{k^3}. \tag{203}$$

This is exactly the same expression we found in the development of the Rayleigh–Taylor instability in Chapter X, § 97, equation (235).

The roots of equation (202) are (cf. equation (30))

$$n = -k_x(\alpha_1 U+\alpha_2 U_2)\pm \pm\left[gk(\alpha_1-\alpha_2)+k_x^2\frac{\mu H^2}{2\pi(\rho_1+\rho_2)}-k_x^2\alpha_1\alpha_2(U_1-U_2)^2\right]^{\frac{1}{2}}. \tag{204}$$

From this equation it follows that a uniform *magnetic field acting parallel to the direction of streaming will suppress the Kelvin–Helmholtz instability if*

$$\alpha_1\alpha_2(U_1-U_2)^2 \leqslant \frac{\mu H^2}{2\pi(\rho_1+\rho_2)}, \tag{205}$$

i.e. if the relative speed does not exceed the root-mean-square Alfvén speed in the two media.

(b) The effect of a magnetic field transverse to the direction of streaming

When the impressed magnetic field is transverse to the direction of $\mathbf{U}$, the relevant perturbation equations are

$$i\rho(n+k_x U)u+\rho(DU)w+\frac{\mu H}{4\pi}(ik_x h_y-ik_y h_x) = -ik_x\delta p, \tag{206}$$

$$i\rho(n+k_x U)v = -ik_y\delta p, \tag{207}$$

$$i\rho(n+k_x U)w-\frac{\mu H}{4\pi}(ik_y h_z-Dh_y) = -D\delta p-g\delta\rho, \tag{208}$$

and

$$i(n+k_x U)\mathbf{h} = ik_y H\mathbf{u}+h_z D\mathbf{U}. \tag{209}$$

These equations replace equations (172)–(175); and the remaining equations are unaltered.

Treating the present set of equations in the same manner as in § (*a*) above, we now obtain in place of (196)

$$D\{\rho(n+k_x U)Dw-\rho k_x(DU)w\}$$
$$=k^2\rho(n+k_x U)w+k_y^2\frac{\mu H^2}{4\pi}\left\{D\left(\frac{Dw}{n+k_x U}\right)-\frac{k^2 w}{n+k_x U}\right\}-$$
$$-k_x k_y^2\frac{\mu H^2}{4\pi}D\left\{\frac{DU}{(n+k_x U)^2}w\right\}+gk^2\frac{D\rho}{n+k_x U}w. \quad (210)$$

We observe that equation (210) reduces to hydrodynamic equation (16) of § 100 in case $k_y = 0$. Therefore, *the development of the Kelvin–Helmholtz instability, in the direction of the streaming, is uninfluenced by the presence of the magnetic field in the transverse direction.*

BIBLIOGRAPHICAL NOTES

The following general references may be noted:

1. Sir Horace Lamb, *Hydrodynamics*, §§ 232 and 268, Cambridge, England, 1932.
2. J. Proudman, *Dynamical Oceanography*, chap. xii, §§ 181–3, John Wiley & Sons, Inc., New York, 1953.

§ 101. The first reference to what has come to be known as the Kelvin–Helmholtz instability occurs in the writings of Helmholtz:

3. H. Helmholtz, 'Ueber discontinuirliche Flüssigkeitsbewegungen', *Wissenschaftliche Abhandlungen*, 146–57, J. A. Barth, Leipzig, 1882; translation by Guthrie in *Phil. Mag.*, Ser. 4, **36**, 337–46 (1868).

(The quotation in the text is from Guthrie's translation.)

Helmholtz's discussion was largely qualitative. The same basic question was treated analytically by:

4. Lord Kelvin, 'Hydrokinetic solutions and observations', 'On the motion of free solids through a liquid, 69–75', 'Influence of wind and capillarity on waves in water supposed frictionless, 76–85', *Mathematical and Physical Papers*, iv, *Hydrodynamics and General Dynamics*, Cambridge, England, 1910.

Kelvin's discussion of the stability of superposed fluids in a state of differential streaming is exceptionally complete: there is hardly a formula in this section which cannot be found in Kelvin's paper. The critical wind velocity, 650 cm/sec (or 660 cm/sec as Kelvin gave it), occurs, for example, for the first time in this paper. As stated in the text, the significance of this particular velocity for the generation of waves by wind has been the subject of much discussion; an account of it will be found in Proudman's book (reference 2; chap. xii). The recognition that Kelvin's critical velocity may, indeed, have a bearing on phenomena which are observed is more recent:

5. W. H. Munk, 'A critical wind speed for air-sea boundary processes', *J. Marine Research*, **6**, 203–18 (1947).

For a further discussion of these and related matters, see:

6. J. W. MILES, 'On the generation of surface waves by shear flows', *J. Fluid Mech.* **3,** 185–204 (1957).
7. —— 'On the generation of surface waves by shear flows. II', ibid. **6,** 568–82 (1959).
8. —— 'On the generation of surface waves by shear flows. III. Kelvin–Helmholtz instability', ibid. 583–98 (1959).

The experimental demonstration of the Kelvin–Helmholtz instability illustrated in the text (Fig. 117) is due to:

9. J. R. D. FRANCIS, 'Wave motions and the aerodynamic drag on a free oil surface', *Phil. Mag.*, Ser. 7, **45,** 695–702 (1954).
10. —— 'Wave motions on a free oil surface', ibid., Ser. 8, **1,** 685–8 (1956).

§ 102. The distributions of ρ and U considered in this section have been investigated by:

11. SIR GEOFFREY TAYLOR, 'Effect of variation in density on the stability of superposed streams of fluid', *Proc. Roy. Soc.* (*London*) A, **132,** 499–523 (1931).
12. S. GOLDSTEIN, 'On the stability of superposed streams of fluids of different densities', ibid. 524–48 (1931).

§ 103. The case of heterogeneous fluid with an exponentially decreasing density and a linearly increasing stream velocity was first considered by Sir Geoffrey Taylor (reference 11). The definitive discussion of this case is due to:

13. K. M. CASE, 'Stability of an idealized atmosphere. I. Discussion of results', *The Physics of Fluids*, **3,** 149–54 (1960).
14. F. J. DYSON, 'Stability of an idealized atmosphere. II. Zeros of the confluent hypergeometric function', ibid. 155–7 (1960).

In connexion with Case's investigation (reference 13), see his other papers:

15. K. M. CASE, 'Taylor instability of an inverted atmosphere', ibid. 366–8 (1960).
16. —— 'Stability of inviscid plane Couette flow', ibid. 143–8 (1960).

§ 104. The example treated in this section is due to:

17. P. G. DRAZIN, 'The stability of a shear layer in an unbounded heterogeneous inviscid fluid', *J. Fluid Mech.* **4,** 214–24 (1958).

§ 105. See:

18. S. CHANDRASEKHAR, 'The effect of rotation on the development of the Kelvin–Helmholtz instability', unpublished.

For a different treatment of the same problem see:

19. H. SOLBERG, 'Integrationen der atmosphärischen Störungsgleichungen, I. Wellenbewegungen in rotierenden, inkompressiblen Flüssigkeitsschichten', *Geofysiske Publikasjoner*, **9,** 1–120 (1928).

See also:

20. V. BJERKNES, J. BJERKNES, H. SOLBERG, and T. BERGERON, *Physikalische Hydrodynamik*, chap. 11, Springer, Berlin, 1933.

§ 106. The Kelvin–Helmholtz instability in the framework of hydromagnetics has been considered by:

21. D. H. MICHAEL, 'The stability of a combined current and vortex sheet in a perfectly conducting fluid', *Proc. Camb. Phil. Soc.* **51,** 528–32 (1955).

22. T. G. Northrop, 'Helmholtz instability of a plasma', *Physical Rev.*, **103**, 1150–4 (1956).

Michael's discussion is restricted to a uniform density of the medium; but he allowed for different magnetic field strengths on the two sides. Northrop's discussion is restricted to the case when the impressed magnetic field is transverse to the direction of streaming; but he allowed for different strengths of the magnetic field in the two fluids.

XII

THE STABILITY OF JETS AND CYLINDERS

107. Introduction

In this chapter we turn our attention to a different group of problems: the stability of jets and cylinders. As Lord Rayleigh has remarked in his *Theory of Sound*: 'a large class of phenomena, interesting not only in themselves but also in throwing light upon others yet more obscure, depend for their explanation upon the transformations undergone by a cylindrical body of liquid when slightly displaced from its equilibrium configuration and left to itself.' The classic example in this connexion is, of course, the instability of water issuing from a nozzle as a cylindrical jet. The cause of this instability, as Plateau first showed, is surface tension which makes the infinite cylinder an unstable figure of equilibrium and entails its breaking into separate pieces with a total superficial area which is less than that of the original cylinder. Specifically, Plateau showed that if the equilibrium surface $\varpi = R_0$ is deformed so that it becomes

$$\varpi = R+\epsilon \cos kz, \tag{1}$$

where ϵ is a small constant, z is measured along the axis, and ϖ is the horizontal distance from the axis, the deformation is stable or unstable according as kR is greater than or less than unity, i.e. according as the wavelength of the *varicosity*† is less than or greater than the circumference of the cylinder. Plateau concluded from this result that a cylindrical jet will break-up into pieces of length $2\pi R$. This conclusion, as Rayleigh showed, is incorrect: for, to ascertain the manner in which an unstable system will tend to break-up, one must determine the *mode of maximum instability*. The principle here is the following. If a system is characterized by a number of unstable modes symbolized by $\mathbf{k}_1, \mathbf{k}_2,..., \mathbf{k}_n$, the amplitudes of which increase like $e^{\sigma_1 t}, e^{\sigma_2 t},..., e^{\sigma_n t}$ (where all the σ's have positive real parts), then the mode of maximum

† In introducing this terminology to describe the deformation (1), Lord Rayleigh has said: 'In speaking of this subject, I have often been embarrassed for want of an appropriate word to describe the condition in question. But a few days ago, during a biological discussion, I found that there is a recognized if not a very pleasant word. The cylindrical jet may be said to become *varicose* and that the varicosity goes on increasing until eventually it leads to absolute disruption.' In recent times, 'sausage instability' has been used to describe the same condition; but this is also not a very 'pleasant' description, and varicose instability would seem preferable.

instability is that one which belongs to the σ with the largest real part. It is by this mode that the system will increasingly tend to break-up when an arbitrary initial disturbance is reduced to evanescence: for, if $\mathbf{k}_j$ denotes the mode of maximum instability, then its amplitude, relative to another mode $\mathbf{k}_i$, after a certain time t, will be $\exp[\text{re}(\sigma_j-\sigma_i)t]$; and this tends to infinity as t tends to infinity.

Two principal problems with which we shall be concerned in this Chapter (§§ 108–12) are the *gravitational* and the *capillary* instability of an infinite liquid cylinder. The two problems will be considered in the frameworks of hydrodynamics and hydromagnetics, when dissipative mechanisms are not present and when they are present. In §§ 113 and 114 we shall consider the effect of fluid motions on the hydromagnetic stability of certain cylindrical systems; and in § 115 we shall consider the stability of the so-called pinch configuration.

108. The gravitational instability of an infinite cylinder

Consider a uniform infinite cylinder of an incompressible, inviscid fluid. The equations governing small departures from equilibrium are

$$\frac{\partial \mathbf{u}}{\partial t} = -\text{grad}\,\Pi, \tag{2}$$

$$\text{div}\,\mathbf{u} = 0, \quad \nabla^2\delta\mathfrak{B} = 0, \tag{3}$$

and

$$\Pi = -\delta\mathfrak{B}+\delta p/\rho, \tag{4}$$

where the various symbols have their usual meanings.

Since we are considering departures from an equilibrium right-cylindrical shape of an incompressible fluid, a normal mode can be expressed uniquely in terms of the deformed surface. Suppose then that the deformed surface is described by the equation

$$\varpi = R+\epsilon e^{i(kz+m\varphi)}, \tag{5}$$

where R is the radius of the unperturbed cylinder, ϵ is some function of the time (which we shall presently specify to be of the form $\epsilon_0\, e^{\sigma t}$), k is the wave number of the disturbance in the z-direction, and m is an integer (positive, zero, or negative).

For deformations of the boundary described by equation (5), the change in the gravitational potential can be obtained as follows. The external and the internal gravitational potentials satisfy the Laplace and the Poisson equations

$$\nabla^2\mathfrak{B}_{\text{ext}} = 0 \quad \text{and} \quad \nabla^2\mathfrak{B}_{\text{int}} = -4\pi G\rho. \tag{6}$$

The solutions of these equations appropriate to the deformed boundary (5) are

$$\mathfrak{B}_{\text{ext}} = -2\pi G\rho R^2 \log(\varpi/R) + \epsilon A e^{i(kz+m\varphi)} K_m(k\varpi) + c_0 \tag{7}$$

and
$$\mathfrak{B}_{\text{int}} = -\pi G\rho\varpi^2 + \epsilon B e^{i(kz+m\varphi)} I_m(k\varpi), \tag{8}$$

where I_m and K_m are the Bessel functions of order m for a purely imaginary argument, A and B are constants to be determined, and c_0 is an additive constant with which we need not be further concerned.

The constants A and B are to be determined by the condition that the potential (with a suitable fixed choice of reference) and its derivative are continuous on the boundary (5). To the first order in ϵ, these conditions give

$$AK_m(x) = BI_m(x) \tag{9}$$

and
$$AK'_m(x) - BI'_m(x) = -4\pi G\rho/k, \tag{10}$$

where
$$x = kR. \tag{11}$$

On solving equations (9) and (10), we find†

$$A = 4\pi G\rho R I_m(x) \quad \text{and} \quad B = 4\pi G\rho R K_m(x). \tag{12}$$

The solution for $\delta\mathfrak{B}$ is therefore given by

$$\delta\mathfrak{B} = 4\pi G\rho R\epsilon K_m(x) I_m(k\varpi) e^{i(kz+m\varphi)}. \tag{13}$$

Returning to equation (2) and taking the divergence of this equation, we obtain

$$\nabla^2\Pi = 0. \tag{14}$$

Under the present circumstances the required solution for Π must, therefore, be of the form

$$\Pi = \epsilon\Pi_0 I_m(k\varpi) e^{i(kz+m\varphi)}, \tag{15}$$

where Π_0 is a constant, unspecified for the present.

We shall now specify the dependence on time of the various quantities by assuming that

$$\epsilon = \epsilon_0 e^{\sigma t}, \tag{16}$$

where σ is a constant. Equation (2) then gives

$$\mathbf{u} = -\frac{\epsilon_0 \Pi_0}{\sigma} e^{\sigma t} \operatorname{grad}\{I_m(k\varpi) e^{i(kz+m\varphi)}\}. \tag{17}$$

In particular,
$$u_\varpi = -\frac{\epsilon_0 \Pi_0}{\sigma} k I'_m(k\varpi) e^{i(kz+m\varphi)+\sigma t}. \tag{18}$$

The boundary conditions which must be satisfied by the solution represented by equation (15) are: (1) the radial component of the velocity

† In obtaining these solutions, use has been made of the relation

$$I_m(x)K'_m(x) - K_m(x)I'_m(x) = -1/x.$$

given by (18) must be compatible with the assumed form of the deformed surface, namely,

$$\varpi = R+\epsilon_0\, e^{i(kz+m\varphi)+\sigma t}; \tag{19}$$

and (2) the pressure must vanish on the same boundary. The first of these conditions clearly requires

$$-\frac{\epsilon_0\, \Pi_0}{\sigma} kI'_m(x) = \epsilon_0\, \sigma \tag{20}$$

or

$$\sigma^2 = -\frac{\Pi_0}{R} xI'_m(x). \tag{21}$$

Now the pressure in the unperturbed cylinder is given by

$$p_0 = \pi G\rho^2(R^2-\varpi^2). \tag{22}$$

The vanishing of the pressure, $p_0+\delta p$, in the perturbed cylinder, on the boundary (19) gives

$$-2\pi G\rho R\epsilon_0\, e^{i(kz+m\varphi)+\sigma t}+\left(\frac{\delta p}{\rho}\right)_R = 0; \tag{23}$$

whereas according to equations (4), (13), and (15),

$$-4\pi G\rho R\epsilon_0\, K_m(x)I_m(x)e^{i(kz+m\varphi)+\sigma t}+\left(\frac{\delta p}{\rho}\right)_R = \epsilon_0\, \Pi_0\, I_m(x)e^{i(kz+m\varphi)+\sigma t}. \tag{24}$$

Eliminating $(\delta p/\rho)_R$ from equations (23) and (24), we find

$$4\pi G\rho R[K_m(x)I_m(x)-\tfrac{1}{2}] = -\Pi_0\, I_m(x). \tag{25}$$

From equations (21) and (25) we now obtain

$$\frac{\sigma_m^2}{4\pi G\rho} = \frac{xI'_m(x)}{I_m(x)}[K_m(x)I_m(x)-\tfrac{1}{2}], \tag{26}$$

where we have written σ_m in place of σ to distinguish the modes of different m's.

From the known properties of the Bessel functions it follows that

$$K_m(x)I_m(x) < \tfrac{1}{2} \quad \text{for all } m \neq 0. \tag{27}$$

Hence

$$\sigma_m^2 < 0 \quad \text{for all } m \neq 0. \tag{28}$$

The cylinder is, therefore, stable for all purely non-axisymmetric perturbations.

Considering the case $m = 0$ (when the inequality (27) does not hold for all x), we can write (since $I'_0(x) = I_1(x)$),

$$\frac{\sigma_0^2}{4\pi G\rho} = \frac{xI_1(x)}{I_0(x)}[K_0(x)I_0(x)-\tfrac{1}{2}]. \tag{29}$$

The asymptotic behaviours of the different Bessel functions which appear in equation (29) are

$$I_0(x) \to 1, \qquad I_1(x) \to \tfrac{1}{2}x, \quad \text{and} \quad K_0(x) \to -(\gamma+\log\tfrac{1}{2}x), \quad \text{for } x \to 0, \tag{30}$$

where γ (= 0·5772...) is Euler's constant, and

$$I_0(x) \to \frac{e^x}{\sqrt{(2\pi x)}}, \qquad I_1(x) \to \frac{e^x}{\sqrt{(2\pi x)}}, \quad \text{and} \quad K_0(x) \to e^{-x}\sqrt{\frac{\pi}{2x}}$$

$$\text{for } x \to \infty. \tag{31}$$

Hence, while $I_0(x)K_0(x)$ tends to infinity logarithmically for $x \to 0$, it

TABLE LIII

The dispersion relation for the gravitationally unstable modes of an infinite cylinder

x	$\frac{\sigma_0}{\sqrt{(4\pi G\rho)}}$	x	$\frac{\sigma_0}{\sqrt{(4\pi G\rho)}}$	x	$\frac{\sigma_0}{\sqrt{(4\pi G\rho)}}$
0	0	0·40	0·2275	0·75	0·2291
0·05	0·05718	0·45	0·2362	0·80	0·2176
0·10	0·09825	0·50	0·2420	0·85	0·2023
0·15	0·1315	0·55	0·2450	0·90	0·1826
0·20	0·1590	**0·580**	**0·2455**	0·95	0·1568
0·25	0·1818	0·60	0·2453	1·00	0·1215
0·30	0·2005	0·65	0·2428	1·05	0·06216
0·35	0·2156	0·70	0·2374	1·0668	0

tends to zero as $1/(2x)$ for $x \to \infty$; moreover, $I_0(x)K_0(x)$ is a monotonic decreasing function of x. From these facts it follows that the equation

$$I_0(x)K_0(x) - \tfrac{1}{2} = 0 \tag{32}$$

allows a single positive root. If x_* denotes this root,

$$\sigma_0^2 > 0 \text{ for } 0 < x < x_* \quad \text{and} \quad \sigma_0^2 < 0 \text{ for } x > x_*. \tag{33}$$

This clearly implies that *the cylinder is gravitationally unstable for all varicose deformations with wavelengths exceeding*

$$\lambda_* = 2\pi R/x_*, \tag{34}$$

where x_* is the root of equation (32). Numerically, it is found that (see Table LIII)

$$x_* = 1{\cdot}0668. \tag{35}$$

Table LIII gives the values of $\sigma_0/\sqrt{(4\pi G\rho)}$ for values of x in the range $0 \leqslant x \leqslant 1{\cdot}067$ (see also Fig. 123). It will be observed that in this range of x, σ_0 attains a maximum value ($\sim 0{\cdot}2455\sqrt{(4\pi G\rho)}$) at $x = 0{\cdot}580$. By arguments which have been stated in § 107, it is by this mode of maximum instability that an infinite cylinder will break-up gravitationally when the amplitude of the initial disturbance tends to zero; and the characteristic time for break-up is

$$\tau = \frac{1}{0{\cdot}2455\sqrt{(4\pi G\rho)}}. \tag{36}$$

For a cylinder of radius 250 parsecs ($7{\cdot}7\times10^{20}$ cm) and density $\rho = 2\times10^{-24}$ g/cm³, the wavelength of the mode of maximum instability is $2{\cdot}7\times10^{3}$ parsecs; and the corresponding characteristic time for break-up is $1{\cdot}0\times10^{8}$ years.

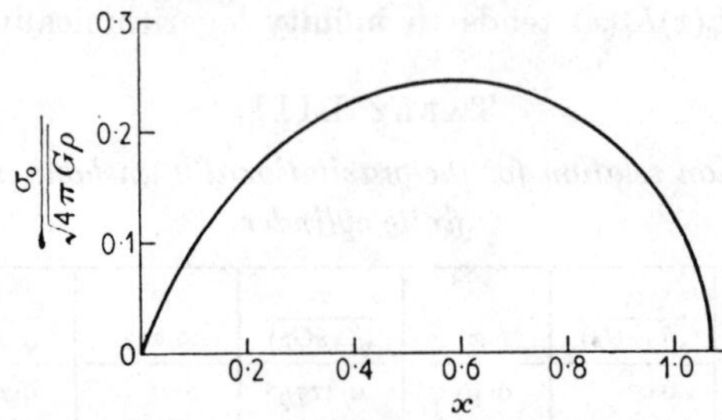

FIG. 123. The dispersion relation for the gravitationally unstable modes ($x < 1{\cdot}067$) of an infinite cylinder. The abscissa measures the wave number in the unit $1/R$ and the ordinate the growth rate in the unit $\sqrt{(4\pi G\rho)}$. The mode of maximum instability occurs for $x = 0{\cdot}580$.

(a) *The origin of the gravitational instability and an alternative method of determining the characteristic frequencies*

The origin of the gravitational instability of an infinite cylinder can be traced directly to the sign of the change in the gravitational potential energy $\Delta\mathfrak{W}$ per unit length resulting from varicose deformations of different wavelengths. We shall find that $\Delta\mathfrak{W} < 0$ for precisely those modes for which the cylinder is unstable according to equations (26) and (29).

Since the potential energy per unit length of an infinite cylinder is infinite, the evaluation of $\Delta\mathfrak{W}$ requires some care. We proceed as follows.

In evaluating $\Delta\mathfrak{W}$, it is convenient to suppose that the equation of the deformed surface is given in the real form:

$$\varpi = R+\epsilon\cos kz\cos m\varphi. \tag{37}$$

Suppose further that the amplitude of the deformation is increased by $\delta\epsilon$. The change $\delta\Delta\mathfrak{W}$ in the potential energy consequent to this infinitesimal increase in the amplitude of the deformation can be determined by evaluating the work done in the redistribution of the matter required to effect the change in ϵ. For evaluating this work it is necessary to specify in a quantitative manner the redistribution which does take place; and we shall now do this.

An arbitrary deformation of an incompressible fluid can be thought of as resulting from a Lagrangian displacement $\boldsymbol{\xi}$ applied to each point of the fluid. The incompressibility of the medium requires that $\operatorname{div}\boldsymbol{\xi} = 0$; and since no loss of generality is implied by supposing that the displacement is carried out irrotationally, we shall write

$$\boldsymbol{\xi} = \operatorname{grad}\psi \tag{38}$$

and require that

$$\nabla^2\psi = 0. \tag{39}$$

A solution of equation (39) which is suitable for considering the deformation of a uniform cylinder into one whose boundary is given by (37) is

$$\psi = AI_m(k\varpi)\cos kz\cos m\varphi, \tag{40}$$

where A is a constant. The corresponding radial component of $\boldsymbol{\xi}$ is

$$\xi_\varpi = \frac{\partial\psi}{\partial\varpi} = AkI'_m(k\varpi)\cos kz\cos m\varphi. \tag{41}$$

Since at $\varpi = R$, ξ_ϖ must reduce to $\epsilon\cos kz\cos m\varphi$,

$$A = \frac{\epsilon}{kI'_m(x)}, \tag{42}$$

where $x = kR$, as before. Hence,

$$\psi = \frac{\epsilon}{kI'_m(x)}I_m(k\varpi)\cos kz\cos m\varphi \tag{43}$$

and

$$\boldsymbol{\xi} = \frac{\epsilon}{kI'_m(x)}\operatorname{grad}\{I_m(k\varpi)\cos kz\cos m\varphi\}. \tag{44}$$

Accordingly, the displacement $\delta\boldsymbol{\xi}$ which must be applied to each point of the fluid to increase the amplitude of the deformation by an amount $\delta\epsilon$ is given by

$$\delta\boldsymbol{\xi} = \frac{\delta\epsilon}{kI'_m(x)}\operatorname{grad}\{I_m(k\varpi)\cos kz\cos m\varphi\}. \tag{45}$$

The change in the potential energy $\delta\Delta\mathfrak{W}$ per unit length involved in the additional deformation $\delta\epsilon$ can be obtained now by integrating, over the whole cylinder, the work done by the displacement $\delta\boldsymbol{\xi}$ in the gravitational potential (cf. equations (8) and (13))

$$\mathfrak{B} = -\pi G\rho\varpi^2+4\pi G\rho R\epsilon K_m(x)I_m(k\varpi)\cos kz\cos m\varphi. \tag{46}$$

Thus,

$$\delta\Delta\mathfrak{W} = -2\pi\rho\left\langle\!\!\left\langle\int\limits_0^{R+\epsilon\cos kz\cos m\varphi}(\delta\boldsymbol{\xi}\cdot\operatorname{grad}\mathfrak{B})\varpi\,d\varpi\right\rangle\!\!\right\rangle, \tag{47}$$

where the angular brackets signify that the quantity enclosed should be averaged over all z and φ.

Substituting for $\mathfrak{V}$ and $\delta\xi$ in (47) and neglecting terms of order ϵ^2 and higher, we find:

$$\delta\Delta\mathfrak{W} = -2\pi\rho\delta\epsilon\left\langle\!\!\left\langle \int\limits_0^{R+\epsilon\cos kz\cos m\varphi} (-2\pi G\rho\varpi)\frac{I'_m(k\varpi)}{I'_m(x)}\cos kz\cos m\varphi\,\varpi\,d\varpi \right\rangle\!\!\right\rangle - $$

$$-\tfrac{1}{4}(1+\delta_{0,m})4\pi G\rho R\epsilon K_m(x)\frac{2\pi\rho\delta\epsilon}{kI'_m(x)}\int\limits_0^R \left\{k^2[I'_m(k\varpi)]^2+\right.$$

$$\left.+\left(k^2+\frac{m^2}{\varpi^2}\right)I_m^2(k\varpi)\right\}\varpi\,d\varpi, \tag{48}$$

where
$$\delta_{0,m} = 1 \text{ if } m = 0 \text{ and zero otherwise.} \tag{49}$$

The reduction of the first of the two integrals in (48) is immediate and we can write

$$\delta\Delta\mathfrak{W} = (1+\delta_{0,m})\pi^2 G\rho^2 R^2\epsilon\,\delta\epsilon -$$

$$-(1+\delta_{0,m})2\pi^2 G\rho^2 R^2\frac{K_m(x)}{xI'_m(x)}\epsilon\,\delta\epsilon\int\limits_0^x \left\{[I'_m(x)]^2+\left(1+\frac{m^2}{x^2}\right)I_m^2(x)\right\}x\,dx; \tag{50}$$

and making use of the identity (which follows from Bessel's equation)

$$\frac{d}{dx}\{xI_m(x)I'_m(x)\} = x\left\{[I'_m(x)]^2+\left(1+\frac{m^2}{x^2}\right)I_m^2(x)\right\}, \tag{51}$$

we obtain

$$\delta\Delta\mathfrak{W} = -(1+\delta_{0,m})2\pi^2 G\rho^2 R^2[K_m(x)I_m(x)-\tfrac{1}{2}]\epsilon\,\delta\epsilon; \tag{52}$$

and integrating this last expression from 0 to ϵ, we finally obtain

$$\Delta\mathfrak{W} = -(1+\delta_{0,m})\pi^2 G\rho^2 R^2[K_m(x)I_m(x)-\tfrac{1}{2}]\epsilon^2. \tag{53}$$

Comparing equation (53) for $\Delta\mathfrak{W}$ with equation (26) for σ_m^2, we observe that the criterion for stability which follows from the conditions $\Delta\mathfrak{W} > 0$ and $\sigma_m^2 < 0$ is the same. It is, of course, necessary on physical grounds that the two conditions lead to the same criterion; that they must is often referred to as the *energy principle*.

The foregoing method of determining the criterion for stability by seeking the sign of the change in $\Delta\mathfrak{W}$ consequent to a particular perturbation can be extended to provide a complete analytical solution to the problem. For this purpose we suppose that the amplitude ϵ of the deformation is a function of time and, considering it as a Lagrangian coordinate, we seek an equation of motion for it by defining a suitable Lagrangian function.

When ϵ is a function of time, each element of the fluid will execute motions; these can be derived from the Lagrangian displacement $\boldsymbol{\xi}$ by

$$\mathbf{u} = \frac{\partial \boldsymbol{\xi}}{\partial t}. \tag{54}$$

With $\boldsymbol{\xi}$ given by equation (44),

$$\mathbf{u} = \frac{R}{xI'_m(x)}[\text{grad}\{I_m(k\varpi)\cos kz \cos m\varphi\}]\frac{d\epsilon}{dt}. \tag{55}$$

The kinetic energy $\mathfrak{T}$ per unit length, associated with the motions specified by (55), is

$$\mathfrak{T} = \tfrac{1}{2}\rho\left(\frac{d\epsilon}{dt}\right)^2 \frac{2\pi R^2}{x^2[I'_m(x)]^2}\int_0^R \langle\!\langle|\text{grad}\{I_m(k\varpi)\cos kz \cos m\varphi\}|^2\rangle\!\rangle \varpi\, d\varpi$$

$$= \tfrac{1}{4}(1+\delta_{0,m})\left(\frac{d\epsilon}{dt}\right)^2 \frac{\pi R^2\rho}{x^2[I'_m(x)^2]}\int_0^x \left\{[I'_m(x)]^2+\left(1+\frac{m^2}{x^2}\right)I_m^2(x)\right\}x\,dx; \tag{56}$$

or, making use of the same identity (51), we obtain

$$\mathfrak{T} = \tfrac{1}{4}(1+\delta_{0,m})\pi R^2\rho\frac{I_m(x)}{xI'_m(x)}\left(\frac{d\epsilon}{dt}\right)^2. \tag{57}$$

Combining this expression for the kinetic energy with the corresponding expression for the potential energy given by (53), we can now define the Lagrangian function

$$\mathfrak{L} = \tfrac{1}{4}(1+\delta_{0,m})\pi R^2\rho\left\{\frac{I_m(x)}{xI'_m(x)}\left(\frac{d\epsilon}{dt}\right)^2+4\pi G\rho[K_m(x)I_m(x)-\tfrac{1}{2}]\epsilon^2\right\}. \tag{58}$$

The equation of motion for ϵ which follows from this Lagrangian is

$$\frac{I_m(x)}{xI'_m(x)}\frac{d^2\epsilon}{dt^2} = 4\pi G\rho[K_m(x)I_m(x)-\tfrac{1}{2}]\epsilon. \tag{59}$$

Therefore,

$$\epsilon = \epsilon_0\, e^{\sigma t}, \tag{60}$$

where σ is given by the same formula (26). While this method of deriving σ is perhaps less direct analytically, it is more direct physically, since the origin of the instability is traced to the energetics of the problem.

109. The effect of viscosity on the gravitational instability of an infinite cylinder

We shall now consider the effect of viscosity on the gravitational instability of an infinite cylinder. Since the principal interest in this connexion is in the unstable modes, we shall restrict the present discussion to axisymmetric perturbations. It will be convenient then to use the representation of axisymmetric solenoidal vectors described in Chapter VI, § 61 (a).

When allowance is made for viscosity, equation (2) of § 108 is replaced by

$$\frac{\partial \mathbf{u}}{\partial t} = -\text{grad}\,\Pi - \nu\,\text{curl}^2\mathbf{u}, \tag{61}$$

where ν denotes the kinematic viscosity; the remaining equations (3) and (4) are unaffected.

Consider the symmetric deformation of the cylinder specified by

$$\varpi = R + \epsilon_0\, e^{ikz+\sigma t}. \tag{62}$$

Then (see equations (13) and (15)),

$$\Pi = \epsilon_0\,\Pi_0\, I_0(k\varpi)e^{ikz+\sigma t}, \tag{63}$$

where

$$\Pi = -\delta\mathfrak{V} + \delta p/\rho = -4\pi G\rho R\epsilon_0\, K_0(x) I_0(k\varpi)e^{ikz+\sigma t} + \delta p/\rho. \tag{64}$$

In the terminology of § 61 (a), grad Π is a poloidal vector with the defining scalar

$$i\epsilon_0\,\Pi_0 \frac{I_1(k\varpi)}{\varpi}\, e^{ikz+\sigma t}. \tag{65}$$

From equations (61) and the poloidal character of grad Π, we conclude that $\mathbf{u}$ is also poloidal; we can accordingly express it in the manner (cf. equation VI (201))

$$\mathbf{u} = \left\{-ik\varpi U \mathbf{1}_{\varpi} + \frac{1}{\varpi}\frac{d}{d\varpi}(\varpi^2 U)\mathbf{1}_z\right\} e^{ikz+\sigma t}, \tag{66}$$

where U is a function of ϖ only.

Remembering that the operation of curl² on a poloidal axisymmetric vector is equivalent to the operation of $-\Delta_5$ on the defining scalar, we obtain from equation (61) the scalar equation

$$\sigma U = -i\epsilon_0\,\Pi_0 \frac{I_1(k\varpi)}{\varpi} + \nu\left(\frac{d^2 U}{d\varpi^2} + \frac{3}{\varpi}\frac{dU}{d\varpi} - k^2 U\right), \tag{67}$$

or, rearranging, we have

$$\frac{d^2 U}{d\varpi^2} + \frac{3}{\varpi}\frac{dU}{d\varpi} - \left(k^2 + \frac{\sigma}{\nu}\right)U = i\frac{\epsilon_0\,\Pi_0}{\nu}\frac{I_1(k\varpi)}{\varpi}. \tag{68}$$

Making use of the relation

$$\left(\frac{d^2}{d\varpi^2} + \frac{3}{\varpi}\frac{d}{d\varpi}\right)\frac{I_1(\alpha\varpi)}{\varpi} = \alpha^2 \frac{I_1(\alpha\varpi)}{\varpi}, \tag{69}$$

we can readily write down the general solution of equation (68) which is free of singularity at the origin; it is given by

$$U = i\left[A\frac{I_1(\kappa\varpi)}{\varpi} - \frac{\epsilon_0\,\Pi_0}{\sigma}\frac{I_1(k\varpi)}{\varpi}\right], \tag{70}$$

where

$$\kappa^2 = k^2 + \sigma/\nu, \tag{71}$$

and A is a constant of integration. The corresponding solutions for the components of the velocity are

$$u_{\varpi} = k\left[AI_1(\kappa\varpi) - \frac{\epsilon_0 \Pi_0}{\sigma} I_1(k\varpi)\right] e^{ikz+\sigma t} \tag{72}$$

and

$$u_z = i\left[A\kappa I_0(\kappa\varpi) - \frac{\epsilon_0 \Pi_0}{\sigma} kI_0(k\varpi)\right] e^{ikz+\sigma t}. \tag{73}$$

The solution represented by equations (72) and (73) must satisfy certain boundary conditions. These are: (1) the radial component of the velocity u_{ϖ} must be compatible with the assumed form of the deformed boundary, namely, (62); (2) the tangential viscous stresses must vanish at $\varpi = R$; and (3) the (ϖ, ϖ)-component of the total stress tensor must vanish on the deformed boundary.

The first of the foregoing conditions gives

$$\epsilon_0 \sigma = k\left[AI_1(y) - \frac{\epsilon_0 \Pi_0}{\sigma} I_1(x)\right], \tag{74}$$

where we have used the abbreviations

$$x = kR \quad \text{and} \quad y = \kappa R. \tag{75}$$

The tangential viscous stress which must vanish at $\varpi = R$ is

$$p_{\varpi z} = \rho\nu\left(\frac{\partial u_{\varpi}}{\partial z} + \frac{\partial u_z}{\partial \varpi}\right). \tag{76}$$

With the representation (66) for $\mathbf{u}$,

$$p_{\varpi z} = \rho\nu\left\{\frac{d}{d\varpi}\left(\frac{1}{\varpi}\frac{d}{d\varpi}\varpi^2 U\right) + k^2\varpi U\right\} e^{ikz+\sigma t}$$

$$= \rho\nu\varpi\left(\frac{d^2U}{d\varpi^2} + \frac{3}{\varpi}\frac{dU}{d\varpi} + k^2 U\right) e^{ikz+\sigma t}; \tag{77}$$

and this must vanish at $\varpi = R$. For U given by (70), the condition gives

$$A(k^2+\kappa^2)I_1(y) - 2\frac{\epsilon_0 \Pi_0}{\sigma} k^2 I_1(x) = 0, \tag{78}$$

or

$$A = \frac{\epsilon_0 \Pi_0}{\sigma} \frac{2k^2}{k^2+\kappa^2} \frac{I_1(x)}{I_1(y)}. \tag{79}$$

Inserting this value of A in equation (74), we obtain

$$\sigma^2 = \frac{\Pi_0}{R} \frac{k^2-\kappa^2}{k^2+\kappa^2} x I_1(x). \tag{80}$$

It remains to satisfy the last of the boundary conditions which requires the vanishing of the (ϖ, ϖ)-component of the total stress tensor on (62). The required component of the stress tensor is

$$p_{\varpi\varpi} = p_0 + \delta p - 2\nu\rho \frac{\partial u_\varpi}{\partial \varpi}, \tag{81}$$

where p_0 is the pressure in the unperturbed configuration given by (22). Retaining only terms of the first order in ϵ, we find that the vanishing of $p_{\varpi\varpi}$ on (62) requires

$$-2\pi G\rho R\epsilon_0 e^{ikz+\sigma t} + \left(\frac{\delta p}{\rho}\right)_R - 2\nu\left(\frac{\partial u_\varpi}{\partial \varpi}\right)_R = 0. \tag{82}$$

On the other hand, from equations (63) and (64),

$$-4\pi G\rho R\epsilon_0 K_0(x)I_0(x)e^{ikz+\sigma t} + \left(\frac{\delta p}{\rho}\right)_R = \epsilon_0 \Pi_0 I_0(x)e^{ikz+\sigma t}. \tag{83}$$

Eliminating $(\delta p/\rho)_R$ between these last two equations, we obtain

$$\{4\pi G\rho R[K_0(x)I_0(x) - \tfrac{1}{2}] + \Pi_0 I_0(x)\}\epsilon_0 e^{ikz+\sigma t} = 2\nu\left(\frac{\partial u_\varpi}{\partial \varpi}\right)_R. \tag{84}$$

Now from equation (72) we find

$$\left(\frac{\partial u_\varpi}{\partial \varpi}\right)_R = k^2\left[A\frac{\kappa}{k}I_1'(y) - \frac{\epsilon_0 \Pi_0}{\sigma}I_1'(x)\right]e^{ikz+\sigma t}. \tag{85}$$

Using this result in equation (84), we obtain

$$2\nu k^2\left[A\frac{\kappa}{k}I_1'(y) - \frac{\epsilon_0 \Pi_0}{\sigma}I_1'(x)\right] = \epsilon_0 \Pi_0 I_0(x) + 4\pi G\rho R\epsilon_0[K_0(x)I_0(x) - \tfrac{1}{2}]; \tag{86}$$

and substituting for A in this equation its value (79), we obtain

$$\Pi_0 I_0(x)\left\{1 + \frac{2\nu k^2}{\sigma}\frac{I_1'(x)}{I_0(x)}\left[1 - \frac{2k\kappa}{k^2+\kappa^2}\frac{I_1(x)}{I_1(y)}\frac{I_1'(y)}{I_1'(x)}\right]\right\} = -4\pi G\rho R[K_0(x)I_0(x) - \tfrac{1}{2}]. \tag{87}$$

Finally, eliminating Π_0 between equations (80) and (87), we obtain the characteristic equation

$$\sigma^2\left\{1 + \frac{2\nu k^2}{\sigma}\frac{I_1'(x)}{I_0(x)}\left[1 - \frac{2k\kappa}{k^2+\kappa^2}\frac{I_1(x)}{I_1(y)}\frac{I_1'(y)}{I_1'(x)}\right]\right\} = -4\pi G\rho\frac{xI_1(x)}{I_0(x)}[K_0(x)I_0(x) - \tfrac{1}{2}]\frac{k^2-\kappa^2}{k^2+\kappa^2}, \tag{88}$$

where according to equations (71) and (75)

$$\sigma R^2/\nu = y^2 - x^2. \tag{89}$$

Using this last relation, we can rewrite equation (88) in the form

$$(x^2-y^2)^2\left\{1-\frac{2x^2}{x^2-y^2}\frac{I_1'(x)}{I_0(x)}\left[1-\frac{2xy}{x^2+y^2}\frac{I_1(x)}{I_1(y)}\frac{I_1'(y)}{I_1'(x)}\right]\right\}$$

$$= -J\frac{xI_1(x)}{I_0(x)}[K_0(x)I_0(x)-\tfrac{1}{2}]\frac{x^2-y^2}{x^2+y^2}, \quad (90)$$

where
$$J = \frac{4\pi G\rho}{(\nu/R^2)^2} = \frac{4\pi G\rho R^4}{\nu^2}. \quad (91)$$

Equation (90) can be written more conveniently in the form

$$2x^2(x^2+y^2)\frac{I_1'(x)}{I_0(x)}\left[1-\frac{2xy}{x^2+y^2}\frac{I_1(x)}{I_1(y)}\frac{I_1'(y)}{I_1'(x)}\right]-(x^4-y^4)$$

$$= J\frac{xI_1(x)}{I_0(x)}[K_0(x)I_0(x)-\tfrac{1}{2}]. \quad (92)$$

One general result which follows at once from equation (92) is that *the instability of the modes for* $0 < x < x_*$ $(= 1{\cdot}0668)$ *is not affected by viscosity.*

(a) The case when the effects of viscosity are dominant

Before we consider the characteristic equation (92) quite generally, it will be useful to examine the case when $\nu \to \infty$ and the effects of viscosity dominate inertia. In this limit, we may write

$$y = x+\delta \text{ and treat } \delta \text{ as small.} \quad (93)$$

Supposing that this is the case, we find that to the first order in δ, the quantity on the left-hand side of equation (92) becomes

$$4x^3\delta-4x^4\delta\frac{I_1'}{I_0}\left(\frac{I_1''}{I_1'}-\frac{I_1'}{I_1}\right), \quad (94)$$

where the argument of all the Bessel functions is now x. By making use of the known properties of the Bessel functions, we can reduce (94) to the form

$$4x^3\delta\left\{1-\frac{x}{I_0 I_1}\left[\left(1+\frac{1}{x^2}\right)I_1^2-I_0 I_1'\right]\right\} = 4x^4\delta\left\{\frac{I_0}{I_1}-\left(1+\frac{1}{x^2}\right)\frac{I_1}{I_0}\right\}. \quad (95)$$

Since in the present approximation

$$2x\delta = \sigma R^2/\nu, \quad (96)$$

the characteristic equation reduces in this limit to

$$\frac{\sigma R^2}{\nu} = J\frac{K_0 I_0-\frac{1}{2}}{2[x^2I_0^2/I_1^2-(1+x^2)]}; \quad (97)$$

or substituting for J its value (91), we have

$$\sigma = \frac{4\pi G\rho R^2}{\nu} \frac{K_0(x)I_0(x)-\frac{1}{2}}{2[x^2I_0^2(x)/I_1^2(x)-(1+x^2)]}, \tag{98}$$

where the arguments of the Bessel functions have been restored.

TABLE LIV

The dispersion relation in the limit of high viscosity for the gravitationally unstable modes of an infinite cylinder

x	$\dfrac{\sigma\nu}{4\pi G\rho R^2}$	x	$\dfrac{\sigma\nu}{4\pi G\rho R^2}$	x	$\dfrac{\sigma\nu}{4\pi G\rho R^2}$
0	∞				
0·01	0·7036	0·35	0·1284	0·75	0·03318
0·02	0·5881	0·40	0·1099	0·80	0·02651
0·05	0·4360	0·45	0·09411	0·85	0·02048
0·10	0·3222	0·50	0·08048	0·90	0·01500
0·15	0·2569	0·55	0·06859	0·95	0·01002
0·20	0·2117	0·60	0·05813	1·00	0·005474
0·25	0·1776	0·65	0·04886	1·05	0·001313
0·30	0·1506	0·70	0·04059	1·0668	0

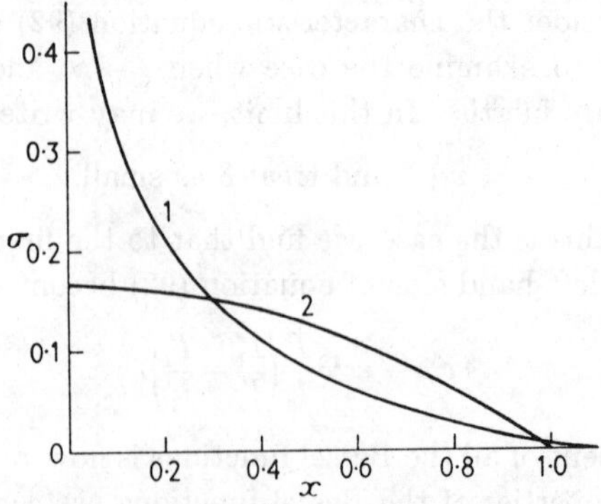

FIG. 124. The dispersion relations in the limit of high viscosity for the modes of gravitational (curve labelled 1) and capillary (curve labelled 2) instability of an infinite cylinder. The abscissa measures the wave number in the unit $1/R$ and the ordinate measures the growth rate in the unit $(4\pi G\rho R^2)/\nu$ for the gravitational modes and in the unit $T/(\rho\nu R)$ for the capillary modes.

In the limit $\nu \to \infty$, the function of x which occurs on the right-hand side of equation (98) gives σ in the unit $4\pi G\rho R^2/\nu$. Values of this function are given in Table LIV for some values of x (see also Fig. 124). We observe that in contrast to the case when viscosity is absent, there is, in this limit, no finite mode of maximum instability: σ is, in fact, a monotonic decreasing function and, moreover,

$$\sigma \to -\frac{2\pi G\rho R^2}{3\nu}(\gamma+\tfrac{1}{2}+\log\tfrac{1}{2}x) \quad \text{as } x \to 0, \tag{99}$$

where γ ($= 0{\cdot}5772\ldots$) is Euler's constant. Thus, when viscosity is paramount, the cylinder will not have a tendency to break up at points separated by distances comparable to the circumference of the cylinder; rather, it will have a tendency to just give way at a few and distant places.

TABLE LV

The function $\Phi(x) = 2[x^2I_0^2(x)/I_1^2(x)-(1+x^2)]$

x	$\Phi(x)$	x	$\Phi(x)$
0	6·00000	0·60	6·00517
0·10	6·00000	0·70	6·00942
0·20	6·00007	0·80	6·01578
0·30	6·00033	0·90	6·02478
0·40	6·00104	1·00	6·03694
0·50	6·00252	1·05	6·04437

Values of the function

$$2[x^2I_0^2(x)/I_1^2(x)-(1+x^2)] \tag{100}$$

which occurs in the denominator of the expression for σ on the right-hand side of equation (98), are given in Table LV. It is seen that for $0 \leqslant x \leqslant x_*$, the function does not depart from its value at $x = 0$ (namely, 6) by more than one per cent. Therefore, to a good approximation, we can write

$$\sigma = \frac{2\pi G\rho R^2}{3\nu}[K_0(x)I_0(x)-\tfrac{1}{2}]. \tag{101}$$

(*b*) *The general case*

We shall now show how one can explicitly obtain the dispersion relation implicitly expressed by equations (89) and (92). For an assigned value of x ($< 1{\cdot}0668$), we can derive a (σ, J)-relation by evaluating σ by equation (89) and J by equation (92) for a sequence of values of y. Relations derived in this manner for a number of assigned values of x are given in Table LVI and exhibited in Fig. 125. The way in which the (σ, J)-curves belonging to different values of x cross each other is noteworthy. By reading the curves in Fig. 125, or by interpolation among the tabulated values, we can deduce the values of σ for different values of x and a given value of J. Dispersion relations derived in this manner for different values of J are illustrated in Fig. 126. The progressive shifting of the mode of maximum instability towards smaller values of x as J decreases is apparent.

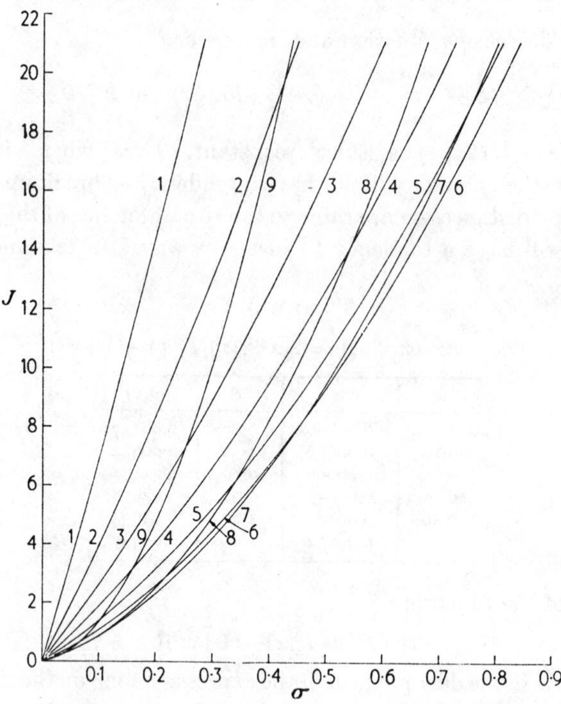

Fig. 125. The effect of viscosity on the gravitationally unstable modes of an infinite cylinder. The (σ, J)-curves for different initially assigned values of the wave number x. The curves labelled 1, 2,..., 9 are for values of $x = 0{\cdot}9$, $0{\cdot}8$, $0{\cdot}7$, $0{\cdot}6$, $0{\cdot}5$, $0{\cdot}4$, $0{\cdot}3$, $0{\cdot}2$, $0{\cdot}1$, respectively.

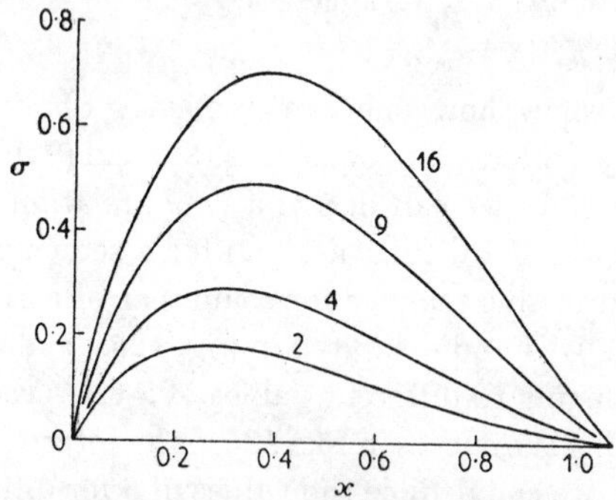

Fig. 126. The effect of viscosity on the gravitational instability of an infinite cylinder. The abscissa measures the wave number in the unit $1/R$ and the ordinate the growth rate in the unit ν/R^2. The different curves are labelled by the values of J $(= 4\pi G\rho R^4/\nu^2)$ to which they refer.

TABLE LVI

The (σ, J)-relations for various initially assigned values of x (σ is given in the unit ν/R^2)

$x = 0·1$		$x = 0·2$		$x = 0·3$		$x = 0·4$		$x = 0·5$	
σ	J	σ	J	σ	J	σ	J	σ	J
0	0	0	0	0	0	0	0	0	0
0·0125	0·054983	0·0225	0·12630	0·0325	0·24211	0·0425	0·42162	0·0525	0·69942
0·0300	0·18634	0·0500	0·33507	0·0700	0·58676	0·0900	0·97551	0·1100	1·5736
0·0525	0·44846	0·0825	0·65894	0·1125	1·0619	0·1425	1·6892	0·1725	2·6519
0·0800	0·91126	0·1200	1·1365	0·1600	1·6994	0·2000	2·5933	0·2400	3·9665
0·1125	1·6602	0·1625	1·8122	0·2125	2·5345	0·2625	3·7209	0·3125	5·5519
0·1500	2·7963	0·2100	2·7365	0·2700	3·6065	0·3300	5·1085	0·3429	6·2702
0·1925	4·4361	0·2625	3·9658	0·3325	4·9582	0·4025	6·7951	0·3900	7·4454
0·2400	6·7116	0·3200	5·5624	0·4000	6·6364	0·4800	8·8228	0·4389	8·7458
0·2925	9·7705	0·3825	7·5945	0·4725	8·6914	0·5289	10·222	0·4725	9·6867
0·3500	13·776	0·4500	10·136	0·5500	11·177	0·5625	11·237	0·5069	10·690
0·4125	18·906	0·5225	13·268	0·5824	12·305	0·5969	12·321	0·5244	11·216
0·4800	25·356	0·5684	15·466	0·6325	14·152	0·6321	13·478	0·5600	12·318
		0·6000	17·075	0·6844	16·197	0·6864	15·356	0·6525	15·384
		0·6489	19·722	0·7381	18·453	0·7425	17·417	0·7500	18·933
		0·6825	21·651	0·7749	20·083	0·8400	21·288	0·8316	22·152

$x = 0·6$		$x = 0·7$		$x = 0·8$		$x = 0·9$	
σ	J	σ	J	σ	J	σ	J
0	0	0	0	0	0	0	0
0·0625	1·1402	0·0429	1·0896	0·0161	0·61281	0·0181	1·2163
0·1024	1·9362	0·1184	3·1664	0·0324	1·2445	0·0364	2·4662
0·1300	2·5178	0·1500	4·0959	0·0489	1·8954	0·0549	3·7504
0·2025	4·1666	0·1824	5·0859	0·0656	2·5660	0·0736	5·0695
0·2484	5·3010	0·2156	6·1391	0·0996	3·9679	0·0925	6·4243
0·2961	6·5542	0·2496	7·2584	0·1344	5·4538	0·1116	7·8155
0·3289	7·4600	0·2844	8·4466	0·1521	6·2293	0·1309	9·2436
0·3625	8·4250	0·3200	9·7069	0·1700	7·0271	0·1504	10·710
0·3969	9·4519	0·3564	11·042	0·2064	8·6915	0·1701	12·214
0·4321	10·544	0·3936	12·455	0·2436	10·451	0·1900	13·757
0·4864	12·308	0·4316	13·950	0·2816	12·309	0·2304	16·965
0·5236	13·574	0·4704	15·529	0·3204	14·269	0·2716	20·337
0·5809	15·614	0·5100	17·196	0·3600	16·336		
0·6400	17·833	0·5504	18·954	0·4004	18·513		
0·7216	21·087	0·5916	20·807	0·4416	20·805		

110. The effect of a uniform axial magnetic field on the gravitational instability of an infinite cylinder

We shall now consider the effect of a uniform magnetic field along the axis of an infinite cylinder on its gravitational instability. We shall limit this present discussion to the case when the fluid is inviscid and of infinite electrical conductivity. The relevant perturbation equations are (cf. equations IV (10) and (31))

$$\frac{\partial \mathbf{u}}{\partial t} = -\operatorname{grad} \Pi + \frac{\mu H}{4\pi\rho}\frac{\partial \mathbf{h}}{\partial z}, \tag{102}$$

$$\frac{\partial \mathbf{h}}{\partial t} = \text{curl}(\mathbf{u}\times\mathbf{H}), \tag{103}$$

$$\text{div}\,\mathbf{u} = 0, \quad \text{div}\,\mathbf{h} = 0, \tag{104}$$

and
$$\Pi = -\delta\mathfrak{B}+\frac{\delta p}{\rho}+\frac{\mu H h_z}{4\pi\rho}, \tag{105}$$

where the various symbols have their usual meanings.

Since in the absence of a magnetic field the cylinder is unstable only for certain axisymmetric modes, we shall restrict our consideration of equations (102)–(105) to the axisymmetric case.

As in § 109, let the deformation of the surface be described by

$$\varpi = R+\epsilon_0\, e^{ikz+\sigma t}. \tag{106}$$

The corresponding solution for Π is (cf. equations (63) and (64))

$$\Pi = -4\pi G\rho R\epsilon_0\, K_0(x) I_0(k\varpi)e^{ikz+\sigma t}+\frac{\delta p}{\rho}+\frac{\mu H h_z}{4\pi\rho} = \epsilon_0\,\Pi_0\, I_0(k\varpi)e^{ikz+\sigma t}, \tag{107}$$

where Π_0 is a constant; and the defining scalar of grad Π is

$$i\epsilon_0\,\Pi_0\frac{I_1(k\varpi)}{\varpi}\, e^{ikz+\sigma t}. \tag{108}$$

We shall express $\mathbf{u}$ in terms of a scalar function $U(\varpi)$ in the manner (66). Then $\mathbf{u}\times\mathbf{H}$ is a toroidal vector with the defining scalar

$$ikHU(\varpi)e^{ikz+\sigma t}; \tag{109}$$

and curl$(\mathbf{u}\times\mathbf{H})$ is a poloidal vector with the same defining scalar. From equation (103) it now follows that the defining scalar of $\mathbf{h}$ is

$$\frac{ikH}{\sigma}\, U(\varpi)e^{ikz+\sigma t}; \tag{110}$$

and the components of $\mathbf{h}$ are given by

$$\mathbf{h}^{(\text{in})} = \frac{ikH}{\sigma}\left\{-ik\varpi U\mathbf{1}_{\varpi}+\frac{1}{\varpi}\frac{d}{d\varpi}(\varpi^2 U)\mathbf{1}_z\right\}e^{ikz+\sigma t}. \tag{111}$$

In writing equation (111), we have distinguished $\mathbf{h}$ by $\mathbf{h}^{(\text{in})}$ to emphasize that this is the perturbation in the magnetic field in the *interior* of the fluid. There will be a corresponding perturbation in the field *exterior* to the fluid; and since this latter perturbation is in a vacuum, it can be derived from a potential (in the same way as an irrotational velocity field) and can be expressed in the form

$$\mathbf{h}^{(\text{ex})} = A\{-K_1(k\varpi)\mathbf{1}_{\varpi}+iK_0(k\varpi)\mathbf{1}_z\}e^{ikz+\sigma t}, \tag{112}$$

where A is a constant to be determined.

Returning to equation (102) and replacing each term in the equation by its defining scalar, we obtain

$$\sigma U = -i\epsilon_0 \Pi_0 \frac{I_1(k\varpi)}{\varpi} - \frac{\mu k^2 H^2}{4\pi\rho\sigma} U. \tag{113}$$

Hence,
$$U = -\frac{i\epsilon_0 \Pi_0}{\sigma(1+\Omega_A^2/\sigma^2)} \frac{I_1(k\varpi)}{\varpi}, \tag{114}$$

where
$$\Omega_A = \sqrt{\frac{\mu H^2 k^2}{4\pi\rho}} = x\sqrt{\frac{\mu H^2}{4\pi\rho R^2}} \tag{115}$$

is the Alfvén frequency.

The solution represented by equations (111), (112), and (114) must satisfy certain boundary conditions. These are: (1) the radial component of the velocity at $\varpi = R$ must be compatible with the assumed form of the boundary; (2) the normal component of **h**, namely h_ϖ, must be continuous across the boundary; and (3) the normal component of the total stress tensor must be similarly continuous across the deformed boundary.

The first of the foregoing conditions gives (cf. equations (66) and (114))

$$u_\varpi(R) = -\frac{\epsilon_0 \Pi_0 k}{\sigma(1+\Omega_A^2/\sigma^2)} I_1(x) e^{ikz+\sigma t} = \epsilon_0 \sigma e^{ikz+\sigma t}, \tag{116}$$

where $x = kR$. Alternatively, we can write

$$\sigma^2+\Omega_A^2 = -\frac{\Pi_0}{R} x I_1(x). \tag{117}$$

From equations (111), (112), and (114), we find that the continuity of h_ϖ at $\varpi = R$ gives

$$h_\varpi^{(\mathrm{in})}(R) = -\frac{i\epsilon_0 \Pi_0 k^2 H}{\sigma^2+\Omega_A^2} I_1(x) e^{ikz+\sigma t} = -A K_1(x) e^{ikz+\sigma t} = h_\varpi^{(\mathrm{ex})}(R). \tag{118}$$

Hence,
$$A = \frac{i\epsilon_0 \Pi_0 H}{(\sigma^2+\Omega_A^2)R^2} x^2 \frac{I_1(x)}{K_1(x)}. \tag{119}$$

Therefore (see equation (112)),

$$h_z^{(\mathrm{ex})}(R) = -\frac{\epsilon_0 \Pi_0 H}{(\sigma^2+\Omega_A^2)R^2} x^2 \frac{I_1(x) K_0(x)}{K_1(x)} e^{ikz+\sigma t}. \tag{120}$$

It remains to satisfy the last of the boundary conditions, namely, the continuity of the total normal stress on the deformed boundary (106). This condition gives

$$-2\pi G\rho R\epsilon_0 e^{ikz+\sigma t} + \left(\frac{\delta p}{\rho}\right)_R + \frac{\mu H h_z^{(\mathrm{in})}(R)}{4\pi\rho} = \frac{\mu H h_z^{(\mathrm{ex})}(R)}{4\pi\rho}, \tag{121}$$

where the first term on the left-hand side is the value of the unperturbed pressure (22) on the boundary displaced by the amount $\epsilon_0 e^{ikz+\sigma t}$. On the other hand, from equation (107) it follows that

$$-4\pi G\rho R\epsilon_0 K_0(x)I_0(x)e^{ikz+\sigma t}+\left(\frac{\delta p}{\rho}\right)_R+\frac{\mu H h_z^{(\mathrm{in})}(R)}{4\pi\rho} = \epsilon_0 \Pi_0 I_0(x)e^{ikz+\sigma t}. \tag{122}$$

Eliminating $(\delta p/\rho)_R$ from equations (121) and (122) and making use of equation (120), we obtain

$$-4\pi G\rho R[K_0(x)I_0(x)-\tfrac{1}{2}] = \Pi_0 I_0(x)+\frac{\mu\Pi_0 H^2x^2}{4\pi\rho R^2(\sigma^2+\Omega_A^2)}\frac{I_1(x)K_0(x)}{K_1(x)}, \tag{123}$$

or, recalling the definition of Ω_A^2 and rearranging, we have

$$\frac{\Pi_0}{R}I_0(x)\left[1+\frac{\Omega_A^2}{\sigma^2+\Omega_A^2}\frac{I_1(x)K_0(x)}{I_0(x)K_1(x)}\right] = -4\pi G\rho[K_0(x)I_0(x)-\tfrac{1}{2}]. \tag{124}$$

Equations (117) and (124) now give

$$\sigma^2+\Omega_A^2\left\{1+\frac{I_1(x)K_0(x)}{I_0(x)K_1(x)}\right\} = 4\pi G\rho\frac{xI_1(x)}{I_0(x)}[K_0(x)I_0(x)-\tfrac{1}{2}], \tag{125}$$

or, making use of the relation

$$I_0(x)K_1(x)+I_1(x)K_0(x) = \frac{1}{x}, \tag{126}$$

we finally obtain

$$\sigma^2 = 4\pi G\rho\frac{xI_1(x)}{I_0(x)}[K_0(x)I_0(x)-\tfrac{1}{2}]-\frac{\mu H^2}{4\pi\rho R^2}\frac{x}{I_0(x)K_1(x)}. \tag{127}$$

Defining

$$H_G = 4\pi\rho R\sqrt{(G/\mu)}, \tag{128}$$

we can rewrite the characteristic equation (127) in the form

$$\sigma^2 = 4\pi G\rho\left\{\frac{xI_1(x)}{I_0(x)}[K_0(x)I_0(x)-\tfrac{1}{2}]-\left(\frac{H}{H_G}\right)^2\frac{x}{I_0(x)K_1(x)}\right\}. \tag{129}$$

The functions

$$\frac{xI_1(x)}{I_0(x)}[K_0(x)I_0(x)-\tfrac{1}{2}] \quad \text{and} \quad \frac{x}{I_0(x)K_1(x)} \tag{130}$$

which occur in equations (129) are listed in Table LVII. Also, it may be noted in this connexion that for $R = 250$ parsecs and $\rho = 2\times10^{-24}$ g/cm³, $H_G = 5\times10^{-6}$ gauss; this is the order of magnitude of the magnetic field believed to prevail in the spiral arms of galaxies.

(a) *The nature of the stabilizing effect of the axial magnetic field*

From the sign of the term in H^2 in equation (129) it is apparent that the axial magnetic field will have the effect of increasing the wavelength at which instability occurs; but *no magnetic field, however strong, can stabilize the cylinder for disturbances of all wavelengths*: *the gravitational instability of sufficiently long waves will persist.* The reason for this lies, of course, in the logarithmic singularity of $\Delta\mathfrak{W}$ for wavelengths tending to infinity; and we can never compensate for this by any increase in the magnetic energy which follows a deformation.

In general, the wavelength at which instability occurs will be determined by the root of the equation

$$\frac{xI_1(x)}{I_0(x)}[K_0(x)I_0(x)-\tfrac{1}{2}] = \left(\frac{H}{H_G}\right)^2 \frac{x}{I_0(x)K_1(x)}. \tag{131}$$

From the tabulation of the functions of x which occur on the two sides of this equation in Table LVII, it is clear that the equation allows, for

TABLE LVII

The auxiliary functions

$$\psi(x) = \frac{xI_1(x)}{I_0(x)}[K_0(x)I_0(x)-\tfrac{1}{2}] \quad \text{and} \quad \chi(x) = \frac{x}{I_0(x)K_1(x)}$$

x	ψ	χ	x	ψ	χ
0	0	0			
0·05	0·003269	0·002510	0·55	0·06004	0·34888
0·10	0·009654	0·01012	0·60	0·06017	0·42172
0·15	0·01729	0·02303	0·65	0·05894	0·50259
0·20	0·02528	0·04146	0·70	0·05638	0·59175
0·25	0·03305	0·06569	0·75	0·05250	0·68941
0·30	0·04021	0·09600	0·80	0·04734	0·79580
0·35	0·04650	0·13267	0·85	0·04094	0·91110
0·40	0·05174	0·17601	0·90	0·03334	1·03550
0·45	0·05579	0·22630	0·95	0·02460	1·1692
0·50	0·05858	0·28383	1·00	0·01475	1·3122

an assigned value of H/H_G, a single positive root (exclusive of the root $x = 0$). If x_* denotes this root, then

$$\sigma^2 < 0 \text{ for } x > x_* \quad \text{and} \quad \sigma^2 > 0 \text{ for } 0 < x < x_*. \tag{132}$$

The cylinder is, therefore, unstable for all varicose deformations with wavelengths exceeding $2\pi R/x_*$, where x_* now depends on the strength of the impressed magnetic field through H/H_G. And, as in the absence of the field, there is a mode of maximum instability which occurs at a wave number, x_m, where the quantity on the right-hand side of equation (129) attains its maximum value.

In Table LVIII, the values of x_*, x_m, and $\sigma/\sqrt{(4\pi G\rho)}$ at x_m are listed for a few values of H/H_G. This table exhibits the strong tendency towards stabilization which the magnetic field exerts. This is shown, especially, by the very rapid decrease in σ, at its maximum, as the strength of the impressed magnetic field is increased. The correspondingly rapid increase in the characteristic times for break-up which this entails is really an exponential one, as we shall now show.

Since $x_m = 0{\cdot}055$ already for $H = H_G$, for $H > H_G$ we can replace the Bessel functions which occur in equation (131) by the dominant terms in their respective series expansions for $x \to 0$. Thus, we obtain

$$\tfrac{1}{2}+\gamma+\log\tfrac{1}{2}x = -2\left(\frac{H}{H_G}\right)^2. \tag{133}$$

Hence, $$x_* = 2\exp\{-(\gamma+\tfrac{1}{2}+2H^2/H_G^2)\}, \tag{134}$$

or, numerically, $$x_* = 0{\cdot}6811e^{-2(H/H_G)^2} \quad (H > H_G). \tag{135}$$

TABLE LVIII

The wave numbers x_* and x_m at which instability occurs and at which it is a maximum

H/H_G	x_*	x_m	$\dfrac{\sigma(x_m)}{\sqrt{(4\pi G\rho)}}$
0	1·067	0·580	0·246
0·25	0·7899	0·452	0·204
0·50	0·4460	0·266	0·129
0·75	0·2205	0·134	0·0669
1·00	0·0910	0·055	0·0278

The corresponding solution for x_m can be obtained by a similar consideration of equation (129). For $x \ll 1$, this equation gives

$$\frac{\sigma^2}{4\pi G\rho} = -\tfrac{1}{2}x^2(\gamma+\tfrac{1}{2}+\log\tfrac{1}{2}x)-x^2\frac{H^2}{H_G^2}. \tag{136}$$

The expression on the right-hand side attains its maximum value when

$$\gamma+\tfrac{1}{2}+\log\tfrac{1}{2}x+\tfrac{1}{2}+2(H/H_G)^2 = 0. \tag{137}$$

Hence $$x_m = 2\exp\{-(\gamma+1)-2H^2/H_G^2\} = 0{\cdot}4131e^{-2(H/H_G)^2}. \tag{138}$$

We find from equation (136) that the corresponding expression for $\sigma_{\max}$ is

$$\frac{\sigma_{\max}}{\sqrt{(4\pi G\rho)}} = \tfrac{1}{2}x_m = 0{\cdot}2065e^{-2(H/H_G)^2}. \tag{139}$$

Equations (135), (138), and (139) (valid for $H > H_G$) emphasize the facts which are already apparent from Table LVIII.

In illustration of the foregoing results, we may consider the following numerical example. An infinite cylinder of radius 250 parsecs and density 2×10^{-24} g/cm^3, will, in the absence of a magnetic field, break-up into pieces of lengths of the order of $2{\cdot}7\times10^3$ parsecs in a time of the order of $1{\cdot}0\times10^8$ years (see p. 520). If there should be an axial magnetic field of strength $7{\cdot}5\times10^{-6}$ gauss ($=1{\cdot}5H_G$), the corresponding figures are $3{\cdot}4\times10^5$ parsecs and $1{\cdot}1\times10^{10}$ years. The fact that the characteristic times for break-up can be prolonged by such prodigious factors is of considerable astronomical interest.

111. The capillary instability of a liquid jet

We shall now consider the stability of a liquid jet held together in its equilibrium state by capillary forces. This is the classical problem of Plateau.

The analysis of the preceding sections can be adapted to this problem with only slight modifications. Thus, in the inviscid case, equations (2) and (3) governing departures from the equilibrium state are, of course, of general validity; only, in place of equation (4), we now have

$$\Pi = \delta p/\rho, \tag{140}$$

since in the present case there is no additional contribution derived from a potential.

A normal mode of perturbation of the jet can again be specified by defining the shape of the boundary as by equation (5). And it is clear, also, that the arguments leading to the solutions (15) and (18) and the result (21) are unaffected. However, the analysis following equation (21) which seeks to determine Π_0 is not valid now: for, in the unperturbed state the pressure p_0 inside the jet is a constant and is

$$p_0 = T/R, \tag{141}$$

where T denotes the surface tension. (We are restricting our present discussion to the case when the jet is in free space and the external pressure can be ignored; in § (*b*) below we consider a case in which the situation is exactly the reverse.) According to equation X (24), the condition we must satisfy in the perturbed state is that on the deformed boundary,

$$p_0+\delta p = T\left(\frac{1}{R_1}+\frac{1}{R_2}\right), \tag{142}$$

where R_1 and R_2 are the principal radii of curvature of the surface defined by equation (5). Clearly,

$$\begin{aligned}\frac{1}{R_1}+\frac{1}{R_2} &= \frac{1}{R+\epsilon e^{i(kz+m\varphi)}}+\epsilon\left(k^2+\frac{m^2}{R^2}\right)e^{i(kz+m\varphi)}\\ &= \frac{1}{R}-\frac{1}{R^2}(1-m^2-x^2)\epsilon e^{i(kz+m\varphi)}.\end{aligned} \tag{143}$$

Hence,

$$\left(\frac{\delta p}{\rho}\right)_R = -\frac{T}{R^2\rho}(1-m^2-x^2)\epsilon e^{i(kz+m\varphi)}, \tag{144}$$

whereas according to equations (15) and (140)

$$\left(\frac{\delta p}{\rho}\right)_R = \epsilon \Pi_0 I_m(x) e^{i(kz+m\varphi)}. \tag{145}$$

Therefore,
$$\frac{\Pi_0}{R} = -\frac{T}{R^3\rho}\frac{1-m^2-x^2}{I_m(x)}. \tag{146}$$

Inserting this expression for Π_0/R in equation (21), we obtain

$$\sigma_m^2 = \frac{T}{R^3\rho}\frac{xI'_m(x)}{I_m(x)}(1-m^2-x^2); \tag{147}$$

in particular,
$$\sigma_0^2 = \frac{T}{R^3\rho}\frac{xI_1(x)}{I_0(x)}(1-x^2). \tag{148}$$

These formulae were first derived by Rayleigh.

TABLE LIX

The dispersion relation for the modes of capillary instability of an infinite cylinder

x	$\frac{\sigma_0}{\sqrt{(T/R^3\rho)}}$	x	$\frac{\sigma_0}{\sqrt{(T/R^3\rho)}}$	x	$\frac{\sigma_0}{\sqrt{(T/R^3\rho)}}$
0	0				
0·05	0·03531	0·40	0·2567	0·70	0·3433
0·10	0·07031	0·45	0·2807	0·75	0·3393
0·15	0·10472	0·50	0·3016	0·80	0·3269
0·20	0·1382	0·55	0·3189	0·85	0·3036
0·25	0·1705	0·60	0·3321	0·90	0·2647
0·30	0·2012	0·65	0·3406	0·95	0·1992
0·35	0·2301	**0·697**	**0·3433**	1·00	0

From equations (147) and (148) it is apparent that

$$\begin{gathered}\sigma_m^2 < 0 \quad \text{for all } m \neq 0 \text{ and } x > 0 \\ \sigma_0^2 < 0 \text{ for } x > 1 \quad \text{and} \quad \sigma_0^2 > 0 \text{ for } 0 < x < 1.\end{gathered} \tag{149}$$

In other words: *the liquid jet is stable for all purely non-axisymmetric deformations; but it is unstable for symmetric varicose deformations with wavelengths exceeding the circumference of the cylinder.* This last result is due to Plateau.

The values of $\sigma_0/\sqrt{(T/R^3\rho)}$ for $0 \leqslant x \leqslant 1$ are given in Table LIX. It is seen that in this case the mode of maximum instability occurs for $x = 0{\cdot}697$; the corresponding wavelength of the varicose deformation is $2\pi R/0{\cdot}697 = 4{\cdot}51\times 2R$. And the value of σ_0 at its maximum is $0{\cdot}3433\sqrt{(T/R^3\rho)}$. According to this last result, a water jet of diameter 1 cm has a characteristic time of break-up of about one-eighth sec.

(a) The origin of the capillary instability

As in § 108 (*a*), we can trace the origin of the capillary instability of a liquid jet to the sign of the change in the potential energy consequent to a deformation.

For the deformation specified by equation (37), the volume per unit length of the cylinder is given by

$$\text{volume per unit length} = \pi\langle\langle\varpi^2\rangle\rangle = \pi\{R^2+\tfrac{1}{4}(1+\delta_{0,m})\epsilon^2\}, \quad (150)$$

where $\delta_{0,m}$ has the meaning given in equation (49). From equation (150) it is clear that to the second order in ϵ the radius R_0 of the unperturbed cylinder cannot be the same as the mean radius R of the deformed cylinder: the two radii must, in fact, be connected by the relation

$$R_0^2 = R^2+\tfrac{1}{4}(1+\delta_{0,m})\epsilon^2,$$

or

$$R = R_0-\tfrac{1}{8}(1+\delta_{0,m})\frac{\epsilon^2}{R}. \quad (151)$$

Now the potential energy of a system arising from capillary forces is simply proportional to the total superficial area. For the deformation given by equation (37), this area per unit length is:

superficial area per unit length

$$= 2\pi\left\langle\left\langle\varpi\left\{1+\left(\frac{\partial\varpi}{\partial z}\right)^2+\left(\frac{\partial\varpi}{\varpi\partial\varphi}\right)^2\right\}^{\frac{1}{2}}\right\rangle\right\rangle$$

$$= 2\pi R+\tfrac{1}{4}\pi(1+\delta_{0,m})(k^2R^2+m^2)\frac{\epsilon^2}{R}; \quad (152)$$

or, expressing R in terms of R_0, we have:

$$\text{superficial area per unit length} = 2\pi R_0+\tfrac{1}{4}\pi(1+\delta_{0,m})(x^2+m^2-1)\frac{\epsilon^2}{R}, \quad (153)$$

where $x = kR$. The change in the surface energy $\Delta\mathfrak{S}$ per unit length consequent on the deformation is, therefore, given by

$$\Delta\mathfrak{S} = \tfrac{1}{4}\pi T(1+\delta_{0,m})(x^2+m^2-1)\frac{\epsilon^2}{R}. \quad (154)$$

(It is not necessary to distinguish any longer between R and R_0.)

Comparing equation (154) for $\Delta\mathfrak{S}$ with equation (147) for σ_m^2, we observe that the criterion for stability which follows from the conditions $\Delta\mathfrak{S} > 0$ and $\sigma_m^2 < 0$ is the same. Indeed, by combining (154) with the kinetic energy of the associated motions given by (57), we can define a Lagrangian which will lead to the same formula for σ_m^2 as (147).

(b) The capillary instability of a hollow jet

In formulating the boundary condition in the manner (142), we are neglecting the inertia of the fluid which may be outside the jet. This is clearly permissible in the case of a liquid jet in air; but there are cases in which this is not permissible and it is the inertia of the fluid outside, rather than that of the fluid inside, that is important. An example is provided by a jet of air injected into water. The stability of such a *hollow jet* can be discussed very similarly to solid jets. Indeed, it is clear that the corresponding formula for σ_m^2 will involve the K-functions in place of the I-functions. Thus, for the important axisymmetric modes, $m = 0$, we will have

$$\sigma_0^2 = \frac{T}{R^3\rho}\frac{xK_1(x)}{K_0(x)}(1-x^2) \tag{155}$$

TABLE LX

The dispersion relation for the modes of capillary instability of a hollow jet

x	$\dfrac{\sigma_0}{\sqrt{(T/R^3\rho)}}$	x	$\dfrac{\sigma_0}{\sqrt{(T/R^3\rho)}}$	x	$\dfrac{\sigma_0}{\sqrt{(T/R^3\rho)}}$
0	0	0·20	0·7233	0·60	0·8021
0·001	0·3774	0·25	0·7548	0·65	0·7822
0·002	0·3975	0·30	0·7797	0·70	0·7534
0·005	0·4298	0·35	0·7985	0·75	0·7144
0·01	0·4601	0·40	0·8115	0·80	0·6626
0·02	0·4979	0·45	0·8186	0·85	0·5942
0·05	0·5647	**0·484**	**0·8201**	0·90	0·5017
0·10	0·6340	0·50	0·8197	0·95	0·3665
0·15	0·6840	0·55	0·8144	1·00	0

instead of (148). The square root of the function of x which occurs on the right-hand side of equation (155) is listed in Table LX (see also Fig. 127). We observe that the mode of maximum instability occurs at a somewhat longer wavelength than for a solid jet of the same diameter; and the characteristic time for break-up is about $2\frac{1}{2}$ times shorter.

(c) *The effect of viscosity*

The effect of viscosity on the capillary instability of a liquid jet can be discussed in the manner of § 109: in fact, a large part of the analysis carries over without any modifications. Thus, except for the definition of Π in equation (64), the analysis up to and including equation (80) holds. The condition that $p_{\varpi\varpi}$ given by equation (81) should vanish on the deformed boundary is now replaced by (cf. equation (142))

$$p_{\varpi\varpi}^{(R)} = \left(p+\delta p-2\nu\rho\frac{\partial u_\varpi}{\partial\varpi}\right)_R = T\left(\frac{1}{R_1}+\frac{1}{R_2}\right). \tag{156}$$

Making use of equation (144) (for the case $m = 0$) and equation (140), we find that equation (84) is replaced by

$$\left\{\frac{T}{R^2\rho}(1-x^2)+\Pi_0\, I_0(x)\right\}\epsilon_0\, e^{ikz+\sigma t} = 2\nu\left(\frac{\partial u_\varpi}{\partial \varpi}\right)_R. \tag{157}$$

Continuing the analysis beyond this point exactly as in § 109 (using equation (157) instead of equation (84)), we shall finally obtain (cf. equations (91) and (92))

$$2x^2(x^2+y^2)\frac{I_1'(x)}{I_0(x)}\left[1-\frac{2xy}{x^2+y^2}\frac{I_1(x)}{I_1(y)}\frac{I_1'(y)}{I_1'(x)}\right]-(x^4-y^4) = J\frac{xI_1(x)}{I_0(x)}(1-x^2), \tag{158}$$

where, now,

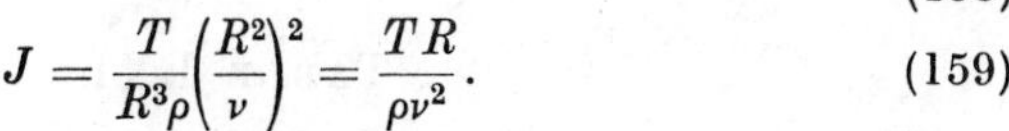

$$J = \frac{T}{R^3\rho}\left(\frac{R^2}{\nu}\right)^2 = \frac{TR}{\rho\nu^2}. \tag{159}$$

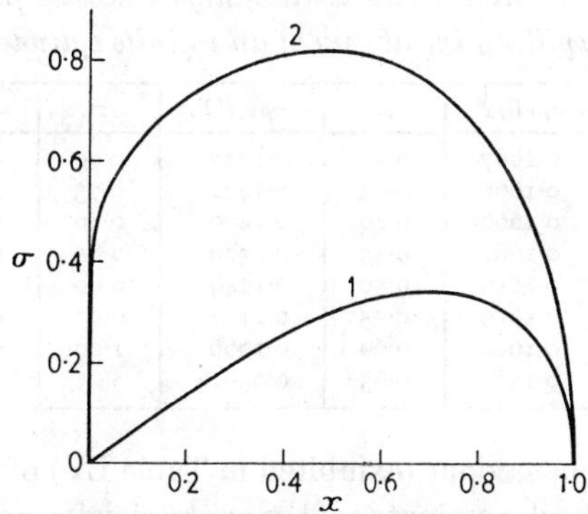

Fig. 127. The dispersion relations for the modes of capillary instability ($x < 1$) of a solid (curve labelled 1) and a hollow (curve labelled 2) jet. The abscissa measures the wave number in the unit $1/R$ and the ordinate the growth rate in the unit $\sqrt{(T/R^3\rho)}$. The mode of maximum instability occurs at $x = 0{\cdot}697$ for a solid jet and at $x = 0{\cdot}484$ for a hollow jet.

We observe that equation (158) differs from equation (92) only by a redefinition of J and the replacement of $(K_0\, I_0-\frac{1}{2})$ (which is proportional to $\Delta\mathfrak{W}$) by $(1-x^2)$ (which is proportional to $\Delta\mathfrak{S}$). This identity between the two equations, apart from the differences mentioned, is the present analogue (in the context of a cylindrical geometry) of the theorem of § 99.

When viscosity is paramount, the equation for σ which is analogous to (98) is

$$\sigma = \frac{T}{\rho\nu R}\,\frac{1-x^2}{2[x^2 I_0^2(x)/I_1^2(x)-(1+x^2)]}. \tag{160}$$

The dependence of σ on x predicted by this equation is exhibited in Table LXI. We again observe that in this limit there is no finite mode of maximum instability (see Fig. 124); and this means that under the circumstances, the jet will not break-up into 'drops' at mutual distances comparable to the circumference of the cylinder; rather, it will have a tendency to just give way at a few and distant places. This is in agreement with the observed behaviour of highly viscous threads of glass or treacle. A separation of a liquid jet into numerous drops, or varicosity suggesting such a separation, may be taken as an indication that 'fluidity has been sufficient to bring inertia into play'.

TABLE LXI

The dispersion relation in the limit of high viscosity for the modes of capillary instability of an infinite cylinder

x	$\sigma\rho\nu R/T$	x	$\sigma\rho\nu R/T$	x	$\sigma\rho\nu R/T$
0	0·1667	0·30	0·1517	0·70	0·08487
0·01	0·1666	0·35	0·1462	0·75	0·07277
0·02	0·1666	0·40	0·1400	0·80	0·05984
0·05	0·1662	0·45	0·1329	0·85	0·04610
0·10	0·1650	0·50	0·1249	0·90	0·03154
0·15	0·1629	0·55	0·1162	0·95	0·01617
0·20	0·1600	0·60	0·1066	1·00	0
0·25	0·1562	0·65	0·09614		

In view of the behaviour (exhibited in Table LV) of the function of x which occurs in the denominator of the expression for σ given by equation (160), we may, to a sufficient approximation, write

$$\sigma = \frac{T}{6\rho\nu R}(1-x^2). \tag{161}$$

More generally, the dispersion relation implicitly expressed by equation (158) can be exhibited in the manner described in § 109 in the case of gravitational instability. The (σ, J)-relations which follow from equations (89) and (158) for different assigned values of x are given in Table LXII and illustrated in Fig. 128; and the dispersion relations for different J's are shown in Fig. 129.

112. The effect of a uniform axial magnetic field on the capillary instability of a liquid jet

We shall now consider the effect of a uniform magnetic field along the axis of a liquid jet on its capillary instability. If we limit ourselves to the case when the liquid is of zero viscosity and resistivity, the effect of

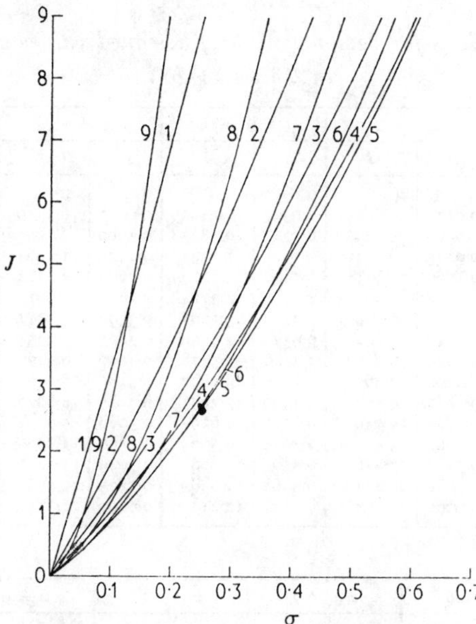

Fig. 128. The effect of viscosity on the modes of capillary instability of an infinite cylinder. The (σ, J)-curves for different initially assigned values of the wave number x. The curves labelled 1, 2,..., 9 are for values of $x = 0{\cdot}9, 0{\cdot}8, 0{\cdot}7, 0{\cdot}6, 0{\cdot}5, 0{\cdot}4, 0{\cdot}3, 0{\cdot}2, 0{\cdot}1$, respectively.

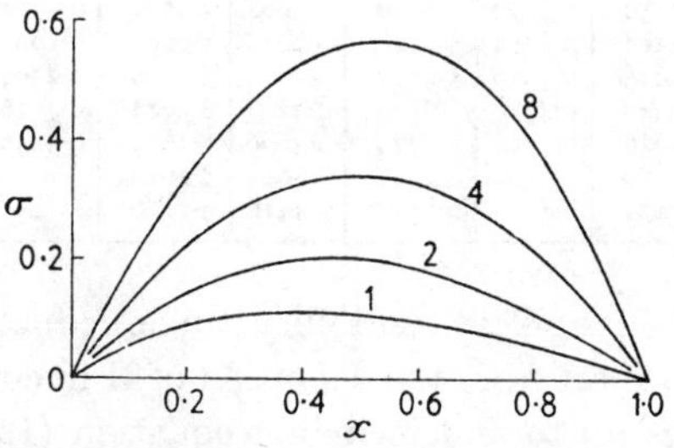

Fig. 129. The effect of viscosity on capillary instability of an infinite cylinder. The abscissa measures the wave number in the unit $1/R$ and the ordinate the growth rate in the unit ν/R^2. The different curves are labelled by the values of J $(= TR/\rho\nu^2)$ to which they refer.

Table LXII

The (σ, J)-relations for various initially assigned values of x (σ is given in the unit ν/R^2)

x = 0·1		x = 0·2		x = 0·3		x = 0·4		x = 0·5	
σ	J	σ	J	σ	J	σ	J	σ	J
0	0	0	0	0	0	0	0	0	0
0·0125	0·10736	0·0225	0·16712	0·0325	0·24039	0·0425	0·33105	0·0525	0·45052
0·0300	0·36386	0·0500	0·44336	0·0700	0·58258	0·0900	0·76596	0·1100	1·0136
0·0525	0·87570	0·0825	0·87190	0·1125	1·0544	0·1425	1·3264	0·1725	1·7082
0·0800	1·7794	0·1200	1·5038	0·1600	1·6873	0·2000	2·0362	0·2400	2·5550
0·0989	2·5779	0·1625	2·3979	0·2125	2·5164	0·2625	2·9216	0·3125	3·5762
0·1125	3·2418	0·2100	3·6209	0·2700	3·5808	0·3300	4·0111	0·3429	4·0389
0·1269	4·0264	0·2625	5·2475	0·3325	4·9229	0·4025	5·3354	0·3900	4·7959
0·1500	5·4602	0·2849	6·0293	0·3589	5·5478	0·4800	6·9276	0·4389	5·6336
0·1749	7·2475	0·3200	7·3601	0·4000	6·5892	0·5289	8·0260	0·4725	6·2395
0·1925	8·6622	0·3569	8·8982	0·4429	7·7651	0·5625	8·8229	0·5069	6·8857
0·2400	13·105	0·3825	10·049	0·4725	8·6295	0·5969	9·6742	0·5244	7·2245
0·2925	19·078	0·4500	13·412	0·5029	9·5623	0·6500	11·059	0·5600	7·9344
0·3500	26·900	0·5225	17·556	0·5500	11·098	0·7049	12·581	0·6525	9·9095
		0·5684	20·464	0·5661	11·648	0·7425	13·676	0·7500	12·195
		0·6000	22·594	0·5824	12·217	0·8400	16·715		

x = 0·6		x = 0·7		x = 0·8		x = 0·9	
σ	J	σ	J	σ	J	σ	J
0	0	0	0	0	0	0	0
0·0625	0·62194	0·0429	0·52118	0·0161	0·27148	0·0181	0·57866
0·1024	1·0561	0·0725	0·89906	0·0324	0·55131	0·0364	1·1733
0·1300	1·3734	0·1184	1·5146	0·0489	0·83967	0·0549	1·7843
0·2025	2·2727	0·1500	1·9592	0·0656	1·1368	0·0736	2·4119
0·2484	2·8915	0·1824	2·4327	0·0825	1·4427	0·0925	3·0565
0·2961	3·5751	0·2156	2·9365	0·0996	1·7578	0·1116	3·7184
0·3289	4·0691	0·2496	3·4718	0·1344	2·4161	0·1309	4·3979
0·3625	4·5955	0·2844	4·0402	0·1700	3·1131	0·1504	5·0953
0·3969	5·1556	0·3200	4·6430	0·2064	3·8504	0·1701	5·8110
0·4144	5·4488	0·3564	5·2816	0·2436	4·6298	0·1900	6·5454
0·4500	6·0625	0·3936	5·9577	0·2816	5·4528	0·2304	8·0713
0·5049	7·0540	0·4316	6·6726	0·3204	6·3212	0·2716	9·6760
0·5425	7·7646	0·4704	7·4279	0·3600	7·2368	0·3136	11·362
0·5809	8·5169	0·5100	8·2252	0·4004	8·2013		
0·6400	9·7271	0·5916	9·9524	0·4416	9·2166		

the magnetic field on capillary instability can be deduced from the analysis of § 110. Thus, apart from the definition of Π in equations (105) and (107), the analysis up to and including equation (120) holds. But in place of equations (121) and (122) we now have

$$\left(\frac{\delta p}{\rho}+\frac{\mu H h_z^{(\mathrm{in})}}{4\pi\rho}\right)_R = \left(\frac{\mu H h_z^{(\mathrm{ex})}}{4\pi\rho}\right)_R - \epsilon_0 \frac{T}{R^2\rho}(1-x^2)e^{ikz+\sigma t} \tag{162}$$

and

$$\left(\frac{\delta p}{\rho}+\frac{\mu H h_z^{(\mathrm{in})}}{4\pi\rho}\right)_R = \epsilon_0\,\Pi_0\,I_0(x)e^{ikz+\sigma t}; \tag{163}$$

using these equations, we find

$$\frac{\Pi_0}{R} I_0(x)\left[1+\frac{\Omega_A^2}{\sigma^2+\Omega_A^2}\frac{I_1(x)K_0(x)}{I_0(x)K_1(x)}\right] = -\frac{T}{R^3\rho}(1-x^2) \tag{164}$$

instead of equation (124). Continuing the analysis beyond this point exactly as in § 110, we shall finally obtain (in place of equations (128) and (129))

$$\sigma^2 = \frac{T}{R^3\rho}\left\{\frac{xI_1(x)}{I_0(x)}(1-x^2)-\left(\frac{H}{H_s}\right)^2\frac{x}{I_0(x)K_1(x)}\right\}, \tag{165}$$

where

$$H_s = \sqrt{\left(\frac{T}{R^3\rho}\frac{4\pi\rho R^2}{\mu}\right)} = \sqrt{\frac{4\pi T}{R\mu}}. \tag{166}$$

The magnetic field has clearly the effect of increasing the wavelength at which capillary instability occurs; but unlike in the case of gravitational instability, the capillary instability can be suppressed completely by a magnetic field of sufficient strength. This difference in the two cases arises from the different behaviours of $\Delta\mathfrak{S}$ (which is proportional to $1-x^2$) and $\Delta\mathfrak{W}$ (which is proportional to $I_0 K_0-\frac{1}{2}$) in the limit of long wavelengths. Specifically, since

$$I_1(x)K_1(x) \leqslant \tfrac{1}{2} \quad \text{for } x \geqslant 0, \tag{167}$$

magnetic fields of strengths

$$H > \frac{1}{\sqrt{2}}H_s = \sqrt{\frac{2\pi T}{R\mu}} \tag{168}$$

will stabilize the jet for varicose deformations of all wavelengths.

With the known surface tension of mercury ($T = 470$ dynes/cm), the minimum field according to (168) is $77/\sqrt{R}$ gauss: a very small field, indeed. However, we shall see below that the finite electrical conductivity of mercury alters this expectation completely.

(a) *The effect of finite electrical conductivity*

Since the resistivity η of metallic liquids such as mercury is very high ($\sim 10^4$) under normal terrestrial conditions, it is important to ascertain how the finiteness of the resistivity affects the possibility of suppressing the capillary instability of a liquid jet by an axial magnetic field.

When we allow for finite resistivity, the relevant perturbation equations are:

$$\frac{\partial \mathbf{u}}{\partial t} = -\text{grad}\,\Pi + \frac{\mu H}{4\pi\rho}\frac{\partial \mathbf{h}}{\partial z} \tag{169}$$

and

$$\frac{\partial \mathbf{h}}{\partial t} = \text{curl}(\mathbf{u}\times\mathbf{H}) - \eta\,\text{curl}^2\mathbf{h}, \tag{170}$$

where

$$\Pi = \frac{\delta p}{\rho}+\frac{\mu H h_z}{4\pi\rho}, \tag{171}$$

and **u** and **h** are solenoidal vectors. In seeking solutions of the foregoing equations, we shall again restrict ourselves to axisymmetric perturbations and consider the normal mode of deformation specified by equation (106).

The solution for Π is again given by equation (63); and in the representation of axisymmetric solenoidal vectors used in the preceding sections, the defining scalar of grad Π is (again, as before)

$$i\epsilon_0\,\Pi_0\frac{I_1(k\varpi)}{\varpi}e^{ikz+\sigma t}. \tag{172}$$

We shall now represent **u** and **h** in terms of two functions $U(\varpi)$ and $P(\varpi)$ in the manner (cf. equation (66))

$$\mathbf{u} = \left\{-ik\varpi U\mathbf{1}_\varpi+\frac{1}{\varpi}\frac{d}{d\varpi}(\varpi^2 U)\mathbf{1}_z\right\}e^{ikz+\sigma t} \tag{173}$$

and

$$\mathbf{h}^{(\text{in})} = \left\{-ik\varpi P\mathbf{1}_\varpi+\frac{1}{\varpi}\frac{d}{d\varpi}(\varpi^2 P)\mathbf{1}_z\right\}e^{ikz+\sigma t}, \tag{174}$$

where we have distinguished **h** by a superscript '(in)' to indicate that this is the perturbation in the field in the interior of the fluid. (We must also consider the perturbation in the field exterior to the fluid; and we shall do this presently.)

Returning to equations (169) and (170) and replacing each of the terms in the equations by its respective defining scalar, we obtain

$$\sigma U = -i\epsilon_0\,\Pi_0\frac{I_1(k\varpi)}{\varpi}+\frac{ik\mu H}{4\pi\rho}P \tag{175}$$

and

$$\sigma P = ikHU+\eta\Delta_5 P, \tag{176}$$

where, for functions with the chosen dependence on z,

$$\Delta_5 = \frac{d^2}{d\varpi^2}+\frac{3}{\varpi}\frac{d}{d\varpi}-k^2. \tag{177}$$

Eliminating U between equations (175) and (176), we obtain

$$\eta\Delta_5 P-\sigma P = -\frac{ikH}{\sigma}\left\{-i\epsilon_0\,\Pi_0\frac{I_1(k\varpi)}{\varpi}+\frac{ik\mu H}{4\pi\rho}P\right\}. \tag{178}$$

Inserting for Δ_5 its expression (175) and rearranging, we have

$$\left(\frac{d^2}{d\varpi^2}+\frac{3}{\varpi}\frac{d}{d\varpi}\right)P-\left\{k^2+\frac{1}{\sigma\eta}(\sigma^2+\Omega_A^2)\right\}P = -\epsilon_0\,\Pi_0\frac{kH}{\sigma\eta}\frac{I_1(k\varpi)}{\varpi}, \tag{179}$$

where Ω_A denotes the Alfvén frequency defined in equation (115).

Making use of the relation (69), we can at once write down the solution of equation (179) which is free of singularity at the origin; it is

$$P = \epsilon_0 H\left\{A\frac{I_1(\kappa\varpi)}{\varpi}+\frac{k\Pi_0}{\sigma^2+\Omega_A^2}\frac{I_1(k\varpi)}{\varpi}\right\}, \tag{180}$$

where A is a constant and

$$\kappa^2 = k^2+\frac{1}{\sigma\eta}(\sigma^2+\Omega_A^2). \tag{181}$$

From equation (175), we find that the corresponding solution for U is

$$U = -i\frac{\epsilon_0\,\Pi_0\,\sigma}{\sigma^2+\Omega_A^2}\frac{I_1(k\varpi)}{\varpi}+i\epsilon_0 A\frac{\mu H^2 k}{4\pi\rho\sigma}\frac{I_1(\kappa\varpi)}{\varpi}. \tag{182}$$

With P given by equation (180) the components of $\mathbf{h}^{(\mathrm{in})}$ are:

$$h_\varpi^{(\mathrm{in})} = -ik\epsilon_0 H\left\{AI_1(\kappa\varpi)+\frac{k\Pi_0}{\sigma^2+\Omega_A^2}I_1(k\varpi)\right\}e^{ikz+\sigma t}, \tag{183}$$

$$h_z^{(\mathrm{in})} = \epsilon_0 H\left\{A\kappa I_0(\kappa\varpi)+\frac{k^2\Pi_0}{\sigma^2+\Omega_A^2}I_0(k\varpi)\right\}e^{ikz+\sigma t}. \tag{184}$$

The corresponding perturbation in the magnetic field exterior to the fluid is derivable from a potential (being a vacuum field) and we can write (cf. equation (112))

$$h_\varpi^{(\mathrm{ex})} = i\epsilon_0\,BHK_1(k\varpi)e^{ikz+\sigma t} \tag{185}$$

and

$$h_z^{(\mathrm{ex})} = \epsilon_0\,BHK_0(k\varpi)e^{ikz+\sigma t}, \tag{186}$$

where B is a constant.

Finally, we may note that the radial component of $\mathbf{u}$ is given by (cf. equations (173) and (182))

$$u_\varpi = \epsilon_0\left\{-\frac{k\sigma\Pi_0}{\sigma^2+\Omega_A^2}I_1(k\varpi)+A\frac{\Omega_A^2}{\sigma}I_1(\kappa\varpi)\right\}e^{ikz+\sigma t}. \tag{187}$$

The solution represented by equations (183)–(187) must satisfy certain boundary conditions. These are: (1) the velocity u_ϖ (given by equation (187)) must be compatible, at $\varpi = R$, with the assumed form of the deformed boundary; (2) the field $\mathbf{h}$ must be continuous across the boundary; and (3) the normal component of the total stress must also be continuous across the boundary.

The first of the foregoing conditions gives

$$\frac{k\Pi_0}{\sigma^2+\Omega_A^2}I_1(x)-A\frac{\Omega_A^2}{\sigma^2}I_1(y) = -1, \tag{188}$$

where

$$x = kR \quad\text{and}\quad y = \kappa R. \tag{189}$$

And the continuity of $\mathbf{h}$ across the boundary requires the components of $\mathbf{h}$ given by equations (183) and (184) to agree with those given by equations (185) and (186) at $\varpi = R$. These conditions give

$$B = -\frac{1}{K_1(x)}\left\{AkI_1(y)+\frac{k^2\Pi_0}{\sigma^2+\Omega_A^2}I_1(x)\right\}$$

$$= +\frac{1}{K_0(x)}\left\{A\kappa I_0(y)+\frac{k^2\Pi_0}{\sigma^2+\Omega_A^2}I_0(x)\right\}. \tag{190}$$

From the equality of the two different expressions for B in (190), we find

$$A\{\kappa I_0(y)K_1(x)+kI_1(y)K_0(x)\}$$

$$= -\frac{k^2\Pi_0}{\sigma^2+\Omega_A^2}\{I_1(x)K_0(x)+I_0(x)K_1(x)\} = -\frac{k^2\Pi_0}{x(\sigma^2+\Omega_A^2)}. \tag{191}$$

Thus,

$$A = -\frac{k\Pi_0}{\Phi(\sigma^2+\Omega_A^2)}, \tag{192}$$

where

$$\Phi = yI_0(y)K_1(x)+xK_0(x)I_1(y). \tag{193}$$

With A given by equation (192), equation (188) becomes

$$\frac{k\Pi_0}{\sigma^2+\Omega_A^2}\left[I_1(x)+\frac{\Omega_A^2}{\sigma^2\Phi}I_1(y)\right] = -1. \tag{194}$$

Hence,

$$\frac{\Pi_0}{R} = -\frac{\sigma^2+\Omega_A^2}{xI_1(x)}\frac{1}{1+(\Omega_A^2/\sigma^2)I_1(y)/\Phi I_1(x)}. \tag{195}$$

It remains to satisfy the last of the boundary conditions, namely, the continuity of the normal component of the total stress on the deformed boundary. This condition leads to the same equations (162) and (163); from these equations we obtain (by subtraction)

$$\left(\frac{\mu H h_z^{(\mathrm{ex})}}{4\pi\rho}\right)_R - \epsilon_0\frac{T}{R^2\rho}(1-x^2)e^{ikz+\sigma t} = \epsilon_0\,\Pi_0 \cdot I_0(x)e^{ikz+\sigma t}. \tag{196}$$

Now substituting for $h_z^{(\mathrm{ex})}$ $(= h_z^{(\mathrm{in})})$ from equation (184), and rearranging, we obtain

$$\frac{T}{R^2\rho}(1-x^2) = \frac{\mu H^2}{4\pi\rho}\left\{A\kappa I_0(y)+\frac{k^2\Pi_0}{\sigma^2+\Omega_A^2}I_0(x)\right\}-\Pi_0 I_0(x)$$

$$= -\frac{\sigma^2\Pi_0}{\sigma^2+\Omega_A^2}I_0(x)+\frac{\kappa}{k^2}A\Omega_A^2 I_0(y). \tag{197}$$

Inserting for A in this last equation its value (192), we obtain

$$\frac{T}{R^3\rho}(1-x^2) = -\frac{\Pi_0}{R}\frac{\sigma^2 I_0(x)}{\sigma^2+\Omega_A^2}\left[1+\frac{\Omega_A^2}{\sigma^2\Phi}\frac{yI_0(y)}{xI_0(x)}\right]. \tag{198}$$

Finally, eliminating Π_0/R between equations (195) and (198), we find

$$\sigma^2\left[1+\frac{\Omega_A^2}{\sigma^2\Phi}\frac{yI_0(y)}{xI_0(x)}\right] = \frac{T}{R^3\rho}\left[1+\frac{\Omega_A^2}{\sigma^2\Phi}\frac{I_1(y)}{I_1(x)}\right]\frac{xI_1(x)}{I_0(x)}(1-x^2), \tag{199}$$

where it may be recalled that (see equation (181))

$$y^2 = x^2+\frac{R^2}{\eta\sigma}(\sigma^2+\Omega_A^2) \tag{200}$$

and

$$\Omega_A^2 = \frac{\mu H^2}{4\pi\rho R^2}x^2. \tag{201}$$

In the limit of zero resistivity,

$$y\to\infty,\quad \Phi\to yI_0(y)K_1(x)\quad (\eta\to 0) \tag{202}$$

and equation (199) tends to

$$\sigma^2 = \frac{T}{R^3\rho}\frac{xI_1(x)}{I_0(x)}(1-x^2)-\Omega_A^2\frac{1}{xI_0(x)K_1(x)}\qquad (\eta\to 0), \tag{203}$$

in agreement with equation (165). On the other hand, in the limit of infinite resistivity,

$$y\to x,\qquad \Phi\to 1, \tag{204}$$

and

$$\sigma^2\to\frac{T}{R^3\rho}\frac{xI_1(x)}{I_0(x)}(1-x^2)\quad (\eta\to\infty); \tag{205}$$

in this limit the magnetic field has no effect on the capillary instability as should, indeed, be the case.

(b) The case of high resistivity

When the resistivity η is high but not infinite, we may write

$$y = x+\delta \quad\text{and treat } \delta \text{ as a small quantity.} \tag{206}$$

To the first order in δ, equation (199) becomes

$$\sigma^2\left[1+\delta\frac{\Omega_A^2}{\sigma^2+\Omega_A^2}\frac{d}{dx}\log(xI_0)\right] = \frac{T}{R^3\rho}\frac{xI_1(x)}{I_0(x)}(1-x^2)\left[1+\delta\frac{\Omega_A^2}{\sigma^2+\Omega_A^2}\frac{d}{dx}\log I_1\right], \tag{207}$$

where all the Bessel functions have now the argument x. An equivalent form of equation (207) is

$$\sigma^2\left[1+\delta\frac{\Omega_A^2}{\sigma^2+\Omega_A^2}\left\{\frac{d}{dx}\log(xI_0)-\frac{d}{dx}\log I_1\right\}\right] = \frac{T}{R^3\rho}\frac{xI_1}{I_0}(1-x^2). \tag{208}$$

By making use of the recurrence relations satisfied by the Bessel functions, we find

$$\frac{d}{dx}\log(xI_0)-\frac{d}{dx}\log I_1 = \frac{I_0+xI_1}{xI_0}-\frac{I_1'}{I_1} = \frac{I_1}{I_0}-\frac{xI_1'-I_1}{xI_1} = \frac{I_1}{I_0}-\frac{I_2}{I_1}. \tag{209}$$

Moreover, in the present approximation (cf. equations (200) and (206))

$$2x\delta = \frac{R^2}{\eta\sigma}(\sigma^2+\Omega_A^2). \tag{210}$$

We thus obtain

$$\sigma^2\left[1+\frac{\mu H^2}{4\pi\rho\eta}\frac{x(I_1^2-I_0 I_2)}{2I_1 I_0}\frac{1}{\sigma}\right] = \frac{T}{R^3\rho}\frac{xI_1}{I_0}(1-x^2). \tag{211}$$

In considering equation (211), it is convenient to measure σ in the unit

$$\sigma_S = \sqrt{\frac{T}{R^3\rho}}\,\text{sec}^{-1}; \tag{212}$$

TABLE LXIII

The function $f(x) = \dfrac{x(I_1^2-I_0 I_2)}{4I_0 I_1}$

x	$f(x)$	x	$f(x)$
0	0		
0·1	0·000624	0·6	0·02092
0·2	0·002479	0·7	0·02776
0·3	0·00552	0·8	0·03524
0·4	0·00968	0·9	0·04321
0·5	0·01485	1·0	0·05155

equation (211) then takes the form

$$\sigma^2\left(1+x\frac{I_1^2-I_0 I_2}{2I_1 I_0}\frac{\mathscr{Q}}{\sigma}\right) = \frac{xI_1}{I_0}(1-x^2), \tag{213}$$

where

$$\mathscr{Q} = \frac{\mu H^2}{4\pi\rho\eta}\sqrt{\frac{R^3\rho}{T}} = \frac{\mu H^2}{4\pi\eta}\sqrt{\frac{R^3}{\rho T}}. \tag{214}$$

For the validity of the present approximation, it is clearly necessary that the second term in parentheses on the left-hand side of equation (213) is small compared to unity. Assuming that this is the case, we obtain from equation (213) the following approximate formula for the roots

$$\sigma = \pm\sqrt{\frac{xI_1(1-x^2)}{I_0}} - x\frac{I_1^2-I_0 I_2}{4I_0 I_1}\mathscr{Q}. \tag{215}$$

The function of x which occurs as the factor of $\mathscr{Q}$ in this equation is listed briefly in Table LXIII. This function, multiplied by $\mathscr{Q}$, must be subtracted from the corresponding entries in Table LIX. From the relative values of the two functions listed in Tables LIX and LXIII, we find that a value of $\mathscr{Q} \sim 6$ is needed to shift the critical wave number at which capillary instability occurs from 1·0 to 0·9. From this it would appear that values of $\mathscr{Q}$ as high as 20 or more will be needed to overcome significantly the paramount effects of finite resistivity.

With the values of the physical constants appropriate for mercury and liquid sodium, we find

$$\mathcal{Q} = 1{\cdot}33\times 10^{-7}H^2R^{\frac{3}{2}} \quad \text{for Hg with } T = 470 \text{ dyne/cm},$$
$$\eta = 7{\cdot}5\times 10^3 \text{ cm}^2/\text{sec}, \quad \text{and} \quad \rho = 13{\cdot}5 \text{ g/cm}^3; \tag{216}$$

and

$$\mathcal{Q} = 7{\cdot}4\times 10^{-5}H^2R^{\frac{3}{2}} \quad \text{for Na with } T = 200 \text{ dyne/cm},$$
$$\eta = 7{\cdot}6\times 10^2 \text{ cm}^2/\text{sec}, \quad \text{and} \quad \rho = 1 \text{ g/cm}^3. \tag{217}$$

We might conclude from these values that for experiments with mercury, for example, magnetic field strengths of the order of 10^4 gauss, or more, will be needed to demonstrate the stabilizing effect of an axial magnetic field.

(c) *The general case*

While the first order formulae of § (*b*) would suggest that under ordinary laboratory conditions it will be difficult to overcome the effects of finite resistivity, it may still be useful to outline how one might treat equations (199) and (200) generally.

Measuring H and σ in the units specified in equations (166) and (212), we first rewrite equations (199) and (200) in the non-dimensional forms

$$\sigma^2\left[1+\frac{H^2}{\sigma^2}\frac{xyI_0(y)}{\Phi I_0(x)}\right] = \frac{xI_1(x)}{I_0(x)}(1-x^2)\left[1+\frac{H^2}{\sigma^2}\frac{x^2I_1(y)}{\Phi I_1(x)}\right] \tag{218}$$

and

$$y^2 = x^2+\frac{S}{\sigma}(\sigma^2+x^2H^2), \tag{219}$$

where

$$S = \frac{R^2}{\eta}\sqrt{\frac{T}{R^3\rho}} = \frac{1}{\eta}\sqrt{\frac{TR}{\rho}}. \tag{220}$$

For a given set of values of H, x, and y, equation (218) can be solved as a quadratic for σ^2; and equation (219) can then be used to determine S. Keeping H and x fixed and allowing y to run through a sequence of values, we can derive a (σ, S)-relation (to be associated with the chosen values of H and x). Interpolating among the (σ, S)-relations for different values of x, but the same value of H, we can obtain the (σ, x)-relation appropriate to some fixed value of S and the chosen value of H. In this manner, the dispersion relation for any set of initial conditions can be deduced.

113. The stability of the simplest solution of the equations of hydromagnetics

Some interesting examples in the stability of cylindrical systems arise in connexion with the simplest solution (cf. Chapter IV, § 40 (*a*)),

$$u_j = \pm H_j\sqrt{\frac{\mu}{4\pi\rho}} \quad \text{and} \quad \Pi = \frac{p}{\rho}+\mathfrak{B}+\frac{\mu|\mathbf{H}|^2}{8\pi\rho} = \text{constant}, \tag{221}$$

of the equations governing an incompressible inviscid fluid of infinite electrical conductivity. Before we consider these examples, we shall first establish the stability of the solution (221) generally.

Writing
$$h_j = \pm\left(\frac{\mu}{4\pi\rho}\right)^{\frac{1}{2}} H_j, \tag{222}$$
we have the equations
$$\frac{\partial u_j}{\partial t}+\frac{\partial}{\partial x_k}(u_j u_k - h_j h_k) = -\frac{\partial \Pi}{\partial x_j} \tag{223}$$
and
$$\frac{\partial h_j}{\partial t}+\frac{\partial}{\partial x_k}(h_j u_k - u_j h_k) = 0. \tag{224}$$

In addition we have the equations expressing the solenoidal characters of **u** and **h**.

Let
$$u_j = h_j = U_j, \quad \frac{\partial U_j}{\partial x_j} = 0, \quad \text{and} \quad \Pi = \text{constant} \tag{225}$$
represent the initial stationary state; and let
$$U_j + u_j, \quad U_j + h_j, \quad \text{and} \quad \Pi + P \tag{226}$$
represent a slightly perturbed state. The equations governing the perturbations u_j, h_j, and P are
$$\frac{\partial u_j}{\partial t}-\frac{\partial}{\partial x_k}(U_k \xi_j + U_j \xi_k) = -\frac{\partial P}{\partial x_j} \tag{227}$$
and
$$\frac{\partial h_j}{\partial t}+\frac{\partial}{\partial x_k}(U_k \xi_j - U_j \xi_k) = 0, \tag{228}$$
where
$$\xi_j = h_j - u_j. \tag{229}$$
From equations (227) and (228) we obtain, by subtraction,
$$\frac{\partial \xi_j}{\partial t}+2U_k\frac{\partial \xi_j}{\partial x_k} = \frac{\partial P}{\partial x_j}. \tag{230}$$
In addition,
$$\frac{\partial \xi_j}{\partial x_j} = 0. \tag{231}$$

For solutions with a dependence on time given by
$$e^{i\sigma t}, \tag{232}$$
equation (230) becomes
$$i\sigma\xi_j = -2U_k\frac{\partial \xi_j}{\partial x_k}+\frac{\partial P}{\partial x_j}. \tag{233}$$

By multiplying equation (233) by ξ_j^* (and summing over j as the notation implies) and integrating over the volume V occupied by the fluid, we obtain

$$i\sigma \int\limits_V |\xi_j|^2\,dV = -2\int\limits_V U_k\,\xi_j^*\frac{\partial \xi_j}{\partial x_k}\,dV + \int\limits_V \frac{\partial}{\partial x_j}(P\xi_j^*)\,dV$$

$$= -2\int\limits_V U_k\,\xi_j^*\frac{\partial \xi_j}{\partial x_k}\,dV + \int\limits_S P\xi_j^*\,dS_j, \tag{234}$$

where S is the surface bounding V. We shall suppose that the conditions on the boundary are such that the surface integral in (234) vanishes. (It will suffice if either P or the normal component of $\boldsymbol{\xi}$ vanishes on S.) Equation (234) will then give

$$i\sigma \int\limits_V |\xi_j|^2\,dV = -2\int\limits_V U_k\,\xi_j^*\frac{\partial \xi_j}{\partial x_k}\,dV. \tag{235}$$

The complex conjugate of this equation is

$$-i\sigma^* \int\limits_V |\xi_j|^2\,dV = -2\int\limits_V U_k\,\xi_j\frac{\partial \xi_j^*}{\partial x_k}\,dV. \tag{236}$$

Adding equations (235) and (236), we obtain

$$i(\sigma-\sigma^*)\int\limits_V |\xi_j|^2\,dV = -2\int\limits_V \frac{\partial}{\partial x_k}(U_k|\xi_j|^2)\,dV = -2\int\limits_S |\xi_j|^2 U_k\,dS_k. \tag{237}$$

Again, we shall suppose that the initial conditions and/or the boundary conditions are such that the surface integral in (237) vanishes. (It will suffice if the normal component of $\mathbf{U}$ vanishes on S.) Equation (237) will then give

$$\mathrm{im}(\sigma)\int\limits_V |\xi_j|^2\,dV = 0. \tag{238}$$

It follows that σ is real; and this establishes the stability of the solution (225).

With the knowledge that σ is real, we can derive a variational basis for its determination. For this purpose, we write

$$\xi_j = \xi_j^{(1)} + i\sigma\xi_j^{(2)} \quad \text{and} \quad P = P^{(1)} + i\sigma P^{(2)} \tag{239}$$

and separate the real and the imaginary parts of equation (233). We then obtain the pair of equations

$$\xi_j^{(1)} = -2U_k\frac{\partial \xi_j^{(2)}}{\partial x_k} + \frac{\partial P^{(2)}}{\partial x_j} \tag{240}$$

and

$$\sigma^2\xi_j^{(2)} = +2U_k\frac{\partial \xi_j^{(1)}}{\partial x_k} - \frac{\partial P^{(1)}}{\partial x_j}. \tag{241}$$

(Both $\boldsymbol{\xi}^{(1)}$ and $\boldsymbol{\xi}^{(2)}$ are, of course, solenoidal.)

By multiplying equation (241) by $\xi_j^{(2)}$ and integrating over the volume of the fluid, we obtain

$$\sigma^2 \int\limits_V |\xi_j^{(2)}|^2\,dV$$
$$= 2\int\limits_V \xi_j^{(2)}\frac{\partial}{\partial x_k}(U_k\,\xi_j^{(1)})\,dV - \int\limits_V \frac{\partial}{\partial x_j}(P^{(1)}\xi_j^{(2)})\,dV$$
$$= -2\int\limits_V \xi_j^{(1)}\,U_k\frac{\partial\xi_j^{(2)}}{\partial x_k}\,dV + 2\int\limits_S \xi_j^{(1)}\,\xi_j^{(2)}\,U_k\,dS_k - \int\limits_S P^{(1)}\xi_j^{(2)}\,dS_j$$
$$= \int\limits_V |\xi_j^{(1)}|^2\,dV - \int\limits_V \frac{\partial}{\partial x_j}(P^{(2)}\xi_j^{(1)})\,dV$$
$$= \int\limits_V |\xi_j^{(1)}|^2\,dV - \int\limits_S P^{(2)}\xi_j^{(1)}\,dS_j = \int\limits_V |\xi_j^{(1)}|^2\,dV, \tag{242}$$

if we suppose that the conditions on the boundary are such that the various surface integrals which have occurred during the reductions vanish. Thus,

$$\sigma^2 = \frac{\int\limits_V |\xi_j^{(1)}|^2\,dV}{\int\limits_V |\xi_j^{(2)}|^2\,dV}. \tag{243}$$

It can be shown in the usual manner that equation (243) provides the basis for a variational formulation of the problem. However, it is important to note that the extremal property of σ^2 is true only if, when performing the variation of the quantity on the right-hand side of equation (243), we treat equation (240) as a constraint which must be satisfied by the variations as well.

(a) An example

To illustrate in a concrete manner how the stability of the equipartition solutions arises, we shall consider an example for which σ^2 can be explicitly found. In the example, the lines of magnetic force and the streamlines of the fluid particles are helices of pitch $2\pi p$.

The stationary solution we consider, in cylindrical polar coordinates, is

$$\mathbf{u} = \mathbf{H}\left(\frac{\mu}{4\pi\rho}\right)^{\frac{1}{2}} = \mathbf{U} = U\left(0,\frac{\varpi}{p},1\right), \tag{244}$$

where U is a constant. We shall further suppose that the fluid is confined in a cylinder of radius R.

We have seen that in general the perturbation equations can be reduced to the form (230). In cylindrical coordinates this equation becomes:

$$\frac{\partial \xi_\varpi}{\partial t}+2(\mathbf{U}\,.\,\mathrm{grad})\xi_\varpi-2\frac{U_\varphi \xi_\varphi}{\varpi} = \frac{\partial P}{\partial \varpi}, \tag{245}$$

$$\frac{\partial \xi_\varphi}{\partial t}+2(\mathbf{U}\,.\,\mathrm{grad})\xi_\varphi+2\frac{U_\varphi \xi_\varpi}{\varpi} = \frac{1}{\varpi}\frac{\partial P}{\partial \varphi}, \tag{246}$$

and
$$\frac{\partial \xi_z}{\partial t}+2(\mathbf{U}\,.\,\mathrm{grad})\xi_z = \frac{\partial P}{\partial z}, \tag{247}$$

where
$$\mathbf{U}\,.\,\mathrm{grad} = U_\varpi\frac{\partial}{\partial \varpi}+\frac{U_\varphi}{\varpi}\frac{\partial}{\partial \varphi}+U_z\frac{\partial}{\partial z}. \tag{248}$$

In addition we have the divergence condition

$$\frac{\partial \xi_\varpi}{\partial \varpi}+\frac{\xi_\varpi}{\varpi}+\frac{1}{\varpi}\frac{\partial \xi_\varphi}{\partial \varphi}+\frac{\partial \xi_z}{\partial z} = 0. \tag{249}$$

Analysing the disturbance into normal modes, we seek solutions whose (t, φ, z)-dependence is given by

$$e^{i(\sigma t+m\varphi+kz)}. \tag{250}$$

For $\mathbf{U}$ given by (244) and solutions having the chosen dependence on the coordinates,

$$\frac{\partial}{\partial t}+2\mathbf{U}\,.\,\mathrm{grad} = i\left\{\sigma+2U\left(\frac{m}{p}+k\right)\right\} = i\omega \quad \text{(say)}, \tag{251}$$

and the perturbation equations become

$$i\omega\xi_\varpi-2\frac{U}{p}\xi_\varphi = DP, \tag{252}$$

$$i\omega\xi_\varphi+2\frac{U}{p}\xi_\varpi = \frac{im}{\varpi}P, \tag{253}$$

$$i\omega\xi_z = ikP, \tag{254}$$

and
$$D_*\xi_\varpi+\frac{im}{\varpi}\xi_\varphi+ik\xi_z = 0, \tag{255}$$

where
$$D = d/d\varpi \quad \text{and} \quad D_* = D+1/\varpi. \tag{256}$$

From equations (252) and (253), we find

$$\left(\omega^2-4\frac{U^2}{p^2}\right)\xi_\varpi = -i\left(\omega DP+2\frac{mU}{p}\frac{P}{\varpi}\right) \tag{257}$$

and
$$\left(\omega^2-4\frac{U^2}{p^2}\right)\xi_\varphi = +\left(2\frac{U}{p}DP+m\omega\frac{P}{\varpi}\right). \tag{258}$$

Substituting for the components of $\boldsymbol{\xi}$ (in accordance with equations (254), (257), and (258)) in equation (255) and making use of the identity

$$D_*\frac{1}{\varpi} \equiv \frac{1}{\varpi}D, \tag{259}$$

we obtain
$$D_*DP+\left\{\frac{k^2}{\omega^2}\left(\frac{4U^2}{p^2}-\omega^2\right)-\frac{m^2}{\varpi^2}\right\}P = 0. \tag{260}$$

The solution of equation (260) which is free of singularity at the origin is

$$P = AJ_m(\kappa\varpi), \tag{261}$$

where A is a constant, J_m is Bessel's function of order m, and

$$\kappa^2 = k^2\left(\frac{4U^2}{p^2\omega^2}-1\right). \tag{262}$$

The boundary condition requires that the total pressure Π be continuous on the deformed surface. In this instance, since Π is a constant, the condition† requires that the perturbation in Π (namely, P) vanishes for $\varpi = R$. Therefore,

$$J_m(\kappa R) = 0. \tag{263}$$

Accordingly, if $x_{m,j}$ denotes the jth root of $J_m(x)$,

$$\kappa^2R^2 = x^2\left(\frac{4U^2}{p^2\omega^2}-1\right) = x^2_{m,j}, \tag{264}$$

where $x = kR$. From equation (264) we obtain (cf. equation (251))

$$\frac{\omega}{2U} = \frac{\sigma}{2U}+\left(\frac{m}{p}+k\right) = \pm\frac{x}{p\sqrt{(x^2+x^2_{m,j})}} \tag{265}$$

or
$$\sigma = 2\frac{U}{R}\left\{\pm\frac{lx}{\sqrt{(x^2+x^2_{m,j})}}-(x+lm)\right\}, \tag{266}$$

where
$$l = R/p \tag{267}$$

is a measure of the reciprocal pitch of the helix in the unit $1/R$.

The frequencies of oscillation which follow from equation (266) are clearly real; and this establishes the stability of the system considered.

114. The effect of fluid motions on the stability of helical magnetic fields

In § 113 we have seen how the equipartition solution is stable under very general conditions. The essential characteristic of the solution is

† We are passing over complications which might arise from a detailed consideration of the perturbation in the field outside the cylinder. Such complications can be formally avoided by considering U to be of the form $(0, \varpi g(\varpi), 1)$ where $g(\varpi) = 1/p$ for $0 \leqslant \varpi \leqslant R$ and $= R^2/p\varpi^2$ for $\varpi \geqslant R$. This latter form for U has been considered by Trehan and Reid (and by Roberts in a different connexion; see references (15) and (17) in the Bibliographical Notes at the end of the chapter). And the results of Trehan and Reid do not differ in their qualitative aspects from the solution obtained in the text.

that in the stationary state both magnetic fields and fluid motions are present. It is, therefore, of interest to see, somewhat more generally than in the case of the equipartition solution, how fluid motions affect the stability of magnetic field configurations. A simple example of such more general configurations is when **u** and **H** are parallel to each other (as in the equipartition solution) but the constant of proportionality is such that equipartition of energy does not obtain.

When **u** and **h** are stationary and are parallel to each other, equation (224) governing the magnetic field is satisfied identically; but equation (223) cannot be satisfied so simply except when equipartition prevails. However, equation (223) can be satisfied under special circumstances; thus

$$\mathbf{u} = U\left(0, \frac{\varpi}{p}, 1\right) \quad \text{and} \quad \mathbf{h} = H\left(0, \frac{\varpi}{p}, 1\right) \tag{268}$$

(where U and H are constants) satisfy equations IX (1)–(3) (with the terms in ν suppressed) provided

$$\frac{d\Pi}{d\varpi} = \frac{\varpi}{p^2}\left(U^2 - \frac{\mu H^2}{4\pi\rho}\right) \tag{269}$$

or

$$\Pi = \text{constant} + \frac{\varpi^2}{2p^2}\left(U^2 - \frac{\mu H^2}{4\pi\rho}\right). \tag{270}$$

The stability of these general helical fields was first considered by Trehan.

Returning to equations (223) and (224), we obtain by addition and subtraction the symmetrical pair of equations

$$\frac{\partial Y_j}{\partial t} - X_k \frac{\partial Y_j}{\partial x_k} = -\frac{\partial \Pi}{\partial x_j} \tag{271}$$

and

$$\frac{\partial X_j}{\partial t} + Y_k \frac{\partial X_j}{\partial x_k} = +\frac{\partial \Pi}{\partial x_j}, \tag{272}$$

where

$$X_j = h_j - u_j \quad \text{and} \quad Y_j = h_j + u_j. \tag{273}$$

Let equations (271) and (272) allow a stationary solution of the form

$$X_j = \alpha_2 U_j \quad \text{and} \quad Y_j = \alpha_1 U_j, \tag{274}$$

where α_1 and α_2 are constants. (We shall presently identify **U** with a helical field of the type (268).) Let ξ_j, η_j, and P represent a first order perturbation of X_j, Y_j, and Π, respectively. The equations governing the perturbations are

$$\frac{\partial \xi_j}{\partial t} + \alpha_1 U_k \frac{\partial \xi_j}{\partial x_k} + \alpha_2 \eta_k \frac{\partial U_j}{\partial x_k} = +\frac{\partial P}{\partial x_j} \tag{275}$$

and

$$\frac{\partial \eta_j}{\partial t} - \alpha_2 U_k \frac{\partial \eta_j}{\partial x_k} - \alpha_1 \xi_k \frac{\partial U_j}{\partial x_k} = -\frac{\partial P}{\partial x_j}. \tag{276}$$

These equations are general. Now suppose that

$$\mathbf{U} = U\left(0, \frac{\varpi}{p}, 1\right), \tag{277}$$

where U is a constant and the condition corresponding to (270) is satisfied.

Writing equations (275) and (276) in cylindrical polar coordinates† and seeking solutions whose (t, φ, z)-dependence is given by (250), we obtain:

$$i\omega_1 \xi_\varpi - \alpha_1 \frac{U}{p} \xi_\varphi - \alpha_2 \frac{U}{p} \eta_\varphi = DP, \tag{278}$$

$$i\omega_1 \xi_\varphi + \alpha_1 \frac{U}{p} \xi_\varpi + \alpha_2 \frac{U}{p} \eta_\varpi = \frac{im}{\varpi} P, \tag{279}$$

$$i\omega_2 \eta_\varpi + \alpha_2 \frac{U}{p} \eta_\varphi + \alpha_1 \frac{U}{p} \xi_\varphi = -DP, \tag{280}$$

$$i\omega_2 \eta_\varphi - \alpha_2 \frac{U}{p} \eta_\varpi - \alpha_1 \frac{U}{p} \xi_\varpi = -\frac{im}{\varpi} P, \tag{281}$$

$$i\omega_1 \xi_z = ikP, \quad \text{and} \quad i\omega_2 \eta_z = -ikP, \tag{282}$$

where
$$\omega_1 = \sigma + \alpha_1 U\left(\frac{m}{p} + k\right) \quad \text{and} \quad \omega_2 = \sigma - \alpha_2 U\left(\frac{m}{p} + k\right). \tag{283}$$

Solving equations (278)–(281) for ξ_ϖ and ξ_φ in terms of DP and P, we readily find

$$\left[\omega_1^2 \omega_2^2 - \frac{U^2}{p^2}(\alpha_2 \omega_1 - \alpha_1 \omega_2)^2\right] \xi_\varpi = i\omega_2 \left[-\omega_1 \omega_2 DP + \frac{U}{p}(\alpha_2 \omega_1 - \alpha_1 \omega_2) \frac{mP}{\varpi}\right] \tag{284}$$

and

$$\left[\omega_1^2 \omega_2^2 - \frac{U^2}{p^2}(\alpha_2 \omega_1 - \alpha_1 \omega_2)^2\right] \xi_\varphi = -\omega_2 \left[\frac{U}{p}(\alpha_2 \omega_1 - \alpha_1 \omega_2) DP - \omega_1 \omega_2 \frac{mP}{\varpi}\right] \tag{285}$$

We also have
$$\xi_z = \frac{k}{\omega_1} P. \tag{286}$$

Now substituting for the components of $\boldsymbol{\xi}$, in accordance with the foregoing equations, in

$$D_* \xi_\varpi + \frac{im}{\varpi} \xi_\varphi + ik\xi_z = 0 \tag{287}$$

† This can be most readily accomplished by noting that in cylindrical polar coordinates, the vector $(\mathbf{A}.\text{grad})\mathbf{B}$ has the components

$$(\mathbf{A}.\text{grad})B_\varpi - A_\varphi B_\varphi/\varpi, \quad (\mathbf{A}.\text{grad})B_\varphi + A_\varphi B_\varpi/\varpi, \quad \text{and} \quad (\mathbf{A}.\text{grad})B_z$$

where $(\mathbf{A}.\text{grad})$ is defined in the manner (248). For $\mathbf{U}$ given by (277) and for solutions having the chosen (t, φ, z)-dependence, $(\mathbf{U}.\text{grad}) \equiv iU(k + m/p)$. Also, it may be noted that the components of $(\mathbf{q}.\text{grad})\mathbf{U}$, where $\mathbf{q}$ is an arbitrary vector, are $(-q_\varphi U/p,\ q_\varpi U/p,\ 0)$. With these relations in mind, equations (278)–(282) are self-evident.

(which expresses the solenoidal character of $\boldsymbol{\xi}$), we obtain (on making use of the identity (259))

$$D_* DP+\left\{k^2\left[\frac{U^2}{p^2}\frac{(\alpha_2\omega_1-\alpha_1\omega_2)^2}{\omega_1^2\omega_2^2}-1\right]-\frac{m^2}{\varpi^2}\right\}P = 0. \tag{288}$$

The solution of this equation which is free of singularity at the origin is

$$P = AJ_m(\kappa\varpi), \tag{289}$$

where A is a constant, J_m is Bessel's function of order m, and

$$\kappa^2 = k^2\left[\frac{U^2}{p^2}\frac{(\alpha_2\omega_1-\alpha_1\omega_2)^2}{\omega_1^2\omega_2^2}-1\right]. \tag{290}$$

We shall suppose that the boundary conditions† require the vanishing of P at $\varpi = R$. Then

$$\kappa R = x_{m,j}, \tag{291}$$

where $x_{m,j}$ is the jth zero of $J_m(x)$. From equation (290) it now follows that

$$\omega_1\omega_2 = q\frac{U}{p}(\alpha_2\omega_1-\alpha_1\omega_2), \tag{292}$$

where $$q = \pm\frac{x}{\sqrt{(x^2+x_{m,j}^2)}} \quad\text{and}\quad x = kR. \tag{293}$$

(Notice the double sign in the definition of q.) Substituting for ω_1 and ω_2 from (283) in (292), we obtain after some rearrangement

$$\sigma^2+(\alpha_1-\alpha_2)U\left(k+\frac{m}{p}+\frac{q}{p}\right)\sigma-\alpha_1\alpha_2U^2\left[\left(k+\frac{m}{p}\right)^2+2\left(k+\frac{m}{p}\right)\frac{q}{p}\right] = 0. \tag{294}$$

The roots of this last equation are given by

$$\sigma = -\tfrac{1}{2}(\alpha_1-\alpha_2)U\left(k+\frac{m}{p}+\frac{q}{p}\right)\pm\tfrac{1}{2}U\left\{(\alpha_1+\alpha_2)^2\left(k+\frac{m}{p}+\frac{q}{p}\right)^2-4\alpha_1\alpha_2\frac{q^2}{p^2}\right\}^{\frac{1}{2}}. \tag{295}$$

It is now convenient to express the equation for σ in terms of the fraction of the energy which is in the magnetic field. Since in the unperturbed state

$$h_j-u_j = \alpha_2 U_j \quad\text{and}\quad h_j+u_j = \alpha_1 U_j, \tag{296}$$

$$f = \frac{\text{magnetic energy}}{\text{total energy}} = \frac{(\alpha_1+\alpha_2)^2}{2(\alpha_1^2+\alpha_2^2)}. \tag{297}$$

Further, let (cf. equation (267))

$$l = R/p; \tag{298}$$

† In general the boundary conditions may be more complex (see footnote on page 556 and subsection (*b*) below). The use of the boundary condition $P = 0$ at $\varpi = R$ is to be considered no more than as illustrative of the results which might follow.

and measure σ in the unit

$$\frac{U}{R}\sqrt{\tfrac{1}{2}(\alpha_1^2+\alpha_2^2)}\ \text{sec}^{-1}. \tag{299}$$

With these definitions, equation (295) takes the form

$$\sigma = -\frac{(\alpha_1-\alpha_2)}{|\alpha_1-\alpha_2|}[x+(q+m)l]\sqrt{(1-f)}\pm\sqrt{\{[x+(q+m)l]^2f+(1-2f)q^2l^2\}}. \tag{300}$$

[Note that when $f = \frac{1}{2}$, this equation reduces to equation (266) derived in § 113 (a).]

Since for any given value of x and m we can associate with q the two values $\pm|q|$, there are, in general, four distinct modes of oscillation corresponding to the four roots for σ which follow from equation (300).

From equation (300) it is apparent that σ cannot have a complex root if $f \leqslant \frac{1}{2}$. Accordingly, *the system is stable if the energy in the magnetic field does not exceed the energy in the velocity field*. This result is due to Trehan.

Before we consider the full implications of equation (300), we shall examine the two limiting cases, $f = 0$ and $f = 1$, when the energy in the field is entirely of one kind or another.

(a) *The case* $f = 0$

This is the case of a pure velocity field which can be obtained by setting $\alpha_1 = -\alpha_2$; and on the assumption that α_1 is positive, equation (300) gives

$$\sigma = -[x+(q+m)l]\pm ql; \tag{301}$$

or

$$\sigma = -[x+(2q+m)l] \quad \text{or} \quad -(x+m)l. \tag{302}$$

However, a direct treatment of this problem leads only to the two modes represented by the first of the two solutions included in (302). To see how this arises and to clarify, also, the possibility of other boundary conditions (besides the one we have used), we shall treat this case *de novo*.

The perturbation equation in the case of a pure velocity field is

$$\frac{\partial u_j}{\partial t}+U_k\frac{\partial u_j}{\partial x_k}+u_k\frac{\partial U_j}{\partial x_k} = -\frac{\partial P}{\partial x_j}. \tag{303}$$

For solutions having the (t, φ, z)-dependence given by (250) and for the same form of U as (277), equation (303) gives (see the footnote on p. 558)

$$i\omega u_\varpi-2\frac{U}{p}u_\varphi = -DP, \tag{304}$$

$$i\omega u_\varphi+2\frac{U}{p}u_\varpi = -\frac{im}{\varpi}P, \tag{305}$$

and $$i\omega u_z = -ikP, \tag{306}$$

where $$\omega = \sigma + U\left(\frac{m}{p}+k\right). \tag{307}$$

From the formal identity of the set of equations (304)–(306) with the set (252)–(254), we can write (cf. equations (261) and (262))

$$P = AJ_m(\kappa\varpi), \tag{308}$$

where $$\kappa^2 = k^2\left(\frac{4U^2}{p^2\omega^2}-1\right). \tag{309}$$

If we impose on P the same boundary condition as we did on the solution (289), we shall obtain

$$\omega = \frac{2U}{p}q, \tag{310}$$

where q has the same meaning as in (293). The corresponding solution for σ is

$$\sigma = -U\left(\frac{m}{p}+k\right)+\frac{2U}{p}q. \tag{311}$$

An equivalent form of this equation is

$$\frac{\sigma R}{U} = -(x+lm)+2lq \tag{312}$$

which (since q can be $\pm|q|$ for a given x and m) is the same as the first of the two solutions included in (302).

The preceding analysis shows how, in this case, the particular result (312) is a consequence of the chosen boundary conditions. If we should suppose, instead, that the fluid is confined by a rigid wall at $\varpi = R$, the proper boundary condition would be that u_ϖ vanishes at $\varpi = R$. Since (cf. equation (257))

$$\left(\omega^2-4\frac{U^2}{p^2}\right)u_\varpi = i\left(\omega DP+\frac{2mU}{p\varpi}P\right), \tag{313}$$

the vanishing of u_ϖ at $\varpi = R$ will lead to the condition

$$yJ'_m(y)+\frac{2mU}{p\omega}J_m(y) = 0, \tag{314}$$

where $$y^2 = x^2\left(\frac{4U^2}{p^2\omega^2}-1\right). \tag{315}$$

Eliminating $2U/p\omega$ from equation (314) by making use of equation (315), we obtain $$yJ'_m(y)\pm m(1+y^2/x^2)^{\frac{1}{2}}J_m(y) = 0. \tag{316}$$

We observe that equation (316) is the same as the characteristic equation which governs the periods of oscillation of a rotating column of liquid (Chapter VII, § 68 (*a*)). The dispersion relations illustrated in Fig. 63 (p. 286) are, therefore, applicable to this problem with minor re-identifications.

If we denote by $y_{m,j}$ the jth zero of the transcendental equation (314), the required expression for σ is

$$\frac{\sigma R}{U} = -(x+lm)\pm 2\frac{xl}{\sqrt{(x^2+y_{m,j}^2)}} \tag{317}$$

which apart from the replaeement of $x_{j,m}$ by $y_{j,m}$ is the same as equation (312). The main point is, however, that the results which follow from the use of the two different boundary conditions ($P = 0$ or $u_\varpi = 0$ at $\varpi = R$) are substantially the same: thus, for the modes $m = 0$, the change consists in the substitution of the zeros of $J_0(x)$ by the zeros of $J_1(x)$; and this is not very consequential.

(*b*) *The case* $f = 1$

This is the case of a pure magnetic field which can be obtained by setting $\alpha_1 = \alpha_2$. Equation (300) gives for this case

$$\sigma = \pm\sqrt{\{[x+(\pm|q|+m)l]^2-q^2l^2\}}. \tag{318}$$

From this equation for σ it is evident that the condition for instability to occur is

$$|x+(\pm|q|+m)l| < |q|l. \tag{319}$$

From the definition of q in equation (293), it is apparent that $|q|$ is a monotonic increasing function of x and, moreover,

$$|q| \leqslant 1, \tag{320}$$

the value 1 being attained only for $x \to \infty$.

In view of (320), the inequality (319) can never be met for $m \geqslant 2$; therefore, *the system is stable for all modes belonging to* $m \geqslant 2$. On the other hand if m should be negative, then with the choice of the negative sign in front of $|q|$ on the left-hand side of (319), the inequality which must be satisfied for instability to occur is

$$|x-(|q|+|m|)l| < |q|l, \tag{321}$$

or

$$(2|q|+|m|)l > x > (|q|+|m|)l. \tag{322}$$

In the finite range of x specified by (322) the system is certainly unstable; therefore, *there exist ranges of wave numbers for which the system is unstable for modes belonging to any negative m.*

It remains to consider the modes belonging to $m = 0$ and 1. When $m = 0$, the inequality (319) becomes

$$|x\pm|q|l| < |q|l. \tag{323}$$

This inequality cannot be met with the positive sign on the left-hand side; therefore, it reduces to

$$x < 2|q|l = \frac{2xl}{\sqrt{(x^2+x_{0,j}^2)}} \tag{324}$$

or

$$l > \tfrac{1}{2}\sqrt{(x^2+x_{0,j}^2)} > \tfrac{1}{2}x_{0,1}. \tag{325}$$

A necessary condition that for some x and j a mode belonging to $m = 0$ is unstable is

$$l = R/p > \tfrac{1}{2}x_{0,1} = 1{\cdot}202. \tag{326}$$

Finally, considering the case $m = 1$, we will have instability if

$$|x+(\pm|q|+1)l| < |q|l, \tag{327}$$

a condition which cannot again be met with the positive sign in front of $|q|$. Therefore, the condition reduces to (in view of (320))

$$x+(1-|q|)l < |q|l \tag{328}$$

or

$$x < (2|q|-1)l. \tag{329}$$

Inserting for $|q|$ its value from (293), the condition becomes

$$x < \left[\frac{2x}{\sqrt{(x^2+x_{1,j}^2)}}-1\right]l. \tag{330}$$

The right-hand side of (330) is negative if

$$x < x_{1,j}/\sqrt{3} < x_{1,1}/\sqrt{3} = 2{\cdot}212. \tag{331}$$

Hence, for $x < 2{\cdot}212$ the modes belonging to $m = 1$ are certainly stable; however, when this value is exceeded, instability is possible if l is sufficiently small.

The main conclusion to be drawn from the foregoing discussion is that the helical magnetic field is unstable. The remarkable fact is that fluid motions parallel to the field, if of sufficient intensity, can stabilize it.

(c) *The general case*

We have seen that while equation (300) predicts four distinct modes for a general value of f, there are only two modes for the particular values $f = 0$, $\tfrac{1}{2}$, and 1. The way in which the different modes coalesce and separate again as f passes through these values is of interest.

When $f = 1$, we have the modes given by equation (318). In this case, we have only two distinct modes, since we need not distinguish the $\pm$ sign which occurs as a factor in this equation. Let the two modes for $f = 1$ be distinguished by 1 and 2. When f decreases, each of these two modes splits into two modes; we shall distinguish them by 1*a* and 1*b*, and 2*a* and 2*b*, respectively. In the case of a complex σ, the imaginary part of σ decreases, while the real part increases as f decreases. The imaginary part vanishes for some $f > \tfrac{1}{2}$.

When f tends to the equipartition value $\frac{1}{2}$, the values of σ for the modes $2a$, and $1a$ or $1b$, tend to zero; and for $f = \frac{1}{2}$ there exist only the two modes of oscillation given by (266).

When f decreases below $\frac{1}{2}$, we recover the four modes once again and

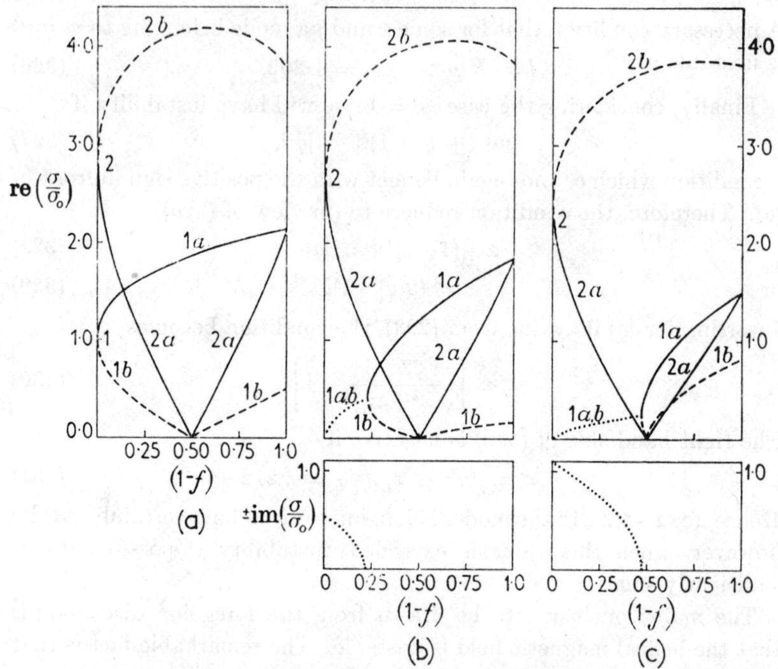

Fig. 130. The dependence of the frequency of oscillation on the fraction of the total energy in the velocity field for the mode $m = 0$, and for $kR = 1$. The curves in (a), (b), and (c) are for $R/p = 1, \sqrt{2}$, and 2, respectively. The 'hydromagnetic' and the 'velocity' modes are shown by the solid and the dashed lines, respectively, while the unstable modes are distinguished by the dotted lines. For $R/p = \sqrt{2}$ and 2, we have stability for $1-f \geqslant 0{\cdot}226$ and $0{\cdot}476$, respectively.

now all these modes are necessarily real. As f tends to zero, the modes $1a$ and $2a$ coalesce while the remaining two modes tend to finite limits; and we obtain the three modes included in (302). However, as we have seen, one of the modes included in (302), namely, $\sigma = -(x+m)l$, is spurious; and one can, indeed, verify that in this limit the proper function belonging to this mode vanishes identically.

The dependence on f of the four roots which follow from equation (300) is shown in Figs. 130 (for the case $m = 0$) and 131 (for the case $m = -1$).

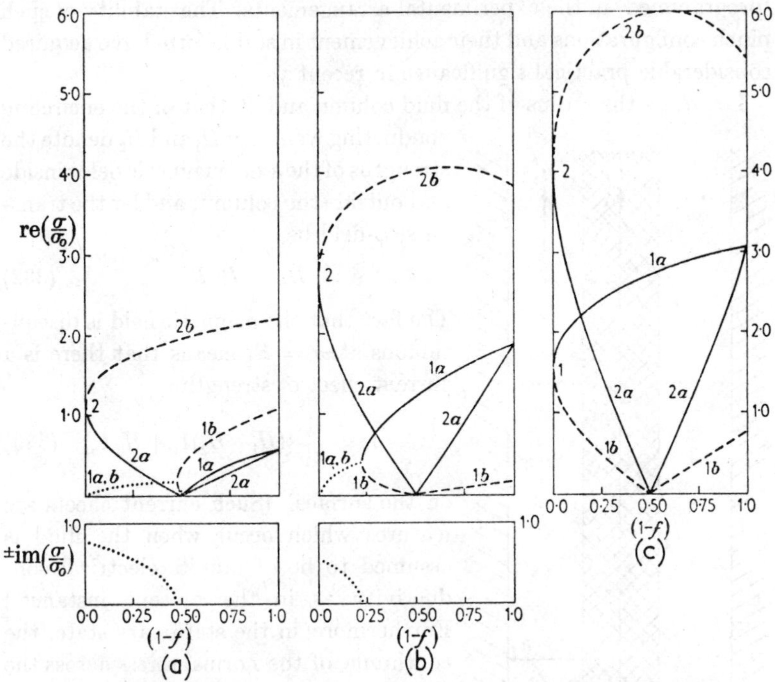

Fig. 131. The dependence of the frequency of oscillation on the fraction of the total energy in the velocity field for the mode $m = -1$, and for $kR = 0{\cdot}85$. The curves in (a), (b), and (c) are for $R/p = 1$, $\sqrt{2}$, and 2, respectively. The 'hydromagnetic' and 'velocity' modes are shown by the solid and the dashed lines, respectively, while the modes which are unstable are distinguished by dotted lines. For $R/p = 1$ and $\sqrt{2}$, we have stability for $1-f \geqslant 0{\cdot}475$ and $0{\cdot}225$, respectively.

115. The stability of the pinch

The simplest of the so-called *pinch configurations* consists of a cylindrical column of fluid (the 'plasma') inside of which a uniform axial magnetic field is present, while outside there is a similar axial field together with a transverse φ-field falling off as ϖ^{-1}. In the usual arrangements, the column of fluid is circled by a concentric conducting wall (see Fig. 132). Configurations of this kind are achieved in the laboratory by sending a high current through a gas by means of a discharge. The axial current produces a transverse magnetic field which 'pinches' the column of gas into a configuration which we have idealized in the description; and the axial fields of different strengths inside and outside the column of fluid, which we have postulated, are 'programmed' in the experimental arrangements. The stability of such pinch configurations and their achievement in stable form have acquired considerable practical significance in recent years.

Let R_1 be the radius of the fluid column and R_2 that of the encircling conducting wall. Let H_1 and H_2 denote the strengths of the axial magnetic fields inside and outside the column; and let the transverse φ-field be

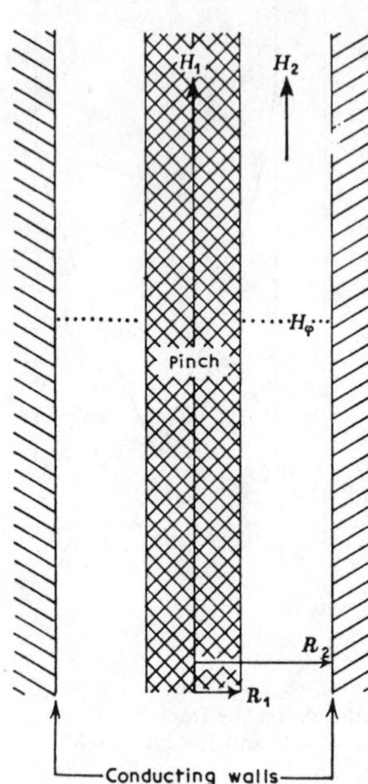

FIG. 132. The pinch configuration.

$$H_\varphi = H_0 R_1/\varpi. \tag{332}$$

The fact that the magnetic field is discontinuous at $\varpi = R_1$ means that there is a current sheet of strength

$$\mathbf{J}_s = \frac{1}{4\pi}\{(H_1-H_2)\mathbf{1}_\varphi + H_0\mathbf{1}_z\} \tag{333}$$

on the surface. (Such current sheets are features which occur when the fluid is assumed to be of infinite electrical conductivity as in the present instance.) Furthermore, in the stationary state, the continuity of the normal stress across the surface of the fluid requires that the constant pressure p_0 inside the fluid is given by

$$p_0 = \frac{\mu}{8\pi}(H_0^2+H_2^2-H_1^2). \tag{334}$$

It will be convenient to express H_1 and H_2 in terms of H_0. Let

$$H_1 = \beta_1 H_0 \quad \text{and} \quad H_2 = \beta_2 H_0. \tag{335}$$

An inequality which must obtain in virtue of (334) is

$$1+\beta_2^2 \geqslant \beta_1^2. \tag{336}$$

The equations governing small perturbations of the fluid column are the same as equations (102)–(104). However, Π has now the meaning

$$\Pi = \frac{\delta p}{\rho} + \frac{\mu H h_z^{(\text{in})}}{4\pi\rho}, \tag{337}$$

where we have distinguished $\mathbf{h}$ by $\mathbf{h}^{(\text{in})}$ to denote that it signifies the perturbation in the field in the interior of the fluid. (We shall similarly distinguish the perturbation in the field in the exterior of the fluid by $\mathbf{h}^{(\text{ex})}$.)

By considering a normal mode of disturbance by specifying that the deformed surface be governed by the equation

$$\varpi = R_1 + \epsilon_0 \, e^{i(kz+m\varphi)+\sigma t}, \tag{338}$$

we seek solutions of the perturbation equations whose (t, φ, z)-dependence is given by

$$e^{i(kz+m\varphi)+\sigma t}. \tag{339}$$

The corresponding solution for Π is

$$\Pi = \epsilon_0 \, \Pi_0 \, I_m(k\varpi) e^{i(kz+m\varphi)+\sigma t}. \tag{340}$$

The integral of equation (103) when $\mathbf{H}$ is axial is

$$\mathbf{h}^{(\text{in})} = \frac{ikH_1}{\sigma} \mathbf{u}. \tag{341}$$

The equation of motion for $\mathbf{u}$ then gives

$$\sigma\left(1 + \frac{\mu H_1^2 k^2}{4\pi\rho\sigma^2}\right)\mathbf{u} = -\text{grad}\,\Pi. \tag{342}$$

In particular,

$$u_\varpi = -\frac{\epsilon_0 \, \sigma k \Pi_0}{\sigma^2 + \mu H_1^2 k^2/4\pi\rho} I'_m(k\varpi) e^{i(kz+m\varphi)+\sigma t}. \tag{343}$$

The perturbation in the magnetic field outside of the fluid is a vacuum field and, accordingly, under the present circumstances can be derived from a potential in the manner

$$\mathbf{h}^{(\text{ex})} = \text{grad}\{[C_1 \, I_m(k\varpi) + C_2 \, K_m(k\varpi)] e^{i(kz+m\varphi)+\sigma t}\}, \tag{344}$$

where C_1 and C_2 are constants. For the components of $\mathbf{h}^{(\text{ex})}$ we have

$$h_\varpi^{(\text{ex})} = k[C_1 \, I'_m(k\varpi) + C_2 \, K'_m(k\varpi)] e^{i(kz+m\varphi)+\sigma t}, \tag{345}$$

$$h_\varphi^{(\text{ex})} = \frac{im}{\varpi}[C_1 \, I_m(k\varpi) + C_2 \, K_m(k\varpi)] e^{i(kz+m\varphi)+\sigma t}, \tag{346}$$

$$h_z^{(\text{ex})} = ik[C_1 \, I_m(k\varpi) + C_2 \, K_m(k\varpi)] e^{i(kz+m\varphi)+\sigma t}. \tag{347}$$

The solutions represented by equations (340), (341), (343), and (345)–(347) must satisfy a number of boundary conditions. These are (1) the velocity u_ϖ at $\varpi = R_1$ must be compatible with the assumed form (338) of the deformed surface; (2) the radial component $h_\varpi^{(\text{ex})}$ must vanish on the conducting wall at $\varpi = R_2$; (3) the component of the magnetic field normal to the deformed surface must be continuous; and (4) the normal component of the total stress tensor must also be continuous on the deformed surface.

The first of the foregoing conditions requires

$$\epsilon_0 \, \sigma e^{i(kz+m\varphi)+\sigma t} = u_\varpi(R_1), \tag{348}$$

or, using equation (343), we obtain

$$\sigma^2 + \frac{\mu H_1^2 x_1^2}{4\pi\rho R_1^2} = -\frac{\Pi_0}{R_1} x_1 I'_m(x_1), \tag{349}$$

where
$$x_1 = kR_1. \tag{350}$$

From equations (341) and (348) we obtain the relation

$$h_{\varpi}^{(\text{in})} = ik\epsilon_0 H_1 e^{i(kz+m\varphi)+\sigma t} \quad (\varpi = R_1). \tag{351}$$

The vanishing of $h_{\varpi}^{(\text{ex})}$ at $\varpi = R_2$ gives (cf. equation (345))

$$\frac{C_2}{C_1} = -\frac{I'_m(x_2)}{K'_m(x_2)} \quad \text{where } x_2 = kR_2. \tag{352}$$

Since the tangential components of the magnetic field at the boundary of the fluid, in the unperturbed state, are not continuous, we must allow for this fact in applying the third of the enumerated boundary conditions. When discontinuities in the tangential magnetic field are present in the stationary state, the boundary condition which must be applied in the perturbed state is

$$\mathbf{N}^{(0)}.\Delta(\mathbf{h}) + \delta\mathbf{N}.\Delta(\mathbf{H}^{(0)}) = 0, \tag{353}$$

where $\Delta(f)$ signifies the jump which a quantity f experiences at the boundary, $\mathbf{N}^{(0)}$ is the unit outward normal to the unperturbed boundary, $\delta\mathbf{N}$ the change in $\mathbf{N}^{(0)}$ consequent to the perturbation, $\mathbf{H}^{(0)}$ the unperturbed magnetic field, and $\mathbf{h}$ the perturbation in it. In the case under consideration

$$\mathbf{N}^{(0)} = \mathbf{1}_{\varpi}; \quad \delta\mathbf{N} = -\left(\frac{im}{R_1}\mathbf{1}_{\varphi} + ik\mathbf{1}_z\right)\epsilon_0 e^{i(kz+m\varphi)+\sigma t};$$

and
$$\Delta(\mathbf{H}^{(0)}) = H_0 \mathbf{1}_{\varphi} + (H_2 - H_1)\mathbf{1}_z. \tag{354}$$

The application of the condition (353) therefore gives

$$[h_{\varpi}^{(\text{ex})} - h_{\varpi}^{(\text{in})}]_{R_1} - \frac{\epsilon_0}{R_1}[imH_0 + ix_1(H_2 - H_1)]e^{i(kz+m\varphi)+\sigma t} = 0. \tag{355}$$

Making use of equations (345) and (351), we obtain

$$k[C_1 I'_m(x_1) + C_2 K'_m(x_1)] = \frac{i\epsilon_0 H_0}{R_1}(\beta_2 x_1 + m), \tag{356}$$

where it may be recalled that $\beta_2 = H_2/H_0$. In combination with equation (352), equation (356) gives

$$C_1 = i\epsilon_0 H_0 \frac{(\beta_2 x_1 + m)K'_m(x_2)}{x_1[K'_m(x_2)I'_m(x_1) - I'_m(x_2)K'_m(x_1)]}. \tag{357}$$

It remains to satisfy the last of the four boundary conditions enumerated. It requires

$$\epsilon_0 \Pi_0 I_m(x_1) e^{i(kz+m\varphi)+\sigma t} = \delta\left(\frac{\mu \mathbf{H}^2}{8\pi\rho}\right)_{R_1+\epsilon_0 e^{i(kz+m\varphi)+\sigma t}}$$

$$= \frac{\epsilon_0 \mu H_0}{4\pi\rho}\left(\frac{dH_\varphi}{d\varpi}\right)_{\varpi=R_1} e^{i(kz+m\varphi)+\sigma t} + \frac{\mu}{4\pi\rho}(H_0 h_\varphi^{(\mathrm{ex})} + H_2 h_z^{(\mathrm{ex})})_{\varpi=R_1}. \tag{358}$$

Substituting for the components of $\mathbf{h}^{(\mathrm{ex})}$ in accordance with equations (346), (347), (352), and (357), and remembering that $H_\varphi = H_0 R_1/\varpi$, we obtain after simplifications

$$\frac{\Pi_0}{R_1} x_1 I_m(x_1) = -\frac{\mu H_0^2}{4\pi\rho R_1^2}\left[x_1 + (\beta_2 x_1 + m)^2 \frac{K'_m(x_2)I_m(x_1) - I'_m(x_2)K_m(x_1)}{K'_m(x_2)I'_m(x_1) - I'_m(x_2)K'_m(x_1)}\right]. \tag{359}$$

Now eliminating Π_0 between equations (349) and (359), we finally obtain the dispersion relation

$$\left(\beta_1^2 + \frac{\sigma^2}{\Omega_A^2}\right)\frac{x_1^2 I_m(x_1)}{I'_m(x_1)}$$

$$= x_1 + (\beta_2 x_1 + m)^2 \frac{K'_m(x_2)I_m(x_1) - I'_m(x_2)K_m(x_1)}{K'_m(x_2)I'_m(x_1) - I'_m(x_2)K'_m(x_1)}, \tag{360}$$

where Ω_A is the Alfvén frequency defined in terms of H_0.

(a) *On stable pinch configurations*

The dispersion relation (360) involves the three parameters

$$\beta_1 = H_1/H_0, \quad \beta_2 = H_2/H_0, \quad \text{and} \quad q = x_2/x_1 = R_2/R_1. \tag{361}$$

Since β_1 occurs as β_1^2 in (360), its sign is clearly immaterial; and we shall suppose that it is positive. On the other hand, the sign of β_2 does matter; however, since a reversal of the sign of β_2 is equivalent to a reversal of the sign of m and since both signs of m are, moreover, allowed, there is again no loss of generality in supposing that β_2 is also positive.

The parameters β_1^2 and β_2^2 are not unrestricted in view of (336); also it is apparent that $q \geqslant 1$. The principal question of interest in connexion with the present problem is therefore: *in what parts of the space*

$$q \geqslant 1, \quad \beta_1 \geqslant 0, \quad \beta_2 \geqslant 0, \quad \text{and} \quad 1+\beta_2^2 \geqslant \beta_1^2, \tag{362}$$

of the parameters do we have stability?

A necessary and sufficient condition for stability can be given at once. From equation (360) it is apparent that the principle of the exchange of stabilities is valid for the present problem and that *a necessary and sufficient condition for stability is*

$$x_1+(\beta_2 x_1 \pm m)^2 \frac{K'_m(qx_1)I_m(x_1)-I'_m(qx_1)K_m(x_1)}{K'_m(qx_1)I'_m(x_1)-I'_m(qx_1)K'_m(x_1)} < \beta_1^2 \frac{x_1^2 I_m(x_1)}{I'_m(x_1)}, \quad (363)$$

for all x_1 (> 0) *and integral* m ($\geqslant 0$); for equation (360) cannot, then, allow a solution with $\sigma^2 > 0$.

The states of marginal stability can be derived by considering (363) with the equality sign in place of the inequality sign.

Two conclusions follow directly from (363). The first, and from the practical standpoint of achieving stable pinch configurations an important, conclusion is that *stability is impossible if* $\beta_1 = 0$. For when $\beta_2 x_1 = m$, the left-hand side of the inequality (363) (with the negative sign for m chosen) is positive, which contradicts (363). The instability, in the absence of an internal z-field when $\beta_2 x_1 + m = 0$, occurs when the lines of force of the perturbed field are helices congruent to the one described by the unperturbed external field on the surface $\varpi = R_1$.

It is also clear that, other things being equal, stability for a negative m implies stability for the corresponding positive m: for the additional term on the left-hand side of (363) on passing from m to $-m$ is

$$-4|m|\beta_2 x_1 \frac{K'_m(qx_1)I_m(x_1)-I'_m(qx_1)K_m(x_1)}{K'_m(qx_1)I'_m(x_1)-I'_m(qx_1)K'_m(x_1)}; \quad (364)$$

and this is negative.

There is one further elementary result which we can derive. This concerns the case $m = 0$. When $m = 0$, the inequality (363) becomes

$$1-\beta_2^2 x_1 \frac{I_1(qx_1)K_0(x_1)+K_1(qx_1)I_0(x_1)}{I_1(qx_1)K_1(x_1)-K_1(qx_1)I_1(x_1)} < \beta_1^2 \frac{x_1 I_0(x_1)}{I_1(x_1)}. \quad (365)$$

From the known properties of the Bessel functions, it can be shown that if the inequality (365) holds for some value of x_1, then it holds for all larger values of x_1. Therefore, in this case, a necessary and sufficient condition that (365) hold for all $x_1 > 0$ can be obtained by considering the limiting form of the inequality for $x_1 \to 0$. We thus obtain

$$\frac{1}{2}-\frac{\beta_2^2}{q^2-1} < \beta_1^2. \quad (366)$$

The foregoing is a necessary and sufficient condition that the modes belonging to $m = 0$ are stable.

An analytical discussion of the equation governing marginal stability for values of m besides zero is possible. However, since in this particular instance exhaustive numerical results are available, we shall content ourselves by giving an account of them. The particular results which we shall describe and illustrate are those of Tayler.

When there is no encircling wall, $q \to \infty$, and the equation for marginal stability reduces to

$$x_1+(\beta_2 x_1 \pm m)^2 \frac{K_m(x_1)}{K'_m(x_1)} = \beta_1^2 \frac{x_1^2 I_m(x_1)}{I'_m(x_1)}. \tag{367}$$

From this equation it appears that by letting β_1 exceed a certain determinate value (depending on β_2 except, as we shall see, in the case $m = 0$), one can achieve stability for all modes except those belonging to $m = -1$.

TABLE LXIV

Critical values of β_1 for the stability of the modes belonging to different m's in case there is no encircling conducting wall

$-m$ \ β_2	0·0	0·5	1·0	1·5	2·0
0	0·707	0·707	0·707	0·707	0·707
2	0·259	0·676	1·122	1·592	2·073
3	0·169	0·492	0·801	1·130	1·468
4	0·126	0·409	0·660	0·926	1·202

Thus, the modes belonging to $m = 0$ can be stabilized by the choice (see equation (366))

$$\beta_1^2 > 0·5 \tag{368}$$

independently of β_2. There are similar results for other values of $-|m|$ (depending, however, on β_2). These limits are given in Table LXIV and are exhibited in Fig. 133. In Fig. 133 the curve $\beta_1^2 = 1+\beta_2^2$ and the loci of the marginally stable configurations for $m = 0$ and $m = -2$ bound a region (shown shaded in the figure) in which all modes belonging to $m \neq -1$ are stable. In the same shaded region, disturbances belonging to $m = -1$ and wave numbers not exceeding certain determinate values are unstable. Table LXV gives, for the configurations represented by points on the upper and the lower boundaries of the shaded region in Fig. 133, the wave numbers below which they are unstable for disturbances belonging to $m = -1$.

If an encircling wall is allowed, it is found that the modes belonging to $m = -1$ can also be stabilized if q (> 1) is less than a certain critical value. Table LXVI gives the maximum values of q which will ensure stability for given values of β_2 when β_1 has the maximum permissible value, namely, $\sqrt{(1+\beta_2^2)}$. We observe that for $\beta_2 = 0$, the maximum value of q compatible with stability is 5; when β_2 increases, the allowed range of q decreases very rapidly. For a given β_2, when β_1 decreases

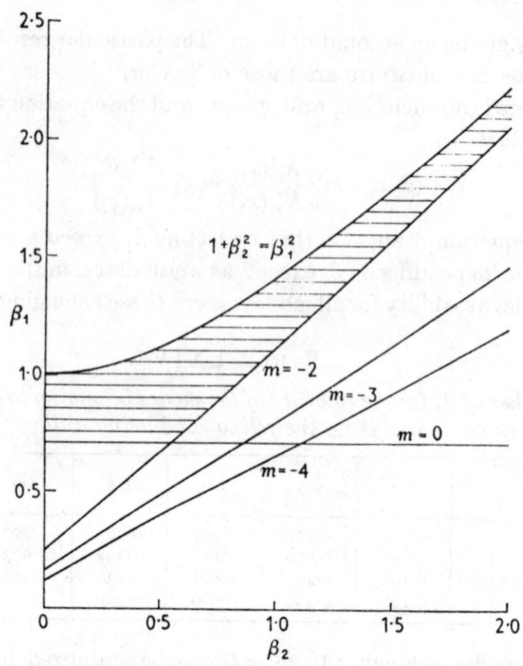

FIG. 133. Stability diagram for pinch configurations when there are no conducting walls. The modes belonging to $\pm m$ are stable when the representative point in the plane is above the curve labelled by $m = -|m|$ and below the locus $1+\beta_2^2 = \beta_1^2$. Complete stability for all modes except those belonging to $m = -1$ is obtained for values of β_2 and β_1 in the shaded region.

TABLE LXV

Wave numbers below which the modes belonging to $m = -1$ are unstable (no encircling conducting wall)

β_2	0·0	0·500	1·000	1·500	2·000
β_1	1·000	1·118	1·414	1·803	2·236
x_1	0·450	0·869	0·736	0·571	0·455
β_1	0·707	0·707	1·122	1·592	2·073
x_1	1·025	1·886	0·991	0·661	0·497

TABLE LXVI

Critical wall radii for the stability of the modes belonging to $m = -1$

β_2	0·0	0·100	0·200	0·500	1·000
β_1	1·000	1·005	1·020	1·118	1·414
q	5·00	3·31	2·62	1·81	1·39

below $\sqrt{(1+\beta_2^2)}$, the maximum allowed value of q, for stability, also decreases. For $q = 1$, the minimum values of β_1 required for the stability of the modes belonging to $m = -1$ are given in Table LXVII. The results for this and other values of q are included in Fig. 134; this figure contains the answer to the principal question raised at the beginning of this section.

TABLE LXVII

Minimum values of β_1 for the stability of the modes belonging to $m = -1$ when $q = 1$

β_2	0·0	0·200	0·500	1·000	1·500	2·000
β_1	0·000	0·429	0·683	1·114	1·580	2·061

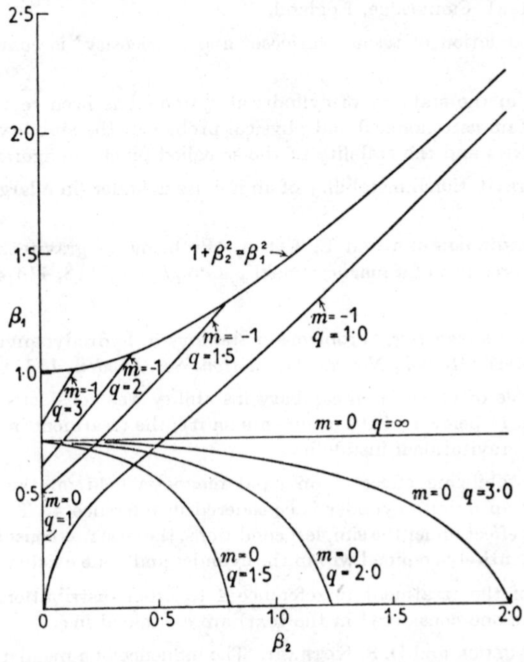

FIG. 134. Stability diagram for pinch configurations when conducting walls are present at a radius $R_2 = qR_1$. For a given value of q, the modes belonging to $\pm m$ are stable when the representative point in the plane is above the curve labelled by $m = -|m|$ and the given q, and below the locus $1+\beta_2^2 = \beta_1^2$. Complete stability for a given q is obtained when the representative point lies in the region appropriate for the stability of the modes belonging to both $m = 0$ and -1.

BIBLIOGRAPHICAL NOTES

Interest in the stability of liquid jets was stimulated by some beautiful experiments of Savart and their interpretation by Plateau in terms of capillarity. And Rayleigh laid the foundations for the theoretical treatment of these and similar problems in:

1. Lord Rayleigh, 'On the instability of jets', *Scientific Papers*, i, **361–71**, Cambridge, England, 1899.

In this paper, Rayleigh developed the important concept of the mode of maximum instability. An account of all this early work will be found in:

2. Lord Rayleigh, *Theory of Sound*, 357–64, Dover Publications, New York, 1945.

A delightful lecture including some beautiful illustrations is:

3. Lord Rayleigh, 'Some applications of photography', *Scientific Papers*, iii, 441–51, Cambridge, England, 1902.

The recommendation of terms 'varicose' and 'varicosity' is contained in this lecture.

The interest in the stability of cylindrical systems has been revived in recent years by certain astronomical and physical problems: the stability of the spiral arms of galaxies and the stability of the so-called pinch configurations.

§ 108. The gravitational instability of an infinite cylinder (in a larger context) is considered in:

4. S. Chandrasekhar and E. Fermi, 'Problems of gravitational stability in the presence of a magnetic field', *Astrophys. J.* **118,** 116–41 (1953).

See also:

5. S. Chandrasekhar, 'Problems of stability in hydrodynamics and hydromagnetics', *Monthly Notices Roy. Astron. Soc. London,* **113,** 667–78 (1953).

§ 109. The role of viscosity in capillary instability was considered by Rayleigh (see reference 11 below and the comments on it); the treatment in the text is an extension to gravitational instability.

§ 110. The stabilizing effect of an axial magnetic field on the gravitational instability of an infinite cylinder is considered in reference 4. To illustrate the nature of the effect under the simplest conditions, the case was considered in which the field was entirely confined within the cylinder and none existed outside it.

Extensions of the treatment in reference 4 to other distributions of the field (including the one considered in the text) are contained in:

6. F. C. Auluck and D. S. Kothari, 'The influence of a magnetic field on the longitudinal stability of a gravitating cylinder', *Z. f. Astrophysik,* **42,** 101–13 (1957).
7. S. K. Trehan, 'The stability of an infinitely long cylinder with a prevalent force-free magnetic field', *Astrophys. J.* **127,** 436–45 (1958).
8. R. Simon, 'The hydromagnetic oscillations of an incompressible cylinder', ibid. **128,** 375–83 (1958).
9. —— 'Les oscillations hydromagnétiques d'un cylindre incompressible', *Ann. d'Astrophys.* **22,** 712–26 (1959).

§ 111. As we have stated, the classical investigations on capillary instability are those of Savart, Plateau, and Rayleigh.

§ 111 (*b*). The capillary instability of a hollow jet is treated in:

10. LORD RAYLEIGH, 'On the instability of cylindrical fluid surfaces,' *Scientific Papers*, iii, 594–6, Cambridge, England, 1902.

§ 111 (*c*). The effect of viscosity was also treated by Rayleigh:

11. LORD RAYLEIGH, 'On the instability of a cylinder of viscous liquid under capillary force', *Scientific Papers*, iii, 585–93, Cambridge, England, 1902.

However, Rayleigh restricted himself to the case when the effect of viscosity was paramount. The general case does not seem to have been considered before.

§ 112. The problem considered in this section does not seem to have been investigated though some beautiful experiments bearing on this topic have been reported by:

12. A. DATTNER, B. LEHNERT, and S. LUNDQUIST, 'Liquid conductor model of instabilities in a pinched discharge', *Second UN Geneva Conference*, 325–7, Pergamon Press, London, 1959.

§ 113. The stability of the simplest solution of the equations of hydromagnetics is proved in:

13. S. CHANDRASEKHAR, 'On the stability of the simplest solution of the equations of hydromagnetics', *Proc. Nat. Acad. Sci.* **42**, 273–6 (1956).

The example considered in § (*a*) is due to:

14. S. K. TREHAN, 'The hydromagnetic oscillations of twisted magnetic fields, I', *Astrophys. J.* **127**, 446–53 (1958).

See also:

15. —— and W. H. REID, 'The hydromagnetic oscillations of twisted magnetic fields. II', ibid. 454–8 (1958).

§ 114. The general problem treated in this section was first considered by:

16. S. K. TREHAN, 'The effect of fluid motions on the stability of twisted magnetic fields,' ibid. **129**, 475–82 (1959).

The case when no motions are present was treated by:

17. P. H. ROBERTS, 'Twisted magnetic fields', ibid. **124**, 430–42 (1956).

See also:

18. S. LUNDQUIST, 'On the stability of magneto-hydrostatic fields', *Physical Review*, **83**, 307–11 (1951).
19. J. W. DUNGEY and R. E. LOUGHHEAD, 'Twisted magnetic fields in conducting fluids', *Australian J. Phys.* **7**, 5–13 (1954).

§ 115. The literature on the stability of pinch configurations is an extensive one. The following is a very brief list of those papers which bear on the very special problem treated in the text:

20. M. KRUSKAL and J. L. TUCK, 'The instability of a pinched fluid with a longitudinal magnetic field', *Proc. Roy. Soc.* (*London*) A, **245**, 222–37 (1958).
21. M. ROSENBLUTH, 'Stability of the pinch', *Los Alamos Scientific Laboratory Report*, No. 2030, April 1956.

22. R. J. Tayler, 'A note on hydromagnetic stability problems', *Phil. Mag.*, Ser. 8, **2**, 33–36 (1957).

23. —— 'Hydromagnetic instabilities of an ideally conducting fluid', *Proc. Phys. Soc. London*, B, **70**, 31–48 (1957).

24. —— 'The influence of an axial magnetic field on the stability of a constricted gas discharge', ibid. 1049–63 (1957).

25. V. D. Shafranov, 'On the stability of a plasma column in the presence of a longitudinal magnetic field and a conduction casing', *Plasma Physics and the Problem of Controlled Thermonuclear Reactions*, ii, 197–215, edited by M. A. Leontovich, Pergamon Press, 1959.

26. T. F. Volkov, 'On stability of a plasma cylinder in an external magnetic field', ibid. 216–22, edited by M. A. Leontovich, Pergamon Press, 1959.

27. K. Hain, R. Lüst, and A. Schlüter, 'Zur Stabilität eines Plasmas', *Z. f. Naturforschung*, **12a**, 833–41 (1957).

28. —— —— 'Zur Stabilität zylindersymmetrischer Plasmakonfigurationen mit Volumenströmen', ibid. **13a**, 936–40 (1958).

29. S. Chandrasekhar, A. N. Kaufman, and K. M. Watson, 'The stability of the pinch', *Proc. Roy. Soc.* (*London*) A, **245**, 435–55 (1958).

The numerical results described and illustrated in the text are those of Tayler (reference 24).

XIII

GRAVITATIONAL EQUILIBRIUM AND GRAVITATIONAL INSTABILITY

116. Introduction

In this chapter we shall consider some general theorems on gravitational equilibrium and gravitational instability. These matters clearly have astronomical overtones; indeed, certain aspects of them have been extensively investigated in the astronomical contexts. However, our concern with them will be only to the extent they bear on the general theory of hydrodynamic and hydromagnetic stability.

117. The virial theorem

Consider an inviscid fluid of zero electrical resistivity in which a magnetic field $\mathbf{H}(\mathbf{x})$ prevails. Suppose that the fluid is a perfect gas and that the ratio of the specific heats is γ. Suppose further that apart from the prevailing magnetic field and gas pressure the only other force acting on the medium is that derived from its own gravitation. Under these circumstances the equation of motion governing the fluid velocities is

$$\rho\frac{du_i}{dt} = -\frac{\partial}{\partial x_i}\left(p+\frac{|\mathbf{H}|^2}{8\pi}\right)+\rho\frac{\partial \mathfrak{B}}{\partial x_i}+\frac{1}{4\pi}\frac{\partial}{\partial x_j}H_i H_j, \tag{1}$$

where

$$\frac{d}{dt} = \frac{\partial}{\partial t}+u_j\frac{\partial}{\partial x_j} \tag{2}$$

is the total time derivative. In equation (1) $\mathfrak{B}$ denotes the gravitational potential, and the rest of the symbols have their usual meanings. (We are setting $\mu = 1$ in the present analysis.)

(a) *Some definitions and relations*

Since we have supposed that the gravitational potential $\mathfrak{B}(\mathbf{x})$ is derived from the distribution of the matter present,

$$\mathfrak{B}(\mathbf{x}) = G\int_V \frac{\rho(\mathbf{x}')}{|\mathbf{x}-\mathbf{x}'|}\,d\mathbf{x}', \tag{3}$$

where for the sake of brevity we have written

$$d\mathbf{x}' = dx_1\,dx_2\,dx_3 \tag{4}$$

and abridged three integral signs into one. In (3) the integration is effected over the entire volume V occupied by the fluid.

We shall find it convenient to define the symmetric tensor,

$$\mathfrak{V}_{ik}(\mathbf{x}) = G \int_V \rho(\mathbf{x}') \frac{(x_i - x_i')(x_k - x_k')}{|\mathbf{x} - \mathbf{x}'|^3} \, d\mathbf{x}', \tag{5}$$

which represents a generalization of (3) and to which it reduces on contraction:

$$\mathfrak{V}_{ii} = \mathfrak{V}. \tag{6}$$

Similarly, in generalization of the usual definition of the gravitational potential energy, we shall define

$$\mathfrak{W}_{ik} = -\tfrac{1}{2} G \int_V \int_V \rho(\mathbf{x}) \rho(\mathbf{x}') \frac{(x_i - x_i')(x_k - x_k')}{|\mathbf{x} - \mathbf{x}'|^3} \, d\mathbf{x} d\mathbf{x}'; \tag{7}$$

and the contraction of this tensor gives the gravitational potential energy

$$\mathfrak{W}_{ii} = \mathfrak{W} = -\tfrac{1}{2} G \int_V \int_V \frac{\rho(\mathbf{x}) \rho(\mathbf{x}')}{|\mathbf{x} - \mathbf{x}'|} \, d\mathbf{x} d\mathbf{x}'. \tag{8}$$

An identity which follows from the foregoing definitions is

$$\mathfrak{W}_{ik} = \int_V \rho(\mathbf{x}) x_i \frac{\partial \mathfrak{V}}{\partial x_k} \, d\mathbf{x} = \int_V \rho(\mathbf{x}) x_k \frac{\partial \mathfrak{V}}{\partial x_i} \, d\mathbf{x}. \tag{9}$$

This can be established as follows: by definition,

$$\begin{aligned} \int_V d\mathbf{x} \rho(\mathbf{x}) x_i \frac{\partial \mathfrak{V}}{\partial x_k} &= G \int_V d\mathbf{x} \rho(\mathbf{x}) x_i \frac{\partial}{\partial x_k} \int_V d\mathbf{x}' \frac{\rho(\mathbf{x}')}{|\mathbf{x} - \mathbf{x}'|} \\ &= -G \int_V \int_V d\mathbf{x} \, d\mathbf{x}' \rho(\mathbf{x}) \rho(\mathbf{x}') \frac{x_i (x_k - x_k')}{|\mathbf{x} - \mathbf{x}'|^3}. \end{aligned} \tag{10}$$

By transposing the primed and the unprimed variables of integrations in (10) and taking the average of the two equivalent expressions, we verify that the double integral on the last line of (10) is the same as $\mathfrak{W}_{ik}$ defined in equation (7).

In addition to $\mathfrak{V}_{ik}$ and $\mathfrak{W}_{ik}$, we shall find it useful to define the further tensors

$$\mathfrak{T}_{ik} = \tfrac{1}{2} \int_V \rho u_i u_k \, d\mathbf{x} \tag{11}$$

and

$$\mathfrak{M}_{ik} = \frac{1}{8\pi} \int_V H_i H_k \, d\mathbf{x}; \tag{12†}$$

† The specification of the volume V over which this integration is effected requires some care; we return to this matter in § (*b*) below.

and the contraction of these tensors gives the kinetic and the magnetic energies of the system

$$\mathfrak{T} = \mathfrak{T}_{ii} = \tfrac{1}{2}\int\limits_V \rho|\mathbf{u}|^2\,d\mathbf{x} \tag{13}$$

and

$$\mathfrak{M} = \mathfrak{M}_{ii} = \frac{1}{8\pi}\int\limits_V |\mathbf{H}|^2\,d\mathbf{x}. \tag{14}$$

The analogous expression for the internal heat energy of the system is

$$\mathfrak{U} = \frac{1}{\gamma-1}\int\limits_V p\,d\mathbf{x}. \tag{15}$$

Even as $\mathfrak{W}_{ik}$, $\mathfrak{T}_{ik}$, $\mathfrak{M}_{ik}$, and $\mathfrak{U}$ characterize, in terms of a few parameters, the distribution of the different forms of energy in the configuration, so does the *inertia tensor*

$$I_{ik} = \int\limits_V \rho x_i x_k\,d\mathbf{x} \tag{16}$$

similarly characterize the distribution of the density in the configuration. The contraction of I_{ik} gives the moment of inertia

$$I = I_{ii} = \int\limits_V \rho|\mathbf{x}|^2\,d\mathbf{x}. \tag{17}$$

(*b*) *The general form of the virial theorem*

Returning to equation (1), multiply it by x_k and integrate it over the *entire* volume V in which the fluid *and* the field pervade. The left-hand side of the equation can be reduced in the manner

$$\begin{aligned}\int\limits_V \rho x_k \frac{du_i}{dt}\,d\mathbf{x} &= \int\limits_V \rho x_k \frac{d^2x_i}{dt^2}\,d\mathbf{x}\\ &= \int\limits_V \rho\frac{d}{dt}\left(x_k\frac{dx_i}{dt}\right)d\mathbf{x} - \int\limits_V \rho\frac{dx_k}{dt}\frac{dx_i}{dt}\,d\mathbf{x}\\ &= \int\limits_V \rho\frac{d}{dt}\left(x_k\frac{dx_i}{dt}\right)d\mathbf{x} - 2\mathfrak{T}_{ik}.\end{aligned} \tag{18}$$

The terms on the right-hand side, similarly, give

$$\begin{aligned}-\int\limits_V x_k\frac{\partial}{\partial x_i}\left(p+\frac{|\mathbf{H}|^2}{8\pi}\right)d\mathbf{x}\\ &= -\int\limits_S \left(p+\frac{|\mathbf{H}|^2}{8\pi}\right)x_k\,dS_i + \delta_{ik}\int\limits_V \left(p+\frac{|\mathbf{H}|^2}{8\pi}\right)d\mathbf{x}\\ &= -\int\limits_S \left(p+\frac{|\mathbf{H}|^2}{8\pi}\right)x_k\,dS_i + \delta_{ik}\{(\gamma-1)\mathfrak{U}+\mathfrak{M}\},\end{aligned} \tag{19}$$

$$\int_V \rho x_k \frac{\partial \mathfrak{B}}{\partial x_i}\, d\mathbf{x} = \mathfrak{W}_{ik}, \tag{20}$$

and

$$\frac{1}{4\pi}\int_V x_k \frac{\partial}{\partial x_j} H_i H_j\, d\mathbf{x} = \frac{1}{4\pi}\int_S x_k H_i H_j\, dS_j - \frac{1}{4\pi}\int_V H_i H_k\, d\mathbf{x}$$

$$= \frac{1}{4\pi}\int_S x_k H_i H_j\, dS_j - 2\mathfrak{M}_{ik}, \tag{21}$$

where in equations (19) and (21) the surface integrals are extended over the surface S bounding V.

Combining equations (18)–(21), we obtain

$$\int_V \rho \frac{d}{dt}\left(x_k \frac{dx_i}{dt}\right) d\mathbf{x} = 2\mathfrak{T}_{ik} + \delta_{ik}\{(\gamma-1)\mathfrak{U} + \mathfrak{M}\} +$$

$$+\mathfrak{W}_{ik} - 2\mathfrak{M}_{ik} + \frac{1}{8\pi}\int_S x_k(2H_i H_j\, dS_j - |\mathbf{H}|^2\, dS_i) - \int_S p x_k\, dS_i. \tag{22}$$

We shall suppose that the volume V, over which the integrations are extended, includes the *whole* system so that all the variables, p, ρ, and $\mathbf{H}$ may be assumed to vanish on S. It is important to note that this definition of V may require us to include in it volumes which one may normally consider as external to the 'natural' boundary of the system, namely, the surface on which the density ρ and the material pressure p vanish. The assumption that $\mathbf{H}$ vanishes on S may, indeed, require us to place S at infinity. This latter possibility arises because magnetic fields can extend far beyond the conventional limits of a material object; but to the extent the object is the seat of the magnetic field, there is justification in including all portions of space into which the field extends as parts of the system. And since the field of an object isolated in space must decrease at least as rapidly as that of a dipole (i.e. as r^{-3}), the surface integrals over the components of $\mathbf{H}$ in equation (22) will vanish under these circumstances when $S \to \infty$. However, it may sometimes be convenient to let S coincide with the natural boundary, in which case the surface integrals over $\mathbf{H}$ in equation (22) must be retained; and, moreover, $\mathfrak{M}_{ik}$ will then refer to only that part of the field which is interior to S.

In the present analysis we shall suppose that V includes (as we have already remarked) all parts of space in which the fluid *and* the field pervade. With this explicit understanding, equation (22) becomes

$$\int_V \rho \frac{d}{dt}\left(x_k \frac{dx_i}{dt}\right) d\mathbf{x} = 2\mathfrak{T}_{ik} - 2\mathfrak{M}_{ik} + \mathfrak{W}_{ik} + \delta_{ik}\{(\gamma-1)\mathfrak{U} + \mathfrak{M}\}. \tag{23}$$

Since all the tensors on the right-hand side of this equation are symmetric, the tensor on the left-hand side of the equation must also be symmetric. Therefore,

$$\int_V \rho \frac{d}{dt}\left(x_k \frac{dx_i}{dt}\right) d\mathbf{x} = \int_V \rho \frac{d}{dt}\left(x_i \frac{dx_k}{dt}\right) d\mathbf{x}. \tag{24}$$

An immediate consequence of this result is

$$\int_V \rho \frac{d}{dt}\left(x_k \frac{dx_i}{dt} - x_i \frac{dx_k}{dt}\right) d\mathbf{x} = \frac{d}{dt} \int_V \rho \left(x_k \frac{dx_i}{dt} - x_i \frac{dx_k}{dt}\right) d\mathbf{x} = 0, \tag{25}$$

where in taking d/dt outside of the integral sign, we have made use of the constancy of $\rho\, d\mathbf{x}$ assured by the equation of continuity. *Equation* (25) *expresses the constancy of the total angular momentum of the system.* It is worth noting that the existence of this integral of the equations of motion has not been affected by the presence of the magnetic field.

A further consequence of the identity (24) is that the quantity on the left-hand side of equation (23) can be replaced by

$$\frac{1}{2} \int_V \rho \frac{d}{dt}\left(x_k \frac{dx_i}{dt} + x_i \frac{dx_k}{dt}\right) d\mathbf{x} = \frac{1}{2} \frac{d^2}{dt^2} \int_V \rho x_i x_k \, d\mathbf{x} = \frac{1}{2} \frac{d^2 I_{ik}}{dt^2}. \tag{26}$$

With this replacement, equation (23) gives

$$\frac{1}{2} \frac{d^2 I_{ik}}{dt^2} = 2\mathfrak{T}_{ik} - 2\mathfrak{M}_{ik} + \mathfrak{W}_{ik} + \delta_{ik}\{(\gamma-1)\mathfrak{U} + \mathfrak{M}\}. \tag{27}$$

This equation represents the complete statement of *the virial theorem for hydromagnetics*.

When $i \neq k$, equation (27) gives

$$\frac{1}{2} \frac{d^2 I_{ik}}{dt^2} = 2\mathfrak{T}_{ik} - 2\mathfrak{M}_{ik} + \mathfrak{W}_{ik} \quad (i \neq k); \tag{28}$$

and by contracting the indices in equation (27), we obtain

$$\frac{1}{2} \frac{d^2 I}{dt^2} = 2\mathfrak{T} + \mathfrak{M} + \mathfrak{W} + 3(\gamma-1)\mathfrak{U}. \tag{29}$$

(c) *The virial theorem for equilibrium configurations*

For configurations in equilibrium and in a steady state, equations (27) and (29) give

$$2\mathfrak{T}_{ik} - 2\mathfrak{M}_{ik} + \mathfrak{W}_{ik} + \delta_{ik}\{(\gamma-1)\mathfrak{U} + \mathfrak{M}\} = 0 \tag{30}$$

and

$$2\mathfrak{T} + \mathfrak{M} + \mathfrak{W} + 3(\gamma-1)\mathfrak{U} = 0. \tag{31}$$

An alternative way of stating the result expressed by equation (30) is that *the tensor* $2(\mathfrak{T}_{ik} - \mathfrak{M}_{ik}) + \mathfrak{W}_{ik}$ *is isotropic.*

Consider first the case when there are no fluid motions or magnetic fields. In this case, equation (30) gives

$$\mathfrak{W}_{ik} = \int_V \rho(\mathbf{x}) x_i \frac{\partial \mathfrak{B}}{\partial x_k} \, d\mathbf{x} = -(\gamma-1)\mathfrak{U}\delta_{ik}. \tag{32}$$

This relation is compatible with a spherical symmetry of the underlying density distribution but does not require it.

When $\mathfrak{T}_{ik}$ and $\mathfrak{M}_{ik}$ are not zero, $\mathfrak{W}_{ik}$ will not, in general, be diagonal. Consequently, *a spherical symmetry of the configuration is, in general, incompatible with the presence of fluid motions and magnetic fields.* An exception is possible if $\mathfrak{T}_{ik} = \mathfrak{M}_{ik}$—an identity which is satisfied, for example, by the equipartition solution discussed in Chapter XII, § 113; for, when $\mathfrak{T}_{ik} = \mathfrak{M}_{ik}$, equation (30) gives

$$\mathfrak{W}_{ik} = \int_V \rho(\mathbf{x}) x_i \frac{\partial \mathfrak{B}}{\partial x_k} \, d\mathbf{x} = -\{(\gamma-1)\mathfrak{U}+\mathfrak{M}\}\delta_{ik}; \tag{33}$$

and this equation is not incompatible with spherical symmetry.

In the special case when only magnetic fields are present,

$$\mathfrak{W}_{ik} = 2\mathfrak{M}_{ik} - \delta_{ik}\{(\gamma-1)\mathfrak{U}+\mathfrak{M}\}; \tag{34}$$

and the impossibility of spherical symmetry, in general, follows from the relation

$$\mathfrak{W}_{ik} = 2\mathfrak{M}_{ik} \neq 0 \quad \text{for } i \neq k. \tag{35}$$

Consider next the contracted form (31) of the general relation. If $\mathfrak{E}$ denotes the total energy of the configuration,

$$\mathfrak{E} = \mathfrak{T}+\mathfrak{U}+\mathfrak{M}+\mathfrak{W}. \tag{36}$$

From equations (31) and (36) it would appear that, as far as these scalar equations go, the magnetic energy can be considered together with the gravitational potential energy. Thus, by eliminating $\mathfrak{M}+\mathfrak{W}$ between the two equations, we obtain the relation,

$$\mathfrak{E} = -\mathfrak{T}-(3\gamma-4)\mathfrak{U}, \tag{37}$$

which is independent of $\mathfrak{M}$. Alternatively, by eliminating $\mathfrak{U}$, we obtain

$$\mathfrak{E} = \frac{1}{3(\gamma-1)}\{(3\gamma-4)(\mathfrak{M}+\mathfrak{W})+(3\gamma-5)\mathfrak{T}\}. \tag{38}$$

In case $\mathfrak{T} = 0$, equations (37) and (38) give

$$\mathfrak{E} = -(3\gamma-4)\mathfrak{U} \tag{39}$$

and

$$\mathfrak{E} = \frac{3\gamma-4}{3(\gamma-1)}(\mathfrak{M}+\mathfrak{W}). \tag{40}$$

An immediate consequence of equation (39) is this: in a configuration for which $\gamma = \frac{4}{3}$, $\mathfrak{E} = 0$ in a steady state; and this is true independently of the radius of the configuration. A small radial expansion during which the configuration changes from one equilibrium state to an adjacent equilibrium state without any expenditure of energy is, therefore, possible. Moreover, since a gravitationally stable configuration must have a total energy which is negative, it is clearly necessary that *for stability the ratio of the specific heats* γ *must be greater than* $\frac{4}{3}$.

A further consequence of equation (40) is that for the total energy $\mathfrak{E}$ to be negative it is necessary that

$$\mathfrak{M}+\mathfrak{W} < 0 \quad \text{or} \quad \mathfrak{M} < |\mathfrak{W}|. \tag{41}$$

Therefore, *for the existence of a stable equilibrium, it is necessary that the total magnetic energy of a system does not exceed its negative gravitational potential energy.*

The inequality (41) enables one to place a useful upper limit on the possible strengths of the magnetic fields of cosmical bodies. Thus, by writing

$$\mathfrak{M} = \tfrac{1}{6}R^3\langle H^2\rangle_{\text{av}} \quad \text{and} \quad |\mathfrak{W}| = q\frac{GM^2}{R}, \tag{42}$$

where R is a measure of the linear dimension of the system, M is its mass, and q is a numerical constant of order unity, we obtain

$$\sqrt{\langle H^2\rangle_{\text{av}}} < \frac{M}{R^2}\sqrt{(6qG)}. \tag{43}$$

118. The virial theorem for small oscillations about equilibrium

We shall now consider small oscillations about equilibrium of a configuration in which magnetic fields are present. We shall, however, suppose that in the stationary state there are no fluid motions; and we shall seek the form which the virial theorem takes under these circumstances.

Considering periodic oscillations with a gyration frequency σ, let $\boldsymbol{\xi}e^{i\sigma t}$ denote the Lagrangian displacement of an element of mass, $dm = \rho\,d\mathbf{x}$, from its equilibrium position at $\mathbf{x}$. Let $\delta p e^{i\sigma t}$, $\delta\rho e^{i\sigma t}$, and $\delta\mathbf{H}e^{i\sigma t}$ denote the corresponding changes in the other physical variables as we follow the fluid element with its motion. The equation of continuity ensures the constancy of dm for such Lagrangian displacements; and if we suppose that the oscillations take place adiabatically, then

$$\delta p/p = \gamma\delta\rho/\rho, \tag{44}$$

where

$$\delta\rho/\rho = -\operatorname{div}\boldsymbol{\xi}. \tag{45}$$

If $\delta I_{ik}\, e^{i\sigma t}$, $\delta \mathfrak{M}_{ik}\, e^{i\sigma t}$, etc., denote the first-order changes in the various integrals representing these quantities in the stationary state, then equation (27) gives

$$-\tfrac{1}{2}\sigma^2 \delta I_{ik} = \delta \mathfrak{W}_{ik} + \delta_{ik}\{(\gamma-1)\delta\mathfrak{U} + \delta\mathfrak{M}\} - 2\delta\mathfrak{M}_{ik}; \tag{46}$$

the term in $\mathfrak{T}_{ik}$ makes no contribution in this order since we have supposed that there are no zero-order fluid motions.

We shall now consider, in turn, the various quantities occurring in equation (46). Clearly,

$$\delta I_{ik} = \int_V \rho(\xi_i x_k + \xi_k x_i)\, d\mathbf{x}. \tag{47}$$

Also, according to the definitions of $\mathfrak{B}_{ik}$ and $\mathfrak{W}_{ik}$ given in equations (5) and (7),

$$\begin{aligned} \delta \mathfrak{W}_{ik} &= -G \int_V \int_V dm\, dm' \xi_j \frac{\partial}{\partial x_j} \frac{(x_i - x_i')(x_k - x_k')}{|\mathbf{x}-\mathbf{x}'|^3} \\ &= -G \int_V d\mathbf{x} \rho(\mathbf{x}) \xi_j \frac{\partial}{\partial x_j} \int_V d\mathbf{x}' \rho(x') \frac{(x_i - x_i')(x_k - x_k')}{|\mathbf{x}-\mathbf{x}'|^3} \\ &= -\int_V d\mathbf{x} \rho(\mathbf{x}) \xi_j \frac{\partial \mathfrak{B}_{ik}}{\partial x_j}. \end{aligned} \tag{48}$$

This last relation is a generalization of the familiar formula,

$$\delta \mathfrak{W} = -\int_V d\mathbf{x} \rho(\mathbf{x}) \xi_j \frac{\partial \mathfrak{B}}{\partial x_j}, \tag{49}$$

for the first-order change in the gravitational potential energy consequent to a slight redistribution of the matter in the system.

Turning next to the change in the internal energy, we have

$$(\gamma-1)\delta\mathfrak{U} = \int_V \delta\!\left(\frac{p}{\rho}\right) \rho\, d\mathbf{x}, \tag{50}$$

or, making use of the relations (44) and (45), we have

$$(\gamma-1)\delta\mathfrak{U} = (\gamma-1) \int_V \frac{p}{\rho} \frac{\delta\rho}{\rho} \rho\, d\mathbf{x} = -(\gamma-1) \int_V p \frac{\partial \xi_j}{\partial x_j}\, d\mathbf{x}. \tag{51}$$

Letting
$$\Pi = p + |\mathbf{H}|^2/8\pi, \tag{52}$$
we can write

$$\delta\mathfrak{U} = -\int_V \Pi \frac{\partial \xi_j}{\partial x_j}\, d\mathbf{x} + \frac{1}{8\pi} \int_V |\mathbf{H}|^2 \frac{\partial \xi_j}{\partial x_j}\, d\mathbf{x}. \tag{53}$$

Integrating by parts the first of the two integrals on the right-hand side, we obtain

$$\int\limits_V \Pi \frac{\partial \xi_j}{\partial x_j}\, d\mathbf{x} = \int\limits_S \Pi\, \xi_j\, dS_j - \int\limits_V \xi_j \frac{\partial \Pi}{\partial x_j}\, d\mathbf{x}. \tag{54}$$

We shall suppose that S is so placed that all the physical variables and their perturbations vanish on S. (As we have remarked in § 117, this may require that S be placed at infinity.) Then

$$\int\limits_V \Pi \frac{\partial \xi_j}{\partial x_j}\, d\mathbf{x} = -\int\limits_V \xi_j \frac{\partial \Pi}{\partial x_j}\, d\mathbf{x}, \tag{55}$$

or, making use of the relation,

$$\frac{\partial \Pi}{\partial x_j} = \rho \frac{\partial \mathfrak{B}}{\partial x_j} + \frac{1}{4\pi} \frac{\partial}{\partial x_l} H_l H_j, \tag{56}$$

which obtains in equilibrium, we have

$$\int\limits_V \Pi \frac{\partial \xi_j}{\partial x_j}\, d\mathbf{x} = -\int\limits_V \rho \xi_j \frac{\partial \mathfrak{B}}{\partial x_j}\, d\mathbf{x} - \frac{1}{4\pi} \int\limits_V \xi_j \frac{\partial}{\partial x_l} H_l H_j\, d\mathbf{x}. \tag{57}$$

Integrating by parts the second of the two integrals on the right-hand side of this equation, we obtain

$$\int\limits_V \Pi \frac{\partial \xi_j}{\partial x_j}\, d\mathbf{x} = -\int\limits_V \rho \xi_j \frac{\partial \mathfrak{B}}{\partial x_j}\, d\mathbf{x} + \frac{1}{4\pi} \int\limits_V H_l H_j \frac{\partial \xi_j}{\partial x_l}\, d\mathbf{x}; \tag{58}$$

the integrated part, again, making no contribution. Now combining equations (53) and (58), we have

$$\delta \mathfrak{U} = \int\limits_V \rho \xi_j \frac{\partial \mathfrak{B}}{\partial x_j}\, d\mathbf{x} + \frac{1}{8\pi} \int\limits_V \left(|\mathbf{H}|^2 \frac{\partial \xi_j}{\partial x_j} - 2 H_l H_j \frac{\partial \xi_j}{\partial x_l} \right) d\mathbf{x}. \tag{59}$$

Finally, considering $\delta \mathfrak{M}_{ik}$, we have

$$\delta \mathfrak{M}_{ik} = \frac{1}{8\pi} \int\limits_V (H_k\, \delta H_i + H_i\, \delta H_k)\, d\mathbf{x} + \frac{1}{8\pi} \int\limits_V H_i H_k \frac{\partial \xi_j}{\partial x_j}\, d\mathbf{x}, \tag{60}$$

where the second term arises from allowing for the variation of the volume element $d\mathbf{x}$, following the motion in accordance with equation (45), and the constancy of $dm = \rho\, d\mathbf{x}$.

Now the change in the magnetic field $\delta \mathbf{H}$, as we follow the motion of the fluid element, is given by

$$\delta \mathbf{H} = \operatorname{curl}(\boldsymbol{\xi} \times \mathbf{H}) + (\boldsymbol{\xi} \cdot \operatorname{grad}) \mathbf{H}. \tag{61}$$

The first term on the right-hand side gives the Eulerian change in $\mathbf{H}$ at a *fixed point* caused by the redistribution of the matter, while the second term is the allowance for the fact that, as we follow the fluid element, it finds itself in a slightly displaced location. In the notation of Cartesian tensors, equation (61) becomes

$$\delta H_i = H_j \frac{\partial \xi_i}{\partial x_j} - H_i \frac{\partial \xi_j}{\partial x_j}. \tag{62}$$

Making use of this relation, we find that equation (60) reduces to

$$\delta \mathfrak{M}_{ik} = \frac{1}{8\pi} \int_V \left\{ H_j \left(H_k \frac{\partial \xi_i}{\partial x_j} + H_i \frac{\partial \xi_k}{\partial x_j} \right) - H_i H_k \frac{\partial \xi_j}{\partial x_j} \right\} d\mathbf{x}; \tag{63}$$

and by contracting, we find

$$\delta \mathfrak{M} = -\frac{1}{8\pi} \int_V \left(|\mathbf{H}|^2 \frac{\partial \xi_j}{\partial x_j} - 2 H_j H_l \frac{\partial \xi_l}{\partial x_j} \right) d\mathbf{x}. \tag{64}$$

Now inserting in equation (46) for δI_{ik}, $\delta \mathfrak{W}_{ik}$, etc., the expressions we have derived for them, we find

$$-\tfrac{1}{2}\sigma^2 \int_V \rho(\xi_i x_k + \xi_k x_i)\, d\mathbf{x} = -\int_V \rho \xi_j \frac{\partial \mathfrak{B}_{ik}}{\partial x_j}\, d\mathbf{x} +$$

$$+\delta_{ik} \left\{ (\gamma - 1) \int_V \rho \xi_j \frac{\partial \mathfrak{B}}{\partial x_j}\, d\mathbf{x} + \frac{\gamma - 2}{8\pi} \int_V \left(|\mathbf{H}|^2 \frac{\partial \xi_j}{\partial x_j} - 2 H_j H_l \frac{\partial \xi_l}{\partial x_j} \right) d\mathbf{x} \right\} +$$

$$+\frac{1}{4\pi} \int_V \left\{ H_i H_k \frac{\partial \xi_j}{\partial x_j} - H_j \left(H_k \frac{\partial \xi_i}{\partial x_j} + H_i \frac{\partial \xi_k}{\partial x_j} \right) \right\} d\mathbf{x}. \tag{65}$$

By contracting this equation, we obtain

$$-\sigma^2 \int_V \rho \xi_i x_i\, d\mathbf{x}$$

$$= (3\gamma - 4) \left\{ \int_V \rho \xi_j \frac{\partial \mathfrak{B}}{\partial x_j}\, d\mathbf{x} + \frac{1}{8\pi} \int_V \left(|\mathbf{H}|^2 \frac{\partial \xi_j}{\partial x_j} - 2 H_j H_l \frac{\partial \xi_l}{\partial x_j} \right) d\mathbf{x} \right\}. \tag{66}$$

(a) A characteristic equation for determining the periods of oscillation

Equations (65) and (66) can be used to obtain estimates for σ^2 by inserting in them suitable 'trial' functions for $\boldsymbol{\xi}$. Thus, the simplest assumption

$$\boldsymbol{\xi} = \text{constant } \mathbf{x} \tag{67}$$

together with equation (66), yields the formula

$$\sigma^2 = -(3\gamma - 4) \frac{\mathfrak{W} + \mathfrak{M}}{I}. \tag{68}$$

However, since a configuration with a prevalent magnetic field cannot be spherically symmetric, the use of a trial function which corresponds to a uniform radial expansion would appear unsuitable; indeed, the substitution of (67) in the tensor equation (65) will lead to gross inconsistencies unless the departures from spherical symmetry of the equilibrium configuration are small. If the latter should not be the case, a more consistent procedure would appear to be the following.

Assume that
$$\boldsymbol{\xi} = \mathbf{X}\mathbf{x}, \tag{69}$$
where $\mathbf{X}$ is a symmetric matrix:
$$X_{ij} = X_{ji}. \tag{70}$$
Further, define the super-matrix
$$\mathfrak{W}_{lj;ik} = \int_V \rho(\mathbf{x}) x_l \frac{\partial \mathfrak{V}_{ik}}{\partial x_j}\, d\mathbf{x}. \tag{71}$$
By contracting this super-matrix with respect to i and k, we obtain
$$\mathfrak{W}_{lj;ii} = \int_V \rho(\mathbf{x}) x_l \frac{\partial \mathfrak{V}}{\partial x_j}\, d\mathbf{x} = \mathfrak{W}_{lj}. \tag{72}$$
Also, let
$$\boldsymbol{\mathfrak{M}} = (\mathfrak{M}_{ik}), \quad \boldsymbol{\mathfrak{W}} = (\mathfrak{W}_{ik}), \quad \text{and} \quad \mathbf{I} = (I_{ik}). \tag{73}$$
Clearly,
$$\mathfrak{M} = \mathfrak{M}_{ii} = \operatorname{tr}(\boldsymbol{\mathfrak{M}}) \quad \text{and} \quad \mathfrak{W} = \mathfrak{W}_{ii} = \operatorname{tr}(\boldsymbol{\mathfrak{W}}), \tag{74}$$
where 'tr' stands for trace (meaning diagonal sum). Similarly,
$$\frac{\partial \xi_j}{\partial x_j} = \operatorname{tr}(\mathbf{X}). \tag{75}$$

With the foregoing definitions, the substitution of the trial function (69) in equation (65) leads to the result
$$\begin{aligned} -\tfrac{1}{2}\sigma^2(\mathbf{IX}+\mathbf{XI})_{ik} = {} & -X_{jl}\,\mathfrak{W}_{lj;ik} + \\ & +\delta_{ik}\{(\gamma-1)\operatorname{tr}(\mathbf{X}\boldsymbol{\mathfrak{W}})+(\gamma-2)[\operatorname{tr}(\mathbf{X})\operatorname{tr}(\boldsymbol{\mathfrak{M}})-2\operatorname{tr}(\mathbf{X}\boldsymbol{\mathfrak{M}})]\}+ \\ & +2\{\operatorname{tr}(\mathbf{X})\mathfrak{M}_{ik}-[\mathbf{X}\boldsymbol{\mathfrak{M}}+\boldsymbol{\mathfrak{M}}\mathbf{X}]_{ik}\}. \end{aligned} \tag{76}$$
Equation (76) represents a system of linear homogeneous equations for the six coefficients of the (assumed) symmetric linear transformation $\mathbf{X}$. The determinant of the system must, therefore, vanish; and the resulting characteristic equation will not only determine σ^2 but also the transformation $\mathbf{X}$ (apart from a constant of proportionality).

Finally, we may note that by contracting the tensor equation (76) we obtain the scalar equation
$$-\sigma^2 \operatorname{tr}(\mathbf{IX}) = (3\gamma-4)\{\operatorname{tr}(\mathbf{X}\boldsymbol{\mathfrak{W}})+\operatorname{tr}(\mathbf{X})\operatorname{tr}(\boldsymbol{\mathfrak{M}})-2\operatorname{tr}(\mathbf{X}\boldsymbol{\mathfrak{M}})] \tag{77}$$
which is a generalization of (68).

119. The gravitational instability of an infinite homogeneous medium; Jeans's criterion

A particularly simple example of gravitational instability was discovered by Jeans. In this example, we start from an infinite homogeneous medium at rest and consider the velocity of propagation of a small fluctuation in the density. If the gravitational effects of the fluctuation are ignored, the problem reduces to the classical one of the propagation of sound; and as is well known, the velocity of sound c is independent of the wave number and is given by

$$c^2 = \gamma p/\rho. \tag{78}$$

If the change in the gravitational potential consequent to the fluctuation $\delta\rho$ in the density is taken into account, the velocity of wave propagation depends, as we shall presently see, on the wave number k, and is, indeed, imaginary for all wave numbers less than a certain value k_J. The instability which follows for $k < k_J$ is the gravitational instability discovered by Jeans.

The demonstration of the gravitational instability of an infinite homogeneous medium is simple. The relevant perturbation equations are

$$\rho\frac{\partial \mathbf{u}}{\partial t} = -\operatorname{grad}\delta p+\rho\operatorname{grad}\delta\mathfrak{V}, \tag{79}$$

$$\frac{\partial}{\partial t}\delta\rho = -\rho\operatorname{div}\mathbf{u}, \tag{80}$$

and

$$\nabla^2\delta\mathfrak{V} = -4\pi G\delta\rho. \tag{81}$$

Further, we shall suppose that the fluctuations in the pressure and the density take place adiabatically so that

$$\delta p/p = \gamma\delta\rho/\rho \tag{82}$$

or

$$\delta p = c^2\delta\rho. \tag{83}$$

Inserting this last expression for δp in equation (79) and taking the divergence of the resulting equation, we obtain

$$\rho\frac{\partial}{\partial t}\operatorname{div}\mathbf{u} = -c^2\nabla^2\delta\rho+\rho\nabla^2\delta\mathfrak{V}. \tag{84}$$

Now making use of equations (80) and (81), we find

$$\frac{\partial^2}{\partial t^2}\delta\rho = c^2\nabla^2\delta\rho+4\pi G\rho\delta\rho. \tag{85}$$

This equation allows solutions of the form

$$\delta\rho = \text{constant}\, e^{i(\mathbf{k}.\mathbf{x}+\sigma t)} \tag{86}$$

provided

$$\sigma^2 = c^2k^2-4\pi G\rho. \tag{87}$$

The velocity of propagation V_J of these waves is, therefore, given by

$$V_J = \frac{\sigma}{k} = c\sqrt{\left(1-\frac{4\pi G\rho}{c^2k^2}\right)}. \tag{88}$$

We may call V_J the *Jeans velocity*. From equation (87) or (88) it follows that *there is instability for all wave numbers*

$$k < k_J \quad where \quad k_J = \frac{1}{c}\sqrt{(4\pi G\rho)}. \tag{89}$$

This is Jeans's result; and the condition (89) for instability is often referred to as *Jeans's criterion.*

The origin of the gravitational instability associated with the long waves lies in the circumstance that, while the generation of a pressure wave requires the expenditure of energy proportional to c^2, it simultaneously results in a release of gravitational potential energy of an amount proportional to $4\pi G\rho/k^2$; and when Jeans's criterion is satisfied, the system can spontaneously go to states of lower energy with the liberation of thermal energy; and this, of course, means instability.

120. The effect of a uniform rotation and a uniform magnetic field on Jeans's criterion

We shall prove that *Jeans's criterion for the gravitational instability of an infinite homogeneous medium is unaffected by the presence, separately, or simultaneously, of a uniform rotation and a uniform magnetic field.*

(a) *The effect of a uniform rotation*

The effect of a uniform rotation $\mathbf{\Omega}$ on Jeans's criterion is readily determined. In a frame of reference rotating with an angular velocity $\mathbf{\Omega}$, equation (79) is replaced by

$$\rho\frac{\partial \mathbf{u}}{\partial t} = -\text{grad}\,\delta p+\rho\,\text{grad}\,\delta\mathfrak{B}+2\rho\mathbf{u}\times\mathbf{\Omega} \tag{90}$$

and equations (80) and (81) are unaffected.

We shall consider solutions of equations (80), (81), and (90) which correspond to the propagation of waves in the z-direction (say). For such solutions $\partial/\partial z$ is the only non-vanishing component of the gradient; and in a system of coordinates so oriented that

$$\mathbf{\Omega} = (0,\Omega_y,\Omega_z), \tag{91}$$

the perturbation equations become:

$$\frac{\partial u_x}{\partial t}+2\Omega_y u_z-2\Omega_z u_y = 0, \tag{92}$$

$$\frac{\partial u_y}{\partial t}+2\Omega_z u_x = 0, \tag{93}$$

$$\frac{\partial u_z}{\partial t}+\frac{c^2}{\rho}\frac{\partial}{\partial z}\delta\rho-\frac{\partial}{\partial z}\delta\mathfrak{B}-2\Omega_y u_x = 0, \tag{94}$$

$$\frac{\partial}{\partial t}\delta\rho+\rho\frac{\partial u_z}{\partial z} = 0, \tag{95}$$

and

$$\frac{\partial^2}{\partial z^2}\delta\mathfrak{B}+4\pi G\delta\rho = 0. \tag{96}$$

Seeking solutions of these equations whose (z, t)-dependence is given by

$$e^{i(kz+\sigma t)}, \tag{97}$$

we are led to the following system of linear homogeneous equations:

$$\begin{vmatrix} 2i\Omega_z & \sigma & -2i\Omega_y & 0 & 0 \\ \sigma & -2i\Omega_z & 0 & 0 & 0 \\ 0 & +2i\Omega_y & \sigma & c^2k/\rho & -k \\ 0 & 0 & \rho k & \sigma & 0 \\ 0 & 0 & 0 & 4\pi G & -k^2 \end{vmatrix} \begin{vmatrix} u_y \\ u_x \\ u_z \\ \delta\rho \\ \delta\mathfrak{B} \end{vmatrix} = 0. \tag{98}$$

The determinant of this system of equations must vanish; and on expanding the determinant and reducing, we obtain the characteristic equation

$$\sigma^4-(4\Omega^2+c^2k^2-4\pi G\rho)\sigma^2+4\Omega^2(c^2k^2-4\pi G\rho)\cos^2\vartheta = 0, \tag{99}$$

where ϑ is the inclination of the direction of $\mathbf{\Omega}$ to the direction of wave propagation.

From equation (99) it follows that there are in general two modes in which a wave can be propagated in the medium. If σ_1 and σ_2 are the gyration frequencies of the two modes, then,

$$\sigma_1^2+\sigma_2^2 = 4\Omega^2+c^2k^2-4\pi G\rho \tag{100}$$

and

$$\sigma_1^2\sigma_2^2 = 4\Omega^2(c^2k^2-4\pi G\rho)\cos^2\vartheta. \tag{101}$$

It is also clear that both the roots σ_1^2 and σ_2^2 are real.

From equation (101) we can now conclude that if Jeans's condition,

$$c^2k^2-4\pi G\rho < 0, \tag{102}$$

is satisfied, then one of the two roots σ_1^2 or σ_2^2 must be negative; and, accordingly, one of the two modes must be unstable. An exception arises when the direction of wave propagation is at right angles to the direction of $\mathbf{\Omega}$; for, in this case, $\vartheta = \frac{1}{2}\pi$ and equation (99) gives

$$\sigma^2 = 4\Omega^2+c^2k^2-4\pi G\rho; \tag{103}$$

and the condition for the instability of these waves, namely,

$$4\Omega^2+c^2k^2-4\pi G\rho < 0, \tag{104}$$

cannot be met if $\Omega^2 > \pi G\rho$. We have thus shown: *for waves propagated in a direction at right angles to the direction of* $\mathbf{\Omega}$, *gravitational instability cannot occur if* $\Omega^2 > \pi G\rho$; *for propagation in every other direction gravitational instability will occur if Jeans's condition,* $c^2k^2-4\pi G\rho < 0$, *is satisfied.*

In considering equation (99) generally, it is convenient to measure

$$\begin{gathered}\sigma \text{ in the unit } \sqrt{(4\pi G\rho)}\ \text{sec}^{-1}, \\ k \text{ in the unit } \frac{1}{c}\sqrt{(4\pi G\rho)}\ \text{cm}^{-1},\end{gathered} \tag{105}$$

and $\qquad \sigma/k$ in the unit c cm/sec.

In these units, Jeans's criterion is, for example, $k < 1$.

When σ and k are measured in the units specified in (105), equation (99) takes the form

$$\sigma^4-(\Lambda^2+k^2-1)\sigma^2+\Lambda^2(k^2-1)\cos^2\vartheta = 0, \tag{106}$$

where
$$\Lambda = \Omega/\sqrt{(\pi G\rho)} \tag{107}$$

is the non-dimensional combination in which Ω enters this problem. The roots of equation (106) are given by

$$2\sigma^2 = \Lambda^2+k^2-1\pm\sqrt{\{(\Lambda^2+k^2-1)^2-4\Lambda^2(k^2-1)\cos^2\vartheta\}}. \tag{108}$$

The dispersion relations derived from this equation for $\Lambda^2 = 0{\cdot}5$, $1{\cdot}0$, and $2{\cdot}0$ and $\vartheta = 0°$, $45°$, and $90°$ are illustrated in Fig. 135. The manner in which the two modes cross one another and fill restricted parts of the σ, k-plane is curious and worth noting.

(b) *The effect of a uniform magnetic field*

When a uniform magnetic field of intensity $\mathbf{H}$ pervades the whole medium, the relevant perturbation equations are

$$\rho\frac{\partial \mathbf{u}}{\partial t} = \frac{1}{4\pi}\operatorname{curl}\mathbf{h}\times\mathbf{H}-c^2\operatorname{grad}\delta\rho+\rho\operatorname{grad}\delta\mathfrak{B}, \tag{109}$$

$$\frac{\partial \mathbf{h}}{\partial t} = \operatorname{curl}(\mathbf{u}\times\mathbf{H}), \tag{110}$$

$$\frac{\partial}{\partial t}\delta\rho = -\rho\operatorname{div}\mathbf{u}, \quad \text{and} \quad \nabla^2\delta\mathfrak{B} = -4\pi G\delta\rho. \tag{111}$$

We shall seek solutions of equations (109)–(111) which correspond to the propagation of waves in the z-direction. Then from the solenoidal character of $\mathbf{h}$ it follows that $h_z = 0$. And by further choosing the orientation of the coordinate system such that

$$\mathbf{H} = (0, H_y, H_z), \tag{112}$$

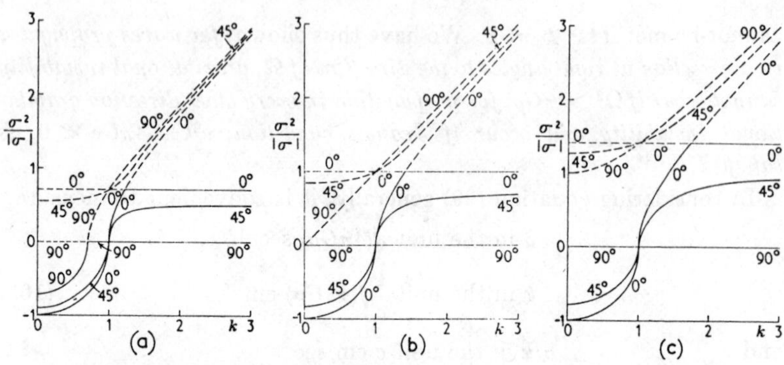

FIG. 135. The effect of rotation on the gravitational instability of an infinite homogeneous medium. The dispersion relations derived from equation (106) are illustrated for values of Λ^2 $(= \Omega^2/\pi G\rho) = 0\cdot5$, $1\cdot0$, and $2\cdot0$ in figures (a), (b), and (c), respectively. The curves are labelled by the angle which the direction of wave propagation makes with the direction $\mathbf{\Omega}$. The modes belonging to the same branch are drawn in the same way: either full-line or dashed. The abscissa gives the wave number in the unit $c^{-1}\sqrt{(4\pi G\rho)}$ cm^{-1} while the ordinate gives $\sigma^2/|\sigma|$ with σ measured in the unit $\sqrt{(4\pi G\rho)}$ sec^{-1}. (Negative values of the ordinate correspond to instability.)

we find that the equations which follow from equations (109)–(111) break up into the two non-combining systems:

$$\rho\frac{\partial u_x}{\partial t} = \frac{H_z}{4\pi}\frac{\partial h_x}{\partial z}, \qquad \frac{\partial h_x}{\partial t} = H_z\frac{\partial u_x}{\partial z}, \tag{113}$$

and

$$\rho\frac{\partial u_y}{\partial t} - \frac{H_z}{4\pi}\frac{\partial h_y}{\partial z} = 0, \tag{114}$$

$$\rho\frac{\partial u_z}{\partial t} + \frac{H_y}{4\pi}\frac{\partial h_y}{\partial z} + c^2\frac{\partial}{\partial z}\delta\rho - \rho\frac{\partial}{\partial z}\delta\mathfrak{B} = 0, \tag{115}$$

$$\frac{\partial h_y}{\partial t} + H_y\frac{\partial u_z}{\partial z} - H_z\frac{\partial u_y}{\partial z} = 0, \tag{116}$$

$$\frac{\partial}{\partial t}\delta\rho + \rho\frac{\partial u_z}{\partial z} = 0, \tag{117}$$

$$\frac{\partial^2}{\partial z^2}\delta\mathfrak{B} + 4\pi G\delta\rho = 0. \tag{118}$$

Equations (113) can be combined to give

$$\frac{\partial^2 h_x}{\partial t^2} = \frac{H_z^2}{4\pi\rho}\frac{\partial^2 h_x}{\partial z^2} \quad \text{and} \quad \frac{\partial^2 u_x}{\partial t^2} = \frac{H_z^2}{4\pi\rho}\frac{\partial^2 u_x}{\partial z^2}. \tag{119}$$

These equations are the same as those leading to the ordinary hydromagnetic waves of Alfvén propagated with the velocity $H_z/\surd(4\pi\rho)$ (cf. Chapter IV, § 39).

Turning next to equations (114)–(118) and seeking solutions having a (z, t)-dependence given by (97), we are led to the following system of linear homogeneous equations:

$$\begin{vmatrix} \rho\sigma & -kH_z/4\pi & 0 & 0 & 0 \\ 0 & +kH_y/4\pi & \rho\sigma & kc^2 & -k\rho \\ -kH_z & \sigma & kH_y & 0 & 0 \\ 0 & 0 & k\rho & \sigma & 0 \\ 0 & 0 & 0 & 4\pi G & -k^2 \end{vmatrix} \begin{Vmatrix} u_y \\ h_y \\ u_z \\ \delta\rho \\ \delta\mathfrak{B} \end{Vmatrix} = 0. \tag{120}$$

The vanishing of the determinant of this system leads to the characteristic equation

$$\sigma^4 - \left(\frac{H^2k^2}{4\pi\rho} + c^2k^2 - 4\pi G\rho\right)\sigma^2 + \frac{H^2k^2}{4\pi\rho}(c^2k^2 - 4\pi G\rho)\cos^2\vartheta = 0, \tag{121}$$

where ϑ is now the inclination of the direction of $\mathbf{H}$ to the direction of wave propagation.

From equation (121) it now follows that there are in general two modes of wave propagation besides the pure Alfvén modes described by equations (119). If σ_1 and σ_2 are the gyration frequencies of the two modes described by equation (121), then

$$\sigma_1^2 + \sigma_2^2 = \frac{Hk^2}{4\pi\rho} + c^2k^2 - 4\pi G\rho \tag{122}$$

and

$$\sigma_1^2\sigma_2^2 = \frac{H^2k^2}{4\pi\rho}(c^2k^2 - 4\pi G\rho)\cos^2\vartheta. \tag{123}$$

It is also clear that both the roots σ_1^2 and σ_2^2 are real.

From equation (123) we can now conclude that if Jeans's criterion (102) is satisfied, then one of the two roots σ_1^2 or σ_2^2 is negative; and accordingly, one of the two modes must be unstable. Thus, *Jeans's condition for gravitational instability is unaffected by the presence of a uniform magnetic field.* The physical reason for this is evident for a mode in which the density waves are perpendicular to the lines of force; for, in that case the motion of the particles are parallel to the lines of force and are, therefore, unaffected by the magnetic field.

In considering equation (121) generally, it is convenient to measure σ and k in the units specified in (105). Equation (121) then becomes

$$\sigma^4 - \left(\frac{V_A^2}{c^2}k^2 + k^2 - 1\right)\sigma^2 + \frac{V_A^2}{c^2}k^2(k^2-1)\cos^2\vartheta = 0, \tag{124}$$

where

$$V_A = H/\sqrt{(4\pi\rho)} \tag{125}$$

is the Alfvén speed.

The dispersion relations derived from equation (124) for $V_A^2/c^2 = 0{\cdot}5$, $1{\cdot}0$, and $2{\cdot}0$ and $\vartheta = 0°$, $45°$, and $90°$ are illustrated in Fig. 136. The

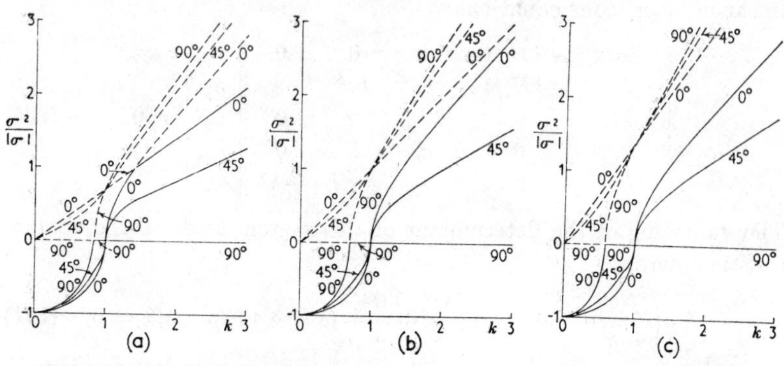

FIG. 136. The effect of a uniform magnetic field on the gravitational instability of an infinite homogeneous medium. The dispersion relation derived from equation (124) is illustrated for values of V_A^2/c^2 ($= H^2/(4\pi\rho c^2)$) $= 0{\cdot}5, 1{\cdot}0$, and $2{\cdot}0$ in figures (*a*), (*b*), and (*c*), respectively. The curves are labelled by the angles which the direction of wave propagation makes with the direction of the field. The modes belonging to the same branch are drawn in the same way: either full-line or dashed. The abscissa gives the wave number in the unit $c^{-1}\sqrt{(4\pi G\rho)}$ cm^{-1} while the ordinate gives $\sigma^2/|\sigma|$ with σ measured in the unit $\sqrt{(4\pi G\rho)}$sec^{-1}. (Negative values of the ordinate correspond to instability.)

manner in which the two modes cross one another and fill restricted parts of the (σ, k)-plane is worth noting. The principal fact concerning these crossings is that the modes which, for $H \to 0$, are the pure Jeans and the pure Alfvén modes become, when $H \to \infty$, *modified Alfvén* and *modified Jeans* modes, respectively: the 'modification' consists in the replacement of H_z by H in the expression for the Alfvén velocity in the former case and in the addition of a factor $\cos\vartheta$ to the Jeans velocity in the latter case. This crossing of the two modes is not relevant to the unstable modes which always occur for $k < 1$ (in the chosen units).

(c) *The effect of the simultaneous presence of rotation and magnetic field*

Finally, we shall consider the joint effects of a uniform rotation and a uniform magnetic field on the gravitational instability of an infinite homogeneous medium. In this general case, equation (109) is replaced by

$$\rho\frac{\partial \mathbf{u}}{\partial t} = \frac{1}{4\pi}\operatorname{curl}\mathbf{h}\times\mathbf{H}+2\rho\mathbf{u}\times\mathbf{\Omega}-c^2\operatorname{grad}\delta p+\rho\operatorname{grad}\delta\mathfrak{B}; \tag{126}$$

but the remaining equations (110) and (111) are unchanged.

Choosing the orientation of the coordinate axes such that

$$\mathbf{H} = (0, H_y, H_z) \quad\text{and}\quad \mathbf{\Omega} = (\Omega_x, \Omega_y, \Omega_z), \tag{127}$$

and seeking solutions which are independent of x and y and whose dependence on z and t is given by (97), we are led to the following system of linear homogeneous equations:

$$\begin{vmatrix} \sigma & 0 & -kH_z & 0 & 0 & 0 & 0 \\ 0 & \sigma & 0 & -kH_z & kH_y & 0 & 0 \\ -kH_z/4\pi\rho & 0 & \sigma & +2i\Omega_z & -2i\Omega_y & 0 & 0 \\ 0 & -kH_z/4\pi\rho & -2i\Omega_z & \sigma & +2i\Omega_x & 0 & 0 \\ 0 & +kH_y/4\pi\rho & +2i\Omega_y & -2i\Omega_x & \sigma & c^2k/\rho & -k \\ 0 & 0 & 0 & 0 & \rho k & \sigma & 0 \\ 0 & 0 & 0 & 0 & 0 & 4\pi G & -k^2 \end{vmatrix} \begin{Vmatrix} h_x \\ h_y \\ u_x \\ u_y \\ u_z \\ \delta\rho \\ \delta\mathfrak{B} \end{Vmatrix} = 0. \tag{128}$$

The vanishing of the determinant of this system leads to the characteristic equation

$$\sigma^6-\sigma^4\{(4\Omega^2+\Omega_A^2)+(\Omega_A^2+\Omega_B^2+\Omega_J^2)\}+$$
$$+\sigma^2\{\Omega_J^2(4\Omega_z^2+\Omega_A^2)+\Omega_A^2(4\Omega_x^2+\Omega_A^2+\Omega_B^2+\Omega_J^2)+4(\Omega_y\Omega_A-\Omega_z\Omega_B)^2\}-$$
$$-\Omega_A^4\Omega_J^2 = 0, \tag{129}$$

where
$$\Omega_A^2 = \frac{k^2H_z^2}{4\pi\rho}, \quad \Omega_B^2 = \frac{k^2H_y^2}{4\pi\rho}, \quad \Omega^2 = |\mathbf{\Omega}|^2,$$

and
$$\Omega_J^2 = c^2k^2-4\pi G\rho. \tag{130}$$

From equation (129) it follows that there are in general three modes in which a wave can be propagated in the medium; and if σ_1, σ_2, and σ_3 denote the gyration frequencies of the three modes, then

$$\sigma_1^2+\sigma_2^2+\sigma_3^2 = 4\Omega^2+2\Omega_A^2+\Omega_B^2+\Omega_J^2 = 4\Omega^2+\frac{(|\mathbf{H}|^2+H_z^2)k^2}{4\pi\rho}+c^2k^2-4\pi G\rho \tag{131}$$

and
$$\sigma_1^2\sigma_2^2\sigma_3^2 = \Omega_A^4\Omega_J^2 = \frac{k^4H_z^4}{(4\pi\rho)^2}(c^2k^2-4\pi G\rho). \tag{132}$$

From equation (132) (or, more directly from equation (129)) it follows that if Ω_J^2 is negative, then one of the roots of σ^2 must be negative; and this means that one of the three modes is unstable. The theorem stated at the beginning of the section follows.

BIBLIOGRAPHICAL NOTES

For an exceptionally complete and penetrating analysis of the problems bearing on questions of stability as they occur in astrophysics, see:

1. P. Ledoux, 'Stellar stability', *Handbuch der Physik*, **51**, 605–88 (1958), see § 17.

Related problems on stellar variability and stellar pulsation are treated in:

2. P. Ledoux and T. Walraven, 'Variable stars', ibid. 353–604 (1958), see §§ 59 and 82.

§ 117. The scalar form of the virial theorem for an assembly of particles interacting according to some general law of force is well known in statistical mechanics; see, for example:

3. R. H. Fowler, *Statistical Mechanics*, § 9.7, Cambridge, England, 1936.

Essentially the same theorem has been known in celestial mechanics and classical dynamics for a very long time: it was discovered by Lagrange as a useful integral formula for the theory of the three-body problem; and it was generalized to n bodies by Jacobi:

4. J. L. Lagrange, *Oeuvres*, vi, 240; ix, 836, Paris, 1873.
5. C. G. J. Jacobi, *Vorlesungen über Dynamik*, pp. 22–23, Druck & Verlag von Georg Reimer, Berlin, 1866.

Lagrange's identity, as it is referred to in books on celestial mechanics, plays an important role in the modern developments of the subject:

6. C. L. Siegel, *Vorlesungen über Himmelsmechanik*, § 6, Springer-Verlag, Berlin, 1956.

The usefulness of the virial theorem for stellar dynamics was recognized by

7. H. Poincaré, *Leçons sur les hypothèses cosmogoniques*, § 74, Paris, 1911.
8. A. S. Eddington, 'The kinetic energy of a star cluster', *Mon. Not. Roy. Astr. Soc.* **76**, 525–8 (1916).

For an account of these latter applications see:

9. S. Chandrasekhar, *Principles of Stellar Dynamics*, chap. v, University of Chicago Press, 1942.

An extension of the virial theorem to include frictional terms proportional to the velocity was given by Milne:

10. E. A. Milne, 'An extension of the theorem of the virial', *Phil. Mag.*, Ser. 6, **50**, 409–14 (1925).

The suggestion that one might formulate the virial theorem in tensor form seems to have been made, first, by Rayleigh:

11. Lord Rayleigh, 'On a theorem analogous to the virial theorem', *Scientific Papers*, iv, 491–3, Cambridge, England, 1903.

More recently, this aspect has been revived by:

12. E. N. Parker, 'Tensor virial equations', *Physical Rev.* **96**, 1686–9 (1954).

The extension of the virial theorem to hydromagnetics was given by:

13. S. Chandrasekhar and E. Fermi, 'Problems of gravitational stability in the presence of a magnetic field', *Astrophys. J*, **118**, 116–41 (1953).

13a. S. CHANDRASEKHAR, 'The virial theorem in hydromagnetics', *Journal of Mathematical Analysis and Applications*, **1,** 240–52 (1960).

See also:

14. S. CHANDRASEKHAR, 'Problems of stability in hydrodynamics and hydromagnetics', *Mon. Not. Roy. Astr. Soc.* **113,** 667–78 (1953).

In reference 13 the scalar form of the virial theorem (equation (29) in the text) was derived; and its consequences for gravitational equilibrium (in particular the inequality, $\mathfrak{M} < |\mathfrak{W}|$) were discussed. The general treatment of the virial theorem given in the text follows reference 13a.

§ 118. The first use of the virial theorem to obtain an integral formula for σ^2 is contained in:

15. P. LEDOUX, 'On the radial pulsation of gaseous stars', *Astrophys. J.* **102,** 143–53 (1945).

In this paper Ledoux derives the formula $\sigma^2 = -(3\gamma-4)\mathfrak{W}/I$. In this particular instance, Ledoux was able to show that the same formula follows, also, from a different integral relation provided by a variational treatment of the underlying characteristic value problem for σ^2. For this latter treatment see:

16. P. LEDOUX and C. L. PEKERIS, 'Radial pulsations of stars', ibid. **94,** 124–35 (1941).

The generalization of Ledoux's method to allow for prevalent magnetic fields is contained in:

17. S. CHANDRASEKHAR and D. NELSON LIMBER, 'On the pulsation of a star in which there is a prevalent magnetic field', ibid. **119,** 10–13 (1954).

In this paper the formula $\sigma^2 = -(3\gamma-4)(\mathfrak{W}+\mathfrak{M})/I$ is derived. The inconsistencies in the assumption of a trial function for $\boldsymbol{\xi}$, which corresponds to a uniform radial expansion, are avoided by the substitution $\boldsymbol{\xi} = \mathbf{X}\mathbf{x}$ considered in the text.

See also:

18. E. N. PARKER, 'The gross dynamics of a hydromagnetic gas cloud', *Astrophys. J. Supp., Ser* **3,** 51–76 (1957).

§ 119. As stated in the text, the gravitational instability of an infinite homogeneous medium was discovered by Jeans. His first statement of the theorem occurs in:

19. J. H. JEANS, 'The stability of a spherical nebula', *Phil. Trans. Roy. Soc.* (*London*), **199,** 1–53 (1902), see particularly, § 46.

For a later account by the same author see:

20. J. H. JEANS, *Astronomy and Cosmogony*, p. 313, Cambridge, England, 1929.

The effect of turbulence on Jeans's criterion is considered by:

21. S. CHANDRASEKHAR, 'The gravitational instability of an infinite homogeneous turbulent medium', *Proc. Roy. Soc.* (*London*) A, **210,** 26–29 (1951).

§ 120. An excellent and an original review of the effects of rotation and magnetic field on gravitational instability has been written by:

22. A. G. PACHOLCZYK and J. S. STODÓŁKIEWICZ, 'On the gravitational

instability of some magnetohydrodynamical systems of astrophysical interest', *Acta Astronomica* (Polska Akademia Nauk) **10,** 1–29 (1960).

§ 120 (*a*). The effect of rotation on Jeans's criterion was considered by:

23. S. CHANDRASEKHAR, 'The gravitational instability of an infinite homogeneous medium when a Coriolis acceleration is acting', *Vistas in Astronomy I*, 344–7, edited by A. Beer, Pergamon Press, 1955.

The illustration of the dispersion relations in the manner of Fig. 135 follows a suggestion of Pacholczyk and Stodółkiewicz (reference 22). In their account these latter authors include the effects of viscosity and thermal conductivity.

The extension of Jeans's criterion to nonuniform rotation is considered by:

24. E. SCHATZMAN and N. BEL, 'Instabilité d'une masse fluide étendue', *Comptes rendus des séances de l'Académie des Sciences,* **241,** 20–22 (1955).

25. N. BEL, 'Instabilité d'une masse fluide étendue', ibid. 163–4 (1955).

See also:

26. P. LEDOUX, 'Sur la stabilité gravitationnelle d'une nébuleuse isotherme', *Ann. d'Astrophys.* **14,** 438–47 (1951).

27. W. FRICKE, 'On the gravitational stability in a rotating isothermal medium', *Astrophys. J.* **120,** 356–9 (1954).

§ 120 (*b*). The effect of a uniform magnetic field on Jeans's criterion was considered by Chandrasekhar and Fermi (reference 13). Again, the illustration of the dispersion relations in the manner of Fig. 136 follows a suggestion of Pacholczyk and Stodółkiewicz (reference 22). These latter authors have discussed, at the same time, the effects of finite electrical conductivity.

§ 120 (*c*). The joint effects of rotation and magnetic field are considered by:

28. S. CHANDRASEKHAR, 'The gravitational instability of an infinite homogeneous medium when Coriolis force is acting and a magnetic field is present', *Astrophys. J.* **119,** 7–9 (1954).

XIV

A GENERAL VARIATIONAL PRINCIPLE

121. Introduction

In the preceding chapters we have considered instabilities of hydrodynamic and hydromagnetic systems arising from many causes: from adverse distributions of temperature and density, of angular velocity and angular momentum, and from shear, and from gravity and capillarity. In this chapter we shall enunciate a general variational principle which can well form the starting point for further developments.

122. A variational principle for treating the stability of hydromagnetic systems

To be specific, suppose that the initial state is one in which an incompressible fluid of infinite electrical conductivity is confined in a volume V bounded by a surface S. Let the space surrounding the fluid, either finite or infinite, be a vacuum. Suppose further that in the stationary state a magnetic field $\mathbf{H}$ prevails.

We shall distinguish the variables pertaining to the interior and to the exterior of the volume V by superscripts 'in' and 'ex', if there should be an ambiguity in the context as to which is meant. When there is no such ambiguity we shall dispense with the distinguishing superscripts.

In the stationary state the equations which must be satisfied inside V are

$$\frac{\partial \Pi}{\partial x_i} = \frac{1}{4\pi}\frac{\partial}{\partial x_k} H_i H_k, \tag{1}$$

where

$$\Pi = p + |\mathbf{H}|^2/8\pi. \tag{2}$$

Outside of V, the field, being that of a vacuum, must be current free; accordingly,

$$\operatorname{curl} \mathbf{H}^{(\mathrm{ex})} = 0. \tag{3}$$

On the bounding surface S the continuity of the normal stress requires

$$p^{(\mathrm{in})} + \frac{|\mathbf{H}^{(\mathrm{in})}|^2}{8\pi} = \frac{|\mathbf{H}^{(\mathrm{ex})}|^2}{8\pi}. \tag{4}$$

Also, we shall restrict our present considerations to the case when

$$\mathbf{H}.d\mathbf{S} = 0 \quad \text{everywhere on } S. \tag{5}$$

This condition states that the lines of force of the magnetic field lie on the surface: a necessary condition if infinite accelerations on the surface are not to arise.

The equations governing a first order perturbation of the initial state are

$$\rho\frac{\partial u_i}{\partial t} = -\frac{\partial \varpi}{\partial x_i}+\frac{1}{4\pi}\frac{\partial}{\partial x_j}(H_j h_i+h_j H_i) \tag{6}$$

and

$$\frac{\partial h_i}{\partial t} = \frac{\partial}{\partial x_j}(H_j u_i-u_j H_i), \tag{7}$$

where **h** denotes the perturbation in the field and

$$\varpi = \delta p+\frac{1}{4\pi}H_j h_j. \tag{8}$$

It is convenient to define the displacement $\boldsymbol{\xi}$ by the equation

$$u_i = \frac{\partial \xi_i}{\partial t}, \tag{9}$$

so that equation (7) immediately integrates to give

$$h_i = \frac{\partial}{\partial x_j}(H_j \xi_i-\xi_j H_i). \tag{10}$$

For an incompressible fluid $\boldsymbol{\xi}$ must, of course, be solenoidal.

Suppose now that all the quantities describing the perturbed state have the time dependence $e^{i\sigma t}$. Then, the equation of motion (6) gives

$$\sigma^2\rho\xi_i = \frac{\partial \varpi}{\partial x_i}-\frac{1}{4\pi}\frac{\partial}{\partial x_j}(H_j h_i+h_j H_i). \tag{11}$$

Also, the perturbation in the field outside V (i.e. in $V^{(\mathrm{ex})}$) must satisfy the equations

$$\operatorname{curl}\mathbf{h}^{(\mathrm{ex})} = 0 \quad \text{and} \quad \operatorname{curl}\mathbf{E}^{(\mathrm{ex})} = -i\sigma\mathbf{h}^{(\mathrm{ex})}. \tag{12}$$

Solutions of equations (10)–(12) must be sought which satisfy certain boundary conditions. These conditions which, strictly, must be satisfied on the displaced boundary, $\mathbf{S}+\delta\mathbf{S}$, are

$$\Delta(p+|\mathbf{H}|^2/8\pi) = 0, \tag{13†}$$

$$\mathbf{N}.\mathbf{H} = 0, \tag{14}$$

and

$$\mathbf{N}\times\Delta(\mathbf{E}) = (\mathbf{N}.\mathbf{u})\Delta(\mathbf{H}) = i\sigma(\mathbf{N}.\boldsymbol{\xi})\Delta(\mathbf{H}), \tag{15}$$

where $\Delta(f)$ is the jump a quantity f experiences on $\mathbf{S}+\delta\mathbf{S}$, and $\mathbf{N}$ is the unit outward normal to this surface. The first of these conditions ensures the continuity of the normal stress, the second, the vanishing of

† In equations (13)–(15), **H** stands for the total field, the perturbed plus the unperturbed.

the normal component of **H** (in accordance with the assumption (5)), and the third derives from the relation

$$\mathbf{E}+\mathbf{u}\times\mathbf{H} = 0. \tag{16}$$

In applying the conditions (13)–(15) one reduces them to conditions on **S** by expanding all the quantities, consistently, to the first order.

Equations (11) and (12) together with the boundary conditions (13)–(15) constitute a characteristic value problem for σ^2. If $\sigma^{(\lambda)}$ is a characteristic value, we shall distinguish the proper solutions belonging to it by the same superscript.

Consider equation (11) belonging to $\sigma^{(\lambda)}$ and after multiplication by $\xi_i^{(\mu)}$ (belonging to a different characteristic value $\sigma^{(\mu)}$) integrate over the volume V occupied by the fluid. We obtain

$$(\sigma^{(\lambda)})^2 \int\limits_V \rho\xi_i^{(\lambda)}\,\xi_i^{(\mu)}\,dV$$
$$= \int\limits_V \frac{\partial}{\partial x_i}(\varpi^{(\lambda)}\xi_i^{(\mu)})\,dV - \frac{1}{4\pi}\int\limits_V H_j\frac{\partial h_i^{(\lambda)}}{\partial x_j}\xi_i^{(\mu)}\,dV - \frac{1}{4\pi}\int\limits_V h_j^{(\lambda)}\frac{\partial H_i}{\partial x_j}\xi_i^{(\mu)}\,dV. \tag{17}$$

Integrating by parts the first two integrals on the right-hand side of (17), we obtain

$$(\sigma^{(\lambda)})^2 \int\limits_V \rho\xi_i^{(\lambda)}\,\xi_i^{(\mu)}\,dV$$
$$= \int\limits_S \varpi^{(\lambda)}\xi_i^{(\mu)}\,dS_i + \frac{1}{4\pi}\int\limits_V \left(H_j h_i^{(\lambda)}\frac{\partial \xi_i^{(\mu)}}{\partial x_j} - h_j^{(\lambda)}\,\xi_i^{(\mu)}\frac{\partial H_i}{\partial x_j}\right)dV, \tag{18}$$

the integrated part from the second integral vanishing on account of (5). Now inserting for **h** from equation (10), we can transform the volume integral on the right-hand side of equation (18) as follows:

$$\int\limits_V H_j\left(H_k\frac{\partial \xi_i^{(\lambda)}}{\partial x_k} - \xi_k^{(\lambda)}\frac{\partial H_i}{\partial x_k}\right)\frac{\partial \xi_i^{(\mu)}}{\partial x_j}\,dV - \int\limits_V \frac{\partial H_i}{\partial x_j}\left(H_k\frac{\partial \xi_j^{(\lambda)}}{\partial x_k} - \xi_k^{(\lambda)}\frac{\partial H_j}{\partial x_k}\right)\xi_i^{(\mu)}\,dV$$
$$= \int\limits_V \left(H_k\frac{\partial \xi_i^{(\lambda)}}{\partial x_k}\right)\left(H_j\frac{\partial \xi_i^{(\mu)}}{\partial x_j}\right)dV + \int\limits_V \xi_k^{(\lambda)}\,\xi_i^{(\mu)}\frac{\partial H_i}{\partial x_j}\frac{\partial H_j}{\partial x_k}\,dV -$$
$$- \int\limits_V \left(H_j\frac{\partial H_i}{\partial x_k}\xi_k^{(\lambda)}\frac{\partial \xi_i^{(\mu)}}{\partial x_j} + H_k\frac{\partial H_i}{\partial x_j}\frac{\partial \xi_j^{(\lambda)}}{\partial x_k}\xi_i^{(\mu)}\right)dV. \tag{19}$$

Interchanging the suffixes j and k in the second term inside the integral on the last line of (19), we obtain

$$-\int\limits_V H_j\frac{\partial H_i}{\partial x_k}\frac{\partial}{\partial x_j}(\xi_k^{(\lambda)}\,\xi_i^{(\mu)})\,dV = \int\limits_V \xi_k^{(\lambda)}\,\xi_i^{(\mu)}\,H_j\frac{\partial^2 H_i}{\partial x_j\,\partial x_k}\,dV. \tag{20}$$

Combining this with the second integral on the second line of (19), we have

$$\int_V \xi_k^{(\lambda)} \xi_i^{(\mu)} \left(\frac{\partial H_i}{\partial x_j}\frac{\partial H_j}{\partial x_k} + H_j \frac{\partial^2 H_i}{\partial x_j \partial x_k}\right) dV$$

$$= \int_V \xi_k^{(\lambda)} \xi_i^{(\mu)} \frac{\partial}{\partial x_k}\left(H_j \frac{\partial H_i}{\partial x_j}\right) dV = 4\pi \int_V \xi_k^{(\lambda)} \xi_i^{(\mu)} \frac{\partial^2 \Pi}{\partial x_i \partial x_k} dV, \quad (21)$$

where, in the last step, we have made use of equation (1). Thus equation (18) now becomes

$$(\sigma^{(\lambda)})^2 \int_V \rho \xi_i^{(\lambda)} \xi_i^{(\mu)} dV = \int_S \varpi^{(\lambda)} \xi_i^{(\mu)} dS_i +$$

$$+\frac{1}{4\pi} \int_V \left\{\left(H_k \frac{\partial \xi_i^{(\lambda)}}{\partial x_k}\right)\left(H_j \frac{\partial \xi_i^{(\mu)}}{\partial x_j}\right) + 4\pi \xi_k^{(\lambda)} \xi_i^{(\mu)} \frac{\partial^2 \Pi}{\partial x_i \partial x_k}\right\} dV. \quad (22)$$

It remains to consider the surface integral in this last equation. We now make use of the condition which follows from (13) requiring the continuity of the total pressure on $\mathbf{S}+\delta\mathbf{S}$; we have (cf. equation XII (358))

$$(\varpi^{(\lambda)})_S = \frac{1}{4\pi} \mathbf{H}^{(\text{ex})} . \mathbf{h}^{(\text{ex};\lambda)} + (\mathbf{N}^{(0)} . \boldsymbol{\xi}^{(\lambda)})[\mathbf{N}^{(0)} . \text{grad}\, \Delta_S(\Pi)], \quad (23)$$

where $\Delta_S(f)$ now denotes the jump which a quantity f experiences on S, and $\mathbf{N}^{(0)}$ is the unit outward normal to S. Substituting this expression for $(\varpi^{(\lambda)})_S$ in the surface integral occurring in equation (22), we have

$$\int_S \varpi^{(\lambda)} \xi_i^{(\mu)} dS_i = \int_S (\mathbf{N}^{(0)} . \boldsymbol{\xi}^{(\lambda)})(\mathbf{N}^{(0)} . \boldsymbol{\xi}^{(\mu)})[\mathbf{N}^{(0)} . \text{grad}\, \Delta_S(\Pi)]\, dS +$$

$$+\frac{1}{4\pi} \int (\mathbf{H}^{(\text{ex})} . \mathbf{h}^{(\text{ex};\lambda)})(\mathbf{N}^{(0)} . \boldsymbol{\xi}^{(\mu)})\, dS. \quad (24)$$

Now from equation (15) it follows that

$$\mathbf{N}^{(0)} \times \mathbf{E}^{(\text{ex};\mu)} = i\sigma^{(\mu)} (\mathbf{N}^{(0)} . \boldsymbol{\xi}^{(\mu)}) \mathbf{H}^{(\text{ex})}. \quad (25)$$

The second integral on the right-hand side of equation (24) thus becomes

$$\frac{1}{4\pi i \sigma^{(\mu)}} \int \mathbf{h}^{(\text{ex};\lambda)} . (\mathbf{N}^{(0)} \times \mathbf{E}^{(\text{ex};\mu)})\, dS = \frac{-1}{4\pi i \sigma^{(\mu)}} \int_S \epsilon_{ijk}\, dS_i\, h_j^{(\text{ex};\lambda)} E_k^{(\text{ex};\mu)}. \quad (26)$$

The surface integral can now be transformed by Gauss's theorem into a volume integral over $V^{(\text{ex})}$ *exterior* to V; thus,

$$\frac{1}{4\pi i \sigma^{(\mu)}} \int_{V^{(\text{ex})}} \epsilon_{ijk} \frac{\partial}{\partial x_i} (h_j^{(\lambda)} E_k^{(\mu)})\, dV$$

$$= -\frac{1}{4\pi i \sigma^{(\mu)}} \int_{V^{(\text{ex})}} \mathbf{h}^{(\lambda)} . \text{curl}\, \mathbf{E}^{(\mu)}\, dV = \frac{1}{4\pi} \int_{V^{(\text{ex})}} h_i^{(\lambda)} h_i^{(\mu)}\, dV. \quad (27)$$

The final form of equation (22) is, therefore,

$$(\sigma^{(\lambda)})^2 \int\limits_{V^{(\text{in})}} \rho \xi_i^{(\lambda)} \xi_i^{(\mu)}\, dV = \int\limits_S (\mathbf{N}^{(0)}.\boldsymbol{\xi}^{(\lambda)})(\mathbf{N}^{(0)}.\boldsymbol{\xi}^{(\mu)})[\mathbf{N}^{(0)}.\operatorname{grad}\Delta_S(\Pi)]\, dS +$$

$$+\frac{1}{4\pi} \int\limits_{V^{(\text{in})}} \left\{\left(H_k \frac{\partial \xi_i^{(\lambda)}}{\partial x_k}\right)\left(H_j \frac{\partial \xi_i^{(\mu)}}{\partial x_j}\right) + 4\pi \xi_k^{(\lambda)} \xi_i^{(\mu)} \frac{\partial^2 \Pi}{\partial x_i \partial x_k}\right\} dV + \frac{1}{4\pi} \int\limits_{V^{(\text{ex})}} h_i^{(\lambda)} h_i^{(\mu)}\, dV. \tag{28}$$

We observe that the expression on the right-hand side of equation (28) is symmetrical in λ and μ. Therefore

$$\int\limits_{V^{(\text{in})}} \rho \xi_i^{(\lambda)} \xi_i^{(\mu)}\, dV = 0 \quad (\lambda \neq \mu). \tag{29}$$

This clearly establishes the self-adjoint character of the underlying characteristic value problem. Also, by writing in equation (28) the complex conjugates of $\xi_i^{(\lambda)}$ and $h_i^{(\lambda)}$ in place of $\xi_i^{(\mu)}$ and $h_i^{(\mu)}$ and suppressing the superscripts, we obtain

$$\sigma^2 \int\limits_{V^{(\text{in})}} \rho |\boldsymbol{\xi}|^2\, dV = \int\limits_S |\mathbf{N}^{(0)}.\boldsymbol{\xi}|^2[\mathbf{N}^{(0)}.\operatorname{grad}\Delta_S(\Pi)]\, dS +$$

$$+\frac{1}{4\pi} \int\limits_{V^{(\text{in})}} \left\{\left|H_j \frac{\partial \boldsymbol{\xi}}{\partial x_j}\right|^2 + 4\pi \xi_i \xi_k^* \frac{\partial^2 \Pi}{\partial x_i \partial x_k}\right\} dV + \frac{1}{4\pi} \int\limits_{V^{(\text{ex})}} |\mathbf{h}|^2\, dV = \Sigma \text{ (say)}. \tag{30}$$

All the quantities on the right-hand side of this equation are clearly real. Therefore σ^2 must be real and can be either positive or negative. In the former case we will have stability; and in the latter case, instability:

$$\Sigma > 0 \text{ implies stability} \qquad \text{and} \qquad \Sigma < 0 \text{ implies instability.} \tag{31}$$

From the self-adjoint character of this problem, it is apparent that *equation* (30) *provides the basis for a variational treatment of the problem.*

(*a*) *The effect of viscosity*

The preceding analysis can be readily modified to allow for effects of viscosity. The modification that is necessary consists only in replacing equation (11) by (see Chapter II, § 7 (*b*), equations (15)–(17))

$$\sigma^2 \rho \xi_i = \frac{\partial \varpi_{ij}}{\partial x_j} - \frac{1}{4\pi}\frac{\partial}{\partial x_j}(H_j h_i + h_j H_i), \tag{32}$$

where
$$\varpi_{ij} = \left(\delta p + \frac{H_j h_j}{4\pi}\right)\delta_{ij} - p_{ij} \tag{33}$$

and
$$p_{ij} = \rho\nu\left(\frac{\partial u_i}{\partial x_j} + \frac{\partial u_j}{\partial x_i}\right) \tag{34}$$

is the viscous stress tensor. The equation governing the magnetic field is unchanged; but the boundary condition (13) must now be replaced by

$$N_i \Delta(P_{ij}) = 0. \tag{35}$$

It is easy to trace the effect of replacing the scalar pressure ϖ by the tensor pressure ϖ_{ij} in the preceding analysis: in place of the simple surface integral on the right-hand side of equation (18) we now have

$$\int\limits_S \varpi_{ij}^{(\lambda)}\, \xi_i^{(\mu)}\, dS_j + \int\limits_V p_{ij}^{(\lambda)} \frac{\partial \xi_i^{(\mu)}}{\partial x_j}\, dV. \tag{36}$$

By virtue of the changed boundary condition (35), the further reduction of the surface integral in (36) proceeds exactly as before and leads to the same final result. The net change in equation (28) is then the addition of the volume integral

$$\int\limits_{V^{(\mathrm{in})}} p_{ij}^{(\lambda)} \frac{\partial \xi_i^{(\mu)}}{\partial x_j}\, dV = \rho\nu(i\sigma^{(\lambda)}) \int\limits_{V^{(\mathrm{in})}} \left(\frac{\partial \xi_i^{(\lambda)}}{\partial x_j} + \frac{\partial \xi_j^{(\lambda)}}{\partial x_i}\right) \frac{\partial \xi_i^{(\mu)}}{\partial x_j}\, dV \tag{37}$$

to the terms on the right-hand side. The corresponding form of equation (30) is therefore

$$\sigma^2 \int\limits_{V^{(\mathrm{in})}} \rho |\boldsymbol{\xi}|^2\, dV = \Sigma + \rho\nu(i\sigma) \int\limits_{V^{(\mathrm{in})}} \left(\frac{\partial \xi_i}{\partial x_j} + \frac{\partial \xi_j}{\partial x_i}\right) \frac{\partial \xi_i^*}{\partial x_j}\, dV. \tag{38}$$

The volume integral on the right-hand side represents the dissipation due to viscosity (cf. equation II (29)). Thus letting

$$\Phi = \tfrac{1}{2}\rho\nu \int\limits_V \left|\frac{\partial \xi_i}{\partial x_j} + \frac{\partial \xi_j}{\partial x_i}\right|^2 dV \tag{39}$$

and

$$I = \int\limits_V \rho |\boldsymbol{\xi}|^2\, dV, \tag{40}$$

we can rewrite equation (38) in the form

$$(i\sigma)^2 I + (i\sigma)\Phi + \Sigma = 0. \tag{41}$$

The roots of this equation are

$$i\sigma = \frac{1}{2I}\{-\Phi \pm \sqrt{(\Phi^2 - 4\Sigma)}\}. \tag{42}$$

From this equation it follows that *a necessary and sufficient condition for stability is that* $\Sigma > 0$; *this condition is the same as in the absence of viscosity*. This last result is in agreement with what was found in Chapter XII, §§ 109 and 111 (*c*).

While the system is stable for $\Sigma > 0$, the modes will be either damped-periodic or damped-aperiodic depending on the magnitude of Φ, i.e. on the magnitude of the viscous dissipation.

123. Extension of the variational principle to allow for compressibility

The variational principle derived in the preceding section can be extended to allow for compressibility. Incompressibility simplifies the analysis by $\boldsymbol{\xi}$ being solenoidal; this latter property cannot be used if the fluid is compressible. On examining the details of the derivation in § 122, we find that the solenoidal property of $\boldsymbol{\xi}$ was used only in two places: first, in reducing the integral

$$\int_V \frac{\partial \varpi^{(\lambda)}}{\partial x_i} \xi_i^{(\mu)}\, dV \tag{43}$$

to the simple surface integral in (18); and again in substituting from (10) for $\mathbf{h}$ in the volume integral on the right-hand side of equation (18) and writing it as on the left-hand side of equation (19). The terms which have been neglected on these two accounts and which we must now include are

$$I = -\int_V \varpi^{(\lambda)} \frac{\partial \xi_i^{(\mu)}}{\partial x_i}\, dV \tag{44}$$

and

$$II = \frac{1}{4\pi} \int_V \frac{\partial \xi_k^{(\lambda)}}{\partial x_k} \left(H_j \frac{\partial H_i}{\partial x_j} \xi_i^{(\mu)} - H_i H_j \frac{\partial \xi_i^{(\mu)}}{\partial x_j} \right) dV. \tag{45}$$

The reduction of the surface integral which remains in (18), as well as the others, does not depend on the solenoidal character of $\boldsymbol{\xi}$ and can be carried out as before. The net result, then, is that on the right-hand side of equation (28) we must now include the terms I and II with the others already present.

Considering I, we rewrite it in the form

$$I = -\int_V \left(\delta p^{(\lambda)} + \frac{H_k h_k^{(\lambda)}}{4\pi} \right) \frac{\partial \xi_i^{(\mu)}}{\partial x_i}\, dV, \tag{46}$$

where $\delta p^{(\lambda)}$ is the perturbation in the material pressure. We must now express (46) like the others, in terms of $\boldsymbol{\xi}$; to accomplish this, a physical hypothesis is needed to relate δp with the displacement $\boldsymbol{\xi}$; and it is customary to suppose that the variations of density and pressure take place adiabatically as we follow fluid elements with their motions. In the Eulerian framework this means

$$\frac{\partial}{\partial t} \delta p + u_j \frac{\partial p}{\partial x_j} = \gamma \frac{p}{\rho} \left(\frac{\partial}{\partial t} \delta \rho + u_j \frac{\partial \rho}{\partial x_j} \right), \tag{47}$$

where γ denotes the ratio of the specific heats and p is the pressure on the unperturbed configuration. In addition to (47) we also have the equation of continuity

$$\frac{\partial}{\partial t}\delta\rho+u_j\frac{\partial \rho}{\partial x_j} = -\rho\frac{\partial u_j}{\partial x_j}. \tag{48}$$

In terms of $\boldsymbol{\xi}$ equations (47) and (48) can be integrated to give

$$\delta\rho = -\xi_k\frac{\partial \rho}{\partial x_k}-\rho\frac{\partial \xi_k}{\partial x_k} \tag{49}$$

and

$$\delta p = -\xi_j\frac{\partial p}{\partial x_j}-c^2\rho\frac{\partial \xi_k}{\partial x_k}, \tag{50}$$

where c denotes the velocity of sound. Inserting accordingly for $\delta p^{(\lambda)}$ in (46), we have

$$\int\limits_V \left(\xi_j^{(\lambda)}\frac{\partial p}{\partial x_j}+c^2\rho\frac{\partial \xi_k^{(\lambda)}}{\partial x_k}\right)\frac{\partial \xi_i^{(\mu)}}{\partial x_i}\,dV = \int\limits_V c^2\rho\frac{\partial \xi_k^{(\lambda)}}{\partial x_k}\frac{\partial \xi_i^{(\mu)}}{\partial x_i}\,dV+\int\limits_V \xi_j^{(\lambda)}\frac{\partial \xi_i^{(\mu)}}{\partial x_i}\frac{\partial p}{\partial x_j}\,dV. \tag{51}$$

Similarly substituting for $\mathbf{h}^{(\lambda)}$ from equation (10) in the other term in (46), we obtain

$$-\frac{1}{4\pi}\int\limits_V H_k\left(H_j\frac{\partial \xi_k^{(\lambda)}}{\partial x_j}-H_k\frac{\partial \xi_j^{(\lambda)}}{\partial x_j}-\xi_j^{(\lambda)}\frac{\partial H_k}{\partial x_j}\right)\frac{\partial \xi_i^{(\mu)}}{\partial x_i}\,dV$$

$$= \frac{1}{4\pi}\int\limits_V |\mathbf{H}|^2\frac{\partial \xi_j^{(\lambda)}}{\partial x_j}\frac{\partial \xi_i^{(\mu)}}{\partial x_i}\,dV+\int\limits_V \xi_j^{(\lambda)}\frac{\partial \xi_i^{(\mu)}}{\partial x_i}\frac{\partial}{\partial x_j}\frac{|\mathbf{H}|^2}{8\pi}\,dV-$$

$$-\frac{1}{4\pi}\int\limits_V H_j H_k\frac{\partial \xi_k^{(\lambda)}}{\partial x_j}\frac{\partial \xi_i^{(\mu)}}{\partial x_i}\,dV. \tag{52}$$

We thus obtain

$$I = \int\limits_V\left(c^2\rho+\frac{|\mathbf{H}|^2}{4\pi}\right)\frac{\partial \xi_j^{(\lambda)}}{\partial x_j}\frac{\partial \xi_i^{(\mu)}}{\partial x_i}\,dV+\int\limits_V \xi_j^{(\lambda)}\frac{\partial \xi_i^{(\mu)}}{\partial x_i}\frac{\partial}{\partial x_j}\left(p+\frac{|\mathbf{H}|^2}{8\pi}\right)dV-$$

$$-\frac{1}{4\pi}\int\limits_V H_j H_k\frac{\partial \xi_k^{(\lambda)}}{\partial x_j}\frac{\partial \xi_i^{(\mu)}}{\partial x_i}\,dV. \tag{53}$$

Now making use of equation (1) governing the stationary state, we can rewrite equation (53) in the manner (cf. equation (45))

$$I = \int\limits_V\left(c^2\rho+\frac{|\mathbf{H}|^2}{4\pi}\right)\frac{\partial \xi_j^{(\lambda)}}{\partial x_j}\frac{\partial \xi_i^{(\mu)}}{\partial x_i}\,dV+\frac{1}{4\pi}\int\limits_V\left(H_k\frac{\partial H_j}{\partial x_k}\xi_j^{(\lambda)}-H_j H_k\frac{\partial \xi_k^{(\lambda)}}{\partial x_j}\right)\frac{\partial \xi_i^{(\mu)}}{\partial x_i}\,dV. \tag{54}$$

Now adding the terms I and II given by equations (45) and (54) to the right-hand side of equation (28), we finally obtain

$$(\sigma^{(\lambda)})^2 \int\limits_{V^{(\text{in})}} \rho\xi_i^{(\lambda)}\,\xi_i^{(\mu)}\,dV = \int\limits_S (\mathbf{N}^{(0)}.\,\boldsymbol{\xi}^{(\lambda)})(\mathbf{N}^{(0)}.\,\boldsymbol{\xi}^{(\mu)})[\mathbf{N}^{(0)}.\,\text{grad}\,\Delta_S(\Pi)]\,dS +$$

$$+\frac{1}{4\pi}\int\limits_{V^{(\text{in})}} \left\{\left(H_k\frac{\partial\xi_i^{(\lambda)}}{\partial x_k}\right)\left(H_j\frac{\partial\xi_i^{(\mu)}}{\partial x_j}\right)+4\pi\xi_k^{(\lambda)}\,\xi_i^{(\mu)}\frac{\partial^2\Pi}{\partial x_i\,\partial x_k}\right\}dV+$$

$$+\int\limits_{V^{(\text{in})}}\left(c^2\rho+\frac{|\mathbf{H}|^2}{4\pi}\right)\frac{\partial\xi_j^{(\lambda)}}{\partial x_j}\,\frac{\partial\xi_i^{(\mu)}}{\partial x_i}\,dV+\frac{1}{4\pi}\int\limits_{V^{(\text{ex})}} h_i^{(\lambda)}\,h_i^{(\mu)}\,dV+$$

$$+\frac{1}{4\pi}\int\limits_{V^{(\text{in})}}\left\{H_k\frac{\partial H_j}{\partial x_k}\left(\xi_j^{(\lambda)}\frac{\partial\xi_i^{(\mu)}}{\partial x_i}+\xi_j^{(\mu)}\frac{\partial\xi_i^{(\lambda)}}{\partial x_i}\right)-\right.$$

$$\left.-H_j\,H_k\left(\frac{\partial\xi_k^{(\lambda)}}{\partial x_j}\,\frac{\partial\xi_i^{(\mu)}}{\partial x_i}+\frac{\partial\xi_k^{(\mu)}}{\partial x_j}\,\frac{\partial\xi_i^{(\lambda)}}{\partial x_i}\right)\right\}dV. \quad (55)$$

We observe that, as in the incompressible case, the expression on the right-hand side of equation (55) is symmetrical in λ and μ; and, therefore,

$$\int\limits_{V^{(\text{in})}} \rho\xi_i^{(\lambda)}\,\xi_i^{(\mu)}\,dV = 0 \quad (\lambda \neq \mu). \quad (56)$$

This clearly establishes the self-adjoint character of the problem. Also by writing the complex conjugates of $\xi_i^{(\lambda)}$ and $h_i^{(\lambda)}$ in place of $\xi_i^{(\mu)}$ and $h_i^{(\mu)}$ and suppressing the superscripts, we obtain

$$\sigma^2 \int\limits_{V^{(\text{in})}} \rho|\boldsymbol{\xi}|^2\,dV = \int\limits_S |\mathbf{N}^{(0)}.\,\boldsymbol{\xi}|^2[\mathbf{N}^{(0)}.\,\text{grad}\,\Delta_S(\Pi)]\,dS+$$

$$+\frac{1}{4\pi}\int\limits_{V^{(\text{in})}}\left\{\left|H_j\frac{\partial\boldsymbol{\xi}}{\partial x_j}\right|^2+4\pi\xi_i\,\xi_j^*\frac{\partial^2\Pi}{\partial x_i\,\partial x_j}\right\}dV+$$

$$+\int\limits_{V^{(\text{in})}}\left(c^2\rho+\frac{|\mathbf{H}|^2}{4\pi}\right)\left|\frac{\partial\xi_i}{\partial x_i}\right|^2 dV+\frac{1}{4\pi}\int\limits_{V^{(\text{ex})}}|\mathbf{h}|^2\,dV+$$

$$+\frac{1}{2\pi}\int\limits_{V^{(\text{in})}}\left\{H_k\frac{\partial H_j}{\partial x_k}\text{re}\left(\xi_j^*\frac{\partial\xi_i}{\partial x_i}\right)-H_j\,H_k\,\text{re}\left(\frac{\partial\xi_k^*}{\partial x_j}\,\frac{\partial\xi_i}{\partial x_i}\right)\right\}dV. \quad (57)$$

From this equation it follows that σ^2 *is necessarily real and that a necessary and sufficient condition for stability is that the right-hand side of equation* (57) *be positive.*

From the self-adjoint character of the problem it is apparent that equation (57) provides the basis for a variational treatment of the problem.

BIBLIOGRAPHICAL NOTES

The variational principles discussed in this chapter are derived from:

1. I. B. BERNSTEIN, E. A. FRIEMAN, M. D. KRUSKAL, and R. M. KULSRUD, 'An energy principle for hydromagnetic stability problems', *Proc. Roy. Soc.* (*London*) A, **244,** 17–40 (1958).
2. K. HAIN, R. LÜST, and A. SCHLÜTER, 'Zur Stabilität eines Plasmas', *Z. f. Naturforschung*, **12a,** 833–41 (1957).
3. A. HARE, 'The effect of viscosity on the stability of incompressible magnetohydrodynamic systems', *Phil. Mag.*, Ser. 8, **48,** 1305–10 (1959).
4. T. G. COWLING, 'Magneto-hydrodynamic oscillations of a rotating fluid globe', *Proc. Roy. Soc.* (*London*) A, **233,** 319–22 (1955).

APPENDIX I

INTEGRAL RELATIONS GOVERNING STEADY CONVECTION

124. Introduction

A GENERAL result which emerges from the study of thermal instability in Chapters II–VI is that the onset of stationary convection is merely the signal that a steady balance can be maintained between the energy dissipated by irreversible processes and the energy released by the buoyancy force. This balance is clearly necessary whenever a state of steady convection prevails, not only at marginal stability when such a balance first becomes possible and the amplitude of the convection is infinitesimal. There is, however, one important difference: under the more general conditions we now envisage, the variation of the average temperature with height will no longer be that deduced from the equation of heat conduction in the absence of motions; and we must allow for the effects arising from the finite amplitudes of the prevailing motions.

In this appendix we shall derive certain integral relations which must govern a general state of steady convection under the circumstances of the simple Bénard problem beyond the onset of instability; and we shall indicate how these relations can be made the basis for estimating the amplitudes of the perturbations just past the state of marginal stability.

125. The integral relations

We shall start from the equations of motions in the Boussinesq approximation. These equations are (equations II (43) and (46)):

$$\frac{\partial u_i}{\partial t}+\frac{\partial}{\partial x_j}(u_i u_j) = -\frac{\partial \Pi}{\partial x_i}-g\left(1+\frac{\delta\rho}{\rho_0}\right)\lambda_i+\nu\nabla^2 u_i, \tag{1}$$

$$\frac{\partial T}{\partial t}+\frac{\partial}{\partial x_j}(Tu_j) = \kappa\nabla^2 T, \tag{2}$$

and

$$\frac{\partial u_i}{\partial x_i} = 0, \tag{3}$$

where $\Pi = p/\rho_0$ and the rest of the symbols have their standard meanings.

We shall now suppose that in a state of steady convection (beyond marginal stability) the circumstances are such that

$$\langle u_i \rangle = 0, \quad \langle \Pi \rangle = \Pi_0(z), \quad \text{and} \quad \langle T \rangle = T_0(z), \tag{4}$$

where the angular brackets signify that the quantity enclosed has been averaged over the horizontal plane; and further, that before the averaging,

$$\Pi = \Pi_0(z) + \varpi(x_i, t) \quad \text{and} \quad T = T_0(z) + \theta(x_i, t). \tag{5}$$

With the usual assumption concerning the 'equation of state',

$$\delta\rho = -\alpha\rho_0[T_0(z) - T_0(0) + \theta], \tag{6}$$

where α is the coefficient of volume expansion. In virtue of these definitions,

$$\langle \varpi \rangle = 0, \quad \langle \theta \rangle = 0, \quad \text{and} \quad \langle \delta\rho \rangle = -\alpha\rho_0[T_0(z) - T_0(0)]. \tag{7}$$

On the premisses stated, the result of averaging equation (1) over the horizontal plane is

$$\left\langle \frac{\partial}{\partial x_j}(u_i u_j) \right\rangle = -\frac{\partial \Pi_0}{\partial x_i} - g\lambda_i\{1 - \alpha[T_0(z) - T_0(0)]\}. \tag{8}$$

The only non-vanishing component of equation (8) is in the direction of the vertical (namely, λ_i); and in this direction the equation gives

$$\frac{d}{dz}(\Pi_0 + \langle w^2 \rangle) = -g\{1 - \alpha[T_0(z) - T_0(0)]\}. \tag{9}$$

This equation signifies that on the average the equilibrium is hydrostatic.

In the same way, the result of averaging equation (2) over the horizontal plane is

$$\frac{d}{dz}\langle \theta w \rangle = \kappa \frac{d^2 T_0}{dz^2}. \tag{10}$$

This equation can be integrated to give

$$\kappa T_0(z) = \int_0^z \langle \theta w \rangle \, dz + c_1 z + c_2, \tag{11}$$

where c_1 and c_2 are constants. The constants are related to the temperatures, $T_0(0)$ and $T_0(d)$, which are steadily maintained at the bottom ($z = 0$) and at the top ($z = d$) surfaces, respectively. Expressing c_1 and c_2 in terms of these temperatures, we can rewrite equation (11) in the form

$$\kappa T_0(z) = \kappa T_0(0) + \int_0^z \langle \theta w \rangle \, dz - \left\{ \kappa[T_0(0) - T_0(d)] + \int_0^d \langle \theta w \rangle \, dz \right\} \frac{z}{d}. \tag{12}$$

From this relation we derive

$$\kappa\frac{dT_0}{dz} = -\beta\kappa + \langle\theta w\rangle - \frac{1}{d}\int_0^d \langle\theta w\rangle\,dz, \tag{13}$$

where

$$\beta = [T_0(0) - T_0(d)]/d \tag{14}$$

is the *mean* adverse temperature gradient which is prevailing.

Equation (13) exhibits how a linear temperature gradient, in the absence of motions, is modified by the presence of motions.

(a) *The integral relations*

Rewriting equation (2) in the form

$$\frac{\partial\theta}{\partial t} + w\frac{dT_0}{dz} + \frac{\partial}{\partial x_j}(\theta u_j) = \kappa\frac{d^2T_0}{dz^2} + \kappa\nabla^2\theta, \tag{15}$$

we multiply it by θ and, after averaging over the horizontal plane, integrate over the range of z. We then find

$$\frac{1}{2}\frac{d}{dt}\int_0^d \langle\theta^2\rangle\,dz + \int_0^d \langle\theta w\rangle\frac{dT_0}{dz}\,dz = \kappa\int_0^d \langle\theta\nabla^2\theta\rangle\,dz. \tag{16}$$

Now substituting for dT_0/dz in accordance with equation (13), we obtain

$$\frac{1}{2}\frac{d}{dt}\int_0^d \langle\theta^2\rangle\,dz = \beta\int_0^d \langle\theta w\rangle\,dz + \kappa\int_0^d \langle\theta\nabla^2\theta\rangle\,dz - \\ -\frac{1}{\kappa}\left\{\int_0^d \langle\theta w\rangle^2\,dz - \frac{1}{d}\left(\int_0^d \langle\theta w\rangle\,dz\right)^2\right\}. \tag{17}$$

Similarly, by multiplying equation (1) scalarly by u_i and integrating over the range of z after averaging over the horizontal plane, we obtain

$$\frac{1}{2}\frac{d}{dt}\int_0^d \langle u_i^2\rangle\,dz = g\alpha\int_0^d \langle\theta w\rangle\,dz + \nu\int_0^d \langle u_i\nabla^2 u_i\rangle\,dz. \tag{18}$$

If the conditions should be such that on the average the state is a steady one, then equations (17) and (18) lead to the integral relations

$$\beta\int_0^d \langle\theta w\rangle\,dz + \kappa\int_0^d \langle\theta\nabla^2\theta\rangle\,dz = \frac{1}{\kappa}\left\{\int_0^d \langle\theta w\rangle^2\,dz - \frac{1}{d}\left(\int_0^d \langle\theta w\rangle\,dz\right)^2\right\} \tag{19}$$

and

$$g\alpha\int_0^d \langle\theta w\rangle\,dz + \nu\int_0^d \langle u_i\nabla^2 u_i\rangle\,dz = 0. \tag{20}$$

Equation (20) expresses merely the equality between ϵ_ν and ϵ_g (cf.

equations II (175) and (178), p. 33); and as we have seen on several occasions, it is this equality which underlies the variational principles characteristic of the subject.

126. The amplitude of the steady convection past marginal stability

It is known empirically that, in problems in which instability sets in as a stationary secondary flow, the pattern of motions which first appears at marginal stability continues to manifest itself long past the critical conditions. This is the case, for example, in the simple Bénard problem; and it is also the case when the flow between two cylinders rotating in the same direction becomes unstable. These empirical facts suggest that one might use the integral relations (19) and (20) to estimate the amplitudes of the convective motions on the assumption that, just past the marginal state, the motions continue to be of the same pattern as at the onset of instability.

We shall suppose, then, that in the state of finite amplitude convection which immediately follows instability the solutions for w and θ are of the forms

$$w = AW(z)f(x,y) \quad \text{and} \quad \theta = A\Theta(z)f(x,y), \tag{21}$$

where $W(z)$ and $\Theta(z)$ are suitably normalized solutions appropriate for marginal stability; $f(x,y)$ represents a two-dimensional periodic wave with the wave number k $(= a/d)$ which is manifested *at* marginal stability; and A is the amplitude which is left undetermined by the linear stability theory.

In terms of the solution for w, the horizontal components of the velocity are given by (cf. equation II (171))

$$u = \frac{1}{a^2}\frac{\partial^2 w}{\partial x \partial z} = A\frac{DW}{a^2}f_x \quad \text{and} \quad v = \frac{1}{a^2}\frac{\partial^2 w}{\partial y \partial z} = A\frac{DW}{a^2}f_y, \tag{22}$$

if linear distances are measured in units of the depth (d) of the layer of fluid. (The unit of distance just mentioned will be used in the rest of this discussion.)

Since in the linear theory A is an undetermined constant of proportionality, there is no loss of generality in supposing that

$$\langle f^2 \rangle = 1. \tag{23}$$

And from the fact that f satisfies the equation

$$\frac{\partial^2 f}{\partial x^2} + \frac{\partial^2 f}{\partial y^2} = -a^2 f, \tag{24}$$

we can conclude, moreover, that

$$\langle f_x^2\rangle+\langle f_y^2\rangle = a^2\langle f^2\rangle = a^2. \tag{25}$$

Inserting the assumed forms for the solutions in equations (19) and (20) (rewritten with z expressed in the unit d) and making use of equations (23) and (25), we obtain

$$\beta\int_0^1 \Theta W\,dz+\frac{\kappa}{d^2}\int_0^1 \Theta(D^2-a^2)\Theta\,dz = \frac{A^2}{\kappa}\left\{\int_0^1 \Theta^2W^2\,dz-\left(\int_0^1 \Theta W\,dz\right)^2\right\} \tag{26}$$

and

$$g\alpha\int_0^1 \Theta W\,dz = -\frac{\nu}{d^2}\int_0^1 \{\langle w(D^2-a^2)w\rangle+\langle u(D^2-a^2)u\rangle+\langle v(D^2-a^2)v\rangle\}\,dz$$

$$= -\frac{\nu}{d^2}\int_0^1 \left\{W(D^2-a^2)W+\frac{1}{a^2}DW(D^2-a^2)\,DW\right\}dz. \tag{27}$$

On further reductions, equations (26) and (27) take the forms

$$\beta\int_0^1 \Theta W\,dz-\frac{\kappa}{d^2}\int_0^1 [(D\Theta)^2+a^2\Theta^2]\,dz = \frac{A^2}{\kappa}\left\{\int_0^1 \Theta^2W^2\,dz-\left(\int_0^1 \Theta W\,dz\right)^2\right\} \tag{28}$$

and
$$g\alpha\int_0^1 \Theta W\,dz = \frac{\nu}{a^2d^2}\int_0^1 [(D^2-a^2)W]^2\,dz. \tag{29}$$

Since by equation II (181) (p. 34)

$$\Theta = \frac{\nu}{g\alpha a^2d^2}F = \frac{\nu}{g\alpha a^2d^2}(D^2-a^2)^2W, \tag{30}$$

equation (29) is in reality an identity; for, in virtue of the boundary conditions of the problem,

$$\int_0^1 W(D^2-a^2)^2W\,dz \equiv \int_0^1 [(D^2-a^2)W]^2\,dz. \tag{31}$$

Turning to equation (28) and expressing Θ in terms of F, we have

$$\frac{A^2}{\kappa}\left\{\int_0^1 W^2F^2\,dz-\left(\int_0^1 WF\,dz\right)^2\right\} = \frac{g\alpha\beta}{\nu}a^2d^2\int_0^1 WF\,dz-$$

$$-\frac{\kappa}{d^2}\int_0^1 [(DF)^2+a^2F^2]\,dz, \tag{32}$$

or, alternatively,

$$A^2\left\{\int_0^1 W^2F^2\,dz-\left(\int_0^1 WF\,dz\right)^2\right\}$$
$$=\frac{\kappa^2a^2}{d^2}\left\{\frac{g\alpha\beta}{\kappa\nu}d^4-\frac{\int_0^1[(DF)^2+a^2F^2]\,dz}{a^2\int_0^1[(D^2-a^2)W]^2\,dz}\right\}\int_0^1 WF\,dz. \tag{33}$$

Recalling that the critical Rayleigh number for the onset of instability is given by (see equation II (185))

$$R_c=\frac{\int_0^1[(DF)^2+a^2F^2]\,dz}{a^2\int_0^1[(D^2-a^2)W]^2\,dz}, \tag{34}$$

we can rewrite equation (33) in the form

$$A^2=\frac{\kappa^2a^2}{d^2}R_c\frac{\int_0^1 WF\,dz}{\int_0^1 W^2F^2\,dz-\left(\int_0^1 WF\,dz\right)^2}\left(\frac{R}{R_c}-1\right). \tag{35}$$

According to this equation, *the amplitude of the disturbance past the marginal state increases like* $(R-R_c)^{\frac{1}{2}}$.

In terms of the exact solutions for W and F given in § 15, we can explicitly evaluate the coefficient of (R/R_c-1) in equation (35) for the different cases considered. Thus, for the case when both bounding surfaces are free,

$$W=\sin\pi z,\qquad F=(\pi^2+a^2)^2\sin\pi z,$$
$$a^2=\tfrac{1}{2}\pi^2,\quad\text{and}\quad R_c=\tfrac{27}{4}\pi^4; \tag{36}$$

and equation (35) gives

$$A^2=6\pi^2\frac{\kappa^2}{d^2}\left(\frac{R}{R_c}-1\right)=59{\cdot}22\frac{\kappa^2}{d^2}\left(\frac{R}{R_c}-1\right). \tag{37}$$

For the other two sets of boundary conditions considered in § 15, the amplitudes associated with the solutions given in Table II (p. 40) can be most readily determined by evaluating the necessary integrals numerically. Thus, we find†

$$\int_0^{\frac{1}{2}} F_eW_e\,dz=0{\cdot}229732(R_ca^2)^{\frac{2}{3}},\quad \int_0^{\frac{1}{2}} F_e^2W_e^2\,dz=0{\cdot}178585(R_ca^2)^{\frac{4}{3}}$$
$$(R_c=1707{\cdot}76\text{ and }a=3{\cdot}117), \tag{38}$$

† With the normalization appropriate to the tabulated solutions.

and

$$\int_0^{\frac{1}{2}} F_o W_o\, dz = 0{\cdot}239680(R_c a^2)^{\frac{2}{3}}, \quad \int_0^{\frac{1}{2}} F_o^2 W_o^2\, dz = 0{\cdot}184027(R_c a^2)^{\frac{2}{3}}$$

$$(R_c = 1100{\cdot}65 \text{ and } a = 2{\cdot}682); \quad (39)$$

and inserting these values in equation (35), we find that the squares of the respective amplitudes are given by

$$A^2 = 80{\cdot}23\frac{\kappa^2}{d^2}\left(\frac{R}{R_c}-1\right) \quad \text{(when both bounding surfaces are rigid)}, \quad (40)$$

and

$$A^2 = 69{\cdot}11\frac{\kappa^2}{d^2}\left(\frac{R}{R_c}-1\right) \quad \begin{array}{l}\text{(when one of the bounding surfaces is rigid} \\ \quad \text{and the other is free).}\end{array} \quad (41)$$

It is clear that the basic ideas underlying the preceding considerations are of wider generality than the simple context in which they have been described. Their extension to the case when a magnetic field is present, for example, is immediate: we have only to include with ϵ_ν in equation (20) the term ϵ_σ giving the rate of dissipation of energy by Joule heating. However, the extension to cases when overstability is possible requires some care; but these are matters which go considerably beyond the scope of this book.

BIBLIOGRAPHICAL NOTES

It was Landau who first suggested from very general considerations that the amplitudes of the disturbance must increase like $(\text{R} - \text{R}_c)^{\frac{1}{2}}$ (where R is some suitably defined 'Reynolds number') beyond the onset of instability at R_c:

1. L. D. Landau, 'On the problem of turbulence', *C.R. Doklady Acad. Sci. URSS* **44**, 311–14 (1944).

See also:

2. L. D. Landau and E. M. Lifshitz, *Fluid Mechanics*, § 27, Pergamon Press, 1959.

Specifically in connexion with the stability of Couette flow, Stuart developed Landau's arguments to yield a quantitative estimate for the amplitude.

3. J. T. Stuart, 'On the non-linear mechanics of hydrodynamic stability', *J. Fluid Mech.* **4**, 1–21 (1958).

The discussion in § 126 is modelled on Stuart's paper. The non-linear aspects of the Bénard problem have been further discussed by:

4. W. V. R. Malkus, 'The heat transport and spectrum of thermal turbulence', *Proc. Roy. Soc.* (*London*) A, **225**, 196–212 (1954).
5. L. P. Gor'kov, 'Stationary convection in a plane liquid layer near the critical heat transfer point', *Soviet Physics, JETP*, **6**, 311–15 (1958).

6. W. V. R. Malkus and G. Veronis, 'Finite amplitude cellular convection', *J. Fluid Mech.* **4**, 225–60 (1958).

7. G. Veronis, 'Cellular convection with finite amplitude in a rotating fluid', ibid. **5**, 401–35 (1959).

8. Y. Nakagawa, 'Heat transport by convection', *Physics of Fluids*, **3**, 82–86 (1960).

9. —— 'Heat transport by convection in presence of an impressed magnetic field', ibid. **3**, 87–93 (1960).

These papers go into details of non-linear stability theories.

APPENDIX II

THE VARIATIONAL FORMULATION OF THE PROBLEM CONSIDERED IN CHAPTER V

127. The general equation governing the energy-balance

FROM the theorem stated in Chapter III (on p. 134) and its extension in Chapter IV (on p. 169) to include modes of dissipation of energy besides viscosity, it is clear that the general problem considered in Chapter V must equally allow a variational formulation. In particular, in case the onset of instability is as stationary convection, the variational principle cannot signify anything more than the equality

$$\epsilon_g = \epsilon_\nu + \epsilon_\sigma, \tag{1}$$

where

$$\epsilon_g = g\alpha\rho \int_0^1 \langle \theta w \rangle \, dz \tag{2}$$

is the average rate of liberation of energy by the buoyancy force, and

$$\epsilon_\nu = -\frac{\rho\nu}{d^2} \int_0^1 \langle u_i \nabla^2 u_i \rangle \, dz \tag{3}$$

and

$$\epsilon_\sigma = \frac{\mu\eta}{4\pi d^2} \int_0^1 \langle |\mathrm{curl}\,\mathbf{h}|^2 \rangle \, dz \tag{4}$$

are the average rates of dissipation of energy in a unit column of the fluid by viscosity and by Joule heating, respectively (cf. IV, § 43 (*b*)).

When rotation is present, the z-components of the vorticity and the current density do not vanish and the necessity to allow for them is the only feature which distinguishes the present problem from that considered in Chapter IV. In Chapter III (§ 33) we have already allowed for a non-vanishing z-component of the vorticity in the expression for ϵ_ν; and the expression derived there (equation (286), p. 131) is of equal applicability here. Thus,

$$\epsilon_\nu = \frac{\rho\nu}{4a^2 d^2} \int_0^1 \{[(D^2-a^2)W]^2 + d^2[(DZ)^2 + a^2 Z^2]\} \, dz. \tag{5}$$

In deriving the analogous expression for ϵ_σ, we must not forget that the expressions for the horizontal components of the magnetic field used for a similar purpose in Chapter IV do not allow for a non-vanishing z-component of the current density; and that we must now use the general relations (equations IV (126) and (127))

$$h_x = \frac{1}{a^2}\left(\frac{\partial^2 h_z}{\partial z \partial x} + d\frac{\partial \xi}{\partial y}\right) \quad \text{and} \quad h_y = \frac{1}{a^2}\left(\frac{\partial^2 h_z}{\partial z \partial y} - d\frac{\partial \xi}{\partial x}\right), \tag{6}$$

where $\xi/(4\pi)$ is the z-component of the current density.

Without any loss of generality, we may now suppose that (cf. equation IV (112))

$$\xi = X(z)\cos a_x x \cos a_y y \tag{7}$$

and

$$h_z = K(z)\cos a_x x \cos a_y y. \tag{8}$$

The corresponding expressions for h_x and h_y are:

$$h_x = -\frac{1}{a^2}(a_x DK \sin a_x x \cos a_y y + a_y dX \cos a_x x \sin a_y y), \tag{9}$$

$$h_y = -\frac{1}{a^2}(a_y DK \cos a_x x \sin a_y y - a_x dX \sin a_x x \cos a_y y). \tag{10}$$

With the components $\mathbf{h}$ given by the foregoing equations, we find that

$$(\text{curl}\,\mathbf{h})_x = +\frac{1}{a^2}\{a_y(D^2-a^2)K \cos a_x x \sin a_y y - a_x d\,DX \sin a_x x \cos a_y y\}, \tag{11}$$

$$(\text{curl}\,\mathbf{h})_y = -\frac{1}{a^2}\{a_x(D^2-a^2)K \sin a_x x \cos a_y y + a_y d\,DX \cos a_x x \sin a_y y\}, \tag{12}$$

and

$$(\text{curl}\,\mathbf{h})_z = dX \cos a_x x \cos a_y y. \tag{13}$$

Now evaluating ϵ_σ in accordance with equation (4), we find

$$\epsilon_\sigma = \frac{\mu\eta}{16\pi d^2 a^2}\int_0^1 \{[(D^2-a^2)K]^2 + d^2[(DX)^2 + a^2 X^2]\}\,dz. \tag{14}$$

Equations (5) and (14) exhibit the completely parallel roles of W and Z and of K and X in determining the respective dissipations.

Eliminating K from equation (14) by means of the relation (equation V (46))

$$(D^2-a^2)K = -\frac{Hd}{\eta}DW, \tag{15}$$

we obtain

$$\epsilon_\sigma = \frac{\rho\nu}{4a^2 d^2}Q\int_0^1 (DW)^2\,dz + \frac{\mu\eta}{16\pi a^2}\int_0^1 [(DX)^2 + a^2 X^2]\,dz. \tag{16}$$

This expression for ϵ_σ should be contrasted with the one derived in Chapter IV (equation (158), p. 169).

Now combining equations (5) and (16), we have

$$\epsilon_\nu+\epsilon_\sigma = \frac{\rho\nu}{4a^2d^2}\left[\int_0^1 \{[(D^2-a^2)W]^2+Q(DW)^2+d^2[(DZ)^2+a^2Z^2]\}\,dz+ \right.$$
$$\left. +\frac{\mu\eta}{4\pi\rho\nu}\,d^2\int_0^1 [(DX)^2+a^2X^2]\,dz\right]. \quad (17)$$

The expression for ϵ_g is the same as in the other similar connexions in which we have considered it; it is given by (cf. equations II (182), III (291), and IV (151))

$$\epsilon_g = \frac{\rho\kappa\nu^2}{4g\alpha\beta a^4d^6}\int_0^1 [(DF)^2+a^2F^2]\,dz, \quad (18)$$

where

$$F = \frac{g\alpha a^2d^2}{\nu}\Theta. \quad (19)$$

By equating (17) and (18), we obtain

$$Ra^2\left[\int_0^1 \{[(D^2-a^2)W]^2+Q(DW)^2+d^2[(DZ)^2+a^2Z^2]\}\,dz+\right.$$
$$\left.+\frac{\mu\eta}{4\pi\rho\nu}\,d^2\int_0^1 [(DX)^2+a^2X^2]\,dz\right] = \int_0^1 [(DF)^2+a^2F^2]\,dz. \quad (20)$$

Equation (20) expresses R as the ratio of two positive integrals; and to verify that it can be made the basis of a variational treatment, we shall derive it directly from the relevant equations of the problem.

With F defined as in equation (19), equations V (35) and (39) (with σ set equal to zero) become

$$(D^2-a^2)F = -Ra^2W \quad (21)$$

and

$$(D^2-a^2)^2W+\frac{\mu Hd}{4\pi\rho\nu}D(D^2-a^2)K-\frac{2\Omega d^3}{\nu}DZ = F. \quad (22)$$

We can eliminate K from equation (22) by making use of equation (15); and we obtain

$$[(D^2-a^2)^2-QD^2]W-\frac{2\Omega d^3}{\nu}DZ = F. \quad (23)$$

Equations (21) and (23) must be considered together with the equations (cf. equations V (37) and (38))

$$(D^2-a^2)Z+\frac{\mu Hd}{4\pi\rho\nu}DX = -\frac{2\Omega d}{\nu}DW \quad (24)$$

and
$$(D^2-a^2)X = -\frac{Hd}{\eta}DZ, \tag{25}$$
and the boundary conditions V (49)–(51).

Now multiply equation (21) by F and integrate over the range of z; we obtain
$$\int_0^1 [(DF)^2+a^2F^2]\,dz = Ra^2\int_0^1 WF\,dz$$
$$= Ra^2\int_0^1 W\Big\{(D^2-a^2)^2W-QD^2W-\frac{2\Omega d^3}{\nu}DZ\Big\}\,dz. \tag{26}$$
After one or more integrations by parts, the integral on the right-hand side of equation (26) can be brought to the form
$$\int_0^1 \{[(D^2-a^2)W]^2+Q(DW)^2\}\,dz+\frac{2\Omega d^3}{\nu}\int_0^1 Z\,DW dz. \tag{27}$$
By making use of equation (24), we can transform the second integral in (27) in the manner
$$\frac{2\Omega d^3}{\nu}\int_0^1 Z\,DWdz = -d^2\int_0^1 Z\Big\{(D^2-a^2)Z+\frac{\mu Hd}{4\pi\rho\nu}DX\Big\}\,dz$$
$$= d^2\int_0^1 [(DZ)^2+a^2Z^2]\,dz+\frac{\mu Hd^3}{4\pi\rho\nu}\int_0^1 X\,DZdz. \tag{28}$$
By making similar use of the remaining equation (25), we can transform the second integral on the right-hand side of equation (28) to give
$$\frac{\mu Hd^3}{4\pi\rho\nu}\int_0^1 X\,DZdz = -\frac{\mu\eta}{4\pi\rho\nu}\,d^2\int_0^1 X(D^2-a^2)X\,dz$$
$$= \frac{\mu\eta}{4\pi\rho\nu}\,d^2\int_0^1 [(DX)^2+a^2X^2]\,dz. \tag{29}$$

By combining the results of the foregoing reductions we recover equation (20); and it can be readily shown that equation (20) does indeed provide a basis for a variational determination of R. However, the present derivation of equation (20) shows that in the variational treatment, equations (23)–(25) must be considered as *subsidiary conditions* on the characteristic value problem specified by equation (21). In other words, for a chosen form of F (compatible with the boundary conditions that it vanish for $z = 0$ and 1) we must *solve* equations (23)–(25) for W, Z, and X and arrange that the required boundary conditions on them are satisfied. For these last purposes it will be convenient to combine equations (24) and (25) in the form

$$[(D^2-a^2)^2-QD^2]Z = -\frac{2\Omega d}{\nu}(D^2-a^2)DW, \tag{30}$$

and use this equation to eliminate Z from equation (23) to obtain

$$\{[(D^2-a^2)^2-QD^2]^2+T(D^2-a^2)D^2\}W = [(D^2-a^2)^2-QD^2]F. \tag{31}$$

This is an equation of order eight for W and the two pairs of boundary conditions on W and the pair each on Z and X will suffice to determine the solution uniquely. However, it is evident that a variational treatment of the problem based on the foregoing equations and for the various permissible sets of boundary conditions will be very laborious; it was for this reason that the treatment in the text was confined to the one case (*albeit*, an artificial case) for which an explicit solution is possible.

The generalization of the preceding analysis to include the case of overstability is straightforward and will be omitted.

BIBLIOGRAPHICAL NOTE

See:

1. S. Chandrasekhar, 'The thermodynamics of thermal instability in liquids', *Max Planck Festschrift 1958*, 103–14, Veb Deutscher Verlag der Wissenschaften, Berlin, 1958.

APPENDIX III

TOROIDAL AND POLOIDAL VECTOR FIELDS

128. A general characterization of solenoidal vector fields. The fundamental basis

THE vector fields defined by the equations

$$\mathbf{T} = \operatorname{curl}\left(\frac{\Psi}{r}\mathbf{r}\right) \tag{1}$$

and

$$\mathbf{S} = \operatorname{curl}\left[\operatorname{curl}\left(\frac{\Phi}{r}\mathbf{r}\right)\right], \tag{2}$$

where Ψ and Φ are arbitrary scalar functions of position, are clearly solenoidal. A field derivable in the manner (1) is said to be *toroidal* with the *defining scalar* Ψ; and a field derivable in the manner (2) is said to be *poloidal* with the defining scalar Φ. Alternative ways of writing $\mathbf{T}$ and $\mathbf{S}$ are

$$\mathbf{T} = \operatorname{grad}\frac{\Psi}{r}\times\mathbf{r} \tag{3}$$

and

$$\mathbf{S} = \operatorname{curl}\left(\operatorname{grad}\frac{\Phi}{r}\times\mathbf{r}\right). \tag{4}$$

In a system of spherical coordinates, (r,ϑ,φ), the components of $\mathbf{S}$ and $\mathbf{T}$ are:

$$T_r = 0, \qquad T_\vartheta = \frac{1}{r\sin\vartheta}\frac{\partial\Psi}{\partial\varphi}, \qquad T_\varphi = -\frac{1}{r}\frac{\partial\Psi}{\partial\vartheta}, \tag{5}$$

and

$$S_r = \frac{1}{r^2}L^2\Phi, \qquad S_\vartheta = \frac{1}{r}\frac{\partial^2\Phi}{\partial r\partial\vartheta}, \qquad S_\varphi = \frac{1}{r\sin\vartheta}\frac{\partial^2\Phi}{\partial r\partial\varphi}, \tag{6}$$

where L^2 stands for the operator (defined in equation VI (25))

$$L^2 = -\frac{1}{\sin\vartheta}\frac{\partial}{\partial\vartheta}\sin\vartheta\frac{\partial}{\partial\vartheta}-\frac{1}{\sin^2\vartheta}\frac{\partial^2}{\partial\varphi^2}. \tag{7}$$

Since $\mathbf{T}$ is solenoidal,

$$(\operatorname{curl}^2\mathbf{T})_i = -\frac{\partial^2 T_i}{\partial x_s^2} = \frac{\partial^2}{\partial x_s^2}\epsilon_{ijk}\,x_j\frac{\partial}{\partial x_k}\frac{\Psi}{r} = \epsilon_{ijk}\,x_j\frac{\partial}{\partial x_k}\nabla^2\frac{\Psi}{r}; \tag{8}$$

thus,

$$\operatorname{curl}^2\mathbf{T} = -\operatorname{grad}\left(\nabla^2\frac{\Psi}{r}\right)\times\mathbf{r}. \tag{9}$$

Similarly,

$$\operatorname{curl}\mathbf{S} = -\operatorname{grad}\left(\nabla^2\frac{\Phi}{r}\right)\times\mathbf{r}. \tag{10}$$

From equations (3), (4), (9), and (10) it is apparent that curl **T** is a poloidal field with the same defining scalar Ψ, while curl²**T** is a toroidal field with the defining scalar

$$\tilde{\Psi} = -r\nabla^2 \frac{\Psi}{r}; \tag{11}$$

correspondingly, curl **S** is a toroidal field with the defining scalar

$$\tilde{\Phi} = -r\nabla^2 \frac{\Phi}{r}, \tag{12}$$

while curl²**S** is a poloidal field with the same defining scalar $\tilde{\Phi}$.

We obtain a *fundamental basis* (on a sphere) for these toroidal and poloidal fields by expanding Ψ and Φ in spherical harmonics with coefficients which are allowed to be functions of r. Thus, the elements of the basis are the fields derived from the scalars

$$\Psi = T(r)Y_l^m(\vartheta,\varphi) \quad \text{and} \quad \Phi = S(r)Y_l^m(\vartheta,\varphi), \tag{13}$$

where

$$Y_l^m(\vartheta,\varphi) = e^{im\varphi}P_l^{|m|}(\cos\vartheta). \tag{14}$$

The components of the fields defined by these special scalars are:

$$T_r = 0, \qquad T_\vartheta = \frac{T(r)}{r\sin\vartheta}\frac{\partial Y_l^m}{\partial\varphi}, \qquad T_\varphi = -\frac{T(r)}{r}\frac{\partial Y_l^m}{\partial\vartheta}, \tag{15}$$

and

$$S_r = \frac{l(l+1)}{r^2}S(r)Y_l^m, \quad S_\vartheta = \frac{1}{r}\frac{dS}{dr}\frac{\partial Y_l^m}{\partial\vartheta}, \quad S_\varphi = \frac{1}{r\sin\vartheta}\frac{dS}{dr}\frac{\partial Y_l^m}{\partial\varphi}, \tag{16}$$

where in simplifying the expression for S_r, we have made use of the identity (equation VI (26))

$$L^2 Y_l^m = l(l+1)Y_l^m. \tag{17}$$

The defining scalars of the associated fields curl²**T** and curl **S** are readily found. We have (cf. equation VI (32))

$$\tilde{\Psi} = -r\nabla^2\left(\frac{T}{r}Y_l^m\right) = -rY_l^m\mathscr{D}_l\left(\frac{T}{r}\right) = -rY_l^m\left[\frac{1}{r^2}\frac{d}{dr}r^2\frac{d}{dr} - \frac{l(l+1)}{r^2}\right]\frac{T}{r}$$

$$= Y_l^m\left[\frac{l(l+1)}{r^2}T - \frac{d^2T}{dr^2}\right]; \tag{18}$$

and similarly,

$$\tilde{\Phi} = Y_l^m\left[\frac{l(l+1)}{r^2}S - \frac{d^2S}{dr^2}\right]. \tag{19}$$

129. The orthogonality properties of the basic toroidal and poloidal fields

It is convenient to define on the unit sphere the vector operator

$$\boldsymbol{\nabla}_s = \mathbf{1}_\vartheta\frac{\partial}{\partial\vartheta} + \mathbf{1}_\varphi\frac{1}{\sin\vartheta}\frac{\partial}{\partial\varphi}, \tag{20}$$

where $\mathbf{1}_\vartheta$ and $\mathbf{1}_\varphi$ are unit vectors, along the principal meridian and the latitude circle, through the point considered. The effect of ∇_s on any scalar function $\Psi(r,\vartheta,\varphi)$ is to yield the vector

$$\nabla_s \Psi = \mathbf{1}_\vartheta \frac{\partial \Psi}{\partial \vartheta} + \mathbf{1}_\varphi \frac{1}{\sin\vartheta}\frac{\partial \Psi}{\partial \varphi}. \tag{21}$$

On the other hand, if

$$\boldsymbol{\xi} = \mathbf{1}_\vartheta \xi_\vartheta + \mathbf{1}_\varphi \xi_\varphi \tag{22}$$

is a two-dimensional vector point-function (on the unit sphere), then we *define*

$$\operatorname{div}\boldsymbol{\xi} = \nabla_s \cdot \boldsymbol{\xi} = \frac{1}{\sin\vartheta}\frac{\partial}{\partial\vartheta}(\xi_\vartheta \sin\vartheta) + \frac{1}{\sin\vartheta}\frac{\partial \xi_\varphi}{\partial\varphi}. \tag{23}$$

Accordingly,

$$\nabla_s^2 \Psi = \nabla_s \cdot (\nabla_s \Psi) = \nabla_s \cdot \left(\mathbf{1}_\vartheta \frac{\partial\Psi}{\partial\vartheta} + \mathbf{1}_\varphi \frac{1}{\sin\vartheta}\frac{\partial\Psi}{\partial\varphi}\right)$$

$$= \frac{1}{\sin\vartheta}\frac{\partial}{\partial\vartheta}\left(\sin\vartheta \frac{\partial\Psi}{\partial\vartheta}\right) + \frac{1}{\sin^2\vartheta}\frac{\partial^2\Psi}{\partial\varphi^2} = -L^2\Psi. \tag{24}$$

Hence, in particular,

$$\nabla_s^2 Y_l^m = -L^2 Y_l^m = -l(l+1) Y_l^m. \tag{25}$$

Two elementary consequences of the foregoing definitions may be noted. *First*, if $\boldsymbol{\xi}$ is a single-valued vector point-function on the unit sphere, then

$$\iint \nabla_s \cdot \boldsymbol{\xi}\, d\Sigma = \int_0^\pi \int_0^{2\pi} \left\{\frac{1}{\sin\vartheta}\frac{\partial}{\partial\vartheta}(\xi_\vartheta \sin\vartheta) + \frac{1}{\sin\vartheta}\frac{\partial\xi_\varphi}{\partial\varphi}\right\} \sin\vartheta\, d\vartheta d\varphi$$

$$= \int_0^{2\pi} d\varphi [\xi_\vartheta \sin\vartheta]_0^\pi + \int_0^\pi d\vartheta [\xi_\varphi]_0^{2\pi} = 0. \tag{26}$$

Thus, in the present geometry Gauss's theorem takes the form

$$\iint \nabla_s \cdot \boldsymbol{\xi}\, d\Sigma = 0. \tag{27}$$

A *second* consequence which directly follows from the definitions is

$$\nabla_s \cdot (\Psi\boldsymbol{\xi}) = \Psi(\nabla_s \cdot \boldsymbol{\xi}) + \boldsymbol{\xi} \cdot (\nabla_s \Psi). \tag{28}$$

Making use of the results expressed in equations (27) and (28), we deduce the following important identity:

$$0 = \iint \nabla_s \cdot (Y_{l'}^{m'} \nabla_s Y_l^m)\, d\Sigma$$

$$= \iint \{Y_{l'}^{m'} \nabla_s^2 Y_l^m + (\nabla_s Y_l^m) \cdot (\nabla_s Y_{l'}^{m'})\}\, d\Sigma$$

$$= -l(l+1) \iint Y_{l'}^{m'} Y_l^m\, d\Sigma + \iint (\nabla_s Y_l^m) \cdot (\nabla_s Y_{l'}^{m'})\, d\Sigma. \tag{29}$$

From the known orthogonality properties of the spherical harmonics, we can infer that

$$\iint (\nabla_s Y_l^m).(\nabla_s Y_{l'}^{m'})\,d\Sigma = l(l+1)N_l^{|m|}\delta_{ll'}\,\delta_{m,-m'}, \tag{30}$$

where
$$N_l^{|m|} = \frac{4\pi}{2l+1}\frac{(l+|m|)!}{(l-|m|)!}. \tag{31}$$

The various orthogonality relations among the basic toroidal and poloidal fields enumerated in § 56 (*d*) (p. 226) are now seen to be direct consequences of equation (30). Thus, considering two toroidal fields, $\mathbf{T}_{1;l}^m$ and $\mathbf{T}_{2;l'}^{m'}$ associated, respectively, with the spherical harmonics Y_l^m and $Y_{l'}^{m'}$, we have

$$\begin{aligned} r^2 \iint \mathbf{T}_{1;l}^m . \mathbf{T}_{2;l'}^{m'}\,d\Sigma &= T_1 T_2 \iint \left(\frac{1}{\sin^2\vartheta}\frac{\partial Y_l^m}{\partial\varphi}\frac{\partial Y_{l'}^{m'}}{\partial\varphi} + \frac{\partial Y_l^m}{\partial\vartheta}\frac{\partial Y_{l'}^{m'}}{\partial\vartheta}\right) d\Sigma \\ &= T_1 T_2 \iint (\nabla_s Y_l^m).(\nabla_s Y_{l'}^{m'})\,d\Sigma \\ &= l(l+1)N_l^{|m|}T_1 T_2\,\delta_{ll'}\,\delta_{m,-m'}; \end{aligned} \tag{32}$$

and similarly,

$$\begin{aligned} r^2 \iint \mathbf{S}_{1;l}^m . \mathbf{S}_{2;l'}^{m'}\,d\Sigma &= l(l+1)l'(l'+1)\frac{S_1 S_2}{r^2}\iint Y_l^m Y_{l'}^{m'}\,d\Sigma + \\ &\qquad + \frac{dS_1}{dr}\frac{dS_2}{dr}\iint (\nabla_s Y_l^m).(\nabla_s Y_{l'}^{m'})\,d\Sigma \\ &= l(l+1)N_l^{|m|}\left\{\frac{l(l+1)}{r^2}S_1 S_2 + \frac{dS_1}{dr}\frac{dS_2}{dr}\right\}\delta_{ll'}\,\delta_{m,-m'}. \end{aligned} \tag{33}$$

And finally, the orthogonality of every poloidal field to every toroidal field can be established as follows:

$$\begin{aligned} r^2 \iint \mathbf{S}_l^m . \mathbf{T}_{l'}^{m'}\,d\Sigma &= T\frac{dS}{dr}\int_0^{\pi}\int_0^{2\pi} \left(\frac{\partial Y_l^m}{\partial\vartheta}\frac{\partial Y_{l'}^{m'}}{\partial\varphi} - \frac{\partial Y_l^m}{\partial\varphi}\frac{\partial Y_{l'}^{m'}}{\partial\vartheta}\right) d\vartheta d\varphi \\ &= T\frac{dS}{dr}2\pi i m'\delta_{m,-m'}\int_0^{\pi}\left(P_{l'}^{|m|}\frac{dP_l^{|m|}}{d\vartheta} + P_l^{|m|}\frac{dP_{l'}^{|m|}}{d\vartheta}\right) d\vartheta \\ &= T\frac{dS}{dr}2\pi i m'\delta_{m,-m'}[P_l^{|m|}(\vartheta)P_{l'}^{|m|}(\vartheta)]_0^{\pi}. \end{aligned} \tag{34}$$

The quantity in square brackets in the last line of (34) clearly vanishes if l and l' are both even or both odd; but if only one of them, say l, is odd, then, $P_l^{|m|}$ ($|m| \neq 0$) vanishes at $\vartheta = 0$ and $\vartheta = \pi$; hence, in all cases the quantity in square brackets vanishes; and the required orthogonality property is established.

BIBLIOGRAPHICAL NOTES

To the references given in the Bibliographical Notes for Chapter VI (Nos. 13 and 14, p. 269) we may add the following:

1. J. M. Blatt and V. F. Weisskopf, *Theoretical Nuclear Physics*, Appendix B, John Wiley & Sons, New York, 1952.
2. G. E. Backus, 'A class of self-sustaining dissipative spherical dynamos', *Annals of Physics*, **4**, 372–447 (1958).

APPENDIX IV

VARIATIONAL METHODS BASED ON ADJOINT DIFFERENTIAL SYSTEMS

130. Adjointness of differential systems. An example

IN this appendix we shall provide a variational basis for a method of solution of characteristic value problems to which we have had frequent recourse, namely, when the equations and the boundary conditions are such as not to allow a variational formulation in terms of an expression for the characteristic value as a ratio of two integrals which are (generally) positive definite. We shall illustrate the principal ideas in the particular context of the problem treated in § 71 (a) though the ideas themselves are of wider generality.

The problem treated in § 71 (a) is the following: to solve the equations

$$(D^2-a^2)^2u = (1+\alpha\zeta)v \tag{1}$$

and

$$(D^2-a^2)v = -\lambda u, \tag{2}$$

together with the boundary conditions

$$u = Du = v = 0 \quad \text{for } \zeta = 0 \text{ and } 1. \tag{3}$$

The characteristic value parameter is λ (denoted by Ta^2 in the text), a and α are real constants, and $D = d/d\zeta$.

The method by which the characteristic value problem presented by equations (1)–(3) was solved is the following.

We expand v in a sine series,

$$v = \sum_n A_n \sin n\pi\zeta, \tag{4}$$

and express u as the sum

$$u = \sum_n A_n u_n, \tag{5}$$

where u_n is the unique solution of the equation

$$(D^2-a^2)^2u_n = (1+\alpha\zeta)\sin n\pi\zeta \tag{6}$$

which satisfies the boundary conditions

$$u_n = Du_n = 0 \quad \text{for } \zeta = 0 \text{ and } 1. \tag{7}$$

Having determined u_n in this fashion, we insert the expansions (4) and (5) in equation (2) to obtain

$$\sum_m A_m(m^2\pi^2+a^2)\sin m\pi\zeta = \lambda \sum_m A_m u_m. \tag{8}$$

Multiplying equation (8) by $\sin n\pi\zeta$ and integrating over the range of ζ, we obtain an infinite set of linear homogeneous equations for the A_m's which in turn leads to the secular equation

$$\left\|\frac{1}{2\lambda}(n^2\pi^2+a^2)\delta_{mn}-(m|n)\right\| = 0, \tag{9}$$

where

$$(m|n) = \int_0^1 u_m \sin n\pi\zeta \, d\zeta. \tag{10}$$

Now consider the slightly different system:

$$(D^2-a^2)^2u^\dagger = v^\dagger, \tag{11}$$

$$(D^2-a^2)v^\dagger = -\lambda^\dagger(1+\alpha\zeta)u^\dagger, \tag{12}$$

and

$$u^\dagger = Du^\dagger = v^\dagger = 0 \quad \text{for } \zeta = 0 \text{ and } 1. \tag{13}$$

Let us seek the solution of this system by the same method which we used for the solution of the system (1)–(3). Accordingly, we express $v^\dagger$ and $u^\dagger$ as series in the forms

$$v^\dagger = \sum_n B_n \sin n\pi\zeta \quad \text{and} \quad u^\dagger = \sum_n B_n u_n^\dagger, \tag{14}$$

where $u_n^\dagger$ is now the unique solution of the equation

$$(D^2-a^2)^2u_n^\dagger = \sin n\pi\zeta \tag{15}$$

which satisfies the boundary conditions

$$u_n^\dagger = Du_n^\dagger = 0 \quad \text{for } \zeta = 0 \text{ and } 1. \tag{16}$$

Equation (12) will then lead to the secular equation

$$\left\|\frac{1}{2\lambda^\dagger}(n^2\pi^2+a^2)\delta_{nm}-(m|n)^\dagger\right\| = 0, \tag{17}$$

where

$$(m|n)^\dagger = \int_0^1 u_m^\dagger(1+\alpha\zeta)\sin n\pi\zeta \, d\zeta. \tag{18}$$

We shall now show that *the matrices* $(m|n)$ *and* $(m|n)^\dagger$ *are the transposed forms of one another*. To show this, insert for $(1+\alpha\zeta)\sin n\pi\zeta$ in the integrand for $(m|n)^\dagger$ according to equation (6); we then obtain

$$(m|n)^\dagger = \int_0^1 u_m^\dagger(D^2-a^2)^2u_n \, d\zeta. \tag{19}$$

After two integrations by parts, equation (19) can be brought to the form

$$(m|n)^\dagger = \int_0^1 [(D^2-a^2)u_m^\dagger][(D^2-a^2)u_n] \, d\zeta. \tag{20}$$

Similarly, by inserting for $\sin n\pi\zeta$ in the integrand for $(m|n)$, in accordance with equation (15), we obtain

$$(m|n) = \int_0^1 u_m(D^2-a^2)^2 u_n^\dagger \, d\zeta \tag{21}$$

which after two integrations by parts becomes

$$(m|n) = \int_0^1 [(D^2-a^2)u_m][(D^2-a^2)u_n^\dagger] \, d\zeta. \tag{22}$$

A comparison of equations (20) and (22) shows that

$$(m|n) = (n|m)^\dagger. \tag{23}$$

A consequence of this last relation is that the secular equations (9) and (17) are derived from matrices which are the transposed forms of one another. Therefore, *the characteristic equations for λ and $\lambda^\dagger$ are identical.* We have thus shown that *the systems* (1)–(3) *and* (11)–(13) *determine the same set of characteristic values.* For this reason we shall call the two systems *the adjoints of one another.*

Quite generally, *two systems can be said to be the adjoints of one another if they determine the same set of characteristic values.* (It should be pointed out in this connexion that more than two systems can be the adjoints of one another.)

131. The dual relationship and the variational principle

From the identity of the characteristic values of the systems (1)–(3) and (11)–(13) we can derive a *relationship of duality* between the proper solutions of the two systems. To elucidate the nature of this duality, consider the solutions u_j and v_j belonging to λ_j and the solutions $u_k^\dagger$ and $v_k^\dagger$ belonging to a different characteristic value λ_k. The equations satisfied by these solutions are

$$(D^2-a^2)^2 u_j = (1+\alpha\zeta)v_j, \tag{24}$$

$$(D^2-a^2)v_j = -\lambda_j u_j, \tag{25}$$

and

$$(D^2-a^2)^2 u_k^\dagger = v_k^\dagger, \tag{26}$$

$$(D^2-a^2)v_k^\dagger = -\lambda_k(1+\alpha\zeta)u_k^\dagger. \tag{27}$$

Now consider

$$\lambda_j \int_0^1 u_j v_k^\dagger \, d\zeta = -\int_0^1 v_k^\dagger (D^2-a^2)v_j \, d\zeta. \tag{28}$$

By two integrations by parts carried out in succession, we find

$$\lambda_j \int_0^1 u_j v_k^\dagger \, d\zeta = \int_0^1 [(Dv_j)(Dv_k^\dagger) + a^2 v_j v_k^\dagger] \, d\zeta$$

$$= -\int_0^1 v_j (D^2 - a^2) v_k^\dagger \, d\zeta. \tag{29}$$

On the other hand, by substituting for $v_k^\dagger$ from equation (26), we obtain

$$\lambda_j \int_0^1 u_j v_k^\dagger \, d\zeta = \lambda_j \int_0^1 u_j (D^2 - a^2)^2 u_k^\dagger \, d\zeta$$

$$= \lambda_j \int_0^1 [(D^2 - a^2) u_j][(D^2 - a^2) u_k^\dagger] \, d\zeta. \tag{30}$$

Therefore, defining

$$G_j = (D^2 - a^2) u_j \quad \text{and} \quad G_k^\dagger = (D^2 - a^2) u_k^\dagger, \tag{31}$$

we have the relation

$$\lambda_j \int_0^1 G_j G_k^\dagger \, d\zeta = \int_0^1 [(Dv_j)(Dv_k^\dagger) + a^2 v_j v_k^\dagger] \, d\zeta. \tag{32}$$

Considering the second alternative form for the integral on the right-hand side of equation (32) (given in (29)) and substituting for $(D^2 - a^2) v_k^\dagger$ from equation (27), we obtain

$$\int_0^1 [(Dv_j)(Dv_k^\dagger) + a^2 v_j v_k^\dagger] \, d\zeta = \lambda_k \int_0^1 v_j (1 + \alpha\zeta) u_k^\dagger \, d\zeta. \tag{33}$$

On the right-hand side of this equation, we can replace $(1+\alpha\zeta)v_j$ by $(D^2 - a^2)^2 u_j$ (in accordance with equation (24)). We thus find

$$\int_0^1 [(Dv_j)(Dv_k^\dagger) + a^2 v_j v_k^\dagger] \, d\zeta = \lambda_k \int_0^1 u_k^\dagger (D^2 - a^2)^2 u_j \, d\zeta$$

$$= \lambda_k \int_0^1 [(D^2 - a^2) u_j][(D^2 - a^2) u_k^\dagger] \, d\zeta$$

$$= \lambda_k \int_0^1 G_j G_k^\dagger \, d\zeta. \tag{34}$$

Now combining the results expressed by equations (32) and (34), we have the relation

$$(\lambda_j - \lambda_k) \int_0^1 G_j G_k^\dagger \, d\zeta = 0. \tag{35}$$

Hence
$$\int_0^1 G_j\, G_k^\dagger\, d\zeta = 0 \quad \text{if } j \neq k. \tag{36}$$

The proper solutions G_j and $G_k^\dagger$ are therefore in a dual relationship of orthogonality.

When $j = k$, equations (32) and (34) yield the same relation

$$\lambda = \frac{\int_0^1 [(Dv)(Dv^\dagger)+a^2vv^\dagger]\,d\zeta}{\int_0^1 [(D^2-a^2)u][(D^2-a^2)u^\dagger]\,d\zeta} = \frac{I_1}{I_2} \quad \text{(say)}, \tag{37}$$

where the distinguishing subscripts have been suppressed. We shall now show that equation (37) provides the basis for a variational formulation of the underlying characteristic value problems.

Consider, then, the effect on λ (evaluated in accordance with equation (37)) due to infinitesimal variations δv and $\delta v^\dagger$ (in v and $v^\dagger$) which are arbitrary except for the requirement that they vanish at $\zeta = 0$ and 1. The variations δu and $\delta u^\dagger$ (in u and $u^\dagger$) consequent to the variations in v and $v^\dagger$ are to be determined as the unique solutions of the equations

$$(D^2-a^2)^2\,\delta u = (1+\alpha\zeta)\,\delta v \tag{38}$$

and
$$(D^2-a^2)^2\,\delta u^\dagger = \delta v^\dagger \tag{39}$$

which satisfy the boundary conditions

$$\delta u = D\,\delta u = 0 \quad \text{and} \quad \delta u^\dagger = D\,\delta u^\dagger = 0 \quad \text{for } \zeta = 0 \text{ and } 1. \tag{40}$$

In other words, the variations are to be carried out subject to equations (1) and (11) as *constraints*.

Denoting by $\delta\lambda$ the resulting first order change in λ, we have

$$\delta\lambda = \frac{1}{I_2}(\delta I_1 - \lambda\,\delta I_2), \tag{41}$$

where
$$\delta I_1 = \int_0^1 \{(D\,\delta v)(D\,v^\dagger)+(Dv)(D\,\delta v^\dagger)+a^2(v\,\delta v^\dagger+v^\dagger\,\delta v)\}\,d\zeta \tag{42}$$

and

$$\delta I_2 = \int_0^1 \{[(D^2-a^2)\,\delta u][(D^2-a^2)u^\dagger]+[(D^2-a^2)u][(D^2-a^2)\,\delta u^\dagger]\}\,d\zeta \tag{43}$$

are the corresponding variations in I_1 and I_2. Making use of boundary conditions imposed on the various increments, we can reduce the expressions for δI_1 and δI_2 by one or more integrations by parts to obtain

$$\delta I_1 = -\int_0^1 \{\delta v^\dagger(D^2-a^2)v+\delta v(D^2-a^2)v^\dagger\}\,d\zeta \tag{44}$$

and
$$\delta I_2 = \int_0^1 \{u^\dagger(D^2-a^2)^2\,\delta u + u(D^2-a^2)^2\,\delta u^\dagger\}\,d\zeta. \tag{45}$$

Since δu and $\delta u^\dagger$ are subject to equations (38) and (39), an equivalent expression for δI_2 is

$$\delta I_2 = \int_0^1 \{\delta v(1+\alpha\zeta)u^\dagger + \delta v^\dagger u\}\,d\zeta. \tag{46}$$

With δI_1 and δI_2 given by equations (44) and (46), the required first order change in λ is

$$\delta\lambda = -\frac{1}{I_2}\int_0^1 \{\delta v^\dagger[(D^2-a^2)v+\lambda u]+\delta v[(D^2-a^2)v^\dagger+\lambda(1+\alpha\zeta)u^\dagger]\}\,d\zeta. \tag{47}$$

From this last equation it follows that $\delta\lambda = 0$ *to the first order for all small variations* δv *and* $\delta v^\dagger$ (*which vanish at* $\zeta = 0$ *and* 1) *provided*

$$(D^2-a^2)v+\lambda u = 0 \quad \text{and} \quad (D^2-a^2)v^\dagger+\lambda(1+\alpha\zeta)\,u^\dagger = 0; \tag{48}$$

i.e. provided equations (2) and (12) governing u and v and $u^\dagger$ and $v^\dagger$ are simultaneously satisfied. It is evident that the converse of this statement is also true. Therefore, equation (37) does provide the basis for a variational treatment of the problem.

Finally, we shall show that the method described at the outset for the solution of the characteristic value problems presented by equations (1)–(3) and (11)–(13) is equivalent to a variational procedure (based on equation (37)) in which the coefficients A_n and B_n, in the expansions for v and $v^\dagger$, are treated as variational parameters. Thus, with the chosen forms for v and $v^\dagger$, the expression for λ becomes

$$\lambda = \frac{\frac{1}{2}\sum_m A_m\,B_m(m^2\pi^2+a^2)}{\sum_n\sum_m A_m\,B_n\int_0^1 [(D^2-a^2)u_m][(D^2-a^2)u_n^\dagger]\,d\zeta}. \tag{49}$$

The matrix element which occurs in the denominator of this expression is, by equation (22), none other than $(m|n)$. Accordingly,

$$\lambda = \frac{\frac{1}{2}\sum_m A_m\,B_m(m^2\pi^2+a^2)}{\sum_m\sum_n A_m(m|n)B_n}. \tag{50}$$

We must now seek an extremum of this expression for λ considered as a function of the parameters A_m and B_n; and the simplest way to accomplish this is to treat λ as a Lagrangian undetermined multiplier and seek directly the extremum of the expression

$$J = \frac{1}{2\lambda}\sum_m A_m B_m(m^2\pi^2+a^2)-\sum_m\sum_n A_m(m|n)B_n. \tag{51}$$

By requiring that

$$\frac{\partial J}{\partial A_m} = \frac{\partial J}{\partial B_n} = 0 \quad (m,n = 1,2,...), \tag{52}$$

we clearly obtain the same secular equations as (9) and (17). This establishes the equivalence of the procedure outlined in § 130 with a variational treatment based on equation (37).

The ideas described in the context of the system (1)–(3) are clearly of wider generality: for example, the methods described for the solution of the various characteristic value problems in §§ 60, 73, 76, 87, and 88 can be similarly provided with variational bases.

BIBLIOGRAPHICAL NOTES

The ideas described in this appendix derive from the following paper by Roberts, though the treatment and the approach are different.

1. P. H. Roberts, 'Characteristic value problems posed by differential equations arising in hydrodynamics and hydromagnetics', *J. Math. Analysis and Applications*, **1**, 195–214 (1960).

See also:

2. S. Chandrasekhar, 'Adjoint differential systems in the theory of hydrodynamic stability', *J. Math. and Mech.* **10** (in press).

APPENDIX V

ORTHOGONAL FUNCTIONS WHICH SATISFY FOUR BOUNDARY CONDITIONS

132. Introduction

In the solution of hydrodynamic problems it is often convenient to be able to expand the prevailing velocity fields in a complete set of orthogonal functions. However, the usual expansions in terms of trigonometric or Bessel functions are not useful (if not impossible) in problems involving viscous flow since, in these problems, the boundary conditions require that on a rigid wall, for example, not only the normal component of the velocity, but also its normal derivative, vanishes (the latter condition ensuring the vanishing of the parallel component of the velocity). In these problems, expansions must therefore be sought in terms of orthogonal functions which together with their first derivatives vanish at the ends of the chosen interval. In this appendix, we shall give an account of the functions which have been constructed for these applications, though their completeness for the purposes of expansion requires a more careful analysis than has been given.

133. Functions suitable for problems with plane boundaries

Consider the characteristic value problem defined by the equation

$$\frac{d^4y}{dx^4} = \alpha^4 y \tag{1}$$

and the boundary conditions

$$y = 0 \quad \text{and} \quad dy/dx = 0 \quad \text{for} \quad x = \pm\tfrac{1}{2}. \tag{2}$$

Let α_m denote a characteristic value and let the proper solution belonging to it be distinguished by the subscript m. Then multiplying the equation governing y_m by y_n (belonging to α_n) and integrating over the range of x, we obtain

$$\alpha_m^4 \int_{-\frac{1}{2}}^{+\frac{1}{2}} y_m y_n \, dx = \int_{-\frac{1}{2}}^{+\frac{1}{2}} y_n \frac{d^4 y_m}{dx^4} \, dx. \tag{3}$$

Transforming the integral on the right-hand side by two successive integrations by parts, we are left with

$$\alpha_m^4 \int_{-\frac{1}{2}}^{+\frac{1}{2}} y_m y_n \, dx = \int_{-\frac{1}{2}}^{+\frac{1}{2}} \frac{d^2 y_n}{dx^2} \frac{d^2 y_m}{dx^2} \, dx, \tag{4}$$

the integrated parts vanishing (both times) on account of the boundary conditions on y_n. From the symmetry of the integral on the right-hand side of equation (4) in n and m, it follows that

$$(\alpha_m^4 - \alpha_n^4) \int_{-\frac{1}{2}}^{+\frac{1}{2}} y_m y_n \, dx = 0. \tag{5}$$

Hence
$$\int_{-\frac{1}{2}}^{+\frac{1}{2}} y_m y_n \, dx = 0 \quad \text{if } m \neq n. \tag{6}$$

The solutions y_m, therefore, form an orthogonal set; and we shall suppose that they are complete for the purposes of expanding an arbitrary continuous function in the interval $(+\frac{1}{2}, -\frac{1}{2})$ which together with its first derivative vanishes at the ends.

The proper solutions of equations (1) and (2) are readily found. They fall into two non-combining groups of even and odd solutions with cosine-like and sine-like behaviours. As standard forms of these solutions we shall take

$$C_m(x) = \frac{\cosh \lambda_m x}{\cosh \frac{1}{2}\lambda_m} - \frac{\cos \lambda_m x}{\cos \frac{1}{2}\lambda_m} \tag{7}$$

and
$$S_m(x) = \frac{\sinh \mu_m x}{\sinh \frac{1}{2}\mu_m} - \frac{\sin \mu_m x}{\sin \frac{1}{2}\mu_m}. \tag{8}$$

Defined in this manner, the functions clearly vanish at $x = \pm\frac{1}{2}$; that their derivatives also vanish at these points requires that λ_m and μ_m are the roots of the characteristic equations

$$\tanh \tfrac{1}{2}\lambda + \tan \tfrac{1}{2}\lambda = 0 \tag{9}$$

and
$$\coth \tfrac{1}{2}\mu - \cot \tfrac{1}{2}\mu = 0, \tag{10}$$

respectively.

It can be readily verified that the functions $C_m(x)$ and $S_m(x)$ are already normalized so that

$$\int_{-\frac{1}{2}}^{+\frac{1}{2}} C_m(x) C_n(x) \, dx = \int_{-\frac{1}{2}}^{+\frac{1}{2}} S_m(x) S_n(x) \, dx = \delta_{mn}. \tag{11}$$

Also, it is evident that every function C_m is orthogonal to every function S_n; thus

$$\int_{-\frac{1}{2}}^{+\frac{1}{2}} C_m(x) S_n(x) \, dx = 0. \tag{12}$$

In Table LXVIII the first few roots of equations (9) and (10) and some related constants are listed.

TABLE LXVIII

The characteristic roots λ_m and μ_m and related constants

m	λ_m	μ_m	$\tanh \frac{1}{2}\lambda_m$	$\coth \frac{1}{2}\mu_m$
1	4·73004074	7·85320462	0·98250221	1·00077731
2	10·99560784	14·13716549	0·99996645	1·00000145
3	17·27875966	20·42035225	0·99999994	1·00000000
4	23·56194490	26·70353756	1·00000000	1·00000000

For $m > 4$ the asymptotic formulae

$$\lambda_m \to (2m-\tfrac{1}{2})\pi \quad \text{and} \quad \mu_m \to (2m+\tfrac{1}{2})\pi \quad (m \to \infty) \tag{13}$$

give the roots correct to ten decimals.

The functions C_m and S_m for $m = 1$, 2, 3, and 4 (together with their first three derivatives) have been tabulated to nine decimals by Harris and Reid for $x = 0(0{\cdot}005)0{\cdot}5$.

(*a*) *Integrals involving $C_n(x)$ and $S_m(x)$*

In the applications of the C- and the S-functions, we have often to evaluate matrix elements involving them. Simple integrals of the types one generally encounters (as in §§ 86–88) can be most readily evaluated by raising by four the order of the derivative with which $C_n(x)$ (or $S_m(x)$) occurs in the integrand (by making use of the differential equation (1) which they satisfy) and then following by a sequence of four integrations by parts. As illustrative of this method we shall consider two typical examples.

(1) Consider
$$(S_m|C_n''') = \int_{-\frac{1}{2}}^{+\frac{1}{2}} S_m(x)C_n'''(x)\,dx. \tag{14}$$

We first rewrite this integral as

$$\mu_m^4(S_m|C_n''') = \int_{-\frac{1}{2}}^{+\frac{1}{2}} S_m^{\text{iv}}(x)C_n'''(x)\,dx = (S_m^{\text{iv}}|C_n'''). \tag{15}$$

An integration by parts now gives

$$\mu_m^4(S_m|C_n''') = 2S_m'''(\tfrac{1}{2})C_n'''(\tfrac{1}{2})-\lambda_n^4(S_m'''|C_n). \tag{16}$$

By a further sequence of integration by parts, we find

$$(S_m'''|C_n) = -(S_m''|C_n') = +(S_m'|C_n'') = -(S_m|C_n'''). \tag{17}$$

Using this last result in equation (16), we obtain

$$(S_m|C_n''') = \frac{2S_m'''(\frac{1}{2})C_n'''(\frac{1}{2})}{\mu_m^4-\lambda_n^4}. \tag{18}$$

(2) Consider the matrix element

$$(S_m|x|C_n) = \int_{-\frac{1}{2}}^{+\frac{1}{2}} S_m\, x C_n\, dx. \tag{19}$$

In the same way as in the preceding example, we find

$$\lambda_n^4(S_m|x|C_n) = (S_m|x|C_n^{\text{iv}}) = -(S_m|C_n''')-(S_m'|x|C_n''')$$
$$= -(S_m|C_n''')+(S_m'|C_n'')-(S_m''|C_n')+(S_m'''|C_n)+(S_m^{\text{iv}}|x|C_n). \tag{20}$$

We thus find (making further use of the relations included in (17))

$$(S_m|x|C_n) = -\frac{4(S_m|C_n''')}{\lambda_n^4-\mu_m^4} = \frac{8S_m'''(\tfrac{1}{2})C_n'''(\tfrac{1}{2})}{(\lambda_n^4-\mu_m^4)^2}. \tag{21}$$

134. Functions suitable for problems with cylindrical and spherical boundaries

For problems with cylindrical and spherical boundaries, the required functions can be derived as the proper solutions of the characteristic value problem defined by the equation

$$\mathscr{D}_\nu^2 y = \left(\frac{d^2}{dr^2}+\frac{1}{r}\frac{d}{dr}-\frac{\nu^2}{r^2}\right)^2 y = \alpha^4 y, \tag{22}$$

and the boundary conditions

$$y = 0 \quad\text{and}\quad dy/dr = 0 \quad \text{for } r = 1 \text{ and } r = \eta\ (< 1). \tag{23}$$

First we may observe that if f is any continuous and bounded function in the interval $(\eta, 1)$ and y satisfies the boundary conditions (23), then

$$\int_\eta^1 ry\mathscr{D}_\nu f\, dr = \int_\eta^1 y\left\{\frac{d}{dr}\left(r\frac{df}{dr}\right)-\frac{\nu^2}{r}f\right\} dr$$
$$= -\int_\eta^1 r\left(\frac{dy}{dr}\frac{df}{dr}+\frac{\nu^2}{r^2}yf\right) dr = \int_\eta^1 rf\mathscr{D}_\nu y\, dr. \tag{24}$$

If y_m and y_n are proper solutions belonging to two different characteristic values α_m and α_n, then it follows from the equation governing y_m that

$$\alpha_m^4 \int_\eta^1 ry_n y_m\, dr = \int_\eta^1 ry_n \mathscr{D}_\nu^2 y_m\, dr. \tag{25}$$

Now making use of the result expressed in (24), we obtain

$$\alpha_m^4 \int_\eta^1 ry_n y_m\, dr = \int_\eta^1 r(\mathscr{D}_\nu y_n)(\mathscr{D}_\nu y_m)\, dr. \tag{26}$$

From the symmetry of the integral on the right-hand side of equation (26) in n and m we conclude that

$$(\alpha_m^4-\alpha_n^4)\int_\eta^1 ry_n\,y_m\,dr = 0. \tag{27}$$

Hence
$$\int_\eta^1 ry_n\,y_m\,dr = 0 \quad \text{if } n \neq m. \tag{28}$$

The proper solutions y_n, therefore, form an orthogonal set in the interval $(1, \eta)$ with the weight function r.

Returning to equation (22), we may write its general solution as a linear combination of the Bessel functions of the different kinds and arguments of order ν; thus

$$y = AJ_\nu(\alpha r)+BY_\nu(\alpha r)+CI_\nu(\alpha r)+DK_\nu(\alpha r), \tag{29}$$

where A, B, C, and D are arbitrary constants. (In writing the solution in the form (29) we have assumed that ν is an integer; if ν should not be an integer we have only to write $J_{-\nu}$ in place of Y_ν.)

From the recurrence relations satisfied by the Bessel functions, it follows that

$$\frac{1}{\alpha r^\nu}\frac{d}{dr}(r^\nu y) = AJ_{\nu-1}(\alpha r)+BY_{\nu-1}(\alpha r)+CI_{\nu-1}(\alpha r)-DK_{\nu-1}(\alpha r). \tag{30}$$

The application of the boundary conditions (23) to the solution (29), therefore, leads to the characteristic equation

$$\begin{Vmatrix} J_\nu(\alpha) & Y_\nu(\alpha) & I_\nu(\alpha) & K_\nu(\alpha) \\ J_\nu(\alpha\eta) & Y_\nu(\alpha\eta) & I_\nu(\alpha\eta) & K_\nu(\alpha\eta) \\ J_{\nu-1}(\alpha) & Y_{\nu-1}(\alpha) & I_{\nu-1}(\alpha) & -K_{\nu-1}(\alpha) \\ J_{\nu-1}(\alpha\eta) & Y_{\nu-1}(\alpha\eta) & I_{\nu-1}(\alpha\eta) & -K_{\nu-1}(\alpha\eta) \end{Vmatrix} = 0. \tag{31}$$

If α_m is a characteristic root and $(1, B_m, C_m, D_m)$ is the vector which is annihilated by the matrix in (31) (when $\alpha = \alpha_m$), then

$$\mathscr{C}_{\nu;m}(r) = J_\nu(\alpha_m r)+B_m Y_\nu(\alpha_m r)+C_m I_\nu(\alpha_m r)+D_m K_\nu(\alpha_m r) \tag{32}$$

is a proper solution belonging to α_m.

(a) The normalization integral

Let
$$u_{\nu;m}(r) = J_\nu(\alpha_m r)+B_m Y_\nu(\alpha_m r)$$

and
$$v_{\nu;m}(r) = C_m I_\nu(\alpha_m r)+D_m K_\nu(\alpha_m r), \tag{33}$$

so that
$$\mathscr{C}_{\nu;m}(r) = u_{\nu;m}(r)+v_{\nu;m}(r). \tag{34}$$

In virtue of the boundary conditions satisfied by $\mathscr{C}_{\nu;m}$,

$$u_{\nu;m}(1) = -v_{\nu;m}(1); \qquad u'_{\nu;m}(1) = -v'_{\nu;m}(1);$$
$$u_{\nu;m}(\eta) = -v_{\nu;m}(\eta); \qquad u'_{\nu;m}(\eta) = -v'_{\nu;m}(\eta); \tag{35}$$

and in particular,

$$u_{\nu;m}\,v'_{\nu;m} - u'_{\nu;m}\,v_{\nu;m} = 0 \quad \text{at} \quad r = 1 \text{ and } \eta. \tag{36}$$

The functions $u_{\nu;m}$ and $v_{\nu;m}$ each being a linear combination of Bessel functions with like (i.e. real or imaginary) arguments, they satisfy the differential equations

$$\frac{1}{r}\frac{d}{dr}\left(r\frac{du_{\nu;m}}{dr}\right) - \frac{\nu^2}{r^2}u_{\nu;m} = -\alpha_m^2\,u_{\nu;m} \tag{37}$$

and
$$\frac{1}{r}\frac{d}{dr}\left(r\frac{dv_{\nu;m}}{dr}\right) - \frac{\nu^2}{r^2}v_{\nu;m} = +\alpha_m^2\,v_{\nu;m}. \tag{38}$$

Multiplying equations (37) and (38) by $rv_{\nu;m}$ and $ru_{\nu;m}$, respectively, and then subtracting one from another, we find

$$2\alpha_m^2\,ru_{\nu;m}\,v_{\nu;m} = u_{\nu;m}\frac{d}{dr}\left(r\frac{dv_{\nu;m}}{dr}\right) - v_{\nu;m}\frac{d}{dr}\left(r\frac{du_{\nu;m}}{dr}\right)$$
$$= \frac{d}{dr}\left(ru_{\nu;m}\frac{dv_{\nu;m}}{dr} - rv_{\nu;m}\frac{du_{\nu;m}}{dr}\right); \tag{39}$$

and integrating this equation over the range of r, we obtain

$$2\alpha_m^2\int_\eta^1 ru_{\nu;m}\,v_{\nu;m}\,dr = [r(u_{\nu;m}\,v'_{\nu;m} - v_{\nu;m}\,u'_{\nu;m})]_\eta^1. \tag{40}$$

The quantity on the right-hand side of equation (40) vanishes by virtue of equation (36). Hence,

$$\int_\eta^1 ru_{\nu;m}\,v_{\nu;m}\,dr = 0. \tag{41}$$

Making use of this property, we can write

$$\int_\eta^1 r\mathscr{C}_{\nu;m}^2\,dr = \int_\eta^1 r(u_{\nu;m}^2 + v_{\nu;m}^2)\,dr. \tag{42}$$

To evaluate the integrals on the right-hand side, consider, first, the differential equation satisfied by $v_{\nu;m}$; after multiplication by $2r^2v'_{\nu;m}$ the equation can be written in the form

$$\frac{d}{dr}(rv'_{\nu;m})^2 = \alpha_m^2\,r^2\frac{dv_{\nu;m}^2}{dr} + \nu^2\frac{dv_{\nu;m}^2}{dr}. \tag{43}$$

On integrating this equation over the range of r, we obtain

$$[r^2(v'_{\nu;m})^2-\nu^2 v^2_{\nu;m}]^1_\eta = \alpha^2_m \int\limits_\eta^1 r^2\frac{dv^2_{\nu;m}}{dr}\,dr = [\alpha^2_m r^2 v^2_{\nu;m}]^1_\eta - 2\alpha^2_m \int\limits_\eta^1 r v^2_{\nu;m}\,dr. \tag{44}$$

We thus have

$$2\alpha^2_m \int\limits_\eta^1 r v^2_{\nu;m}\,dr = [-r^2(v'_{\nu;m})^2+(\alpha^2_m r^2+\nu^2)v^2_{\nu;m}]^1_\eta. \tag{45}$$

From the equation satisfied by $u_{\nu;m}$ we similarly find

$$2\alpha^2_m \int\limits_\eta^1 r u^2_{\nu;m}\,dr = [+r^2(u'_{\nu;m})^2+(\alpha^2_m r^2-\nu^2)u^2_{\nu;m}]^1_\eta. \tag{46}$$

Now adding equations (45) and (46), we have

$$2\alpha^2_m \int\limits_\eta^1 r(u^2_{\nu;m}+v^2_{\nu;m})\,dr = [r^2\{(u'_{\nu;m})^2-(v'_{\nu;m})^2\}+ \\ +\alpha^2_m r^2(u^2_{\nu;m}+v^2_{\nu;m})-\nu^2(u^2_{\nu;m}-v^2_{\nu;m})]^1_\eta. \tag{47}$$

In view of the relations given in (35), equation (47) reduces to

$$\int\limits_\eta^1 r(u^2_{\nu;m}+v^2_{\nu;m})\,dr = [r^2 u^2_{\nu;m}]^1_\eta = [r^2 v^2_{\nu;m}]^1_\eta. \tag{48}$$

Returning to equation (42) and remembering that the functions $\mathscr{C}_{\nu;m}$ and $\mathscr{C}_{\nu;n}$ for $m \neq n$ are orthogonal (in the interval $(1, \eta)$ and with the weight function r), we can write

$$\int\limits_\eta^1 r\mathscr{C}_{\nu;m}\mathscr{C}_{\nu;n}\,dr = N_{\nu;m}\,\delta_{mn}, \tag{49}$$

where $$N_{\nu;m} = u^2_{\nu;m}(1)-\eta^2 u^2_{\nu;m}(\eta) = v^2_{\nu;m}(1)-\eta^2 v^2_{\nu;m}(\eta). \tag{50}$$

(*b*) *The cylinder functions of order* 1

Axisymmetric viscous flows between rotating cylinders can be expanded in terms of the functions of order 1; such an expansion was used for example in Chapter VII, § 73. In view of the special importance of the functions of order 1, we list in Table LXIX a few of the characteristic roots and certain related constants for three different values of η.

For the particular case $\eta = \frac{1}{2}$, the solutions belonging to the first three roots have been tabulated by Chandrasekhar and Elbert.

(*c*) *The spherical functions of half-odd integral orders*

The functions of half-odd integral orders are useful in problems of viscous flow in spheres and in spherical shells. For flows in a sphere,

TABLE LXIX

The characteristic roots and related constants for $\nu = 1$

	$\eta = \frac{1}{4}$			$\eta = \frac{1}{2}$			$\eta = \frac{3}{4}$
	$m = 1$	$m = 2$	$m = 3$	$m = 1$	$m = 2$	$m = 3$	$m = 1$
α_m	$+6{\cdot}390336$	$+10{\cdot}551651$	$+14{\cdot}729000$	$+9{\cdot}49896$	$+15{\cdot}739746$	$+22{\cdot}017821$	$+18{\cdot}93440$
B_m	$-5{\cdot}870144$	$-0{\cdot}4281099$	$+0{\cdot}7125113$	$-9{\cdot}076420$	$-17{\cdot}3573$	$-22{\cdot}4667$	$-10{\cdot}2780$
C_m	$-1{\cdot}497055\times10^{-2}$	$+4{\cdot}167599\times10^{-5}$	$-7{\cdot}144507\times10^{-7}$	$+1{\cdot}006354\times10^{-3}$	$+3{\cdot}67581\times10^{-6}$	$+8{\cdot}86448\times10^{-9}$	$-8{\cdot}91136\times10^{-8}$
D_m	$-10{\cdot}77988$	$-5{\cdot}986757$	$-19{\cdot}90159$	$+4{\cdot}396374\times10^{2}$	$-1{\cdot}95381\times10^{4}$	$+5{\cdot}91172\times10^{5}$	$-6{\cdot}65593\times10^{6}$
$u_m(1) = -v_m(1)$	$+1{\cdot}330279$	$-0{\cdot}1884864$	$+0{\cdot}1803302$	$-1{\cdot}681248$	$-2{\cdot}46940$	$-2{\cdot}70290$	$+1{\cdot}34981$
$u_m(\eta) = -v_m(\eta)$	$+2{\cdot}619034$	$+0{\cdot}3728952$	$+0{\cdot}3581128$	$-2{\cdot}371094$	$+3{\cdot}48706$	$-3{\cdot}81917$	$+1{\cdot}55828$
$u'_m(1) = -v'_m(1)$	$+7{\cdot}807580$	$-1{\cdot}903352$	$+2{\cdot}570776$	$-14{\cdot}93868$	$-37{\cdot}7250$	$-58{\cdot}2068$	$+24{\cdot}4679$
$u'_m(\eta) = -v'_m(\eta)$	$-23{\cdot}29673$	$-4{\cdot}83996$	$-6{\cdot}10685$	$+24{\cdot}82964$	$-58{\cdot}7097$	$+88{\cdot}1452$	$-30{\cdot}0835$
$N_{1:m}$	$+1{\cdot}340934$	$+0{\cdot}02683645$	$+0{\cdot}02450369$	$+1{\cdot}42107$	$+3{\cdot}05805$	$+3{\cdot}65915$	$+0{\cdot}456104$

the requirement that the solutions have no singularity at the origin implies that in the solution (30) we cannot include the terms in Y_ν (or $J_{-\nu}$) and K_ν. Therefore, in this case we can write

$$\mathscr{C}_{l+\frac{1}{2}}(r) = \frac{J_{l+\frac{1}{2}}(\alpha r)}{J_{l+\frac{1}{2}}(\alpha)} - \frac{I_{l+\frac{1}{2}}(\alpha r)}{I_{l+\frac{1}{2}}(\alpha)}. \tag{51}$$

Defined in this manner, the function $\mathscr{C}_{l+\frac{1}{2}}(r)$ clearly vanishes at $r = 1$; and the condition that its derivative also vanishes at $r = 1$ leads to the characteristic equation

$$J_{l+\frac{1}{2}}(\alpha)I_{l-\frac{1}{2}}(\alpha) - J_{l-\frac{1}{2}}(\alpha)I_{l+\frac{1}{2}}(\alpha) = 0. \tag{52}$$

The first three roots of this equation for $l = 1$, 3, and 5 are listed in Table LXX.

The roots of the equation

$$J_{l+\frac{1}{2}}(\alpha)I'_{l+\frac{1}{2}}(\alpha) - I_{l+\frac{1}{2}}(\alpha)J'_{l+\frac{1}{2}}(\alpha) = \alpha J_{l+\frac{1}{2}}(\alpha)I_{l+\frac{1}{2}}(\alpha) \tag{53}$$

needed in a related connexion (§ 62) are also included in Table LXX.

Table LXX

The roots of equations (52) *and* (53)

m	*Roots of equation* (52)			*Roots of equation* (53)		
	$l = 1$	$l = 3$	$l = 5$	$l = 1$	$l = 3$	$l = 5$
1	5·267568	7·748590	10·10830	4·180881	6·813002	9·232055
2	8·506951	11·19090	13·73380	7·573795	10·309365	12·881869
3	11·68768	14·47661	17·12845	10·802354	13·618716	16·289137

BIBLIOGRAPHICAL NOTES

The usefulness of the functions described in this appendix for the solution of hydrodynamical problems was pointed out by:

1. S. Chandrasekhar and W. H. Reid, 'On the expansion of functions which satisfy four boundary conditions', *Proc. Nat. Acad. Sci.* **43,** 521–7 (1957).

§ 133. The functions which we have denoted by C_m and S_m occur in the theory of vibrating beams and in this connexion they have been considered by Rayleigh:

2. Lord Rayleigh, *Theory of Sound,* p. 278, Dover Publications, New York, 1945.

In the hydrodynamical context these functions have been studied most extensively by:

3. W. H. Reid, 'On the stability of viscous flow in a curved channel', *Proc. Roy. Soc.* (*London*), A, **244,** 186–98 (1958).
4. D. L. Harris and W. H. Reid, 'On orthogonal functions which satisfy four boundary conditions. I. Tables for use in Fourier-type expansions', *Astrophys. J. Supp. Ser.* **3,** 429–47 (1958).
5. W. H. Reid and D. L. Harris, 'On orthogonal functions which satisfy

four boundary conditions. II. Integrals for use with Fourier-type expansions', *Astrophys. J. Supp. Ser.* **3,** 448–52 (1958).

In reference 4 Harris and Reid have tabulated the functions C_m and S_m and their first three derivatives for $m = 1, \ldots, 4$ to 9 decimals for $x = 0\ (0{\cdot}005)\ 0{\cdot}5$. Less extensive tables have been published earlier by:

6. R. E. D. Bishop and D. C. Johnson, *Vibration Analysis Tables*, Cambridge University Press, New York, 1956.

In reference 5 Reid and Harris have provided a very useful compilation of integrals involving the C- and the S-functions.

§ 134. The functions denoted by $\mathscr{C}_{\nu:m}$ seem to have been considered for the first time in these connexions in reference 1. See also:

7. S. Chandrasekhar, 'The stability of viscous flow between rotating cylinders', *Proc. Roy. Soc.* (*London*) A, **246,** 301–11 (1958).

The numerical results included in Table LXIX are taken from:

8. —— and Donna D. Elbert, 'On orthogonal functions which satisfy four boundary conditions. III. Tables for use in Fourier–Bessel–type expansions', *Astrophys. J. Supp. Ser.* **3,** 453–8 (1958).

The functions $\mathscr{C}_{1:m}$ for $m = 1$, 2, and 3 and $\eta = \frac{1}{2}$ are tabulated in reference 8.

The characteristic roots listed in Table LXX are taken from:

9. F. E. Bisshopp, 'On the thermal instability of a rotating fluid sphere', *Phil. Mag.* Ser. 8, **3,** 1342–60 (1958).

SUBJECT INDEX

INDEX OF DEFINITIONS

Theorems are enunciated on the following pages

Section number	*Page number*	*Section number*	*Page number*	*Section number*	*Page number*
3	6	58	229	103	491
13	29	60	240	105	507
	31		251	106	511
14	34	61	252		512
20	77	66	273	108	518
	79	67	280		519
	80	70	298	109	527
22	83	71	313	110	534
	84	72	316	111	538
23	86	78	361	112	545
26	93		362	113	553
27	104		366	114	560
28	108		367		562
	109		370	115	570
29	116	81	389	117	581
	118	83	395		582
	119	84	398		583
30	126	86	413	118	587
33	134	87	420	119	589
38	152	92	434	120	589
	153		435		591
39	156		436		593
40	159	93	440	122	603
43	167	94	447		604
45	177	95	456	123	607
	179	96	458	126	614
	181		464	130	628
46	183	99	475		629
47	189	101	485	131	631
56	226		486		632